Geophysical Monograph Series

Including
IUGG Volumes
Maurice Ewing Volumes
Mineral Physics Volumes

Geophysical Monograph Series

Geophysical Monograph 196

Extreme Events and Natural Hazards: The Complexity Perspective

A. Surjalal Sharma
Armin Bunde
Vijay P. Dimri
Daniel N. Baker
Editors

American Geophysical Union
Washington, DC

Published under the aegis of the AGU Books Board

Library of Congress Cataloging-in-Publication Data

Extreme events and natural hazards : the complexity perspective / A. Surjalal Sharma ... [et al.] editors.
p. cm. – (Geophysical monograph, ISSN 0065-8448 ; 196)
Includes bibliographical references and index.
ISBN 978-0-87590-486-3
1. Geophysical prediction. 2. Natural disasters. 3. Computational complexity. 4. Hazard mitigation. I. Sharma, A. Surjalal, 1951-
QC807.E98 2012
551.01'511352–dc23
2012018176

ISBN: 978-0-87590-486-3
ISSN: 0065-8448

Cover Image: Extreme events and natural hazards originate from many sources, from the Sun to the Earth's interior, and can occur together. (top left) A solar prominence eruption, captured here by the NASA Solar Dynamics Observatory, is the main cause of severe space weather events. Photo credit NASA/GSFC/Solar Dynamics Observatory's AIA Instrument. (top right) Hurricane Katrina powered a category 5 storm (August 2005) and is the most destructive hurricane to date to strike the United States. Photo by NOAA Gulfstream IV-SP aircraft. (bottom right) On 18 August 2005 a tornado touched down in Stoughton, Wisconsin. Photo credit C. McDermott, National Weather Service, NOAA. (bottom left) A village is devastated near the coast of Sumatra following the tsunami that struck Southeast Asia is shown. Photo credit U.S. Navy Photographer's Mate 2nd Class P. A. McDaniel.

Printed in the United States of America.

CONTENTS

PREFACE

Understanding extreme natural events and their societal consequences is among the most pressing scientific challenges of our time. Like most of the major scientific challenges in the Earth and space sciences, there is increasing recognition that an integrated approach involving multiple disciplines will be needed to advance the science underlying extreme events that lead to natural hazards. Complexity science, with its multidisciplinary nature and its focus on dynamical instability and sudden changes, provides a natural framework for this effort. The main goal of this volume is to bring together the research in extreme events, complexity science, and natural hazards and to explore new ways to advance our knowledge and develop strategies.

The need for an integrated approach to the understanding of extreme events and the resulting natural hazards was highlighted by the devastating consequences of the March 2011 Japan earthquake of magnitude 9.0 and the resulting tsunami. The severe damage done by the tsunami was further compounded by the exposure of the nuclear power plants to potential accidents with unprecedented consequences. The possibility of such a potential confluence of events also brings into focus the effects of other extreme events such as severe space weather. The technological infrastructure we depend on, such as telecommunications and electric power networks, are susceptible to disruption in severe space weather, and the scenarios of such conditions during other natural hazards need serious study.

The low-probability, high-impact nature of extreme events makes their understanding a continuing imperative, and complexity science with its systems-based approach provides an important complement to the traditional first-principles studies. The nature of the distribution function of the events is essential to the characterization of extreme events, and recent studies have shown many interesting results. This could lead to better quantification of their likelihood. The distributed nature of the components and the strong interaction among them is another feature common to systems exhibiting extreme events. This is evident in many branches of geosciences, e.g., atmospheric, hydrologic, oceanic, and space sciences, and has led to the characterization of the long-range nature of the correlations among the components. The combination of the frequency of extreme events and the strong correlation among them is an important feature in assessing their potential hazard to society. Such advances make a strong case for pursuing the approaches based on the developments in complexity science.

Responding to extreme events and natural hazards depends strongly on timely warning with specified likelihoods. The development of the capability to provide such warnings is a major objective of the research efforts. The preparedness of our society for natural disasters depends on the accomplishments of such research, and we hope that the insights gained from this volume will stimulate new initiatives.

This volume derives from the Chapman Conference on Complexity and Extreme Events in Geosciences, held in Hyderabad, India, in February 2010. This conference was truly inter-disciplinary, as is evident from the coverage of the many disciplines in this volume, and provided a forum for exploring new research ideas and collaborations. The National Geophysical Research Institute, Hyderabad, hosted the conference, and the National Science Foundation supported it with a travel grant.

The editors would like to thank the colleagues who participated in the evaluation of the papers submitted to this volume. We owe the high quality of the articles in this volume to the diligence, expertise, and rigorous standards of these reviewers.

A. Surjalal Sharma
University of Maryland

Armin Bunde
Justus Liebig University Giessen

Vijay P. Dimri
National Geophysical Research Institute

Daniel N. Baker
University of Colorado

Extreme Events and Natural Hazards: The Complexity Perspective
Geophysical Monograph Series 196

10.1029/2012GM001257

Complexity and Extreme Events in Geosciences: An Overview

A. Surjalal Sharma,[1] Daniel N. Baker,[2] Archana Bhattacharyya,[3] Armin Bunde,[4] Vijay P. Dimri,[5] Harsh K. Gupta,[5] Vijay K. Gupta,[6] Shaun Lovejoy,[7] Ian G. Main,[8] Daniel Schertzer,[9] Hans von Storch,[10,11] and Nicholas W. Watkins[12,13,14]

Extreme events are an emergent property of many complex, nonlinear systems in which various interdependent components and their interaction lead to a competition between organized (interaction dominated) and irregular (fluctuation dominated) behavior. Recent advances in nonlinear dynamics and complexity science provide a new approach to the understanding and modeling of extreme events and natural hazards. The main connection of extreme events to nonlinear dynamics arises from the recognition that they are not isolable phenomena but must be understood in terms of interactions among different components, within and outside the specific system. A wide range of techniques and approaches of complexity science are directly relevant to geosciences, e.g., nonlinear modeling and prediction, state space reconstruction, statistical self-similarity and its dynamical origins, stochastic cascade models, fractals and multifractals, network theory, self-organized criticality, etc. The scaling of processes in geosciences has been one of the most active areas of studies and has the potential to provide better tools for risk assessment and analysis. Many studies of extreme events in geosciences are also contributing to the basic understanding of their inherent properties, e.g., maximum entropy production and criticality, space-time cascades, and fractional Lévy processes. The need for better data for extreme events is evident in the necessity for detailed statistical analysis, e.g., in marine storms, nonlinear correlations, etc. The Chapman Conference on Complexity and Extreme Events held (2010) in Hyderabad,

[1]Department of Astronomy, University of Maryland, College Park, Maryland, USA.
[2]Laboratory for Atmospheric and Space Physics, University of Colorado, Boulder, Colorado, USA.
[3]Indian Institute of Geomagnetism, Navi Mumbai, India.
[4]Institut für Theoretische Physik III, Justus-Liebig-Universität Giessen, Giessen, Germany.
[5]National Geophysical Research Institute, Hyderabad, India.

Extreme Events and Natural Hazards: The Complexity Perspective
Geophysical Monograph Series 196

10.1029/2012GM001233

[6]Department of Civil, Environmental and Architectural Engineering and Cooperative Institute for Research in Environmental Science, Boulder, Colorado, USA.
[7]Department of Physics, McGill University, Montreal, Quebec, Canada.
[8]School of GeoScience, University of Edinburgh, Edinburgh, UK.
[9]LEESU, Ecole des Ponts ParisTech, Universite Paris-Est, Paris, France.
[10]Institute of Coastal Research, Helmholtz-Zentrum Geesthacht, Geesthacht, Germany.
[11]Meteorological Institute, University of Hamburg, Hamburg, Germany.
[12]British Antarctic Survey, Cambridge, UK.
[13]Centre for the Analysis of Time Series, London School of Economics and Political Science, London, UK.
[14]Centre for Fusion, Space and Astrophysics, University of Warwick, Coventry, UK.

India, was focused on the understanding of natural hazards mainly from the perspective of complexity science. The emerging theme from the conference was the recognition of complexity science as the interdisciplinary framework for the understanding of extreme events and natural hazards.

1. INTRODUCTION

Complexity refers to the behavior of systems with many interdependent components that lead to organized as well as irregular (irreducibly stochastic) features. In such systems the knowledge of the parts does not necessarily lead to the predictable behavior of the entire system. The coupling among the components is essentially nonlinear, and this leads to a rich variety of dynamical behavior, geometrical patterns, and statistical distributions that are seen in virtually all disciplines. In geosciences the studies of nonlinear dynamical process and complexity have been an active field of research for many decades [*Lovejoy et al.*, 2009]. The basic ideas underlying complexity science have now matured to the point that they are generating new approaches to addressing problems in many different disciplines, including geosciences. One such area is the nature of extreme events, in particular the natural hazards, whose connection to complexity arises from the recognition that they are not isolable phenomena but must be understood in terms of interactions among different components, inside and outside the specific system.

Extreme events are of both natural and anthropogenic origin and are of widespread concern mainly because of their damaging consequences [*Lubchenco and Karl*, 2012]. Although there is no single definition, at least in the physical sciences, of extreme events [*Bunde et al.*, 2002; *Jentsch et al.*, 2006; *Moffat and Shuckburgh*, 2011], there *is* a significant, widely accepted body of work [*Coles*, 2001], since that of *Fisher and Tippett* [1928], which informs much statistical practice and goes by the name of extreme value theory. In science and applications, however, the practical interpretation of the degree of "extremeness" often mixes the statistical attributes of infrequent occurrence and low probability, the physics- or prediction-related property of being unexpected, and the societal or economic aspect of strong impact, etc. In general, it is not so clear that extreme events can be characterized or "marked" by only one or even a few measures. However, it is, nonetheless, clear that extreme events are typically rare, and in the distribution of events of all magnitudes they are identified as those outside the bulk; that is, they occur in the tail of the distribution. A main objective in the analysis of extreme events thus relates directly to the understanding of the distribution functions of the events, in particular the values in the tail. Even when a single probability distribution suffices to capture a system's amplitude variation, however, we may still distinguish at least two scenarios. One is the case where the distribution is relatively "short tailed" in its fluctuations, the model for this being the Gaussian. In this framework, extreme events really are rare and, in the Gaussian example, will only very rarely exceed three standard deviations from the mean. Conversely, if the underlying distribution is actually heavy tailed, with examples being the power law, lognormal, and stretched exponential families, the mean will be a much less adequate characterization of the expected behavior, and we will see effects like the 80–20 rule of thumb from actuarial science, where 80% of losses come from only 20% of claims [e.g., *Embrechts et al.*, 1997].

Another feature of extreme events is that they occur suddenly, often without clear warning, and on large spatial scales compared to the system size. Such long-range order, i.e., the value of a physical variable at an arbitrary point is correlated with its value at a point located far away, is also a property of thermodynamic phase transitions. In the case of second-order phase transitions the correlation length reaches the system size, giving rise to arbitrarily large extreme events at the critical point because of the competition between random fluctuations and cooperative dynamical interactions. This leads to the recognition that long-range correlations are important indicators of the emergence of extreme events. In view of these features the dynamical and statistical approaches of complexity science provide a natural framework for the study of extreme events [*Sharma et al.*, 2010]. An aspect of correlation of particular importance to extremes, and not always recognized, is that while a low probability of occurrence must indeed imply that such an event will be, on average, rare, the correlations in time between extreme events can mean that several such "black swan" tail events may still follow each other in close succession (dubbed "bunched black swans" by *Watkins et al.* [2011]). The recurrence time distribution is not simply prescribed by the frequency distribution, a point which becomes progressively more significant as temporal correlations become longer ranged.

A widely known property of many natural phenomena is the dynamical instability of nonlinear systems, which leads to limits of predictability because of sensitivity to the initial conditions, even for systems with few degrees of freedom. This behavior in deterministic systems such as the Lorenz equations [*Lorenz*, 1963] is a key dynamical origin for irregularity in nature and is now known as *chaos*. The Lorenz attractor, perhaps the best known case, has contributed immensely to the understanding of dynamical systems studies but has weaker ties to atmospheric circulation, of which it is a model. Such dynamical systems, known as strange attractors, also have the interesting geometrical property of being a fractal with self-similar characteristics. This property introduced by *Mandelbrot* [1967] arose from its preponderance in nature.

For systems with many degrees of freedom, thermodynamics or information theory can often be used to provide significant constraints on the dynamics of a population. In a thermodynamic system at equilibrium a formal entropy can be defined as a conjugate variable to temperature, and the scaling properties can be determined by maximizing the Boltzmann-Gibbs entropy subject to known constraints. Near-equilibrium states can also be modeled using this approach, if the rate of entropy production (by an irreversible process) is constrained to be a minimum, and provides an explanation for spontaneous self-organization in space and time for many open chemical and physical systems (an idea dating back to *Prigogine* [1945]). For such irreversible systems it is actually difficult to define or interpret the meaning of a macroscopic entropy in a rigorous way. As a consequence such systems are often analyzed using the more general concept of information theory [*Shannon*, 1948]. Here we also maximize an entropy-like function (with no Boltzmann pre-factor), also subject to constraints, and determine the maximum expectation value of the log (probability) of states [*Jaynes*, 1957]. In this formulation a power law distribution of event sizes may result physically from the geometric degeneracy of the system's energy states, with an exponent constrained by a geometric mean of the energy (see, e.g., *Main and Burton* [1984] for the case of earthquake populations).

For some systems, under certain constraints, the maximum entropy solution is also one of maximum entropy production [*Dewar*, 2003; *Martyushev and Seleznev*, 2006]. For example, the longitudinal temperature profile of the Earth is close to that predicted by a state of maximum entropy production [*Paltridge*, 2005]. Physically this state provides a mechanism for maintaining the atmosphere in a permanent state of balance between stable laminar flow and turbulence at intermediate Rayleigh number over long time scales.

Tsallis [1995] has proposed an alternate explanation for the common observation of power law statistics in complex systems, based on a nonextensive generalization of the Boltzmann-Gibbs entropy. The main disadvantages of this approach are that simple constraints such as system energy per se cannot be applied and there is no direct connection to information theory because the Tsallis entropy is not additive for independent sources of uncertainty.

The interplay of dynamics, geometry, and information and their effects on the scaling properties of the population and the predictability of individual events is an important feature of complexity science in general and in the studies of extreme events in particular. This chapter is an overview of recent advances in the understanding of extreme events in the natural world, in particular those made through the use of nonlinear dynamics, statistical physics, and related approaches. The topics covered include the recent studies of earthquakes, river flows, and climate variations, which have demonstrated long-range dependences (clustering) in the appearance of extreme events. These have important practical implications for climate change, natural hazard forecasting, and risk management. Frequency-magnitude statistics of many natural hazards follow power laws, exhibiting features of complex behavior, but robust power law statistics for a population often occur hand in hand with restricted predictability of individual events. This poses significant challenges in developing and communicating operationally useful forecasts of such events. The interdisciplinary nature of the research leads naturally to strong connections with a wide range of applications. Here we focus on the relationship between the science underlying the behavior of a complex system and the emergence of intermittent, perhaps clustered, extreme events, and identify approaches that may lead to significant advances in understanding in future.

2. NONLINEAR DYNAMICS, COMPLEXITY, AND EXTREME EVENTS

The complex behavior in deterministic systems with a few degrees of freedom, such as the Lorenz attractor, led to models of low-dimensional chaos for dynamical systems. In mathematical terms such systems are represented by a small number of first-order ordinary differential equations. These were complemented by approaches that include more complex spatiotemporal dynamics, namely, in the form of partial differential equations such as Ginzburg-Landau and Kuramoto-Sivashinsky equations. While these models were successful in representing many laboratory systems, they faced many difficulties in describing large-scale open natural systems. The next advance came from the recognition of the nonlinear coupling and the dissipative nature of many dynamical systems, responsible for the contraction of phase space. These two properties underlie the elucidation of strange attractors as the hallmark of chaotic dynamics, leading to many new developments in the dynamical systems theory. One application of dynamical systems with a few degrees of freedom to understanding of the Hurst effect is given by *Mesa et al.* [this volume]. Another set of applications require the assumption of an effectively low-dimensional nature of large-scale systems and of the applicability of the "embedding theorem," thus enabling the reconstruction of dynamical models from time series data. The converse assumption, of high dimensionality, has been termed "stochastic chaos" and is the assumption used in multifractal and other stochastic approaches [see, e.g., *Lovejoy and Schertzer*, this volume]. The use of low dimensionality as a paradigm has stimulated many new approaches to the study of seemingly complicated behavior in many natural systems, including in geosciences. In the studies of the dynamics of the geospace environment this approach provided the first predictive models of geomagnetic activity, enabled by the extensive data from ground-based and spaceborne instruments reviewed by *Sharma* [1995].

Various studies reveal that many of the Earth's processes show fractal statistics, where the relevant distributions follow power laws with noninteger ("fractal") exponents. Fractal structures that show self-similarity and are governed by a noninteger dimension are ubiquitous in geosciences. For example, they occur in the frequency size statistics of events and in the behavior of various physical properties such as density, susceptibility, electrical resistivity, and thermal conductivity. Nonlinearity and nonuniqueness are nearly always encountered in the inverse problem in geophysical exploration, including the components of data acquisition, processing, and not least interpretation (often by expert elicitation). Fractal dimension plays a vital role in the design of survey networks with proper sampling and gridding [*Dimri*, 1998; *Srivastava et al.*, 2007]. The interpretation of geophysical data such as gravity and magnetic data has shown good results assuming a fractal distribution of sources. Theoretical relation between fractal sources and their geophysical response has been used to derive source parameters [*Maus and Dimri*, 1994]. The fractal theory has led to the development of a wide variety of physical models of seismogenesis including nonlinear dynamics that can be used to characterize the seismicity pattern of a region [*Kagan and Knopoff*, 1980; *Sunmonu and Dimri*, 2000]. Similarly, the "coda" or tail of a seismogram can be modeled well with a fractal distribution of scatterers that emerge in a growing fracture population [*Vlastos et al.*, 2007].

The late Per Bak and colleagues sought to explain the widespread appearance of spatial fractals and temporal "$1/f$" noise by proposing a new class of avalanche-like models [*Bak et al.*, 1987] that exhibited fractally structured growing instabilities ("avalanches"), obeying heavy-tailed (algebraic) and thus extreme probability distributions. In fact, Bak most tersely expressed his intent a little later (with K. Chen [*Bak and Chen*, 1989, p. 5]) in the memorably short abstract: "Fractals in nature originate from self-organized critical dynamical processes." These self-organized critical (SOC) processes not only have extremes but also entire structures determined by them. There are now many models of self-organized criticality and significant developments toward an analytical theory. The applications of the SOC paradigm range from geosciences (including earthquakes, forest fires, solar activity, and rainfall) to the financial markets, while the renewed emphasis on the heavy tails generated by SOC processes has had a direct influence on modern network science and its applications. Although generally the SOC approach does not provide dynamical predictions, it describes the statistical features and thus yields the probabilities of events. For example, SOC provides a single unifying theory for much of what had previously been separate empirical observations of earthquake phenomena [*Main*, 1996]. While this provides a useful basis for seismic hazard calculation based on the population [*Main*, 1995], the proximity to criticality and the inherent stochastic element inevitably degrades any predictive power for *individual* events (see http://www.nature.com/nature/debates/earthquake/equake_frameset.html) [*Kagan*, 2006]. Similarly, in those (common) cases where the dynamics act over large ranges of scale, multifractal processes provide an interesting paradigm for extremes. In these cascade type processes the variability at any given resolution (i.e., the process averaged at the given scale) is precisely the consequence of the wide scale range of the dynamics. This includes not only the range from the largest scale to the resolution scale, but somewhat surprisingly, it also depends on the smaller scales whose variability is not completely "smoothed out" and which leads to extreme variability in the form of "fat-tailed" (power law) probability distributions. Since they involve both fractal structures and power law distributions, multifractal processes thus provide a "nonclassical" route to SOC.

More generally, the advances in the studies of complexity in many areas of geosciences have led to new predictive and diagnostic approaches that exploit how patterns, processes, and probabilities are mutually coupled. For example, nonlinear dynamics and complexity can be exploited to reduce forecast error, to understand atmospheric flow transitions, to test climate models by analyzing their atmospheric variability [*Govindan et al.*, 2002; *Vyushin et al.*, 2004; *Rybski et al.*, 2008], and to explain atmospheric and oceanic teleconnections [*Tsonis*, this volume]. The nonlinear aspects of the climate can regulate El Niño's background state, while the related approaches to data analysis, which are becoming increasingly popular, can be used to check the consistency of general circulation models in the representation of fundamental statistical properties of the atmosphere. In other areas of geosciences the role of nonlinear dynamics and complexity in complex landscape patterns of Earth's surface and hydrologic processes are well recognized, as presented in this volume.

Recent advances in the studies of extreme events from the viewpoint of nonlinear dynamics and complexity have demonstrated the existence and role of long-term memory in many complex systems [*Bunde et al.*, 2005; *Sharma and Veeramani*, 2011; *Mesa et al.*, this volume]. For example, when the memory can be described by a linear autocorrelation function that decays algebraically with an exponent γ, then the probability density function of the return intervals between events above some threshold no longer show the exponential decay typical for uncorrelated data but a stretched exponential decay characterized by the same exponent γ as the autocorrelation function [*Bunde et al.*, 2005]. Also, the return intervals themselves are long-term correlated, again characterized by the same exponent. This approach provides a new way to understand the clustering of extreme events. When the linear correlations vanish, and long-term memory exists only in the form of nonlinear correlations, such as the volatility bunching seen in finance [e.g., *Mantegna and Stanley*, 2000], the effect becomes even stronger, and both the probability distribution functions of the return intervals and their autocorrelation function decay as a power law [*Bogachev et al.*, 2007].

When considering complexity in the atmosphere and ocean, and to some extent space plasmas (especially when modeled using magnetohydrodynamics), we must recall the strong historical and theoretical links with the field of fully developed turbulence, in particular the classical turbulence laws associated with the pioneers L. F. Richardson, A. N. Kolmogorov, A. Obukhov, S. Corrsin, and R. Bolgiano. These classical laws were proposed as emergent laws in the field of sufficiently strong hydrodynamic turbulence where "developed turbulence as a new macroscopic state of matter" [*Manneville*, 2010] appears at high Reynolds number (*Re*). This new state has been considered as a form of matter with properties that cannot simply be reduced to, or simply deduced from, the governing Navier-Stokes equations. Although fluid (and plasma) geosystems certainly differ from incompressible hydrodynamics in several important respects, we may, nevertheless, expect higher-level laws to emerge and that they will share at least some of the features of fully developed turbulence.

In the atmosphere, key obstacles to applying classical turbulence are the strong anisotropy (especially stratification) and intermittency. However, over the years, new types of models and new symmetry principles have been developed in order to generalize the classical laws so as to handle these issues. For weather the key generalizations are from isotropic to anisotropic notions of scale and from smooth, quasi-Gaussian variability to strong, cascade-generated multifractal intermittency. *Lovejoy and Schertzer* [this volume, and references therein] argue that this leads to a model of atmospheric and oceanic dynamics as a system of coupled anisotropic cascade processes. They go on to indicate how the same approach can be extended to much longer (climate) time scales.

Extreme events in the Earth's near-space environment are driven by the solar wind, which brings the energetic plasma and fields from the solar eruptive events such as coronal mass ejections (CME) to geospace. Forecasting of space weather is an active field of research, and recently many new techniques and approaches have been developed. The nonlinear dynamical approach to space weather forecasting played a pioneering role by showing the predictability of space weather. Research in this area is now quite advanced and has provided techniques for dynamical and statistical forecasting [*Ukhorskiy et al.*, 2004]. Many extreme space weather events in the recent past have caused serious damage to technological systems such as satellites, power transmission, etc. [*National Research Council* (*NRC*), 2008]. Although these events may not seem devastating by themselves, a confluence of natural hazards in the different regions of the Earth's environment can make our society and its technological systems highly vulnerable because of the interconnectedness [*Baker and Allen*, 2000]. In this aspect the nonlinear dynamical framework for the study of the clustering of events, described above, becomes directly relevant to the extended Earth and space system. The studies of these extreme events in nature are summarized in the following sections.

3. EARTH SCIENCES: EARTHQUAKES AND LANDSCAPE DYNAMICS

One aspect of near-critical dynamics anticipated by *Bak et al.* [1987] and others is that earthquakes can also occur or be induced in the interior of the plates not just at the boundaries. For example, the Bhuj earthquake (M_w 7.7) of 26 January 2001 is recognized to be the deadliest among the recorded intraplate earthquakes [*Gupta et al.*, 2001], and many aspects, including its recurrence time, have been studied extensively. This earthquake has attracted further interest as an analog for the New Madrid earthquakes (M_w 7.5–8.0) that struck the central United states almost two centuries ago [*Ellis et al.*, 2001]. From the standpoint of extreme event studies the recurrence time of such earthquakes is of key interest. In this case the recurrence time is estimated to be ~1500 years. The bounds on recurrence times are naturally limited by the available data, and this estimate suffers from the lack of sufficient data. However, this also highlights the need for integrating the different types of available data with the models to develop better estimates for key features such as the recurrence times, aftershock distributions, etc. The studies of more than 500 aftershocks ($M > 2.0$) of the Bhuj earthquake using the data of 3-D velocity, gravity, magnetic,

GPS, and satellite observations have shown its features such as rupture propagation along two trends in India [*Kayal and Mukhopadhyay*, 2006]. The studies of the postseismic deformation of this earthquake [*Rastogi et al.*, this volume], focused on the changes in the seismicity of the Gujarat region in space and time, have identified the Kachchh and the Saurastra regions as more vulnerable, consistent with the observed increase in seismicity. They modeled the postseismic stress changes due to the earthquake and interpreted the deformation in these regions due to the migration of the stress pulse via viscoelastic process in lower crust and upper mantle resulting from the 20 MPa stress drop of the 2001 Bhuj earthquake.

The Koyna Dam located in western India is the most outstanding example of reservoir-triggered seismicity (RTS) where triggered earthquakes have been occurring since the impoundment in 1962 in a restricted area of 20 × 30 km^2. These include the largest RTS earthquake of M 6.3 on 10 December 1967, 22 earthquakes of $M > 5$, and about 200 $M \sim 4$ earthquakes [*Narain and Gupta*, 1968; *Gupta*, 2002]. The maximum credible earthquake estimated for the Koyna region is M 6.8, and it is reasonable to infer that the Koyna region was stressed close to critical before the relatively small perturbation to the stress field on impoundment of the reservoir. The impoundment could serve only as a trigger, leading to many small events as well as the extreme, as well as fluctuations in the subsurface pore fluid pressure. So far about 60% of the energy of an M 6.8 earthquake has been released, and the rest of the energy could be released over the next 3 to 4 decades [*Gupta et al.*, 2002]. The occurrence of $M > 5$ earthquakes is governed by factors like the rate of loading, highest water levels reached, duration of retention of the high water levels, and whether the previous water maxima has been exceeded or not (Kaiser effect). In a very exciting development a borehole to probe the physical state and mechanical behavior of the Koyna fault parameters at the hypocentral depths of 7 km is under advanced stages of planning [*Gupta et al.*, 2011]. Downhole measurements complemented by observations on cores and cuttings, analyses of fluids and gas samples, and geological and geophysical characterization studies would help answer questions related to the genesis of stable continental region earthquakes in general with a special emphasis on RTS.

If Earth's brittle lithosphere is in a state of self-organized criticality, as implied by these and other observations, that, in turn, begs the question of how the system evolved to such a marginally stable, far-from-equilibrium, stationary state in the first place. Using the information theoretic approach outlined above, *Dewar* [2003] showed that the most likely (maximum entropy) state for such systems was one where the entropy production rate was also a maximum. This is not a general conclusion (maximum entropy production is not a principle as such, though it is often referred to as one) and depends on key assumptions being met for specific systems, for example, that there exists a single, stable, attractor steady state solution. *Main and Naylor* [2008, 2010] tested the hypothesis of maximum entropy production (MEP) on synthetic earthquake populations generated by a dissipative cellular automaton model. By tuning the dissipation (entropy production), MEP occurs when the global elastic strain is near critical, with small relative fluctuations in macroscopic strain energy expressed by a low seismic efficiency and broad-bandwidth power law scaling of frequency and rupture area. These phenomena, all as observed in natural earthquake populations, are hallmarks of the broad conceptual definition of SOC, though the MEP state is near but strictly subcritical. In the MEP state the strain field retains some memory of past events, expressed as coherent "domains," implying a degree of predictability, albeit strongly limited in practice by the proximity to criticality and our inability to map the natural stress field at an equivalent resolution to the numerical model. The resulting theoretical and practical limitations have led to the debate on earthquake predictability moving on to what can be done (or not) with the data in real time, in a low-probability, high-impact forecasting environment, and what the consequences might be for operational forecasting, including quantifying the consistency, quality, and utility of any forecast and issues of communication between scientists, the relevant authorities, and the general public in an uncertain, complex, and nonlinear world [*Jordan et al.*, 2011].

4. ATMOSPHERIC AND OCEAN SCIENCES

Obviously, it is difficult to accurately define what an extreme event in atmospheric, oceanic, hydrologic, cryospheric, or others disciplinary contexts of the Earth system are. When "climate" is defined as the statistics of atmospheric, oceanic, cryospheric, and hydrologic states and "climate change" is defined as the change of these statistics, this helps, but when enlarging the time scales, other components add to this system [cf. *van Andel*, 1994]. However, the difficulty of defining extreme events is largely a consequence of the subjective nature of this usual yet vague definition of climate. When the weather and the climate are objectively defined in terms of their type of scaling variability [e.g., *Lovejoy and Schertzer*, this volume], then precise definitions of extremes are indeed possible. *Lovejoy and Schertzer* [this volume] argue that an objective scaling approach is needed to clarify the key distinction, occulted in the usual approaches, between low-frequency weather and the climate and that this is a prerequisite for objective definitions of climate states and climate change.

For the time being, we refer to extremes in the present context mostly to short-term events, which extend over a few hours, maybe a few days, but hardly more than a few months. Some of them are caused by mechanisms, which are considered mostly external to the climate system, namely, earthquakes or landslides, which may accompany tsunamis and other catastrophic events. Others are due to the internal dynamics of mainly the oceanic and atmospheric systems, such as extreme rainfall events associated with river flooding, but not all flooding is associated with extreme rainfall because of the role of river networks in aggregating flows as explained in section 5. Marine storms, such as tropical or midlatitude cyclones cause havoc not only because they are associated with rainfall but because of related storm surges and coastal inundations [*von Storch and Woth*, 2008]. Such events go along with massive societal losses, in particular with loss of life sometimes of the order of 100,000 lives and more, the last time in 2006 when tropical storm Nargis struck the coast of Myanmar [*Fritz et al.*, 2009]. Other important examples are the Sumatra-Andaman M_w 9.2 earthquake of 26 December 2004 and the resultant tsunami that claimed over 250,000 human lives in south and southeast Asian countries and caused immense financial losses [*Dimri and Srivastava*, 2007; *Gupta*, 2008; *Swaroopa Rani et al.*, 2011] and the Tohoku, Japan, M_w 9.0 earthquake of 11 March 2011 and the resultant tsunami that caused nuclear accidents [*Lay and Kanamori*, 2011]. The Japan earthquake has given rise to a global debate on the anticipated maximum size of an earthquake in a given region and the safety of nuclear power plants in the coastal region. As a matter of fact, in the first 11 years of the 21st century the number of human lives lost because of earthquakes has exceeded the total number of human lives lost in the entire 20th century because of earthquakes.

To what extent such extreme events really go along with catastrophes with societal losses depends very much on the vulnerability of the society and of the degree of adaptation. Thus, a tropical cyclone may cause much more damage when hitting the coast of the United States than when the same storm hits the coast of Cuba.

Since climate is presently changing, and will likely continue to do so as a response to ever increasing greenhouse gas concentrations in the atmosphere (also the characteristics of some extreme events are changing and will change in future), the state of knowledge has been assessed recently in the 2011 report of the *Intergovernmental Panel on Climate Change* (*IPCC*) [2011]. For the present change the *IPCC* [2011, p. 7] asserts

> There is evidence that some extremes have changed as a result of anthropogenic influences, including increases in atmospheric concentrations of greenhouse gases. It is likely that anthropogenic influences have led to warming of extreme daily minimum and maximum temperatures on the global scale. There is medium confidence that anthropogenic influences have contributed to intensification of extreme precipitation on the global scale. It is likely that there has been an anthropogenic influence on increasing extreme coastal high water due to increase in mean sea level. The uncertainties in the historical tropical cyclone records, the incomplete understanding of the physical mechanisms linking tropical cyclone metrics to climate change and the degree of tropical cyclone variability provide only low confidence for the attribution of any detectable changes in tropical cyclone activity to anthropogenic influences.

For the future, the *IPCC* [2011, pp. 10, 11, 12] finds the following changes "likely": ". . . frequency of heavy precipitation or the proportion of total rainfall from heavy falls will increase in the 21st century over many areas of the globe"; "Average tropical cyclone maximum wind speed is . . . to increase, although increases may not occur in all ocean basins. . . . the global frequency of tropical cyclones will either decrease or remain essentially unchanged"; and "It is very likely that mean sea level rise will contribute to upward trends in extreme coastal high water levels in the future."

Although the usual approach to assessing possible anthropogenic climate change is to use numerical models, this can also be done by first understanding the natural variability and then developing statistical tests to assess the probability of any observed changes occurring naturally. This approach is discussed in detail by *Lennartz and Bunde* [this volume]. By providing probability information as a function of space-time scales, scaling approaches to atmospheric variability thus provide a different, complementary path to studying anthropogenic effects. *Pielke et al.* [2009, p. 413] give evidence in support of the hypothesis that

> Although the natural causes of climate variations and changes are undoubtedly important, the human influences are significant and involve a diverse range of first-order climate forcings, including, but not limited to, the human input of carbon dioxide. Most, if not all, of these human influences on regional and global climate will continue to be of concern during the coming decades.

When dealing with the issue of real-time warning and prediction as well as the issue of determining present statistics and possible future changes of these statistics, mostly dynamical models are applied: models of the form $\Delta Y/\Delta t = \Sigma$ processes, where some processes are explicitly described by first principles such as mass conservation and others (such as boundary layer turbulence) are parameterized; that is, their net effect on the resolved scales and parameters is semiempirically closed [cf. *Müller and von Storch*, 2004]. Such models usually exhibit chaotic behavior, so that the predictability is limited often to only a few days (e.g., in case of the extratropical atmosphere) or even hours (convective rainfall). However, this chaotic feature does not inhibit the model from skillfully representing the statistics of the system properly, as well as their dependence on certain external factors; the best example represents the annual cycle, as a response to solar radiation. Thus, contemporary climate models can simulate thousands of years, exhibiting a realistic chaotic behavior and realistic long-term memory [cf. *Rybski et al.*, 2008]. (For a broader discussion about models, their specifics and their role in a societal context, refer to *von Storch et al.* [2011].) At present, there is a discussion if the low-frequency variability of global circulation models is too low or not compared with paleodata [*Lovejoy and Schertzer*, this volume, Figures 1a and 1b].

The model simulations include the formation of extreme events, for instance, mesoscale marine storms, such as polar lows in the Atlantic [*Zahn and von Storch*, 2010] or medicanes (rare hurricane-like small storms in Mediterranean Sea) [*Cavicchia and von Storch*, 2012]. Their formation is realistic in terms of numbers, across-scales link to large-scale conditions, and dynamical features; they allow the derivation of scenarios of future conditions associated with elevated greenhouse gas concentrations in the atmosphere.

On the longer time scales, there is an emerging realization that abrupt and prevalent temperature variability with significant impacts has recurred in the past when the Earth system was forced beyond threshold, although scaling approaches suggest that they could be the result of internal variability mechanisms repeating scale after scale over wide ranges (e.g., cascades and multifractals). The emerging evidence supports the recognition that Earth's climate has characteristics of a critical phenomenon and also is sensitive to small changes in solar output on the centennial time scale. Quasi-cyclic manifestations of "critical forcing parameter" like solar cycles appear in the spectra because of their imprint at the time of major changes.

The reconstructed proxy record of temperature variability decoded from tree rings provides suitable data for the study of abrupt changes in temperature. The study by *Tiwari et al.* [this volume] is based on the spectra of empirical orthogonal functions of a newly reconstructed tree ring temperature variability record decoded from the western Himalaya for the period spanning 1227–2000 A.D., addressing the frequency resolution of interdecadal and interannual oscillatory modes. The spectral analysis of first principal component (PC1) with about 61.46% variance reveals the dominance of some significant solar cycles notably peaking around 81, 32, 22, and 8–14 years. This analysis in the light of the recent ocean-atmospheric model results suggests that even small variation in solar output in conjunction with atmospheric-ocean system and other related feedback processes could have caused abrupt temperature variability at the time of "criticality" through the triggering mechanism. Identification of different natural frequency modes from a complex noisy temperature record is essential for a better understanding of the climate response to internal/external forcing.

5. HYDROLOGIC SCIENCE: FLOODS AND DROUGHTS

The *NRC* [1991] classified hydrologic science as a distinct geoscience. The importance of the global water cycle and its nonlinear interactions with the natural Earth system and the engineered Earth system was instrumental in devoting an entire chapter to nonlinearity in the *NRC* [1991] volume titled "Hydrology and Applied Mathematics." The *NRC* [2012] has revisited the challenges and opportunities. Our focus here is on hydrologic extremes that include high-frequency flood events and low-frequency drought events. Analyses of paleo-hydrologic and paleoclimate time series for understanding droughts have a long history in hydrology that dates back to the classic work of *Hurst* [1951]. Indeed, the "Hurst effect" has become quite well known in the nonlinear science literature [*Feder*, 1988], and Hurst's scaling approach has since been greatly generalized by *Mandelbrot* [1999] and others into a multifractal approach with the implications for the extremes mentioned above, including power law tails on the rain rate and river flow distributions [e.g., *Bunde et al.*, this volume]. *Mesa et al.* [this volume] give a brief overview of the pertinent literature from hydrologic science, climate science, and dynamical systems theory. Simulations from the Daisyworld model with and without the hydrologic cycle as a simple climate model reveal complex nonstationary behavior that inhibits consistent interpretation of the Hurst exponent using three well-known estimation methods. Challenging problems for future investigations are suggested that offer a broad context for nonlinear geophysics in understanding droughts.

Accurate estimates of the magnitude and frequency of floods are needed for the design of water-use and water-control projects, for floodplain definition and management, and for the design of transportation infrastructure such as bridges and roads. These practical engineering needs have led to the development of regional flood frequency analysis that *Dawdy et al.* [2012] have summarized. For example, *Fuller* [1914] analyzed peak flows from around the world, but particularly from the United States, and observed that the mean of the maximum annual flood can be related to the drainage area as a power law with an exponent of 0.8. Similar empirical relationships have since been employed to relate discharge to drainage basin characteristics.

Unfortunately, the accuracy of flood quantile estimates is constrained by the data available at a stream gauging site: record lengths are often limited to 100 years and are typically less than 30 years. To overcome these limitations, numerous statistical methods to estimate flood quantiles have been developed through the years. Furthermore, quantile estimates are often needed for ungauged sites at which no historical stream flow data are available. The challenging problem of prediction in ungauged basins (PUB) led to an international decadal initiative [*Sivapalan et al.*, 2003].

Global warming, and other human influences like large-scale deforestation in the Amazon River basin, is changing the global and regional water cycle and the climate system. Owing to strong nonlinear coupling between climate and the water cycle, future floods are not expected to be statistically similar to the past [*Milly et al.*, 2002; *Dawdy*, 2007]. As a result, estimates of flood quantiles using historic data would be biased. Can the results in the regional flood frequency analyses be understood in terms of physical mechanisms producing floods? This is a key question. A geophysical understanding of regional flood frequencies is anchored in spatial power law statistics or scaling as summarized below.

How the scaling behavior in the quantiles of annual peak discharge arises and how the slopes and intercepts can be predicted from physical processes require understanding of scaling relations in individual rainfall-runoff events. Experimental research along this line was initiated in the small 21 km^2 Goodwin Creek experimental watershed in Mississippi [*Ogden and Dawdy*, 2003]. Event scaling was a major shift in focus from the study of regional annual flood statistics that is well established in the literature. *Ogden and Dawdy* [2003] first observed scaling relations in peak discharges for 226 individual rainfall-runoff events that spanned hourly to daily time scales and found that the scaling slopes and intercepts vary from one event to another. The mean of 226 event slopes (0.82) was close to the common slope of mean annual and 20 year return peak discharges (0.77), which suggested that it should be possible to predict scaling in annual flood frequencies from event scaling.

Key results in the last 20 years have established the theoretical and observational foundations for developing a nonlinear geophysical theory or the scaling theory of floods in river basins. It has the explicit goal of linking the statistics of space-time rainfall input and the physics of flood-generating processes and self-similar branching patterns of drainage networks with spatial power law statistical relations between floods and drainage areas across multiple scales of space and time. A substantial literature in the last 30 years has developed around stochastic point process models and multifractal approaches to space-time rainfall intensities. *Gupta et al.* [2007] reviewed the developments in the scaling flood theory. Published results have shown that the spatial power law statistical relations emerge asymptotically from conservation equations of mass and momentum in self-similar (self-affine) channel networks as drainage area goes to infinity. These results have led to a key hypothesis that the physical basis of power laws in floods has its origin in the self-similarity (self-affinity) of channel networks. Self-similarity is also the basis for the widely observed fractal structure and Horton relations in river networks [*Rodriguez-Iturbe and Rinaldo*, 1997; *Gupta et al.*, 2007; *McConnell and Gupta*, 2008]. Observed power laws in floods range from hours and days of individual flood events to an annual time scale of flood frequencies. They serve as the foundation for developing a new diagnostic framework to test different assumptions governing multi-scale spatiotemporal variability in physical processes that arise in predicting power law statistical relations between floods and drainage areas [*Gupta et al.*, 2010].

Important new developments are taking place toward generalizing the scaling theory of floods to medium and large basins involving annual flood frequencies. For example, *Poveda et al.* [2007] discovered a link between mean annual runoff that is estimated from annual water balance and annual flood scaling statistics. Their study included many large river basins in Colombia that have varying hydrology and climates ranging from arid to humid. *Lima and Lall* [2010] found scaling in annual flood quantiles in large basins of Brazil (~800,000 km^2). They developed a Bayesian approach to estimate scaling parameters on inter-annual time scales. Generalizations to large basins have important relevance to flood prediction under climate change [*Milly et al.*, 2002]. Future work requires an understanding of how scaling slopes and intercepts in annual flood quantiles are modified in a changing climate, which can serve as a basis for making future flood predictions. All these developments and many others not given here have made a substantial contribution to solving the PUB problem [*Sivapalan et al.*, 2003].

6. SPACE WEATHER: SOLAR ACTIVITY, MAGNETOSPHERE, IONOSPHERE, AND SOCIETAL IMPACTS

Extreme events in space weather occur during periods when the magnetosphere is strongly driven by the solar wind, bringing energetic plasma and fields from the solar eruptive events such as coronal mass ejections to geospace. Many extreme space weather events have caused serious damage to technological systems such as satellites, power transmission systems, etc. Some well-known examples are the collapse of the Hydro Quebec power grid during the great geomagnetic storm of March 1989, the Canadian telecommunication satellite outage during a period of enhanced energetic electron fluxes at geosynchronous orbit in January 1994, the electrical breakdowns and satellite malfunctions during the magnetic cloud event of July 2000 (Bastille Day event), the disabling of

GPS-based aviation system during the severe space weather events of October–November 2003 (Halloween storms), the disturbances in commercial airline traffic during several days of enhanced geomagnetic activity in January 2005, etc. [*NRC*, 2008]. Although these events may not seem devastating by themselves, a confluence of natural hazards in the different regions of the environment of the Earth can make our society and its technological systems highly vulnerable because of their interconnectedness [*Baker and Allen*, 2000; *NRC*, 2008]. In this aspect the nonlinear dynamical approach, with its ability to integrate many interacting components, becomes a natural framework for the study of the extreme events in the extended Earth and space system.

The modeling of space weather events relies strongly on the availability of good geospace data, and among the most widely used data are the geomagnetic indices. Geomagnetic field data from ground magnetometer stations around the globe have been recorded for more than one and half centuries, and these data have been used to compute the geomagnetic indices. Among the many indices the auroral electrojet indices (*AE*, *AL*, and *AU*) characterize the substorms, and the ring current index *Dst* represents the geomagnetic or space storms. The substorms, with a characteristic time of ~1 hour, are episodic in nature and are the essential elements of magnetospheric dynamics. The auroral electrojet indices provide the detailed dynamical features of the global aspects of substorms. On the other hand, the geomagnetic storms, with a typical time scale of ~10 hours, are the more global space weather disturbances during which intense substorms occur. The auroral electrojet indices computed from the horizontal component of the magnetic field disturbances reflect the strengths of the large-scale ionospheric currents driven by the reconfiguration of the magnetosphere during substorms. They are highly variable during strongly disturbed periods, with peak values of 1000–2000 nT during extreme events cited earlier.

The substorms with *AL* index values less than −1000 nT are considered strong disturbances and can be characterized as extreme events. The geomagnetic storms with *Dst* values less than −100 nT are referred to as intense storms. The substorms occur during the storms, and their episodic nature and high variability are evident in the sharp peaks of the electrojet indices. As is the case with extreme events in general, there is no single measure of the extreme events in space weather. For example, the *Dst* index for the well-known "Carrington" storm of 1–2 September 1859 is estimated to have reached a value of −1760 nT [*Tsurutani et al.*, 2003], and its effects were felt across the globe. The more recent Bastille Day event of 14–16 July 2000 with a *Dst* minimum of −300 nT was an extreme space weather event that led to significant damage to satellites and other technological infrastructure. It should be emphasized here that the main objective in the studies of extreme events is the nature of their distribution.

From observed solar-terrestrial dynamics and detailed studies of the complexity of the magnetospheric system, many studies (see reviews of *Sharma* [1995], *Klimas et al.* [1996], *Vassiliadis* [2006], and *Baker* [2012, this volume]) have shown evidence that the solar wind–magnetosphere-ionosphere realm clearly is characterized by nonlinear dynamics. It is argued that methods borrowed from other branches of physics, chemistry, and engineering can offer useful, if imperfect, analogies on how to deal with nonlinearity in the Sun-Earth system. Ultimately, space weather prediction probably must depart from idealized local "stability" analyses of various plasma domains to consider true global system responses and must incorporate realistic, nonlinear aspects. Forecasting of space weather using data-based nonlinear dynamical models of solar wind–magnetosphere coupling [*Vassiliadis et al.*, 1995; *Valdivia et al.*, 1996; *Ukhorskiy et al.*, 2004; *Chen and Sharma*, 2006] would be useful not only at high latitudes where geomagnetically induced currents pose a risk to power transmission grids but also at equatorial and low latitudes where the development of irregular structure in the ionosphere can cause degradation and even disruption in the operation of satellite-based communication and navigation systems such as the GPS. A disturbance dynamo set up by enhanced currents in the auroral ionosphere during magnetic substorms can after a few hours give rise to conditions in the equatorial ionosphere that are conducive for the development of such irregular structure [*Bhattacharyya*, this volume].

In the light of such understandings, there is strong motivation to go away from the usual plasma physics techniques to consider more aggressive and comprehensive approaches. It may be that the magnetotail has a distributed set of interacting control agents, some of which are local plasma conditions and some of which are remote in the ionosphere or solar wind. Thus, it may appear that instabilities are spontaneous or seemingly random. We need to use new analysis tools to address such issues. Incorporating both traditional forecasting methods for solar storms and also using the array of methods embodied in the most modern tools of nonlinear dynamics offers the best hope for dealing with extreme space weather events [*Sharma and Veeramani*, 2011; *Baker*, this volume].

The vision of a space weather forecasting plan begins with understanding what is, in effect, space "climate." It is known that the Sun undergoes an approximately 11 year cycle of increasing and then decreasing sunspot activity. Some solar cycles have been relatively weak, while others in the historical record have been very strong [*Baker*, this volume]. The stronger maxima tend to have more episodes of strong solar flares, intense outbursts of ultraviolet light, and (perhaps) greater chances of solar particle "superstorms." Conversely, weaker maxima may exhibit fewer of these extreme characteristics. However, even these general patterns of behavior are not absolute. Some of the largest historical events, for example, occurred during a relatively weak sunspot maximum period [see *Baker*, this volume]. On even shorter temporal scales it is known that forecasting space weather events will require continuous and effective observations of the Sun and its corona. Being able to observe the initiation of a powerful solar flare and detecting an earthward bound CME at its early stages gives society the best chance of predicting a powerful geomagnetic storm. Such CME observations can provide perhaps 18 hour (or longer) alerts that a solar-induced geomagnetic storm is imminent.

By using models developed in the research community and driving such models with space-based and ground-based solar observations, the natural hazard community has moved an important step closer to providing tens of hours warning of impending geomagnetic storms. Both ground-based and space-based observations of the Sun can be fed into highly capable models to forecast solar wind conditions near Earth. In the magnetosphere-ionosphere system, models of the complete, coupled geospace system can be used to provide forecasts ranging from tens of minutes to several days. The present hazard response plan still has a long way to go to achieve the levels of accuracy and reliability that are desired by the U.S. industrial, military, and commercial sectors. Moreover, knowing with accuracy the near-Earth solar and solar wind drivers of space weather still demands further understanding of the nonlinear responses of the geospace system that are described here [*Baker*, this volume]. The complex feedbacks within the solar wind–magnetosphere-ionosphere system may ultimately limit our ability to predict the size, duration, and precise location of magnetospheric current systems that can wreak havoc on human technologies, but we must understand these better to avert severe consequences for humans.

7. OVERARCHING ISSUES: INTERDISCIPLINARY APPROACH, DATA, AND LARGE-SCALE SIMULATIONS

As we have seen above, at least three research threads have all made significant contributions to fundamental complexity science and to the geosciences over the last 40 or more years. One thread has been the use of low-dimensional deterministic models, particularly chaotic ones, based on ordinary differential equations or maps [*Lorenz*, 1963; *May*, 1976]. This has been accompanied by data analysis methods such as the recovery of attractors by embedding and has led to many applications [*Abarbanel et al.*, 1993; *Kantz and Schrieber*, 1999]. In space weather this approach has provided the earliest predictive models and is now the basis for near-real-time forecasting. The irregular behavior of geomagnetic activity, however, cannot be described entirely by complexity in low-dimensional dynamics, and there are strong multiscale features that exhibit long-range behavior.

A second thread, described as stochastic chaos above, has involved the use of additive stochastic models that can exhibit heavy tails and/or long-range dependence. Described as H self-similar by applied mathematicians, these models include Mandelbrot's Lévy (alpha stable) motion for heavy tails, his fractional Brownian motion for long-range dependence, and his subsequent fractional hyperbolic model of 1969 that combined both effects. Here H is a single self-similarity exponent, which is the same as the more familiar Hurst exponent in the absence of heavy tails [e.g., *Mercik et al.*, 2003]. Related models like autoregressive fractionally integrated moving average, which are not completely self-similar but do show long-range dependence, have been extensively studied in applied statistics [e.g., *Beran*, 1994]. A third thread has developed from the recognition that the "wild" variability seen in some natural systems, particularly features such as nonlinear correlation in amplitudes that gives rise to observed "volatility bunching," can be modeled by multiplicative cascade multifractal models rather than additive monofractal ones. These generalize the single scaling exponent to a continuous spectrum of exponents.

Clearly, exactly what sort of extremes a given natural system exhibits may reflect the extent to which the above descriptions apply to it, as, for example, discussed in the contribution by *Watkins et al.* [this volume], where the implications of monofractal and multifractal models of integrated extreme "bursts" above a threshold are compared. Focusing on the burst problem, we note that several interdisciplinary stochastic approaches exist that provide quantitative information about extreme bursts. Some are more focused on the probability of single large events, while others are more concerned with extended dwell times above a given spatiotemporal threshold. As mentioned above, a statistical extreme value theory is available. In addition, there is a mathematical theory of record breaking [*Nevzorov*, 2001]; while image processing and fluid dynamics contribute the idea of level sets, stochastic processes offer sojourn times [*Berman*, 1992]; and as noted above, complexity science has contributed models of "avalanches" in the space-time activity of nonequilibrium systems, where the term "avalanche" refers here to a cascading, spatiotemporal, coherent energy release event. It can be seen that even without historical and disciplinary factors, it would be completely natural for knowledge relevant to a given burst problem to be siloed and so not yet be "joined up." In practice, the choice of approach has also differed between and within application domains, sometimes reflecting the data's own limitations and also reflecting the differing cultures and assumptions of theoretical physics, complexity science, the geosciences, applied stochastic modeling, and statistical inference. Although such knowledge is not yet integrated, particularly for Earth and sciences applications, nonetheless, significant efforts have started along this direction.

The rise of complexity science in geophysics has given currency to the expression "emergent" for higher-level laws, i.e., properties that cannot be predicted easily from the elementary interactions. For example, processes which appear complicated at one scale often produce emergent laws in a suitable limit at a higher scale that are relatively simple. A great deal of complexity science has so far been precisely the elaboration of various nonlinear models or, even paradigms, which allow one to reduce "complexity" to "simplicity" in this way. In many, even most, geophysical cases, these nonlinear models capture a good deal of the phenomenology and physics of processes that had hitherto been impervious to realistic modeling. It is particularly true for cases where ubiquitous assumptions of homogeneity and linearity had reigned. For example, the prototypical example of self-organized criticality, the classical sandpile model, has been widely applied across the geosciences including studies of the atmosphere where cascades can generate a "nonclassical" SOC. By generically describing some phenomena that are so extreme that even their mean behaviors are determined by avalanche-like events, it is frequently the only type of model that is even remotely realistic, e.g., earthquakes and forest fires.

The space-time variability of physical processes on multiple scales is common to many disciplines of geosciences. But this feature presents a huge scientific challenge in developing physically based models. Let us briefly explain this cross-cutting and overarching issue. The first challenge is suitable parameterizations of processes at "appropriate scales" that are required in specifying the coupled conservation equations of mass, momentum, energy, etc. Typically, observations are not available at the scales of interest, which require statistical parameterizations. For example, the coupled conservation equations underlying the scaling theory of floods [*Gupta et al.*, 2010] need how much rainfall becomes runoff involving surface and subsurface processes in each hillslope (~0.05 km^2) of a river basin. But each basin has a large number of hillslopes. For example, a basin of area A has 20A hillslopes, so a 100 km^2 "small basin" has about 2000 hillslopes. There is no known measurement technology that can be used to measure runoff in each hillslope even for a single experimental basin in the world, because runoff varies both in space and time among hillslopes. Furthermore, no known theory is available that can be used to parameterize runoff from hillslopes, so new theories are needed to solve this problem. As new and diverse approaches are developed from different fields of geosciences, a major issue for future conferences and workshops would be to identify if parameterization models have some features in common that can apply across disciplines.

In taking stock, it is clear that complexity science has been most effective in quantifying the dynamical features and scaling laws of spatiotemporal data of various types and in providing an explanation for such data in a unified framework, sometimes when there is no competing theory. It has helped us solve the puzzle of why such scaling laws in a population of events can be robust, while at the same time leading in many cases to pessimism (some would say realism) on the potential for the predictability of the timing of individual extreme events. Principles of self-organization have been shown to have a thermodynamic explanation in terms of entropy production (minimum for a near-equilibrium, linear system, leading to Euclidean geometry, or maximum for a far-from-equilibrium nonlinear one, leading to fractal geometry), though this may not be universal for all systems and is a subject of much debate. Scaling laws are extremely useful in quantifying hazard and risk in an operational sense, but much needs to be done to quantify the uncertainties in forecasting future risk based on a limited sample of past extreme events [*Main et al.*, 2008].

8. CONCLUSION

The long-term dependencies in geophysical systems and their implications for catastrophic extreme events have direct applications to other areas such as life sciences. The elucidation of long-range (scaling) dependencies can be used in forecasting extreme events [*Bunde et al.*, this volume]. Indeed, over the last 30 years, work on general scaling processes has shown that the extremes are precisely the consequence of variability that builds up scale after scale over huge ranges.

An important aspect of extreme events research is the extensive demands for computational resources. Enhanced computational resources have led to investigations of complexity in a wide range of Earth and space systems, and this cyber-enabled discovery will lead to more complex but reliable forward models. The ensemble of model results can then yield statistical properties that reflect the uncertainties and their propagation, providing the scaling behaviors with respect to the external forcing parameters. These statistical approaches are essential for characterizing extreme events accurately or their uncertainties realistically if this is not possible.

The different geoscience disciplines have many important problems that need to be studied using approaches beyond those commonly used within the specific discipline. The underlying features of many of these problems arise from the nonlinear nature, and complexity science provides a framework for studying the seemingly unconnected phenomena. This has been evident from the progress made in the Earth and space sciences based on the understanding of nonlinear dynamics, chaos, fractals, multifractals, scaling, etc.

Currently, the first principle or theoretical models are limited in providing predictable models of extreme events. Among the many reasons for this is the disparate space and time scales associated with extreme events. A promising approach is the data-driven modeling in which the inherent features of the many phenomena contributing to an extreme event can be used together, independent of a priori assumptions, to build an integrative model. We are also entering a new age of data-intensive research, where key information can be extracted and new objects and patterns identified from very large, continuously recorded data sets, a process known as "data assimilation" [*Hey et al.*, 2009]. It remains to be seen whether predictive power can be improved by such models or such improved short-term databases and methods of assimilation, but this can only be done by applying rigorous tests that reflect the epistemic (model dependent) and aleatoric (statistical) uncertainties involved.

Acknowledgments. The Chapman Conference was hosted by CSIR-National Geophysical Research Institute, Hyderabad, and supported by the NSF under grant AGS-1036473.

REFERENCES

Abarbanel, H. D. I., R. Brown, J. J. Sidorowich, and L. S. Tsimiring (1993), The analysis of observed chaotic data in physical systems, *Rev. Mod. Phys.*, *65*, 1331–1392.

Bak, P., and K. Chen (1989), The physics of fractals, *Physica D*, *38*, 5–12.

Bak, P., C. Tang, and K. Wiesenfeld (1987), Self-organized criticality: An explanation of $1/f$ noise, *Phys. Rev. Lett.*, *59*, 381–384.

Baker, D. N. (2012), Extreme space weather: Forecasting behavior of a nonlinear dynamical system, in *Complexity and Extreme Events in Geoscience*, *Geophys. Monogr. Ser.*, doi:10.1029/2011GM001075, this volume.

Baker, D. N., and J. H. Allen (2000), Confluence of natural hazards: A possible scenario, *Eos Trans. AGU*, *81*(23), 254.

Beran, J. (1994), *Statistics for Long-Memory Processes*, Chapman and Hall, New York.

Berman, S. (1992), *Sojourns and Extremes of Stochastic Processes*, Chapman and Hall, New York.

Bhattacharyya, A. (2012), Development of intermediate-scale structure in the nighttime equatorial ionosphere, in *Complexity and Extreme Events in Geoscience*, *Geophys. Monogr. Ser.*, doi:10.1029/2011GM001078, this volume.

Bogachev, M. I., J. F. Eichner, and A. Bunde (2007), Effect of nonlinear correlations on the statistics of return intervals in multifractal data sets, *Phys. Rev. Lett.*, *99*, 240601, doi:10.1103/PhysRevLett.99.240601.

Bunde, A., J. Kropp, and H. J. Schellnhuber (2002), *The Science of Disasters*, Springer, Berlin.

Bunde, A., J. F. Eichner, J. W. Kantelhardt, and S. Havlin (2005), Long-term memory: A natural mechanism for the clustering of extreme events and anomalous residual times in climate records, *Phys. Rev. Lett.*, *94*, 048701, doi:10.1103/PhysRevLett.94.048701.

Bunde, A., M. I. Bogachev, and S. Lennartz (2012), Precipitation and river flow: Long-term memory and predictability of extreme events, in *Complexity and Extreme Events in Geoscience*, *Geophys. Monogr. Ser.*, doi:10.1029/2011GM001112, this volume.

Cavicchia, L., and H. von Storch (2012), The simulation of medicanes in a high-resolution regional climate model, *Clim. Dyn.*, doi:10.1007/s00382-011-1220-0, in press.

Chen, J., and A. S. Sharma (2006), Modeling and prediction of the magnetospheric dynamics during intense geospace storms, *J. Geophys. Res.*, *111*, A04209, doi:10.1029/2005JA011359.

Coles, S. (2001), *An Introduction to Statistical Modeling of Extreme Values*, Springer, New York.

Dawdy, D. (2007), Prediction versus understanding, *J. Hydrol. Eng.*, *12*(1), 1–3.

Dawdy, D. R., V. W. Griffis, and V. K. Gupta (2012), Regional flood frequency analysis: How we got here and where we are going, *J. Hydrol. Eng.*, in press.

Dewar, R. (2003), Information theory explanation of the fluctuation theorem, maximum entropy production and self-organized criticality in non-equilibrium stationary states, *J. Phys. A Math. Gen.*, *36*, 631–641, doi:10.1088/0305-4470/36/3/303.

Dimri, V. P. (1998), Fractal behavior and detectibility limits of geophysical surveys, *Geophysics*, *63*, 1943–1947.

Dimri, V. P., and K. Srivastava (2007), Tsunami propagation of the 2004 Sumatra earthquake and the fractal analysis of the aftershock activity, *Indian J. Mar. Sci.*, *36*(2), 128–135.

Ellis, M., J. Gomberg, and E. Schweig (2001), Indian earthquake may serve as analog for New Madrid earthquakes, *Eos Trans. AGU*, *82*(32), 345.

Embrechts, P. M., C. Kluppelberg, and T. Mikosch (1997), *Modelling Extremal Events for Insurance and Finance*, Springer, Berlin.

Feder, J. (1988), *Fractals*, Plenum, New York.

Fisher, R. A., and L. H. C. Tippett (1928), Limiting forms of the frequency distribution of the largest or smallest member of a sample, *Proc. Cambridge Philos. Soc.*, *24*, 180–190.

Fritz, H. M., C. D. Blount, S. Thwin, M. K. Thu, and N. Chan (2009), Cyclone Nargis storm surge in Myanmar, *Nat. Geosci.*, *2*, 448–449, doi:10.1038/ngeo558.

Fuller, W. E. (1914), Flood flows, *Trans. Am. Soc. Civ. Eng.*, *77*, 567–617.

Govindan, R. B., D. Vjushin, A. Bunde, S. Brenner, S. Havlin, and H.-J. Schellnhuber (2002), Global climate models violate scaling of the observed atmospheric variability, *Phys. Rev. Lett.*, *89*, 028501, doi:10.1103/PhysRevLett.89.028501.

Gupta, H. K. (2002), A review of recent studies of triggered earthquakes by artificial water reservoirs with special emphasis on earthquakes in Koyna, India, *Earth Sci. Rev.*, *58*, 279–310.

Gupta, H. K. (2008), India's initiative in mitigating tsunami and storm surge hazard, *J. Earthquake Tsunami*, *2*(4), 287–295.

Gupta, H. K., N. P. Rao, B. K. Rastogi, and D. Sarkar (2001), The deadliest intraplate earthquake, *Science*, *291*, 2101–2102.

Gupta, H. K., P. Mandal, and B. K. Rastogi (2002), How long will triggered earthquakes at Koyna, India, continue?, *Curr. Sci.*, *82*, 202–210.

Gupta, H. K., S. Nayak, and Y. J. Bhaskar Rao (2011), Planning a deep drilling project in the Koyna region of India, *Eos Trans. AGU*, *92*(34), 283.

Gupta, V. K., B. Troutman, and D. R. Dawdy (2007), Towards a nonlinear geophysical theory of floods in river networks: An overview of 20 years of progress, in *Nonlinear Dynamics in Geosciences*, edited by A. A. Tsonis and J. B. Elsner, chap. 8, pp.121–151, Springer, New York.

Gupta, V. K., R. Mantilla, B. M. Troutman, D. Dawdy, and W. F. Krajewski (2010), Generalizing a nonlinear geophysical flood theory to medium-sized river networks, *Geophys. Res. Lett.*, *37*, L11402, doi:10.1029/2009GL041540.

Hey, A. J. G., S. Tansley, and K. Tolle (Eds.) (2009), *The Fourth Paradigm: Data-Intensive Scientific Discovery*, 286 pp., Microsoft Res., Redmond, Wash.

Hurst, H. (1951), Long term storage capacity of reservoirs, *Trans. Am. Soc. Civ. Eng.*, *116*, 776–808.

Intergovernmental Panel on Climate Change (IPCC) (2011), Summary for policymakers, in *Intergovernmental Panel on Climate Change Special Report on Managing the Risks of Extreme Events and Disasters to Advance Climate Change Adaptation*, edited by C. B. Field et al., pp. 1–19, Cambridge Univ. Press, New York.

Jaynes, E. T. (1957), Information theory and statistical mechanics, *Phys. Rev.*, *106*, 620–630.

Jentsch, V., H. Kantz, and S. Albeverio (2006), Extreme events: Magic, mysteries, and challenges, in *Extreme Events in Nature and Society*, edited by S. Albeverio, V. Jentsch, and H. Kantz, pp. 1–18, Springer, Berlin.

Jordan, T., Y. Chen, P. Gasparini, R. Madariaga, I. Main, W. Marzocchi, G. Papadopoulos, G. Sobolev, K. Yamaoka, and J. Zschau (2011), Operational earthquake forecasting: State of knowledge and guidelines for utilization, *Ann. Geophys.*, *54*(4), 361–391, doi:10.4401/ag-5350.

Kagan, Y. Y. (2006), Why does theoretical physics fail to explain and predict earthquake occurrence?, in *Modeling Critical and Catastrophic Phenomena in Geoscience: A Statistical Approach*, edited by P. Bhattacharyya and B. K. Chakrabarti, *Lect. Notes Phys.*, *705*, 303–359.

Kagan, Y. Y., and L. Knopoff (1980) Spatial distribution of earthquakes: The two-point correlation function, *Geophys. J. R. Astron. Soc.*, *62*(2), 303–320, doi:10.1111/j.1365-246X.1980.tb04857.x.

Kantz, H., and T. Schrieber (1999), *Nonlinear Time Series Analysis*, Cambridge Univ. Press, Cambridge, U. K.

Kayal, J. R., and S. Mukhopadhyay (2006), Seismotectonics of the 2001 Bhuj earthquake (*Mw* 7.7) in western India: Constraints from aftershocks, *J. Ind. Geophys. Union*, *10*, 45–57.

Klimas, A. J., D. Vassiliadis, D. N. Baker, and D. A. Roberts (1996), The organized nonlinear dynamics of the magnetosphere, *J. Geophys. Res.*, *101*(A6), 13,089–13,113.

Lay, T., and H. Kanamori (2011), Insights from the great 2011 Japan earthquake, *Phys. Today*, *64*, 33–39, http://dx.doi.org/10.1063/PT.3.1361.

Lennartz, S., and A. Bunde (2012), On the estimation of natural and anthropogenic trends in climate records, in *Complexity and Extreme Events in Geoscience*, *Geophys. Monogr. Ser.*, doi:10.1029/2011GM001079, this volume.

Lima, C. H. R., and U. Lall (2010), Spatial scaling in a changing climate: A hierarchical Bayesian model for nonstationary multisite annual maximum and monthly streamflow, *J. Hydrol.*, *383* (3–4), 307–318, doi:10.1016/j.jhydrol.

Lorenz, E. N. (1963), Deterministic nonperiodic flow, *J. Atmos. Sci.*, *20*, 130–141.

Lovejoy, S., and D. Schertzer (2012), Low-frequency weather and the emergence of the climate, in *Complexity and Extreme Events in Geoscience*, *Geophys. Monogr. Ser.*, doi:10.1029/2011GM001087, this volume.

Lovejoy, S., et al. (2009), Nonlinear geophysics: Why we need it, *Eos Trans. AGU*, *90*(48), 455.

Lubchenco, J., and T. R. Karl (2012), Predicting and managing extreme weather events, *Phys. Today*, *65*, 31–36, http://dx.doi.org/10.1063/PT.3.1475.

Main, I. G. (1995), Earthquakes as critical phenomena: Implications for probabilistic seismic hazard analysis, *Bull. Seismol. Soc. Am.*, *85*, 1299–1308.

Main, I. (1996), Statistical physics, seismogenesis, and seismic hazard, *Rev. Geophys.*, *34*, 433–462.

Main, I. G., and P. W. Burton (1984), Information theory and the earthquake frequency-magnitude distribution, *Bull. Seismol. Soc. Am.*, *74*, 1409–1426.

Main, I. G., and M. Naylor (2008), Maximum entropy production and earthquake dynamics, *Geophys. Res. Lett.*, *35*, L19311, doi:10.1029/2008GL035590.

Main, I. G., and M. Naylor (2010), Entropy production and self-organized (sub) criticality in earthquake dynamics, *Philos. Trans. R. Soc. A*, *368*, 131–144, doi:10.1098/rsta.2009.0206.

Main, I. G., L. Li, J. McCloskey, and M. Naylor (2008), Effect of the Sumatran mega-earthquake on the global magnitude cut-off and event rate, *Nat. Geosci.*, *1*, 142, doi:10.1038/ngeo141.

Mandelbrot, B. B. (1967), How long is the coast of Britain? Statistical self-similarity and fractional dimension, *Science*, *155*, 636–638.

Mandelbrot, B. B. (1999), *Multifractals and 1/f Noise: Wild Self-Affinity in Physics (1963-1976): Selecta Volume N*, Springer, New York.

Manneville, P. (2010), *Instabilities, Chaos and Turbulence*, 2nd ed., Imperial College Press, London.

Mantegna, R. N., and H. E. Stanley (2000), *Econophysics*, Cambridge Univ. Press, Cambridge, U. K.

Martyushev, L. M., and V. D. Seleznev (2006), Maximum entropy production principle in physics, chemistry and biology, *Phys. Rep.*, *426*, 1–45.

Maus, S., and V. P. Dimri (1994), Scaling properties of potential fields due to scaling sources, *Geophys. Res. Lett.*, *21*, 891–894.

May, R. M. (1976), Simple mathematical models with very complicated dynamics, *Nature*, *261*, 459–467.

McConnell, M., and V. Gupta (2008), A proof of the Horton law of stream numbers for the Tokunaga model of river networks, *Fractals*, *16*(3), 227–233.

Mercik, S., K. Weron, K. Burnecki, and A. Weron (2003), Enigma of self-similarity of fractional Levy stable motions, *Acta Phys. Pol. B.*, *34*(7), 3773–3791.

Mesa, O. J., V. K. Gupta, and P. E. O'Connell (2012), Dynamical system exploration of the Hurst phenomenon in simple climate models, in *Complexity and Extreme Events in Geoscience*, *Geophys. Monogr. Ser.*, doi:10.1029/2011GM001081, this volume.

Milly, P. C. D., R. T. Wetherald, K. A. Dunne, and T. L. Delworth (2002), Increasing risk of great floods in a changing climate, *Nature*, *415*, 514–517.

Moffat, H. K., and E. Shuckburgh (Eds.) (2011), *Environmental Hazards: The Fluid Dynamics and Geophysics of Extreme Events*, World Sci., Singapore.

Müller, P., and H. von Storch (2004), *Computer Modelling in Atmospheric and OceanicSciences: Building Knowledge*, 304 pp., Springer, New York.

Narain, H., and H. Gupta (1968), Koyna earthquake, *Nature*, *217*, 1138–1139.

National Research Council (NRC) (1991), *Opportunities in the Hydrologic Sciences*, Natl. Acad. Press, Washington, D. C.

National Research Council (NRC) (2008), *Severe Space Weather Events: Understanding Societal and Economic Impacts: A Workshop Report*, Natl. Acad. Press, Washington, D. C.

National Research Council (NRC (2012), *Challenges and Opportunities in the Hydrologic Sciences*, Natl. Acad. Press, Washington, D. C.

Nevzorov, V. B. (2001), *Records: Mathematical Theory*, Am. Math. Soc., Providence, R. I.

Ogden, F. L., and D. R. Dawdy (2003), Peak discharge scaling in a small Hortonian watershed, *J. Hydrol. Eng.*, *8*(2), 64–73.

Paltridge, G. W. (2005), Stumbling into the MEP racket: An historical perspective, in *Non-equilibrium Thermodynamics and the Production of Entropy: Life, Earth, and Beyond*, edited by A. Kleidon and R. D. Lorenz, chap. 3, pp. 33–40, Springer, New York.

Pielke, R., et al. (2009), Climate change: The need to consider human forcings besides greenhouse gases, *Eos Trans. AGU*, *90* (45), 413, doi:10.1029/2009EO450008.

Poveda, G., et al. (2007), Linking long-term water balances and statistical scaling to estimate river flows along the drainage network of Colombia, *J. Hydrol. Eng.*, *12*, 4–13.

Prigogine, I. (1945), Modération et transformations irreversibles des systemes ouverts, *Bull. Cl. Sci. Acad. R. Belg.*, *31*, 600–606.

Rastogi, B. K., P. Choudhury, R. Dumka, K. M. Sreejith, and T. J. Majumdar (2012), Stress pulse migration by viscoelastic process for long-distance delayed triggering of shocks in Gujarat, India, after the 2001 M_w 7.7 Bhuj earthquake, in *Complexity and Extreme Events in Geoscience*, *Geophys. Monogr. Ser.*, doi:10.1029/2011GM001061, this volume.

Rodriguez-Iturbe, I., and A. Rinaldo (1997), *Fractal River Basins*, Cambridge Univ. Press, Cambridge, U. K.

Rybski, D., A. Bunde, and H. von Storch (2008), Long-term memory in 1000-year simulated temperature records, *J. Geophys. Res.*, *113*, D02106, doi:10.1029/2007JD008568.

Shannon, C. E. (1948), A mathematical theory of communication, *Bell Syst. Tech. J.*, *27*, 379–423, 623–656.

Sharma, A. S. (1995), Assessing the magnetosphere's nonlinear behavior: Its dimension is low, its predictability high, *U.S. Natl. Rep. Int. Union Geod. Geophys. 1991–1994*, *Rev. Geophys.*, *33*, 645–650.

Sharma, A. S., and T. Veeramani (2011), Extreme events and long-range correlations in space weather, *Nonlinear Processes Geophys.*, *18*, 719–725, doi:10.5194/npg-18-719-2011.

Sharma, A. S., D. N. Baker, V. P. Dimri, and A. Bunde (2010), Complexity and extreme events: Interdisciplinary science of natural hazards, *Eos Trans. AGU*, *91*(30), 265.

Sivapalan, M., et al. (2003), IAHS decade on predictions in ungauged basins (PUB), 2003-2012: Shaping an exciting future for the hydrologic sciences, *Hydrol. Sci. J.*, *48*, 857–880.

Srivastava, R. P., N. Vedanti, and V. P. Dimri (2007), Optimum design of a gravity survey network and its application to delineate the Jabera-Damoh structure in the Vindhyan Basin, central India, *Pure Appl. Geophys.*, *164*, 1–14.

Sunmonu, L. A., and V. P. Dimri (2000), Fractal geometry of faults and seismicity of Koyna-Warna region west India using Landsat images, *Pure Appl. Geophys.*, *157*, 1393–1405.

Swaroopa Rani, V., K. Srivastava, and V. P. Dimri (2011), Tsunami propagation and inundation due to tsunamigenic earthquakes in the Sumatra-Andaman subduction zone: Impact at Vishakhapatnam, *Mar. Geod.*, *34*, 48–58, doi:10.1080/01490411.2011.547802.

Tiwari, R. K., R. R. Yadav, and K. P. C. Kaladhar Rao (2012), Empirical orthogonal function spectra of extreme temperature variability decoded from tree rings of the western Himalayas, in *Complexity and Extreme Events in Geoscience*, *Geophys. Monogr. Ser.*, doi:10.1029/2011GM001133, this volume.

Tsallis, C. (1995) Some comments on Boltzmann-Gibbs statistical mechanics, *Chaos Solitons Fractals*, *6*, 539–559.

Tsonis, A. A. (2012), Climate subsystems: Pacemakers of decadal climate variability, in *Complexity and Extreme Events in Geoscience*, *Geophys. Monogr. Ser.*, doi:10.1029/2011GM001053, this volume.

Tsurutani, B. T., W. D. Gonzalez, G. S. Lakhina, and S. Alex (2003), The extreme magnetic storm of 1–2 September 1859, *J. Geophys. Res.*, *108*(A7), 1268, doi:10.1029/2002JA009504.

Ukhorskiy, A. Y., M. I. Sitnov, A. S. Sharma, and K. Papadopoulos (2004), Global and multi-scale features of solar wind-magnetosphere coupling: From modeling to forecasting, *Geophys. Res. Lett.*, *31*, L08802, doi:10.1029/2003GL018932.

Valdivia, J. A., A. S. Sharma, and K. Papadopoulos (1996), Prediction of magnetic storms by nonlinear models, *Geophys. Res. Lett.*, *23*, 2899–2902.

van Andel, T. H. (1994), *New Views on an Old Planet: A History of Global Change*, 2nd ed., 439 pp., Cambridge Univ. Press, Cambridge, U. K.

Vassiliadis, D. (2006), Systems theory for geospace plasma dynamics, *Rev. Geophys.*, *44*, RG2002, doi:10.1029/2004RG000161.

Vassiliadis, D., A. J. Klimas, D. N. Baker, and D. A. Roberts (1995), A description of solar wind-magnetosphere coupling based on nonlinear filters, *J. Geophys. Res.*, *100*(A3), 3495–3512.

Vlastos, S., E. Liu, I. G. Main, and C. Narteau (2007), Numerical simulation of wave propagation in 2-D fractured media: Scattering attenuation at different stages of the growth of a fracture population, *Geophys. J. Int.*, *171*, 865–880.

von Storch, H., and K. Woth (2008), Storm surges, perspectives and options, *Sustainability Sci.*, *3*, 33–44, doi:10.1007/s11625-008-0044-2.

von Storch, H., A. Bunde, and N. Stehr (2011), The physical sciences and climate politics, in *The Oxford Handbook of Climate Change and Society*, edited by J. S. Dyzek, D. Schlosberg, and R. B. Norgaard, pp. 113–128, Oxford Univ. Press, Oxford, U. K.

Vyushin, D., I. Zhidkov, S. Havlin, A. Bunde, and S. Brenner (2004), Volcanic forcing improves Atmosphere-Ocean Coupled General Circulation Model scaling performance, *Geophys. Res. Lett.*, *31*, L10206, doi:10.1029/2004GL019499.

Watkins, N., S. Rosenberg, S. Chapman, M. Naylor, and M. Freeman (2011), When black swans come in bunches: Modelling the impact of temporal correlations on the return periods of heavy tailed risk, *Geophys. Res. Abstr.*, *13*, EGU2011-12404.

Watkins, N. W., B. Hnat, and S. C. Chapman (2012), On self-similar and multifractal models for the scaling of extreme bursty fluctuations in space plasmas, in *Complexity and Extreme Events in Geoscience*, *Geophys. Monogr. Ser.*, doi:10.1029/2011GM001084, this volume.

Zahn, M., and H. von Storch (2010), Decreased frequency of North Atlantic polar lows associated with future climate warming, *Nature*, *467*, 309–312.

D. N. Baker, Laboratory for Atmospheric and Space Physics, University of Colorado, Boulder, CO 80309, USA.

A. Bhattacharyya, Indian Institute of Geomagnetism, Navi Mumbai 410 218, India.

A. Bunde, Institut für Theoretische Physik III, Justus-Liebig-Universität Giessen, Giessen D-35390, Germany.

V. P. Dimri and H. K. Gupta, National Geophysical Research Institute, Hyderabad 500606, India.

V. K. Gupta, Department of Civil, Environmental and Architectural Engineering, Boulder, CO 80309, USA.

S. Lovejoy, Department of Physics, McGill University, Montreal, QC H3A 2T8, Canada.

I. G. Main, School of GeoScience, University of Edinburgh, Edinburgh EH9 3JN, UK

D. Schertzer, LEESU, Ecole des Ponts ParisTech, Universite Paris-Est, F-77455 Paris, France.

A. S. Sharma, Department of Astronomy, University of Maryland, College Park, MD 20742, USA. (ssh@astro.umd.edu)

H. von Storch, Institute of Coastal Research, Helmholtz-Zentrum Geesthacht, D-21502 Geesthacht, Germany.

N. W. Watkins, British Antarctic Survey, Cambridge CB3 0ET, UK.

Earthquakes: Complexity and Extreme Events

M. R. Yoder

Department of Physics, University of California, Davis, California, USA

D. L. Turcotte

Department of Geology, University of California, Davis, California, USA

J. B. Rundle

Departments of Physics and Geology, University of California, Davis, California, USA

Santa Fe Institute, Santa Fe, New Mexico, USA

Earthquakes are clearly complex phenomena; they are chaotic, and they are widely considered to be an example of self-organized criticality. Despite the complexity, earthquakes satisfy several scaling laws to a good approximation. The best known is Gutenberg-Richter (GR) frequency-magnitude scaling. This scaling is valid under a wide range of conditions, including global seismicity. GR scaling is important in seismic hazard assessment because it can be used to estimate the risk of large earthquakes from the rate of occurrence of small earthquakes. Also important in seismic hazard assessment is the concept of characteristic earthquakes (CEs) on mapped faults. In this paper, we address the alternative GR and CE behaviors for faults. We use the sequence of CEs that have occurred on the Parkfield segment of the San Andreas fault. We conclude that the data tend to support the CE hypothesis, but the GR hypothesis cannot be ruled out on the basis of currently available data. We also use numerical simulations to study the CE hypothesis.

1. INTRODUCTION

Earthquakes are one of the most feared of the natural hazards because they incur extensive costs in both life and property and because they occur without warning. Absent reliable precursory phenomena [*Turcotte*, 1991], the seismic hazard is mitigated by enforcing construction standards and facilitating public preparedness in general. Three earthquakes in 2010 illustrate the effectiveness of properly hardening infrastructure against the seismic hazard. The M = 7.0 Haiti earthquake, on 12 January 2010, occurred 25 km west of Port au Prince and caused an estimated 230,000 deaths. The M = 7.1 New Zealand earthquake on 4 September 2010 occurred 40 km west of Christchurch; there were no deaths. The great difference in fatalities can be attributed largely to enforced standards of construction in New Zealand. The M = 8.8 Chile earthquake on 27

Extreme Events and Natural Hazards: The Complexity Perspective
Geophysical Monograph Series 196

10.1029/2011GM001071

February 2010 caused 250 deaths. The relatively low toll, again, can be attributed to construction codes, earthquake safety education, and preparedness among the general public, enforced following the $M = 9.5$ 1960 Chile earthquake, which is the strongest earthquake to occur since seismographs have been available to quantify earthquake magnitudes. Earthquakes of concern generally have magnitudes $M > 6$. These earthquakes occur on preexisting faults, but only a small fraction of the relevant faults have been mapped. Many have no surface expression and are not defined by seismic activity.

It is the purpose of this chapter to discuss the complex nature of large earthquakes and methods by which the earthquake hazard can be assessed. Slider block models are considered as simple analogs for distributed seismicity. A coupled pair of slider blocks have been shown to exhibit deterministic chaos [*Huang and Turcotte*, 1990], and multiple slider blocks exhibit self-organized critical (SOC) behavior [*Carlson and Langer*, 1989]. In the slider block model, blocks are pulled over a surface by a driver plate. The blocks are connected to the driver plate and to each other by springs. The frictional interaction between the blocks and the surface leads to stick-slip behavior; sequences of slip events occur. The frequency-area statistics of the SOC behavior are power law (fractal)

$$N(A) \sim A^{D/2}, \quad (1)$$

where N is the number of slip events of area A, specifically the number of events involving A discrete block elements, and D is the fractal dimension of the system.

Despite the complexity of seismicity, it is accepted that the frequency-magnitude distribution of earthquakes satisfies Gutenberg-Richter (GR) scaling under a wide variety of conditions. The applicable relation is [*Gutenberg and Ritcher*, 1954]

$$\log[N(>M)] = a - bM, \quad (2)$$

where $N(>M)$ is the cumulative number of earthquakes in a specified spatial area and time window with magnitudes greater than M, $b \approx 1$, and the constant a is a measure of seismic intensity. *Aki* [1981] showed that GR scaling, equation (2), is equivalent to the fractal scaling given in equation (1) if A is the earthquake rupture area and $D = 2b$. The association of earthquakes with SOC behavior has been discussed in some detail by *Turcotte* [1997, 1999]. The association of this behavior with statistical physics has been outlined by *Rundle et al.* [2003]. *Pavlos et al.* [2007] and *Iliopoulos and Pavlos* [2010] have discussed the association of SOC behavior with low-dimensional chaos.

2. GR SCALING VERSUS CHARACTERISTIC EARTHQUAKES (CES)

It is recognized that GR scaling is generally applicable to regional and global seismicity. There are, however, two limiting hypotheses that can explain this scaling. In the GR hypothesis, each fault has a GR distribution of earthquakes on it. The largest earthquake would be limited by the size of the fault, but earthquakes occurring would satisfy GR frequency-magnitude scaling. This is the behavior associated with the SOC slider block model. The array of blocks has fractal (GR) frequency-magnitude scaling limited by the size of the array.

The second limiting hypothesis for GR scaling of regional seismicity introduces the concept of CEs. Each fault has CEs, of approximately equal magnitude, that rupture the entire fault. Large faults have large earthquakes and small faults have small earthquakes. It is recognized that faults have a fractal number-area distribution [*Turcotte*, 1997], and this distribution is responsible for the GR scaling of regional seismicity. An example of a CE would be the 1906 San Francisco earthquake on the northern San Andreas fault. Very few earthquakes have occurred on or near this fault since 1906. However, we do not have a sufficiently long record to study the distributions during an entire earthquake cycle. A focus of this chapter will be to consider the alternative GR and CE hypotheses for the sequence of earthquakes that have occurred on the Parkfield segment of the San Andreas fault.

Clearly, actual seismicity will not satisfy either of the extreme hypotheses, but will have characteristics of both. The relative importance of GR scaling versus CEs has important implications for earthquake hazard assessment. GR scaling provides a seismicity-based approach to earthquake hazard assessment. The GR hypothesis is that large earthquakes ($M > 6$) occur where large numbers of small earthquakes ($2 < M < 6$) occur. The GR scaling given in equation (2) extrapolates the numbers of small earthquakes to forecast the probabilities of occurrences of large earthquakes. The direct use of GR scaling for risk assessment is known as the relative intensity method [*Holliday et al.*, 2006; *Shcherbakov et al.*, 2010]. Modifications of the GR approach form the basis of the regional earthquake likelihood models that are currently being tested for California [*Field*, 2007a].

The second approach to the understanding of the earthquake hazard is to consider sequences of large ($M > 6$) CEs that occur on mapped faults. The fault-based CE hypothesis is that a sequence of large earthquakes occurs on mapped faults with a specified statistical distribution of recurrence times. This approach is the basis of the Working Groups on California Earthquake Probabilities hazard assessment [*Field*,

2007b], in which rates and magnitudes of earthquakes, mean slip rates, measured strain, and similar data are evaluated with respect to known fault maps.

The alternative GR and CE approaches have also been applied to earthquakes on specified faults or fault segments. It is accepted that small earthquakes on or near a fault obey GR scaling. In the GR hypothesis, all earthquakes on or near the fault satisfy GR scaling. In the competing CE hypothesis, the large "characteristic" earthquakes lie above GR scaling, which is to say they occur more frequently than predicted by the GR distribution of small events in the region. Arguments favoring the CE hypothesis have been given by *Wesnousky* [1994], *Hofmann* [1996], *Ishibe and Shimazaki* [2009], and others. Arguments favoring the GR hypothesis have been made by *Parsons and Geist* [2009] and *Naylor et al.* [2009], among others. There are several problems associated with comparisons of the two hypotheses. First, sequences of CEs generally involve long intervals so that the data are relatively poor. Paleoseismic studies can be used, but again, there are concerns about data quality. In addition, it is necessary to specify the region to be considered in a comparison; in general, an arbitrary area must be specified.

A primary focus of this chapter is to consider the alternative GR and CE hypotheses in terms of seven $M \approx 6$ CEs that occurred on the Parkfield section of the San Andreas fault between 1857 and 2004. We will concentrate our attention on the CE cycle associated with the 1966 and 2004 events. We will consider the seismicity on and adjacent to the fault during the period 1971 (5 years after the 1966 earthquake) to 2009 (5 years after the 2004 earthquake). We determine the GR statistics of aftershocks and background seismicity during the period. In the GR hypothesis, the 2004 main shock should lie on the GR scaling of the aftershocks and background seismicity; we will test this hypothesis.

Instrumental records of characteristics of CEs are rare. Historical and paleoseismic records are helpful, but their reliability can be questioned. An alternative approach is to use numerical simulations of CE occurrences to study their statistical behavior. We will use the Virtual California (VC) earthquake simulator to generate synthetic catalogs of earthquakes in Northern California. We will concentrate our attention on the Hayward fault and compare its simulated behavior with the behavior of the Parkfield segment of the San Andreas fault.

3. GLOBAL SEISMICITY

Earthquake frequency-magnitude statistics have been recognized to satisfy log-linear scaling both globally and regionally. Because of the many problems associated with the magnitudes of large earthquakes, the preferred approach to global seismicity is to use the Global Centroid Moment Tensor catalog. Using this catalog, the cumulative number of earthquakes per year $N(> M) \cdot \mathrm{yr}^{-1}$ with magnitudes greater than M for the period 1977 to 2009 is given in Figure 1. There is a clear change in slope at $M = 7.55$. The best fit GR scaling from equation (2) in the range $5.5 < M < 7.55$ is $a = 8.09$ and $b = 1.0$. The best fit GR scaling from equation (2) for $M > 7.55$ is $a = 11.68$ and $b = 1.48$. This change in slope has been previously observed [*Pacheco et al.*, 1992]; *Rundle* [1989] predicted the transition from $b = 1.0$ to $b = 1.5$ in the context of self-similarity. In both papers, the break at $M \approx 7.5$ is associated with the thickness of the brittle lithosphere and the transition from a 2-D to an approximately 1-D rupture. The largest earthquake during the last 100 years was the 1960 Chile earthquake, with $M \approx 9.5$. For this earthquake, the large-magnitude GR scaling, as shown in Figure 1, implies a recurrence time $\Delta t =$ 219 years. Within the context of GR scaling, this earthquake was not unexpected.

From Figure 1, traditional GR scaling, where $b = 1$, appears to be valid within the magnitude domain $5.5 < M <$ 7.55; GR scaling with $b \approx 1.5$ persists for $7.55 < b < 9.5$. Studies of seismicity in a deep gold mine in South Africa show GR scaling for magnitudes as small as $M = -4.4$ [*Kwiatek et al.*, 2010], which implies that the lower-magnitude rollover in Figure 1 is due to the sensitivity limit of the global seismic network. For the parameter values, we have obtained within the domain where $b = 1.0$ and $a = 8.09$, that the earthquake magnitude M is related to the rupture area A_r by *Aki* [1981]

$$A_r = 64 \cdot 10^M \ \mathrm{m}^2 \tag{3}$$

with A_r is in meters squared (m^2). Thus, the power law (fractal) scaling for global seismicity ($-4.4 < M < 7.5$) is valid for rupture areas from $2.55 \cdot 10^{-3}\ \mathrm{m}^2 \leq A_r \leq 2024\ \mathrm{km}^2$.

4. CHARACTERISTIC EARTHQUAKES AT PARKFIELD

In the previous section, we considered the earthquakes that occur on many faults. We now turn to the frequency-magnitude statistics of earthquakes associated with a specific fault or fault segment. As previously noted, this is a controversial subject; there are two competing hypotheses. The first hypothesis is that the earthquakes associated with a fault or fault segment obey the GR scaling (equation (2)). In the competing CE hypothesis, large CEs on a fault, or fault segment, occur more frequently than power law scaling of smaller earthquakes would predict. There are several problems associated with comparisons of the two hypotheses.

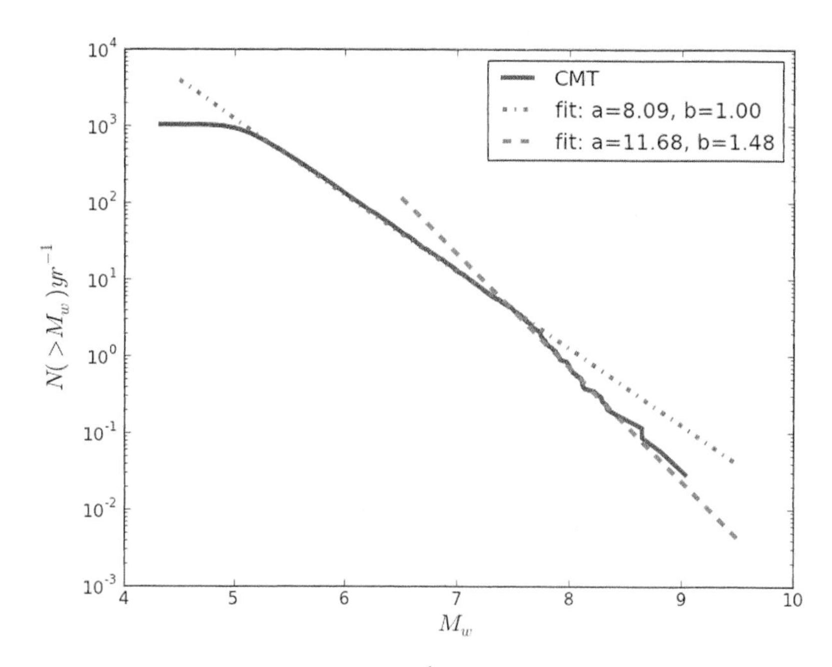

Figure 1. Worldwide number of earthquakes per year $N \cdot \text{yr}^{-1}$ with magnitudes greater than M_w. The solid line represents the cumulative distribution of moment magnitudes from the Harvard Centroid Moment Tensor Catalog, averaged over the period January 1977 to September 2010. The dash-dot line represents a least squares fit to equation (2) for $5.5 < M < 7.55$ with $a = 8.09$, $b = 1.0$. Fitting the data for $M > 7.55$ separately, we find $a = 11.68$, $b = 1.48$, as shown in the dashed line. These values for b and the break at $M \approx 7.5$ are consistent with observed measurements by *Pacheco et al.* [1992] and a theoretical treatment by *Rundle* [1989].

First, sequences of CEs involve long temporal recurrence intervals, so the data are sparse. Additionally, it is necessary to specify the region adjacent to a fault or fault segment where earthquakes are considered.

The best documented sequence of CEs occurred on the Parkfield segment of the San Andreas fault in California. Evidence suggests that earthquakes with $M \approx 6$ occurred in 1857, 1881, 1901, 1922, 1934, 1966, and 2004 [*Bakun et al.*, 2005]. Based on seismograms, the 1922, 1934, 1966, and 2004 events were remarkably similar in intensity. However, the recurrence times, 12, 32, and 38 years, varied considerably.

The main focus of this chapter is to study the seismicity on the Parkfield segment of the San Andreas fault during a full earthquake cycle. Specifically, the cycle associated with the 28 June 1966 $M_s = 6.0$ and the 28 September 2004 $M_w = 5.96$ earthquakes. To take full advantage of the improvements of the seismic network, we will consider the cycle for the period 1972 to 2010. The quality of the data for the aftershocks of the 2004 earthquake is much better than the quality of the data for the aftershocks of the 1966 earthquake.

To isolate seismicity associated with the Parkfield section of the San Andreas fault, we confine our study to the region where the aftershocks of the 2004 earthquake were concentrated. This region is well defined, and we select the specific geometry from the work of *Shcherbakov et al.* [2006]. The region is elliptical, centered at 35.9°N and −120.5°W with semimajor and semiminor axes of 0.4° and 0.15°, respectively, oriented at 137°NW. Both the aftershocks and the elliptical region are shown in Figure 2.

We consider the full seismic cycle between 28 June 1971 (5 years after the 1966 Parkfield earthquake) and 28 September 2009 (5 years after the 2004 Parkfield earthquake). The cumulative frequency magnitude distribution for this time period and region is given in Figure 3; the $M = 5.96$ 2004 Parkfield event is included. The best fit GR scaling from equation (2) is also shown; assuming $b = 1.0$, we find $a = 5.65$. If the Parkfield earthquake was part of the GR scaling, we would require $a = 5.96$. Thus, the observed seismicity for the seismic cycle is a factor of 2 less than that required for full GR scaling. This would indicate that the 2004 Parkfield event was a CE relative to both aftershocks and background seismicity in the aftershock region.

Based on these scaling relations, invoking equation (2), there were $N_{\text{CE}} = 10^{3.65} = 4467$ $M > 2$ earthquakes during the CE cycle. If the GR hypothesis is valid, there would have been $N_{\text{GR}} = 10^{3.96} = 9120$ $M > 2$ earthquakes during the cycle. The earthquakes that occurred include the 2004 main

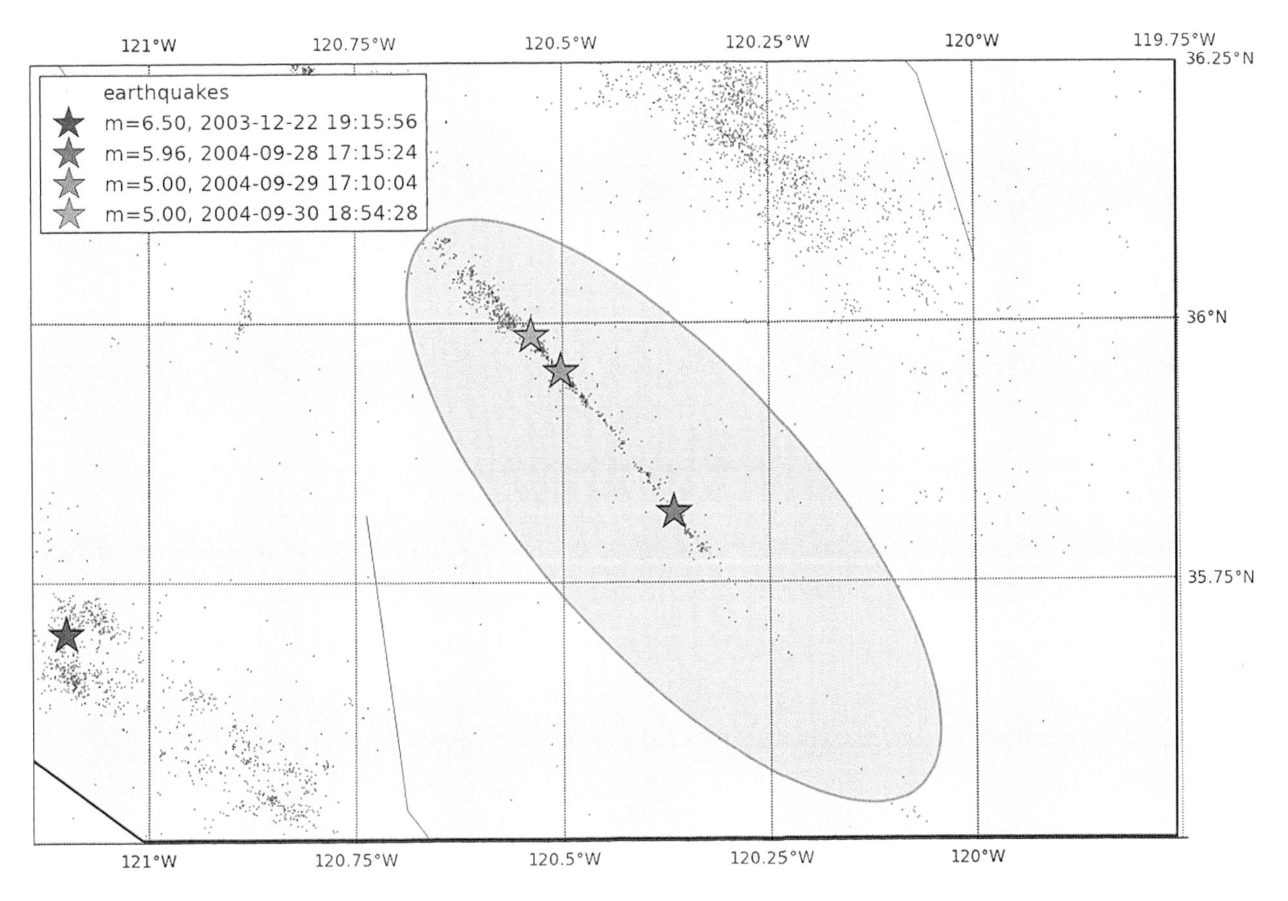

Figure 2. Seismicity in central California for the period 28 September 1999 to 28 September 2009. The aftershocks of the 18 September 2004 Parkfield earthquake are shown in the elliptical area. Aftershocks of the 2004 San Simeon earthquake and residual aftershocks of the 1983 Coalinga earthquake are clearly seen to the southwest and northeast of the study region.

shock, aftershocks of this earthquake, and background seismicity. Based on the aftershock study of *Shcherbakov et al.* [2006], there were $N_{as} = 450$ $M > 2$ aftershocks. This leaves $N_{BG} = 4200$ $M > 2$ background earthquakes during the cycle. The background earthquakes constituted 44% of the number required by the GR hypothesis. The 2004 Parkfield main shock appears to be a well-defined CE, but there was considerable background seismicity.

Characteristic earthquakes are associated with quasiperiodicity, but can also have considerable variability. A measure of the variability of recurrence times on a fault or fault segment is the coefficient of variation C_v, the ratio of the standard deviation σ to the mean μ. For strictly periodic earthquakes on a fault or fault segment, we would have $\sigma = C_v = 0$. For the random (i.e., exponential with no memory) distribution of recurrence times, we would have $\sigma = \mu$ and $C_v = 1$. *Ellsworth et al.* [1999] analyzed 37 series of recurrent earthquakes and suggested a provisional generic value of the coefficient of variation, $C_v \approx 0.5$. A number of alternative statistical distributions have been proposed for this purpose. These include the exponential (Poisson), the lognormal, Brownian passage time (inverse Gaussian), and Weibull distributions [*Davis et al.*, 1989; *Sornette and Knopoff*, 1997; *Mathews et al.*, 2002]. We will primarily consider the Weibull distribution. The cumulative distribution function (CDF) for the Weibull distribution is

$$P(t) = 1 - \exp\left[-\left(\frac{t}{\tau}\right)^{\beta}\right], \tag{4}$$

where $P(t)$ is the fraction of the recurrence times that are shorter than t, and β and τ are fitting parameters. If $\beta = 1$, this is the Poisson (random) distribution. Reasons for preferring the Weibull distribution have been discussed by *Abaimov et al.* [2008].

The CDF of Parkfield recurrence times (t = 12, 20, 21, 24, 32, and 38 years) is given in Figure 4. The mean,

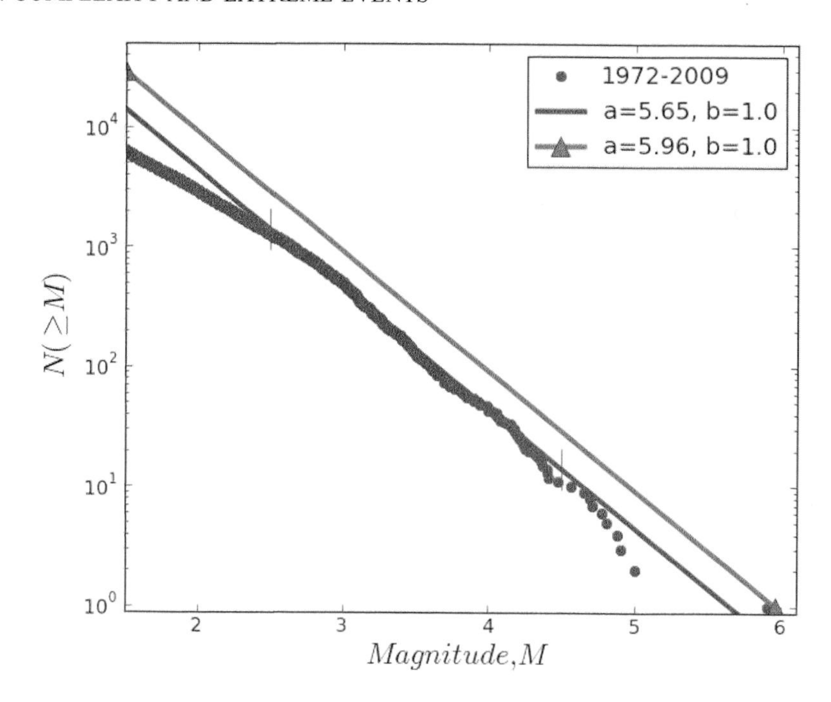

Figure 3. Cumulative numbers of earthquakes N with magnitude greater than M as a function of M for the Parkfield earthquake cycle 1972 to 2009. Least squares best fit to Gutenberg-Richter (GR) scaling (equation (2)) assuming $b = 1.0$, between $2.5 < M < 4.5$ yields $a = 5.65$. GR scaling consistent with the $M = 5.96$ Parkfield earthquake requires $a = 5.96$, as shown.

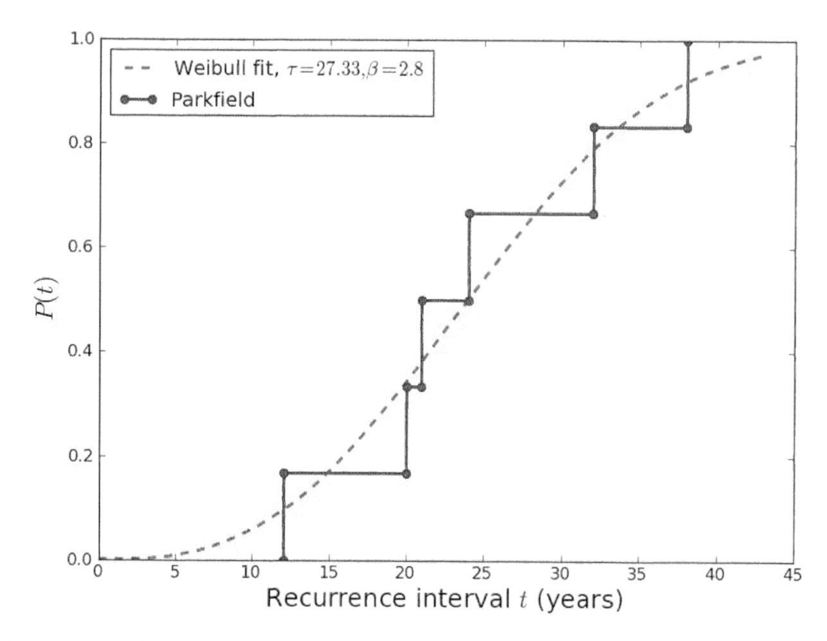

Figure 4. Cumulative distribution of recurrence times of Parkfield earthquakes. The stair-step solid line is the distribution of the six actual recurrence intervals. The continuous dot-dash line is the best fit Weibull distribution. The mean value statistics are $\mu = 24.5$ years, $\sigma = 9.25$ years, and $C_v = 0.377$. The Weibull fit parameters are $\tau = 27.33$ and $\beta = 2.80$.

standard deviation, and coefficient of variation of these recurrence times are $\mu = 24.5$ years, $\sigma = 9.25$ years, and $C_v = \frac{\sigma}{\mu} = 0.377$, respectively. Taking these values, the corresponding fitting parameters for the Weibull distributions are $\tau = 27.33$ years and $\beta = 2.80$. Using these values, the CDF from equation (4) is also shown in Figure 4; good agreement is observed.

5. SIMULATIONS

An alternative approach to the statistics of CEs is to use simulations. A number of fault-based simulations of seismicity have been conducted. A limited simulation model for earthquakes on specified strike-slip faults in California called VC was introduced by *Rundle* [1988] and was subsequently updated [*Rundle et al.*, 2002, 2004, 2006]. VC is a 3-D simulator code that includes static stress interactions between faults using dislocation theory. Vertical fault segments are embedded in an elastic half-space. Stress accumulates by "back slip," a linearly increasing negative displacement across each fault segment is applied at a prescribed velocity. A model earthquake occurs when the resulting stress exceeds a prescribed static coefficient of friction. Related models have been proposed by *Ward* [1992, 1996, 2000], by *Robinson and Benites* [1995, 1996], and by *Dietrich and Richards-Dinger* [2010].

As an example, we will consider results obtained for Northern California by *Yikimaz et al.* [2011] using VC simulations. Twenty strike-slip faults are considered. The faults have a depth of 12 km and four layers of 3 km × 3 km elements are used in the cellular grid computation. These are a total of 3348 elements. Mean slip rates based on geological and geodetic data and mean interval times between "characteristic" earthquakes are specified for each fault segment. The interval times are used to calculate static coefficients of friction on the faults. Without fault interactions, periodic earthquakes would occur on each fault. The fault interactions generate a complex, chaotic statistical behavior.

We will consider a 100,000 year simulation and will focus our attention on the Hayward fault. This fault crosses the University of California Berkeley campus, posing a severe hazard for the Berkeley-Oakland-Hayward region. The fault has a mean specified slip rate $v = 9$ mm yr^{-1} and a mean recurrence interval of $\langle \Delta t_0 \rangle = 171$ years. The probability distribution function (PDF) of earthquake magnitudes on the Hayward fault for the 100,000 year simulation is given in Figure 5. The distribution is dominated by CEs with $M \approx 7.0$. The CDF of recurrence intervals of earthquakes on the Hayward fault with $M > 6.5$ is shown in Figure 6. The observed mean recurrence interval is 141 years with coefficient of variation $C_v = 0.42$. Although there is little variability of magnitudes on the Hayward fault, there is considerable variability of recurrence intervals. This is quite similar to the behavior of CEs on the Parkfield segment of the San Andreas fault. Recently, *Lienkaemper et al.* [2010] studied 12 historical earthquakes on the southern Hayward fault that occurred in the past 1900 years and found a mean recurrence interval of 161 ± 10 years and $C_v = 0.4$, which is in good agreement with our findings.

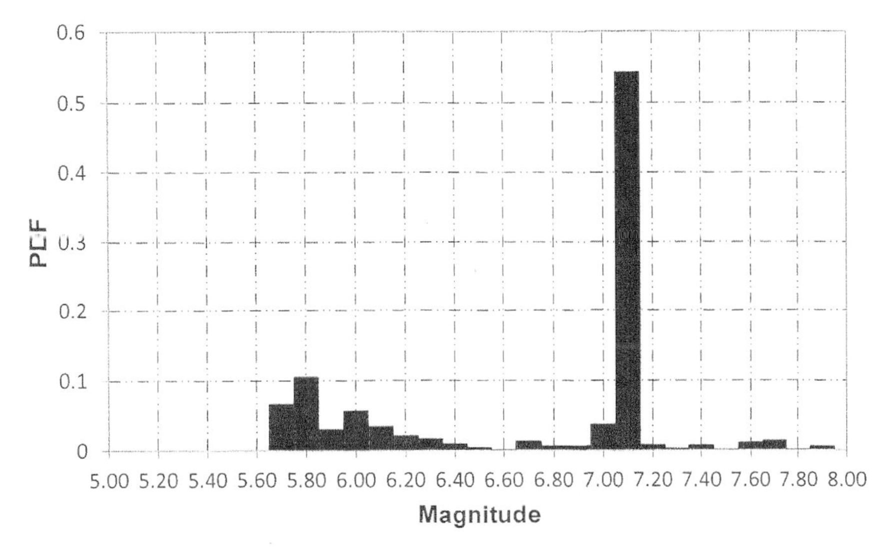

Figure 5. Probability distribution function of earthquake magnitudes on the Hayward fault during the 100,000 year VC simulation. Bins are 0.1 magnitude units.

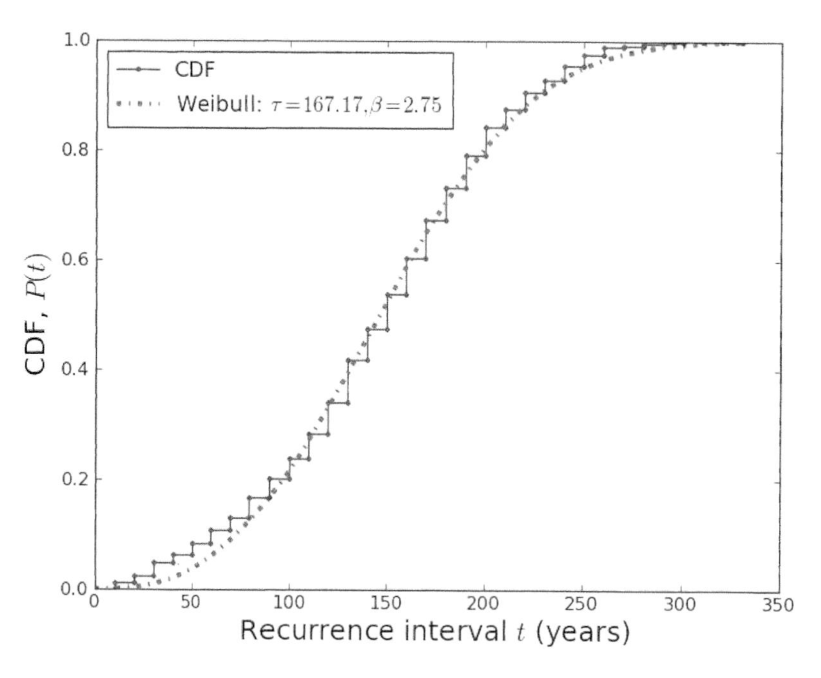

Figure 6. Cumulative distribution function of recurrence intervals for earthquakes with $M > 6.5$ on the Hayward fault during the 100,000 year simulation (dots). The mean recurrence time is 141 years with $C_v = 0.42$; the bin width is 10 years. The dash-dot line shows the least squares best fit to a Weibull distribution (equation (4)) with $\tau = 167.17$ and $\beta = 2.75$.

6. DISCUSSION

The principal focus of this chapter has been on the relative role of GR and CE statistics. Both play important roles in probabilistic seismic hazard analysis. GR statistics are used to estimate the risk of large earthquakes from the measured occurrence of smaller earthquakes. CE statistics are used to estimate the recurrence of earthquakes on mapped faults.

The two hypotheses are related to the basic physics of earthquake occurrence. Two explanations can be given for the applicability of GR statistics. The first is that every fault has a GR distribution of earthquakes on it. The second is that every fault has only CEs on it, and GR statistics result from the fractal distributions of fault areas. Observations indicate that both hypotheses are partially valid.

In order to study the alternative CE hypotheses, we have considered the sequence of CEs that have occurred on the Parkfield segment of the San Andreas fault. A sequence of $M \approx 6$ CEs is well documented. We have examined in detail the GR statistics of earthquakes on and near the fault segment during a CE cycle. The question is whether the CE lies on this GR distribution of seismicity.

We find that the sum of the aftershock activity and background seismicity lies somewhat below the extrapolated GR statistics based on the CE. However, this single example could be a statistical exception; further studies are clearly necessary.

We also use a simulation model of Northern California seismicity to study the statistics of CEs. We have focused on the behavior of the Hayward fault. In terms of the sequence of near equal magnitude CEs and the statistics of recurrence times, the simulation results for the Hayward fault are similar to the observations for the Parkfield segment of the San Andreas fault.

Clearly, more studies are desirable concerning the relative roles of GR versus CE statistics. At the present time, it is clearly desirable to use both in probabilistic seismic hazard analyses.

Acknowledgments. This work has been partially supported by JPL subcontract 1291967 and DOE DE-FG02-04ER15568.

REFERENCES

Abaimov, S., D. Turcotte, R. Shcherbakov, J. B. Rundle, G. Yakovlev, C. Goltz, and W. I. Newman (2008), Earthquakes: Recurrence and interoccurrence times, *Pure Appl. Geophys.*, *165*, 777–795.

Aki, K. (1981), A probabilistic synthesis of precursory phenomena, in *Earthquake Prediction, Maurice Ewing Ser.*, vol. 4, edited by D. W. Simpson and P. G. Richards, pp. 566–574, AGU, Washington D. C.

Bakun, W., et al. (2005), Implications for prediction and hazard assessment from the 2004 Parkfield earthquake, *Nature*, *437*, 969–974.

Carlson, J., and J. Langer (1989), Mechanical model of an earthquake fault, *Phys. Rev. A*, *40*, 6470–6484.

Davis, P. M., D. D. Jackson, and Y. Y. Kagan (1989), The longer it has been since the last earthquake, the longer the expected time till the next, *Bull. Seismol. Soc. Am.*, *79*(5), 1439–1456.

Dietrich, J. H., and K. B. Richards-Dinger (2010), Earthquake recurrence in simulated fault systems, *Pure Appl. Geophys.*, *167*(8–9), 1087–1104.

Ellsworth, W. L., M. V. Mathews, R. M. Nadeau, S. P. Nishenko, P. A. Reasenberg, and R. W. Simpson (1999), A physically based earthquake recurrence model for estimation of long-term earthquake probabilities, *U.S. Geol. Surv. Open File Rep.*, *99–522*.

Field, E. H. (2007a), A summary of previous working groups on California earthquake probabilities., *Bull. Seismol. Soc. Am.*, *97* (4), 1033–1053.

Field, E. H. (2007b), Overview of the working group for the development of regional earthquake likelihood models (RELM), *Seismol. Res. Lett.*, *78*(1), 7–16.

Gutenberg, B., and C. F. Richter (1954), *Seismicity of the Earth and Associated Phenomenon*, 2nd ed., Princeton Univ. Press, Princeton, N. J.

Hofmann, R. B. (1996), Individual faults can't produce a Gutenberg-Richter earthquake recurrence, *Eng. Geol.*, *43*(1), 5–9.

Holliday, J. R., J. B. Rundle, K. F. Tiampo, and D. L. Turcotte (2006), Using earthquake intensities to forecast earthquake occurrence times, *Nonlinear Processes Geophys.*, *13*, 585–593.

Huang, J., and D. L. Turcotte (1990), Are earthquakes an example of deterministic chaos?, *Geophys. Res. Lett.*, *17*(3), 223–226.

Iliopoulos, A., and G. Pavlos (2010), Global low dimensional seismic chaos in the Hellenic region, *Int. J. Bifurcation Chaos*, *20*(7), 2071–2095.

Ishibe, T., and K. Shimazaki (2009), Seismicity in source regions of large interplate earthquakes around Japan and the characteristic earthquake model, *Earth Planets Space*, *61*(9), 1041–1052.

Kwiatek, G., K. Plenkers, N. Nakatani, Y. Yabe, G. Dresen, and JAGUARS-Group (2010), Frequency-magnitude characteristics down to magnitude -4.4 for induced seismicity recorded at Mponeng gold mine, South Africa, *Bull. Seismol. Soc. Am.*, *100*(3), 1165–1173.

Lienkaemper, J. J., P. L. Williams, and T. Guilderson (2010), Evidence for a twelfth large earthquake on the southern Hayward fault in the past 1900 years, *Bull. Seismol. Soc. Am.*, *100*(5A), 2024–2034.

Mathews, M. V., W. L. Ellsworth, and P. A. Reasenberg (2002), A Brownian model for recurrent earthquakes, *Bull. Seismol. Soc. Am.*, *92*(6), 2233–2250.

Naylor, M., J. Greenhough, J. McCloskey, A. F. Bell, and I. G. Main (2009), Statistical evaluation of characteristic earthquakes in the frequency-magnitude distributions of Sumatra and other subduction zone regions, *Geophys. Res. Lett.*, *36*, L20303, doi:10.1029/2009GL040460.

Pacheco, J. F., C. H. Scholz, and L. R. Sykes (1992), Changes in frequency-size relationship from small to large earthquakes, *Nature*, *355*, 71–73.

Parsons, T., and E. L. Geist (2009), Is there a basis for preferring characteristic earthquakes over a Gutenberg-Richter distribution in probabilistic earthquake forecasting?, *Bull. Seismol. Soc. Am.*, *99*(3), 2012–2019.

Pavlos, G. P., A. C. Iliopoulos, and M. A. Athanasiou (2007), Self organized criticality or/and low dimensional chaos in earthquake processes: Theory and practice in Hellenic region, in *Nonlinear Dynamics in Geosciences*, edited by A. A. Tsonis and J. B. Elsner, pp. 235–259, Springer, New York, doi:10.1007/978-0-387-34918-3_14.

Robinson, R., and R. Benites (1995), Synthetic seismicity models of multiple interacting faults, *J. Geophys. Res.*, *100*(B9), 18,229–18,238.

Robinson, R., and R. Benites (1996), Synthetic seismicity models for the Wellington region, New Zealand: Implications for the temporal distribution of large events, *J. Geophys. Res.*, *101* (B12), 27,833–27,844.

Rundle, J. B. (1988), A physical model for earthquakes 2. Application to southern California, *J. Geophys. Res.*, *93*(B6), 6255–6274.

Rundle, J. B. (1989), Derivation of the complete Gutenberg-Richter magnitude-frequency relation using the principle of scale invariance, *J. Geophys. Res.*, *94*(B9), 12,337–12,342.

Rundle, J. B., P. B. Rundle, W. D. Klein, K. F. Tiampo, A. Donnelan, and L. H. Kellogg (2002), GEM plate boundary simulations for the plate boundary observatory: A program for understanding the physics of earthquakes on complex fault networks via observations, theory, and numerical simulation, *Pure Appl. Geophys.*, *159* (10), 2357–2381.

Rundle, J. B., D. L. Turcotte, R. Shcherbakov, W. Klein, and C. Sammis (2003), Statistical physics approach to understanding the multiscale dynamics of earthquake fault systems, *Rev. Geophys.*, *41*(4), 1019, doi:10.1029/2003RG000135.

Rundle, J. B., P. B. Rundle, A. Donnelan, and G. Fox (2004), Gutenberg-Richter statistics in topologically realistic system-level earthquake stress-evolution simulations, *Earth Planets Space*, *56*(8), 761–771.

Rundle, P. B., J. B. Rundle, K. F. Tiampo, A. Donnelan, and D. L. Turcotte (2006), Virtual California: Fault model, frictional parameters, applications, *Pure Appl. Geophys.*, *163*(9), 1819–1846.

Shcherbakov, R., D. L. Turcotte, and J. B. Rundle (2006), Scaling properties of the Parkfield aftershock sequence, *Bull. Seismol. Soc. Am.*, *96*(4B), S376–S384, doi:10.1785/0120050815.

Shcherbakov, R., D. L. Turcotte, J. B. Rundle, K. F. Tiampo, and J. R. Holliday (2010), Forecasting the locations of future large earthquakes: An analysis and verification, *Pure Appl. Geophys.*, *167*(6–7), 743–749.

Sornette, D., and L. Knopoff (1997), The paradox of the expected time until the next earthquake, *Bull. Seismol. Soc. Am.*, *87*(4), 789–798.

Turcotte, D. L. (1991), Earthquake prediction, *Annu. Rev. Earth Planet. Sci.*, *19*, 263–281.

Turcotte, D. L. (1997), *Fractals and Chaos in Geology and Geophysics*, 2nd ed., Cambridge Univ. Press, Cambridge, U. K.

Turcotte, D. L. (1999), Self-organized criticality, *Rep. Prog. Phys.*, *62*, 1377–1429, doi:10.1088/0034-4885/62/10/201.

Ward, S. N. (1992), An application of synthetic seismicity in earthquake statistics: The Middle America trench, *J. Geophys. Res.*, *97*(B5), 6675–6682.

Ward, S. N. (1996), A synthetic seismicity model for southern California: Cycles, probabilities, and hazard, *J. Geophys. Res.*, *101*(B10), 22,393–22,418.

Ward, S. N. (2000), San Francisco bay area earthquake simulations: A step toward a standard physical earthquake model, *Bull. Seismol. Soc. Am.*, *90*(2), 370–386.

Wesnousky, S. G. (1994), The Gutenberg-Richter or characteristic earthquake distribution, which is it?, *Bull. Seismol. Soc. Am.*, *84* (6), 1940–1959.

Yikimaz, M. B., E. M. Heien, D. L. Turcotte, J. B. Rundle, and L. H. Kellogg (2011), A fault and seismicity based composite simulation in northern California, *Nonlinear Processes Geophys.*, *18*, 955–966.

J. B. Rundle and M. R. Yoder, Department of Physics, University of California, Davis, CA 95616, USA. (yoder@physics.ucdavis.edu)

D. L. Turcotte, Department of Geology, University of California, Davis, CA 95616, USA.

Patterns of Seismicity Found in the Generalized Vicinity of a Strong Earthquake: Agreement With Common Scenarios of Instability Development

M. V. Rodkin

Institute of Earthquake Prediction Theory and Mathematical Geophysics, Moscow, Russia

The evolution of cases of development of instability occurring in different systems is believed to obey the universal scenarios. The occurrence of a strong earthquake is ordinarily treated as a typical example of instability development. But, in comparison with a majority of other systems prone to instability, the case of strong earthquake occurrence permits the use of more detailed statistical examination because of a substantial number of strong earthquakes occurring throughout the world. The Harvard seismic moment and U.S. Geological Survey/National Earthquake Information Center catalogs were used to construct a generalized space-time vicinity of strong earthquakes (SEGV) and to investigate the seismicity behavior in SEGV. As a result of this investigation, a few anomalies (besides the expected foreshock and aftershock cascades) were found. Both foreshock and aftershock anomalies increase at the time of approaching the moment of the generalized main shock as a logarithm of the time interval remaining before the main shock occurrence. The anomalies revealed agree with general scenarios of development of instability. Some of these anomalies relevant to the effect of the critical slowing down are discussed in more detail.

1. INTRODUCTION

The evolutions of cases of development of instability occurring in very different systems are believed to obey a few (or even a single one) general scenarios of the development of instability [*Haken*, 1978; *Bowman et al.*, 1998; *Sornette*, 2000; *Turcotte and Malamud*, 2004, and references herein]. The occurrence of a strong earthquake is ordinarily treated as an example of instability development. But, in comparison with a majority of other systems prone to instability, the case of origin of strong earthquakes permits the use of a more detailed statistical examination because of a large number of strong earthquakes occurring throughout the world. Below, we examine the statistical patterns inherent to the origin of strong earthquakes. These patterns are compared with the general features inherent to theoretical scenarios of development of instability.

A precursory process and a large earthquake occurrence are commonly treated as a typical example of instability similar to critical phenomenon [*Bowman et al.*, 1998; *Shebalin et al.*, 2000; *Sornette*, 2000; *Keilis-Borok and Soloviev*, 2003; *Malamud et al.*, 2005]. Many of the earthquake precursors, which are currently used, such as an increase in correlation length, a development of power law foreshock cascades, and an abnormal clustering of earthquakes, are expected to occur in critical processes [*Zöller and Hainzl*, 2002; *Keilis-Borok and Soloviev*, 2003; *Shebalin*, 2006]. Moreover, some of these precursors came into use because of the intentions to treat the strong earthquake occurrence as a sort of a critical phenomenon. In this situation, it was very natural to ask to what extent the suggested critical behavior patterns really typical of evolution of foreshock and aftershock cascades of different strong earthquakes are. *Romashkova and Kosobokov* [2001] have considered the evolution

Extreme Events and Natural Hazards: The Complexity Perspective
Geophysical Monograph Series 196

10.1029/2011GM001060

of foreshock and aftershock activity in the vicinities of 11 strong earthquakes occurring from 1985 until 2000. This analysis has not supported the universality of a power law growth in foreshock activity toward the moment of a large earthquake. It also turned out that aftershock sequences of different strong earthquakes may differ essentially from the Omori law; as a result, it was hypothesized that the class of decay of aftershock cascades does not consist only of sequences obeying the Omori law and that other patterns of aftershock decay can exist.

One can ask whether the observed deviations of a seismic process from the theoretically expected pattern have a stochastic character or a few different types of evolution could take place in the foreshock and aftershock sequences as it was hypothesized by *Romashkova and Kosobokov* [2001]. One can also ask a more general question: Do the universal scenarios take place in the process of strong earthquake origin, or do different regularities take place in different cases? For example, in the examination of seismicity in California [*Shcherbakov et al.*, 2006], no pronounced temporal changes in the correlation length prior to strong earthquakes were found, whereas well-defined variations were found for aftershock sequences and for background seismicity.

An answer to that question can be obtained by revealing common (mean) features of a large number of strong earthquakes fore and aftershock sequences. Strong earthquakes are characterized here by the space-time domains where an evolution of seismicity is significantly influenced by strong earthquake occurrence. In this chapter, the generalized space-time vicinity (SEGV) of strong earthquake is constructed, and the mean anomalies revealed in SEGV are examined. Some of these anomalies were not known earlier. The revealed mean anomalies are compared with the patterns of theoretical scenarios of development of instability. In this chapter, the anomalies similar to those expected in the case of development of the critical slowing-down effect [*Haken*, 1978; *Akimoto and Aizawa*, 2006] are discussed in more detail.

2. THE METHOD AND THE DATA

We use data from the Harvard worldwide seismic moment catalog for 1976–2005 with focal depths $H \leq 70$ km that contains 17,424 earthquakes and the U.S. Geological Survey/National Earthquake Information Center (USGS/NEIC) catalog for 1964–2007 similarly restricted in a depth. Two sets of data from each of the catalogs could be examined. The first one includes all earthquakes from the catalog, and the second includes only completely reported stronger earthquakes. Below, we present the results from processing of the Harvard catalog using all events and the results of examination of USGS/NEIC data using only completely reported events. In the second case, the events with magnitude ≥ 4.7 were used; the total number of such events is 97,615. A similar cutoff of weaker incompletely reported earthquakes from the Harvard catalog would significantly reduce the available information; at such restriction, the needed statistical analysis becomes hardly possible.

It should be noted that a majority of the examined parameters are intensive; that means they do not explicitly depend on the earthquake magnitude; for example, it appears that the difference between body-wave magnitude and moment magnitude $M_b - M_w$ does not significantly depend on the magnitude. For the intensive parameter, the use of a large number of smaller events will hardly bias the estimate from those obtainable only for the case of completely reported larger events. But the increase in a number of events considerably enhances the statistics. When extensive parameters are used (e.g., a half duration of the seismic process), we checked the results obtained by doing the similar examination for norm value of a parameter, namely, the values were divided by the cubic root of the seismic moment $(\text{Mo})^{1/3}$. As a rule, no qualitative distinctions were found.

Both the above catalogs were searched for events falling into the space-time domains surrounding the largest earthquakes, with due account to the seismic moment value in the Harvard catalog and the maximum magnitude in the USGS/NEIC catalog. The generalized vicinity of large earthquakes is understood as a set of events falling into the zone of influence of any of the examined strong earthquakes. The zones of influence were defined as follows. The spatial size of the zone of influence of strong earthquake of magnitude M was calculated from the relationship [*Kasahara*, 1981; *Sobolev and Ponomarev*, 2003] between typical source size L and magnitude M:

$$L\,(\text{km}) = 10^{0.5M-1.9}, \tag{1}$$

where M is a moment magnitude M_w or maximum magnitude reported in the USGS/NEIC catalog. In the examination below, the earthquakes located at distances less than $7 \times L$ and $3 \times L$ from the epicenter of a given large earthquake are taken into account; this choice is argued below.

In constructing a time vicinity of large earthquake, we used the conclusion [*Smirnov and Ponomarev*, 2004] that the duration of cycle of a seismic failure weakly depends from an earthquake magnitude. Hence, in constructing the generalized vicinity of strong earthquakes of different (but close) magnitudes, one can use the simple method of epoch superposition, and this simple method was used below. Negative consequences of this choice are the lower statistics at the edges of the used time interval (there are less data there) and the appearance of a false effect of a systematic growth of a number of quakes toward the central part of the time interval under examination. These negative features were taken into account, so they do not distort the obtained results.

The SEGV zone, which was constructed this way, contained more than 27,000 earthquakes for the Harvard catalog (vicinities of 300 strongest events were summarized) and more than 750,000 earthquakes for the USGS/NEIC catalog (vicinities of 450 strongest $M7+$ earthquakes were summarized). The great number of events available for the analysis resulted from the fact that one and the same earthquake can fall into the space-time vicinity of different strong earthquakes. The great number of events constructing the SEGV has significantly enhanced the possibility of statistical examination. The contribution of an individual strong earthquake into the SEGV data set is determined by the number of earthquakes falling into the corresponding space-time volume, that is, it depends on the magnitude of a particular strong earthquake (which determines the spatial dimension of its zone of influence) and on the density of earthquakes occurring in the vicinity of this earthquake.

The space-time position of each earthquake falling in the SEGV is characterized by the time shift relative to the time of occurrence of the relevant main shock and by the distance from the epicenter of this main shock (norm by the source size L of the relevant strong earthquake). The data of both catalogs (Harvard and USGS/NEIC) were used to study changes in the rate of seismic events and b values in the SEGV. The Harvard catalog was used also to examine the additional parameters available from the seismic moment data. The changes in the mean $M_b - M_w$ values, differences in the depth and time of event occurrence obtained from the first arrivals and from the seismic moments data, changes in the uniformity of orientation of earthquake focal mechanisms, and the change in the apparent stress σ_a values were examined.

Let us now specify the suggested physical sense of some of the used parameters. The difference $M_b - M_w$ characterizes the relative predominance of high and low frequency domains in the seismic source oscillations, which are used to determine the magnitudes M_b and M_w correspondingly. The hypocenter parameters (depth and time) derived from the first arrival data characterize initiation of the seismic rupture; the same parameters (depth and origin time) derived from the seismic moment data characterize the "center of mass" of process of seismic rupture. Hence, the time difference between the two time determinations $\Delta\tau$ characterizes the half duration of seismic wave radiation process, while the difference between two focal depth values ΔH characterizes the vertical half size of the rupture zone and the direction of rupturing (upward or downward). These parameters are irregular for individual earthquakes, but their mean values for an essential number of events were argued to reveal systematic tendencies [*Rodkin*, 2006, 2008]. The change in apparent stress σ_a values characterizes the tendency of change in stress state and rock strength in vicinity of strong earthquakes. The use of the abovementioned rarely used parameters was shown [*Rodkin*, 2006, 2008] to be useful because the tendencies of their change were found to have a reasonable physical interpretation in most cases.

3. CHANGE IN THE RATE OF EARTHQUAKES IN THE GENERALIZED VICINITY OF A STRONG EARTHQUAKE

The best known patterns of seismic behavior in the vicinities of strong earthquakes are the regularities in aftershock and foreshock sequences. In view of this fact, it was tested whether the average activity revealed from the examination of the generalized vicinity of strong earthquake shows the power law behavior, which is believed to be typical of the foreshock and aftershock sequences. Figures 1a and 1b show the foreshock and aftershock sequences in the generalized vicinity of strong earthquake obtained from USGS/NEIC data; there are 304,684 events with $M \geq 4.7$ located at distances less that $3 \times L$ from the epicenter of a corresponding strong earthquake. The time is given in days from the occurrence time of the generalized main event. The rate is presented by time density of clusters consisting of 100 of the subsequent events. The clusters are taken at a step of 50 events; thus, the data points are independent of those next to the adjacent ones.

As can be seen from Figure 1, the evolution of mean foreshock and aftershock sequences is well described by a power law. The Omori-Utsu law [*Utsu*, 1965; *Utsu et al.*, 1995] is known to be a good fit to the aftershock rate. According to this law,

$$n \sim (c + t)^{-p}, \qquad (2)$$

where n is the rate of events, t is the time interval after the main shock occurrence, c is the parameter fitting the rate of earthquakes in the closest vicinity of the main shock, and p is the parameter of the Omori law. The inverse (foreshock) power law cascade can be described in a similar manner; in this case, t is the time before the main shock occurrence, p value is marked as p', and c is suggested to be zero. The lines corresponding to typical $p = 1$ value are given in both Figures 1a and 1b. It can be seen that $p = 1$ value fits the

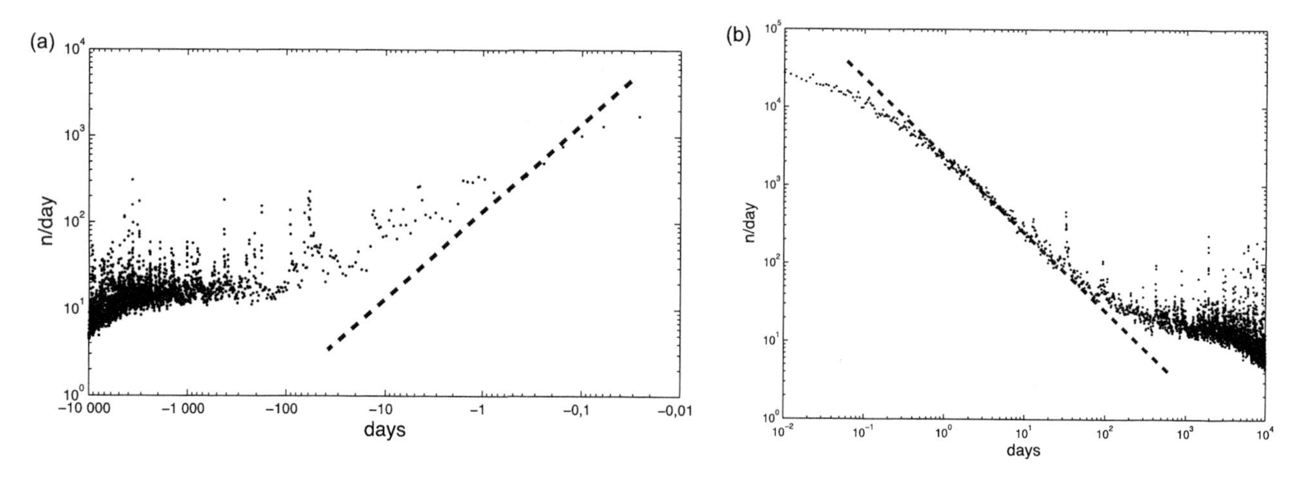

Figure 1. The (a) foreshock and (b) aftershock sequences in the generalized vicinity of strong earthquake. Events flow is given in number of events per day; zero time corresponds to the moment of occurrence of the generalized main shock, the typical $p = 1$ Omori law is shown for comparison by thick dashed line.

mean aftershock decay well for time interval from 0.5 until 100 days. For the case of foreshock cascade, $p' = 1$ fits the data in the narrow time interval from a few hours until 1 day only. The p' value can be seen to be less in general than the p value. Both p and p' values decrease for the distant foreshock and aftershocks. These findings agree well with the previously obtained results [*Ogata et al.*, 1995; *Helmstetter et al.*, 2003; *Faenza et al.*, 2009]. These papers discussed the tendency of a decrease of p value for distant events and the relation $p' < p$.

The aftershock rate during the first hours after the main shock occurrence is of special interest. The deficit of earlier aftershocks described by c parameter in equation (2) is explained sometimes by difficulties in recording numerous weak aftershocks occurring immediately after a large earthquake, but this factor is hardly capable of providing a full explanation of this phenomenon [*Ogata*, 1998; *Shebalin*, 2006; *Lennartz et al.*, 2008]. The deviation toward lower rate of events during the few first hours of the aftershock process is clearly seen in Figure 1b, and the aftershock rate begins to fit the Omori law with $c = 0$ only 2–3 h after the generalized strong earthquake occurrence. At that time, the mean rate of the examined events with $M \geq 4.7$ in the vicinity of a single strong earthquake is a little above one event per hour. Such rate of events cannot cause any difficulty in recording. It seems also unlikely that an earthquake with $M \geq 4.7$ could be missed. Thus, we have got the additional evidence that the effect of a lower rate of earlier aftershocks has a physical nature. This result is not novel [*Shebalin*, 2006; *Lennartz et al.*, 2008; *Lindman et al.*, 2010], but it is of interest as a refinement of the mean character of the aftershock process and as an evidence that the SEGV examination method used here provides a reasonable information. In the work of *Lindman et al.* [2010], the shortage in a number of earlier aftershocks is suggested to be connected with adjustment of stresses in porous-elastic media; moreover, the effect of an increase of pre-power law decay period for stronger earthquakes was mentioned that agrees with a large c parameter value found above in the case of the strongest ($M7+$) earthquakes.

The similar regularities were obtained from the examination of the Harvard catalog, but they are less convincing because of a smaller volume of data.

We now characterize the changes in the rate of seismicity as functions of the distance from the main shock epicenter. Figure 2 shows the distance-time diagram of the rate of a number of events in the vicinity of the generalized main shock. The horizontal axis indicates the time (in days) from the moment of main shock occurrence; an analog of the log-time scale is used for the time intervals adjacent to the main shock occurrence. The vertical axis indicates the distance from the main shock epicenter in units of earthquake source size L from equation (1). For the rate of events, we used a logarithmic scale, $\log(n)$, where n is the density of number of events in a cell of the distance-time diagram. The rate of earthquakes in the vicinity of the main shock begins to increase about 100 days before the main shock, and this activity increase accelerates toward the main shock moment. The foreshock and aftershock processes are found to be symmetrical enough in space and time, despite the strong difference in a number of events. The effect of increase in a seismicity rate from a large earthquake becomes insignificant at a distance of three to four source sizes. This is a reason to examine below the closer part of the generalized vicinity of

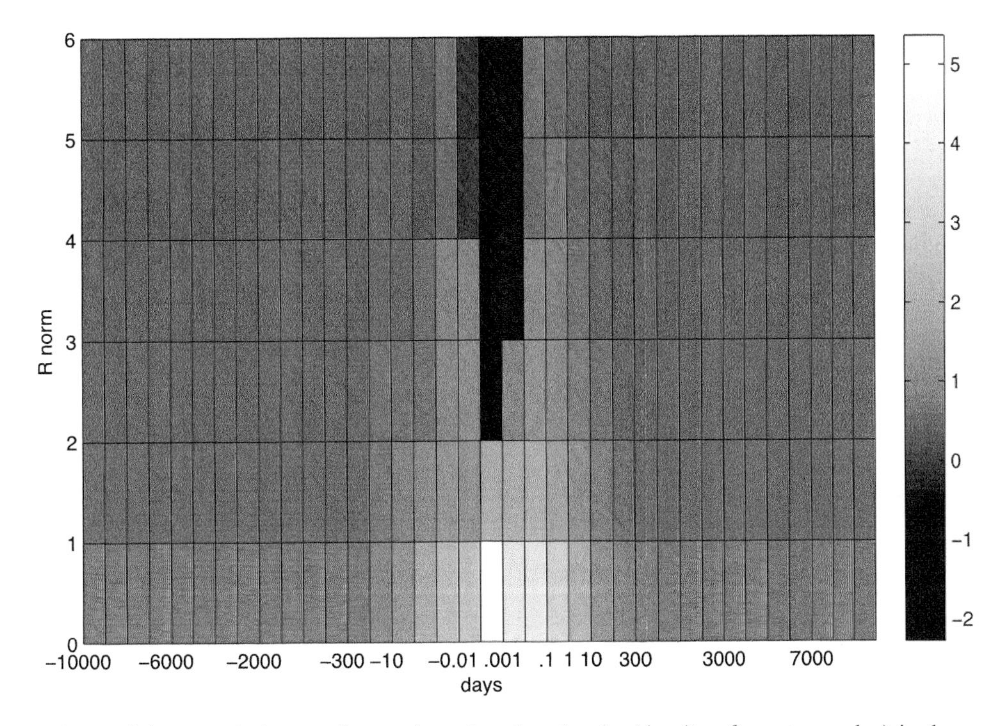

Figure 2. The spatial-temporal change of a number of earthquakes log(density of events per day) in the generalized vicinity of strong earthquake. The distance R norm from the main shock epicenter is given in norm source size units.

strong earthquake limited to three source sizes. The same $3 \times L$ limitation was used in Figure 1 for foreshock and aftershock examination.

Examination of the rate of seismicity in the generalized vicinity of a strong earthquake shows also that the seismic activity outside the zone of three earthquake source sizes decreases in a small time vicinity of the main shock. A small decrease could be noted 1 day before the main shock, and a quite noticeable decrease occurs about a quarter of an hour around the main shock. This effect takes place at distances from three to six to seven earthquake source sizes. It seems possible to explain this effect by development of a process of weakening in the closer vicinity of preparation of a strong earthquake. Below, some other evidence of strength decrease in the vicinity of strong earthquake will be presented. Hence, it can be suggested that the close vicinity of a rupture zone can be treated as becoming "softer" slightly before the earthquake occurrence. In this case, one could expect strain rearrangement from the outer "rigid" region into the inner "soft" region where the large earthquake is about to occur. As a result of stress rearrangement, the probability of earthquake occurrence in the outer "rigid" vicinity of the strong earthquake would become somewhat lower. This very decrease in seismicity can be seen in Figure 2. This effect provides an explanation why the area equal to $7 \times L$ source size from the strong earthquake epicenter was examined.

4. THE b VALUE CHANGE IN THE GENERALIZED VICINITY OF A STRONG EARTHQUAKE

The decrease in b value is used widely in algorithms of an earthquake prediction as an indicator of increase of probability of a strong earthquake occurrence [*Shebalin*, 2006; *Zav'yalov*, 2006]. Thus, it seems reasonable to examine the change in b values in the generalized vicinity of a strong earthquake. The catalog USGS/NEIC was used to present the change in b values (the similar results are obtained in the examination of the Harvard catalog).

The maximum likelihood method was used for b values estimation [*Utsu*, 1965; *Marzocchi and Sandri*, 2003]. By this method, a b value is calculated from

$$b = \log(e)/(M_{av} - M_c), \tag{3}$$

where M_{av} is the average magnitude for each subset of data, and M_c is the lower magnitude limit used in the analysis, $M_c = 4.7$ here. Discreteness of magnitude values because of aggregation in 0.1 bins is small; it influences the b values weakly and uniformly; therefore, it is not taken into account. The maximum likelihood method (equation (3)) gives a quite stable estimation for a number of events exceeding 50. Having this in mind, the groups consisting of 50 subsequent events were used in b value determination. The data points in Figure 3 reflect the b values obtained for such groups with a step of 25 events; thus, the data points are independent of those next to the adjacent ones.

As can be seen in Figure 3, there is an evident tendency of decrease in b values in the time vicinity of the generalized main shock, and this tendency increases strongly with the approaching of the moment of the main shock. A sharp increase in b values in the aftershock sequence takes place during the first several days after the main shock. A slower increase in b values takes place in the following 100 days. These features are fairly expectable and agree with a tendency of lowermost b values in the very beginning of the aftershock sequences and with an increase of b value in the further evolution of aftershock sequences [*Smirnov and Ponomarev*, 2004; *Mandal and Rastogi*, 2005; *Rodkin*, 2008]. A similar tendency was found in the examination of acoustic emission data [*Smirnov and Ponomarev*, 2004]. New findings consist of stronger decrease than it was found before and symmetrical character of this decrease for foreshock and aftershock sequences. The amplitude of b value decrease appears to be proportional to the logarithm of time remaining from the moment of the main shock. Such type of behavior is typical of the critical processes. The new findings were revealed as a result of the use of a huge amount of data available in SEGV.

It may be suggested that the strong decrease in b value observed in the narrow vicinity of a strong earthquake testifies for an increase of stress values in this zone. Different authors [*Wiemer and Wyss*, 1997; *Schorlemmer and Wiemer*, 2005; *Amorese et al.*, 2010] have argued that b values tend to be lower in high-stress zones prone to occurrence of larger earthquakes. This phenomenon has been observed also in laboratory experiments [*Scholz*, 1968; *Wyss*, 1973]. But alternative treatment appears to be more plausible. The change in b value displays the relation between the stress level and the strength of material. Hence, the decrease in b value can display both the increase in stress level and the decrease in strength of the material. We believe that the complex of anomalies revealed in the generalized vicinity of strong earthquake (which were discussed above and will be mentioned below) testifies for the decrease of strength in the area of strong earthquake preparation rather than for an increase in a stress level. Note also in this connection that throughout the Earth's lithosphere, a negative correlation between the density of earthquakes and typical apparent stress values resolutely predominates [*Rodkin*, 1996, 2001a]. One could expect that a positive correlation should take place if the level of seismic activity depends from the stress level, but factually, the negative correlation takes place. The obtained results testify that the role of decrease in

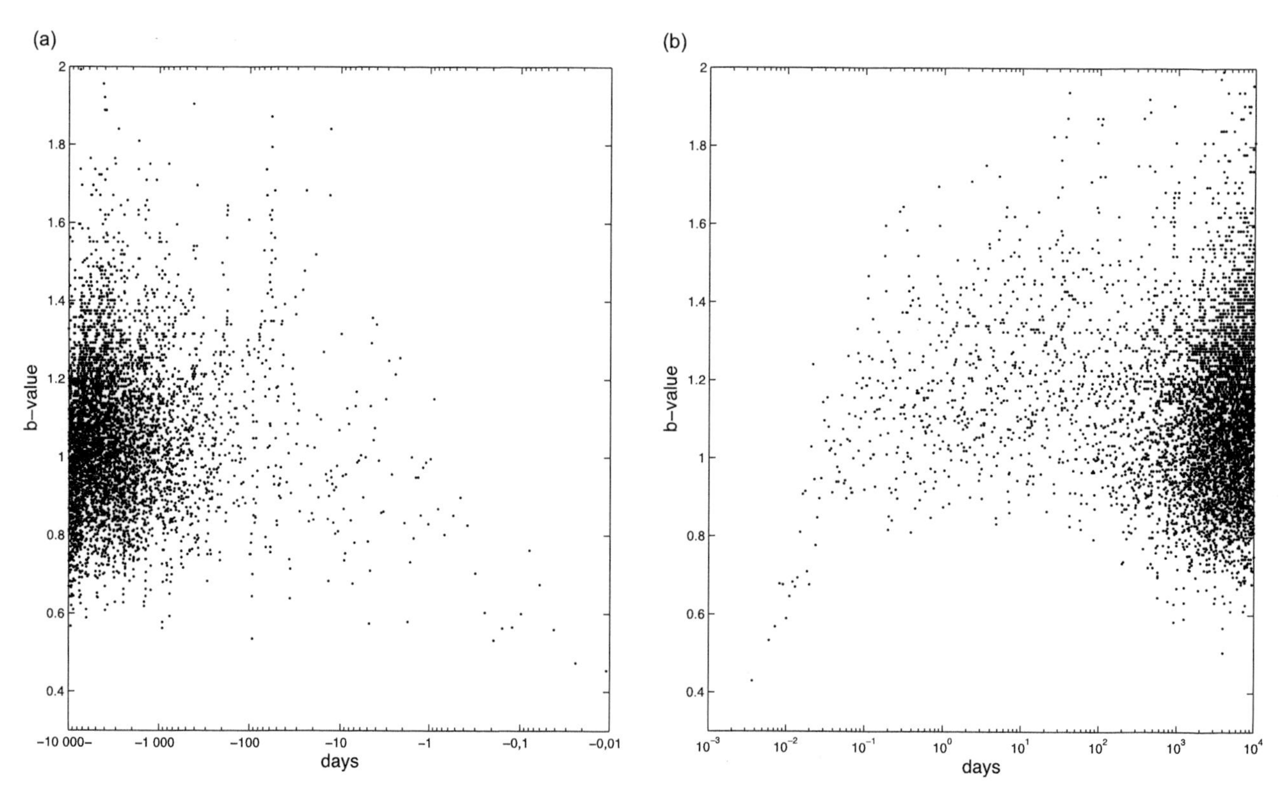

Figure 3. The change of mean b values in the generalized vicinity of strong earthquake: (a) foreshock and (b) aftershock as in Figure 1.

rock strength predominates in a changeability of seismic activity and in decrease of b values in SEGV.

5. CHANGE IN EARTHQUAKE SOURCE CHARACTERISTICS

The data presented in Figures 1 and 2 provide convincing evidence that the evolution of instability inherent to strong earthquake origin manifests itself in the power law growth of a rate of a number of earthquakes. The same power law–like tendency can be seen in a change of the b values (Figure 3) though with a lower statistical reliability.

In this section, the indications of origin of instability in parameters of individual earthquakes will be discussed. The only parameter characterizing the seismic stresses that can be determined for all events described in the Harvard catalog is an apparent stress σ_a value that can be found from the well-known relationship [*Kasahara*, 1981; *Abe*, 1982].

$$\sigma_a = \mu Es/Mo, \tag{4}$$

where μ is the shear modulus, Mo is the seismic moment, and Es is the earthquake energy calculated from the magnitude M_b value as follows:

$$\log\{Es(\text{ergs})\} = 1.5M_b + 11.8. \tag{5}$$

The shear modulus μ in equation (4) was assumed to be depth dependent in accordance with the mean crust and uppermost mantle characteristics [*Bullen*, 1975]. Figure 4 presents the median σ_a values for sets of consecutive earthquakes from the SEGV. The distribution law of individual σ_a values is close to the lognormal distribution [*Rodkin*, 2001b] that means that a rather heavy tail takes place, and the scattering in σ_a values is large. Having this in mind, the number of events in every subset was chosen large and equal to 200 events with a step between neighboring sets of 100 events.

Figure 4 shows that median σ_a values decrease progressively with the approaching of the moment of the main shock. The character of this decrease is close for foreshock and aftershock sequences, and the mean duration of that decrease can be estimated as 100 days from the main shock date. Thin dotted lines show the scatter of median σ_a values evaluated by the boot-strap method. Note that this scatter characterizes only a part of the total scattering of σ_a values; the total scattering in σ_a values could be crudely characterized as twice more. However, despite the essential scattering, the tendency of a prominent exponential law–like decrease in σ_a values seems to be quite valid.

Note that a tendency of a prominent decrease of σ_a values in a close vicinity of a strong earthquake testifies definitely for the strength decrease instead of the alternative possibility of increase of a stress level.

Evidence in support of the strength decrease in the generalized vicinity of strong earthquake can be found also from the examination of the stress-strain state. The relative

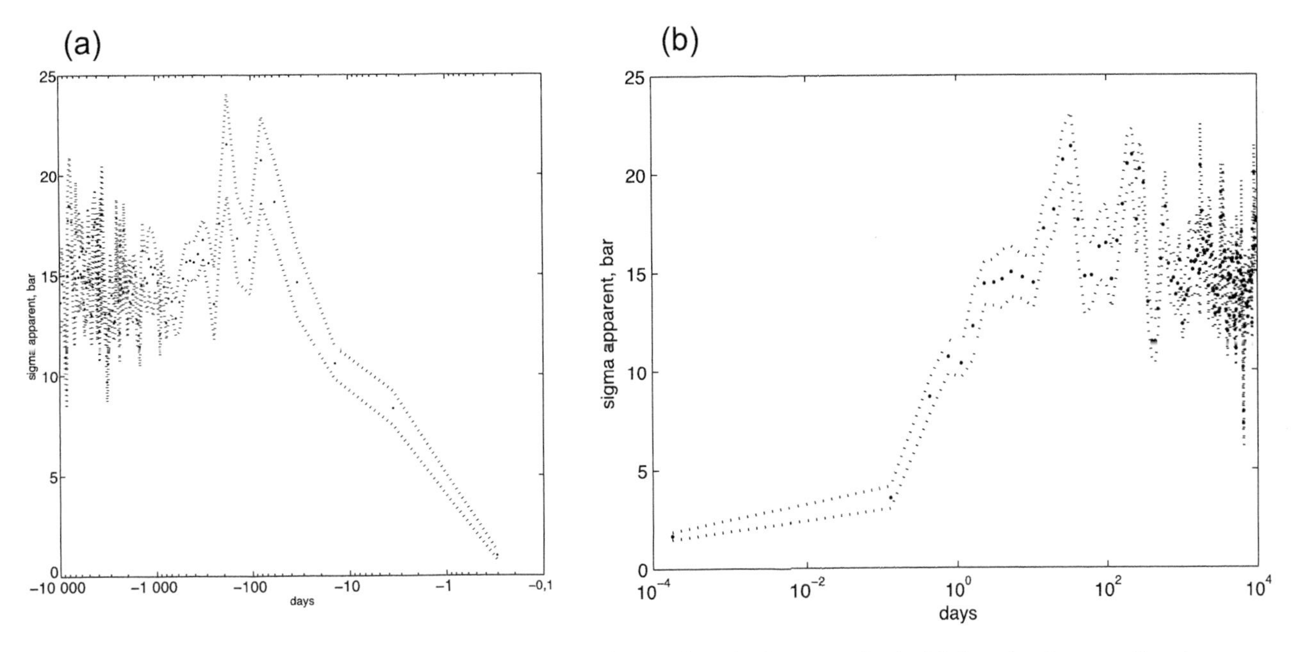

Figure 4. The change of median apparent stress σ_a (bars) values in the generalized vicinity of a strong earthquake: (a) foreshock and (b) aftershock as in Figure 1. Thin dotted lines show the scattering.

orientations of earthquake focal mechanisms were examined. The values of disorientation between the focal mechanism of the main shock (characterizing the dominant strain mode in the area) and that of the weak earthquake occurring in the vicinity of this main shock were examined. Their disorientation is characterized by the scalar product of components of the seismic moment tensor for the main earthquake Mo_{ij} $(j > i)$ and a current weak earthquake M_{ij}. The scalar product (the pairwise product of the components is denoted by an asterisk) is normalized by the product of norms of the respective tensors $||Mo_{ij}||$ and $||M_{ij}||$:

$$K = Mo_{ij} * M_{ij} / \{||Mo_{ij}|| \cdot ||M_{ij}||\}. \quad (6)$$

The case $K = 1$ corresponds to an identical orientation of two tensors, while the condition $K \approx 0$ will be fulfilled (in average, when a sufficient number of pairs of events are examined) when the focal mechanisms' orientations are uncorrelated. The data points in Figure 5 reflect the median K values obtained for groups of 20 consecutive events with a step of 10 events. As it can be seen from Figure 5, the maximum (closer to unity) values of K are typical of the close vicinity of the main shock. The physical nature of this finding is not quite clear, but it could be suggested that the effect is related to a strength decrease. It may be suggested that the strength decrease will first remove the local stresses (an analog of the condition of local equilibrium in nonequilibrium thermodynamics). When the local stresses are relaxed, the overall stress field will approach the regional one responsible for the main shock focal mechanism orientation.

Figure 6 presents the change of the mean $M_b - M_w$ values. Subsets consisting of 50 events with a step of 25 events are

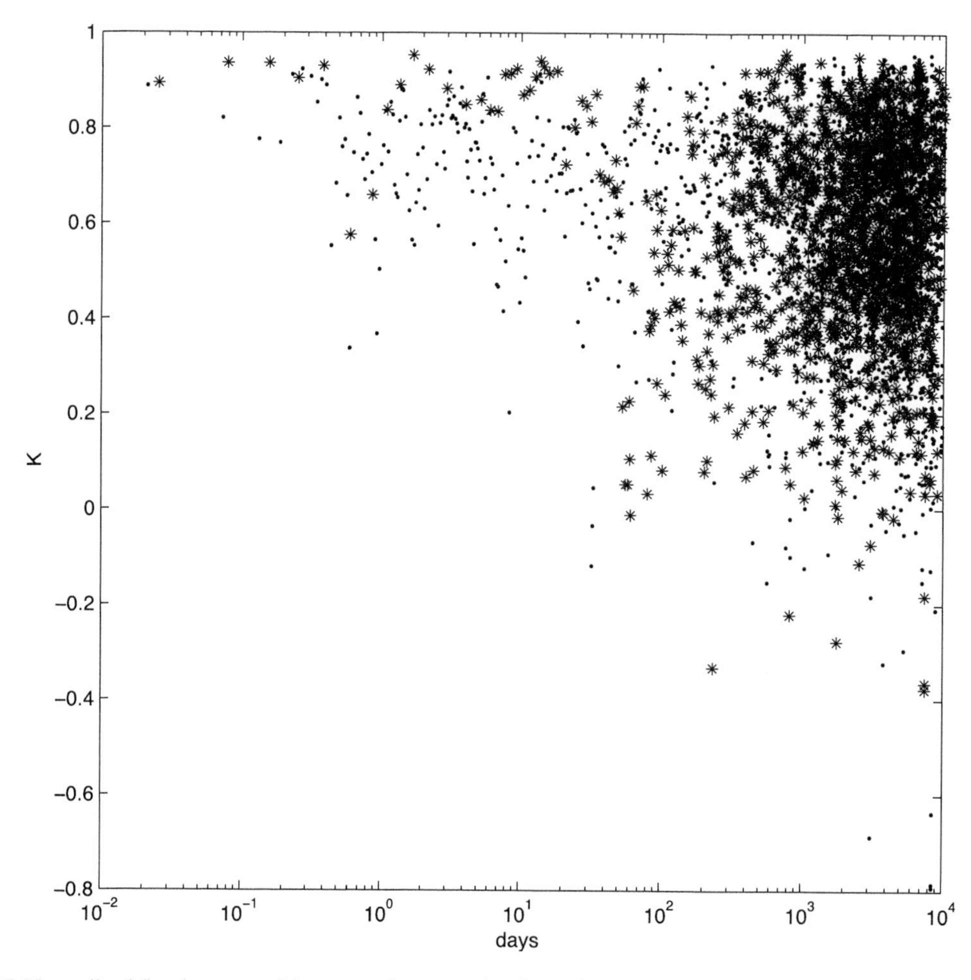

Figure 5. Normalized (by the norm of the respective tensors) values of the scalar product K between the seismic moment tensors for the main shock and a current event in the foreshock (stars) and aftershock (dots) sequences in the general vicinity of strong earthquake. Medians of K values in groups consisting of 20 consecutive events with step 10 events are shown.

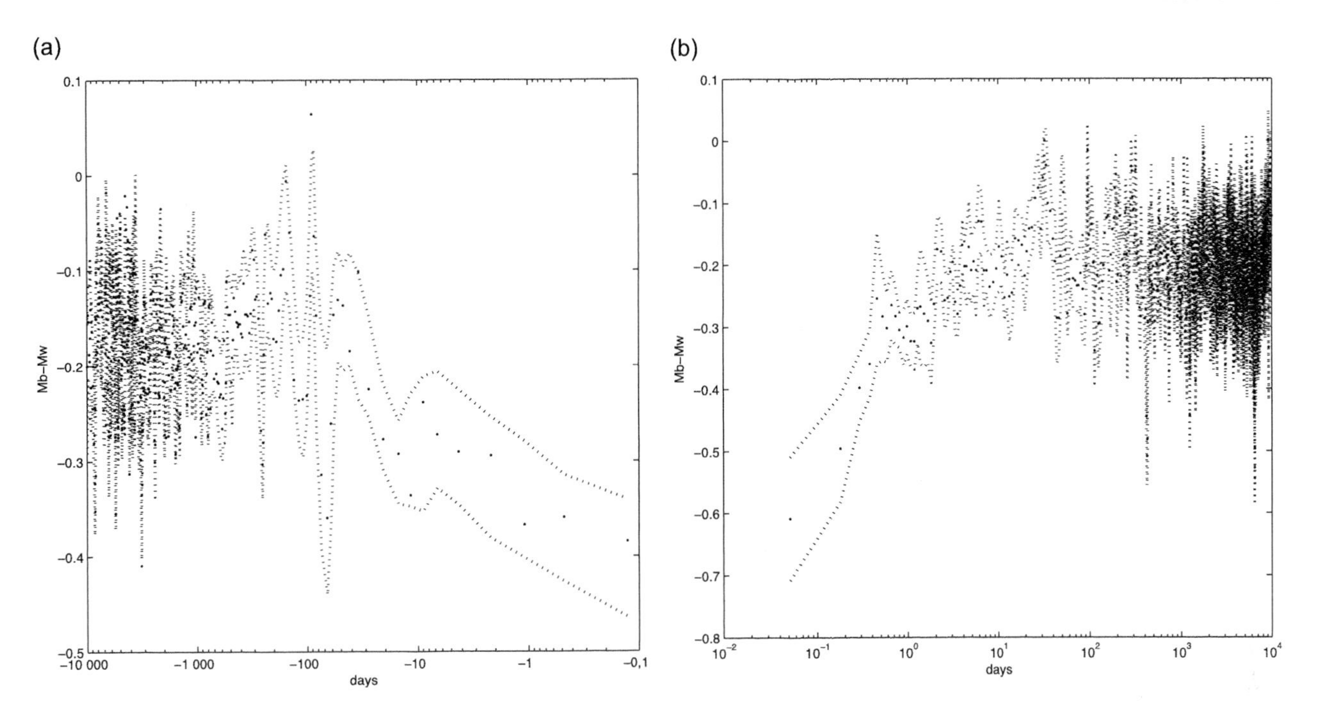

Figure 6. The change of the mean $M_b - M_w$ values in the generalized vicinity of strong earthquake: (a) foreshock and (b) aftershock as in Figures 1 and 4.

used in this case. Figure 6 shows that $M_b - M_w$ values decrease while approaching the strong earthquake moment in a similar manner with the σ_a decrease (Figure 4). This result is expectable because $M_b - M_w$ values closely correlate with apparent stress σ_a values. This relation is evident from the formula for σ_a calculation (equation (4)), where earthquake energy is in the numerator, and the seismic moment is in the denominator.

However, the decrease in $M_b - M_w$ values while approaching the moment of a strong earthquake can have also a different interpretation. As it was mentioned above, the change in $M_b - M_w$ values characterizes the relative input of high- and lower-frequency domains in the seismic oscillations spectra. Thus, the decrease in $M_b - M_w$ values corresponds to the increase of input of the lower-frequency oscillations. The increase of input of a low-frequency domain with an approach of a strong earthquake is not unexpected. The similar results were obtained earlier. It was found [*Levin and Sasorova*, 1999; *Sobolev*, 2003; *Sobolev and Lyubyshin*, 2006] from the examination of variability of microseism spectra and the rate of a number of weak earthquakes that an increase in relative intensity of lower-frequency modes occurs frequently in connection with large earthquake occurrences. This finding appears to agree with a general tendency in the development of instability. It has been shown [*Ma*, 1976; *Haken*, 1978; *Akimoto and Aizawa*, 2006] that an increase in lower-frequency input can be expected during the development of instability. This effect is known in the theory of critical phenomena as the critical slowing-down effect, and it will be discussed below in more detail.

The last parameter of seismicity that is examined here is also connected with an increase of input of a lower-frequency domain. The mean duration $\Delta\tau$ of the earthquake process was evaluated as a difference between the event time determined from the first arrival data and from the seismic moment calculation. Being averaged, this parameter characterizes the typical half duration $\Delta\tau$ of the process of the seismic rupture. The change in mean $\Delta\tau$ values during the foreshock and aftershock sequences is presented in Figure 7. The medium length subsets consisting of 100 events with a step of 50 events are used in this case. As it can be seen in Figure 7, the mean $\Delta\tau$ values have a clear tendency of a linear-like growth with a decrease of a logarithm of time from the main event occurrence. This tendency is similar to the previous ones and is also typical of critical processes: the relaxation time in critical processes is known to increase while approaching the critical point [*Ma*, 1976]. It should be added that the effect of $\Delta\tau$ increase is partly connected with the tendency of magnitude increase at the time of the approaching of a strong earthquake and with a tendency of longer duration of a stronger earthquake. But both these tendencies could explain

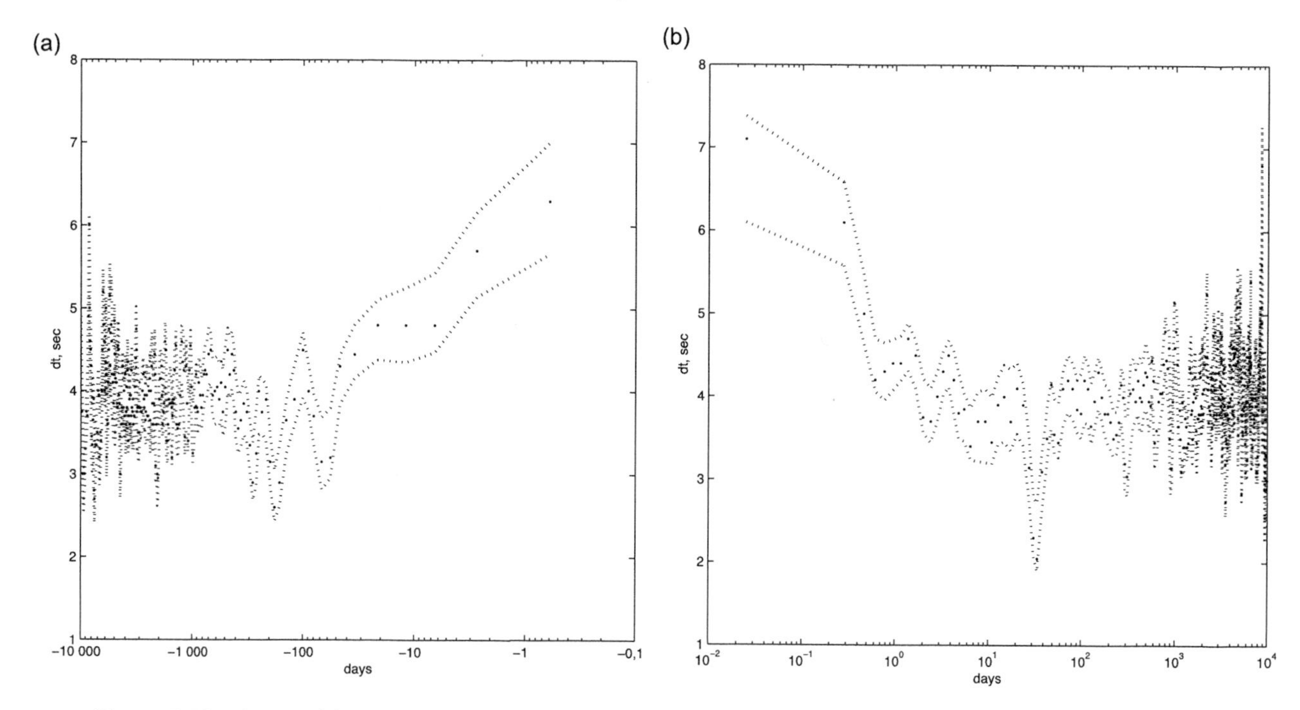

Figure 7. The change of the mean half duration of the seismic process $\Delta\tau$ (seconds) in the generalized vicinity of strong earthquake: (a) foreshock and (b) aftershock as in Figures 1 and 4.

an increase in mean $\Delta\tau$ values up to a few dozen percent only, while an essentially stronger increase takes place in fact (Figure 7).

6. CONNECTION WITH THE CRITICAL PROCESSES AND WITH THE CRITICAL SLOWING-DOWN EFFECT

Critical type of behavior is believed to be typical of different types of natural hazards including major earthquakes [*Sornette*, 2000; *Zöller and Hainzl*, 2002; *Keilis-Borok and Soloviev*, 2003; *Turcotte and Malamud*, 2004]. Thus, critical effects could be expected to be precursors of major earthquakes. Moreover, having in mind that the variance of fluctuations exhibits a power law divergence approaching the bifurcation point, the so-called time-to-failure prognosis could be carried out. This approach was brought to life in a few cases [*Sornette*, 2000; and references herein].

A very simple model [*Haken*, 1978] that illustrates the critical slowing down effect is the following (Figure 8). Let us imagine a ball on a smooth curved surface; the ball undergoes noise oscillations in the vicinity of an equilibrium point (that is, one of local minimums of the surface height). If a barrier between this minimum and the deeper neighbor minimum smoothly decreases, the character of small noise movements of the ball will change: The amplitude of deviations from the local equilibrium point and the period of oscillations of such noise movements of the ball will increase. When the barrier disappears, the returning force acting at the ball will decrease to zero, and the period of corresponding noise oscillations becomes infinite (theoretically, in a linear approximation).

The ball model disagrees in a few important points from the classical model of critical processes, and thus, it can be treated as some sort of critical slowing-down-like model. Moreover, this very simple model appears to be even more relevant to the seismic process than the theory of critical processes and the well-known self-organized criticality model. The fundamental point of the critical processes (and

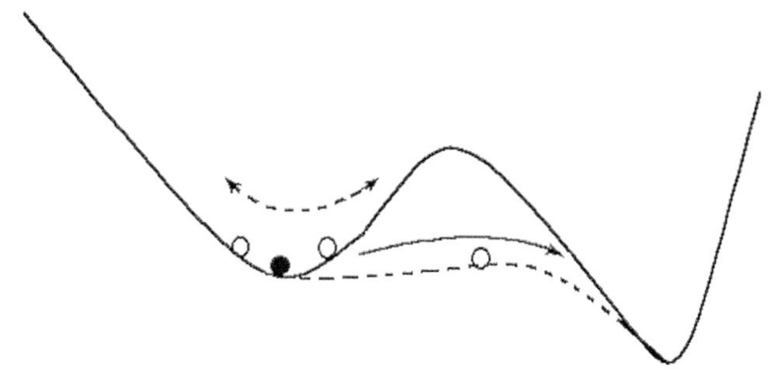

Figure 8. Ball model. The returning force and effective modulus decrease, and the periods of noise oscillations increase while approaching the instability point. Energy surface profile corresponding to a lower stability situation is given by a dashed line.

the second-order phase transitions as a widely used example of such processes) is the absence of energy release (or absorption) during the transformation. But just a huge energy release is a typical feature of a strong earthquake, and the ball model incorporates the effect of energy release.

7. APPLICATION TO EARTHQUAKE PREDICTION

The construction of the SEGV gives possibility to significantly enlarge the volume of data available for statistical examination of seismic behavior connected with strong earthquake occurrence. As a result of SEGV examination, a complex of precursory anomalies was described. The existence of this complex can be treated as an evidence of the fact that the short-term earthquake prediction can be put into effect.

The majority of the described precursory anomalies were known before, and their typical parameters were specified, but other precursor anomalies were not known earlier. The obtained description of a complex of precursor anomalies gives possibility to use it in practice of earthquake prediction in the cases when detailed current seismic information is available.

According to the preliminary interpretation of the precursor anomalies, the material softening instead of the usually suggested increase in a stress level appears to be the predominating factor of seismicity. A similar result was recently described by *Gao and Crampin* [2004].

The power law–like character of change of preshock and postshock anomalies supports the widely used treatment of large earthquakes as a critical-like phenomenon [*Shebalin et al.*, 2000; *Sornette*, 2000; *Zöller and Hainzl*, 2002; *Turcotte and Malamud*, 2004]. It should be pointed out, however, that the mentioned anomalies are not specific, i.e., they do manifest the development of instability, but these anomalies do not contain an explicit indication of the physical mechanism responsible for the development of instability in a particular case of earthquake origin. Thus, these anomalies can hardly supply the needed information that could help to reveal the physical mechanism of development of material softening, which was discussed above.

It seems important to reveal the features that could characterize the processes that produce the strength decrease. We will briefly mention here one such anomaly connected with the difference between the depth of earthquake obtained from the seismic moment solution and the depth of the hypocenter. A negative value of the difference between these depth estimations indicates a rupture propagation moving toward the surface. For the Harvard catalog (for earthquakes with $H < 70$ km), the difference in these depth values for individual earthquakes has a large scattering $\Delta H = -5 \pm 15$ km. But the mean ΔH value obtained by the bootstrap method is mean $(\Delta H) = -5 \pm 0.13$ km, which means that the rupture propagation toward the surface does predominate. The mean ΔH values depend on the earthquake's depth [*Rodkin*, 2008]. The events from the close vicinity of the main shock have greater negative ΔH values; that is, for these events, the tendency of earthquake rupture propagation toward the surface increases. This tendency may be related with an increase in activity of the fluid phase of low density, whose tendency to penetrate into areas of lower pressure may facilitate the rupture development in the direction to the Earth's surface [*Rodkin*, 2008].

8. CONCLUSION

The SEGV is constructed using the data from the Harvard seismic moment catalog and from the USGS/NEIS earthquake catalog. The construction of SEGV gives a possibility to substantially enlarge the volume of data available for statistical examination of preseismic and postseismic phenomena. It was confirmed that the dynamics of the averaged foreshock and aftershock sequences is well described by a power law cascade model. Thus, the disagreement of the evolution of foreshock and aftershock sequences from the power law model found for a number of strong earthquakes [*Romashkova and Kosobokov*, 2001] has a stochastic character. The carried out analysis gives a possibility to specify the characteristic parameters of the inverse (foreshock) cascade and of the aftershock sequence.

Besides the foreshock and aftershock cascades, a few new anomalies were found as a result of the SEGV examination. The apparent stress σ_a and $M_b - M_w$ values and b values were found to decrease, and the half duration of earthquake source oscillation process $\Delta\tau$ was found to increase while approaching the moment of the main shock. The growth of uniformity in the character of stress-strain state was also found. These anomalies were found or described in more detail because of a great amount of data available in the SEGV examination.

Both foreshock and aftershock anomalies increase approaching the main shock moment as a logarithm of time interval from the main shock occurrence. In a majority of cases, the anomalies differ noticeably from the background seismic regime since about 100 days before the main shock occurrence. These results can supplement the practice of earthquake prediction.

Acknowledgments. The work was supported by the Russian Foundation for Basic Research, grant 11-05-00663.

REFERENCES

Abe, K. (1982), Magnitude, seismic moment and apparent stress for major deep earthquakes, *J. Phys. Earth*, *30*, 321–330.

Akimoto, T., and Y. Aizawa (2006), Scaling exponents of the slow relaxation in non-hyperbolic chaotic dynamics, *Nonlin. Phenomena Complex Syst.*, *9*(2), 178–182.

Amorese, D., J. R. Grasso, and P. A. Rydelek (2010), On varying b-values with depth: Results from computer-intensive tests for southern California, *Geophys. J. Int.*, *180*, 347–360.

Bowman, D. D., G. Ouillon, C. G. Sammis, A. Sornette, and D. Sornette (1998), An observational test of the critical earthquake concept, *J. Geophys. Res.*, *103*(B10), 24,359–24,372.

Bullen, K. E. (1975), *The Earth's Density*, 420 pp., Chapman and Hall, London, U. K.

Faenza, L., S. Hainzl, and F. Scherbaum (2009), Statistical analysis of the central-Europe seismicity, *Tectonophysics*, *470*, 195–204.

Gao, Y., and S. Crampin (2004), Observations of stress relaxation before earthquakes, *Geophys. J. Int.*, *157*, 578–582.

Haken, H. (1978), *Synergetics*, 355 pp., Springer, Berlin.

Helmstetter, A., D. Sornette, and J.-R. Grasso (2003), Mainshocks are aftershocks of conditional foreshocks: How do foreshock statistical properties emerge from aftershock laws, *J. Geophys. Res.*, *108*(B1), 2046, doi:10.1029/2002JB001991.

Kasahara, K. (1981), *Earthquake Mechanics*, 272 pp., Cambridge Univ. Press, Cambridge, U. K.

Keilis-Borok, V. I., and A. A. Soloviev (2003), *Nonlinear Dynamics of the Lithosphere and Earthquake Prediction*, 337 pp., Springer, Berlin.

Lennartz, S., A. Bunde, and D. L. Turcotte (2008), Missing data in aftershock sequences: Explaining the deviations from scaling laws, *Phys. Rev. E*, *78*, 041115, doi:10.1103/PhysRevE.78.041115.

Levin, B. V., and E. V. Sasorova (1999), Low frequency seismic signals as regional precursors of earthquakes, *Vulkanol. Seismol.*, *4*, 108–115.

Lindman, M., B. Lund, and R. Roberts (2010), Spatiotemporal characteristics of aftershock sequences in the South Iceland Seismic Zone: Interpretation in terms of pore pressure diffusion and poroelasticity, *Geophys. J. Int.*, *183*, 1104–1118.

Ma, S.-K. (1976), *Modern Theory of Critical Phenomena*, 561 pp., Benjamin, Reading, Mass.

Malamud, B. D., G. Morein, and D. L. Turcotte (2005), Log-periodic behavior in a forest-fire model, *Nonlin. Processes Geophys.*, *12*, 575–585.

Mandal, P., and B. K. Rastogi (2005), Self-organized fractal seismicity and b-value of aftershocks of 2001 Bhuj earthquake in Kutch (India), *Pure Appl. Geophys.*, *162*, 53–72.

Marzocchi, W., and L. Sandri (2003), A review and new insights in the estimation of b-value and its uncertainty, *Ann. Geophys.*, *46*, 1271–1282.

Ogata, Y. (1998), Space–time point-process models for earthquake occurrence, *Ann. Inst. Stat. Math.*, *50*, 379–402.

Ogata, Y., T. Utsu, and K. Katsura (1995), Statistical features of foreshocks in comparison with other earthquake clusters, *Geophys. J. Int.*, *121*, 233–254.

Rodkin, M. V. (1996), Contradictions in the resent seismogenetical notions, *Phys. Chem. Earth.*, *21*(4), 257–260.

Rodkin, M. V. (2001a), The problem of the earthquake source physics: Models and contradictions, *Izv. Russ. Acad. Sci. Phys. Solid Earth, Engl. Transl.*, *37*(8), 653–662.

Rodkin, M. V. (2001b), Statistics of apparent stresses in relation to the origin of earthquake source, *Izv. Russ. Acad. Sci. Phys. Solid Earth, Engl. Transl.*, *37*(8), 663–672.

Rodkin, M. V. (2006), Implications of differences in thermodynamic conditions for the seismic process, *Izv. Russ. Acad. Sci. Phys. Solid Earth, Engl. Transl.*, *42*(9), 745–754.

Rodkin, M. V. (2008) Seismicity in the generalized vicinity of large earthquakes, *J. Volcanol. Seismol.*, *2*(6), 435–445.

Romashkova, L. L., and V. G. Kosobokov (2001), The dynamics of seismic activity before and after great earthquakes of the world, 1985–2000 [in Russian], *Vychisl. Seismol.*, *32*, 162–189.

Scholz, C. H. (1968), The frequency-magnitude relation of microfracturing in rock and its relation to earthquakes, *Bull. Seismol. Soc. Am.*, *58*, 399–415.

Schorlemmer, D., and S. Wiemer (2005), Microseismicity data forecast rupture area, *Nature*, *434*, 1086.

Shcherbakov, R., J. Van Aalsburg, J. B. Rundle, and D. L. Turcotte (2006), Correlations in aftershock and seismicity patterns, *Tectonophysics*, *413*, 53–62.

Shebalin, P. N. (2006), A methodology for prediction of large earthquakes with waiting times less than one year [in Russian], *Vychisl. Seismol.*, *37*, 7–182.

Shebalin, P. N., I. Zaliapin, and V. I. Keilis-Borok (2000), Premonitory rise of the earthquakes' correlation range: Lesser Antilles, *Phys. Earth Planet. Inter.*, *122*, 241–249.

Smirnov, V. B., and A. V. Ponomarev (2004), Seismic regime relaxation properties from in-situ and laboratory data, *Izv. Russ. Acad. Sci. Phys. Solid Earth, Engl. Transl.*, *40*(10), 807–816.

Sobolev, G. A. (2003), Evolution of periodic oscillations of seismic activity before strong earthquakes, *Izv. Russ. Acad. Sci. Phys. Solid Earth, Engl. Transl.*, *11*, 3–15.

Sobolev, G. A., and A. A. Lyubushin (2006), Microseismic impulses as earthquake precursors, *Izv. Russ. Acad. Sci. Phys. Solid Earth,* Engl. Transl., *42*(9), 721–733.

Sobolev, G. A., and A. V. Ponomarev (2003), *Physics of Earthquakes and Precursors* [in Russian], 270 pp., Nauka, Moscow.

Sornette, D. (2000), *Critical Phenomena in Natural Sciences*, 450 pp., Springer, Berlin.

Turcotte, D. L., and B. D. Malamud (2004), Landslides, forest fires, and earthquakes: Examples of self-organized critical behavior, *Physica A*, *340*(4), 580–589.

Utsu, T. (1965), A method for determining the value of b in a formula $\log n = a - bM$ showing the magnitude-frequency relation for earthquakes [in Japanese], *Geophys. Bull. Hokkaido Univ.*, *13*, 99–103.

Utsu, T., Y. Ogata, and R. S. Matsu'ura (1995), The centenary of the Omori formula for a decay law of aftershock activity, *J. Phys. Earth*, *43*, 1–33.

Wiemer, S., and M. Wyss (1997), Mapping the frequency-magnitude distribution in asperities: An improved technique to calculate recurrence times?, *J. Geophys. Res.*, *102*(B7), 15,115–15,128.

Wyss, M. (1973), Towards a physical understanding of the earthquake-frequency distribution, *Geophys. J. R. Astron. Soc.*, *31*, 341–359.

Zav'yalov, A. D. (2006), *Intermediate-Term Earthquake Prediction: Principles, Techniques, Implementation* [in Russian], Nauka, Moscow.

Zöller, G., and S. Hainzl (2002), A systematic spatiotemporal test of the critical point hypothesis for large earthquakes, *Geophys. Res. Lett.*, *29*(11), 1558, doi:10.1029/2002GL014856.

M. V. Rodkin, Institute of Earthquake Prediction Theory and Mathematical Geophysics, Moscow 117997, Russia. (rodkin@-rodkin@mitp.ru)

Characterizing Large Events and Scaling in Earthquake Models With Inhomogeneous Damage

Rachele Dominguez,[1] Kristy Tiampo,[2] C. A. Serino,[3] and W. Klein[4]

Natural earthquake fault systems are highly nonhomogeneous. The inhomogeneities occur because the Earth is made of a variety of materials that hold and dissipate stress differently. In this work, we study scaling in earthquake fault models that are variations of the Olami-Feder-Christensen and Rundle-Jackson-Brown models. We examine the effect of spatial inhomogeneities due to damage and inhomogeneous stress dissipation on the magnitude-frequency scaling relation and the occurrence of large events. Spatially rearranging dead sites on a given lattice affects the numerical distributions of the effective stress dissipation parameters and the scaling behavior of large avalanche events, depending on the homogeneity of the damage and the length scales associated with the clustered dead sites. We find that the scaling depends not only on the amount of damage but also on the spatial distribution of that damage, such that large events require the existence of interconnected regions with lower stress dissipation. In addition, the largest events are more prevalent if the interaction range is long but shorter than the intrinsic damage length scale.

1. INTRODUCTION

The impact to life and property from large earthquakes is potentially catastrophic. In 2010, the *M*7.0 earthquake in Haiti became the fifth most deadly earthquake on record, killing more than 200,000 people and resulting in $8 billion (U.S. dollars) in direct damages [*Cavallo et al.*, 2010]. Direct economic damage from the *M*8.8 earthquake that struck Chile in February of 2010 could reach U.S. $30 billion or 18% of Chile's annual economic output [*Kovacs*, 2010]. As a result of their potential regional and national impact, research into the dynamics of the earthquake fault system, with the goal of better understanding the timing and location of these extreme events, has been ongoing on some level for almost 100 years [*Kanamori*, 1981; *Jordan*, 2006].

While it has long been recognized that temporal and spatial clustering is evident in seismicity data, much of the research associated with these patterns in the early years focused on a relatively small fraction of the events, primarily at the larger magnitudes [*Kanamori*, 1981]. Examples include, but are not limited to, characteristic earthquakes and seismic gaps [*Swan et al.*, 1980; *Haberman*, 1981; *Bakun et al.*, 1986; *Ellsworth and Cole*, 1997], Mogi donuts and precursory quiescence [*Mogi*, 1969; *Yamashita and Knopoff*, 1989; *Wyss et al.*, 1996], temporal clustering [*Frohlich*, 1987; *Press and Allen*, 1995; *Dodge et al.*, 1996; *Eneva and Ben-Zion*, 1997; *Jones and Hauksson*, 1997], aftershock sequences [*Gross and Kisslinger*, 1994; *Nanjo et al.*, 1998], stress transfer and earthquake triggering over large distances [*King et al.*, 1994; *Deng and Sykes*, 1996; *Gomberg*, 1996; *Pollitz and Sacks*, 1997; *Stein*, 1999; *Brodsky*, 2006], scaling relations [*Rundle*,

[1]Department of Physics and Astronomy, Western Kentucky University, Bowling Green, Kentucky, USA.

[2]Department of Earth Sciences, University of Western Ontario, London, Ontario, Canada.

[3]Department of Physics, Boston University, Boston, Massachusetts, USA.

[4]Department of Physics and Center for Computational Science, Boston University, Boston, Massachusetts, USA.

Extreme Events and Natural Hazards: The Complexity Perspective
Geophysical Monograph Series 196

10.1029/2011GM001082

1989; *Pacheco et al.*, 1992; *Romanowicz and Rundle*, 1993; *Saleur et al.*, 1996], pattern recognition [*Keilis-Borok and Kossobokov*, 1990; *Kossobokov et al.*, 1999], and time-to-failure analyses [*Bufe and Varnes*, 1993; *Bowman et al.*, 1998; *Brehm and Braile*, 1998; *Jaume and Sykes*, 1999]. Although this body of research represents important attempts to describe these characteristic patterns using empirical probability density functions, it was hampered by the poor statistics associated with the small numbers of moderate-to-large events either available or considered for analysis.

Despite the multitude of space-time patterns of activity observed, one significant problem associated with studies of the earthquake fault system is forecasting future large events when the underlying dynamics of the system are not observable [*Herz and Hopfield*, 1995; *Rundle et al.*, 2000a]. A second problem, equally serious, is that the nonlinear earthquake dynamics is strongly coupled across a vast range of space and time scales [*Kanamori*, 1981; *Bufe and Varnes*, 1993; *Main*, 1996; *Rundle et al.*, 1999, 2002; *Scholz*, 2002; *Turcotte*, 1997]. Important spatial scales range from the microscopic scale associated with friction to the tectonic plate boundary scale (10^3–10^4 km) associated with the driving force, while important temporal scales range from seconds during dynamic rupture to 10^3–10^4 years for the repeat times of the largest earthquakes. In particular, the relatively small number of the most extreme events occur very rarely, impacting the ability to evaluate the significance of the associated local and regional patterns in the instrumental and historic data using either deterministic or statistical techniques [*Jackson and Kagan*, 2006; *Schorlemmer and Gerstenberger*, 2007; *Vere-Jones*, 1995, 2006; *Zechar et al.*, 2010]. As a result, computational simulations are critical to our understanding of the dynamics and underlying physics of the earthquake system and the occurrence of its largest events [see, e.g., *Rundle et al.*, 2003].

Numerical and experimental models of rock fracture and the earthquake process suggest that spatial inhomogeneities in the fault network play an important role in the occurrence of large events [*Dahmen et al.*, 1998; *Turcotte et al.*, 2003; *Lyakhovsky and Ben-Zion*, 2009]. The spatial arrangement of these fault inhomogeneities is dependent on the geologic history of the fault, and because this history is typically quite complex, the spatial distribution of the various inhomogeneities occurs on many length scales. One way that the inhomogeneous nature of fault systems manifests itself is in the spatial patterns, which emerge in seismicity graphs [*Tiampo et al.*, 2002, 2007].

Despite their inhomogeneous nature, real faults are often modeled as spatially homogeneous systems. One argument for this approach is that earthquake faults have long-range stress transfer [*Klein et al.*, 2007], and if this range is longer than the length scales associated with the inhomogeneities of the system, the dynamics of the system may be unaffected by the inhomogeneities. However, it is not clear that this is the case. Consequently, it is important to investigate the relation between the stress transfer range and the spatial scales associated with the inhomogeneities.

In this work, we study the effect of the spatial heterogeneities and damage on the generation and behavior of extreme events and their effect on the scaling of the system in cellular automaton models of earthquake faults. The first slider block model for studying earthquake faults was introduced by *Burridge and Knopoff* [1967]. Here we use a variation of a model of massless blocks, originally introduced by Rundle, Jackson, and Brown (RJB) and re-introduced independently by Olami, Feder, and Christensen(OFC), to explore the effect of spatial inhomogeneities in earthquake-fault-like systems when stress transfer ranges are long, but not necessarily longer than the length scales associated with the inhomogeneities of the system [*Burridge and Knopoff*, 1967; *Olami et al.*, 1992]. For long-range stress transfer without inhomogeneities, as well as randomly distributed inhomogeneities [*Serino et al.*, 2011], such models have been found to produce scaling similar to Gutenberg-Richter scaling found in real earthquake systems [*Gutenberg and Richter*, 1956]. It has been shown that the scaling found in such models is due to a spinodal in the limit of long-range stress transfer [*Rundle and Klein*, 1993; *Klein et al.*, 2000].

Here we introduce inhomogeneities into the earthquake lattice models through the mechanism through which the stress is dissipated. Stress is dissipated both at the lattice site of failure (site dissipation) and at neighboring sites that are damaged (damage dissipation). Spatial inhomogeneities are incorporated by varying this stress dissipation throughout the system in different spatial arrangements. We find that the scaling for damaged systems depends not only on the amount of damage, but also on the spatial distribution of that damage as well as the relation of the spatial damage or dissipation to the stress transfer range. Studying the effects of various spatial arrangements of site dissipation provides insights into how to construct a realistic model of an earthquake fault, which is consistent with Gutenberg-Richter scaling.

2. MODEL

We use a 2-D cellular automaton model of an earthquake fault, which is a lattice model using simple rules to mimic the evolution of an earthquake avalanche event in discrete time steps. Our version is a variant of the RJB model [*Rundle and Jackson*, 1977; *Rundle and Brown*, 1991] and closely resembles the OFC model [*Olami et al.*, 1992]. We begin with a 2-D lattice, where each site is either dead (damaged) or alive (active). Each live site i contains an internal stress variable, $\sigma_i(t)$, which is a function of time. All stress variables are initially below a given threshold stress, σ^t, and greater than or equal to a residual stress, σ^r (both of which we assume to be spatially homogeneous). Sites transfer stress to z neighbors. Neighbors are defined as all sites within the transfer range, R. Initially, we randomly distribute stress to each site so that $\sigma^r \leq \sigma_i < \sigma^t$. We then increase the stress on all sites equally until one site reaches σ^t At this point, the site at the threshold stress fails. When a site fails, some fraction of that site's stress, given by $\alpha_i(\sigma^t - \sigma^r \mp \eta)$, is dissipated from the system, where α_i is a parameter that characterizes the fraction of stress dissipated from site i, and η is a random flatly distributed noise. The stress of the site is lowered to $\sigma^r \pm \eta$, and the remaining stress is distributed equally to the site's z neighbors.

To model more realistic faults, we use systems which are *damaged*, meaning they have both alive sites, which obey the rules outlined above, and dead sites, which do not hold any stress. Following the work of *Serino et al.* [2010], in addition to the stress dissipation regulated by the site dissipation parameter, α_i, we specify that any stress, which is passed to a neighboring dead site, also gets dissipated from the system. We can therefore regulate the spatial distribution of stress dissipation from the system with the distribution of the α_i and the placement of dead sites on the lattice. After the initial site failure, all live neighbors are then checked to see if their stress has risen above σ^t. If it has, this site goes through the same failure procedure outlined above until all sites have stress below σ^t. The size of the avalanche is the number of failures that stem from the single initiating site. We refer to this whole avalanche process as a plate update.

Because stress is dissipated from the system both at the site of failure (as regulated by α_i) and through dead sites, which may be placed inhomogeneously throughout the system, we may think of each site i as having an effective dissipation parameter, which incorporates both types of dissipation,

$$\gamma_i = 1 - \phi_i(1 - \alpha_i), \tag{1}$$

where ϕ_i is the fraction of live neighbors of site i. The mean value

$$\bar{\gamma} = \sum_i \frac{\gamma_i}{N_a}, \tag{2}$$

where N_a is the number of live sites, is the average fraction of excess stress dissipated from the system per failed site.

We will want to compare the scaling in these systems with the scaling in systems where the damage distribution is uniform. It has been found [*Klein et al.*, 2007; *Serino et al.*, 2011] for these OFC type models with no spatial inhomogeneities (homogeneous damage and constant α_i) that in the mean field limit, the number of avalanche events of size s obeys the scaling form

$$n(s) \sim e^{-\Delta hs}/s^{\tau}. \tag{3}$$

The quantity Δh, which is a function of the fraction of dead sites, is a measure of the distance from the spinodal and $\tau = 3/2$. (Note that $n(s)$ is the number of events of size s, which is the noncumulative distribution, rather than the number of events of size s or smaller, which is the cumulative distribution often discussed in relation to the Gutenberg-Richter law.) We know from the work of *Serino et al.* [2010] that long-range damaged systems, with a fraction ϕ of live sites and constant site dissipation parameter α_i, are equivalent to undamaged systems with site dissipation parameter $\alpha' = 1 - \phi(1 - \alpha)$. These systems approach the spinodal ($\Delta h \to 0$) as the stress dissipation from the system vanishes: $\phi \to 1$ and $\alpha \to 0$. Physically, stress dissipation from the lattice system suppresses large avalanche events.

3. DAMAGE DISSIPATION

We first study the case of damage dissipation only; that is, α_i is constant (we choose $\alpha_i = 0$ for the systems in this section), and damage is distributed throughout the system. Because all stress passed to a dead site is dissipated from the system, the total amount of dissipated stress depends on which live sites are available to pass stress to the dead sites. This depends on the spatial distribution of the dead sites, which we now examine.

3.1. *Effective Dissipation*

In Figure 1, we show 2-D lattices of linear size $L = 256$ and 25% of the sites dead. The lattices have various distributions of the dead sites. The top row shows the placement of dead sites, while the bottom row shows the corresponding values of the effective dissipation parameter γ_i. Roughly, Figure 1 from left to right corresponds to lattices of decreasing homogeneity of dead sites. Figures 1a and 1e show a lattice system with dead sites randomly distributed throughout the system. In the long-range limit, this corresponds to homogenous damage studied in the work of *Serino et al.* [2010]. Figures 1b–1d and 1f–1h incorporate some clustering of dead sites. Figures 1b and 1f show a lattice system with blocks of randomly distributed dead sites with various length scales where the linear block sizes range from 1 to $L/8$. Each block has a fraction p of randomly distributed dead sites, where p

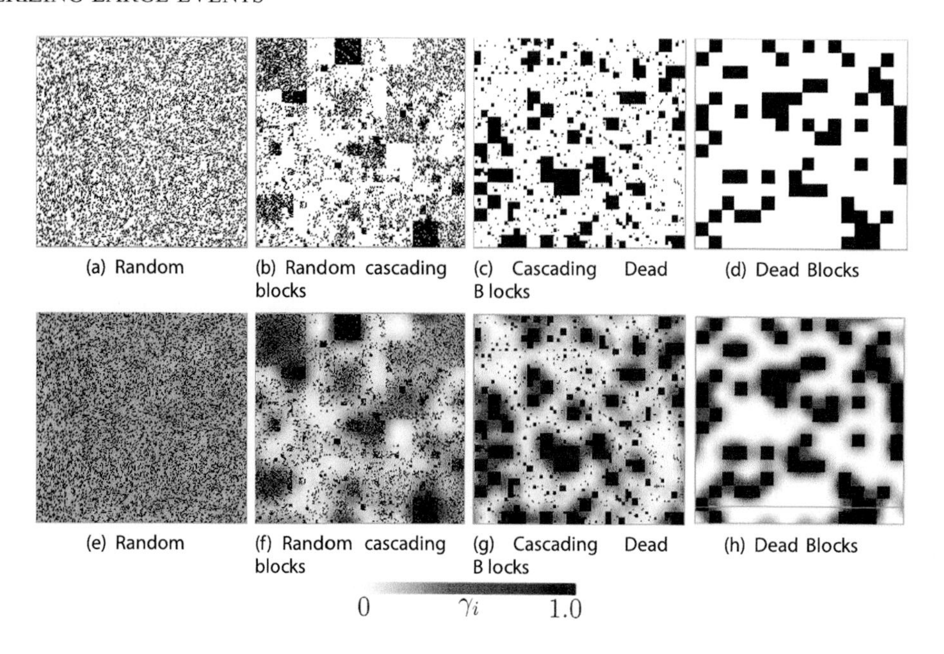

Figure 1. Various configurations of 25% dead sites (in black) for a lattice with linear size $L = 256$. (top) Locations of dead sites. (bottom) Values of the effective dissipation parameter γ_i calculated using equation (1), where lighter colors indicate lower values of γ_i as indicated in the color scale. Lattices contain dead sites distributed randomly (Figures 1a and 1e), blocks of various sizes (Figures 1b and 1f), where each block has p randomly distributed dead sites with p varying for each block, dead blocks of various sizes (Figures 1c and 1g), and dead blocks of a single size (Figures 1d and 1h). The green line drawn in Figure 1h corresponds to the data in green in Figure 4.

varies from block to block. Figures 1c and 1g show a lattice system with dead blocks with various length scales where the linear block sizes also range from 1 to $L/8$. We use various length scales to reflect the variety of length scales present in real earthquake systems. Figures 1d and 1h show a lattice system with randomly distributed dead blocks with blocks of linear size of $L/16$ only. To characterize each configuration in Figure 1, we calculate $\bar{\gamma}$, and the variance of γ_i for a stress transfer range $R = 16$ and $\alpha_i = 0\ \forall i$. The results are summarized in Table 1.

Figure 2 shows $n(s)$, the numerical distribution of avalanche events of size s, corresponding to the various distributions of damage in Figure 1. We find that the scaling behavior of systems with damage depends not only on the total amount of damage to the system but also on the spatial distribution of damage. In particular, large events are suppressed more for lattices with damaged sites distributed more homogeneously. Because these lattices are of equal size, have the same number of damaged sites, and the same stress transfer range ($R = 16$), the differences in the large event behavior are not due to the finite size of the lattice or the finite number of active sites in the lattice. Furthermore, the

Table 1. Averages and Variances of γ_i for the Distributions of Dead Sites Shown in Figure 1[a]

Damage Distribution	$\bar{\gamma}$	Variance
Random	0.2510	2.0×10^{-4}
Random cascading blocks	0.2293	6.9×10^{-3}
Cascading dead blocks	0.2092	8.9×10^{-3}
Dead blocks	0.1803	2.0×10^{-2}

[a]Total number of dead sites is equal to 25% of the lattice for all distributions.

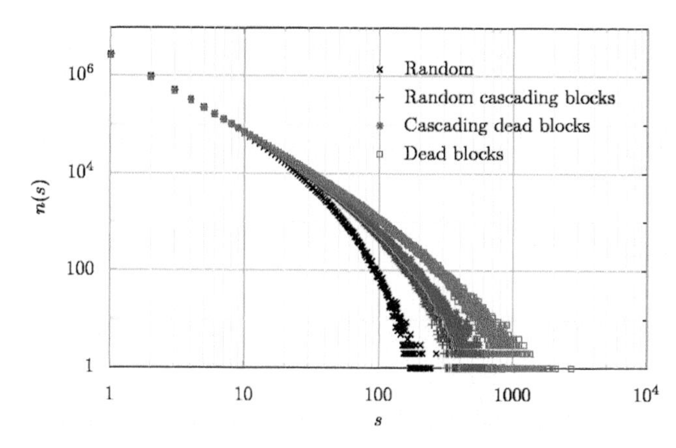

Figure 2. Numerical distribution of avalanche events of size s for various spatial distributions of dead sites. Data corresponds to lattices in Figure 1 with stress transfer range $R = 16$.

results of section 4 indicate that the effect is due to the spatial distribution of γ_is and does not require that the lattice be damaged at all. The calculated quantities in Table 1 would appear to indicate that the large event suppression is correlated with both higher values of the average dissipation parameter $\bar{\gamma}$ and lower values of the variance of $\bar{\gamma}$.

Variations in the *b* value, or slope of the Gutenberg-Richter frequency-magnitude distribution relation for earthquakes, have been studied intensively over the past 20 years [*Imoto et al.*, 1990; *Imoto*, 1991; *Frohlich and Davis*, 1993; *Ogata and Katsura*, 1993; *Cao et al.*, 1996; *Wyss et al.*, 1996; *Keilis-Borok et al.*, 1997; *Rotwain et al.*, 1997; *Wiemer and Wyss*, 1997, 2000, 2002; *Schorlemmer et al.*, 2004, 2005, and many others]. In general, this work demonstrates that the *b* value varies in both space and time and on a wide variety of scales [*Wiemer and Wyss*, 2002; *Schorlemmer et al.*, 2004; *Wiemer and Schorlemmer*, 2007]. These variations have important implications for earthquake hazard assessment because local and regional probabilistic seismic hazard assessment is commonly performed using the Gutenberg-Richter frequency-magnitude distribution [*Field*, 2007; *Wiemer and Schorlemmer*, 2007; *Wiemer et al.*, 2009]. As a result, accurate estimation of the Gutenberg-Richter distribution and its value is critical to our understanding of the magnitude and frequency of the most extreme events. Recent work into regional *b* values has resulted in two important conclusions. First, the *b* value varies with type of faulting mechanism, and the *b* value for thrust events is the smallest ($b \sim 0.7$), for strike-slip events is intermediate ($b \sim 0.9$), for normal events is the greatest ($b \sim 1.1$). This relationship is inversely proportional to the mean stress in each regime [*Schorlemmer et al.*, 2005; *Gulia and Wiemer*, 2010]. In

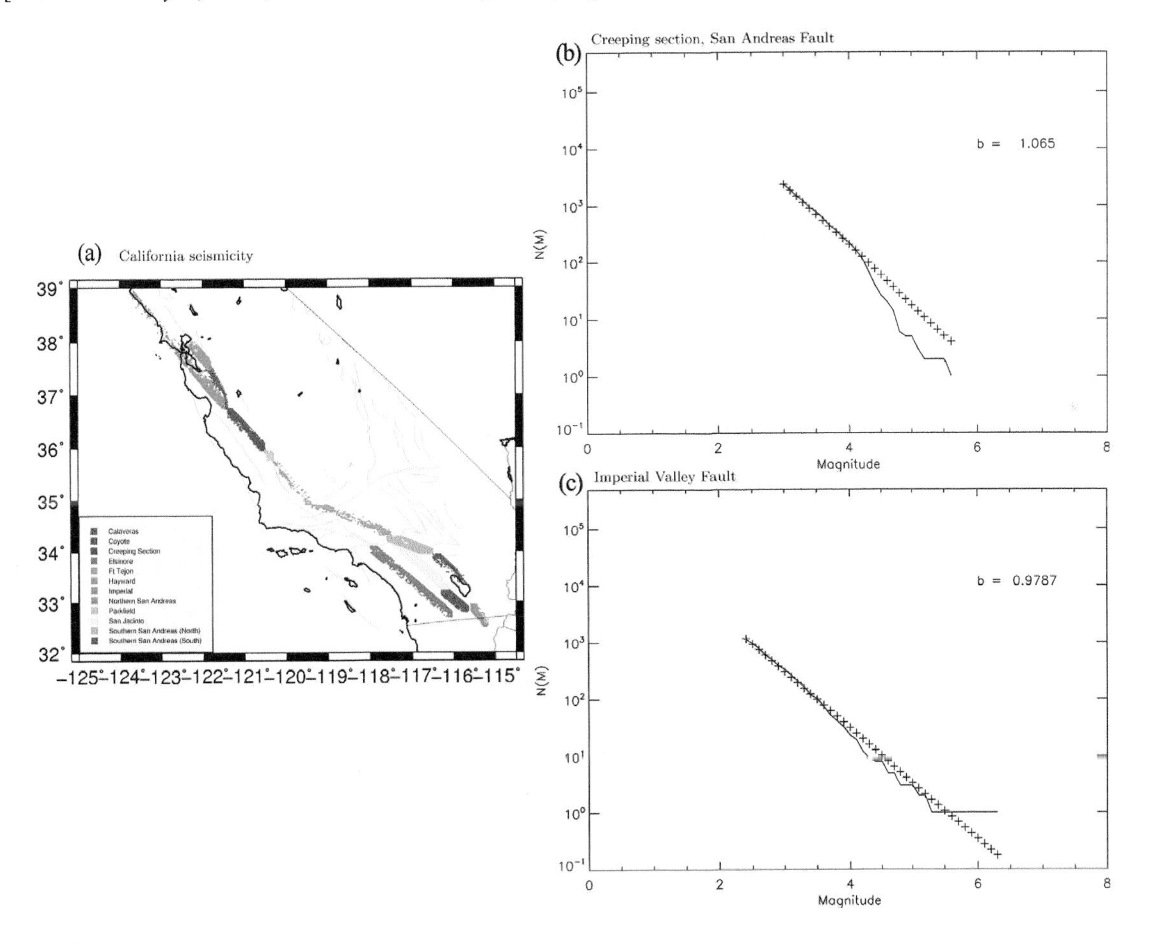

Figure 3. (a) Seismicity in a 20 km swath around major faults of the San Andreas system. Gutenberg-Richter magnitude-frequency relation for (b) the Creeping segment of the San Andreas fault and (c) the Imperial Valley fault. $N(M)$ is the cumulative number of events of magnitude $m \geq M$ for seismicity within 10 km on either side of the referenced fault, for the period 1960 through 2008, inclusive. The crosses show a best fit to a power law with exponents $b = 1.07$ (standard error = 0.020) for the Creeping segment of the San Andreas fault, and $b = 0.98$ (standard error = 0.028) for the Imperial Valley fault.

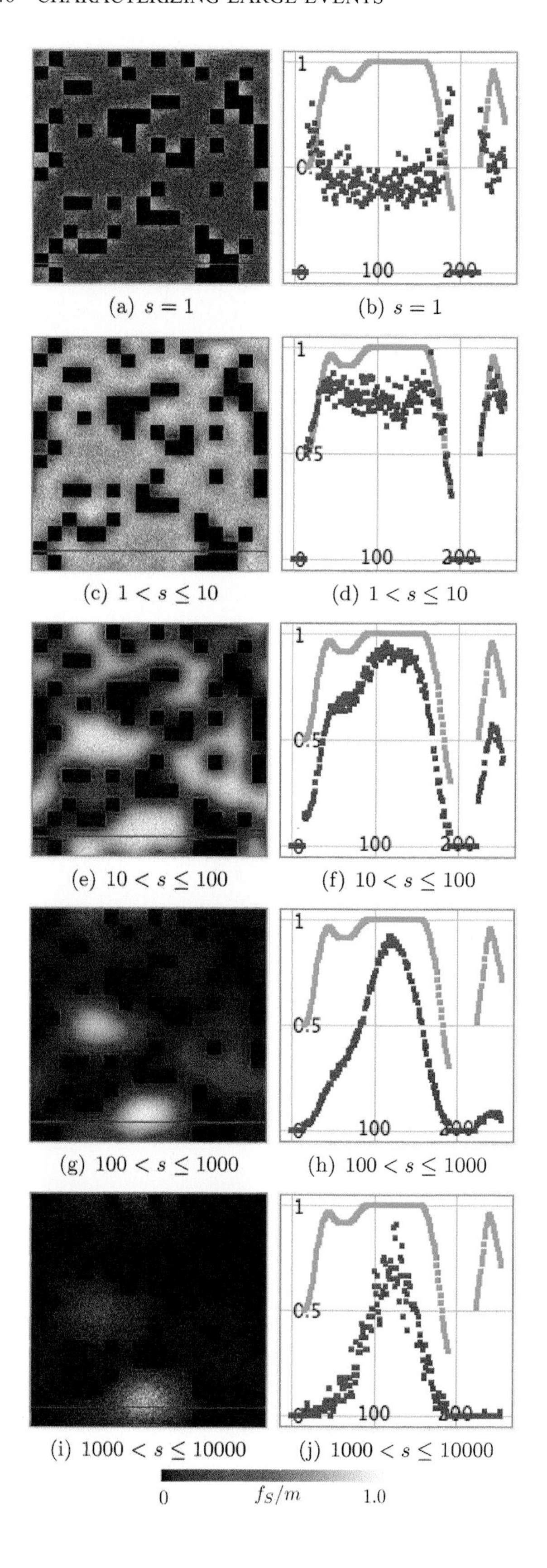

addition, related research suggests that locked patches on faults, or asperities, are characterized by low b values, while creeping faults have higher b values, and that the asperities are related to the occurrence of extreme events [*Wiemer and Wyss*, 1994, 1997, 2002; *Schorlemmer et al.*, 2004, 2005; *Latchman et al.*, 2008]. The Gutenberg-Richter relation and the associated b value identify spatial and temporal variations in stress accumulation and the associated potential for extreme events.

Examples for two faults in California are shown in Figure 3. Here the data are taken from the Advanced National Seismic System (ANSS) seismic catalog (anss.org). Figure 3a shows the seismicity on each fault included in this study, where the color scheme identifies each segment. The minimum magnitude of completeness (M_c) is determined using the method of *Wiemer and Wyss* [2002], while the associated b value is calculated using the maximum likelihood estimate method [*Aki*, 1965] and the standard error of b is computed from the work of *Shi and Bolt* [1982]. Here the minimum magnitude of completion was tested for a sequence of values for both larger and smaller regions around the fault, and a 20 km swath was selected as providing a stable value of M_c. Least-squares estimates for the b value also were computed for the same fault segments. However, the maximum likelihood values provide a better estimate in every case, based on the least-squares residual of the fit to the data.

Figure 3b shows the Gutenberg-Richter relation for the Creeping section of the San Andreas fault in central California from 1960 through 2008, inclusive. $N(M)$ is the cumulative number of events of magnitude $m \geq M$ for seismicity within 10 km on either side of the referenced fault. Figure 3c depicts the Gutenberg-Richter relation for the Imperial Valley fault, located at the southern end of the San Andreas fault system, for the same time period. The form of the relation for the Creeping section (Figure 3b) corresponds to that of the more homogeneous damage (labeled "Random") seen in Figure 2, while that of the Imperial Valley fault (Figure 3c) corresponds to the "Dead Blocks" distribution seen in Figure 2, with an increased number of large

Figure 4. Spatial dependence of avalanche events of different sizes. (left) Shown is f_S/m_S, where f_S is the number of times a site has failed during avalanche events within a size bracket (given below each plot), and m_S is the maximum value of f_S for the given size bracket. Darker colors indicate lower values of f_S/m_S as indicated in the color scale. The total number of plate updates is 6 million. (right) A plot of a slice of the lattice corresponding to the red line drawn on the lattice. The green data on the plots on the right show the values of $1 - \gamma_i$ corresponding to the green line drawn in Figure 1h.

events as a result of the damage inhomogeneities shown in Figures 1d and 1h.

3.2. Spatial Dependence of Large Events

In Figure 4, we present the spatial dependence of avalanche events of different sizes. The left column shows f_S/m_S, where f_S is the number of times a site has failed during avalanche events within a size bracket (which is given in the subfigure captions), and m_S is the maximum value of f_S for the given size bracket. The total number of plate updates is 6 million. The right column shows a plot of a slice of the lattice corresponding to the red line drawn on the lattice. The green data on the plots in the right column show the values of $1 - \gamma_i$ corresponding to the green line drawn in Figure 1h.

Small events occur uniformly throughout the live parts of the lattice, independently of the spatial distribution of γ_i values for live sites. There is an effect at the boundaries of live and dead sites, where events of size $s = 1$ (Figures 4a and 4b) increase in frequency and events of size $1 < s \leq 10$ (Figures 4d and 4c) decrease in frequency. The spatial locations of medium-sized events ($10 < s \leq 100$) generally follow spatial patterns of the γ_i values as can be seen in Figure 4f and by comparing Figures 4e and 1h. However, larger events ($100 < s \leq 1000$ and $1000 < s \leq 10000$) depend not only on the presence of low γ_i values but also on the spatial accumulation of low γ_i values as can be seen in Figures 4g–4j.

3.3. Large Events and Site Connectedness

We now develop the concept of connectedness between lattice sites, which is related to connected nodes in complex networks [*Albert and Barabsi*, 2002]. The data presented, so far, indicates that large avalanche events occur where there is a local accumulation of sites with low effective dissipation parameters. We have assumed that excess stress is transferred uniformly to all sites within a certain stress transfer range, R. We define a site to be connected to another site if it passes excess stress to that site upon failure. Therefore, in our current model, a site is connected to another site if it is within a distance R. However, some sites transfer more stress to others because they have lower site dissipation rates, α_i, and are therefore better connected to nearby sites. (Note that this implies that connectedness is directional since site A may pass a larger fraction of stress to site B than site B passes to site A.) Furthermore, sites are also *indirectly* connected to sites, which are further away than R, because during an avalanche event, stress can pass to a connected neighbor, and then to one of its connected neighbors, and so on. We expect that sites, which are well-connected to other sites in the lattice, be frequent participants in large avalanche events.

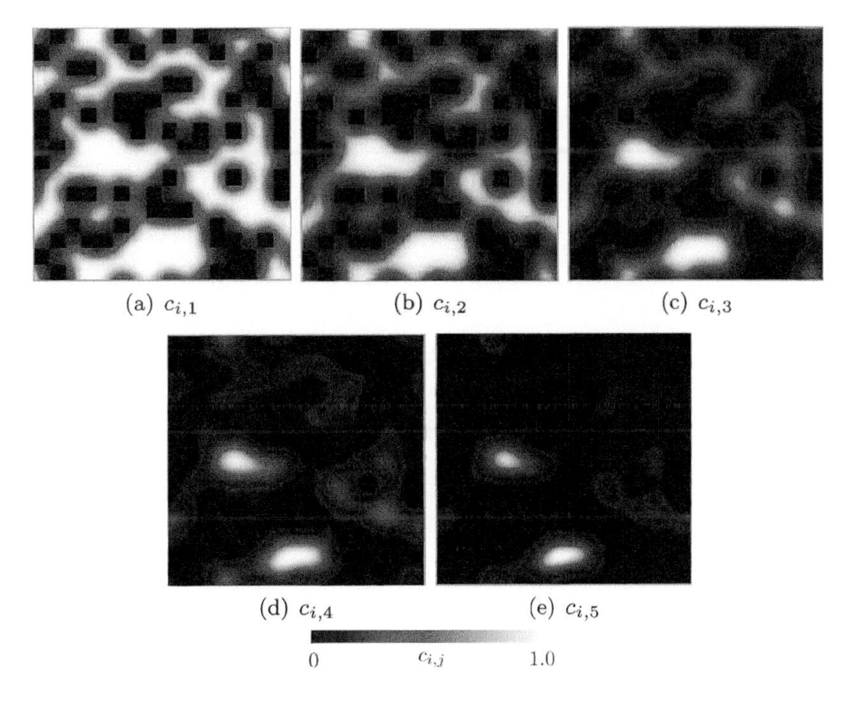

Figure 5. Results for iterative connection parameter calculations using equations (4) and (5) for the lattice with dead block damage arrangement shown in Figures 1d–1h. Darker colors indicate lower values of $c_{i,j}$ as indicated in the color scale.

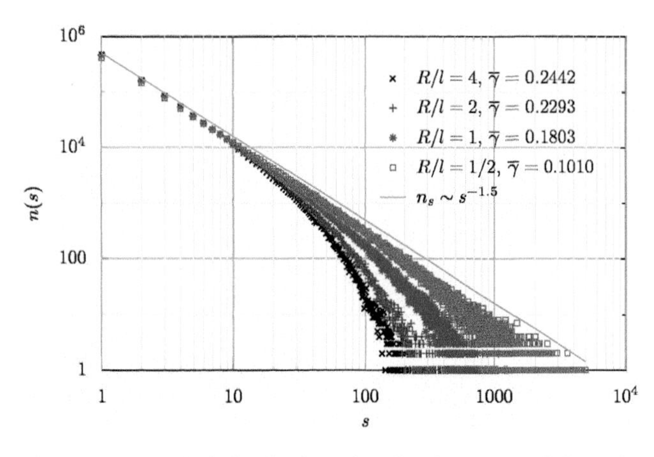

Figure 6. Numerical distribution of avalanche events of size s for blocks of dead sites of linear size l. Systems are characterized by the dimensionless parameter R/l. (Figures 1d and 1h corresponds to $R/l = 1$.) The size of the systems shown here is $L = 256$, and the stress transfer range is $R = 16$. The line is drawn to show that the data is approaching a power law with exponent $-3/2$.

To quantify the total connectedness of individual sites to other sites in the lattice, we define an iterative quantity called the *connection parameter* as follows: the zeroth iteration of the connection parameter for site i depends only on its own γ value:

$$c_{i,0} = 1 - \gamma_i, \tag{4}$$

while the jth iteration is

$$c_{i,j} = c_{i,j-1} \frac{\sum_n c_{n,j-1}}{\sum_n}, \tag{5}$$

where n sums over site i's connected neighbors, i.e., the sites that are within a distance R. Thus, the first iteration of a site's connection parameter incorporates the γ values of the site's connected neighbors, the second iteration incorporates the γ values of the site's connected neighbors' connected neighbors, and so on. The calculation mimics the path that stress may take during an avalanche event.

In Figure 5, we show the results of such a calculation for the system with dead block damage from Figures 1d–1h. The first iteration mimics the spatial arrangement of the γ_i distribution itself, while further iterations retain high values of $c_{i,j}$ only for sites centered within large patches of low γ_i values. These highly connected sites shown in Figure 5e are expected to be participants in large avalanche events. By comparison to Figure 4i, we see that this is the case (though the calculation shown in Figure 5e does not capture all of the spatial details of Figure 4i). Note that the advantage of analyzing lattices in terms of connectedness is that it is generalizable to other systems with more complicated rules of stress transfer.

3.4. Length Scales

For any given distribution of damage, the system will act as if the damage is homogeneous if the stress transfer range is

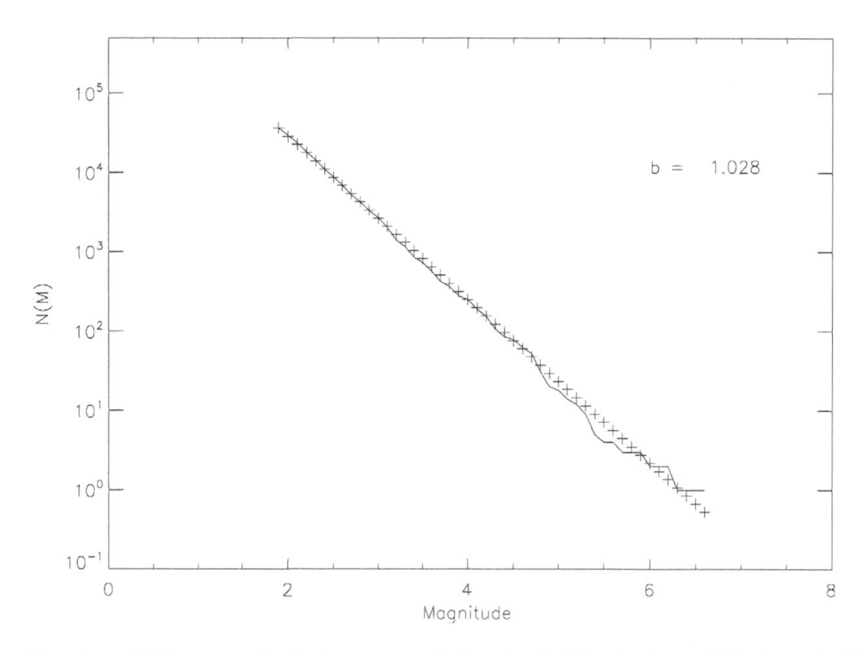

Figure 7. The Gutenberg-Richter magnitude-frequency relation for California from 1980 through 2008, inclusive. The crosses show a best fit to a power law with exponent $b = 1.03$.

long enough compared to the length scales of damage of the lattice. To illustrate the importance of relative length scales, we consider the case of a single length scale associated with damaged areas. We place blocks of damaged sites of size, l, randomly in the system, which has constant $\alpha_i = \alpha$ (see, e.g., Figure 1d). As we vary the physically relevant dimensionless parameter R/l, the measured value of $\bar{\gamma}$ varies from $\bar{\gamma} = \alpha$ for $R/l << 1$ to $\bar{\gamma} = 1 - \phi(1 - \alpha)$ for $R/l >> 1$. In the former case, the live domains of the system appear nearly homogeneous with $\phi_i = 1$ except near the boundaries of dead blocks. (Recall ϕ_i is the total fraction of live sites in the system, and ϕ_i is the fraction of live neighbors at site i.) The latter case is the limit of homogeneously distributed damage. In both limiting cases, the variance of γ_i is small, and the scaling is equivalent to the scaling for an undamaged system with $\alpha' = \bar{\gamma}$.

In Figure 6, we compare systems with randomly distributed dead blocks of various length scales, $R = 16$, $\alpha = 0$, and 25% total damage. As R/l gets small, the values of $\bar{\gamma}$ also get small. The distribution n_s of the corresponding data approaches a power law with the exponent $-3/2$, which is the form of the distribution of a system at the spinodal.

In Figure 7, we plot the Gutenberg-Richter distribution for a set of faults from southern California. Here we designate the major faults or fault segments that account for the greatest percentage of earthquakes in California and identify only that seismicity that occurs within 10 km on either side of those segments. For this study, we include the Calaveras fault, the Coyote Creek fault, the Superstition Hills fault, the Creeping section, the Elsinore fault, the Ft. Tejon segment of the San Andreas, the Hayward fault, the San Andreas north of the Creeping section, the Parkfield segment of the San Andreas fault, the Imperial Valley fault, and the southern San Andreas (see Figure 3a for fault locations). Data used are again from the ANSS catalog (anss.org), from 1980 through 2008, inclusive. We sum up all the seismicity from the individual fault segments and calculate the cumulative Gutenberg-Richter distribution, $N(M)$ is the cumulative number of events of magnitude $m \geq M$, as seen in Figure 7. Here the magnitude-frequency relation is a power law similar to that seen in Figure 6 for the case of a damaged system in which the range of interaction is less than the predominant damage scale.

4. SITE DISSIPATION

The spatial distribution of damaged sites determines the spatial distribution of γ_i values. A more direct way to control the numerical and spatial distributions of γ_i is to use undamaged systems and vary the values of α_i. In this way, we can isolate the effects of spatial redistribution of γ_i values, while holding the numerical distributions of γ_i constant. In this section, we study systems with site dissipation only; that is, they have no damage, and $\gamma_i = \alpha_i$ for each system.

4.1. Numerical Distributions of Site Dissipation Parameters

In Figure 8, we present data for three systems with site dissipation only. We see in Figure 8a that $\bar{\gamma} = 0.5$ for all three systems from the numerical distributions of α_i values, $p(\alpha_i)$. The two systems labeled "Gaussian Split" and "Gaussian Centered" both have a uniform spatial distribution of α_i

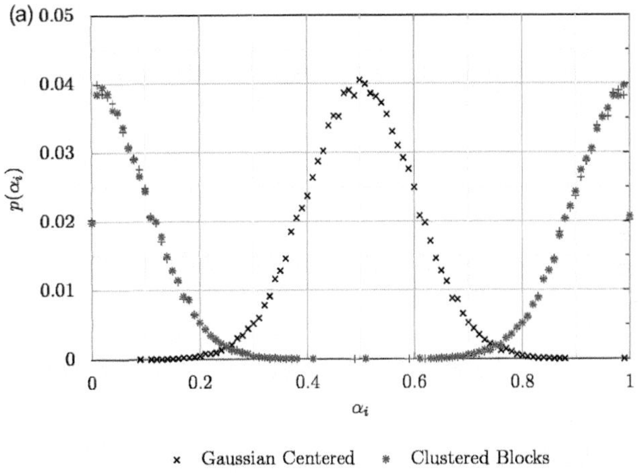

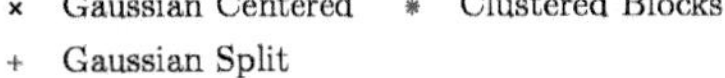

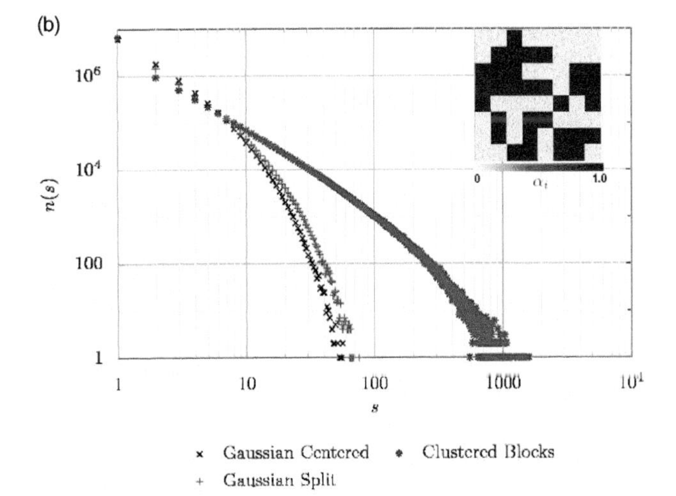

Figure 8. (a) Comparison of three lattice systems with no damage and two different distributions of $p(\alpha_i)$. The lattices labeled by "Gaussian Centered" and "Gaussian Split" are distributed uniformly in space, while the spatial distribution of "Clustered Blocks" is shown in the inset of Figure 8b, where lighter colors indicate lower values of α_i as indicated in the color scale. (b) Numerical distributions of avalanche events of size s.

values. However, the values of α_i for the "Gaussian Centered" system have a Gaussian distribution centered about $\alpha_i = 0.5$, while the values of α_i for the "Gaussian Split" system have partial Gaussian distributions and are clustered near the values of $\alpha_i = 0$ and $\alpha_i = 1$. Figure 8b shows that the numerical distribution of avalanche events, $n(s)$, for these two systems with spatially uniform distributions of α_i are very similar despite the different numerical distributions.

However, we see by studying the "Clustered Blocks" system that spatial distributions of α_i have a robust effect on scaling, even when the averages and variances of α_i are the same. The "Gaussian Split" and "Clustered Blocks" systems have nearly the same numerical distributions of α_i (Figure 8a) and, therefore, have the same value of the variance of α_i. The spatial distributions of these two cases, however, are different: the "Gaussian Split" system has a uniform spatial distribution of α_i values, while the "Clustered Blocks" system has high (and low) α_i values clustered together into blocks as shown in the inset of Figure 8b. Despite having equal values of $\overline{\alpha}_i$ and equal variances of α_i values, the "Clustered Blocks" system experiences much larger events (by an order of magnitude).

Evidently, the larger events depend crucially on the spatial clustering of low dissipation sites. This is because failing sites with low values of γ_i pass along a high percentage of excess stress, encouraging the failure of neighboring sites. Thus, a large earthquake event is more likely to occur if the initial site of failure is well connected to a large number of sites with low dissipation parameters. In our system, connectedness is determined by spatial locality, so we require large clumps of sites with low values of γ_i in order to allow for the occasional large earthquake event.

4.2. Gutenberg-Richter Scaling

The Gutenberg-Richter scaling law states that the cumulative distribution of earthquake sizes is exponential in the magnitude [*Gutenberg and Richter*, 1956]. In terms of the seismic moment, which has succeeded the Richter magnitude as the appropriate measure for earthquake sizes, the law may be reframed to state that the cumulative distribution of earthquake sizes, N_{M_0}, is a power law in the seismic moment, M_0 [*Serino et al.*, 2011].

$$N_{M_0} \sim M_0^{-\beta}, \text{with} \quad \beta \equiv \frac{2b}{3}, \qquad (6)$$

and b is the so-called b value of the Gutenberg-Richter law, which has been measured for many real earthquake systems, where b typically varies from $0.75 \leq b \leq 1.2$ [*Frohlich and Davis*, 1993; *Rundle et al.*, 2003; *Gulia and Wiemer*, 2010]. The seismic moment M_0 is proportional to the size of the earthquake in this model [*Rundle et al.*, 2000b]. Therefore, the relation appropriate for the systems considered in this work is the cumulative distribution of earthquake size:

$$N_s \sim s^{-\beta}, \qquad (7)$$

or the corresponding noncumulative distribution

$$n_s \sim s^{-\tilde{\tau}}, \text{with } \tilde{\tau} = \beta + 1. \qquad (8)$$

Serino et al. [2011] construct a model for an earthquake fault system consisting of an aggregate of lattice models, where each lattice has a fraction q of homogeneously distributed dead sites, and q varies from 0 to 1. The weighting factor D_q gives the fraction of lattices with damage q. For a constant weighting factor in which all values of q are contributing equally to the fault system, they find a value of $\tilde{\tau} = 2$. They also consider a power law distribution of D_q,

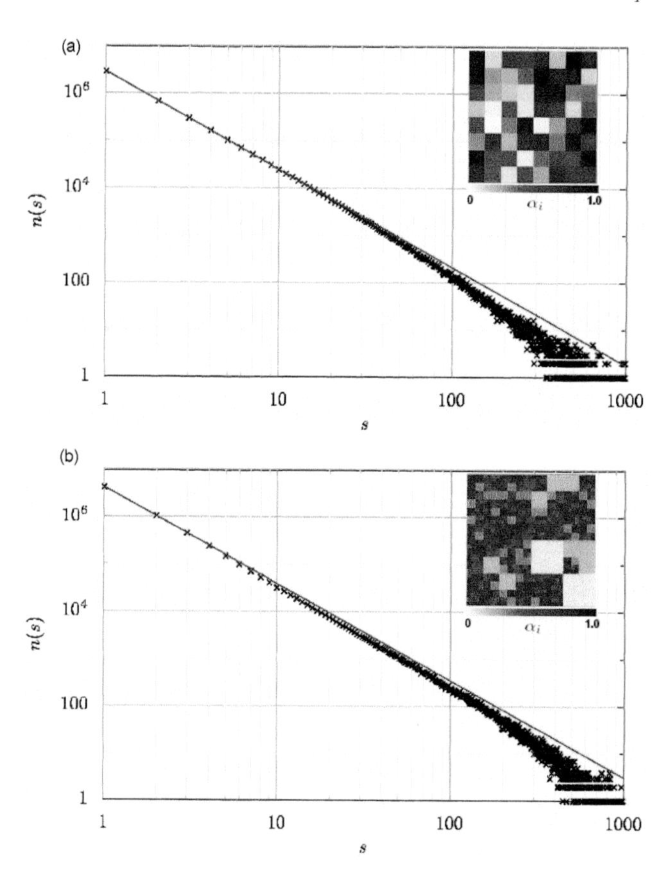

Figure 9. Numerical distribution for avalanche events of size s for systems with uniform numerical distributions for α_i, but nonuniform spatial distributions of α_i, which are shown in the insets. Lighter colors indicate lower values of α_i as indicated in the color scale. Slopes of least squares best fit lines in red are (a) $\tilde{\tau} \simeq 2.0697$ (standard error = 0.0002) and (b) $\tilde{\tau} \simeq 2.0451$ (standard error = 0.0011).

such that $1.5 \leq \tilde{\tau} \leq 2.0$ and $0.75 \leq b \leq 1.5$, corresponding to the Gutenberg-Richter b values found in real earthquake fault systems and as seen in Figure 7.

There are two important differences between the model considered by *Serino et al.* [2011] and our work: In the model treated by Serino et al., (1) the damage is distributed homogeneously and (2) the individual lattices with homogeneous damage q are noninteracting.

To expand upon these results, we investigate both the effect of the spatial arrangement of the damage and its relation to the stress transfer range as well as the effect of stress transfer between regions with different levels of damage.

We construct two lattice systems with a uniform numerical distribution of γ_i, which have scaling consistent with the systems of *Serino et al.* [2011] with constant distribution D_q. The first model essentially pieces together many homogeneous lattice systems: The numerical distribution of α_i values is uniform between 0 and 1, but spatially arranged into N_B blocks of linear size B (see Figure 9a inset), where each block contains a random distribution of α_i values within an interval of size $1/N_B$. There are no dead sites, so that $\alpha_i = \gamma_i$. The effects of the boundaries between the blocks should be negligible if $B << R$. In Figure 9a, we present data from a system with $L = 512$, $R = 16$, and $B = 64$. The straight line shows the least squares best fit to a power law with exponent $\tilde{\tau} \simeq 2.0697$ (standard error = 0.0002), which is consistent with the results for the aggregate lattice system of *Serino et al.* [2011] with $D_q = 1$.

We find that the size of the blocks, B, need not be the same for different values of α_i. It is important that the boxes with lower values of $\bar{\gamma}$ be large enough to accommodate large avalanche events, but blocks with large α_i may be small because they are more likely to seed small avalanches. With this in mind, we construct a lattice system with various length scales of blocks, where the largest blocks have the lowest α_i values and decreasing-sized blocks have increasing values of α_i. The scaling results are shown in Figure 9b, with a least squares best fit power law with exponent $\tilde{\tau} \simeq 2.0451$ (standard error = 0.0011), or a cumulative b value of approximately 1.5, higher than that shown for California in Figure 7. This suggests that not only is the distribution of damage not constant in the natural earthquake fault system (which was previously shown by *Serino et al.* [2011]) but also that it likely is not related to the relative length scale of the damage regions.

5. CONCLUSIONS

We have studied both damage and site dissipation to inform the development of models of realistic earthquake faults with inhomogeneous stress dissipation. Spatially rearranging dead sites on a given lattice affects the numerical distributions of the effective stress dissipation parameters and the scaling behavior of large avalanche events, depending on the homogeneity of the damage and the length scales associated with the clustered dead sites. However, by studying site dissipation, we find that the spatial distribution of the dissipation parameters crucially affects the scaling behavior even when the numerical distributions of dissipation parameters are the same. In addition, the largest events are more prevalent if the stress transfer range is smaller than the intrinsic damage length scale.

Sites with lower stress dissipation, even if only partially distributed throughout the lattice but grouped together, allow for larger avalanche events. Larger events depend crucially on the spatial clustering of low dissipation sites because failing sites with low values of γ_i pass along a high percentage of excess stress, encouraging the failure of neighboring sites. Thus, a large earthquake event is more likely to occur if the initial site of failure is well connected to a large number of sites with low dissipation parameters. In our system, connectedness is determined by spatial locality, so we require large clumps of sites with low values of γ_i in order to allow for the occasional large earthquake event.

We introduce a connectedness parameter, which quantifies the range and direction of stress transfer. Again, results here suggest that it is those sites with the greatest connectedness that participate in the largest events in the system. Finally, we have found models for earthquake fault systems, which have avalanche event size scaling, which is consistent with the new approach for understanding Gutenberg-Richter scaling proposed by *Serino et al.* [2011]. The models studied here go beyond those previously proposed by incorporating inhomogeneities into the lattice and allowing areas with different characteristic dissipation rates to interact.

Acknowledgments. This work was funded by the DOE through grant DE-FG02-95ER14498 and the NSERC and Aon Benfield/ICLR Industrial Research Chair in Earthquake Hazard Assessment.

REFERENCES

Aki, K. (1965), Maximum likelihood estimate of b in the formula log N=a-bM and its confidence limits, *Bull. Earthquake Res. Inst. Univ. Tokyo*, *43*, 237–239.

Albert, R., and A. Barabsi (2002), Statistical mechanics of complex networks, *Rev. Mod. Phys.*, *74*(1), 47–97.

Bakun, W. H., G. C. P. King, and R. S. Cockerham (1986), Seismic slip, aseismic slip, and the mechanics of repeating earthquakes on the Calaveras fault, California, in *Earthquake Source Mechanics*, *Geophys. Monogr. Ser.*, vol. 37, edited by S. Das, J. Boatwright, and C. H. Scholz, pp. 195–207, AGU, Washington, D. C., doi:10.1029/GM037p0195.

Bowman, D., G. Ouillon, C. Sammis, A. Sornette, and D. Sornette (1998), An observational test of the critical earthquake concept, *J. Geophys. Res.*, *103*(B10), 24,359–24,372.

Brehm, D. J., and L. W. Braile (1998), Intermediate-term earthquake prediction using precursory events in the new Madrid seismic zone, *Bull. Seismol. Soc. Am.*, *88*(2), 564–580.

Brodsky, E. E. (2006), Long-range triggered earthquakes that continue after the wave train passes, *Geophys. Res. Lett.*, *33*, L15313, doi:10.1029/2006GL026605.

Bufe, C., and D. Varnes (1993), Predictive modeling of the seismic cycle of the greater San Francisco Bay region, *J. Geophys. Res.*, *98*(B6), 9871–9883.

Burridge, R., and L. Knopoff (1967), Model and theoretical seismicity, *Bull. Seismol. Soc. Am.*, *57*(3), 341–371.

Cao, L., H. Fang, Q. Li, and J. Chen (1996), Forecasting b values for seismic events, *Int. J. Bifurcation Chaos*, *6*, 545–555.

Cavallo, E., A. Powell, and O. Becerra (2010), Estimating the direct economic damage of the 1511 earthquake in Haiti, *Work. Pap. Ser. IDB-WP-163*, Inter-Am. Dev. Bank, Washington, D. C.

Dahmen, K., D. Erta, and Y. Ben-Zion (1998), Gutenberg-Richter and characteristic earthquake behavior in simple mean-field models of heterogeneous faults, *Phys. Rev. E*, *58*(2), 1494–1501.

Deng, J., and L. R. Sykes (1996), Triggering of 1812 Santa Barbara Earthquake by a Great San Andreas Shock: Implications for future seismic hazards in southern California, *Geophys. Res. Lett.*, *23*(10), 1155–1158.

Dodge, D., G. Beroza, and W. Ellsworth (1996), Detailed observations of California foreshock sequences: Implications for the earthquake initiation process, *J. Geophys. Res.*, *101*(B10), 22,371–22,392.

Ellsworth, W. I., and A. T. Cole (1997), A test of the characteristic earthquake hypothesis for the San Andreas Fault in central California, *Seismol. Res. Lett*, *68*, 298.

Eneva, M., and Y. Ben-Zion (1997), Techniques and parameters to analyze seismicity patterns associated with large earthquakes, *J. Geophys. Res.*, *102*(B8), 17,785–17,795.

Field, E. H. (2007), Overview of the working group for the development of Regional Earthquake Likelihood Models (RELM), *Seismol. Res. Lett.*, *78*(1), 7–16.

Frohlich, C. (1987), Aftershocks and temporal clustering of deep earthquakes, *J. Geophys. Res.*, *92*(B13), 13,944–13,956.

Frohlich, C., and S. Davis (1993), Teleseismic b values; or, much ado about 1.0, *J. Geophys. Res.*, *98*(B1), 631–644.

Gomberg, J. (1996), Stress/strain changes and triggered seismicity following the Mw 7.3 Landers, California, earthquake, *J. Geophys. Res.*, *101*(B1), 751–764.

Gross, S. J., and C. Kisslinger (1994), Tests of models of aftershock rate decay, *Bull. Seismol. Soc. Am.*, *84*(5), 1571–1579.

Gulia, L., and S. Wiemer (2010), The influence of tectonic regimes on the earthquake size distribution: A case study for Italy, *Geophys. Res. Lett.*, *37*, L10305, doi:10.1029/2010GL043066.

Gutenberg, B., and F. Richter (1956), Earthquake magnitude, intensity, energy and acceleration, *Ann. Geophys.*, *46*, 105–145.

Haberman, R. E. (1981), Precursory seismicity patterns: Stalking the mature seismic gap, in *Earthquake Prediction: An International Review*, *Maurice Ewing Ser.*, vol. 4, edited by D. W. Simpson and P. G. Richards, pp. 29–42, AGU, Washington, D. C.

Herz, A. V. M., and J. J. Hopfield (1995), Earthquake cycles and neural reverberations: Collective oscillations in systems with pulse-coupled threshold elements, *Phys. Rev. Lett.*, *75*(6), 1222–1225.

Imoto, M. (1991), Changes in the magnitude frequency b value prior to large (M-greater-than-or-equal-to-6.0) earthquakes in Japan, *Tectonophysics*, *193*, 311–325.

Imoto, M., N. Hurukawa, and Y. Ogata (1990), Three-dimensional spatial variations of b-value in the Kanto area, Japan, *J. Seismol. Soc. Jpn.*, *43*, 321–326.

Jackson, D. D., and Y. Y. Kagan (2006), The 2004 Parkfield earthquake, the 1985 prediction, and characteristic earthquakes: Lessons for the future, *Bull. Seismol. Soc. Am.*, *96*(4B), S397–S409.

Jaume, S. C., and L. R. Sykes (1999), Evolving towards a critical point: A review of accelerating seismic moment/energy release prior to large and great earthquakes, *Pure Appl. Geophys.*, *155* (2), 279–305.

Jones, L. M., and E. Hauksson (1997), The seismic cycle in southern California: Precursor or response?, *Geophys. Res. Lett.*, *24*(4), 469–472.

Jordan, T. H. (2006), Earthquake predictability, brick by brick, *Seismol. Res. Lett.*, *77*(1), 3–6.

Kanamori, H. (1981), The nature of seismicity patterns before large earthquakes, in *Earthquake Prediction: An International Review*, *Maurice Ewing Ser.*, vol. 4, edited by D. W. Simpson and P. G. Richards, pp. 1–19, AGU, Washington, D. C..

Keilis-Borok, V. I., and V. G. Kossobokov (1990), Times of increased probability of strong earthquakes ($M \geq 7.5$) diagnosed by algorithm M8 in Japan and adjacent territories, *J. Geophys. Res.*, *95*(B8), 12,413–12,422.

Keilis-Borok, V. I., I. M. Rotwain, and A. A. Soloviev (1997), Numerical modeling of block structure dynamics: Dependence of a synthetic earthquake flow on the structure separateness and boundary movements, *J. Seismol.*, *1*(2), 151–160.

King, G. C., R. S. Stein, and J. Lin (1994), Static stress changes and the triggering of earthquakes, *Bull. Seismol. Soc. Am.*, *84*(3), 935–953.

Klein, W., M. Anghel, C. D. Ferguson, J. B. Rundle, and J. S. Sá Martins (2000), Statistical analysis of a model for earthquake faults with long-range stress transfer, in *Geocomplexity and the Physics of Earthquakes*, *Geophys. Monogr. Ser.*, vol. 120, edited by J. B. Rundle, D. L. Turcotte, and W. Klein, pp. 43–71, AGU, Washington, D. C., doi:10.1029/GM120p0043.

Klein, W., H. Gould, N. Gulbahce, J. B. Rundle, and K. Tiampo (2007), Structure of fluctuations near mean-field critical points and spinodals and its implication for physical processes, *Phys. Rev. E*, *75*(3), 031114, doi:10.1103/PhysRevE.75.031114.

Kossobokov, V. G., L. L. Romashkova, V. I. Keilis-Borok, and J. H. Healy (1999), Testing earthquake prediction algorithms: Statistically significant advance prediction of the largest earthquakes in the Circum-Pacific, 1992–1997, *Phys. Earth Planet. Inter.*, *111*, 187–196.

Kovachs, P. (2010), Reducing the risk of earthquake damage in Canada: Lessons from Haiti and Chile, *ICLR Res. Pap. Ser. 49*, Inst. for Catastrophic Loss Reduct., Toronto, Ont., Canada.

Latchman, J. L., F. D. O. Morgan, and W. P. Aspinall (2008), Temporal changes in the cumulative piecewise gradient of a variant of the Gutenberg-Richter relationship, and the imminence of extreme events, *Earth Sci. Rev.*, *87*(3–4), 94–112.

Lyakhovsky, V., and Y. Ben-Zion (2009), Evolving geometrical and material properties of fault zones in a damage rheology model, *Geochem. Geophys. Geosyst.*, *10*, Q11011, doi:10.1029/2009GC002543.

Main, I. (1996), Statistical physics, seismogenesis, and seismic hazard, *Rev. Geophys.*, *34*(4), 433–462.

Mogi, K. (1969), Some features of recent seismic activity in and near Japan 2, activity before and after large earthquakes, *Bull. Earthquake Res. Inst. Univ. Tokyo*, *47*, 395–417.

Nanjo, K., H. Nagahama, and M. Satomura (1998), Rates of aftershock decay and the fractal structure of active fault systems, *Tectonophysics*, *287*(1–4), 173–186.

Ogata, Y., and K. Katsura (1993), Analysis of temporal and spatial heterogeneity of magnitude frequency distribution inferred from earthquake catalogues, *Geophys. J. Int.*, *113*(3), 727–738.

Olami, Z., H. J. S. Feder, and K. Christensen (1992), Self-organized criticality in a continuous, nonconservative cellular automaton modeling earthquakes, *Phys. Rev. Lett.*, *68*(8), 1244–1247.

Pacheco, J. F., C. H. Scholz, and L. R. Sykes (1992), Changes in frequency-size relationship from small to large earthquakes, *Nature*, *355*(6355), 71–73.

Pollitz, F. F., and I. S. Sacks (1997), The 1995 Kobe, Japan, earthquake: A long-delayed aftershock of the offshore 1944 Tonankai and 1946 Nankaido earthquakes, *Bull. Seismol. Soc. Am.*, *87*(1), 1–10.

Press, F., and C. Allen (1995), Patterns of seismic release in the southern California region, *J. Geophys. Res.*, *100*(B4), 6421–6430.

Romanowicz, B., and J. B. Rundle (1993), On scaling relations for large earthquakes, *Bull. Seismol. Soc. Am.*, *83*(4), 1294–1297.

Rotwain, I. M., V. I. Keilis-Borok, and L. Botvina (1997), Premonitory transformation of steel fracturing and seismicity, *Phys. Earth Planet. Inter.*, *101*, 61–71.

Rundle, J. B. (1989), Derivation of the complete Gutenberg-Richter magnitude-frequency relation using the principle of scale invariance, *J. Geophys. Res.*, *94*(B9), 12,337–12,342.

Rundle, J. B., and S. R. Brown (1991), Origin of rate dependence in frictional sliding, *J. Stat. Phys.*, *65*(1), 403–412.

Rundle, J. B., and D. D. Jackson (1977), Numerical simulation of earthquake sequences, *Bull. Seismol. Soc. Am.*, *67*(5), 1363–1377.

Rundle, J. B., and W. Klein (1993), Scaling and critical phenomena in a cellular automaton slider-block model for earthquakes, *J. Stat. Phys.*, *72*(1), 405–412.

Rundle, J. B., W. Klein, and S. Gross (1999), Physical basis for statistical patterns in complex earthquake populations: Models, predictions and tests, *Pure Appl. Geophys.*, *155*(2), 575–607.

Rundle, J. B., W. Klein, K. Tiampo, and S. Gross (2000a), Linear pattern dynamics in nonlinear threshold systems, *Phys. Rev. E*, *61*(3), 2418–2431.

Rundle, J. B., D. L. Turcotte, and W. Klein (Eds.) (2000b), *Geocomplexity and the Physics of Earthquakes*, *Geophys. Monogr. Ser.*, vol. 120, 284 pp., AGU, Washington, D. C., doi:10.1029/GM120.

Rundle, J. B., K. F. Tiampo, W. Klein, and J. S. S. Martins (2002), Self-organization in leaky threshold systems: The influence of near-mean field dynamics and its implications for earthquakes, neurobiology, and forecasting, *Proc. Natl. Acad. Sci. U. S. A.*, *99*, suppl. 1, 2514–2521.

Rundle, J. B., D. L. Turcotte, R. Shcherbakov, W. Klein, and C. Sammis (2003), Statistical physics approach to understanding the multiscale dynamics of earthquake fault systems, *Rev. Geophys.*, *41*(4), 1019, doi:10.1029/2003RG000135.

Saleur, H., C. G. Sammis, and D. Sornette (1996), Discrete scale invariance, complex fractal dimensions, and log-periodic fluctuations in seismicity, *J. Geophys. Res.*, *101*(B8), 17,661–17,677.

Scholz, C. H. (2002), *The Mechanics of Earthquakes and Faulting*, 496 pp., Cambridge Univ. Press, Cambridge, U. K.

Schorlemmer, D., and M. C. Gerstenberger (2007), RELM testing center, *Seismol. Res. Lett.*, *78*(1), 30–36.

Schorlemmer, D., S. Wiemer, M. Wyss, and D. D. Jackson (2004), Earthquake statistics at Parkfield: 2. Probabilistic forecasting and testing, *J. Geophys. Res.*, *109*, B12308, doi:10.1029/2004JB003235.

Schorlemmer, D., S. Wiemer, and M. Wyss (2005), Variations in earthquake-size distribution across different stress regimes, *Nature*, *437*(7058), 539–542.

Serino, C. A., W. Klein, and J. B. Rundle (2010), Cellular automaton model of damage, *Phys. Rev. E*, *81*(1), 016105, doi:10.1103/PhysRevE.81.016105.

Serino, C. A., K. F. Tiampo, and W. Klein (2011), New approach to Gutenberg-Richter scaling, *Phys. Rev. Lett.*, *106*(10), 108501, doi:10.1103/PhysRevLett.106.108501.

Shi, Y., and B. A. Bolt (1982), The standard error of the magnitude-frequency b-value, *Bull. Seismol. Soc. Am.*, *72*(5), 1667–1687.

Swan, F. H., III, D. P. Schwartz, and L. S. Cluff (1980), Recurrence of moderate to large magnitude earthquakes produced by surface faulting on the Wasatch fault zone, Utah, *Bull. Seismol. Soc. Am.*, *70*(5), 1431–1462.

Stein, R. S. (1999), The role of stress transfer in earthquake occurrence, *Nature*, *402*(6762), 605–609.

Tiampo, K. F., J. B. Rundle, S. McGinnis, S. J. Gross, and W. Klein (2002), Mean-field threshold systems and phase dynamics: An application to earthquake fault systems, *Europhys. Lett.*, *60*(3), 481–488.

Tiampo, K. F., J. B. Rundle, W. Klein, J. Holliday, J. S. SáMartins, and C. D. Ferguson (2007), Ergodicity in natural earthquake fault networks, *Phys. Rev. E*, *75*(6), 066107, doi:10.1103/PhysRevE.75.066107.

Turcotte, D. L. (1997), *Fractals and Chaos in Geology and Geophysics*, 416 pp., Cambridge Univ. Press, Cambridge, U. K.

Turcotte, D. L., W. I. Newman, and R. Shcherbakov (2003), Micro and macroscopic models of rock fracture, *Geophys. J. Int.*, *152* (3), 718–728.

Vere-Jones, D. (1995), Forecasting earthquakes and earthquake risk, *Int. J. Forecasting*, *11*(4), 503–538.

Vere-Jones, D. (2006), The development of statistical seismology: A personal experience, *Tectonophysics*, *413*, 5–12.

Wiemer, S., and D. Schorlemmer (2007), ALM: An asperity-based likelihood model for California, *Seismol. Res. Lett.*, *78*(1), 134–140.

Wiemer, S., and M. Wyss (1994), Seismic quiescence before the landers (M = 7.5) and big bear (M = 6.5) 1992 earthquakes, *Bull. Seismol. Soc. Am.*, *84*(3), 900–916.

Wiemer, S., and M. Wyss (1997), Mapping the frequency-magnitude distribution in asperities: An improved technique to calculate recurrence times?, *J. Geophys. Res.*, *102*(B7), 15,115–15,128.

Wiemer, S., and M. Wyss (2000), Minimum magnitude of completeness in earthquake catalogs: Examples from Alaska, the western United States, and Japan, *Bull. Seismol. Soc. Am.*, *90*(4), 859–869.

Wiemer, S., and M. Wyss (2002), Mapping spatial variability of the frequency-magnitude distribution of earthquakes, *Adv. Geophys.*, *45*, 259–302.

Wiemer, S., D. Giardini, D. Fäh, N. Deichmann, and S. Sellami (2009), Probabilistic seismic hazard assessment of Switzerland: Best estimates and uncertainties, *J. Seismol.*, *13*(4), 449–478.

Wyss, M., K. Shimazaki, and T. Urabe (1996), Quantitative mapping of a precursory seismic quiescence to the Izu-Oshima 1990 (M6.5) earthquake, Japan, *Geophys. J. Int.*, *127*(3), 735–743.

Yamashita, T., and L. Knopoff (1989), A model of foreshock occurrence, *Geophys. J. Int.*, *96*(3), 389–399.

Zechar, J. D., D. Schorlemmer, M. Liukis, J. Yu, F. Euchner, P. J. Maechling, and T. H. Jordan (2010), The collaboratory for the study of earthquake predictability perspective on computational earthquake science, *Concurrency Comput. Pract. Exp.*, *22*(12), 1836–1847.

R. Dominguez, Department of Physics and Astronomy, Western Kentucky University, Bowling Green, KY 42101, USA. (erg.dominguez@wku.edu)

W. Klein, Department of Physics and Center for Computational Science, Boston University, Boston, MA 02215, USA.

C. A. Serino, Department of Physics, Boston University, Boston, MA 02215, USA.

K. Tiampo, Department of Earth Sciences, University of Western Ontario, London, Ontario, N6A 5B7 Canada.

Fractal Dimension and b Value Mapping Before and After the 2004 Megathrust Earthquake in the Andaman-Sumatra Subduction Zone

Sohini Roy

School of Oceanographic Studies, Jadavpur University, Calcutta, India

Uma Ghosh

Department of Mathematics, Lalbaba College, Howrah, India

Sugata Hazra and J. R. Kayal

School of Oceanographic Studies, Jadavpur University, Calcutta, India

About 8000 earthquakes $M_w \geq 4.5$ recorded during the period 1964–2007 in the Andaman-Sumatra subduction zone, relocated by the Engdahl, van der Hilst, and Buland (EHB) method (EHB catalog), are used for fractal dimension and b value mapping. These maps are prepared to examine the state of dynamic evolution of tectonic states before and after the 26 December 2004 megathrust event M_w 9.3. After this megathrust event, one great earthquake M_w 8.6 occurred on 25 March 2005 to its south, and two large/great earthquakes M_w 7.9 and 8.5 occurred in 2007 farther south along the Sumatra subduction zone. We observe fractal correlation dimension (D_c) 0.6–1.6 all along the Andaman-Sumatra-Java subduction zone except for a few patches of lower D_c 0.4–0.5. The lower D_c implies more cluster characteristics of the earthquakes, while the D_c 1.0–1.6 indicates linear to fraction 2-D characteristics of the tectonic features. A significant spatial change in fractal characteristics is observed after the 2004 megathrust event; the aftershock clusters are identified with prominent low D_c (0.4–0.5) at the junction of the West Andaman fault and Andaman Sea Ridge, at the epicenter zone of the 2005 great earthquake, and at the Sumatra fault zone. A similar observation is made with the b value maps. The b value is high (1.4–1.7) all along the subduction zone, but a long stretch of comparatively lower b value (1.1–1.3) zone is observed along the Sumatra trench where all the main shocks occurred, which include the 2004 megathrust event, the 2005 great earthquake, and the two 2007 large/great earthquakes. After the megathrust event, there is a significant spatial change in b value structure, the regional trend remaining the same. A transverse b value structure was noted in between the Sumatra and Java subduction zones before the megathrust event.

1. INTRODUCTION

The Burmese-Andaman-Sumatra-Sunda arc defines a ~5500 km long boundary between the Indo-Asian-Australian plates, from Myanmar to Sumatra and Java to Australia

Extreme Events and Natural Hazards: The Complexity Perspective
Geophysical Monograph Series 196

10.1029/2011GM001072

[*Fitch*, 1970; *Curray et al.*, 1979]. The plate boundary separates the northeast-moving Indian plate from the Southeast Asian plate that includes Burma, Andaman, and Sunda *microplates* [*Curray et al.*, 1979; *Curray*, 2005]. It has been suggested that the Indian plate converges obliquely toward the Asian plate at an average rate of 54 mm yr^{-1} [*DeMetes et al.*, 1994]. The oblique convergence has caused formation of a sliver plate between the subduction zone and the right-lateral Sumatra and Java fault systems in the southern part and the Sagaing fault system in the northern part and opened the Andaman Sea Ridge (ASR) in the Andaman Sea (Figure 1). The nature of convergence varies from continental type in the Burmese arc to oceanic type in the Andaman-Sunda arc [*Kayal*, 2008]. The Andaman-Sunda arc is seismically very active and falls in the category of highest seismic hazard zone (V) at par with the northeast India region with varying degrees of tectonism and volcanic activity along this subducting margin [*Curray*, 2005].

Our study area, the Andaman-Sumatra section of the subduction zone had produced several large and great earthquakes in the past, some of which generated destructive tsunamis [*Billham*, 2005]. The largest among them are the historical earthquakes that occurred in 1833 ($M \sim 8.7$), 1861 ($M \sim 8.5$), 1881 (M_w 7.9), and 1941 (M_w 7.7) (Figure 1). While these large earthquakes ruptured only a few hundreds of kilometers (~200–300 km) of the plate boundary, the 26 December 2004 tsunamigenic Sumatra megathrust earthquake (M_w 9.3) and its large series of aftershocks ruptured more than 1300 km length of the arc, stripping the regions that were ruptured in the past as well as the intervening unbroken patches [*Billham*, 2005] (Figure 1). After this 2004 megathrust event, the Sumatra subduction zone has experienced one great earthquake (M_w 8.6) in 2005 and one great (M_w 8.5) and one large earthquake (M_w 7.9) in 2007 (Figure 1).

About 8000 epicenters, including about 3000 aftershocks of the 2004 megathrust and 2005 great earthquakes recorded during the period 1964–2007, are relocated by the Engdahl, van der Hilst, and Buland (EHB) method [*Engdahl et al.*, 2007; Engdahl, personal communication, 2008] (Figure 2). These events $M_w \geq 4.5$ are much controlled in depth estimation, and magnitude M_w are also well estimated. We have divided these data file into two sets, one before the megathrust earthquake (i.e., from January 1964 to 25 December 2004), which consists of 4748 events (Figure 2a), and the other after the megathrust event (i.e., from 26 December

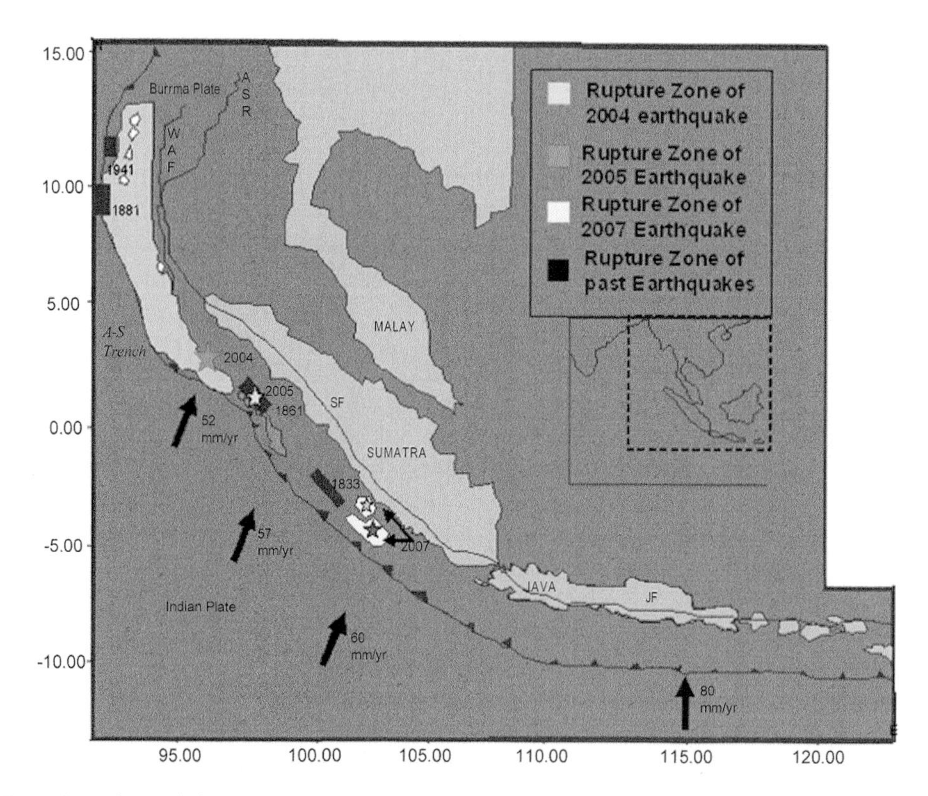

Figure 1. Tectonic setting and plate motion at the Andaman-Sumatra subduction zone. Abbreviations are as follows: WAF, West Andaman fault; ASR, Andaman Sea Ridge; A-S trench, Andaman-Sumatra trench; SF, Sumatra fault; and JF, Java fault systems. Megathrust and large earthquakes with rupture zones are shown. Modified from *Briggs et al.* [2006].

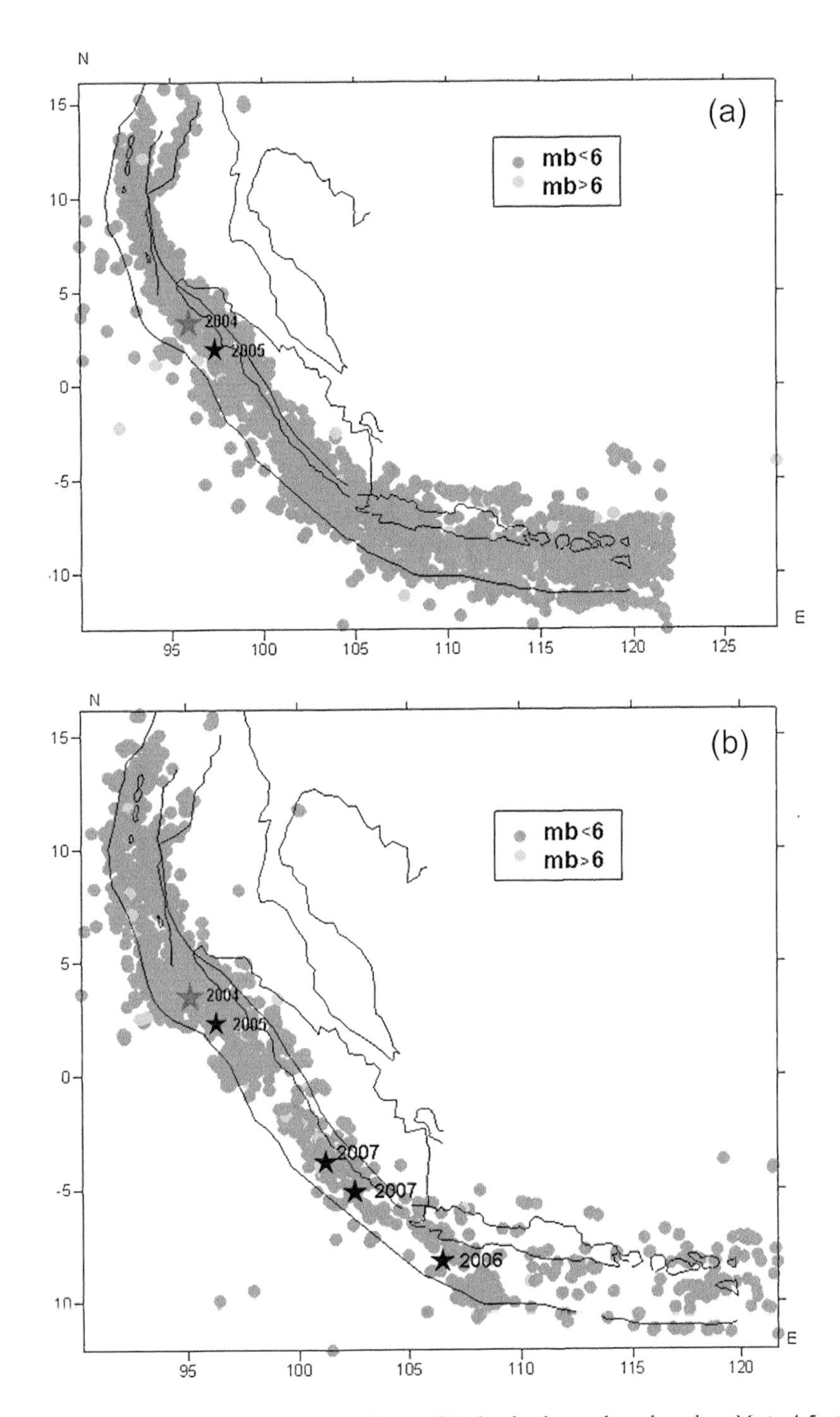

Figure 2. Epicenter maps of the Engdahl, van der Hilst, and Buland relocated earthquakes $M_w \geq 4.5$ at the Andaman-Sumatra subduction zone. (a) Epicenters before the 26 December megathrust earthquake (July 1964 to 25 December 2004) and (b) epicenters of earthquakes after the megathrust event (26 December 2004 to December 2007).

2004 to December 2007) with 3407 events (Figure 2b). These data sets are used to map the fractal dimension and b value characteristics of the ~3000 km long Andaman-Sumatra subduction zone in the region between 15°S–15°N latitude and 90°E–125°E longitude. A comparative study of these seismic characteristic maps is made to understand the effect of the megathrust event on the megathrust subduction zone and its dynamic evolution of the tectonic state.

2. MAPPING SEISMIC CHARACTERISTICS: POWER LAW RELATIONS

The earthquake phenomenon is explained by *power law* relations with respect to magnitude, time, and space. The epicentre distribution of earthquakes is represented by self-similar mathematical construct, the "fractal," and the scaling parameter is called the fractal dimension [*Kagan and Knopoff*, 1980; *Mandelbrot*, 1982]. *Fractal* dimension, the power law relation, is a two-point spatial correlation function for

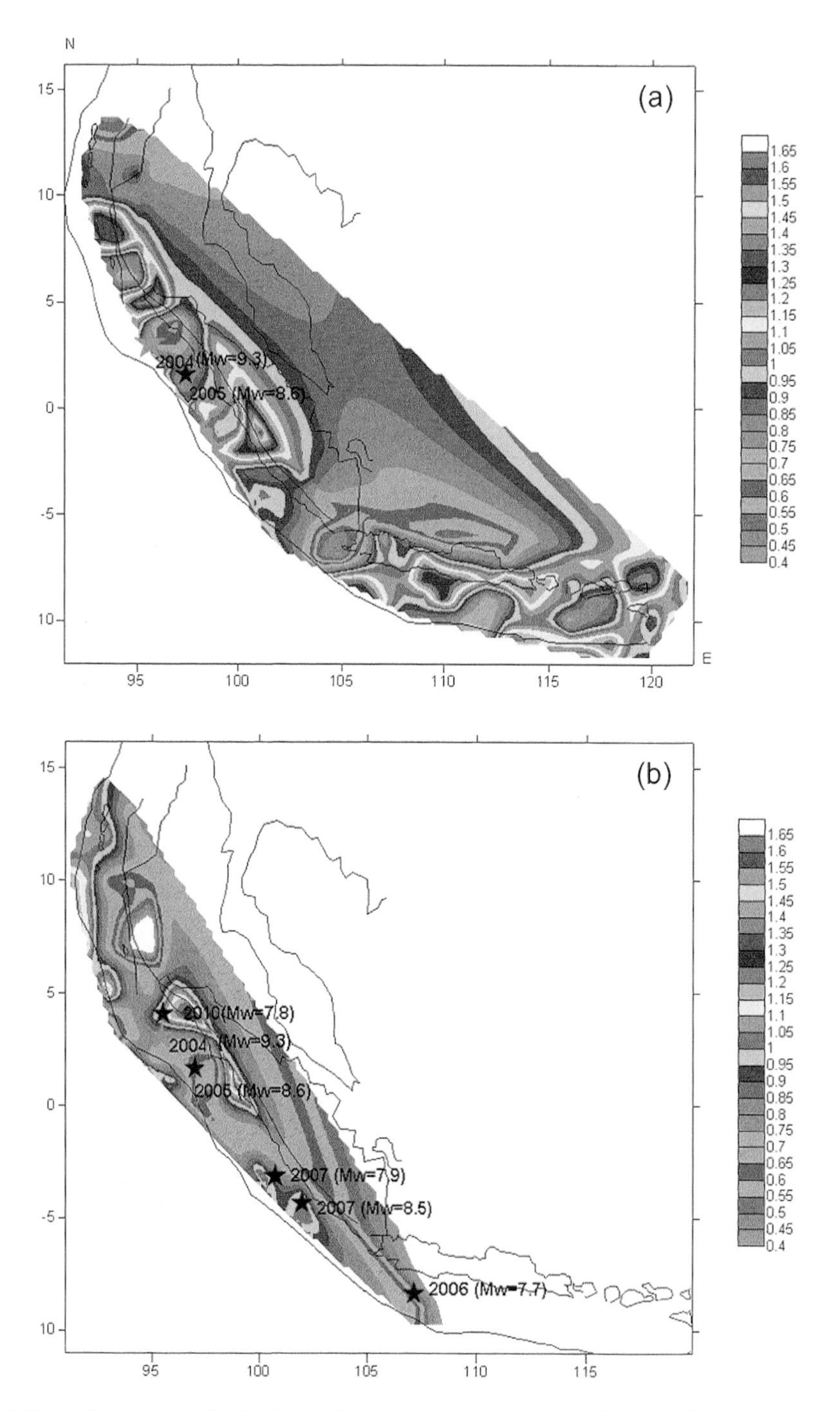

Figure 3. Fractal dimension maps at the Andaman-Sumatra subduction zone: (a) before the December 2004 megathrust earthquake and (b) after the megathrust earthquake.

earthquake epicenters. It is a powerful tool to characterize the fault geometry that has a self-similar structure [*Hirata*, 1989]. As fractal dimension measures the degree of clustering of the seismic events, changes in fractal dimension of the hypocenter distribution data is a good precursor consideration of earthquakes [*Dimri*, 2005].

The b value is another power law relation; it is a frequency-magnitude relation defined by *Gutenberg and Richter* [1944]. The b value of a region reflects the frequency-magnitude characteristics of geotectonic features, stress distribution in space and depth [*Mori and Aberocrombi*, 1997; *Wiemer and Wyss*, 1997; *Wiemer et al.*, 1998]. It represents a statistical measurement of the relative abundance of large and small earthquakes in a group or cluster. If b value is large, large or great earthquakes are relatively rare. There also exists a relationship between the frequency of earthquakes and the length of the faults [*Wang and Lu*, 1997].

For the present analysis to estimate the above power law relations, the study area is gridded by $2° \times 2°$ spacing with an overlapping of 1°. This exercise generated 312 grids. Fractal correlation dimension and the frequency-magnitude relation (b value) are estimated using the events in each grid, each containing more than or equal to 50 events; other grids are discarded. The center of each grid is taken as the plotting point for making contour maps.

2.1. Fractal Dimension

The most commonly used methods to estimate fractal dimension is the box counting method, which measures the *capacity dimension* D_0, and the correlation integral method, which measures the *correlation dimension* D_2 [*Hirata*, 1989; *Bhattacharay et al.*, 2002]. The correlation dimension D_2 is widely applied in seismology, especially to spatial distribution of epicenters. This technique is preferred to box counting algorithm because of its greater reliability and sensitivity to small changes in clustering properties [*Kagan and Knopoff*, 1980; *Hirata*, 1989]. The correlation dimension method measures the spacing between two points, which in this case are the earthquake epicenters [*Grassberger and Procaccia*, 1983].

The epicenter distribution has a fractal structure, and it is given by the following relation:

$$C(r) \sim r^{D_2}, \qquad (1)$$

where D_2 is fractal dimension, more strictly, the correlation dimension [*Grassberger and Procaccia*, 1983]. By plotting $C(r)$ against r on a double logarithmic coordinate, we can practically obtain the fractal dimension D_2 from the slop of the graph. The distance (r) between two events, (θ_1, ϕ_1) and (θ_2, ϕ_2), is calculated by using a spherical triangle as given by *Hirata* [1989]:

$$r = \cos^{-1}(\cos\theta_1\cos\theta_2 + \sin\theta_1\sin\theta_2\cos(\phi_1 - \phi_2)). \qquad (2)$$

The slope is obtained by fitting a least squares line in the scaling region.

The estimated fractal dimensions D_2 in each grid are then used for contouring to prepare the map. Thus, we prepared two maps, one using the events before the 26 December 2004 megathrust earthquake and the other using the events after the megathrust earthquake; these two maps are shown in Figures 3a and 3b.

2.2. Frequency-Magnitude Relation (b Value)

Magnitude of an earthquake indicates its size. The statistical distribution of sizes for a group of earthquakes is very complicated. *Gutenberg and Richter* [1944] provided a simplest frequency-magnitude relation of earthquakes, which describes a power law relation:

$$\log_{10}N = a - bM, \qquad (3)$$

where N is the cumulative number of earthquakes in a group having magnitude larger than M, a is a constant, and b is the slope of the log-linear relation. The estimated slope of the log-linear relation or the coefficient b is known as the b value. The b values are estimated using two methods: *least squares fit method* and *maximum likelihood method*. The maximum likelihood method, which is based on theoretical consideration, is claimed to be a better method [*Aki*, 1965]. We have used the maximum likelihood method to estimate the b value of the earthquakes in each selected grids of the region. In this method, the b value is defined as

$$b = \frac{\log_{10}e}{\overline{M} - M_0},$$

where $\overline{M}$ is the average magnitude, and M_0 is the threshold magnitude.

An estimate of error, standard deviation δb of the b value is given by *Aki* [1965]; then, a modified formulation is given by *Shi and Bolt* [1982] as follows:

$$\delta b = 2.3b^2\sqrt{\frac{\sum(M_i - \overline{M})}{n(n-1)}}.$$

The standard errors of b values are estimated to be $\delta b \pm$ 0.001–0.0015. The estimated b values are considered for contouring to prepare the map. We thus obtained two b value maps (Figures 4a and 4b), one using the data before the 26 December 2004 and the other using the earthquakes recorded after the 26 December 2004 megathrust event, where the data up to December 2007 are used.

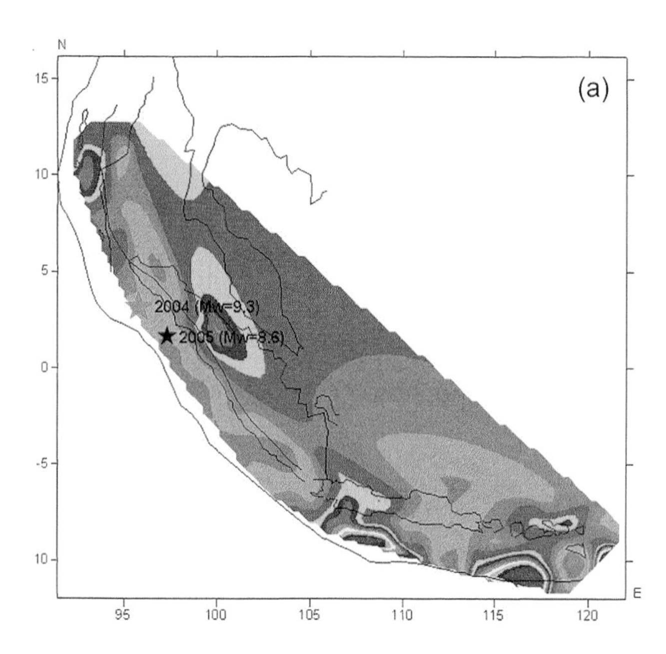

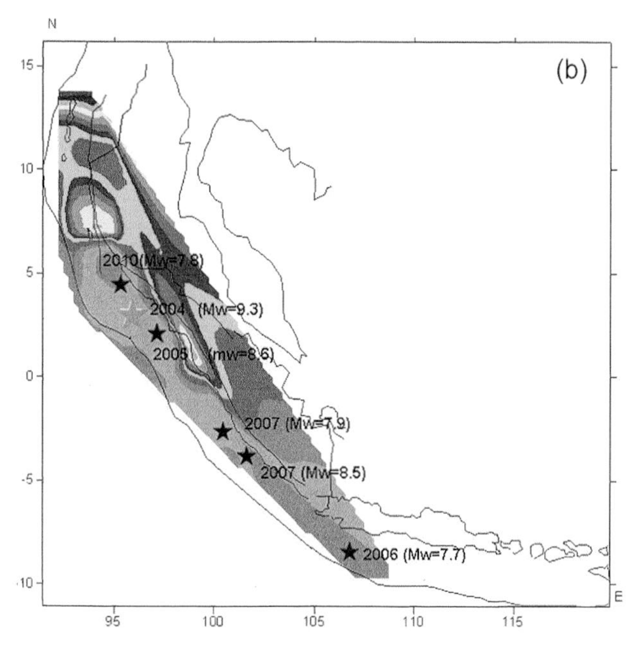

Figure 4. The *b* value maps at the Andaman-Sumatra subduction zone: (a) before the December 2004 megathrust earthquake and (b) after the megathrust earthquake.

3. RESULTS AND DISCUSSION

3.1. Fractal Dimension

The fractal correlation dimension (hereafter called D_c) of the spatial distribution of epicenters may be related to the heterogeneity of the fractured material. A value of D_c close to 3 implies that earthquake fractures are filling up a volume of the crust; a value close to 2 suggests that it is a plane that is being filled up, and a value close to 1 means that line sources are predominant [*Aki*, 1981]. *Tosi* [1998] illustrated that the possible values of D_c for epicenter distribution of earthquakes are bound to range between 0 and 2, which is dependent on the dimension of the embedding space. Interpretation of such limit values is that a set with $D_c \rightarrow 0$ has all events clustered into one point, and at the other end of the scale, $D_c \rightarrow 2$ indicates that the events are randomly or homogeneously distributed over a 2-D embedding space. *Hirata* [1989] demonstrated that active fault systems in Japan possess self-similarity with fractal correlation dimensions of 0.5 to 1.6.

In this study, we observe a spatial variation of D_c all along the subduction zone from north to south and to southeast, Andaman to Sumatra to Java, in both the maps (Figures 3a and 3b). Before the megathrust event, the D_c ranges between 0.6 and 1.6 all along the Andaman-Sumatra-Java subduction zone except for four lowest-value (0.40–0.55) zones, one to the north of Andaman islands, one near the epicenter area of the 2004 megathrust earthquake, and two larger zones in the Java subduction zone (Figure 3a). Among the two longest low-D_c zones, one indicates a transverse structure in between the Sumatra and Java subduction, and the other is along the fore arc of the Java subduction zone. The zones with the lowest D_c 0.40–0.55 indicate that the epicenters are more in cluster characteristics. The zones with D_c 1.0–1.6 indicate linear characteristics to fraction 2-D embedding space. The linear characteristics are observed along the trench and along the known long faults like the Sumatra fault and Java fault systems in the fore arc.

After the 2004 megathrust event, there is a significant change in the fractal structures. The change in the fractal dimension may correspond to the dynamic evolution of the states of the system after the megathrust event. Four lowest-D_c (0.40–0.55) zones are observed, from north to south, one in the West Andaman fault (WAF)-ASR junction, one small circular patch at the Sumatra trench, one near the epicenter of the 2005 great earthquake (M_w 8.7), and the largest zone is in the Sumatra fore arc, a little oblique to the south Sumatra fault system (Figure 3b). These observations are much different from the earlier observations and reflect the dynamic evolution of the tectonic state after the 2004 megathrust and 2005 great events. These structures are compatible with the cluster of aftershocks at the WAF-ASR, the 2005 aftershock cluster, and the intense activity along the Sumatra fault system (Figure 2b). We do not observe aftershocks of the 2004 and 2005 events in the Java subduction zone, and not many regional events are available for the period 2005–2007 after the megathrust event for fractal mapping of this zone (Figure 2b).

3.2. The *b* Value

A statistical measurement of relative abundance of large and small earthquakes in an active zone is well represented by the frequency-magnitude power law relation. In a tectonically active region, a normal *b* value is found to be ~1.0, but it may range from 0.5 to 2.0 depending on the crustal heterogeneities, tectonic stress conditions, thermal structures, etc. [e.g., *Kayal*, 2008]. A higher *b* value means that a smaller fraction of the total earthquakes occur at the higher magnitudes, whereas a lower *b* value implies a larger fraction occur at higher magnitudes. If *b* is large, large and great earthquakes are relatively rare. It has been also reported that the *b* value shows systemic variations in the period preceding a major earthquake [e.g., *Kayal*, 2008]. Hence, it is the most investigated parameter in seismology, observationally as well as theoretically.

In this study, it is observed that before the 2004 megathrust event, the major portions of the subduction zone shows *b* value 1.3–1.7 with a few zones having comparatively lower *b* values 1.1–1.2. Four lower *b* value structures are observed, one to the north of Sumatra island near the epicenter of the 2004 megathrust event, one longer zone along the Sumatra trench, one longer zone across the Java subduction zone, and one small circular zone farther southeast (Figure 4a).

After the megathrust event, there is a significant change in the *b* value structures. We observe higher *b* value (1.6–1.8) structures at the WAF-ASR junction and along the Sumatra fault zone. These zones indicate higher aftershock activity with more lower-magnitude aftershocks (Figure 2b). Two lower *b* value (1.1–1.2) structures are observed, one at the epicenter zone of the 2004/2005 megathrust/great earthquakes and one along the Sumatra trench to the south. These two lower *b* value structures are significantly shifted spatially and extended laterally compared to that were observed before the 2004 megathrust event (Figures 4a and 4b). This observation indicates the state of dynamic evolution with the 2004 megathrust earthquake. It may be mentioned that one large earthquake (M_w 7.8) occurred in 2010 in the low *b* value structure at the 2004/2005 epicenter zone, and two large earthquakes (M_w 8.5 and 7.9) occurred in 2007 and one M_w 7.8 earthquake occurred in 2006 at the southern segment of the lower *b* value structure (Figure 4b). No large earthquake ($M_w > 7.5$) yet occurred in the transverse low *b* value (1.1–1.3) structure at the Java subduction zone, which could be a source zone for an impending large earthquake. The change in this transverse low *b* value structure after the 2004 megathrust event, however, could not be mapped with the insufficient data (2005–2007) (Figures 4a and 4b).

4. CONCLUSIONS

A dynamic evolution of the tectonic states after the December 2004 megathrust earthquake is identified at the Andaman-Sumatra subduction zone. The seismic characteristics, the fractal correlation dimension (D_c), and the *b* value structures show a significant spatial change after the megathrust earthquake. The cluster characteristics of the earthquakes are observed with low D_c (0.4–0.5) and the linear to fraction 2-D fault systems with higher D_c (1.0–1.6). Lower *b* value structures in the subduction zone are found to be the source zones for the large/great earthquakes ($M_w > 7.5$). Spatial shifts of the D_c and *b* value structures are evident after the megathrust event. In general, lower *b* value structures had been the source zones for the large/great earthquakes in the subduction zone. The fractal correlation dimension D_c for these source zones is found to be in the range of 1.0–1.6 with a few low D_c (0.40–0.55) showing the cluster characteristics of the aftershocks.

Acknowledgments. All earthquakes were located by the EHB method, and listings of the events were available from E. R. Engdahl (personal communication, 2008). The Council of Scientific and Industrial Research (CSIR), Government of India, New Delhi, sponsored the research project.

REFERENCES

Aki, K. (1965), Maximum likelihood estimation of *b* in the formula log $N = a - bM$ and its confidence limits, *Bull. Earthquake Res. Inst. Univ. Tokyo*, *43*, 237–239.

Aki, K. (1981), A probabilistic synthesis of precursory phenomena, in *Earthquake Prediction: An International Review, Maurice Ewing Ser.*, vol. 4, edited by W. Simpson and G. Richards, pp. 566–574, AGU, Washington, D. C., doi:10.1029/ME004p0566.

Bhattacharay, P. M., R. K. Majumdar, and J. R. Kayal (2002), Fractal dimension and b-value mapping in northeast, India, *Curr. Sci.*, *82*(12), 1486–1491.

Bilham, R. (2005), A flying start then a slow slip, *Science*, *308*, 1126–1127.

Briggs, R. W., et al. (2006), Deformation and slip along the Sunda megathrust in the great 2005 Nias-Simeulue earthquake, *Science*, *31*, 1897–1901.

Curray, J. R. (2005), Tectonics and history of the Andaman Sea region, *J. Asian Earth Sci.*, *25*, 131–140.

Curray, J. R., D. G. Moore, L. A. Lawver, F. J. Emmel, R. W. Raitt, M. Henry, and R. Kieckhefer (1979), Tectonics of the Andaman Sea and Burma, in *Geological and Geophysical Investigations of Continental Margins*, edited by J. S. Watkins, L. Montadert, and P. Dickenson, *AAPG Mem.*, *29*, 189–198.

DeMets, C., R. G. Gordon, D. F. Argus, and S. Stein (1994), Effect of recent revisions to the geomagnetic reversal time scale on estimates of current plate motions, *Geophys. Res. Lett.*, *21*(20), 2191–2194.

Dimri, V. P. (2005), *Fractal Behaviour of the Earth System*, 224 pp., Springer, New York.

Engdahl, E. R., A. Villasenor, R. H. DeShon, and C. H. Thurber (2007), Teleseismic relocation and assessment of seismicity (1918–2005) in the region of the 2004 M_w Sumatra–Andaman and 2005 M_w 8.6 Nias Island Great Earthquakes, *Bull. Seismol. Soc. Am.*, *97*(1A), S43–S61.

Fitch, T. J. (1970), Earthquake mechanisms in the Himalaya, Burmese, and Andaman regions and continental tectonics in central Asia, *J. Geophys Res.*, *75*, 2699–2709.

Grassberger, P., and I. Procaccia (1983), Measuring the strangeness of strange attractors, *Physica D*, *9*, 189–208.

Gutenberg, R., and C. F. Richter (1944), Frequency of earthquakes in California, *Bull. Seismol. Soc. Am.*, *34*, 185–188.

Hirata, T. (1989), A correlation between the *b* value and fractal dimension of earthquakes, *J. Geophys. Res.*, *94*, 7507–7514.

Kagan, Y. Y., and L. Knopoff (1980), Spatial distribution of earthquakes: The two point correlation function, *Geophys. J. R. Astron. Soc.*, *62*, 303–320.

Kayal, J. R. (2008), *Microearthquake Seismology and Seismotectonics of South Asia*, 282 pp., Springer, Berlin.

Mandelbrot, B. B. (1982), *The Fractal Geometry of Nature*, 460 pp., Freeman, San Francisco, Calif.

Mori, J., and R. E. Aberocrombi (1997), Depth dependence of earthquake frequency-magnitude distribution in California: Implications for the rupture initiation, *J. Geophys. Res.*, *102*, 15,081–15,090.

Shi, Y., and B. Bolt (1982), The standard error of magnitude-frequency *b* value, *Bull. Seismol. Soc. Am.*, *72*, 1677–1687.

Tosi, P. (1998), Seismogenic structure behaviour revealed by spatial clustering of seismocity in the Umbria-Marche region (central Italy), *Ann. Geofis.*, *41*(2), 215–224.

Wang, Z., and H. Lu (1997), Evidence and dynamics for the change of the strike-slip direction of the Changle-Nannao ductile shear zone South East China, *J. Asian Earth Sci.*, *15*(6), 507–515.

Wiemer, S., and M. Wyss (1997), Mapping the frequency-magnitude distribution in asperities: An improved technique to calculate recurrence times, *J. Geophys. Res.*, *102*, 15,115–15,128.

Wiemer, S., S. R. Mcnutt, and M. Wyss (1998), Temporal and three dimensional spatial analysis of the frequency-magnitude distribution near Long Valley Caldera, California, *Geophys. J. Int.*, *134*, 409–421.

U. Ghosh, Department of Mathematics, Lalbaba College, Howrah 711202, India.

S. Hazra, J. R. Kayal, and S. Roy, School of Oceanographic Studies, Jadavpur University, Kolkata 700032, India. (sohini.roy54@gmail.com)

Stress Pulse Migration by Viscoelastic Process for Long-Distance Delayed Triggering of Shocks in Gujarat, India, After the 2001 M_w 7.7 Bhuj Earthquake

B. K. Rastogi, Pallabee Choudhury, and Rakesh Dumka

Institute of Seismological Research, Gandhinagar, India

K. M. Sreejith and T. J. Majumdar

Space Applications Centre, Indian Space Research Organization, Ahmedabad, India

About 200 km × 300 km Kachchh region of Gujarat in western India is seismically one of the most active intraplate regions of the world. It has six major E-W trending faults of the failed Mesozoic rift that are being reactivated by thrusting. Kachchh had earlier experienced earthquakes of *M*7.8 in 1819, *M*6.3 in 1845, and *M*6 in 1956 and a small number of $M < 6$ earthquakes. After the *M*7.7 earthquake in 2001, besides the continuing high seismicity in the rupture zone of 20 km radius, several other faults within distances of 60 km from the rupture zone and even after 5–8 years of the Bhuj earthquake are activated with earthquakes of *M*4–5.7 and associated sequences. Moreover, the seismicity to the *M*3–5 level is triggered along small faults at 20 locations up to 200 km south in the Saurashtra region. The unusually high seismicity along several faults is inferred to be triggered because of a stress pulse migration by viscoelastic processes after a 20 MPa stress drop due to the 2001 Bhuj earthquake.

1. INTRODUCTION

The Gujarat region in western India is seismically one of the most active intraplate regions of the world. The Gujarat region has major E-W trending faults of the failed Mesozoic rifts of Kachchh and Narmada that are being reactivated by thrusting. There are some smaller transverse strike-slip faults. South of Kachchh, in the Deccan Volcanics of Saurashtra, the NW and NE trending smaller strike-slip faults are activated in the form of moderate earthquakes in response to the plate-tectonics stress [*Biswas*, 1987].

The Kachchh part of Gujarat was known to have low seismicity but high hazard in view of the occurrences of several large earthquakes but fewer moderate or smaller shocks. The scenario changed after the Bhuj (23.44°N, 70.31°E) earthquake of *M*7.7 in 2001, as several small to moderate earthquakes occurred (Table 1) along a number of faults in Kachchh. In Saurashtra, south of Kachchh and beyond 30 km wide Gulf of Kachchh also, the number of small to moderate earthquakes increased substantially [*Chopra et al.*, 2008a] (Table 2) when 30 felt shocks (of *M*4 or so) occurred at 20 different locations. In contrast, 20 decades earlier hardly one or two felt shocks per decade were experienced with the exception of the year 1938 when an earthquake of *M*5.7 and five strong foreshocks were felt at Paliyad (22.40°N, 71.80°E) in Saurashtra.

No doubt the detectability of earthquakes and completeness of catalog for different magnitude levels have changed over time for the study region; the space-time changes inferred here are gross and not much influenced by it.

It has been estimated that the catalog of earthquakes for Gujarat region is complete for *M*5 earthquakes for the last

Extreme Events and Natural Hazards: The Complexity Perspective
Geophysical Monograph Series 196

10.1029/2011GM001061

Table 1. Felt and Damaging Earthquakes in Kachchh[a]

Magnitude	Shocks pre-2001 (200 years)	Shocks post-2001 (9 years)
3.5–3.9	46	653
4.0–4.9	25	262
$\geq$ 5.0	11	20

[a]Though post-2001 number of shocks are mostly aftershocks, about 50% of shocks of $M < 5$ and at least one of *M*5.7 are independent main shocks.

120 years [*Yadav et al.*, 2008]. This catalog was homogenized for M_w magnitude by taking instrumentally determined M_w values or converting M_s, m_b, or M_L values to M_w. In the preinstrumental era, reports of strongly felt earthquakes, which could be of magnitude 4 or so, could be relied upon as most parts of the region were inhabited. Local stations started operating in the region from the early 1970s for several river valley projects, enabling detectability of *M*3 earthquakes for last 40 years. From mid-2006 onward a network of 50 broadband seismographs started operating, enabling a detectability of *M*2 earthquakes in the Kachchh active area and *M*2.5 in the rest of Gujarat [*Chopra et al.*, 2008b].

2. SEISMICITY OF THE GUJARAT REGION

The Institute of Seismological Research prepared a catalog of earthquakes in Gujarat from 1668 through 2000 using the available catalogs of *Oldham* [1883], *Chandra* [1977], *Tandon and Srivastava* [1974], and *Malik et al.* [1999] and institutional catalogs of the following: (1) India Meteorological Department, New Delhi, India; (2) Geological Survey of India (seismotectonic atlas of India and its environs, 2000); (3) National Earthquake Information Center, U.S. Geological Survey; (4) International Seismological Centre, Thatcham, United Kingdom; (5) Gujarat Engineering Research Institute, Gujarat, India; and (6) Institute of Seismological Research (ISR), Gandhinagar, Gujarat, India.

The catalog published by ISR in their Annual Report 2007–2008 (available at http://www.isr.gujarat.gov.in/pdf/Ann07.pdf) contains 187 earthquakes of $M \geq 2$ including 32 earthquakes of $M \geq 5.0$. Though the data of many earthquakes especially of the smaller shocks may not be precise, it is included to indicate which faults/areas may have been active.

Because of the large number of aftershocks since 2001 in Kachchh and enhanced activity in Saurashtra, it is difficult to assign main shocks to different faults and to distinguish main shocks from aftershocks in nearby faults of the main shocks. Hence, the catalog during 2001 to 2009 includes only 30 clear main shocks to indicate which faults/areas have been active.

There were thousands of aftershocks of the 2001 Bhuj earthquake. These aftershocks are not included in the catalog. Only the distinct main shocks along different faults are included. Hence, for the period 2001 to 2009, only five shocks of $M \geq 5.0$ and about 25 distinctly felt (*M*2–*M*4.9) are included in the catalog.

Prior to 2001, Kachchh had experienced three large earthquakes: the *M*7.8 Allah Bund (24.00°N, 69.00°E) earthquake in 1819, the *M*6.3 Lakhpat (23.80°N, 68.90°E) earthquake of 1845, and the *M*6 Anjar (23.30°N, 70.00°E) earthquake in 1956. Smaller shocks include seven earthquakes of magnitude 5–5.6 and only 71 of *M*3.5–4.9. Other areas of Gujarat have experienced a few damaging earthquakes of magnitude 6 or less, e.g., the one in 1970 in Bharuch (21.6°N, 72.96°E) along the South Narmada fault. In Saurashtra, the two significant earthquakes were at Ghogha (22°N, 72°E) in 1919 and at Paliyad in 1938. However, the last decade from 2000 onward has witnessed much increased seismicity in Kachchh as well as in Saurashtra as shown in Tables 1 and 2. Additionally, 30 felt main shocks (of *M*4 or so) occurred at 20 different locations in Kachchh and Saurashtra, and numerous smaller ones were felt at different locations. In contrast, the earlier 20 decades experienced hardly one or two felt shocks/decade (barring the year 1938 when four earthquakes of *M*4–*M*5.7 occurred at Paliyad in Saurashtra).

Table 2. Decennial Number of Earthquakes in Saurashtra[a]

Decade	Number of Earthquakes	Magnitudes
1870–1879	1	5.0
1889–1889	4	4.4
1890–1899	1	4.4
1900–1909	-	
1910–1919	1	5.7
1920–1929	1	4.3
1930–1939	8	4.3, 4.1, 5.5, and 5.7
1940–1949	1	5.0
1950–1959	-	
1960–1969	2	4.3 and 4.3
1970–1979	5	4.3, 3.3, 3.1, 3.6, and 3.3
1980–1989	7	3.2, 2.9, 3.2, 3.1, 3.5, 4.3, and 3.8
1990–1999	5	3.1, 4.4, 3.0, 3.2, and 2.5
2000–2009	16	3.6, 4.6, 4.2, 2.5, 2.0, 3.1, 2.5, 3.0, 4.0, 3.3, 3.1, 5.0, 2.9, 2.8, 3.2, and 3.0

[a]An unusually large number of main shocks during the decade starting in 2000 are inferred to be due to triggering caused by stress pulse generated by *M*7.7 earthquake in 2001 at distances of 100–200 km away. Except during 1919 and 1938, there were no significant earthquakes prior to 2000.

3. AFTERSHOCKS OF THE 2001 *M*7.7 EARTHQUAKE AND TRIGGERED EARTHQUAKES IN KACHCHH

The aftershock sequence of the *M*7.7 earthquake in 2001 consisted of 10,000 located aftershocks of $M \geq 1$ (or 3371 located aftershocks of $M \geq 2$) including 20 aftershocks of *M*5–5.7 and about a thousand of *M*3.5–4.9. Aftershocks in the 2001 *M*7.7 earthquake rupture zone in Kachchh continued at the *M*5.6 level until 2006 and at the $M \leq 5$ level subsequently. Annually, on average, six earthquakes of $M \geq 4$ and 60 earthquakes of $M \geq 3$ occur currently. For one year, the activity concentrated along the 2001 rupture zone of 40 km × 40 km with large slip (Figure 1). The total absence of epicenters outside the rupture zone may be noted. Subsequently, the hypocenters expanded to nearby areas along different faults in the E-W direction (more toward east) becoming 70 km × 50 km, 100 km × 75 km, and 125 km × 75 km by 2003, 2004, and 2006, respectively. By 2008, the area further expanded to 200 km × 80 km, covering the South Wagad fault (SWF) and the Banni fault. Additionally, the epicentral area expanded by 60 km toward the NE to the Gedi fault and transverse fault across it by March 2006 (Figure 2). The remarkable contrast from Figure 1 that most of the Kachchh region is full of epicenters may be noted. The EW expansion was as predicted from Coulomb stress change due to the 2001 earthquake but not in other directions. Moreover, the activities along the Allah Bund and the Island Belt faults have also increased, making the north Kachchh area of 250 km × 150 km sparsely active by 2008 with $M < 5$ shocks. Magnitude distribution of felt and damaging earthquakes in Kachch from historic time through 2009 is given in Table 1. About 50% of shocks since 2001 are aftershocks of the 2001 Bhuj earthquake and are located close to the rupture zone of 20 km radius and have focal depths of 10–30 km.

4. TRIGGERED EARTHQUAKES IN SAURASHTRA

The activity had also spread toward south to Saurashtra: 120 km by 2006 and 200 km by 2007 along several small faults in Jamnagar, Junagadh, Porbandar, and Surendranagar districts (Figure 3). At three sites the activity is in the form of sequences with largest shocks of *M* ~4–5 and several hundreds of $M \geq 0.5$ shocks recorded on local networks. Only two such sequences were reported earlier during 1938 at Paliyad in the Bhavnagar district and during 1986 in Valsad, south Gujarat. At some other sites the sequences had fewer shocks. These shocks are shallower than 10 km and are associated with subterranean sounds. Magnitude distribution

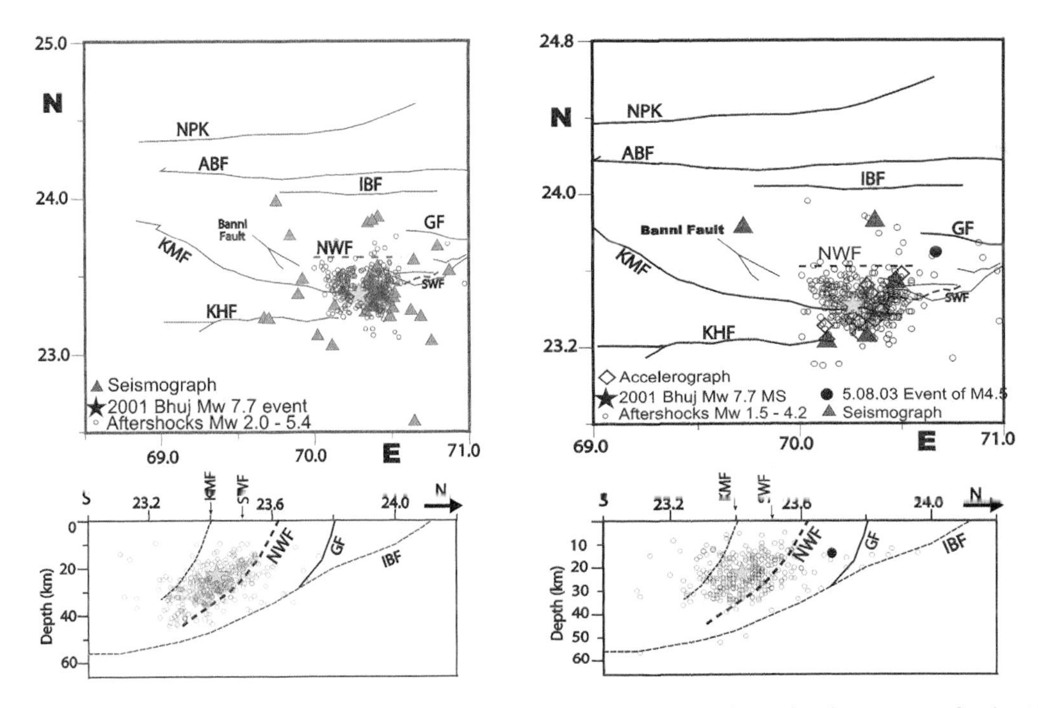

Figure 1. (left) Epicenters and focal depths of Kachchh earthquakes during 2001 shows that they were confined to NWF and 40 km × 40 km area, while (right) during next two years of 2002 to 2003 the other faults became slightly active (National Geophysical Research Institute (NGRI) data). Abbreviations are as follows: NPK, Nagar Parkar fault; ABF, Allah Bund fault; IBF, Island Belt fault; GF, Gedi fault; NWF, North Wagad fault; SWF, South Wagad fault; KMF, Kachchh mainland fault; and KHF, Katrol Hill fault.

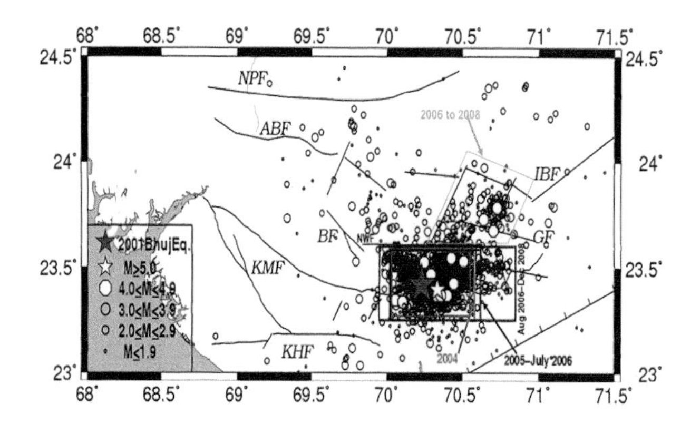

Figure 2. Epicenter of the M7.7 Bhuj earthquake, epicenters of $M > 1$ during 2006–2009, and temporal change in the epicentral area in Kachchh. By 2006 the South Wagad and the Gedi faults were activated, while by 2008 the Allah Bund fault also became active.

of felt and damaging earthquakes in Saurashtra from historic time through 2009 is given in Table 2. Figure 4 shows the seismicity of Kachchh and Saurashtra from 2001 to 2009, and Figure 5 shows the migration pattern of seismicity in these two regions.

5. GPS AND INTERFEROMETRIC SYNTHETIC APERTURE RADAR MEASUREMENTS

Since 2006, ISR is observing seismic deformation with 1 mm yr^{-1} accuracy across geological faults in Gujarat through a network of 25 permanent and 11 campaign GPS stations (Figure 6). Local deformation has been estimated with respect to Gandhinagar/Ahmedabad stations operated more than 200 km east of the epicenter of the main earthquake of 2001. For the period 2001–2009 we combined data of our two stations with that of the Indian Institute of Geomagnetism, which happened to be close by. Near the epicenter, initially the postseismic relaxation was large but exponentially reduced, being 12, 6, 4, and 3 mm for four consecutive 6 month periods of 2001–2002 [*Reddy and Sunil*, 2008; *Chandrasekhar et al.*, 2009]. Presently, the horizontal deformation is found to be very low, i. e., of the order of 2–5 mm yr^{-1} all over Kachchh (Figure 7). However, vertical deformation is found to be quite large, i.e., up to 13 mm yr^{-1} as observed by GPS measurements (Figure 8).

Studies related to differential interferometric synthetic aperture radar (InSAR) aided with corner reflectors have also been carried out. Envisat advanced synthetic aperture radar data sets of 22 June 2008 and 25 October 2009 with a baseline separation of 125 m are utilized for the interferogram

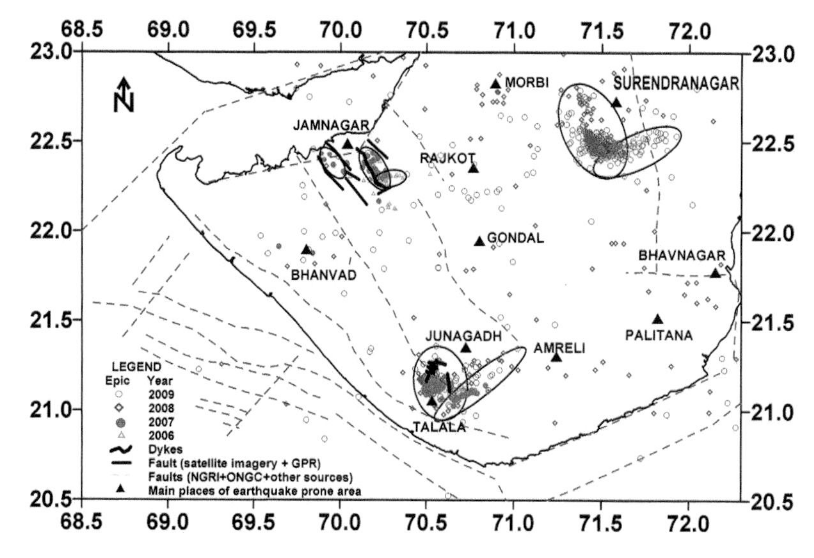

Figure 3. Epicenters in Saurashtra during 2006–2009 of M0.5–4.0. Faults inferred by NGRI and the Oil and Natural Gas Corporation are marked by dashed lines.

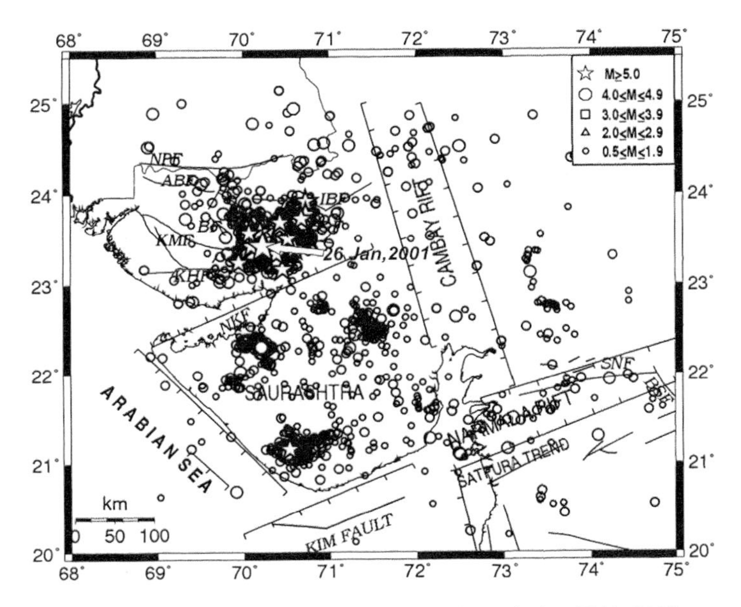

Figure 4. Epicenters of earthquakes in Gujarat during 2001–2009.

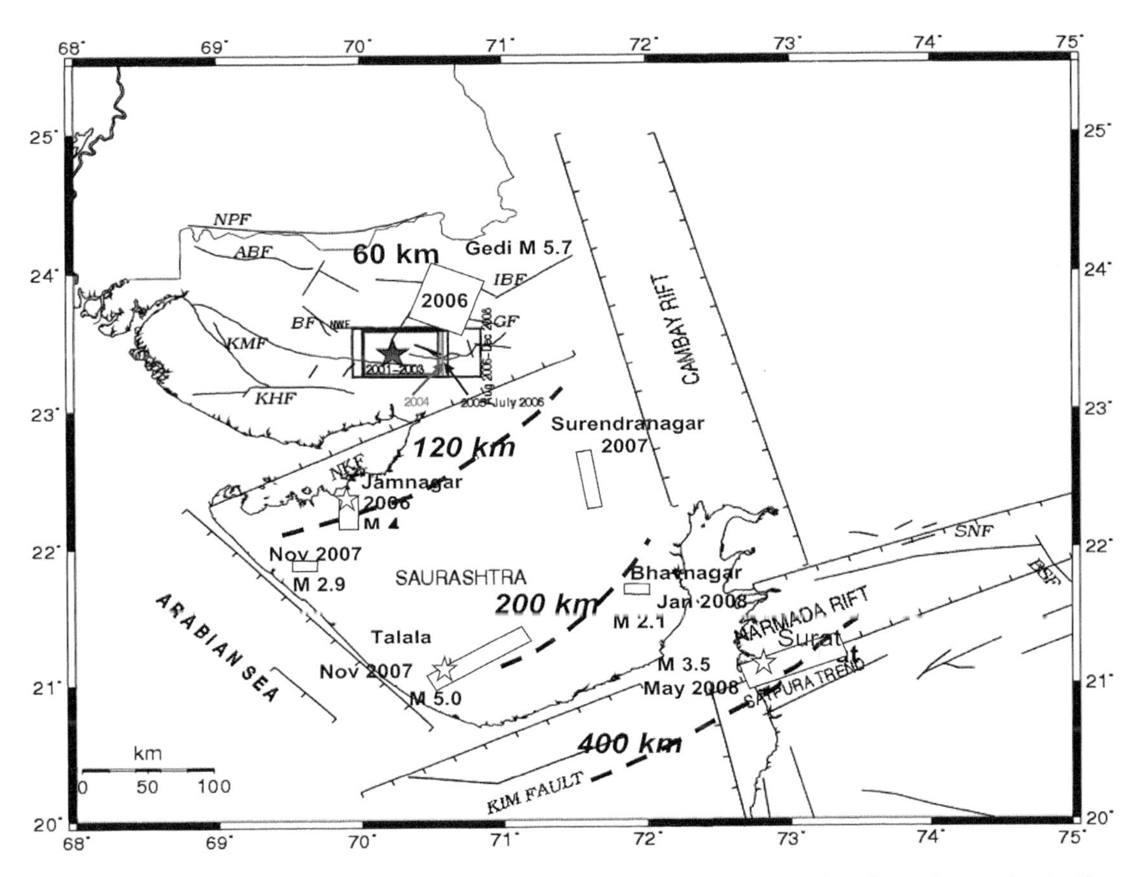

Figure 5. Long-time and delayed triggering of seismicity in Gujarat from 2006 to 2008. Epicentral zones for significant sequences are marked by rectangles. The seismicity migrated to 120 km south to Jamnagar in Saurashtra and to 200 km in the Surendranagar area and Talala in Junagadh.

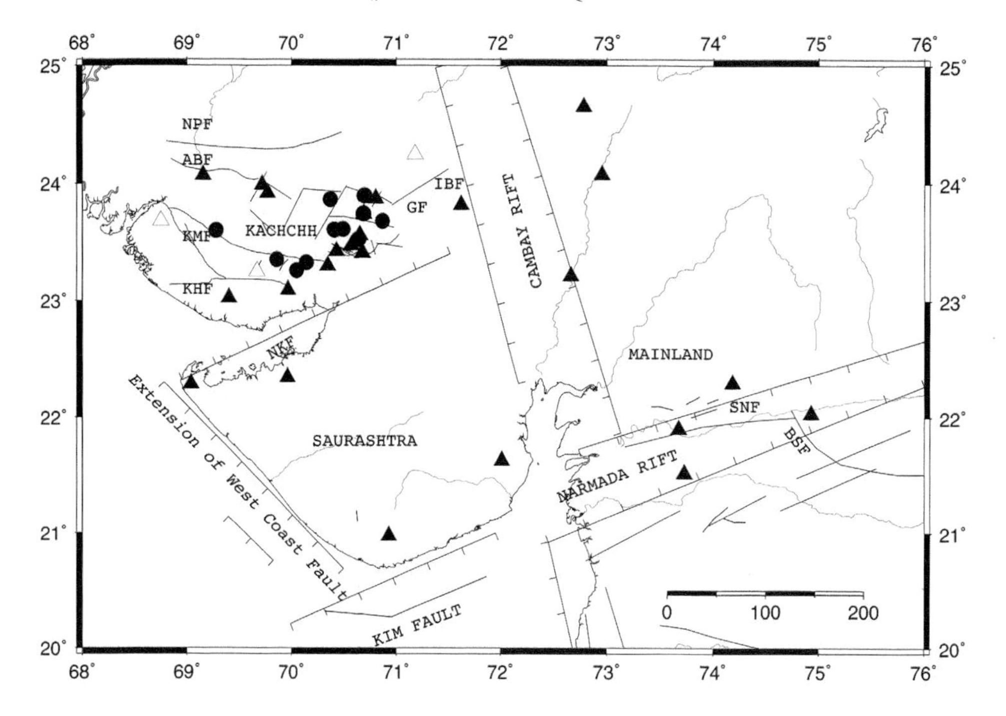

Figure 6. GPS permanent (solid triangles), campaign (circles), and proposed permanent (open triangles) stations of ISR.

generation. The signals related to the deformation are not directly visible in the interferogram because of decorrelation effects. Hence, only areas with coherence >0.2 were considered for the analysis. A differential interferogram was generated by removing the topographic phases, which were then converted to displacement (range change). This indicates deformation rates of 10–40 mm yr^{-1} along the line of sight (LOS) of the satellite in several parts of Kachchh (Figure 9) [*Sreejith et al.*, 2011]. Figure 10 shows the scene covered by the satellite path in the epicentral zone, and Figure 11 shows the range change values along the LOS.

Postseismic horizontal deformation in Wagad shows movement toward the south. The Rann area west of it shows movement toward the west and the Kachchh mainland; that

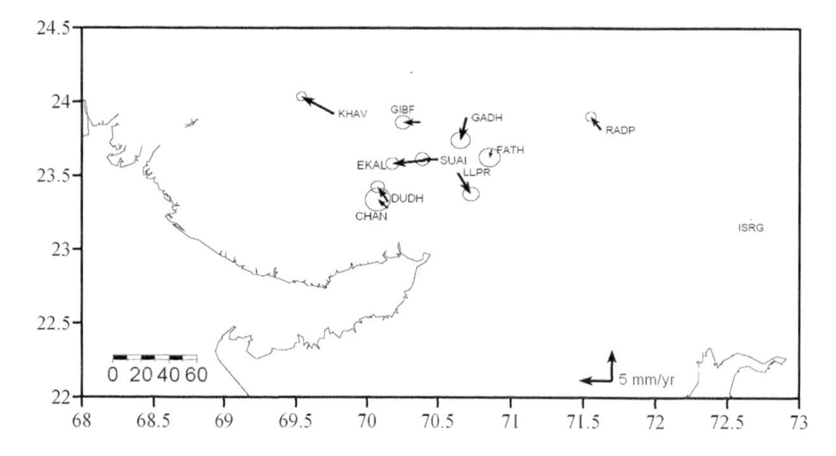

Figure 7. In the Kachchh area, horizontal deformation is found to be very low, i.e., of the order of 2–5 mm yr^{-1}. However, vertical deformation is found to be quite large, i.e., up to 13 mm yr^{-1} by GPS and up to 40 mm yr^{-1} by interferometric synthetic aperture radar along LOS (Figures 8 and 9). Postseismic horizontal deformation in Wagad shows movement toward the south. The Rann area west of it shows movement toward the west, and the Kachchh mainland area shows movement toward the north. Abbreviations are as follows: KHAV, Khavda; GIBF, Amarapar; GADH, Gadadha; EKAL, Ekal; SUAI, Suai; FATH, Fathegadh; RADP, Radhanpur; DUDH, Dudhai; CHAN, Chandrani; and ISRG, Gandhinagar.

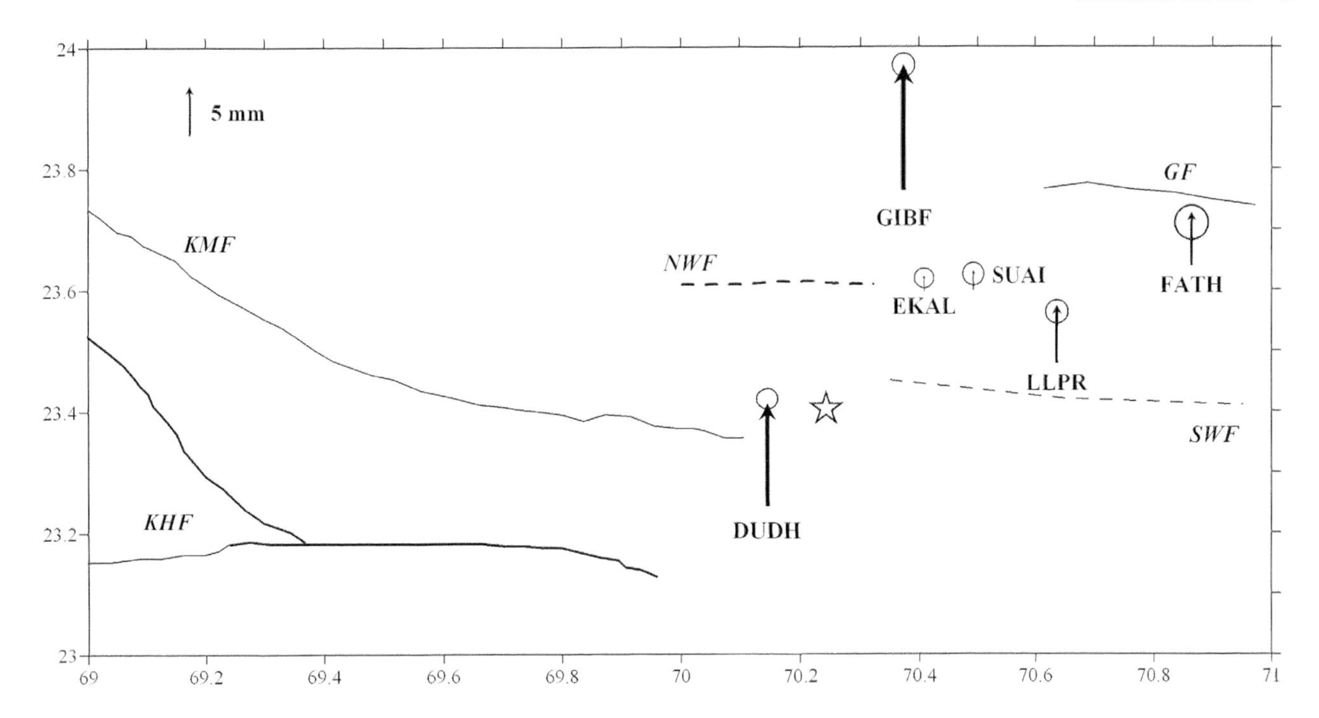

Figure 8. Rate of vertical displacement at campaign sites derived from GPS measurements, Maximum rate of displacement is ~13 mm yr^{-1}. Uncertainty in vertical displacement is 2–3 mm.

is, the hanging wall of the 2001 Bhuj earthquake reverse faulting still shows movement toward the north.

6. POSTSEISMIC DEFORMATION

The shear deformation for adjustment process in the Bhuj earthquake zone is now negligible as deduced from only 2–3 mm yr^{-1} movements of GPS stations. Moreover, the activity after the 2001 earthquake could not be explained by the coseismic Coulomb stress increase, as the NE expansion to Gedi area is in the zone of decreased stress [*Mandal et al.*, 2007], though the EW expansion was as predicted.

Following large earthquakes, coseismic stress changes are further modified by relaxation of a viscous lower crust

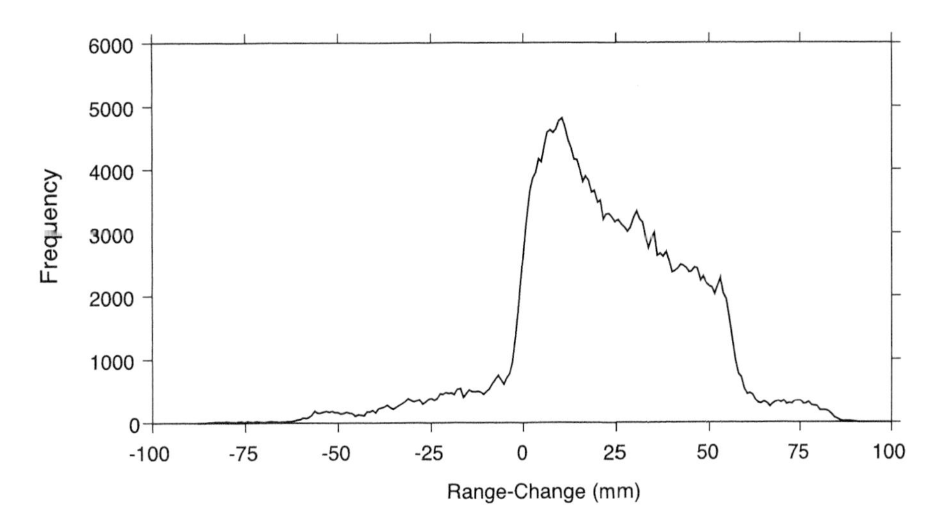

Figure 9. LOS deformation by DInSAR is found to be high: 15 to 50 mm in 1.5 years (between June 2008 and October 2009) or 10–40 mm yr^{-1}.

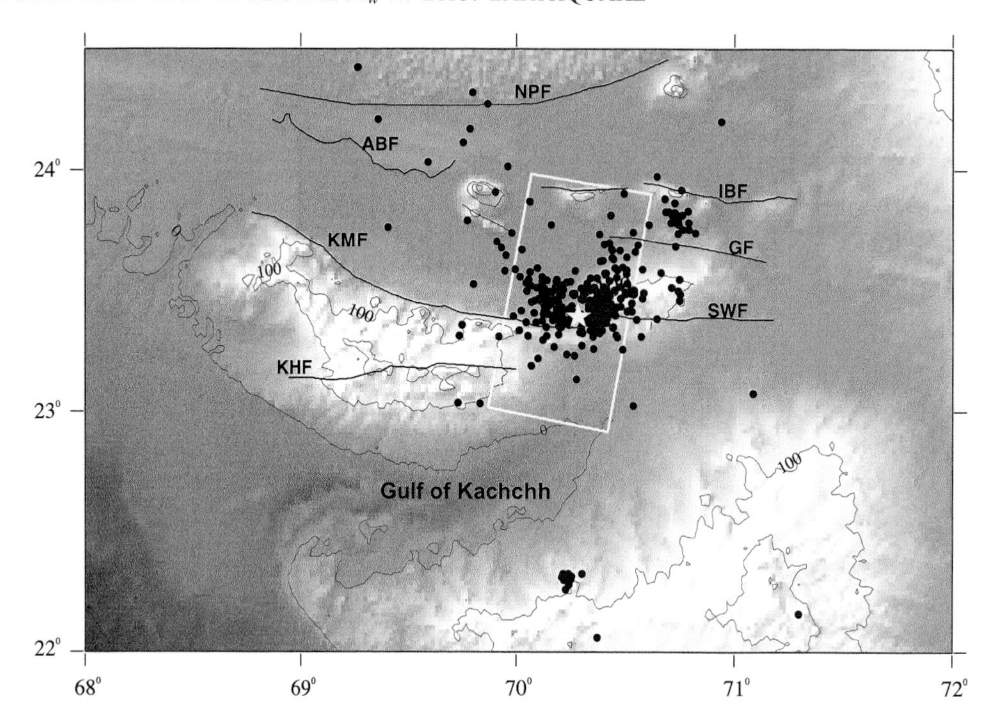

Figure 10. Envisat satellite tracks (white solid line) used for interferogram generation. The star indicates the epicenter of 2001 Bhuj earthquake, and solid circles indicate the epicenters of $M \geq 3$ events during 2006–2009. Major faults (black line) and Shuttle Radar Topography Mission topography contoured at 100 m interval is also shown.

and upper mantle, which serves to transfer stress from these warm regions both upward to the seismogenic crust and outward across a broader region [*Freed et al.*, 2007]. We tried to calculate the postseismic stress change in the Kachchh region due to the 2001 Bhuj earthquake considering a viscoelastic mechanism using EDGRN/EDCMP code developed by *Wang et al.* [2003], which is based on the representation of a layered spherical Earth with elastic-viscoelastic coupling. We adopted the earthquake source parameterization of Y. Yagi and M. Kikuchi (Results of rupture process for January 26, 2001, western India earthquake (M_s 7.9) (revised on March 9, 2001), report, 2001, http://wwweic.eri.u-tokyo.ac.jp, hereinafter referred to as Yagi and Kikuchi, report, 2001). Strike, dip, rake, and seismic moment are 78°, 58°, 81°, and 2.5×10^{20} Nm, respectively. The model rupture is 75 km long and 35 km deep. The variable slip distribution of Yagi and Kikuchi (report, 2001) was taken to calculate the postseismic stress transfer. The velocity model of Kachchh was adopted from *Mandal* [2006]. Viscosity of lower crust/upper mantle (at a depth ~34 km) was taken as 2×10^{19} Pa s [*Chandrasekhar et al.*, 2009].

Figure 12 shows the stress transfer in bars 6 years after the 2001 Bhuj earthquake. The calculated postseismic stress changes (+1 bar) following the Bhuj earthquake seem to correlate with the migration of seismicity toward the east along the SWF and toward NW in the Banni area. The activity along the Gedi fault that started in 2006 is mainly strike-slip activity in the NE direction. This area falls in a zone of stress decrease by −1 bar, which might have decreased the normal stress across the fault and triggered the seismicity. A lobe of positive stress change (0.1 bar) is also observed in the Jamnagar, Junagadh, and Surendranagar regions of Saurashtra (Figure 12), which may be sufficient for triggering seismicity in those regions [*To et al.*, 2004]. Hence, the viscoelastic process appears to be the plausible mechanism for long-distance and delayed triggering of earthquakes by migration of the stress pulse generated by the 2001 earthquake with diffusion rates of 5–30 km yr^{-1} or area growth of 4000 km^2 yr^{-1} contributing stress vertically upward from lower crust and upper mantle to a distance of 200 km in 6 years. The increased stress in the lower crust or upper mantle appears to be transmitted upward to near-surface depths. It may cause more vertical deformation. Unusually high vertical deformation is detected by GPS to the tune of up to 13 mm yr^{-1} and by InSAR measurements up to 40 mm yr^{-1} (along LOS) at several places in Kachchh.

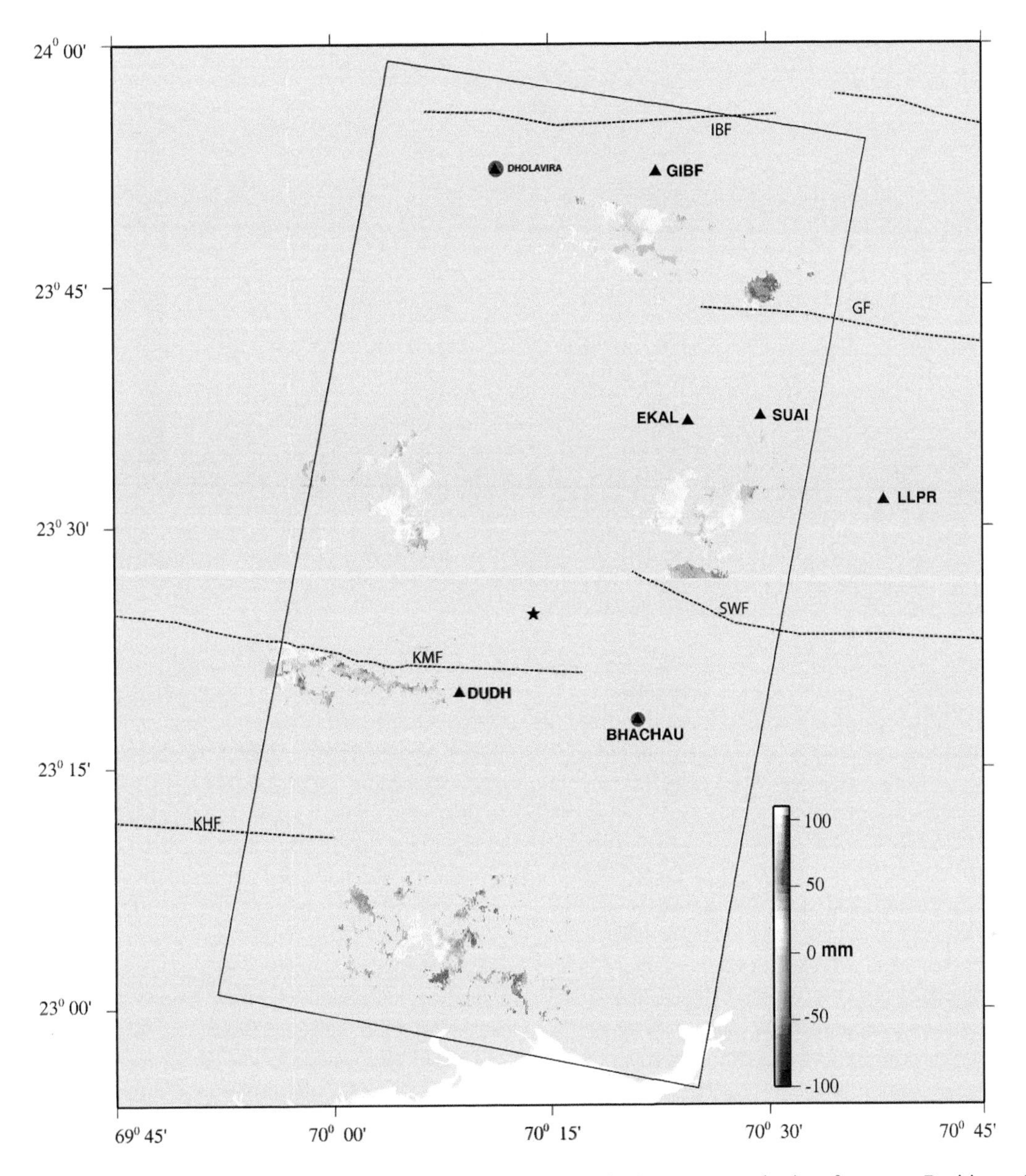

Figure 11. LOS range change generated from Envisat advanced synthetic aperture radar interferogram. Positive values represent displacement toward the satellite (maximum around 50 mm during 1.5 years corresponds to about 40 mm yr^{-1}). Interferogram fringes were identifiable in hilly uplifted areas only but not in large low-lying areas. The star indicates the 2001 Bhuj earthquake.

7. DISCUSSIONS

After the 2001 *M*7.7 Bhuj earthquake we observed migration of active area toward the east and northeast for a distance of about 60 km in 5 years within the Kachchh region. It also migrated to the south for 100 km in 3 years and 200 km in 6 years in the Saurashra region. The Kachchh and the Saurastra regions are separated by the 30 km wide Gulf of Kachchh.

It appears that the stress pulse due to 20 MPa stress drop has migrated via viscoelastic process (rheology change) in lower crust and upper mantle. Unusually all the surrounding faults of the 2001 epicentral zone became activated by 2008. In Saurashtra also, at an unusually high number of places swarm type of activity is associated with earthquakes of *M*3.5–*M*5. This is unusual compared to the catalog of the last 200 years. Though the stress pulse may propagate horizontally through lower crust and upper mantle, it is transferred vertically to

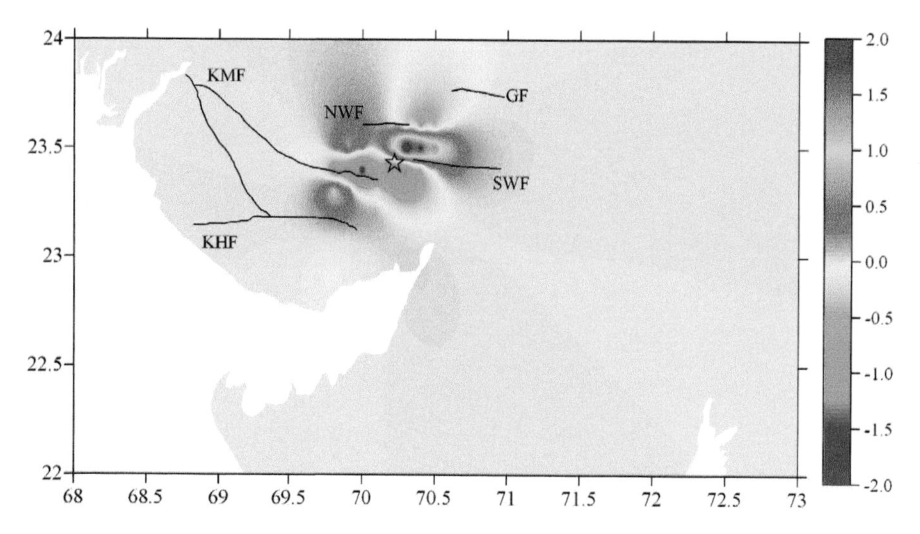

Figure 12. Postseismic stress changes (bars) 6 years after 2001 Bhuj earthquake. The star represents the 2001 Bhuj earthquake. A good correlation is observed between the positive stress region (maximum 1 bar) and the migration of seismicity toward east along the SWF and toward NW in the Banni area. Positive stress changes (0.1 bar) are also observed in Jamnagar, Junagdh, and Surendranagar of Saurashtra.

faults near the surface. The seismicity in Saurashtra is very shallow with 5–6 km focal depth. Some later seismicity in Kachchh too is shallow.

The Gujarat region seems to be critically stressed as evidenced by omnipresent earthquakes along several faults. It is well known that the critically stressed faults can be triggered by small stress perturbation. *To et al.* [2004] estimated that the 0.1 bar Coulomb stress increase near the epicenter of the 2001 Bhuj earthquake has influenced the 2001 earthquake. Viscoelastic modeling indicates stress perturbation up to 1 bar in Kachhh and 0.1 bar in Saurashtra seismic zones. Hence, the sudden appearance of earthquakes along several faults in Kachchh and Saurashtra to 200 km distance from the 2001 Bhuj earthquake epicenter might have been triggered by stress perturbation.

Freymueller [2010] has reviewed the phenomenon of postseismic deformation. Time dependence of postseismic deformation depends on the rheology of the crust and upper mantle [*Burgmann and Dresen*, 2008]. Postseismic deformation can be large and long lasting [*Wang*, 2007]. For example, the post seismic deformation for the 1964 Alaska earthquake is 15–20 mm yr^{-1} even 45 years after its occurrence [*Suito and Freymueller*, 2009]. The seismicity pattern allows estimation of long-term slip rate and viscosity structure [*Johnson et al.*, 2007]. Postseismic deformation results from three factors: after slip of the fault zone, poroelastic rebound, and viscoelastic relaxation of the mantle and lower crust. The shallow after slip is a frictional process [*Smith and Wyss*, 1968] and can lead to an increase in stress in certain regions of the upper crust [*Freed*, 2005]. But after slip may not be the contributor to postseismic stress transfer in the Gujarat region, as very negligible (~2–5 mm yr^{-1}) horizontal deformation is observed through GPS measurements. Poroelastic rebound, which is deformation driven by fluid flow that relieves pressure differences caused by the earthquakes, is generally considered close to the fault and near geometric complexities such as fault step overs and bends [*Freed*, 2005]. However, it appears to act in Gujarat in the same fashion as stress transfer through pore pressure diffusion invoked for reservoir-triggered seismicity [*Talwani et al.*, 2007]. The postseismic models in which the mantle is linear and viscoelastic predict transient deformations that always extend to great distance, even a long time after the earthquake [*Savage and Prescott*, 1978]. Hence, both poroelastic and viscoelastic mechanisms may be potentially important to postseismic deformation in this area.

8. CONCLUSIONS

An unusually large number of at least 30 main shocks occurred during 2003–2009 in an area of 350 km × 300 km along several faults. These earthquakes were strongly felt or had magnitudes of 3–5.7. At least four of these were associated with long sequences of a few years, and some others had sequences of a few months. The shocks are usually shallow (focal depth <10 km) and are accompanied by subterranean sounds. Generation of an unusually large number of main shocks is inferred to be due to triggering caused by a stress pulse generated by the 20 MPa stress drop of the *M*7.7 earthquake in 2001 to distances of 100–200 km, even 6–8 years after the great earthquake. The triggered seismicity may be because of an increase in stress up to 1 bar in Kachhh and 0.1 bar in Saurashtra as estimated by viscoelastic modeling. Poroelastic modeling of the stress perturbation for the Kachchh seismic zone and the Saurashtra region needs to be done to quantify it.

Acknowledgments. This study has been supported by the Ministry of Earth Science and Disaster Management Support Program (R&D) of the Department of Space, Indian Space Research Organization, India. T. J. M. wishes to thank CSIR, New Delhi, for Emeritus Scientist Fellowship since January 2011.

REFERENCES

Biswas, S. K. (1987), Regional framework, structure and evolution of the western marginal basins of India, *Tectonophysics*, *135*, 302–327.

Burgmann, R., and G. Dresen (2008), Rheology of the lower crust and upper mantle: Evidence from rock mechanics, geodesy and field observations, *Annu. Rev. Earth Planet. Sci.*, *36*, 531–567, doi:10.1146/annurev.earth.36.031207.124326.

Chandra, U. (1977), Earthquakes of peninsular India—A seismotectonic study, *Bull. Seismol. Soc. Am.*, *67*(5), 1387–1413.

Chandrasekhar, D. V., R. Burgmann, C. D. Reddy, P. S. Sunil, and D. A. Schmidt (2009), Weak mantle in NW India probed by geodetic measurements following the 2001 Bhuj earthquake, *Earth Planet. Sci. Lett.*, *280*, 229–235.

Chopra, S., K. Madhusudhan Rao, B. Sairam, S. Kumar, A. K. Gupta, H. Patel, M. S. Gadhavi, and B. K. Rastogi (2008a), Earthquake swarm activities after rains in peninsular India and a case study from Jamnagar, *J. Geol. Soc. India*, *72*, 245–252.

Chopra, S., R. B. S. Yadav, H. Patel, S. Kumar, K. M. Rao, B. K. Rastogi, A. Hameed, and S. Srivastava (2008b), The Gujarat (India) seismic network, *Seismol. Res. Lett.*, *79*(6), 806–815.

Freed, A. M. (2005), Earthquake triggering by static, dynamic and postseismic stress transfer, *Annu. Rev. Earth Planet. Sci.*, *33*, 335–367, doi:10.1146/annurev.earth.33.092203.122505.

Freed, A. M., S. T. Ali, and R. Bürgmann (2007), Evolution of stress in Southern California for the past 200 years from coseismic, postseismic and interseismic stress changes, *Geophys. J. Int.*, *169*, 1164–1179, doi:10.1111/j.1365-246X.2007.03391.x.

Freymueller, J. T. (2010), Active tectonics of plate boundary zones and the continuity of plate boundary deformation from Asia to North America, *Curr. Sci.*, *99*(12), 1719–1732.

Johnson, K. M., G. E. Hilley, and R. Bürgmann (2007), Influence of lithosphere viscosity structure on estimates of fault slip rate in the Mojave region of the San Andreas fault system, *J. Geophys. Res.*, *112*, B07408, doi:10.1029/2006JB004842.

Malik, J. N., P. S. Sohoni, R. V. Karanth, and S. S. Merh (1999), Modern and historic seismicity of Kachchh Peninsula, western India, *J. Geol. Soc. India*, *54*, 545–550.

Mandal, P. (2006), Sedimentary and crustal structure beneath Kachchh and Saurashtra regions, Gujarat, India, *Phys. Earth Planet. Inter.*, *155*, 286–299.

Mandal, P., R. K. Chadha, I. P. Raju, N. Kumar, C. Satyamurty, R. Narsaiah, and A. Maji (2007), Coulomb static stress variations in the Kachchh, Gujarat, India: Implications for the occurrences of two recent earthquakes (M_w = 5.6) in the 2001 Bhuj earthquake region, *Geophys. J. Int.*, *169*, 281–285, doi:10.1111/j.1365-246X.2006.03301.x.

Oldham, T. (1883), Catalogue of Indian Earthquakes, *Mem. Geol. Surv. India*, *19*, 163–215.

Reddy, C. D., and P. S. Sunil (2008), Post-seismic crustal deformation and strain rate in Bhuj region, western India, after the 2001 January 26 earthquake, *Geophys. J. Int.*, *172*, 593–606.

Savage, J., and W. Prescott (1978), Asthenosphere readjustment and the earthquake cycle, *J. Geophys. Res.*, *83*, 3369–3376.

Smith, S. W., and M. Wyss (1968), Displacement on the San Andreas fault subsequent to the 1966 Parkfield earthquake, *Bull. Seismol. Soc. Am.*, *58*, 1955–1973.

Sreejith, K. M., T. J. Majumdar, B. K. Rastogi, R. Dumka, P. Choudhury, and F. Bhattacharya (2011), Crustal deformation mapping in Kachchh, India using InSAR and GPS: Initial results, paper presented at AES 2011 International Symposium, Inst. of Seismol. Res., Gandhinagar, India, 22–24 Jan.

Suito, H., and J. T. Freymueller (2009), A viscoelastic and afterslip postseismic deformation model for the 1964 Alaska earthquake, *J. Geophys. Res.*, *114*, B11404, doi:10.1029/2008JB005954.

Talwani, P., L. Chen, and K. Gahalaut (2007), Seismogenic permeability, k_s, *J. Geophys. Res.*, *112*, B07309, doi:10.1029/2006JB004665.

Tandon, A. N., and H. N. Srivastava (1974), *Earthquake Occurrences in India, Jai Krishna Volume*, Sarita Prakashan, Roorkee, India.

To, A., R. Bürgmann, and F. Pollitz (2004), Postseismic deformation and stress changes following the 1819 Rann of Kachchh, India earthquake: Was the 2001 Bhuj earthquake a triggered event?, *Geophys. Res. Lett.*, *31*, L13609, doi:10.1029/2004GL020220.

Wang, K. (2007), Elastic and viscoelastic models of crustal deformation in subduction earthquake cycles, in *The Seismogenic Zone of Subduction Thrust Faults*, edited by T. H. Dixon and J. C. Moore, pp. 540–575, Columbia Univ. Press, New York.

Wang, R., F. L. Martin, and F. Roth (2003), Computation of deformation induced by earthquakes in a multi-layered elastic crust—FORTRAN programs EDGRN/EDCMP, *Comput. Geosci.*, *29*, 195–207.

Yadav, R. B. S., J. N. Tripathi, B. K. Rastogi, and S. Chopra (2008), Probabilistic assessment of earthquake hazard in Gujarat and adjoining region, India, *Pure Appl. Geophys.*, *165*, 1813–1833.

P. Choudhury, R. Dumka, and B. K. Rastogi, Institute of Seismological Research, Gandhinagar 382 009, India. (brastogi@yahoo.com)

T. J. Majumdar and K. M. Sreejith, Space Applications Centre, Indian Space Research Organization, Ahmedabad 380015, India.

Extreme Seismic Events in Models of Lithospheric Block-and-Fault Dynamics

A. T. Ismail-Zadeh

Geophysikalisches Institut, Karlsruher Institut für Technologie, Karlsruhe, Germany
Institut de Physique du Globe de Paris, Paris, France
International Institute of Earthquake Prediction Theory and Mathematical Geophysics, Russian Academy of Sciences, Moscow, Russia

J.-L. Le Mouël

Institut de Physique du Globe de Paris, Paris, France

A. A. Soloviev

International Institute of Earthquake Prediction Theory and Mathematical Geophysics, Russian Academy of Sciences, Moscow, Russia
The Abdus Salam International Centre for Theoretical Physics, Trieste, Italy

Extreme seismic events are manifestations of the complex behavior of the lithosphere structured as a hierarchical system of blocks of different sizes. Driven by mantle convection, these lithospheric blocks are involved in relative movement, resulting in stress localization and earthquakes. We discuss a quantitative approach to simulation of earthquakes in models of block-and-fault dynamics (BAFD), which reproduces basic features of the observed seismicity like the Gutenberg-Richter law, clustering of earthquakes, occurrence of extreme seismic events, aftershocks, and foreshocks. The model provides a link between geodynamical processes and seismicity, allows studying the influence of fault network properties (e.g., fragmentation, block geometry and movement, and direction of driving forces) on seismic patterns and seismic cycles, and assists, in a broader sense, in earthquake forecast modeling. We review the applications of the BAFD model to study earthquake sequences and extreme events in several regions: Vrancea (south-eastern Carpathians), the Tibet-Himalaya region, and the Sunda arc region. Finally, we discuss perspectives in modeling of earthquake occurrences and extreme seismicity.

1. INTRODUCTION

The vulnerability of human civilizations to extreme natural hazards is growing due to the proliferation of high-risk objects, clustering of populations, and destabilization of large cities. Today, a single earthquake may take up to several hundred thousand lives and cause material damage up to several billion dollars. A large earthquake can trigger an ecological catastrophe if it occurs in close vicinity to a nuclear power plant built in an earthquake-prone area (note that the last sentence was written a few weeks before the 2011 Great East Japan earthquake and subsequent tsunami, which damaged the Fukushima Dai-ichi nuclear power plant and resulted in nuclear radiation leaks). About a million earthquakes with

Extreme Events and Natural Hazards: The Complexity Perspective
Geophysical Monograph Series 196

10.1029/2011GM001080

magnitudes greater than 2 are registered each year; about a thousand of them are large enough to be felt; about a hundred earthquakes cause considerable damage, and once in a few decades, an extreme (catastrophic) event occurs.

In a broad sense, an extreme seismic event can be defined as an earthquake occurrence that, with respect to other earthquakes, is notable, rare, unique, and profound or otherwise significant in terms of its impacts, effects, and outcomes. (This definition is adapted for the case of extreme seismicity from the definition of extreme events adopted by the participants of the Extreme Events Workshop held in Boulder, Colorado, 7–9 June 2000; http://www.isse.ucar.edu/extremes.) We shall distinguish two types of extreme seismic events: (1) large-magnitude and rare earthquakes and (2) earthquakes leading to disasters (e.g., the 2010 Haiti earthquake). Extreme seismic events, like the 1755 Lisbon, 1906 San Francisco, 1960 Chile, 2004 Aceh-Sumatra, 2008 Sichuan, and the 2011 Great East Japan earthquakes, belong to both types of extreme seismic events (high-magnitude events and great disasters at the same time). Many great earthquakes around the Pacific seismic belt, fortunately, do not lead to great disasters; nevertheless, they are considered to belong to the first type of extreme seismic events. In this chapter, we deal with the first type of extreme events, that is, the events, which are located in the tail of the frequency-magnitude (FM) distribution curve. Meanwhile, the importance of studies on the second type of the extreme seismic events should not be diminished and should be appreciated.

An extreme seismic event is a key manifestation of the lithosphere dynamics exhibiting a nonlinear system behavior and evolving from stability to a catastrophe over space and time [e.g., *Keilis-Borok*, 1990; *Sornette and Sammis*, 1995; *Turcotte*, 1999; *Keilis-Borok and Soloviev*, 2003]. Driven by mantle convection, lithospheric plates are involved in relative movement resulting in stress localization and earthquakes. The lithosphere presents a hierarchy of blocks where the largest blocks are the major tectonic plates. They are divided into smaller blocks, like shields or mountain chains, and down to the grains of rock. The blocks are separated by less rigid boundary zones by a factor of 10–100 thinner than the corresponding blocks. Each boundary zone presents a similar hierarchical structure: it consists of fracture zones of smaller ranks. The boundary zones bear different names: fault zones (with width of 10^1–10^3 m), cracks (10^{-2}–10^0 m), and microcracks (10^{-4}–10^{-3} m). The blocks (from plates to grains) and faults (from fault zones to microcracks) interact along and across the hierarchy and move relatively to each other under control of lithosphere dynamics. In large part, these movements are realized through formation and subsequent healing of failures on surfaces where displacements are discontinuous as defects, slips, fractures, or earthquakes [e.g., *Rice and Ben-Zion*, 1996; *Keilis-Borok et al.*, 2001; *Ben-Zion*, 2008].

Stress accumulation and its release in earthquakes are governed by nonlinear hierarchical systems, which have a number of degrees of freedom and, therefore, cannot be understood by studying them piece by piece. Since an adequate theoretical base has yet not been well elaborated, theoretical estimation of statistical parameters of earthquake sequences is still a complex problem. Studying seismicity using the statistical and phenomenological analysis of earthquake catalogs has the disadvantage that instrumental observations cover, usually, a too short time interval compared to the duration of the tectonic processes responsible for seismic activity. The patterns of earthquake occurrence identifiable in a catalog may be apparent and yet may not be repeated in the future. Moreover, the historical data on seismicity are usually incomplete. Numerical modeling of seismogenic processes allows generating synthetic earthquake catalogs covering very long time intervals and, therefore, providing a basis for reliable estimates of the parameters of the earthquake occurrences [e.g., *Soloviev and Ismail-Zadeh*, 2003].

It is difficult to detect the impact of a single factor on the dynamics of seismicity by analyzing seismic observations because seismicity is impacted by an assemblage of factors, some of which may be more significant than that under consideration. It is also difficult (if not impossible) to single out the impact of an isolated factor by using seismic observations. This difficulty may be resolved by numerical modeling of processes that generate seismicity and by studying the synthetic earthquake catalogs thus obtained [e.g., *Gabrielov and Newman*, 1994; *Allègre et al.*, 1995]. Quantitative models of lithosphere dynamics are also tools for studying earthquake preparation processes and are useful in seismic hazard and earthquake prediction studies. An adequate model should indicate the physical basis of premonitory patterns determined prior to large events. Note that available data often do not constrain the statistical significance of premonitory patterns. The model can also be used to suggest new premonitory patterns that might exist in catalogs of seismic events.

Advances in understanding of the Earth's dynamics and in computational tools permitting accurate quantitative modeling have already impacted studies of earthquake physics, earthquake rupture, and seismic wave propagations. Modern computer facilities provide useful tools in the modeling of geodynamical processes leading to extreme seismic events and in the monitoring of seismic hazard [e.g., *Soloviev and Ismail-Zadeh*, 2003, *Xing et al.*, 2007; *Ismail-Zadeh et al.*, 2007; *Cui et al.*, 2008; *Bielak et al.*, 2010].

Three types of quantitative models of stress localization and stress drop (earthquakes) have been so far developed:

(1) models of tectonic stress generation, localization, and transfer in a fault zone or its surroundings [e.g., *King et al.*, 1994; *Hirahara*, 2002; *Lin and Stein*, 2004; *Aoudia et al.*, 2007; *Ismail-Zadeh et al.*, 2005, 2010], (2) dynamic systems reproducing "universal" features of seismicity (including large events, main shocks, and aftershocks) common to a wide class of nonlinear systems [e.g., *Burridge and Knopoff*, 1967; *Fukao and Furumoto*, 1985; *Ogata*, 1988; *Narkunskaya and Shnirman*, 1990; *Rundle and Klein*, 1993; *Gabrielov et al.*, 1994; *Gabrielov and Newman*, 1994; *Allègre et al.*, 1995, 1998; *Cochard and Madariaga*, 1996; *Hainzl et al.*, 1999; *Narteau et al.*, 2000; *Zaliapin et al.*, 2003; *Turcotte et al.*, 2007; *Vere-Jones and Zhuang*, 2008; *Lennartz et al.*, 2011], and (3) models specific to the solid Earth, reproducing the features of seismicity specific to a single fault [e.g., *Dieterich*, 1972, 1994; *Ben-Zion and Rice*, 1993; *Schmittbuhl et al.*, 1996; *Lyakhovsky et al.*, 2001; *Zöller et al.*, 2004, 2005; *Ben-Zion and Lyakhovsky*, 2006; *Nodal and Lapusta*, 2010] or to a fault system [e.g., *Wang et al.*, 1983; *Ward*, 1992, 1996, 2000; *Gabrielov et al.*, 1990, 2007; *Soloviev and Ismail-Zadeh*, 2003, *Rundle et al.*, 2006; *Ismail-Zadeh et al.*, 2007; *Zöller and Hainzl*, 2007; *Pollitz*, 2009; *Bielak et al.*, 2010].

The models of the first type provide results on (1) high shear stress localizations in a region and/or (2) static, viscoelastic, and dynamic Coulomb stress changes after an earthquake to determine the potential sites of impending large events. For example, *Ismail-Zadeh et al.* [2005] analyzed stress generation and localization in and around the descending Vrancea slab using a quantitative 3-D model of mantle flow induced by the slab. The model, which was based on a temperature model derived from seismic *P* wave velocity anomalies and surface heat flow, predicted the maximum shear stress localization to coincide with the hypocenters of the intermediate-depth seismicity and stress orientations in a good agreement with the stress regime defined from fault-plane solutions for the intermediate-depth earthquakes. Using the Coulomb failure criterion, *King et al.* [1994] explored how changes in the Coulomb stress conditions associated with an earthquake may trigger subsequent earthquakes (aftershocks). An earthquake alters the shear and normal stress on surrounding faults, and small sudden stress changes cause large changes in seismicity rate. The main aftershock activity is shown to coincide with the area of Coulomb stress increases [*Stein*, 1999]. Increases and decreases in seismicity rate are followed by a time-dependent recovery, which depends on the rheological properties of the crust [*Pollitz and Sacks*, 2002]. Although the models of the first type provide important information on the localization of stresses and stress changes before the next large events, which can be used in hazard assessment, the models do not simulate earthquake occurrences.

The models of the second type feature nonlinear dynamics of fault systems, provide earthquake statistics, and assist in analysis of earthquake pattern, occurrences, clustering, and frequencies. One of the first models to simulate earthquakes was proposed by *Burridge and Knopoff* [1967]. The model connects blocks to each other via springs, and each block interacts with other blocks subject to a force, which is proportional to the distances of the blocks from their equilibrium position and to a friction force. Several models based on spring-block interaction, cellular automata, scaling organization of fracture tectonics, colliding cascades, etc., have been developed later to reproduce general properties of observed seismicity.

Some macroscopic phenomena have their origin in a microscopic organization, which can be transferred to larger scales [*Allègre et al.*, 1982]. This scaling approach was employed to develop the scaling organization of fracture tectonics (SOFT) model for simulation of earthquakes [*Allègre et al.*, 1995, 1998], which links physical and multiblock (like Burridge-Knopoff model) approaches. The SOFT model deals essentially with a single fault zone between two moving sides (e.g., tectonic plates). The fault is considered as a hierarchical system of embedded cells. Each cell interacts with neighboring cells generating a fracture at a larger scale in a larger cell. Fracturing process begins at the lowest level of the hierarchy if the local shear stress exceeds a strength threshold. Each earthquake (new fracture in a given cell) introduces stress heterogeneity in the neighboring cells. This initiates a cascade of stress redistribution by smaller earthquakes: aftershocks. Direct numerical simulations of the model reproduce cycles of seismic activity, foreshocks and aftershocks; temporal decay of aftershocks obeys a power law (Omori law).

Other models using the scaling law to simulate earthquakes and aftershocks are the epidemic-type aftershock sequence model by *Ogata* [1988] and the branching aftershock sequence model by *Turcotte et al.* [2007]. Both models employ the concept of aftershock cascades (when main shock generates aftershocks, the aftershocks generate their aftershocks, etc.) and differ by the law of aftershock generation. Despite the models of the second type describing many features of the earthquake occurrences, only the Earth-specific models (of the third type) simulate earthquakes in a system of crustal/lithospheric faults, e.g., for a large heterogeneous strike-slip fault [*Ben-Zion and Rice*, 1993; *Zöller et al.*, 2005] and for a system of faults [e.g., *Ward*, 1992, 1996, 2000; *Robinson and Benites*, 1996, 2001; *Fitzenz and Miller*, 2001; *Zhou et al.*, 2006].

In this chapter, we review recent advances in modeling of dynamics of lithospheric block-and-fault structures and occurrences of large (extreme) events in the models. A block-

and-fault dynamics (BAFD) model can answer the following questions: (1) how upper crustal (or lithospheric) blocks react to the plate motions and to a flow of the lower crust (or asthenosphere), (2) how earthquakes cluster in the system of major regional faults, (3) at which part of a fault system large (extreme) events can occur and what the occurrence time of the extreme events is, (4) how the properties of the FM relationship change prior extreme events, and (5) how fault zone properties influence the earthquake clustering, its magnitude, and fault slip rates. In section 2, we present the basic principles and mathematical statement of the BAFD model. The basic features of the BAFD model are analyzed in section 3. We review the application of the BAFD model to several earthquake-prone regions: the Vrancea region of the southeastern Carpathians (section 4), the Sunda arc region (section 5), and the Tibet-Himalayan region (section 6), and discuss only the case studies related to preferred numerical experiments (as the most consistent with the observations). We refer to the papers where the full range of the case studies and the sensitivity analysis of the BAFD models are presented. The perspectives in earthquake modeling will be discussed in section 7.

2. BLOCK-AND-FAULT DYNAMICS MODEL

2.1. Model Description

A model of BAFD is used to analyze how the basic features of seismicity and fault slip rates depend on the crust (lithosphere) structure and dynamics. The basic principles of the model have been developed by *Gabrielov et al.* [1990]. The BAFD model considers a seismic region as a structure of perfectly rigid (upper crustal or lithospheric) blocks divided by infinitely thin fault planes. The blocks interact between themselves and with the underlying medium (lower crust or asthenosphere). The structure of blocks moves in response to the prescribed motion of the boundary blocks and of the underlying medium. Because the blocks are perfectly rigid, deformation is localized in the fault zones, and relative block displacements take place along the fault planes. The block motion is defined so that the structure remains in a quasistatic equilibrium state.

The interaction of the blocks along the fault planes is viscoelastic (we refer to it as a normal state) as long as the ratio of the shear stress to the difference between the pore pressure and normal stress remains below a certain strength level. When the critical level is exceeded in some part of a fault plane, a stress drop (failure) occurs resulting in failures in adjacent parts of the fault planes. The failure produces an earthquake. Immediately after the earthquake, the stress-drop-affected parts of the fault planes are in a state of creep. This state differs from the normal state because of a faster growth of inelastic displacements, lasting until the stress falls below a certain level. Thus, the BAFD model generates a catalog of synthetic earthquakes. Using the synthetic catalogs, it is possible to analyze spatial-temporal correlation between earthquakes, their clustering, long-range interaction between the events, and fault slip rates. One can determine model parameters for a particular region, which fit closely the spatial distribution of seismicity, FM relationship in this region, displacement rates of crustal structures, and fault slip rates.

The block-and-fault structure considered in the model is a bounded and simply connected part of a layer of thickness H limited by two horizontal planes (Figure 1). The lateral boundaries of the structure and its subdivision into blocks are formed by portions of planes intersecting the layer; we refer to these planes as *fault planes*. The intersection lines of the fault planes with the upper plane are referred to as *faults*. The fault planes may have arbitrary dip angles, which are specified in the model on the basis of the knowledge of the deep structure of the region under study. A common point of two faults is referred to as a *vertex*. The vertices in the upper and the lower planes are connected by a *rib* of the intersection line of the relevant fault planes (Figure 1). The part of a fault plane between two ribs corresponding to successive vertices on the fault is referred to as a *fault segment*. The upper and the lower surfaces of the blocks are polygons. The common part of a block with the lower plane is referred to as a *block bottom*. The block-and-fault structure is bordered by the infinite confining medium. The motion of the confining medium is defined in areas bounded by two ribs of the structure boundary (refer to each area as a *boundary block*). Boundary blocks and the underlying medium are assumed to move because of applied forces. The movements are assumed to be horizontal, their rates are given, and the rate of movement of the underlying medium can vary with the blocks. Dimensionless time is used in the model.

The state of the block-and-fault structure is considered at discrete values of time $t_i = t_0 + i\Delta t$ ($i = 1, 2, \ldots$), where t_0 is the initial time. At each time step, the translation vectors and

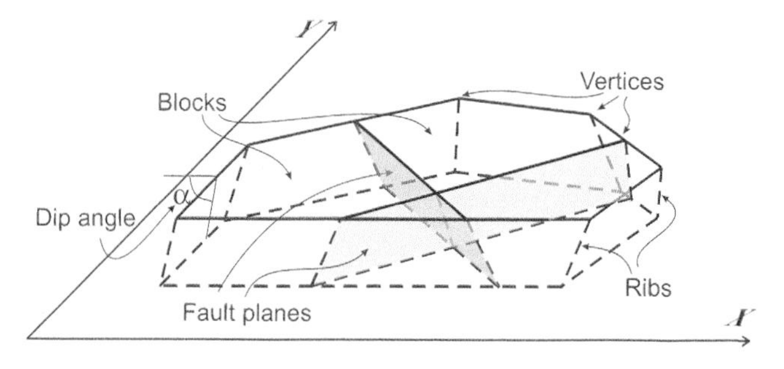

Figure 1. Model geometry: sketch of a block-and-fault structure.

the angles of rotation of the blocks are determined in such a way that the structure is in a quasistatic equilibrium. Relative block displacements take place, as assumed, only along fault planes. All displacements are supposed to be infinitely small compared to the block size. Therefore, the geometry of the block structure does not change during the simulation, and the structure does not move as a whole.

2.2. Interaction of Blocks Along Fault Planes and With the Underlying Medium

Consider a point (X, Y) at the fault plane separating two blocks i and j (for the sake of definiteness, we assume that blocks i and j are on the left and right of the fault with respect to the slip orientation, Figure 2a). The components Δx and Δy of the relative displacement of the blocks are defined as follows:

$$\begin{aligned} \Delta x &= x_i - x_j - (Y - Y_c^i)\varphi_i + (Y - Y_c^j)\varphi_j \\ \Delta y &= y_i - y_j + (X - X_c^i)\varphi_i - (X - X_c^j)\varphi_j, \end{aligned} \quad (1)$$

where X_c^i, Y_c^i, X_c^j, and Y_c^j are the coordinates of the geometric centers of the block bottoms; (x_i, y_i) and (x_j, y_j) are the translational vectors of the blocks; and ϕ_i and ϕ_j are the angles of rotation of the blocks about the geometric centers of their bottoms, all depending on time t. A relationship similar to equations (1) is used to define the relative displacement of the block and the underlying medium.

We assume that the relative block displacements take place only along fault planes (Figure 2a); therefore, the displacement $\Delta = (\Delta_t, \Delta_l)$ along the fault plane is given by

$$\begin{aligned} \Delta_t &= e_x \Delta x + e_y \Delta y \\ \Delta_l &= \Delta_n / \cos\alpha = (e_x \Delta y - e_y \Delta x)/\cos\alpha, \end{aligned} \quad (2)$$

where Δ_t and Δ_l are the displacement components along the fault plane, parallel and normal to the fault line in the upper plane, respectively; (e_x, e_y) are the direction cosines (with respect to the fault slip orientation); α is the dip angle of the fault plane; and Δ_n is the horizontal displacement normal to the fault line in the upper plane (Figure 2b).

Elastic forces arise in the block bases and fault planes in response to the displacement of the blocks relative to the boundary blocks and the underlying medium. At a point (X, Y), the shear stress vector $\sigma = (\sigma_t, \sigma_l)$ (elastic force per unit area acting along the fault plane or the block base) is

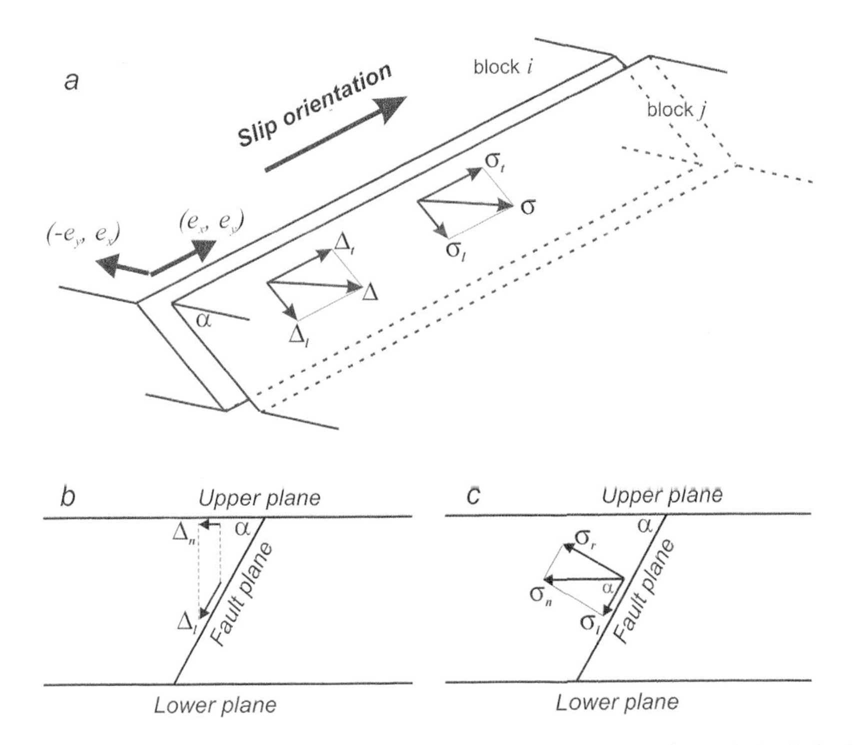

Figure 2. Model geometry: (a) Relative displacements of blocks and the forces per unit area in the fault plane. A vertical section of a block-and-fault structure orthogonal to a fault presenting the (b) relative displacements of blocks and (c) forces per unit area.

related to the strain ε at this point by a viscoelastic (Maxwell) rheological law:

$$\left(\frac{d}{dt}+\frac{1}{\tau}\right)\sigma=\mu\frac{d\varepsilon}{dt}, \quad (3)$$

where τ is the relaxation time ($\tau = \eta/\mu$); μ is the shear modulus; and η is the viscosity. For the convenience of subsequent discussions, we present the relationship (3) in the following equivalent form:

$$\sigma=\xi(\Delta-\delta) \quad \frac{d\delta}{dt}=\zeta\sigma, \quad (4)$$

where $\delta = (\delta_t,\delta_l)$ is the inelastic displacement at a point of the fault planes; $\xi = \mu/a$ and $\zeta = a/\eta$, where a is a characteristic length (a = 3 km is assumed in the model); and $\tau = 1/(\xi\zeta)$.

In addition to the elastic force, there is a reaction force normal to the fault plane; the work of this force is zero because all relative movements are tangent to the fault plane. The normal stress (the reaction force per unit area) σ_r is then defined (Figure 2c)

$$\sigma_r=\sigma_l\tan\alpha. \quad (5)$$

The value of σ_r is positive for a tensional stress. Because of the reaction force introduced in the model, there are no vertical forces acting on the blocks and, hence, no vertical displacement of the blocks. The horizontal movement of the model boundary blocks is prescribed by their translation and rotation about the origin, that is, the coordinates of the geometric center of the block bottom in equation (1) are set to zero for any boundary block (e.g., if the block numbered j is a boundary block, then $X_c^j = Y_c^j = 0$).

The components of the translational vectors of the blocks and the angles of their rotation about the geometric centers of the block bottoms are found from the condition that the total force and the total moment of forces acting on each block must vanish. This is the condition of quasistatic equilibrium of the system and at the same time the condition of minimum energy. The equilibrium equations include only forces caused by specified movements of the underlying medium and of the boundaries of the block-and-fault structure. In fact, it is assumed that the action of all other forces on the structure is ruled out and does not cause displacements of blocks. The space discretization required to carry out numerical simulations of the BAFD model is made by splitting the surfaces, on which the forces act, into small cells of trapezoidal shape, whose linear size does not exceed a specified parameter χ. The coordinates (X, Y), displacement Δ, inelastic displacement δ, and elastic stress σ are supposed to be the same for all the points of a cell.

2.3. Earthquake and Creep

Earthquakes are simulated according to the criterion of the Coulomb failure stress [e.g., *Stein*, 1999] and the dry friction. Namely, for each cell of the fault plane, we introduce the dimensionless number

$$\kappa=\frac{|\sigma|}{P-\sigma_r}, \quad (6)$$

where $|\sigma|$ is the magnitude of the shear stress along the fault plane and P is the difference between the lithostatic and pore (hydrostatic) pressure in the fault zone, which is assumed to be equal to 2×10^8 Pa (the typical value at 15 km depth) for all the faults.

Three critical values of κ, $B > H_f > H_s$, are specified for each fault. The initial conditions for the BAFD model, the translation vectors and angles of rotation of the blocks and the inelastic displacements of the cells, are assumed to satisfy the inequality $\kappa < B$ for all cells of the fault planes. If at any time the value of κ exceeds the level B at one or more cells, a failure (earthquake) occurs. An earthquake in the model is defined as an abrupt decrease of the inelastic displacement δ in the cell. The inelastic displacement is updated after the failure:

$$\tilde{\delta}=\delta+\gamma\sigma, \quad (7)$$

where the coefficient γ is determined from the condition that after the failure, κ is reduced to the level H_f:

$$\gamma=\frac{1}{\xi}\left(1-\frac{P}{|\sigma|/H_f+\sigma_r}\right). \quad (8)$$

Once the new values of the inelastic displacements for all the failed cells are computed, the translation vectors and the angles of rotation of the blocks are determined to satisfy the condition of the quasistatic equilibrium. If after these computations $\kappa > B$ for some cell(s) of the fault planes, the procedure is repeated for this (these) cell(s); otherwise, the numerical simulation is continued in the ordinary way.

On the same fault plane, the cells, in which failure occurs at the same time, form a single earthquake. The coordinates of the earthquake epicenter are defined as the weighted sum of the coordinates of the cells forming the earthquake (the weights are proportional to the areas of the failed cells). The magnitude of the earthquake is estimated using the following formula:

$$M=0.98\log_{10}S+3.93, \quad (9)$$

where S is the total area of the cells forming the earthquake, measured in km^2 [*Utsu and Seki*, 1954]. *Wells and Coppersmith*

[1994] estimated the second term in the right-hand side of equation (9) to be 4.07. Hence, the use of the Wells and Coppersmith's formula will increase the earthquake magnitudes in the models.

Immediately after the earthquake, the cells in which the failure occurred are creeping. It means that the parameter ζ_s ($\zeta_s > \zeta$) in equation (4) is used instead of ζ for these cells. The cells are in the state of creep as long as $\kappa > H_s$. Once $\kappa < H_s$, the cells return to the normal state, and henceforth, the parameter ζ is used in equation (4) for these cells.

3. FEATURES OF THE BAFD MODEL

Several features of the observed seismicity, e.g., clustering of events in time, long-range interaction between events, and transformation of the FM relationship prior to extreme events have been obtained in BAFD models. Different features of the model seismicity are analyzed in detail by *Keilis-Borok et al.* [1997], *Maksimov and Soloviev* [1999], *Soloviev and Ismail-Zadeh* [2003], *Vorobieva and Soloviev* [2005], and *Soloviev* [2008]. Here we review briefly the model results.

3.1. Clustering of Synthetic Earthquakes

Earthquake clustering assists in understanding the dynamics of seismicity and specifically in problems of earthquake forecasting [e.g., *Kagan and Knopoff*, 1978; *Keilis-Borok et al.*, 1980; *Dziewonski and Prozorov*, 1985, *Molchan and Dmitrieva*, 1992]. It is important to clarify whether clustering is caused by specific features of tectonics in an earthquake-prone region or, conversely, whether it is a phenomenon for a wide variety of tectonic conditions and reflects the general features of interacting lithospheric blocks. The clustering of earthquakes in a synthetic catalog obtained from modeling the dynamics of a simple block structure favors the second assumption.

We consider a simple block structure to illustrate how synthetic events cluster in the model: four blocks whose common parts with the upper plane are squares (Figure 3) with a side of 50 km. The thickness of the layer is $H = 20$ km, and the dip angle of 85° is specified for all fault planes. For all faults, the parameters in equation (4) have the following values: $\xi = 1$ (measured in 10^7 Pa m^{-1}) and $\zeta = 0.05$ (measured in 10^{-7} m Pa^{-1} yr^{-1}), which correspond to the shear modulus 3×10^{10} Pa and viscosity 2×10^{19} Pa s; $B = 0.1$, $H_f = 0.085$, $H_s = 0.07$, and $\zeta_s = 2$ (measured in 10^{-7} m Pa^{-1} yr^{-1}). The values of the parameters for the time and space discretization are $\Delta t = 10^{-2}$ yr and $\chi = 5$ km, respectively. The block structure moves as a result of the prescribed boundary movement: translation of fault segments 7 and 8 (Figure 3) with the rate ($v_x = 0.2$, $v_y = -0.05$ m yr^{-1}) and their rotation about the origin with an angular velocity of 10^{-6} rad yr^{-1}. The other parts of the boundary and the underlying medium do not move.

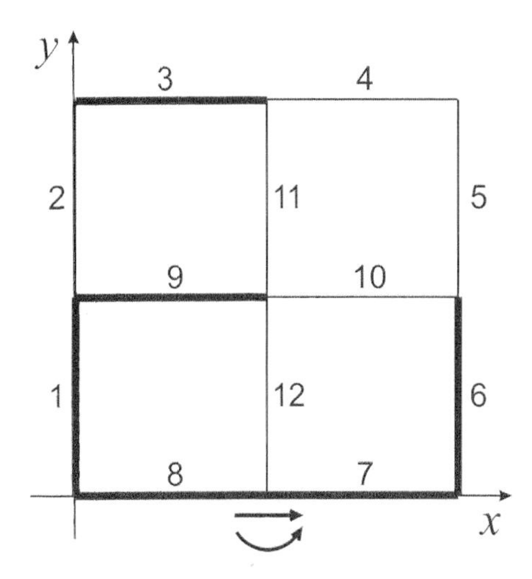

Figure 3. Fault pattern on the upper plane accepted for assessing the clustering of synthetic earthquakes and their long-range interaction. Arrows illustrate movements specified at the boundary consisting of fault segments 7 and 8. Reprinted from *Soloviev and Ismail-Zadeh* [2003] with kind permission of Springer Science and Business Media.

The occurrences of earthquakes (vertical lines) are shown in Figure 4 for individual fault segments and for the whole structure for a time interval of 3 years starting at $t = 480$ years. Earthquakes occur on six fault segments. The respective parts of the faults are marked in Figure 3 by thick lines. Segment 9 has only one earthquake during the period under consideration. Earthquakes cluster clearly on fault segments 1, 3, 6, and 7. Segment 8 has the largest number of earthquakes. Here the clustering appears weaker: the groups of earthquakes are diffuse along the time axis. The pattern for the whole structure looks like that for segment 8, and groups of earthquakes can also be identified. Clustering of earthquakes for other time intervals does not differ significantly from that presented in Figure 4. The obtained model results allow studying the phenomenon of earthquake clustering in earthquake prone regions. In particular, one can ascertain the dependence of clustering on the geometry of a block structure and on the values of its parameters.

3.2. Long-Range Interaction Between Synthetic Earthquakes

An interaction between earthquakes was a long-standing subject of investigations. *Benioff* [1951] investigated seismic strain accumulation and release in the period 1904–1950 and suggested a hypothesis for interdependence of earthquakes

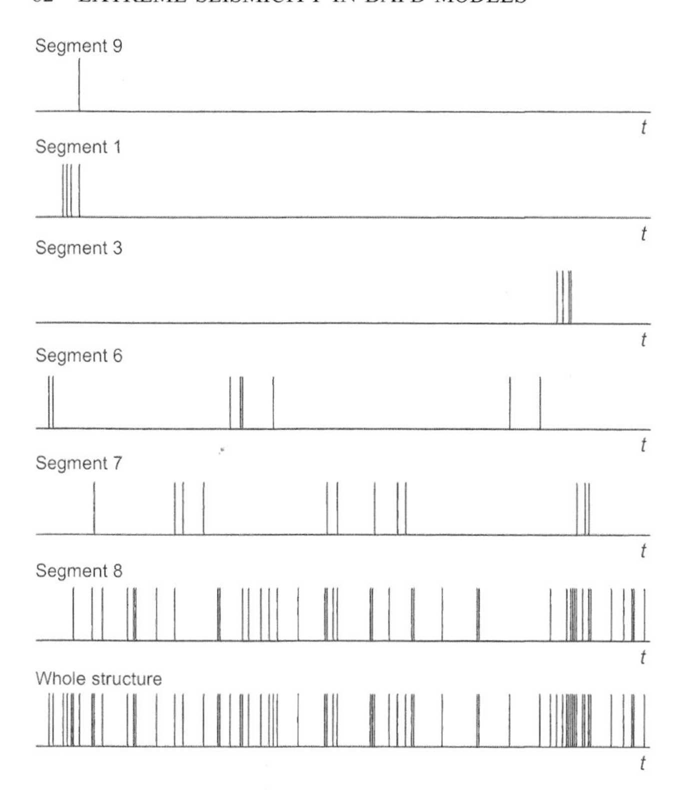

Figure 4. Clustering of synthetic earthquakes. Times of occurrence (vertical lines) are shown in the interval of 3 years for individual segments numbered as in Figure 3 and for the whole structure. Reprinted from *Soloviev and Ismail-Zadeh* [2003] with kind permission of Springer Science and Business Media.

of magnitude $M \geq 8$. *Duda* [1965] argued that this interdependence could be extended to earthquakes of magnitude $M \geq 7$. *Prozorov* [1994a, 1994b] analyzed the long-range interaction between earthquakes separated by distances that exceed considerably their aftershock areas. *Press and Allen* [1995] examined a catalog of Southern California earthquakes using pattern recognition methods and determined traits that characterize earthquakes of the San Andreas (SA) fault system (SA-type earthquakes) and those with non–San Andreas (NSA) attributes (NSA-type earthquakes). The following hypothesis was suggested to explain these traits: earthquakes in Southern California occur within a larger system that includes at least the Great Basin and the Gulf of California; episodes of activity in these adjacent regions alert about subsequent release of the SA-type earthquakes; in the absence of activity in these adjacent regions, the occurrence of SA-type of earthquakes is reduced, and NSA-type earthquakes occur more frequently. In this case, the distance of interaction amounts to several hundreds of kilometers.

Statistical methods were used to analyze the long-range interaction between synthetic events in the BAFD model [*Soloviev and Ismail-Zadeh*, 2003; *Vorobieva and Soloviev*, 2005]. Let us consider the block structure as defined in section 3.1 (Figure 3) with the same model parameters; the model run for 2000 years. We consider earthquakes of $M \geq 6.6$ occurred on fault segments 7 and 8 and earthquakes of $M \geq 6.0$ that occurred on fault segment 9. The analysis shows that the earthquakes in segments 7 and 8 (202 model events) occur more frequently after earthquakes in segment 9 (138 model events) than, on average, for the whole time interval of modeling. Figure 5 shows the $M \geq 6.0$ earthquake occurrences on segment 9 (Figure 5a), and the $M \geq 6.6$ earthquake occurrences on segments 7 and 8 (Figure 5b). The occurrences of the $M \geq 6.6$ earthquakes on segments 7 and 8, when segment 9 is locked ($B = 10$ is specified for this segment), are presented in Figure 5c. A considerable difference can be observed between the earthquake patterns in Figure 5b (segment 9 is unlocked) and Figure 5c (segment 9 is locked). A similar long-range interaction of earthquakes was determined at model fault segments 1 and 6 [*Soloviev and Ismail-Zadeh*, 2003; *Vorobieva and Soloviev*, 2005]. The model results show that the long-range interaction of synthetic earthquakes exists in the BAFD model. It depends on the relative positions of fault segments and on movements specified in the model.

3.3. Transformation of the Frequency-Magnitude Relationship Prior to the Extreme Event

The earthquake FM distribution [*Ishimoto and Iida*, 1939; *Gutenberg and Richter*, 1994] is described by the relation $\log_{10}N = a - bM$, where N is the cumulative number of earthquakes having magnitudes not less than M, a value and b value characterize the total number of earthquakes and size distribution of earthquakes, respectively. Variation of the b value has been found in observable seismicity [e.g., *Smith*, 1986; *Ogata and Katsura*, 1993; *Amelung and King*, 1997; *Wiemer and Wyss*, 1997; *Burroughs and Tebbens*, 2002], in synthetic seismicity [e.g., *Christensen and Olami*, 1992; *Zaliapin et al.*, 2003; *Zöller et al.*, 2006], and for microfractures in laboratory samples [e.g., *Rotwain et al.*, 1997; *Amitrano*, 2003]. Theoretical models have been developed to explain temporal fluctuations in b value [e.g., *Main et al.*, 1989, 1992; *Henderson et al.*, 1992].

The b value change in FM distribution for synthetic earthquake catalogs obtained by means of the BAFD model has been recently studied by *Soloviev* [2008]. Two time periods have been identified in the catalog: D period that includes time intervals preceding strong earthquakes and N period that includes time intervals not preceding strong earthquakes. The separate analysis of these periods shows that the b value becomes smaller before extreme events. Figure 6 shows

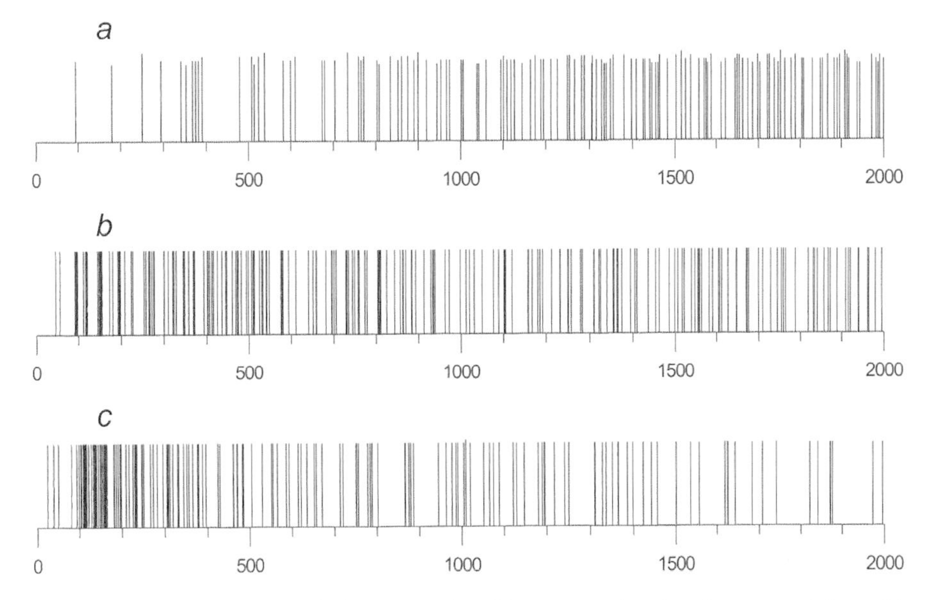

Figure 5. Times of earthquakes that occurred in (a) segment 9 and in segments 7 and 8 when segment 9 was (b) unlocked and (c) locked. Reprinted from *Soloviev and Ismail-Zadeh* [2003] with kind permission of Springer Science and Business Media.

commutative FM plots obtained separately for the D and N periods determined in the catalog of earthquakes that occurred on one of the segments of the block structure studied by *Soloviev* [2008]. One can see that the b value corresponding to the D period is less than that one corresponding to the N period. Decreasing of the b value means that the number of small earthquakes decreases while the number of large earthquakes increases. This phenomenon can be used in earthquake prediction studies as highlighted by *Eneva and Ben-Zion* [1997]. The transformation of the FM relationship before extreme events is incorporated in the earthquake prediction algorithm M8 [*Keilis-Borok and Kossobokov*, 1990].

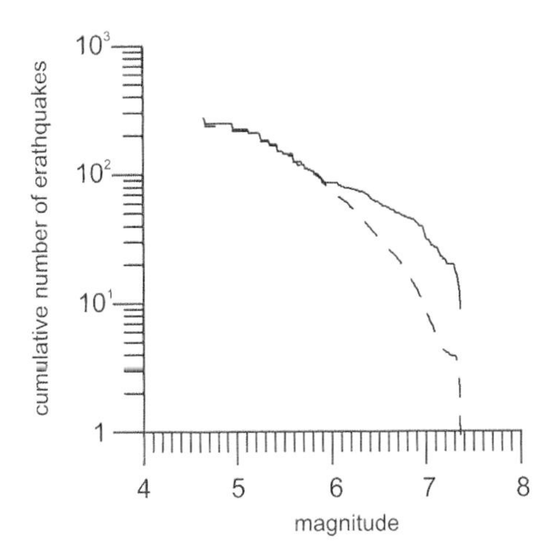

Figure 6. Commutative frequency-magnitude (FM) plots obtained separately for D period (solid curve) and N period (dashed curve) determined for the catalog of earthquakes occurred in one of the segments of the block structure studied by *Soloviev* [2008]. FM relation for N period has been preliminary normalized to obtain the same total number of events as it is for D period.

The BAFD model allows the reproduction of different configurations of blocks and faults and specification of various movements of the underlying medium and boundary blocks and some features of the fault zones. Detecting the transformation of FM plot before strong earthquakes in the model allows the study of how the manifestation of this phenomenon depends on the block structure geometry, the plate movements, and the features of the fault zones.

4. INTERMEDIATE-DEPTH SEISMICITY IN VRANCEA

Large intermediate-depth earthquakes in Vrancea cause destruction in Bucharest (Romania) and shake central and eastern European cities several hundred kilometers away from the hypocenters of the events. The earthquake-prone Vrancea region is situated at the bend of the southeastern Carpathians and is bounded to the north and northeast by the Eastern European platform (EEP), to the east by the Scythian platform, to the southeast by the Dobrogea orogen, to the south and southwest by the Moesian platform (MP), and to the northwest by the Transylvanian basin. The epicenters of the mantle earthquakes in the Vrancea region are concentrated within a very small area (Figure 7a). The projection of the foci on a NW-SE vertical plane across the bend of the eastern Carpathians (section AB in Figure 7b) shows a seismogenic volume about 110 km (deep) × 70 km × 30 km and

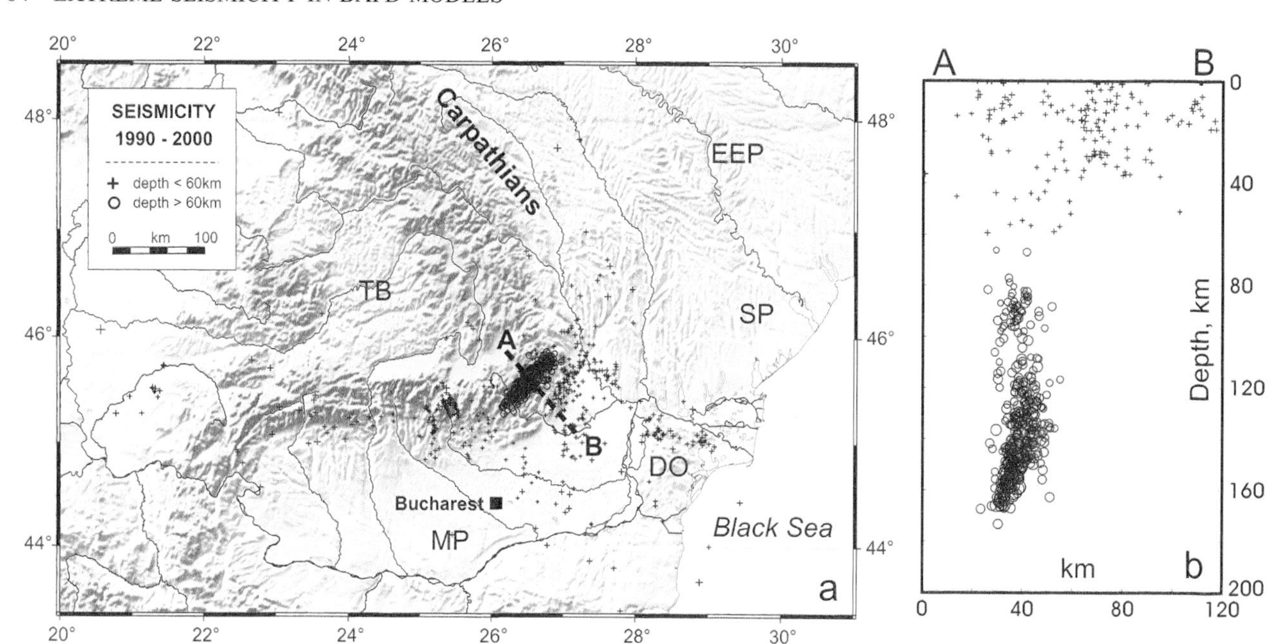

Figure 7. Observed seismicity in Romania for the last decade with magnitude $M_w \geq 3$. (a) Epicenters of Vrancea earthquakes determined by the joint hypocenter method. The background is the topography. (b) Hypocenters of the same earthquakes projected onto the NW-SE vertical plane AB (dashed line in Figure 7a). Abbreviations are DO, Dobrogea orogen; EEP, Eastern European platform; MP, Moesian platform; SP, Scythian platform; TB, Transylvanian basin. Modified from *Ismail-Zadeh et al.* [2005], reprinted with permission from Elsevier (http://www.sciencedirect.com/science/journal/00319201).

extending to a depth of about 180 km. Beyond this depth, the seismicity ends suddenly: one seismic event at 220 km depth represents an exception [*Oncescu and Bonjer*, 1997]. According to the historical catalog of Vrancea events [*Radu*, 1991], strong intermediate-depth shocks with magnitudes $M_w > 6.5$ occur three to five times per century. In the twentieth century, large events at depths d of 70 to 180 km occurred in 1940 (moment magnitude $M_w = 7.7$, $d = 160$ km), in 1977 ($M_w = 7.5$, $d = 100$ km), in 1986 ($M_w = 7.2$, $d = 140$ km), and in 1990 ($M_w = 6.9$, $d = 80$ km) [e.g., *Oncescu and Bonjer*, 1997].

McKenzie [1972] suggested that the intermediate-depth seismicity in Vrancea is associated with a relic part of oceanic lithosphere sinking in the mantle and now overlain by continental crust, and *Fuchs et al.* [1979] suggested that a seismic gap observed at depths of 40–70 km beneath Vrancea can be associated with a lithospheric slab detached from the continental crust. Two principal sets of geodynamic models have been developed so far. One set is based on the assumption that the mantle seismicity in the Vrancea region is associated with a relic oceanic lithosphere (attached to or already detached from the continental crust), which is presently descending beneath the SE Carpathians [e.g., *Linzer*, 1996; *Girbacea and Frisch*, 1998; *Wortel and Spakman*, 2000; *Sperner et al.*, 2001; *Gvirtzman*, 2002]. The other set of models assumes that the Vrancea lithosphere thickened during a continental collision, became unstable, and started to sink in the mantle finally delaminating from the crust and producing intermediate-depth earthquakes [e.g., *Pana and Erdmer*, 1996; *Pana and Morris*, 1999; *Knapp et al.*, 2005; *Fillerup et al.*, 2010]. Seismic tomography imaged a high-velocity body beneath the Vrancea region [e.g., *Martin et al.*, 2006; *Raykova and Panza*, 2006]. This body can be interpreted as a dense and cold lithospheric body sinking in less dense and warm surrounding mantle. The hydrostatic buoyancy forces promote the sinking of the body, but viscous and frictional forces resist the descent. The combination of these forces produces shear stresses at intermediate depths that are high enough to cause earthquakes [*Ismail-Zadeh et al.*, 2000, 2005].

The BAFD model was applied to study the dynamics of the lithosphere and intermediate-depth large earthquakes in the Vrancea region. *Panza et al.* [1997] and *Soloviev et al.* [1999, 2000] considered four main structural elements of Vrancea as the lithospheric blocks comprising the BAFD model: the EEP, the Moesian, the Black Sea, and the Intra-Alpine subplates. We refer to *Ismail-Zadeh et al.* [2012] for the detailed description of the lithospheric configuration in the region. Figure 8 shows the pattern of faults on the upper

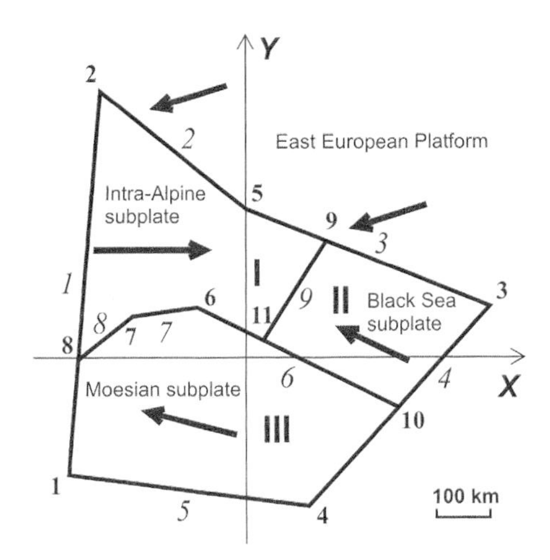

Figure 8. Block structure used for earthquake simulation in the Vrancea region. Vertices are marked by bold numbers (1–11), faults by italic numbers (1–9), and blocks by Roman numerals (I–III). Arrows outside and inside the block structure indicate the movements of blocks and of the sublithospheric mantle, respectively. Modified from *Panza et al.* [1997], reprinted with kind permission of Springer Science and Business Media.

plane of the block structure used to model the Vrancea region. The point with the geographic coordinates 44.2°N and 26.1°E (the Vrancea zone) is chosen as the origin of the reference coordinate system. Although the thickness of the lithosphere varies beneath the region under study, the model thickness was assumed to be 200 km (the depth of the deepest earthquakes in the Vrancea region).

The structure contains nine fault planes. The model fault plane parameters are given by *Panza et al.* [1997]. The movement of the sublithospheric mantle beneath blocks I–III is prescribed (Figure 8): $v_{\rm I} = (0.5, 0)$ cm yr^{-1}, $v_{\rm II} = (-0.3, 0.14)$ cm yr^{-1}, and $v_{\rm III} = (-0.4, 0.1)$ cm yr^{-1}. The model boundary adjacent to fault planes 2 and 3 moves progressively at the rate $v_b = (-0.32, -0.1)$ cm yr^{-1}. The boundary fault planes 1, 4, and 5 do not represent any geologic features and are introduced in the model to limit the block structure (hence $\xi = \zeta = \zeta_s = 0$ at these faults). The time and space discretization are $\Delta t = 5 \times 10^{-2}$ yr and $\chi = 7.5$ km, respectively.

The catalog of synthetic seismicity was computed for the period of 7000 years (after stress stabilization) [see *Panza et al.*, 1997; *Soloviev and Ismail-Zadeh*, 2003]. The synthetic catalog contained 6743 events with magnitudes varying from 5.0 to 7.6. The minimum value of the magnitude corresponds, in accordance with the magnitude relation (9), to the minimum area of one cell. The maximum value of the magnitude in the catalog of model events is 7.6. (Note that the maximum magnitude of model earthquakes is close to the magnitude $M_w = 7.7$ earthquake that occurred in Vrancea in 1940.)

Figure 9 presents the observed seismicity for the period 1900–1995 (Figure 9a) and the distribution of epicenters from the catalog of synthetic events (Figure 9b). The majority of synthetic events occur on fault plane 9 (cluster A), which corresponds to the Vrancea zone, where most of the observed intermediate-depth seismicity is concentrated. Some events cluster on fault plane 6 (cluster B) located southwest of the main seismic area and separated by an aseismic zone. The third cluster of events (cluster C) groups

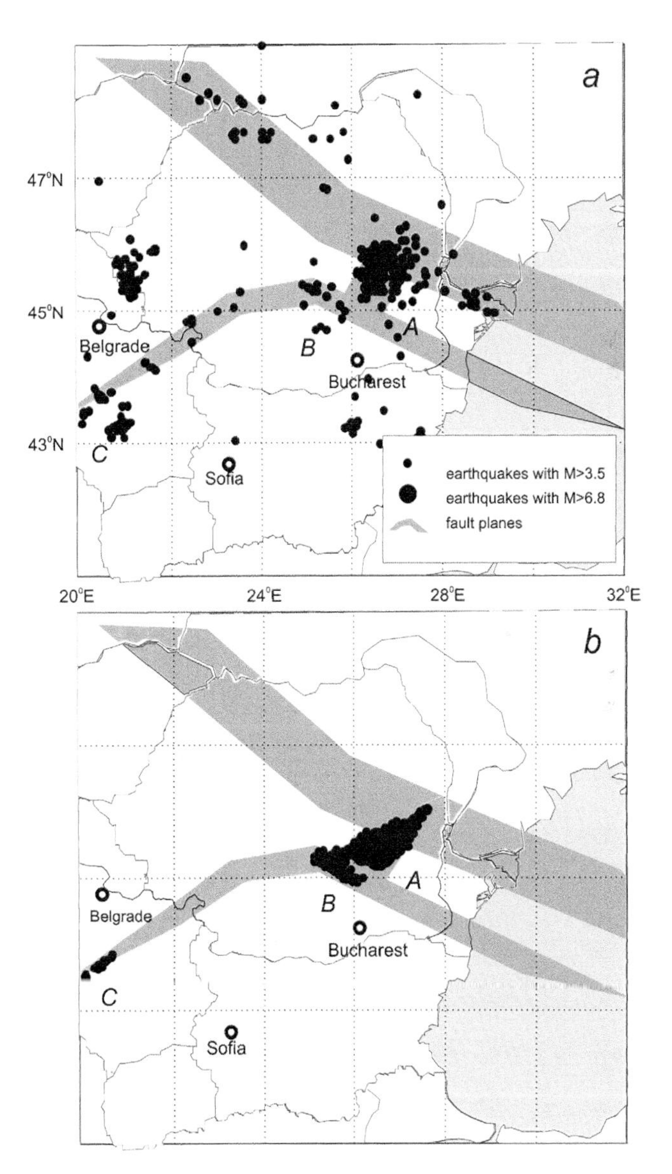

Figure 9. Maps of observed seismicity in Vrancea (a) in the period 1900–1995 and (b) modeled seismicity for 7000 years. Gray areas are the projections of fault planes with $\xi \neq 0$ on the upper plane. Modified from *Soloviev and Ismail-Zadeh* [2003], reprinted with kind permission of Springer Science and Business Media.

on fault plane 8. There are several additional clusters of epicenters on the map of the observed seismicity (Figure 9a) that are absent in the synthetic catalog. This is not surprising because only a few main seismic faults of the Vrancea region are included in the model. The regional earthquakes far from the Vrancea zone are shallow crustal events, and hence, the block structure model should be modified to take into account the upper crustal seismicity. Nevertheless, even a simple structure consisting of only three lithospheric blocks reproduces the main features of the observed seismicity in space.

Figure 10a depicts the Gutenberg-Richter FM plots for the observed seismicity in Vrancea (solid line) and for the synthetic seismicity (dashed line). The FM curve for the synthetic events is close to linear and has approximately the same slope as that for the observed seismicity. Note that the FM curve is shifted upward because the number of modeled seismic events (for 7000 years) is larger than that of observed earthquakes (for about 100 years). The FM distribution for the Vrancea earthquakes is characterized by a deficit of earthquakes around 6.5 magnitude relative to the linear distribution of the modeled FM curve. An analysis of the BAFD model earthquake catalog for the time interval of 7000 years may explain the origin "6.5 magnitude" gap in the Vrancea seismicity as associated with the small time window of relevant observations.

The duration of the earthquake simulations is by a factor of about 70 longer than the total period of the observed catalog (about 1 century long). According to this scale, the synthetic catalog was divided into 70 parts (each 100 years long) in order to find the features similar to the present intermediate-depth seismicity in Vrancea. The catalog of the Vrancea earthquakes contains 71 events of magnitude $M \geq 5.4$ and 1 event of the maximum magnitude 7.4 [*Radu*, 1991]. The number of magnitude $M \geq 5.4$ events in each of the 70 parts of the catalog of synthetic events varies from 53 to 94; the average number of such events is 68, and the maximum magnitude varies from 6.0 to 7.6. If a synthetic event with $M \geq 6.8$ is considered large, then the number of large earthquakes varies from zero to four. There are no large events in 29 parts, one large event in 20 parts, two large events in 16 parts, three large events in 4 parts, and four large events in only 1 part of the total catalog.

The FM plots for periods with and without extreme model events (large synthetic earthquakes) are compared to the FM curve of the observed seismicity. The trend and the intensity of earthquake sequences for the observed (solid line) and the synthetic (dashed line) events (for the period without large model events) are quite similar for the part of the curve accounting for the events of magnitude 6.4 and smaller (Figure 10b). For the period with four large model events, there is a gap in the number of model events of the magnitude range from 6.4 to 6.8 in the FM plot (Figure 10c), similar to that observed in Vrancea. Such a gap is a typical phenomenon for the parts of the catalog of synthetic events with few large shocks. The modeling results lead to a conclusion that presently (for the last 100 years) the Vrancea region experiences an "active seismic

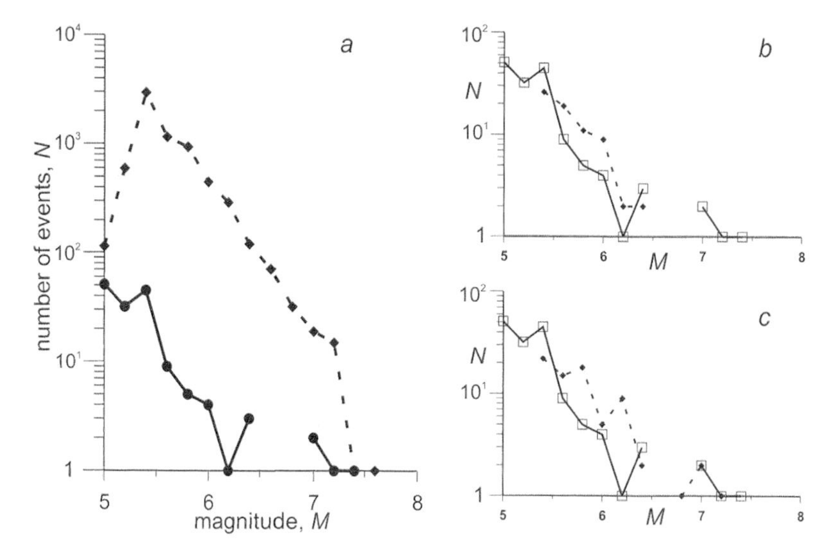

Figure 10. The FM plots for the observed seismicity in Vrancea (solid line) and for three catalogs of synthetic events (dashed lines) of the same experiment: (a) the whole catalog of events for 7000 years, (b) a 100 year long part of the whole catalog *without* large events, and (c) a 100 year long part of the whole catalog *with* four large events. Modified from *Soloviev and Ismail-Zadeh* [2003], reprinted with kind permission of Springer Science and Business Media.

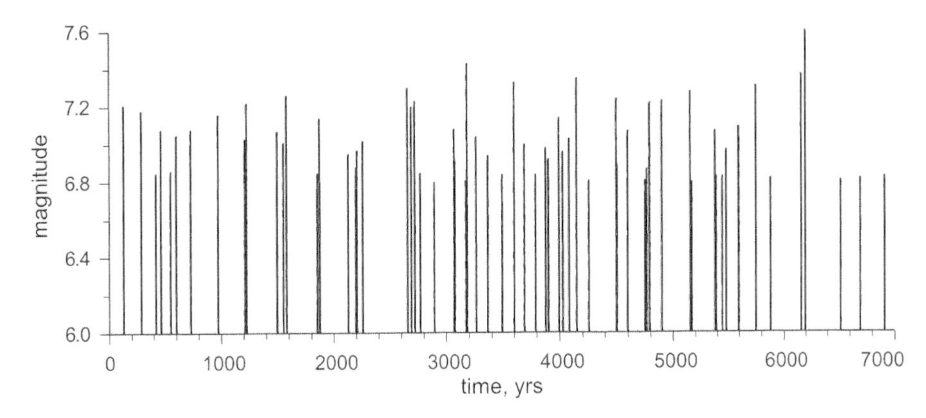

Figure 11. Temporal distribution of large earthquakes in the catalog of synthetic events for Vrancea. Modified from *Soloviev and Ismail-Zadeh* [2003], reprinted with kind permission of Springer Science and Business Media.

period" with four (observed) large earthquakes, but with a deficit of earthquakes at some magnitude range. This gap in magnitudes is likely to be filled with much longer observations.

The temporal distribution of large (M > 6.8) synthetic earthquakes for 7000 years is presented in Figure 11. One can see a strong irregularity in the flow of these events. For example, groups of large earthquakes occur periodically in the time interval from about 500 to 3000 model years, with a return period of about 300–350 years. The unique part of the catalog of modeled events with four large shocks belongs to this time interval. The periodic occurrence of a single large earthquake with a return period of about 100 years is typical of the interval from 3000 to 4000 years. There is no periodicity in the occurrence of large earthquakes in the remaining parts of the catalog of synthetic events. These results demonstrate the importance of a careful estimation of the duration of seismic cycles to predict the occurrence of a future large earthquake.

Soloviev et al. [2000] studied the influence of the variation of the fault plane angle on the model earthquakes. *Ismail-Zadeh et al.* [1999] introduced a mantle flow pattern into a BAFD model. The rate of the motion of the lithospheric blocks was determined from a model of mantle flow induced by a sinking slab beneath the Vrancea region [*Ismail-Zadeh et al.*, 2000]. Several numerical experiments for various model parameters showed that the spatial distribution of synthetic events is significantly sensitive to the directions of the block movements. *Ismail-Zadeh et al.* [1999] showed that changes in modeled seismicity are controlled by small changes in the lithospheric slab's descent (e.g., slab position, dip angle). This is in overall agreement with the study of *Press and Allen* [1995], which showed that small changes in the direction of Pacific plate motion would result in changes of the pattern of seismic release.

5. EARTHQUAKES IN THE SUNDA ISLAND ARC REGION

The Sunda island arc marks an active convergence boundary between the Eurasian plate, which underlies Indonesia with Indian and Australian plates (Figure 12). A chain of volcanoes forms the topographic spine of the islands of Sumatra, Java, and Sunda. The Indian and Australian plates subduct beneath the southwestern part of the Eurasian plate along the Sunda arc. The tectonic deformation and associated stress localization in the Sunda trench caused the 2004 Aceh-Sumatra great megathrust earthquake of magnitude 9.1. Seismic tomography imaging revealed anomalies of seismic waves beneath the Sunda island arc suggesting that the lithospheric slab penetrates into the lower mantle [*Widiyantoro and van der Hilst*, 1996].

Earthquakes in the Sunda arc region were simulated using the BAFD model [*Soloviev and Ismail-Zadeh*, 2003]. The

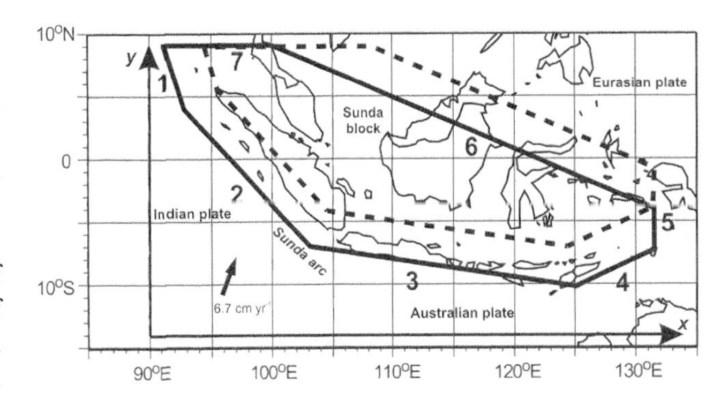

Figure 12. Model geometry of the Sunda arc. The upper and lower planes of the Sunda block are marked by solid and dashed lines, and numbers 1 to 7 mark the faults. Modified from *Soloviev and Ismail-Zadeh* [2003], reprinted with kind permission of Springer Science and Business Media.

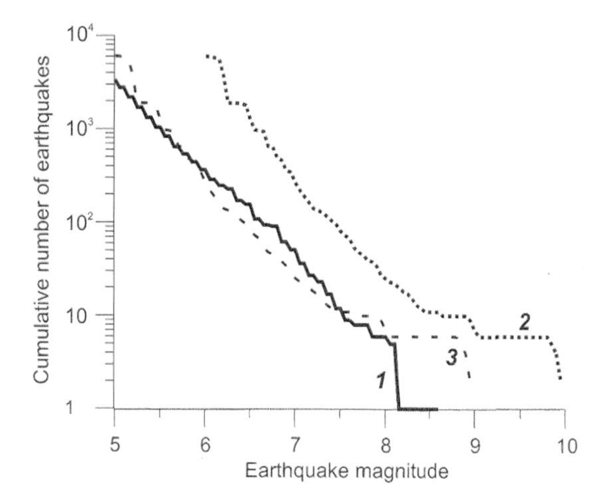

Figure 13. Cumulative FM plots for the observed seismicity (1900–1999) of the Sunda arc (solid curve 1), for the synthetic catalog (dotted curve 2), and for the synthetic catalog with magnitudes reduced by 1 (dashed curve 3). Modified from *Soloviev and Ismail-Zadeh* [2003], reprinted with kind permission of Springer Science and Business Media.

model structure consists of one 130 km thick block bounded by seven fault planes numbered from 1 through 7 (Figure 12). Fault planes 1–4 (forming the boundary between the Sunda and the Indo-Australian plates) have a dip angle of 21°. The other model parameters are given by *Soloviev and Ismail-Zadeh* [2003]. The values $\Delta t = 10^{-3}$ year and $\chi = 15$ km were used for temporal and spatial discretization.

The HS2-NUVEL1 model [*Gripp and Gordon*, 1990] is used to specify the movements. Namely, the model (Sunda) block moves at the translational velocity $v_e = (-0.604, 0.396)$ cm yr^{-1} and the angular velocity 0.06×10^{-8} rad yr^{-1}. The boundary block (Indo-Australian plate) adjacent to fault planes 1–4 moves at the translational velocity $v_a = (1.931, 6.560)$ cm yr^{-1} and the angular velocity 0.42×10^{-8} rad yr^{-1}. In both cases, the coordinate origin is the center of rotation, and its positive direction is counterclockwise. Therefore, the movement of Australia relative to Eurasia is translation at the velocity $v_e = (2.535, 6.164)$ cm yr^{-1} and rotation about the origin at the angular velocity 0.36×10^{-8} rad yr^{-1}. The dynamics of the block structure was simulated for 200 years, starting from zero. The seismicity observed in the region is compared to the stabilized part of the synthetic earthquake catalog from 100 to 200 model years.

Figure 13 presents the cumulative FM plots for the synthetic catalog and the observed seismicity. One can see that the slopes of the curves (b values) are close, but the synthetic curve is shifted to larger magnitudes by about 1 unit of magnitude. The synthetic curve with magnitudes reduced by 1 is also shown in Figure 12. It fits to the curve for the observed seismicity rather well. The results of the BAFD model for the Sunda arc was published in 2003 [*Soloviev and Ismail-Zadeh*, 2003]; that is why the 2004 M9.1 Aceh-Sumatra earthquake and 2005 M8.6 northern Sumatra earthquake are not included in the plot of the observed seismicity (if included, the misfit between the modeled and observed seismicity would be smaller).

The maps of earthquake epicenters of $M \geq 6$ observed in the Sunda arc until 1999 and the synthetic epicenters with $M \geq 7$ are shown in Figure 14. The BAFD model identified two areas prone to the largest earthquakes. The first area is predicted to be located between the Borneo, Sulawesi, Sumbala, Lombok, and Bali islands in the eastern part of the arc. The second area is located in the northwestern part of the Sunda arc, where the 2004 Aceh-Sumatra earthquake occurred (Figure 14). Several experiments have been performed to study the dependence of the cumulative FM plot for the catalog of modeled events on the movements specified in the model [*Soloviev and Ismail-Zadeh*, 2003]. Figure 15 demonstrates the change in the b value of the FM curves of the synthetic events for different experiments. Particularly, Figures 15a–15c show that the increase in the angular velocity of the Indo-Australian plate (when the translation velocity is constant) results in an increase of b values.

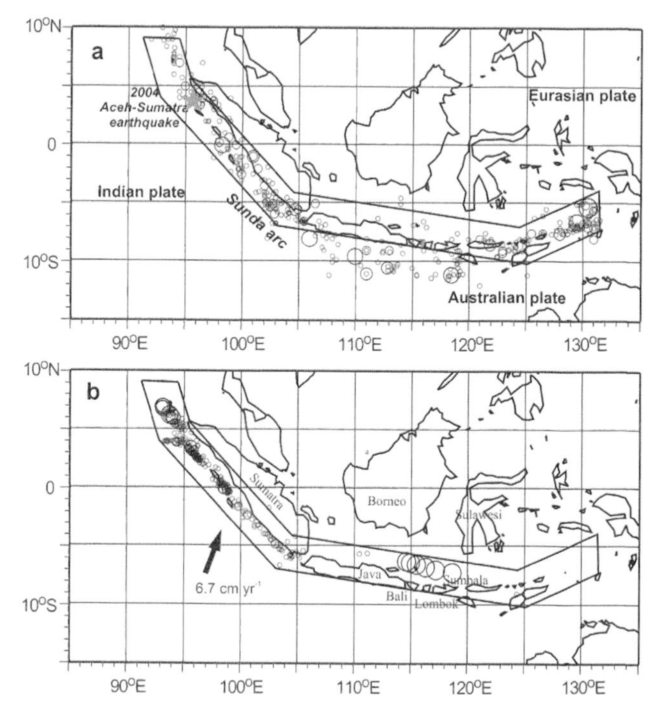

Figure 14. Maps of (a) observed seismicity in the Sunda arc for 1900–1999 and (b) modeled seismicity. Modified from *Soloviev and Ismail-Zadeh* [2003], reprinted with kind permission of Springer Science and Business Media.

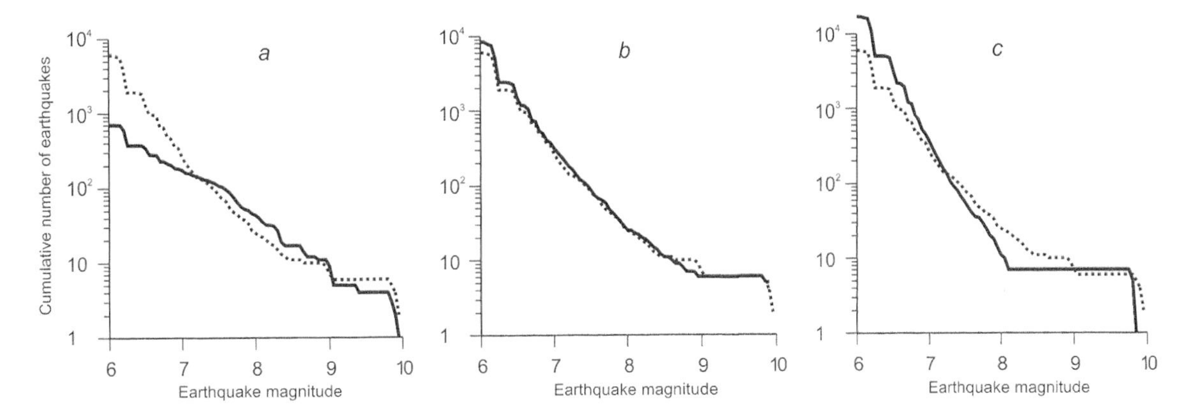

Figure 15. Cumulative FM plots for three experiments, where the angular velocity of the Indo-Australian plate ω_{IA} is varied: (a) $\omega_{IA} = 0.18 \times 10^{-8}$, (b) 0.42×10^{-8}, and (c) 0.54×10^{-8} rad yr^{-1}. Dotted lines illustrate the FM plot for the experiment with $\omega_{IA} = 0.36 \times 10^{-8}$ rad yr^{-1} (the case study presented here). Modified from *Soloviev and Ismail-Zadeh* [2003], reprinted with kind permission of Springer Science and Business Media..

Despite its simplicity, the BAFD model of the Sunda arc shows that the movements specified by the global plate motion yields synthetic seismicity having certain common features with observations, namely, the locations of larger events, their magnitudes, and the slope of the FM plot. An important message is that the BAFD model for the Sunda arc predicted (in advance to the great events of the twenty-first century in the region) a considerable deviation of the FM curve (in the range of magnitudes from 8 to 9+) between the observed (1900–1999) and modeled seismic events. This highlights, again, an importance of earthquake modeling in the analysis of extreme seismic events.

6. EARTHQUAKES IN THE TIBET-HIMALAYAN REGION

Following the closure of the Mesozoic Tethys Ocean, the India-Asia collision initiated the development of the Himalayan range and the Tibetan plateau and induced widespread strain in southeastern Asia and China. The Tibetan plateau is underlain by a thick crust (up to 70 to 80 km) as inferred from gravity anomalies and seismic profiles [*Barazangi and Ni*, 1982; *Le Pichon et al.*, 1992; *Nelson et al.*, 1996]. The Himalayan frontal thrust and the Longmen Shan represent abrupt and steep topographic fronts at the southern and eastern edges of the plateau (Figure 16a).

There are three distinct views about the active deformation in Tibet that dominates the debate on the mechanics of continental deformation. One view is that the deformation is distributed throughout the continental lithosphere [e.g., *Houseman and England*, 1996]. The second view is associated with crustal thinning and deformation due to a channel flow within the mid-to-lower crust [e.g., *Royden et al.*, 1997]. Meanwhile, there is growing evidence supporting the alternative view that a substantial part of the deformation of the continents is localized on long and relatively narrow faults and shear zones separating rigid crustal blocks [e.g., *Tapponnier et al.*, 2001]. Many of these zones cut the base of the crust [e.g., *Vergnes et al.*, 2002], and some extend to the base of the lithosphere [e.g., *Wittlinger et al.*, 2004]. Therefore, such deformations can be described by motions of crustal lithospheric blocks separated by faults.

Using a BAFD model, *Ismail-Zadeh et al.* [2007] developed several sets of numerical experiments to analyze the earthquake clustering, frequency-to-magnitude relationships, earthquake focal mechanisms, and fault slip rates in the Tibet-Himalayan region. The model structure for the region (Figure 16a) is made of six major crustal blocks delineated on the basis of detailed geomorphic and tectonic field studies and large-scale SPOT or Landsat imagery analysis [*Replumaz and Tapponnier*, 2003]. The blocks are separated by thrust and strike-slip faults hundreds of kilometers long. The faults in the model structure do not trace exactly the regional faults but rather represent the main geometrical features of the faults. The model structure contains 41 fault planes, 63 fault segments, 6 crustal blocks, and 6 boundary blocks. The movement of the blocks was specified with the rate constrained by the present rate of convergence between India and Asia [*Bilham et al.*, 1997]. The numerical simulations were performed for 4000 model years starting from zero initial conditions. For further details on this block-and-fault structure and its parameters and on the configuration of the lithospheric block in the Tibet region, we refer to the work of *Ismail-Zadeh et al.* [2007]. Four sets of numerical experiments have been performed to study the dynamics of the block-and-fault structure in the Tibet-Himalayan region. The

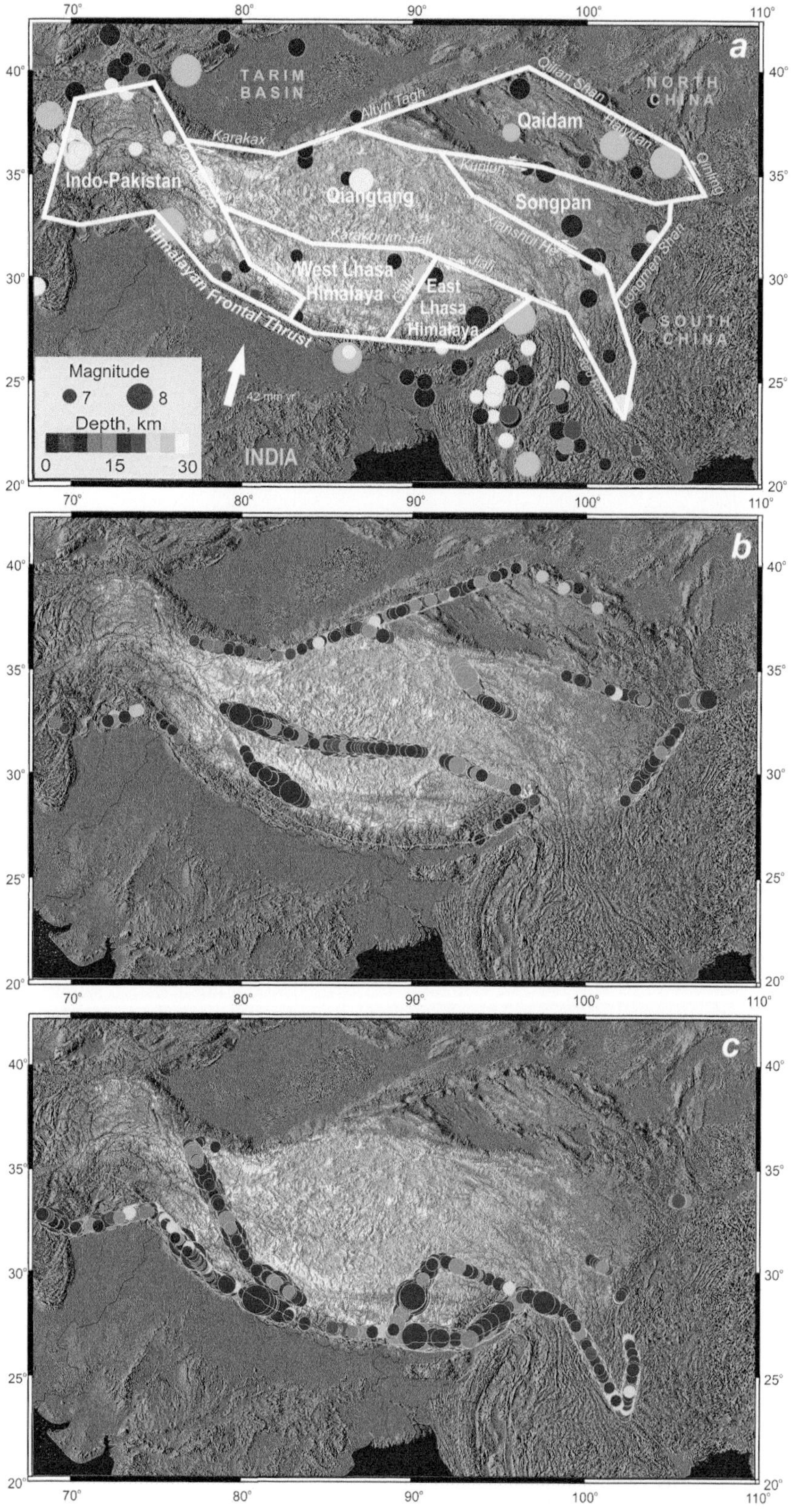
a
b
c
TARIM BASIN
NORTH CHINA
SOUTH CHINA
INDIA
Qaidam
Songpan
Qiangtang
Indo-Pakistan
West Lhasa Himalaya
East Lhasa Himalaya
Altyn Tagh
Qilian Shan
Haiyuan
Qinling
Karakax
Kunlun
Xianshui He
Longmen Shan
Karakorum-Jiali
Himalayan Frontal Thrust
42 mm yr
Magnitude
7
8
Depth, km
0
15
30
70°
80°
90°
100°
110°
20°
25°
30°
35°
40°

Figure 16

Table 1. Large Crustal Earthquakes ($M \geq 7.5$) in the Tibet-Himalayan Region

Date	Time	Latitude (°N)	Longitude (°E)	Depth (km)	Magnitude
22 Aug 1902	03:00:00.0	40.00	77.00	25.0	8.6
4 Apr 1905	00:50:00.0	33.00	76.00	25.0	8.6
22 Dec 1906	18:21:00.0	43.50	85.00	25.0	8.3
21 Oct 1907	04:23:00.0	38.00	69.00	25.0	8.1
3 Jan 1911	23:25:00.0	43.50	77.50	25.0	8.7
18 Feb 1911	18:41:00.0	40.00	73.00	-	7.8
23 May 1912	02:24:00.0	21.00	97.00	25.0	7.9
8 Jul 1918	10:22:00.0	24.50	91.00	-	7.6
16 Dec 1920	12:05:00.0	36.00	105.00	25.0	8.6
22 May 1927	22:32:00.0	36.75	102.00	25.0	8.3
27 Jan 1931	20:09:00.0	25.60	96.80	-	7.6
25 Dec 1932	02:04:00.0	39.25	96.50	-	7.6
15 Jan 1934	08:43:00.0	26.50	86.50	25.0	8.4
30 May 1935	21:32:00.0	29.50	66.75	-	7.5
7 Jan 1937	13:20:00.0	35.50	98.00	-	7.6
12 Sep 1946	15:17:00.0	23.50	96.00	-	7.5
12 Sep 1946	15:20:00.0	23.50	96.00	-	7.8
2 Nov 1946	18:28:00.0	41.50	72.50	-	7.6
17 Mar 1947	08:19:00.0	33.00	99.50	-	7.7
29 Jul 1947	13:43:00.0	28.50	94.00	-	7.9
10 Jul 1949	03:53:00.0	39.00	70.50	-	7.6
15 Aug 1950	14:09:00.0	28.50	96.50	25.0	8.7
18 Nov 1951	09:35:00.0	30.50	91.00	25.0	7.9
17 Aug 1952	16:02:00.0	30.50	91.50	-	7.5
9 Jun 1956	23:13:00.0	35.10	67.50	-	7.6
6 Feb 1973	10:37:00.0	31.40	100.58	-	7.7
23 Aug 1985	12:41:00.0	39.43	75.22	7.0	7.5
19 Aug 1992	02:04:00.0	42.14	73.57	27.0	7.5

experiments analyzed how the following factors influence the seismic clustering and the rates of block displacements and fault slips: the resistance of the model boundary, variations in the movement of the Indian plate with respect to Eurasia, changes in the elastic and viscous properties of the fault zones, and variations in the motion of the lower crust. A lower crust flow along with the Indian plate movement results in earthquakes on the internal fault segments of the Tibetan plateau including segments of the Altyn Tagh, Karakorum, Kunlun, and Longmen Shan faults and the Gulu rift zone (Figure 16b and 16c). Hypocenters of the modeled events are located mainly at depths ranging from 10 to 20 km.

The largest magnitude of the model events was found to be $M = 8.9$. These great events occurred on the fault segment associated with the Gulu rift zone (Figure 16c). It is remarkable to mention that the frequency of occurrence of these extreme events varies considerably: the second of the four events occurred 252 model years after the first event, the new event in 48 years after the second one, and the last event with a time interval of 144 years after the third event. The slope of the FM plots provides additional information on magnitudes of regional earthquakes. The slope of the FM plots for the synthetic events (for two preferred experiments) is rather close to that for observed seismicity in the magnitude range from 6.8 to 8.2 (Figure 17).

Figure 16. (opposite) A block-and-fault structure and spatial distribution of large earthquakes in the Tibet-Himalayan region. White bold lines (model faults) delineate the model blocks (structural geological elements), and the white arrow indicates the motion of India relative to Eurasia. Earthquake epicenters are marked by circles. (a) Observed seismicity ($M > 6.5$) in the region since 1900 (see Table 1 for the large shallow earthquakes of $M \geq 7.5$). (b) and (c) Synthetic seismicity in two numerical experiments. See *Ismail-Zadeh et al.* [2007] for the results of other numerical experiments. From *Ismail-Zadeh et al.* [2007], reprinted with permission from Elsevier (http://www.sciencedirect.com/science/journal/0012821X).

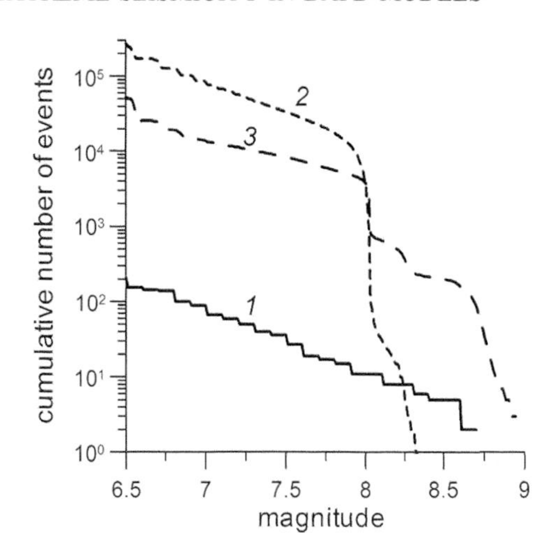

Figure 17. Cumulated FM plots for observed seismicity from 1967 to 2003 (solid curve 1) and synthetic seismicity predicted by the preferred numerical experiment (dashed curves). Curves 2 and 3 correspond to synthetic seismicity presented in Figures 16b and 16c, respectively.

The numerical results demonstrated that the slope of the FM plots and clustering of large earthquakes are sensitive to the changes in the movements of the lower crust and in the rheological properties of fault plane zones in the Tibetan plateau. Large events localize only on some of the faults but not on all of the individual faults where the elastic and viscous coefficients were equally changed. This illustrates the fact that the BAFD model describes the dynamics of a network of crustal blocks and faults rather than the dynamics of individual fault planes. As an example, a cluster of large modeled events along the Longmen Shan fault was identified by the BAFD model (Figure 16b). The 2008 $M = 7.9$ Sichuan (Wenchuan) earthquake occurred along this fault killing about 70,000; in addition, there were about 400,000 injured and about 20,000 missing people.

Ismail-Zadeh et al. [2007] also analyzed the focal mechanisms of the synthetic earthquakes computing the angle between the slip direction (in the fault plane) and the fault line and showed a reasonably good agreement between the focal mechanisms of the synthetic and observed earthquakes. Namely, in both cases (in the model and reality), most thrust faulting events occur on the Himalayan seismic belt and normal faulting events on the Gulu rift.

7. PERSPECTIVES IN EARTHQUAKE SIMULATIONS AND STUDIES OF EXTREME SEISMICITY

In this chapter, we have analyzed a quantitative approach to simulation of earthquakes using the BAFD model. The model provides a link between geodynamical processes and seismicity and allows studying the influence of fault network properties (e.g., fragmentation, block geometry and movement, direction of driving forces) on seismic patterns and cycles. Also, the applications of the BAFD model to study earthquake sequences and extreme seismic events in three earthquake-prone regions have been reviewed. A better integration of knowledge on earthquake physics into BAFD models is required to develop sophisticated realistic models of earthquake occurrences. Seismological (e.g., physical parameters of earthquakes and tectonic stress, fluid migration, etc.), geodetic (e.g., GPS, interferometric synthetic aperture radar, and other measurements of the crustal deformation), and geological (e.g., determination of the time intervals between strongest earthquakes using paleoseismological tools) data can be used and assimilated in numerical simulations of earthquakes to better forecast extreme seismicity [e.g., *Van Aalsburg et al.*, 2007; *Werner et al.*, 2011]. Various properties of the lithosphere, such as spatial heterogeneity, hierarchical block structure, nonlinear rheology, gravitational, thermal and chemical processes, phase transitions, and fluid migration, are likely to influence the characteristics of seismicity, such as a seismic pattern, clustering, activation-quiescence, and significant spatial and temporal variations of seismic activity.

We expect that earthquake simulators could be capable of producing long catalogs of synthetic earthquakes containing hundred thousands to millions of synthetic events of a wide range of magnitudes. This requires high-performance computations to simulate earthquakes on a fault plane with a fault spatial discretization down to 10–100 m and even finer meshes. We consider that the synthetic fault plane geometry should mimic the geometry of the real fault plane as close to reality as possible. Precise geodetic measurements of surface deformation and subplate movements and accurate estimations of lower crustal flow orientation through high-resolution studies of seismic anisotropy could allow determining more accurately fault slip rates, earthquake focal mechanisms, and energy release in earthquake models.

The earthquake simulators have to produce foreshocks and aftershocks, clustering of earthquakes, and long-range interaction and finally to allow forecasting future large events in studied regions. The models should provide an insight into the relationship between the geometrical and rheological properties of fault zones and different types of seismicity patterns in space, time, and magnitude domain. The earthquake simulators should also allow analyzing the model parameters responsible for the power law FM earthquake statistics, the clustering of earthquakes, and the spatial and temporal behavior of extreme events. These models should also be capable of calculating times of increased probability of large events to make effective probabilistic earthquake predictions.

The BAFD model provides a straightforward tool for a broad range of problems: (1) link between geodynamical processes and seismicity, (2) dependence of seismicity on general properties of fault networks (that is, fragmentation of the structure, movement of blocks, direction of driving forces, etc.), (3) study of seismic cycles, and (4) modeling of earthquake forecasting. The model reproduces some basic features of the observed seismicity (like Gutenberg-Richter law, clustering of earthquakes, and dependence of the occurrence of large earthquakes on the fragmentation of the block structure and on the rotation of blocks, etc.). It also enables one to study the relationships between geometry of faults, block movements, and earthquake sequences and to reproduce regional features of seismicity. From the observation of the territorial distribution of seismicity, the BAFD model enables to reconstruct the tectonic driving forces (and evaluate competing geodynamic hypotheses) in such a way as to produce a synthetic seismicity fitting the observed seismicity in an optimal way.

Meanwhile, the BAFD model is still far from a "perfect" model, and several aspects of its performance should be improved. The catalogs of synthetic seismicity obtained from experiments using the BAFD model are used to analyze the sites of large events, the FM (Gutenberg-Richter) relationship for synthetic events, the long-range interaction of earthquakes, the mechanisms of earthquake, the fault slip rates, etc. Meanwhile, the BAFD model does not always reproduce realistic aftershock sequences obeying the Omori law. A relevant modification of the model to allow the realistic aftershock dynamics would improve the model.

Although considerable simplifications have been made in the model, the model provides the possibility to understand the basic features of earthquake occurrences in a given region. For example, the blocks are considered to be rigid, although bounded by normal or thrust faults, they stretch or shrink during geological times. Meanwhile, their deformation for the time of several thousand years (the time scale of BAFD models) is insignificant compared to the block sizes. Perfectly rigid blocks employed in BAFD models allow reproducing a long-range interaction of seismicity and clustering of seismic events, although in real seismicity, the earthquake interaction and clustering cannot be so well pronounced as in the models. Other linked assumptions are that all displacements in the model are infinitely small compared with the typical block size, that the geometry of a block structure does not change during simulation, and that the structure does not move as a whole. Indeed, considering an average annual movement of a crustal block to be about 1–3 cm yr^{-1}, the total average displacement of the blocks can be about 100–300 m for the thousand years.

The interaction of the crustal (lithospheric) blocks along the fault planes is viscoelastic. Despite this rheological behavior is, in general, a good approximation, a nonlinear interaction of blocks should be explored in the model. The relation between earthquake energy and focal area is linear in the BAFD model [Keilis-Borok et al., 1997], which is at variance with the natural proportionality of the energy to the focal volume (or to the focal area raised to the power 3/2). Some elements of the SOFT model [*Allègre et al.*, 1995] were incorporated into the BAFD model to bring the relationship between released energy and focal area closer to the observed one [*Shebalin et al.*, 2002]. *Gabrielov et al.* [2007] further improved the BAFD model by incorporating a fluid flow into the fault system to explore the impact of fluid migration on the dynamics of the lithosphere and earthquake sequences. The BAFD model was also expanded to spherical geometry to simulate the interaction of tectonic plates and to analyze synthetic catalogs of world seismicity [*Rozenberg et al.*, 2005, 2010].

Simulation of realistic earthquake catalogs for a study region is of a great importance and significant challenge at the same time. The catalogs of synthetic events over a large time window can assist in interpreting the seismic cycle behavior or predicting a future extreme event, as the available observations cover only a short time interval (of about a hundred years). If a segment of the catalog of modeled events approximates the observed seismic sequence with a sufficient accuracy, the part of the catalog immediately following this segment might be used to predict the future seismicity and to forecast extreme events.

Acknowledgments. This work was supported by the grants from the French Ministry of Research, the German Research Foundation (DFG), the Russian Academy of Sciences, and the Abdus Salam ICTP. We thank C. Allègre, K. Fuchs, A. Gabrielov, M. Ghil, V. Keilis-Borok, V. Lyakhovsky, G. Panza, I. Vorobieva, and I. Zaliapin, for stimulating discussions on earthquake modeling and extreme seismicity. The authors are very grateful to S. Sharma and two reviewers for constructive comments.

REFERENCES

Allègre, C. J., J.-L. Le Mouël, and A. Provost (1982), Scaling rules in rock fracture and possible implications for earthquake prediction, *Nature*, *297*, 47–49.

Allègre, C. J., J.-L. Le Mouël, H. Duyen, and C. Narteau (1995), Scaling organization of fracture tectonics (S.O.F.T.) and earthquake mechanism, *Phys. Earth Planet. Inter.*, *92*, 215–233.

Allègre, C. J., P. Shebalin, J.-L. Le Mouël, and C. Narteau (1998), Energetic balance in scaling organization of fracture tectonics, *Phys. Earth Planet. Inter.*, *106*, 139–153.

Amelung, F., and G. King (1997), Earthquake scaling laws for creeping and noncreeping faults, *Geophys. Res. Lett.*, *24*, 507–510.

Amitrano, D. (2003), Brittle-ductile transition and associated seismicity: Experimental and numerical studies and relationship with the *b* value, *J. Geophys. Res.*, *108*(B1), 2044, doi:10.1029/2001JB000680.

Aoudia, A., A. T. Ismail-Zadeh, and F. Romanelli (2007), Buoyancy-driven deformation and contemporary tectonic stress in the lithosphere beneath Central Italy, *Terra Nova*, *19*, 490–495.

Barazangi, M., and J. Ni (1982), Velocities and propagation characteristics of Pn and Sn beneath the Himalayan arc and Tibetan plateau: Possible evidence for underthrusting of Indian continental lithosphere beneath Tibet, *Geology*, *10*, 179–185.

Benioff, H. (1951), Global strain accumulation and release as related by great earthquakes, *Bull. Geol. Soc. Am.*, *62*, 331–338.

Ben-Zion, Y. (2008), Collective behavior of earthquakes and faults: Continuum-discrete transitions, evolutionary changes, and corresponding dynamic regimes, *Rev. Geophys.*, *46*, RG4006, doi:10.1029/2008RG000260.

Ben-Zion, Y., and V. Lyakhovsky (2006), Analysis of aftershocks in a lithospheric model with seismogenic zone governed by damage rheology, *Geophys. J. Int.*, *165*, 197–210.

Ben-Zion, Y., and J. R. Rice (1993), Earthquake failure sequences along a cellular fault zone in a three-dimensional elastic solid containing asperity and nonasperity regions, *J. Geophys. Res.*, *98* (B8), 14,109–14,131.

Bielak, J., et al. (2010), The ShakeOut earthquake scenario: Verification of three simulation sets, *Geophys. J. Int.*, *180*, 375–404.

Bilham, R., K. Larson, J. Freymueller, and Project Idylhim members (1997), GPS measurements of present-day convergence across the Nepal Himalayas, *Nature*, *386*, 61–64.

Burridge, R., and L. Knopoff (1967), Model and theoretical seismicity, *Bull. Seismol. Soc. Am.*, *58*, 341–371.

Burroughs, S. M., and S. F. Tebbens (2002), The upper-truncated power law applied to earthquake cumulative frequency-magnitude distributions: Evidence for a time-independent scaling parameter, *Bull. Seismol. Soc. Am.*, *92*, 2983–2993.

Christensen, K., and Z. Olami (1992), Variation of the Gutenberg-Richter *b* values and nontrivial temporal correlations in a spring-block model for earthquakes, *J. Geophys. Res.*, *97*(B6), 8729–8735.

Cochard, A., and R. Madariaga (1996), Complexity of seismicity due to highly rate-dependent friction, *J. Geophys. Res.*, *101*(B11), 25,321–25,336.

Cui, Y., K. Olsen, A. Chourasia, R. Moore, P. Maechling, and T. H. Jordan (2008), The TeraShake computational platform for large-scale earthquake simulations, *Adv. Geocomput.*, *119*, 229–278.

Dieterich, J. H. (1972), Time-dependent friction as a possible mechanism for aftershocks, *J. Geophys. Res.*, *77*(20), 3771–3781.

Dieterich, J. H. (1994), A constitutive law for rate of earthquake production and its application to earthquake clustering, *J. Geophys. Res.*, *99*(B2), 2601–2618.

Duda, S. J. (1965), Secular seismic energy release in the circum Pacific belt, *Tectonophysics*, *2*, 409–452.

Dziewonski, A. M., and A. G. Prozorov (1985), Self-similar determination of earthquake clustering, in *Mathematical Modeling and Interpretation of Geophysical Data I, Comput. Seismol.*, vol. 16, edited by V. I. Keilis-Borok, pp. 7–16, Allerton, New York.

Eneva, M., and Y. Ben-Zion (1997), Techniques and parameters to analyze seismicity patterns associated with large earthquakes, *J. Geophys. Res.*, *102*(B8), 17,785–17,795.

Fillerup, M. A., J. H. Knapp, C. C. Knapp, and V. Raileanu (2010), Mantle earthquakes in the absence of subduction? Continental delamination in the Romanian Carpathians, *Lithosphere*, *2*, 333–340.

Fitzenz, D., and S. Miller (2001), A forward model for earthquake generation on interacting faults including tectonics, fluids, and stress transfer, *J. Geophys. Res.*, *106*(B11), 26,689–26,706.

Fuchs, K., et al. (1979), The Romanian earthquake of March 4, 1977. II. Aftershocks and migration of seismic activity, *Tectonophysics*, *53*, 225–247.

Fukao, Y., and M. Furumoto (1985), Hierarchy in earthquake size distribution, *Phys. Earth Planet. Inter.*, *37*, 149–168.

Gabrielov, A. M., and W. I. Newman (1994), Seismicity modeling and earthquake prediction: A review, in *Nonlinear Dynamics and Predictability of Geophysical Phenomena, Geophys. Monogr. Ser.*, vol. 83, edited by W. I. Newman, A. Gabrielov and D. L. Turcotte, pp. 7–13, AGU, Washington, D. C.

Gabrielov, A. M., T. A. Levshina, and I. M. Rotwain (1990), Block model of earthquake sequence, *Phys. Earth Planet. Inter.*, *61*, 18–28.

Gabrielov, A. M., W. I. Newman, and L. Knopoff (1994), Lattice model of failure: Sensitivity to the local dynamics, *Phys. Rev. E*, *50*, 188–197.

Gabrielov, A. M., V. I. Keilis-Borok, V. Pinsky, O. M. Podvigina, A. Shapira, and V. A. Zheligovsky (2007), Fluid migration and dynamics of a blocks-and-faults system, *Tectonophysics*, *429*, 229–251.

Girbacea, R., and W. Frisch (1998), Slab in the wrong place: Lower lithospheric mantle delamination in the last stage of the Eastern Carpathian subduction retreat, *Geology*, *26*, 611–614.

Gripp, A. E., and R. G. Gordon (1990), Current plate velocities relative to the hotspots incorporating the NUVEL-1 global plate motion model, *Geophys. Res. Lett.*, *17*(8), 1109–1112.

Gutenberg, B., and C. F. Richter (1994), Frequency of earthquakes in California, *Bull. Seismol. Soc. Am.*, *34*, 185–188.

Gvirtzman, Z. (2002), Partial detachment of a lithospheric root under the southeast Carpathians: Toward a better definition of the detachment concept, *Geology*, *30*, 51–54.

Hainzl, S., G. Zöller, and J. Kurths (1999), Similar power laws for foreshock and aftershock sequences in a spring-block model for earthquakes, *J. Geophys. Res.*, *104*(B4), 7243–7253.

Henderson, J., I. Main, P. Meredith, and P. Sammonds (1992), The evolution of seismicity at Parkfield: Observation, experiment and a fracture-mechanical interpretation, *J. Struct. Geol.*, *14*, 905–913.

Hirahara, K. (2002), Interplate earthquake fault slip during periodic earthquake cycles in a viscoelastic medium at a subduction zone, *Pure Appl. Geophys.*, *159*, 2201–2220.

Houseman, G., and P. England (1996), A lithospheric-thickening model for the Indo-Asian collision, in *The Tectonic Evolution of Asia*, edited by A. Yin and T. M. Harrison, pp. 3–17, Cambridge Univ. Press, New York.

Ishimoto, M., and K. Iida (1939), Observations on earthquakes registered with the microseismograph constructed recently, *Bull. Earthquake Res. Inst. Univ. Tokyo*, *17*, 443–478.

Ismail-Zadeh, A. T., V. I. Keilis-Borok, and A. A. Soloviev (1999), Numerical modelling of earthquake flows in the southeastern Carpathians (Vrancea): Effect of a sinking slab, *Phys. Earth Planet. Inter.*, *111*, 267–274.

Ismail-Zadeh, A. T., G. F. Panza, and B. M. Naimark (2000), Stress in the descending relic slab beneath the Vrancea region, Romania, *Pure Appl. Geophys.*, *157*, 111–130.

Ismail-Zadeh, A. T., B. Mueller, and G. Schubert (2005), Three-dimensional modeling of present-day tectonic stress beneath the earthquake-prone southeastern Carpathians based on integrated analysis of seismic, heat flow, and gravity observations, *Phys. Earth Planet. Inter.*, *149*, 81–98.

Ismail-Zadeh, A. T., J.-L. Le Mouël, A. Soloviev, P. Tapponnier, and I. Vorobieva (2007), Numerical modelling of crustal block-and-fault dynamics, earthquakes and slip rates in the Tibet-Himalayan region, *Earth Planet. Sci. Lett.*, *258*, 465–485.

Ismail-Zadeh, A. T., A. Aoudia, and G. F. Panza (2010), Three-dimensional numerical modeling of contemporary mantle flow and tectonic stress beneath the central Mediterranean, *Tectonophysics*, *482*, 226–236.

Ismail-Zadeh, A., L. Matenco, M. Radulian, S. Cloetingh, and G. Panza (2012), Geodynamics and intermediate-depth seismicity in Vrancea (the southeastern Carpathians): Current state-of-the-art, *Tectonophysics*, *530–531*, 50–79.

Kagan, Y., and L. Knopoff (1978), Statistical study of the occurrence of shallow earthquakes, *Geophys. J. R. Astron. Soc.*, *55*, 67–86.

Keilis-Borok, V. I. (1990), The lithosphere of the Earth as non-linear system with implications for earthquake prediction, *Rev. Geophys.*, *28*, 19–34.

Keilis-Borok, V. I., and V. G. Kossobokov (1990), Premonitory activation of earthquake flow: Algorithm M8, *Phys. Earth Planet. Inter.*, *61*, 73–83.

Keilis-Borok, V. I., and A. A. Soloviev (2003), *Nonlinear Dynamics of the Lithosphere and Earthquake Prediction*, Springer, Heidelberg, Germany.

Keilis-Borok, V. I., L. Knopoff, and I. M. Rotwain (1980), Bursts of aftershocks, long-term precursors of strong earthquakes, *Nature*, *283*, 258–263.

Keilis-Borok, V. I., I. M. Rotwain, and A. A. Soloviev (1997), Numerical modeling of block structure dynamics: Dependence of a synthetic earthquake flow on the structure separateness and boundary movements, *J. Seismol.*, *1*, 151–160.

Keilis-Borok, V., A. Ismail-Zadeh, V. Kossobokov, and P. Shebalin (2001), Non-linear dynamics of the lithosphere and intermediate-term earthquake prediction, *Tectonophysics*, *338*, 247–259.

King, G. C. P., R. S. Stein, and J. Lin (1994), Static stress changes and the triggering of earthquakes, *Bull. Seismol. Soc. Am.*, *84*, 935–953.

Knapp, J. H., C. C. Knapp, V. Raileanu, L. Matenco, V. Mocanu, and C. Dinu (2005), Crustal constraints on the origin of mantle seismicity in the Vrancea Zone, Romania: The case for active continental lithospheric delamination, *Tectonophysics*, *410*, 311–323.

Lennartz, S., A. Bunde, and D. L. Turcotte (2011), Modelling seismic catalogues by cascade models: Do we need long-term magnitude correlations?, *Geophys. J. Int.*, *84*, 1214–1222, doi:10.1111/j.1365-246X.2010.04902.x.

Le Pichon, X., M. Fournier, and L. Jolivet (1992), Kinematics, topography, shortening, and extrusion in the India-Eurasia collision, *Tectonics*, *11*(6), 1085–1098.

Lin, J., and R. S. Stein (2004), Stress triggering in thrust and subduction earthquakes and stress interaction between the southern San Andreas and nearby thrust and strike-slip faults, *J. Geophys. Res.*, *109*, B02303, doi:10.1029/2003JB002607.

Linzer, H.-G. (1996), Kinematics of retreating subduction along the Carpathian arc, Romania, *Geology*, *24*, 167–170.

Lyakhovsky, V., Y. Ben-Zion, and A. Agnon (2001), Earthquake cycle, fault zones, and seismicity patterns in a rheologically layered lithosphere, *J. Geophys. Res.*, *106*, 4103–4120, doi:10.1029/2000JB900218.

Main, I. G., P. G. Meredith, and C. Jones (1989), A reinterpretation of the precursory seismic b-value anomaly from fracture mechanics, *Geophys. J. Int.*, *96*, 131–138.

Main, I. G., P. G. Meredith, and P. R. Sammonds (1992), Temporal variations in seismic event rate and b-values from stress corrosion constitutive laws, *Tectonophysics*, *211*, 233–246.

Maksimov, V. I., and A. A. Soloviev (1999), Clustering of earthquakes in a block model of lithosphere dynamics, in *Selected Papers From Volumes 28 and 29 of Vychislitel'naya Seysmologiya, Comput. Seismol. Geodyn.*, vol. 4, edited by D. K. Chowdhury, pp. 124–126, AGU, Washington, D. C.

Martin, M., F. Wenzel, and the CALIXTO working group (2006), High-resolution teleseismic body wave tomography beneath SE-Romania—II. Imaging of a slab detachment scenario, *Geophys. J. Int.*, *164*, 579–595.

McKenzie, D. P. (1972), Active tectonics of the Mediterranean region, *Geophys. J. R. Astron. Soc.*, *30*, 109–185.

Molchan, G. M., and O. E. Dmitrieva (1992), Aftershock identification: Methods and new approaches, *Geophys. J. Int.*, *109*, 501–516.

Narkunskaya, G. S., and M. G. Shnirman (1990), Hierarchical model of defect development and seismicity, *Phys. Earth Planet. Inter.*, *61*, 29–35, doi:10.1016/0031-9201(90)90092-C.

Narteau, C., P. Shebalin, M. Holschneider, J. L. Le Mouël, and C. J. Allègre (2000), Direct simulations of the stress redistribution in the scaling organization of fracture tectonics (SOFT) model, *Geophys. J. Int.*, *141*, 115–135, doi:10.1046/j.1365-246X.2000.00063.x.

Nelson, K. D., et al. (1996), Partially molten middle crust beneath southern Tibet: Synthesis of project INDEPTH results, *Science*, *274*, 1684–1688.

Noda, H., and N. Lapusta (2010), Three-dimensional earthquake sequence simulations with evolving temperature and pore pressure due to shear heating: Effect of heterogeneous hydraulic diffusivity, *J. Geophys. Res.*, *115*, B12314, doi:10.1029/2010JB007780.

Ogata, Y. (1988), Statistical models for earthquake occurrences and residual analysis for point processes, *J. Am. Stat. Assoc.*, *83*, 9–27.

Ogata, Y., and K. Katsura (1993), Analysis of temporal and spatial heterogeneity of magnitude frequency distribution inferred from earthquake catalogues, *Geophys. J. Int.*, *113*, 727–738.

Oncescu, M. C., and K. P. Bonjer (1997), A note on the depth recurrence and strain release of large Vrancea earthquakes, *Tectonophysics*, *272*, 291–302.

Pana, D., and P. Erdmer (1996), Kinematics of retreating subduction along the Carpathian arc, Romania: Comment, *Geology*, *24*, 862–863.

Pana, D., and G. A. Morris (1999), Slab in the wrong place: Lower lithospheric mantle delamination in the last stage of the eastern Carpathian subduction retreat: Comment, *Geology*, *27*, 665–666.

Panza, G. F., A. A. Soloviev, and I. A. Vorobieva (1997), Numerical modeling of block-structure dynamics: Application to the Vrancea region, *Pure Appl. Geophys.*, *149*, 313–336.

Pollitz, F. F. (2009), A viscoelastic earthquake simulator with application to the San Francisco bay region, *Bull. Seismol. Soc. Am.*, *99*, 1760–1785.

Pollitz, F. F., and I. S. Sacks (2002), Stress triggering of the 1999 Hector Mine earthquake by transient deformation following the 1992 Landers earthquake, *Bull. Seismol. Soc. Am.*, *92*, 1487–1496.

Press, F., and C. Allen (1995), Patterns of seismic release in the southern California region, *J. Geophys. Res.*, *100*, 6421–6430.

Prozorov, A. G. (1994a), An earthquake prediction algorithm for the Pamir and Tien Shan region based on a combination of long-range aftershocks and quiescent periods, in *Selected Papers From Volumes 22 and 23 of Vychislitel'naya Seysmologiya, Comput. Seismol. Geodyn.*, vol. 1, edited by D. K. Chowdhury, pp. 31–35, AGU, Washington, D. C.

Prozorov, A. G. (1994b), A new statistics to test the significance of long-range interaction between large earthquakes, in Seismicity and Related Processes in the Environment, vol. 1, edited by V. I. Borok, pp. 38–43, Res. and Coord. Cent. for Seismol. and Eng., Moscow.

Radu, C. (1991), Strong earthquakes occurred on the Romanian territory in the period 1901–1990 [in Romanian], *Vitralii*, *3*, 12–13.

Raykova, R. B., and G. F. Panza (2006), Surface waves tomography and non-linear inversion in the southeast Carpathians, *Phys. Earth Planet. Inter.*, *157*, 164–180.

Replumaz, A., and P. Tapponnier (2003), Reconstruction of the deformed collision zone Between India and Asia by backward motion of lithospheric blocks, *J. Geophys. Res.*, *108*(B6), 2285, doi:10.1029/2001JB000661.

Rice, J. R., and Y. Ben-Zion (1996), Slip complexity in earthquake fault models, *Proc. Natl. Acad. Sci. U. S. A.*, *93*, 3811–3818.

Robinson, R., and R. Benites (1996), Synthetic seismicity models for the Wellington Region, New Zealand: Implications for the temporal distribution of large events, *J. Geophys. Res.*, *101*(B12), 27,833–27,844.

Robinson, R., and R. Benites (2001), Upgrading a synthetic seismicity model for more realistic fault ruptures, *Geophys. Res. Lett.*, *28*(9), 1843–1846.

Rotwain, I., V. Keilis-Borok, and L. Botvina (1997), Premonitory transformation of steel fracturing and seismicity, *Phys. Earth Planet. Inter.*, *101*, 61–71.

Royden, L. H., B. C. Burchfiel, R. W. King, E. Wang, Z. Chen, F. Shen, and Y. Liu (1997), Surface deformation and lower crustal flow in eastern Tibet, *Science*, *276*, 788–790.

Rozenberg, V. L., P. O. Sobolev, A. A. Soloviev, and L. A. Melnikova (2005), The spherical block model: Dynamics of the global system of tectonic plates and seismicity, *Pure Appl. Geophys.*, *162*(1), 145–164, doi:10.1007/s00024-004-2584-4.

Rozenberg, V., A. Ismail-Zadeh, L. Melnikova, and A. Soloviev (2010), A spherical model of lithospheric dynamics and seismicity, *Geophys. Res. Abstr.*, *12*, EGU2010-6485.

Rundle, J. B., and W. Klein (1993), Scaling and critical phenomena in a cellular automaton slider block model for earthquakes, *J. Stat. Phys.*, *72*, 405–412.

Rundle, P. B., J. B. Rundle, K. F. Tiampo, A. Donnellan, and D. L. Turcotte (2006), Virtual California: Fault model, frictional parameters, application, *Pure Appl. Geophys.*, *163*, 1819–1846.

Schmittbuhl, J., J.-P. Vilotte, and S. Roux (1996), A dissipation-based analysis of an earthquake fault model, *J. Geophys. Res.*, *101*(B12), 27,741–27,764, doi:10.1029/96JB02294.

Shebalin, P., A. Soloviev, and J.-L. LeMouël (2002), Scaling organization in the dynamics of blocks-and-faults systems, *Phys. Earth Planet. Inter.*, *131*, 141–153.

Smith, W. D. (1986), Evidence for precursory changes in the frequency-magnitude b-value, *Geophys. J. Int.*, *86*, 815–838.

Soloviev, A. A. (2008), Transformation of frequency-magnitude relation prior to large events in the model of block structure dynamics, *Nonlin. Processes Geophys.*, *15*, 209–220.

Soloviev, A. A., and A. T. Ismail-Zadeh (2003), Models of dynamics of block-and-fault systems, in *Nonlinear Dynamics of the Lithosphere and Earthquake Prediction*, edited by V. I. Keilis-Borok and A. A. Soloviev, pp. 69–138, Springer, Heidelberg, Germany.

Soloviev, A. A., I. A. Vorobieva, and G. F. Panza (1999), Modeling of block-structure dynamics: Parametric study for Vrancea, *Pure Appl. Geophys.*, *156*, 395–420.

Soloviev, A. A., I. A. Vorobieva, and G. F. Panza (2000), Modelling of block structure dynamics for the Vrancea region: Source mechanisms of the synthetic earthquakes, *Pure Appl. Geophys.*, *157*, 97–110.

Sornette, D., and C. G. Sammis (1995), Complex critical exponents from renormalization group theory of earthquakes: Implication for earthquake prediction, *J. Phys. I*, *5*, 607–619.

Sperner, B., F. Lorenz, K. Bonjer, S. Hettel, B. Müller, and F. Wenzel (2001), Slab break-off—Abrupt cut or gradual detachment? New insights from the Vrancea region (SE Carpathians, Romania), *Terra Nova*, *13*, 172–179.

Stein, R. S. (1999), The role of stress transfer in earthquake occurrence, *Nature*, *402*, 605–609.

Tapponnier, P., X. Zhiqin, F. Roger, B. Meyer, N. Arnaud, G. Wittlinger, and Y. Jingsui (2001), Oblique stepwise rise and growth of the Tibet Plateau, *Science*, *294*, 1671–1677.

Turcotte, D. L. (1999), Seismicity and self-organized criticality, *Phys. Earth Planet. Inter.*, *111*, 275–294.

Turcotte, D. L., J. R. Holliday, and J. B. Rundle (2007), BASS, an alternative to ETAS, *Geophys. Res. Lett.*, *34*, L12303, doi:10.1029/2007GL029696.

Utsu, T., and A. Seki (1954), A relation between the area of aftershock region and the energy of main shock, *J. Seismol. Soc. Jpn.*, *7*, 233–240.

Van Aalsburg, J., L. B. Grant, G. Yakovlev, P. B. Rundle, J. B. Rundle, D. L. Turcotte, A. Donnellan (2007), A feasibility study of data assimilation in numerical simulations of earthquake fault systems, *Phys. Earth Planet. Inter.*, *163*, 149–162.

Vere-Jones, D., and J. Zhuang (2008), Distribution of the largest event in the critical epidemic type aftershock-sequence model, *Phys. Rev. E*, *78*, 047102, doi:10.1103/PhysRevE.78.047102.

Vergnes, J., G. Wittlinger, Q. Hui, P. Tapponnier, G. Poupinet, J. Mei, G. Herquel, and A. Paul (2002), Seismic evidence for stepwise thickening of the crust across the N-E Tibetan plateau, *Earth Planet. Sci. Lett.*, *203*, 25–33.

Vorobieva, I. A., and A. A. Soloviev (2005), Long-range interaction between synthetic earthquakes in the block model of lithosphere dynamics, in *Selected Papers From Volume 32 of Vychislitel'naya Seysmologiya, Comp. Seismol. Geodyn.*, vol. 7, edited by D. K. Chowdhury, pp. 170–177, AGU, Washington, D. C.

Wang, R., X. Sun, and Y. Cai (1983), A mathematical simulation of earthquake sequence in north China in the last 700 years, *Sci. Sinica*, *26*, 103–112.

Ward, S. N. (1992), An application of synthetic seismicity in earthquake statistics: The Middle America Trench, *J. Geophys. Res.*, *97*, 6675–6682.

Ward, S. N. (1996), A synthetic seismicity model for southern California: Cycles, probabilities, hazards, *J. Geophys. Res.*, *101*(B10), 22,393–22,418.

Ward, S. N. (2000), San Francisco Bay Area earthquake simulation: A step toward a standard physical earthquake model, *Bull. Seismol. Soc. Am.*, *90*, 370–386.

Wells, D. L., and K. J. Coppersmith (1994), New empirical relationships among magnitude, rupture length, rupture width, rupture area, and surface displacement, *Bull. Seismol. Soc. Am.*, *84*, 974–1002.

Werner, M. J., K. Ide, and D. Sornette (2011), Earthquake forecasting based on data assimilation: Sequential Monte Carlo methods for renewal point processes, *Nonlin. Processes Geophys.*, *18*, 49–70, doi:10.5194/npg-18-49-2011.

Widiyantoro, S., and R. van der Hilst (1996), Structure and evolution of lithospheric slab beneath the Sunda arc, Indonesia, *Science*, *271*, 1566–1570.

Wiemer, S., and M. Wyss (1997), Mapping the frequency-magnitude distribution in asperities: An improved technique to calculate recurrence times?, *J. Geophys. Res.*, *102*(B7), 15,115–15,128.

Wittlinger, G., J. Vergne, and P. Tapponnier (2004), Teleseismic imaging of subducting lithosphere and Moho offsets beneath western Tibet, *Earth Planet. Sci. Lett.*, *221*, 117–130.

Wortel, M. J. R., and W. Spakman (2000), Subduction and slab detachment in the Mediterranean-Carpathian region, *Science*, *290*, 1910–1917.

Xing, H. L., A. Makinouchi, and P. Mora (2007), Finite element modeling of interacting fault systems, *Phys. Earth Planet. Inter.*, *163*, 106–121.

Zaliapin, I., V. Keilis-Borok, and M. Ghil (2003), A Boolean delay model of colliding cascades. II: Prediction of critical transitions, *J. Stat. Phys.*, *111*, 839–861.

Zhou, S., S. Johnston, R. Robinson, and D. Vere-Jones (2006), Tests of the precursory accelerating moment release model using a synthetic seismicity model for Wellington, New Zealand, *J. Geophys. Res.*, *111*, B05308, doi:10.1029/2005JB003720.

Zöller, G., and S. Hainzl (2007), Recurrence time distributions of large earthquakes in a stochastic model for coupled fault systems: The role of fault interaction, *Bull. Seismol. Soc. Am.*, *97*, 1679–1697.

Zöller, G., M. Holschneider, and Y. Ben-Zion (2004), Quasi-static and quasi-dynamic modeling of earthquake failure at intermediate scales, *Pure Appl. Geophys.*, *161*, 2103–2118.

Zöller, G., S. Hainzl, M. Holschneider, and Y. Ben-Zion (2005), Aftershocks resulting from creeping sections in a heterogeneous fault, *Geophys. Res. Lett.*, *32*, L03308, doi:10.1029/2004GL021871.

Zöller, G., S. Hainzl, Y. Ben-Zion, and M. Holschneider (2006), Earthquake activity related to seismic cycles in a model for a heterogeneous strike-slip fault, *Tectonophysics*, *423*, 137–145.

A. T. Ismail-Zadeh, Geophysikalisches Institut, Karlsruher Institut für Technologie, Hertzstr. 16, Karlsruhe D-76187, Germany. (Alik.Ismail-Zadeh@kit.edu)

J.-L. Le Mouël, Institut de Physique du Globe de Paris, 1 rue Jussieu, Paris F-75252, France. (lemouel@ipgp.fr)

A. A. Soloviev, International Institute of Earthquake Prediction Theory and Mathematical Geophysics, Russian Academy of Sciences, Profsoyuznaya str. 84/32, Moscow 117997, Russia. (soloviev@mitp.ru)

Investigation of Past and Future Polar Low Frequency in the North Atlantic

Matthias Zahn

Environmental Systems Science Centre, University of Reading, Reading, UK

Hans von Storch

Institute of Coastal Research, Helmholtz-Zentrum Geesthacht, Geesthacht, Germany
Meteorological Institute, University of Hamburg, Hamburg, Germany

Polar lows are a particular kind of extreme weather event. Polar lows are subsynoptic-scale vigorous cyclones in subpolar maritime region. They are associated with strong winds and severe weather and when developing rapidly can constitute a major threat to human offshore activities. Here we present methods and results of our previous research to investigate long-term frequency changes of polar low occurrences over the North Atlantic in the past and in a projected, anthropogenically warmed future atmosphere. For the past, we downscaled reanalysis data to reproduce, detect, and count polar lows, and for the future, we downscaled global model data from various Intergovernmental Panel on Climate Change scenarios. We found no systematic change of polar low frequency in the past 60 years but a significant decrease of down to half as many cases for the simulated future. We relate this decrease to an increase of vertical atmospheric stability, which is the result of more quickly rising air temperatures compared to the sea surface. The results have recently been published in a series of articles.

1. INTRODUCTION

Apart from the well-known and weather-dominating synoptic-scale depressions, a number of small- to medium-scale (or mesoscale) cyclones, termed mesocyclones, also develop in subpolar maritime waters every year, mainly during the winter seasons [e.g., *Harold et al.*, 1999a, 1999b]. Here the term small to medium scale refers to diameters of up to 1000 km. The intensity spectrum of those mesocyclones ranges from weak swirls via moderate breeze cyclones up to above gale force storms. The extreme events at the upper end of this distribution are commonly referred to as Arctic hurricanes or as polar lows and are treated in this contribution.

General characteristics of polar lows include severe weather, such as strong wind speeds and heavy precipitation as rain, hail, or snow, and their often rapid development within cold polar air masses. Polar lows usually are initiated as disturbances in low-level air, of, e.g., orographic or baroclinic origin [*Harrold and Browning*, 1969; *Klein and Heineman*, 2002; *Rasmussen and Turner*, 2003]. Later during their development, they amplify by convective processes under the influence of large vertical temperature gradients and latent and sensible heat fluxes [*Rasmussen*, 1979; *Emanuel and Rotunno*, 1989; *Rasmussen and Turner*, 2003]. Associated oceanic energy loss elicited by these latent and sensible heat fluxes may influence Greenland Deep Water formation and has been found to influence deep ocean circulation [*Condron et al.*, 2008].

Extreme Events and Natural Hazards: The Complexity Perspective
Geophysical Monograph Series 196

10.1029/2011GM001091

The variety of mechanisms involved in their development makes a characterization difficult, and previously, different characteristics for what a polar low exactly is have been used by different scientists. In our studies, we have used a general definition given by *Rasmussen and Turner* [2003, p. 12], who state:

> A polar low is a small, but fairly intense maritime cyclone that forms poleward of the main baroclinic zone (the polar front or other major baroclinic zone). The horizontal scale of the polar low is approximately between 200 and 1000 kilometers and surface winds near or above gale force.

We will, in the following, describe our recent efforts to investigate long-term changes in the frequency of polar lows, in the past and in an assumed anthropogenically warmed future atmosphere. In section 2, we will explain the methods and techniques developed for our investigations before we present our results in section 3. We published our results in a series of papers [*Zahn et al.*, 2008; *Zahn and von Storch*, 2008a, 2008b, 2010].

This methodology also has a high potential for investigations in other areas. For example, a similar methodology has been developed and tested for the case of East Asian typhoons [*Feser and von Storch*, 2008a, 2008b; *von Storch et al.*, 2010]. The methodology is presently transferred to the analysis and projection of North Pacific polar low statistics and that of medicanes, vigorous below synoptic-scale cyclones in the Mediterranean (The word medicane is a hybrid of Mediterranean and hurricane).

The potential of inferring properties of subsynoptic features, such as polar lows, is a significant added value provided by regional climate models [*Feser et al.*, 2011].

2. DATA AND METHODS DEVELOPED

A number of former and recent studies have addressed climatological issues of polar lows in the North Atlantic by making use of observation data, from ground measurements as well as from satellite data [*Wilhelmsen*, 1985; *Ese et al.*, 1988; *Noer and Ovhed*, 2003; *Blechschmidt*, 2008; *Noer et al.*, 2011]. While those studies are capable of identifying individual cases, preferred regions, or preferred synoptic-scale situations of polar low developments, they are lacking any statement on long-term frequency changes. The periods for which these data are available are too short, and intercomparison between former and recent studies is not possible due to lacking homogeneity. However, homogeneous data in space and time spanning a sufficiently long period are a prerequisite for studies on past long-term changes of the properties of atmospheric features with a certain spatial extension. In atmospheric sciences, this period should usually cover at least a couple of decades. To date, these requirements are only fulfilled by reanalysis, which assimilate a variety of existing observational data into a global model system, resulting in global, but coarsely resolved, fields of a number of atmospheric parameters.

The resolution of global reanalysis data is usually at about 2°, corresponding to about 200 km in the North Atlantic. A proper homogeneous representation of smaller-scale features, such as polar lows, is not possible at this resolution because of insufficient spatial resolution. Anyhow, a number of studies have used reanalysis data to investigate climatological properties of polar lows, but instead of identifying individual cases, they identified particular atmospheric environments favoring polar low development as a proxy [*Kolstad*, 2006; *Kolstad et al.*, 2008, *Bracegirdle and Gray*, 2008; *Kolstad*, 2011]. An advantage of such studies is that they are objective; that is, the selection is not influenced by any subjectivity of the author, and resulting climatologies are thus reproducible. Further, the methods developed can easily be transferred to other data sharing similar technical properties as the past reanalysis data, e.g., model data for global future projections from the Intergovernmental Panel on Climate Change (IPCC) to investigate possible future polar low frequency [*Vavrus et al.*, 2006; *Kolstad and Bracegirdle*, 2008].

Although those proxy-based investigations use homogeneous data, they are not able to resolve and detect individual cases. We therefore have, in the past years, undertaken steps to reproduce individual polar low cases over the past six decades. To do so, we have used National Centers for Environmental Prediction (NCEP)/National Center for Atmospheric Research (NCAR) reanalysis [*Kalnay et al.*, 1996] data at a resolution of about 2°, which we dynamically downscaled over the entire North Atlantic to a grid of ≈0.5° by means of the regional climate model (RegCM) Climate Limited-area Modelling (CLM) [*Rockel et al.*, 2008]. Dynamical downscaling is a method to postprocess coarse gridded data with a finer-scale model to achieve more highly resolved data. Later, in a subsequent study, we also downscaled IPCC global projections of the state of the atmosphere to study possible changes of polar low frequency in a presumed warmed atmosphere. For all downscaling efforts throughout this article, we applied the spectral nudging method of *von Storch et al.* [2000]. In this method, large-scale information of the driving data are nudged into the RegCM simulation above the 850 hPa level to prevent the RegCM deviating from the given and presumably realistic large scale of the forcing data too much.

Counting polar lows manually in several decades of data is infeasible. We thus developed an automated procedure to detect and track polar low cases in the gigabytes of output data from the downscaling experiments. In the following subsection, we here first introduce the benefits of scale separation. At the same time, it is shown that by means of dynamically downscaling reanalysis data, it is indeed possible to reproduce polar low cases. Based on this information, the automated detection procedure is described afterward. Finally, details of the long-term simulations are presented.

2.1. Reproducing an Exemplary Polar Low Case and Its Separation From the Large-Scale Field

A famous polar low case developed in October 1993 at the western side of a synoptic low located north of Scandinavia, which shifted cold air from over the Arctic ice cap over the open ocean. After its initial development on 14 October 1993, it moved southward over the Norwegian Sea until it decayed after hitting the Norwegian mainland in the area around Bergen on 16 Oct. This polar low has been described in several studies [*Nielsen*, 1997; *Claud et al.*, 2004; *Bracegirdle and Gray*, 2009].

We have undertaken a case study to simulate this polar low with CLM in its version 2.4.6, which was the most up-to-date version at the beginning of our studies. We want to know whether we can expect polar lows to be reproduced by long-term downscaling efforts, i.e., independent from the initial conditions. Thus, we have run CLM in climate mode for our case study, which means that the simulation is initialized 2 weeks prior to the actual polar low formation. The result is shown in Figure 1 for 06:00 UTC, 15 October 1993, when the polar low is fully developed in its mature stage. The atmospheric situation at the surface, as given by analyses from the German weather service, Deutscher Wetterdienst (DWD), is shown in Figure 1a. The synoptic low mentioned before is located north of Scandinavia, and the polar low is visible with closed isobars in the mean sea level pressure (MSLP) field off the Norwegian coast, with a core pressure below 995 hPa and adjacent wind speeds above the defined threshold of 13.9 m s^{-1}. The synoptic situation is also included in the NCEP/NCAR data in Figure 1b. However, the polar low is only included insufficiently as a weak surface trough. This trough is considerably deepened after downscaling with CLM (Figure 1c). Although the core pressure is about 10 hPa too high and the location is some hundred kilometers shifted, a small-scale low does develop.

As a first step toward an automated detection procedure, we applied a near-isotropic 2-D digital band-pass filter of *Feser and von Storch* [2005] to the CLM output fields to separate the mesoscale polar low information from the full MSLP fields. The band-pass filter was configured to retain information on scales between 230 and 450 km. Technically, we constructed a digital filter with a footprint of 21 × 21 grid points and the 2-D wavenumber k^*, which we approximated with the response function $\kappa(k^*) = 0$ for all $k^* \leq 6$, $\kappa(k^*) = 1$ for all $k^* \in [8,15]$, and $\kappa(k^*) = 0$ for all $k^* \geq 18$.

The results of applying such a filter to the MSLP fields shown before are visible in Figures 1d–1f. In the filtered DWD field, there is a distinct signature at the length scales of the polar low evident as a local minimum. Also in the filtered MSLP of the CLM simulation, the polar low becomes distinct at the right position (indicated by the dot). In the reanalysis, however, where the surface signature has been weak, no sign of any medium-scale disturbance is visible.

We conclude at this point, that it is principally possible to reproduce polar lows with a RegCM run in climate mode, which considerably deepens the polar low signature compared to the coarse NCEP/NCAR driving data. However, the dynamical details associated to the polar low may not be simulated accurately. That these results also hold for various ensemble members for this case and also for further cases is presented by *Zahn et al.* [2008].

2.2. Automated Detection of Polar Lows in Long-Term RegCM Data

To investigate long-term changes in polar low frequency, first of all, a database of polar low cases is required. We compiled such a database, and to do so automatically detected polar lows in long-term downscaling results at three hourly temporal resolutions. Our tracking algorithm is based on the above-mentioned digital band-pass filter. At first, the mesoscale information of all the three-hourly MSLP output fields are separated. All maritime positions at which the band-pass-filtered MSLP in the simulation output fields are lower than the ad hoc value of −1 hPa and at the same time constitutes a local minimum that is recorded together with information of adjacent atmospheric parameters such as wind speed, sea surface temperature (SST), or temperature at the 500 hPa vertical level. In a second step, all these positions are merged to tracks based on the constraint that the distance between two consecutive positions must not exceed 200 km. Finally, five further constraints are requested along the tracks, accounting for dynamical characteristics of polar lows as well as being technically founded: (1) the band-pass-filtered MSLP must be below the ad hoc value of −2 hPa once along the track, (2) a 10 m wind speed must exceed the threshold of 13.9 m s^{-1} at least 20% of the positions along the track, (3) the vertical temperature difference between the SST and the 500 hPa (T500) must exceed 43°C at least once along the track, (4) the first detected position has to be about 100 km (1°) farther north than the last, and (5) noncoastal grid boxes in more than 50% of the positions are required along the track.

If the last criterion is valid and the band-pass-filtered MSLP in the simulation output fields decreases below

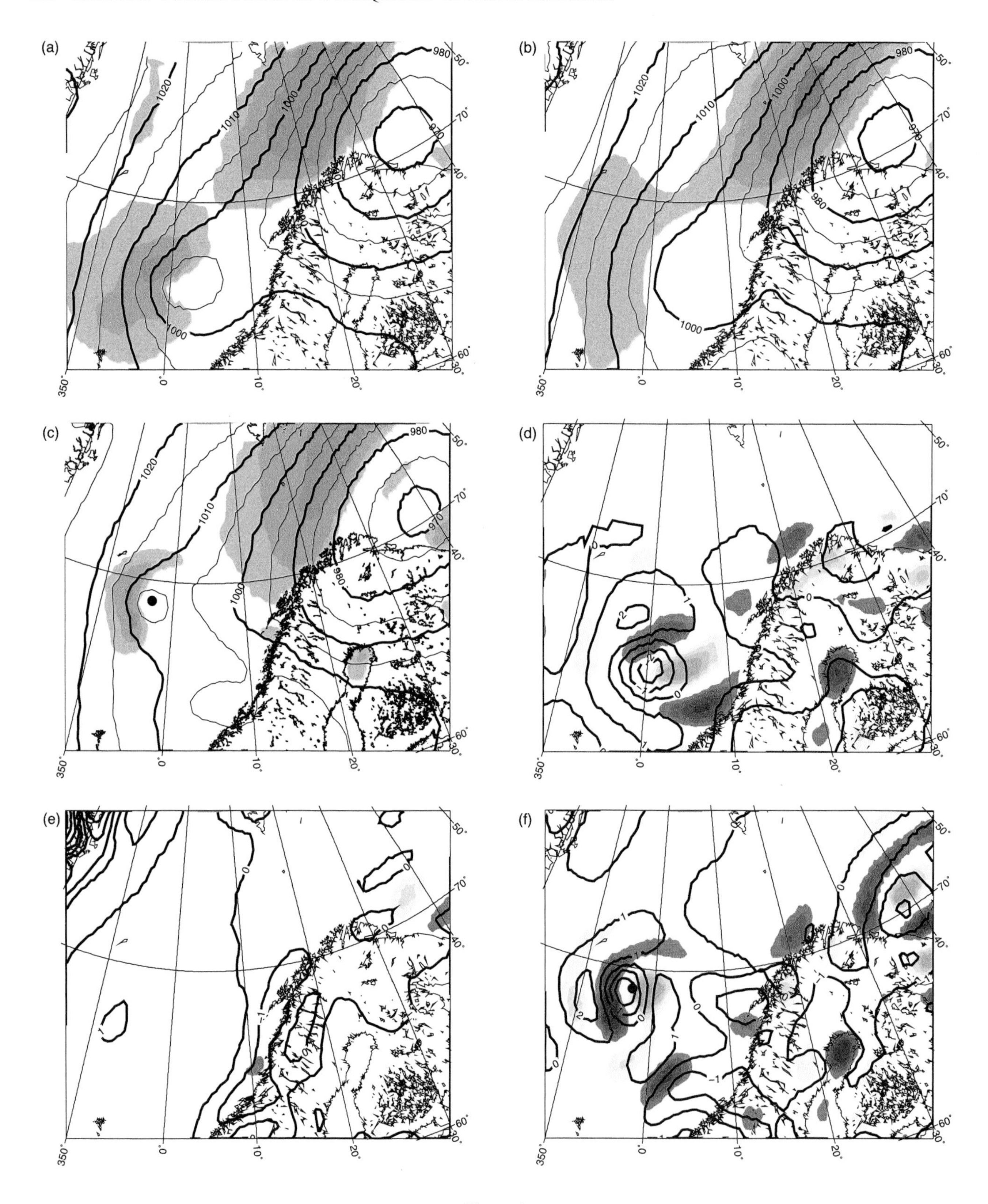

Figure 1

−6 hPa, an exceptional disturbance at the mesoscale is assumed, and all the other criteria are overridden. A more detailed description and justification of these criteria are given by *Zahn and von Storch* [2008a]. They also demonstrated that such an algorithm indeed is capable of detecting polar lows in 2 years lasting CLM simulations.

2.3. Long-Term Simulations Undertaken for Our Study

After concluding in the last two sections that polar lows can be reproduced by means of dynamical downscaling and tracked in the output data, we here describe the setup of the long-term simulations applied in our work, for the past and for future projections.

In our first long-term study, we investigated polar low frequency in the past and gained results as will be described in section 3.1. We used the NCEP/NCAR1 reanalysis [*Kalnay et al.*, 1996] over the period 1 January 1948 through 28 February 2006. Again version 2.4.6 of CLM [*Rockel et al.*, 2008] was used. To maintain consistency, we did not switch the version throughout our studies. We defined a rotated geographical grid with a rotated north pole at 175°E and 21.3°N at a resolution of about 50 km, spanning a rectangle with side lengths of 8987 × 3516 km covering the entire maritime North Atlantic, including regions north of the ice edge in the North, covering land regions of Canada and Russia in the West and East, and reaching as far south as Southern Great Britain to also cover the approximate location of the Polar Front (see also Figure 1 in the work of *Zahn et al.* [2008]). Spectral nudging [*von Storch et al.*, 2000] was applied, and calculations were undertaken at a time step of 240 s. In the following, these simulations are referred to as REA, according to reanalysis based.

For the future projections of polar low frequency as presented in section 3.3, we also downscaled global atmospheric data with the same approach as for REA. However, we excluded parts of the Russian land area to save calculation time and used a somewhat smaller grid (cf. Supplementary Figure 1 in the work of *Zahn and von Storch* [2010]). Again, CLM2.4.6 was used with spectral nudging [*von Storch et al.*, 2000]. As driving data, two periods of transient simulations of the Fourth Assessment Report (AR4) of the IPCC were used. The compilation of AR4 of IPCC global climate models originating from various institutes from all over the world are forced with prescribed Greenhouse Gas (GHG) concentrations. For a possible future, such GHG concentrations are derived from different considerations on how human population and economy may evolve during the twenty-first century. We investigated three different scenarios, the so-called B1, A1B, and A2 scenarios as described in the IPCC—Special Report on Emission Scenarios [*Nakicenovic and Swart*, 2000]. In scenario B1, GHG concentrations and thus atmospheric temperatures increase relatively moderately. In both scenarios A1B and A2, GHG concentrations and temperature increase more rapidly until about 2070. In A1B, this increase is followed by a decline of GHG emissions leading to a reduced increase of temperature, whereas in A2, GHG concentrations and temperature continue to rise quickly.

We used the downscaled realizations of these three scenarios as given by the European Centre/Hamburg 5 (ECHAM5)/Max-Planck-Institut ocean model (MPI-OM) [*Roeckner et al.*, 2003; *Marsland et al.*, 2003] at the end of the twenty-first century, from 2070 until 2099. Global models suffer from a number of systematic differences (biases) to observations and analysis [*Meehl et al.*, 2007]. Therefore, a direct comparison of the scenarios with REA makes little sense. Instead, we used the realization of another scenario applying twentieth century GHG concentrations from 1960 to 1989 calculated by the same model, ECHAM5/MPI-OM. We also downscaled this scenario to use the results as a reference, called C20 in the following. We assume that by using the same model, and thus the same physics and parametrization schemes in the scenario and in the reference simulation, model biases will cancel out when comparing the two.

3. RESULTS

Northern Hemisphere polar lows are phenomena of the cold season. In none of the simulations, neither for the past nor for the future, was any polar low found in the summer months, June and July. To avoid splitting into 2 years the polar lows of the same winter, we aggregate the detected polar lows of each winter and for further discussion invent the term polar low season (PLS). A PLS denotes the period from July to June of the following year, and a given PLS is referred to by the second year.

Figure 1. (opposite) Full mean sea level pressure (hPa) and 10 m wind speed fields (above 13.9 m s^{-1}) from (a) Deutscher Wetterdienst (DWD) analysis data, (b) National Centers for Environmental Prediction (NCEP)/National Center for Atmospheric Research (NCAR), and (c) Community Land Model (CLM) simulations at 06:00 UTC 15 October 1993. Corresponding band-pass-filtered mean sea level air pressure (hPa) and 10 m wind speed from (d) DWD analysis data, (e) NCEP/NCAR, and (f) CLM simulations. From *Zahn et al.* [2008].

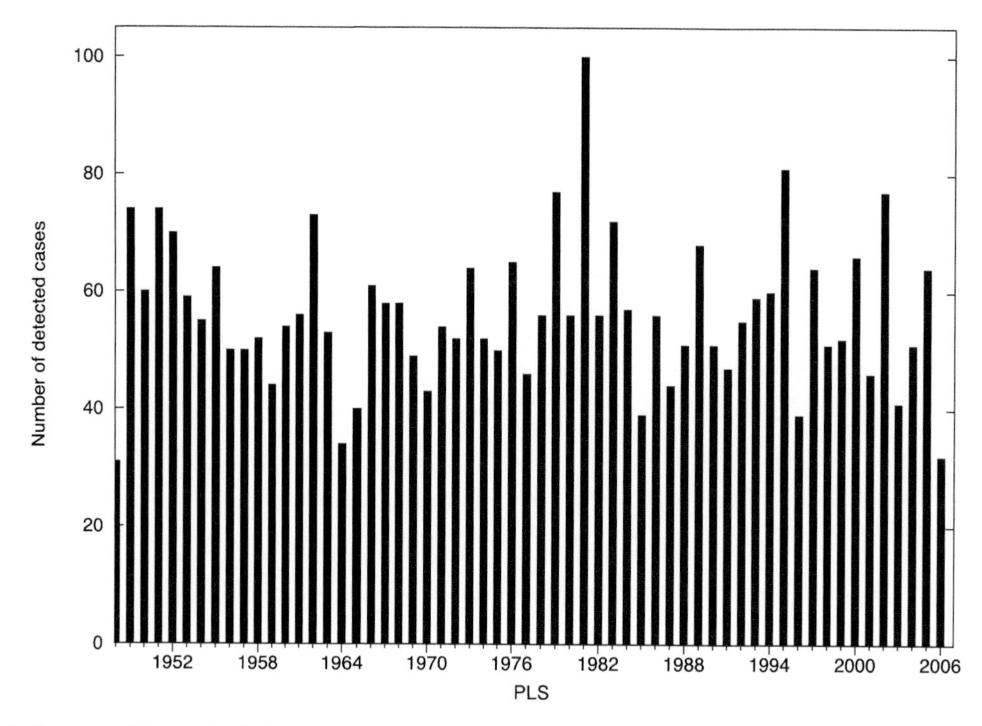

Figure 2. Number of detected polar lows per polar low season (PLS). One PLS is defined as the period starting 1 July and ending 30 June the following year. From *Zahn and von Storch* [2008b].

3.1. Changes of North Atlantic Polar Low Frequency Over the Past Six Decades

From 1948 until 2006, we found an average number of 56 polar lows per PLS (Figure 2) over the entire North Atlantic. A maximum number of 100 cases show up in PLS 1981, a minimum number of 36 cases is found in PLS 1964. We do not consider PLS 1948 and 2006 here, as they both do not consist of the full number of months. There is high year to year variability evident in Figure 2, reflected in a standard deviation of $\sigma \approx 13$, but the yearly average number remains on a constant level over decadal time scales. Accordingly, we

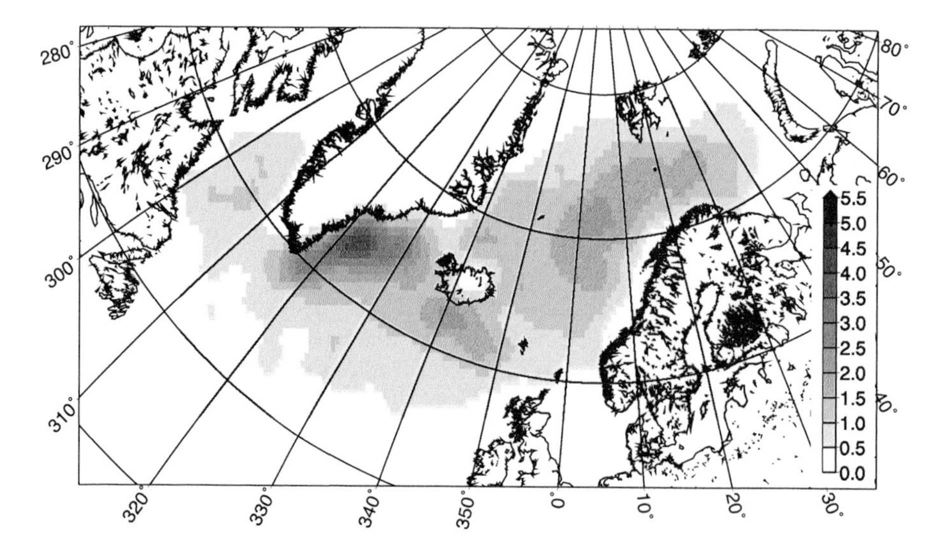

Figure 3. Density distribution of polar low genesis. Unit is number of detected polar low cases per 2500 km^2. From *Zahn and von Storch* [2008b].

could not detect any trend for the number of polar lows per PLS. A least squares fit we applied to the data was not statistically significant, and its sign was highly dependent on the first and last year used.

The thresholds chosen for the algorithm can be modified. To check whether the findings of no trend and high year to year variability are results of the respective threshold chosen, we varied the thresholds for the dynamical criteria and used different criteria for the wind speed and for the vertical temperature gradient. We found a strong effect on the number of detected cases per PLS, but the statistical properties remained the same. The correlations between resulting time series were high.

We conclude that the interannual variability of the frequency of polar lows is high, but there has not been any significant change over the past decades. We published these results in the work of *Zahn and von Storch* [2008b].

3.2. Comparison With Other Studies and Observations

Drawing conclusions for the "real" world from simulated data may give rise to the question of the validity of the results. Checking the validity is usually done by comparing simulated results to available observations. For the case of polar lows, this comparison is restricted by the scarce availability of databases. We here compare the spatial genesis density of polar lows over the North Atlantic as gained from our results with what is known from previous studies (which all considered shorter periods of time) and compare temporal frequency in restricted areas of the North Atlantic with observations mentioned in section 2.

A hot spot of polar low development is the Norwegian Sea off the coast of Norway. Here a warm tongue of the Gulf Stream extends far into the North Atlantic. Frequent outbreaks of cold air can lead to large vertical temperature gradients in this area, destabilizing the air, favoring convection, and providing good environments for polar low development. Also, the regions between Greenland and Iceland are prone to polar low development, as, e.g., detected by *Bracegirdle and Gray* [2008, Figure 5]. Here in the vicinity of high mountain ranges along the coast of Greenland, orographic influences may play a significant role in polar low development. In a recent study, the Labrador Sea has also been found to frequently provide favorable conditions for polar low genesis [*Kolstad*, 2011].

The density distribution gained from the polar lows in our database is shown in Figure 3. Consistent with the main genesis regions described before, we find peaks in polar low intensity between Greenland and Iceland and in the Norwegian Sea. Also, high genesis frequency detected south of Iceland has counterparts as polar lows by *Bracegirdle and Gray* [2008] or as medium sized cyclones by *Condron et al.* [2006]. A weaker peak is found in the Labrador Sea, at the western side of Greenland.

The observational databases are confined to particular areas or are available for very short periods only. We thus compare the number of cases per PLS or per month only. For the comparison, we only used cases of our database that developed in the same areas that were observed. Also, different wind speed thresholds were used in the other studies,

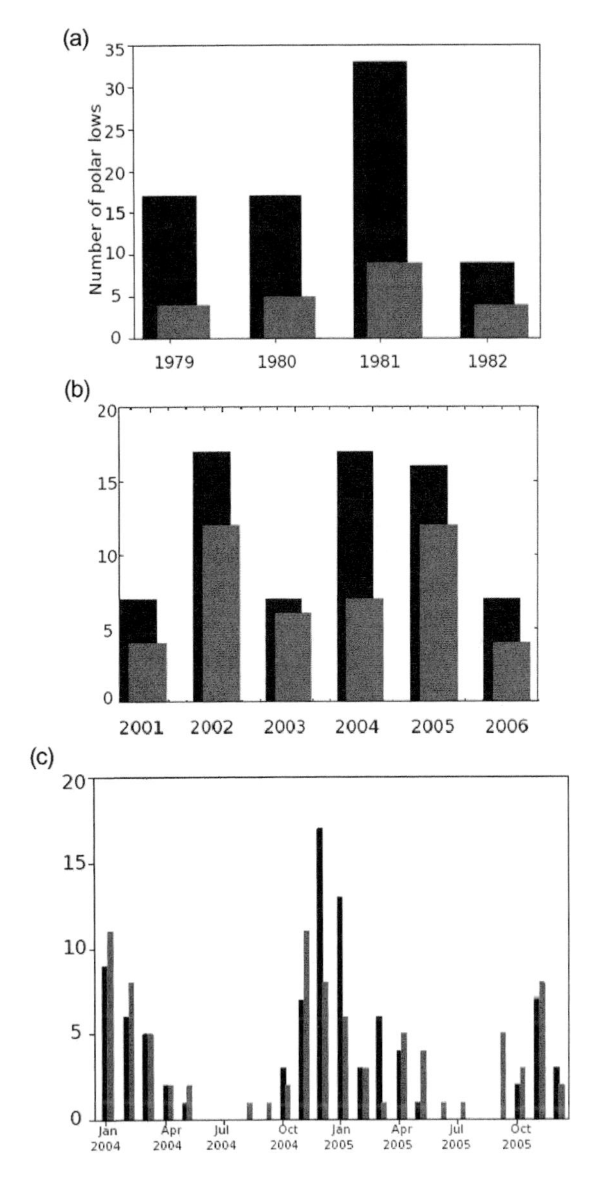

Figure 4. Number of detected polar lows per PLS or per month in our study (black) and in observation-based studies (red) as given by (a) *Wilhelmsen* [1985], (b) *Noer and Ovhed* [2003], and (c) *Blechschmidt* [2008]. Figures 4a and 4c are taken from the supplementary material of *Zahn and von Storch* [2010].

and for comparison, we adapted our threshold to 15 m s^{-1} in the case of *Wilhelmsen* [1985] and *Blechschmidt* [2008] and to 17.2 m s^{-1} in the case of *Noer and Ovhed* [2003]. The database by *Wilhelmsen* [1985] covering the years 1978 through 1982 was compiled mainly using weather charts and observations and only few satellite images. *Noer and Ovhed* [2003] used the wealth of routine observations available at the Norwegian Meteorological Service. Both studies are mainly confined to the area around Norway. *Blechschmidt* [2008] covers the whole northern North Atlantic (north of 60°N), but over a 2 year period only. She used IR advanced very high resolution radiometer quick look images to detect polar low cloud structures by eye and then checked with wind speed from satellite measurements.

The results are shown in Figure 4. Comparison with the work of *Wilhelmsen* [1985] reveals a large offset with about four times more polar lows in our data. We relate the low number of cases given by *Wilhelmsen* [1985] to the scarcity of observation data, especially over the Arctic Ocean, in that time, which probably has led to missing a lot of cases. The main feature, relatively similar numbers of polar lows in three of the four PLSs, and a high peak in PLS 1981, however, is a joint property in both data. The numbers of polar lows in our data are much closer to the one of *Noer and Ovhed* [2003] as shown in Figure 4b. Here in five of six PLS, the course of the number of polar lows is similar. Comparison with the work of *Blechschmidt* [2008] on a yearly basis is meaningless. We thus compared on the basis of the number of cases per month. Although here peak months are not always hit by both data at the same time, correlation between both is high with a coefficient of $c = 0.72$.

We conclude that our simulated results share considerable similarities to other studies and observation-based ones. We thus assume that our methodology of simulating and detecting polar low is capable of generating trustworthy statements of the "real" world.

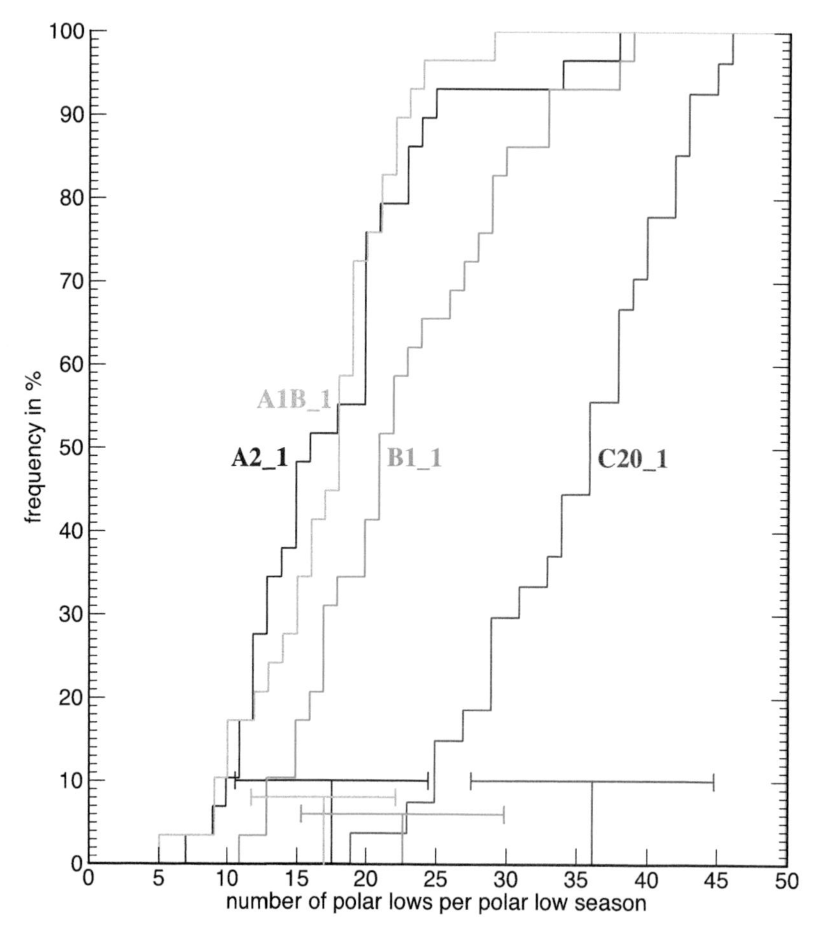

Figure 5. Cumulative frequency of the number of polar lows resulting from the Intergovernmental Panel on Climate Change (IPCC) C20 scenario (red) and future scenarios B1 (green), A1B (yellow), and A2 (blue). Vertical bars denote means, and horizontal error bars denote the standard deviations. From *Zahn and von Storch* [2010].

3.3. Changes of North Atlantic Polar Low Frequency in a Presumed Warmed Atmosphere

After having investigated past changes in polar low frequency, an obvious topic is to investigate systematic changes in their future frequency. We have addressed this question as the last in our series of papers and applied our simulation and tracking methodology to IPCC GHG scenario-driven global model data. Counting polar lows results in the frequencies shown in Figure 5. In the twentieth century reference climate (red lines), we find an average of about 36 polar lows per PLS. This bias is not in opposition to our results for the past reported in section 3.1, where we found 56 cases per PLS. Systematic differences between different models or different data sets are a common feature and are referred to in literature [e.g., *Meehl et al.*, 2007; *Christensen et al.*, 2007; *Jun et al.*, 2008]. They may be caused by different physics involved or different parametrization schemes used to take account for subgrid-scale processes. We will find further examples of systematic differences between models later in this section.

The results for the scenarios B1, A1B, and A2 are shown in Figure 5 as green, yellow, and blue lines. They are gained with the same approach and model and are thus comparable to C20. We find a systematic and statistically significant (at the 99.5% level) decrease in the number of annual polar lows in all the future scenarios. The projected annual number of polar lows is only half as high in the late twenty-first century's climate compared to C20. Among the scenarios, the decrease is more pronounced the higher the GHG concentrations are assumed and the higher the atmospheric temperatures rise.

To find the reason for this dramatic decline, we looked closer at the evolution of temperatures, at the SST, and at the 500 hPa level (T500). Their evolutions, as given by our ECHAM5/MPI-OM driving data, are shown in Figure 6. Variability of yearly mean SSTs averaged over the simulation area is relatively low. Mean values for SST of 7.54°C in C20, 7.96°C in B1, 8.27°C in A1B and also in A2 are relatively close to each other, meaning that average SST is increasing by less than 1°C. The annual mean T500 is more variable. Mean values for T500 read −29.78°C in C20, −27.83°C in B1, −27.11°C in A1B, and −26.79°C in A2. The temperatures increase up to about 3° in T500, resulting in a less pronounced vertical temperature gradient and consequently in a more stable atmospheric stratification, reducing the likeliness of polar low developments. A quicker warming of the atmosphere compared to the ocean is in line with literature. This has especially been found over the Arctic atmosphere, where tropospheric air temperature is expected to warm stronger than the global average in the Arctic [*Meehl et al.*,

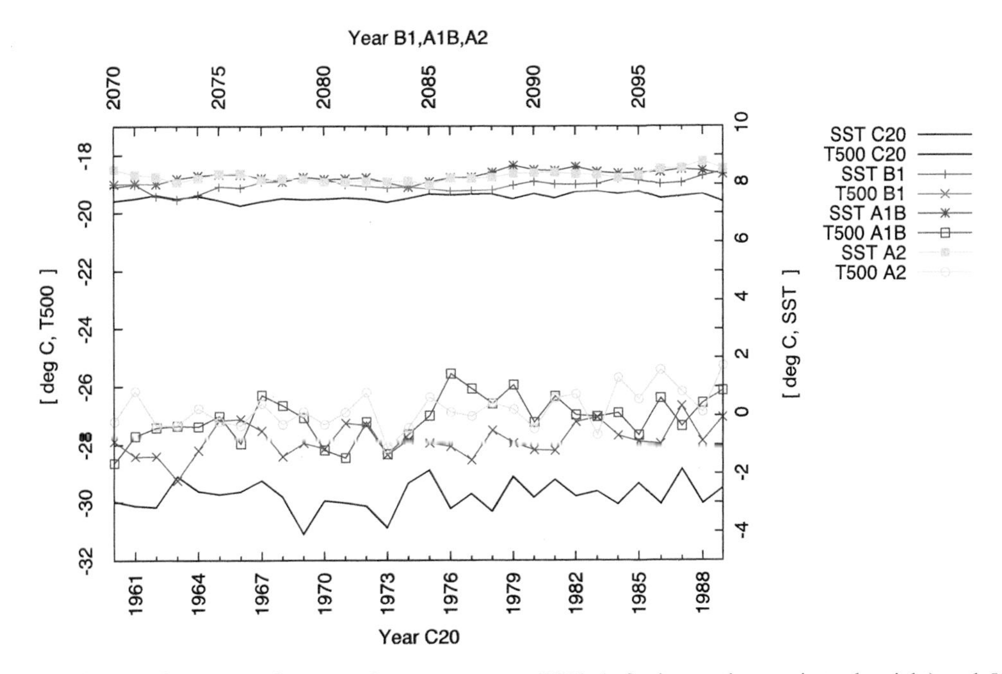

Figure 6. Evolution of mean yearly sea surface temperature (SST) (referring to the *y* axis to the right) and 500 hPa (referring to the *y* axis to the left) in IPCC C20 scenario (black, years as given at the lower *x* axis) and future scenarios (years as given at the upper *x* axis) B1 (red), A1B (blue), and A2 (light blue) as calculated by the European Centre/Hamburg 5/MPI-OM model.

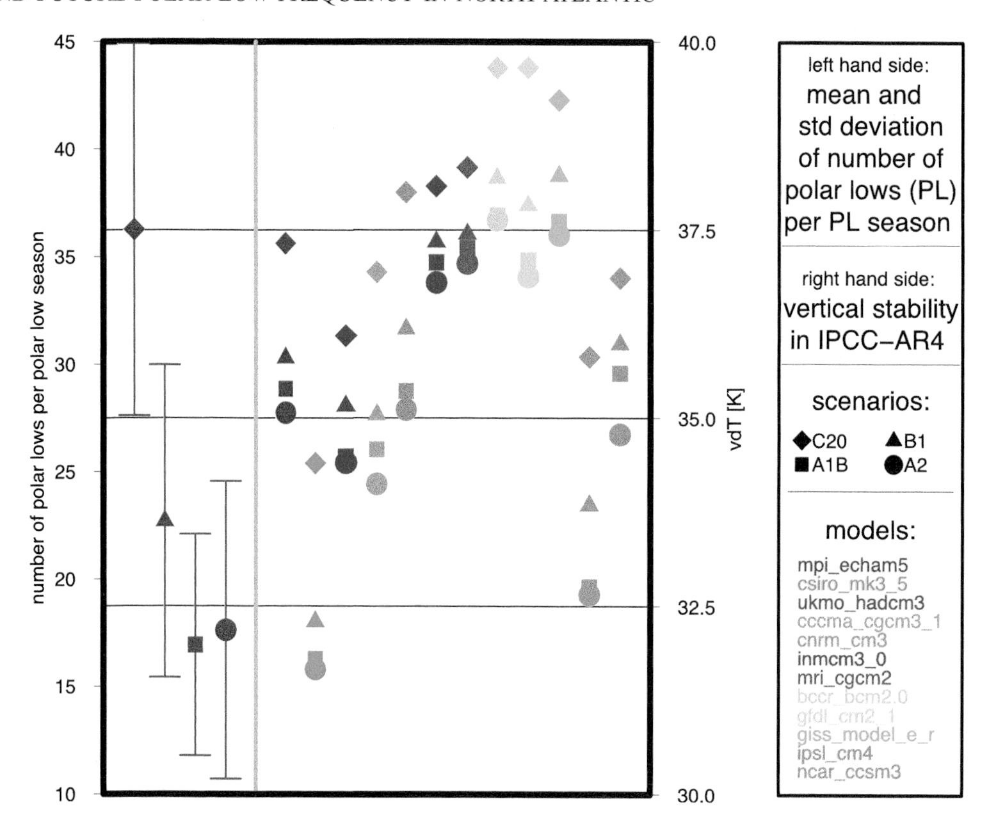

Figure 7. (left) Mean number of polar lows per PLS and standard deviation. (right) Mean vertical difference between the SSTs and temperatures at 500 hPa level from various IPCC models in scenarios C20 (1960–1989), B1, A1B, and A2 (2070–2099). From *Zahn and von Storch* [2010].

2007]. The ocean is warming slower, and some studies have even reported a cooling for some parts, at least over the last decades [*Levitus et al.*, 2000, 2005; *Bindoff et al.*, 2007].

Checking the vertical temperature difference is much less cost intensive in terms of computer power than downscaling all the global data and detect polar lows in the output data. In an additional effort, we tried to verify the robustness of the signal of decreasing polar low frequency. We calculated the vertical temperature difference between SST and T500 as an indicator for the evolution of vertical atmospheric stability in 12 of the IPCC models for the C20, B1, A1B, and A2 scenarios. The results are shown in Figure 7. As discussed when comparing numbers of polar lows in REA and in the IPCC C20 scenario, large intermodel differences of up to 5°C are evident in the vertical temperature gradients. However, comparing C20 and future scenarios delivers the same signals in each of the models: the vertical temperature gradient is strongest in the scenario of the twentieth century and about 2°C or more lower in the end of the twenty-first century. Moreover, we find the same pattern of decline as in the number of polar lows: the higher the increase in GHGs and atmospheric temperature, the stronger the increase in vertical atmospheric stability. This is a robust signal in all IPCC models.

We conclude that the presently available scenarios of possible future climate change all point toward a notable decrease in the frequency of polar lows in the North Atlantic. The dynamical mechanism is plausibly related to the increase in atmospheric stability, which is a robust signal in all the IPCC models used in this study. Our results were published in the work of *Zahn and von Storch* [2010].

4. SUMMARY AND CONCLUSIONS

We have presented the methodology and the main results of our recent work on frequency changes of vigorous North Atlantic small- to medium-scale storms, so-called polar lows. We reproduced polar lows in long-term simulations by downscaling global atmospheric data by means of a RegCM and detected them using an automated tracking algorithm, which is based on a near-isotropic 2-D digital band-pass filter of *Feser and von Storch* [2005]. Global atmospheric data were delivered by reanalysis data for the past and by global models applying IPCC scenarios for the future assessment.

We found high interannual variability of polar low frequency over the past 60 years, but decadal numbers remained on a similar level, and no trend was found. For the future in a presumed anthropogenically warmed atmosphere, we found a significant decrease of polar low frequency, which is more pronounced the stronger the atmosphere warms. We relate this to an increase in atmospheric stability, as the air temperatures rise faster than the SSTs do, thus lowering the vertical temperature gradient and hindering the development of good environments for convection.

To what extent this reduced vertical stability has implications for extratropical future storm activity, in general, remains to be investigated. As a first guess, one would expect increasing atmospheric stability to reduce the number of extratropical large-scale storms, too. There have, indeed, been some studies suggesting a slight, but insignificant, decrease in the number of these storms [*Bengtsson et al.*, 2006; *Ulbrich et al.*, 2009].

Acknowledgments. M.Z. is currently funded by the NERC PREPARE project, NE/G015708/1. H.vS. thanks the research programs REKLIM of HGF and the Center of Excellence CLISAP at the University of Hamburg for support.

REFERENCES

Bengtsson, L., K. I. Hodges, and E. Roeckner (2006), Storm tracks and climate change, *J. Clim.*, *19*, 3518–3543.

Bindoff, N. L., et al. (2007), Observations: Oceanic climate change and sea level, in *Climate Change 2007: The Physical Science Basis: Working Group I Contribution to the Fourth Assessment Report of the IPCC*, edited by S. Solomon et al., pp. 385–432, Cambridge Univ. Press, Cambridge, U. K.

Blechschmidt, A.-M. (2008), A 2-year climatology of polar low events over the Nordic Seas from satellite remote sensing, *Geophys. Res. Lett.*, *35*, L09815, doi:10.1029/2008GL033706.

Bracegirdle, T., and S. Gray (2008), An objective climatology of the dynamical forcing of polar lows in the Nordic seas, *Int. J. Climatol.*, *19*, 1903–1919.

Bracegirdle, T., and S. Gray (2009), The dynamics of a polar low assessed using potential vorticity inversion, *Q. J. R. Meteorol. Soc.*, *135*, 880–893.

Christensen, J., et al. (2007), Regional climate projections, in *Climate Change 2007: The Physical Science Basis: Working Group I Contribution to the Fourth Assessment Report of the IPCC*, edited by S. Solomon et al., pp. 847–940, Cambridge Univ. Press, Cambridge, U. K.

Claud, C., G. Heinemann, E. Raustein, and L. Mcmurdie (2004), Polar low 'Le Cygne': Satellite observations and numerical simulations, *Q. J. R. Meteorol. Soc.*, *130*, 1075–1102.

Condron, A., G. R. Bigg, and I. A. Renfrew (2006), Polar mesoscale cyclones in the northeast Atlantic: Comparing climatologies from ERA-40 and satellite imagery, *Mon. Weather Rev.*, *134*, 1518–1533.

Condron, A., G. R. Bigg, and I. A. Renfrew (2008), Modeling the impact of polar mesocyclones on ocean circulation, *J. Geophys. Res.*, *113*, C10005, doi:10.1029/2007JC004599.

Emanuel, K. A., and R. Rotunno (1989), Polar lows as arctic hurricanes, *Tellus, Ser. A*, *41*, 1–17.

Ese, T., I. Kanestrøm, and K. Pedersen (1988), Climatology of polar lows over the Norwegian and Barents Seas, *Tellus, Ser. A*, *40*, 248–255.

Feser, F., and H. von Storch (2005), A spatial two-dimensional discrete filter for limited area model evaluation purposes, *Mon. Weather Rev.*, *133*, 1774–1786.

Feser, F., and H. von Storch (2008a), Regional modelling of the western Pacific typhoon season 2004, *Meteorol. Z.*, *17*, 519–528.

Feser, F., and H. von Storch (2008b), A dynamical downscaling case study for typhoons in SE Asia using a regional climate model, *Mon. Weather Rev.*, *136*, 1806–1815.

Feser, F., B. Rockel, H. von Storch, J. Winterfeldt, and M. Zahn (2011), Regional climate models add value to global model data a review and selected examples, *Bull. Am. Meteorol. Soc.*, *9*, 1181–1192.

Harold, J. M., G. R. Bigg, and J. Turner (1999a), Mesocyclone activities over the north-east Atlantic. Part 1: Vortex distribution and variability, *Int. J. Climatol.*, *19*, 1187–1204.

Harold, J. M., G. R. Bigg, and J. Turner (1999b), Mesocyclone activities over the north-east Atlantic. Part 2: An investigation of causal mechanisms, *Int. J. Climatol.*, *19*, 1283–1299.

Harrold, T. W., and K. A. Browning (1969), The polar low as a baroclinic disturbance, *Q. J. R. Meteorol. Soc.*, *95*, 710–723.

Jun, M., R. Knutti, and D. W. Nychka (2008), Spatial analysis to quantify numerical model bias and dependence, *J. Am. Stat. Assoc.*, *103*, 934–947.

Kalnay, E., et al. (1996), The NCEP/NCAR reanalysis project, *Bull. Am. Meteorol. Soc.*, *77*, 437–471.

Klein, T., and G. Heinemann (2002), Interaction of katabatic winds and mesocyclones near the eastern coast of Greenland, *Meteorol. Appl.*, *9*, 407–422.

Kolstad, E. W. (2006), A new climatology of favourable conditions for reverse-shear polar lows, *Tellus, Ser. A*, *58*, 344–354, doi:10.1111/j.1600-0870.2006.00171.x.

Kolstad, E. W. (2011), A global climatology of favourable conditions for polar lows, *Q. J. R. Meteorol. Soc.*, *137*, 1749–1761.

Kolstad, E. W., and T. J. Bracegirdle (2008), Marine cold-air outbreaks in the future: An assessment of IPCC AR4 model results for the Northern Hemisphere, *Clim. Dyn.*, *30*, 871–885.

Kolstad, E. W., T. J. Bracegirdle, and I. A. Seierstad (2008), Marine cold-air outbreaks in the North Atlantic: Temporal distribution and associations with large-scale atmospheric circulation, *Clim. Dyn.*, *33*, 187–197.

Levitus, S., J. I. Antonov, T. P. Boyer, and C. Stephens (2000), Warming of the world ocean, *Science*, *287*, 2225–2229.

Levitus, S., J. Antonov, and T. Boyer (2005), Warming of the world ocean, 1955–2003, *Geophys. Res. Lett.*, *32*, L02604, doi:10.1029/2004GL021592.

Marsland, S. J., H. Haak, J. H. Jungclaus, M. Latif, and F. Röske (2003), The Max-Planck-Institute global ocean/sea ice model with orthogonal curvilinear coordinates, *Ocean Modell.*, *5*, 91–127.

Meehl, G., et al. (2007), Global climate projections, in *Climate Change 2007: The Physical Science Basis: Working Group I Contribution to the Fourth Assessment Report of the IPCC*, edited by S. Solomon et al., pp. 747–845, Cambridge Univ. Press, Cambridge, U. K.

Nakicenovic, N., and R. Swart (Eds.) (2000), *IPCC Special Report on Emissions Scenarios*, 612 pp., Cambridge Univ. Press, Cambridge, U. K.

Nielsen, N. (1997), An early autumn polar low formation over the Norwegian Sea, *J. Geophys. Res.*, *102*, 13,955–13,973.

Noer, G., and M. Ovhed (2003), Forecasting of polar lows in the Norwegian and the Barents Sea, paper presented at the 9th Meeting of the EGS Polar Lows Working Group, Eur. Geophys. Soc., Cambridge, U. K.

Noer, G., Ø. Saetra, T. Lien, and Y. Gusdal (2011), A climatological study of polar lows in the Nordic Seas, *Q. J. R. Meteorol. Soc.*, *137*, 1762–1772.

Rasmussen, E. (1979), The polar low as an extratropical CISK disturbance, *Q. J. R. Meteorol. Soc.*, *105*, 531–549.

Rasmussen, E., and J. Turner (2003), *Polar Lows: Mesoscale Weather Systems in the Polar Regions*, 628 pp., Cambridge Univ. Press, Cambridge, U. K.

Rockel, B., A. Will, and A. Hense (2008), The regional climate model COSMO-CLM (CCLM), *Meteorol. Z.*, *17*, 347–348.

Roeckner, E., et al. (2003), The atmospheric general circulation model ECHAM 5. Part I: Model description, *MPI Rep. 349*, Max Planck Inst., Hamburg, Germany.

Ulbrich, U., G. Leckebusch, and J. Pinto (2009), Extra-tropical cyclones in the present and future climate: A review, *Theor. Appl. Climatol.*, *96*, 117–131, doi:10.1007/s00704-008-0083-8.

Vavrus, S., J. E. Walsh, W. L. Chapman, and D. Portis (2006), The behavior of extreme cold air outbreaks under greenhouse warming, *Int. J. Climatol.*, *26*, 1133–1147.

von Storch, H., H. Langenberg, and F. Feser (2000), A spectral nudging technique for dynamical downscaling purposes, *Mon. Weather Rev.*, *128*, 3664–3673.

von Storch, H., F. Feser, and M. Barcikowska (2010), Downscaling tropical cyclones from global re-analysis and scenarios: Statistics of multi-decadal variability of TC activity in E Asia, *Proc. Int. Conf. Coastal Eng.*, *32*, 1–8.

Wilhelmsen, K. (1985), Climatological study of gale-producing polar lows near Norway, *Tellus, Ser. A*, *42A*, 451–459.

Zahn, M., and H. von Storch (2008a), Tracking polar lows in CLM, *Meteorol. Z.*, *17*, 445–453.

Zahn, M., and H. von Storch (2008b), A long-term climatology of North Atlantic polar lows, *Geophys. Res. Lett.*, *35*, L22702, doi:10.1029/2008GL035769.

Zahn, M., and H. von Storch (2010), Decreased frequency of North Atlantic polar lows associated with future climate warming, *Nature*, *467*, 309–312.

Zahn, M., H. von Storch, and S. Bakan (2008), Climate mode simulation of North Atlantic polar lows in a limited area model, *Tellus, Ser. A*, *60*, 620–631.

H. von Storch, Institute of Coastal Research, Helmholtz-Zentrum Geesthacht, D-21502 Geesthacht, Germany.

M. Zahn, Environmental Systems Science Centre, University of Reading, Berkshire, Reading RG6 6AL, UK. (m.zahn@reading.ac.uk)

Variability of North Atlantic Hurricanes: Seasonal Versus Individual-Event Features

Álvaro Corral

Centre de Recerca Matemàtica, Barcelona, Spain

Antonio Turiel

Institut de Ciències del Mar, CSIC, Barcelona, Spain

Tropical cyclones are affected by a large number of climatic factors, which translates into complex patterns of occurrence. The variability of annual metrics of tropical cyclone activity has been intensively studied since the sudden activation of the North Atlantic in the mid-1990s. We provide an overview of work by diverse authors about these annual metrics for the North Atlantic basin, where the natural variability of the phenomenon, the existence of trends, the drawbacks of the records, and the influence of global warming have been the subject of interesting debates. Next, we present an alternative approach that does not focus on seasonal features but on the characteristics of single events. Thus, the individual-storm power dissipation index (PDI) constitutes a natural way to describe each event, with its statistics yielding a robust law for the occurrence of tropical cyclones in terms of a power law distribution. As an important extension, we model the whole range of the PDI density (excluding incompleteness effects at the smallest values) with the gamma distribution, consisting of a power law with an exponential tail. The characteristic scale of this decay, represented by the cutoff parameter, provides very valuable information on the finite size of the basin via the largest values of the PDIs that the basin can sustain. We use the gamma fit to evaluate the positive influence on the occurrence of extreme PDI values of sea surface temperature, the positive phase of the Atlantic multidecadal oscillation, the number of hurricanes in a season, and extreme negative values of the multivariate El Niño–Southern Oscillation index.

1. INTRODUCTION

Tropical cyclones are a rare phenomenon, with less than 100 occurrences per year worldwide (tropical depressions not counted); just compare with 10^4 earthquakes with magnitude larger than 4 per year. Despite this scarcity, the societal impact of these atmospheric systems is huge, as it is perceived year by year by the general public. An important issue for planning of tropical cyclone damage mitigation is the knowledge of the temporal variability of the phenomenon and to try to establish which part of the variability is natural and which part can be due to global warming.

When counting these meteorological monsters, a bit of terminology is useful [*Emanuel*, 2005a]. Tropical cyclones is the generic name encompassing typhoons, hurricanes, tropical storms, and tropical depressions (when the context

Extreme Events and Natural Hazards: The Complexity Perspective
Geophysical Monograph Series 196

10.1029/2011GM001069

allows it, we will also use the vague term "storm"). Typhoons and hurricanes (and severe cyclonic storms) are "mature" tropical cyclones (the difference in name is only of geographical origin), with winds strong enough to achieve a category from 1 to 5 in the Saffir-Simpson scale (T. Schott et al., The Saffir-Simpson hurricane wind scale, 2010, available at http://www.nhc.noaa.gov/sshws.shtml). In contrast, tropical storms do not have so large wind speeds, and for tropical depressions, these are even weaker. The thresholds separating these three classes are 34 knots (17 m s^{-1}) and 64 knots (33 m s^{-1}) for the so-called maximum sustained (1 min) surface (10 m) wind speed (1 knot = 0.5144 m s^{-1}), further thresholds define the Saffir-Simpson category. Then, speed is a way to define intensity, and intensity at a given instant determines the stage of the tropical cyclone (hurricane, tropical storm, etc.) and the category in case of hurricanes or typhoons. The stage corresponding to the maximum lifetime intensity determines how the storm is finally labeled. Generally, the databases in use only contain more or less complete information on tropical storms and hurricane-like systems, and therefore, when talking about tropical cyclones, tropical depressions will be excluded in this text. Note that in some references, tropical depressions are not even considered as tropical cyclones; they are just formative or decaying stages of tropical storms and hurricanes [*Neumann et al.*, 1999].

Although the North Atlantic Ocean (including the Gulf of Mexico and the Caribbean) comprises only about 10% of the tropical cyclones in the world, this is the ocean basin that has been more extensively studied, with routine aircraft reconnaissance since 1944 and an intensive scrutiny of ship log records for previous years (satellite imagery started to be available from the 1960s) [*Neumann et al.*, 1999]. This has yield the longest and most reliable database for tropical cyclones, extending back in time, with obvious limitations, for more than one century and a half.

In this work, we analyze tropical cyclone activity in the North Atlantic. First, previous work on temporal variability is overviewed, from the point of view of direct observations, and the controversy about if a global warming signal can be separated from the natural fluctuations is briefly mentioned. Fully in-depth reviews on this topic, including modeling, theory, and paleoclimatic studies, have been published by *Shepherd and Knutson* [2007], *Knutson et al.* [2010b], and *Knutson et al.* [2010a]. Second (section 3), an approach based on the study of the features of individual tropical cyclones is presented, based on the calculation of their power dissipation index (PDI). Power law fits for the PDI distribution are presented and discussed also, and a superior, more general fitting function (the gamma distribution) is introduced in section 4, taking into account the exponential decay of the tail of the distribution. Sections 5 and 6 analyze the effect of tropical sea surface temperature (SST) and diverse climatic indices (El Niño, North Atlantic oscillation (NAO), Atlantic multidecadal mode (or oscillation) (AMO), number of tropical cyclones) on the PDI distribution.

2. VARIABILITY OF HURRICANE ACTIVITY

The large temporal variability of North Atlantic tropical cyclone frequency is clearly seen in the existing records, with a maximum of 28 occurrences in 2005 (including a subtropical storm) versus only 4 in 1983 (for instance, see Figure 1). Excluding tropical storms, there was a maximum of 15 hurricanes also in 2005 and just 2 in 1982; for major hurricanes, which are those with category 3 or above (96 knots (49 m s^{-1}) threshold), there were 8 in 1950 versus none in 1994 [*Neumann et al.*, 1999; *Gray et al.*, 1992; *Elsner and Kara*, 1999; *Bell et al.*, 2006; C. Landsea, Record number of storms by basin, see http://www.aoml.noaa.gov/hrd/tcfaq/E10.html, 2007].

But high and low tropical cyclone activity comes in multidecadal clusters. *Gray* [1990] reported that the number of major hurricanes decreased to less than one half from the period 1947–1969 to 1970–1987 (3.3 per year versus 1.55), which was interpreted as a regional manifestation of a global-scale climate variation governed by thermohaline processes. If a hypothetical overestimation of speeds previous to 1970 is corrected, the decreasing trend still persists for these periods [*Landsea*, 1993]; nevertheless, see also the work of *Landsea* [2005] with regard to the necessity of this adjustment.

Elsner et al. [2000] and *Goldenberg et al.* [2001] realized that in 1995, a new period of high activity seemed to have started in the North Atlantic. In the first case, the increase was related to the NAO, whereas in the second, it was associated to simultaneous increases in North Atlantic SSTs

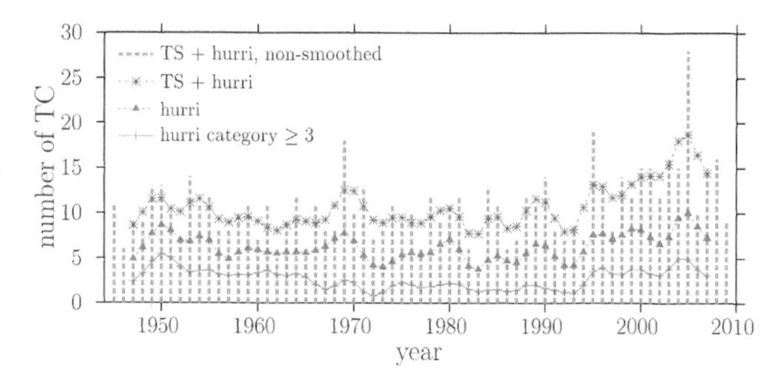

Figure 1. Annual number of tropical cyclones, hurricanes, and major hurricanes in the North Atlantic as a function of time. The data have been smoothed with a 1-2-1 filter applied twice. Unsmoothed data are also shown for the number of tropical cyclones (tropical storms plus hurricanes). Corrections are not performed, neither in the counts nor in the intensities. For the original plot, see the work of *Goldenberg et al.* [2001] or *Webster et al.* [2005].

and decreases in vertical wind shear, governed by variations of natural origin of the AMO [*Kerr*, 2000]. In this way, the period from 1965 to 1994 witnessed no more than three major hurricanes per year, whereas in 1995, there were five of such storms and six in 1996. The return of high levels of activity was labeled as dramatic.

Later, *Webster et al.* [2005] found that an increasing trend for the period 1970–2004 was also significant for hurricanes of at least category 1, in clear contrast to the behavior of other basins worldwide (in which tropical SST also increased). Remarkably, these authors also extended the increase of major tropical cyclones to the other basins (excluding category 3) and noted that this was not inconsistent with climate simulations showing that a doubling of atmospheric CO_2 should lead to an increase in the frequency of the most intense tropical cyclones. However, subsequent analysis have claimed that the sign in the trend of the northeastern Pacific is dependent on the time window selected and that the increase in other basins could be an artifact due to the lack of quality and homogeneity of the records. In any case, the increased activity of North Atlantic hurricanes of categories 4 and 5 was clear and robust [*Klotzbach*, 2006; *Kossin et al.*, 2007].

However, *Landsea* [2007] has shown that the analysis of tropical cyclone variability using counts of events requires enormous caution. Before the advent of aircraft reconnaissance in the Atlantic, in 1944, detection of open-ocean storms was dependent on chance encounters with ships. But even after the 1940s, aircrafts were covering essentially the west half of the basin, so systems developing on the east part were not always observed. The comparison between the number of landfalling storms (which is assumed reliable since 1900) and the total number of storms gives a quite stable ratio of 59% since 1966 (when satellite imagery started to operate), in contrast to a value around 75% for previous years. In addition, new operational tools, as QuikSCAT, that have become available in the turn of this century, have allowed the identification of one additional tropical cyclone per year (on average). This has led Landsea to estimate a deficit in the annual frequency of storms of 3.2 events per year up to 1965 and 1 per year from 1966 to 2002. (One can wonder why the correction to this problem is solved by adding a constant term, rather than multiplying by some factor.)

Obviously, storm counts do not provide a complete characterization of tropical cyclone activity, and other indicators are necessary. (Here we will use the term activity in a broad sense and not as synonymous of frequency or abundance.) In this way, total storm days, defined by *Gray et al.* [1992] as the total time all storms in a year spend in a given storm stage, was found by *Gray* [1990] to quadruple between low and high activity periods for the stage of major hurricanes (a reduction to 2.1 days per year from a previous average of 8.5, for the periods mentioned above). Note that this metric is different from the total duration of major hurricanes (which is the sum of all their durations, in any stage). For the total duration of North Atlantic hurricanes (categories 1 to 5 and including tropical storm stages), *Webster et al.* [2005] reported a significant trend from 1970 to 2004.

Another quantity that has been employed is the net tropical cyclone (NTC) activity, explained in the supplementary information of the work of *Goldenberg et al.* [2001] as an average between storm counts and total storm days, considering tropical storms, nonmajor hurricanes, and major hurricanes, but giving more weight to latter and less weight to the former storms. In this way, information on frequency, duration, and intensity is combined into a single number, resulting in NTC between 1995 and 2000 being twice the value of 1971–1994, which in turn was a factor 1.6 smaller than for 1944–1970.

A significant step forward was taken by *Bell et al.* [2000], who defined the accumulated cyclone energy (ACE) by summing, for all tropical cyclones in a season, the squares of the 6 h maximum sustained wind speed for all records in which the systems were above the tropical storm limit (i.e., speed $\geq$34 knots (17 m s^{-1})). The ACE is a more natural way to combine frequency, intensity, and duration than the NTC, but it should not be interpreted as a kinetic energy (rather, it could be the time integral of something akin to kinetic energy, as its name denotes). It is worth mentioning that previously, *Gray et al.* [1992] had introduced a related quantity, the hurricane destruction potential, with the only difference that tropical storm stages were not taken into account.

Updating the results of Bell et al. for the ACE as an annual indicator, *Trenberth* [2005] noticed that between 1995 and 2004, the North Atlantic activity had been "above normal," with the only exception of the El Niño years of 1997 and 2002, and that this increase was in parallel to the fact that that decade was the one with the highest SST on record in the tropical North Atlantic (by more than 0.1°C) and also to the increased water vapor content over the global oceans. (The results for ACE were confirmed later by *Klotzbach* [2006].) Although many uncertainties remain on how this and other environmental changes can affect the self-organization process of tropical cyclone formation, once formation has taken place, it seems clear that these conditions provide more energy to the storm, which suggests more intense winds and heavier rainfalls under a global warming scenario.

A similar analysis was performed by *Emanuel* [2005b], who introduced what he called PDI, but what we will call annual or accumulated PDI (APDI), defined as the ACE with the main difference that the square of the speeds is replaced by their cube (and also, the time summation was performed for the whole life of the storm and not only for tropical storm and hurricane stages). It is notable that the APDI constitutes a "proxy" estimation of the kinetic energy dissipated by all tropical cyclones in one season (not of their power). Emanuel showed how a well-known formula for the dissipation, used previously by *Bister and Emanuel* [1998], could be converted into the APDI formula under reasonable assumptions. Indeed, the dissipated power is given by the integral over surface of the cube of the velocity field (multiplied by the air density and the drag coefficient, which can be assumed as constants). First, accepting a similar shape for all storms (the same functional form for the velocity field, scaled by the radius of the storm and the maximum-in-space instantaneous speed), this power is proportional to the square of the radius multiplied by the cube of the maximum speed (with the same proportionality constant for all storms). Second, noticing the weak correlation between storm dimensions and speed, one could assign a common radius to all storms (in all stages) and get only random errors in the estimation of the power. As the energy dissipated by a tropical cyclone is given by the time integral (for all its life) of its power, replacing the integral by a discrete summation (as the records come in discrete intervals), one gets that the dissipated energy is roughly proportional then to the sum of the cube of the maximum sustained wind speed, which is (in contrast to the usage of Emanuel), what we call PDI. Further summation for all storms provides the APDI, for which it is expected that the errors coming from the assignment of a common radius will more or less compensate. See the work of *Corral* [2010] for a nonverbal version of this derivation and the appendix here for an estimation of the proportionally constant linking PDI and dissipated energy.

In equations,

$$\mathrm{APDI} = \sum_{i=1}^{n} \mathrm{PDI}_i$$

where i counts the n tropical cyclones in the season under consideration. For a single storm, its PDI is

$$\mathrm{PDI} = \sum_{\forall t} v_t^3 \Delta t,$$

where t labels discrete time, v_t is the maximum sustained surface wind speed at t, and we introduce a time interval Δt that is constant, in principle, and equal to 6 h, just to make the PDI independent on changes on Δt (nevertheless, it is an open question how a better time resolution can influence the stability of the PDI value; this will depend on the smoothness of the time evolution of the speed). Under these assumptions, the SI units of the PDI are m^3 s^{-2} (for instance, hurricane Katrina (2005) yields PDI = 6.5×10^{10} m^3 s^{-2}).

Note that for tropical cyclone i,

$$\mathrm{PDI}_i = \langle v_t^3 \rangle_i T_i,$$

where $\langle v_t^3 \rangle_i$ is the average of the cube of the (6 h) maximum sustained wind speed over storm i, and T_i is its duration. Also, for year y,

$$\mathrm{APDI}_y = n_y \langle \mathrm{PDI} \rangle_y = n_y \langle T \rangle_y \langle v_t^3 \rangle_y$$

where n_y is the number of tropical cyclones in that year, and $\langle \ldots \rangle_y$ is the average of the considered quantity for the same year (the average is performed in a different way for the duration, of which there are n_y data, than for the speed, of which there are $\sum_i T_i/\Delta t$ records).

The previous equations show that the PDI of a tropical cyclone is the product of its duration and its average intensity, if one redefines intensity as the cube of the maximum sustained surface wind speed, whereas the APDI is the product of frequency, average duration, and annually averaged (6 h) intensity. Both the PDI and the APDI turn out to be very natural and convenient ways to estimate tropical cyclone activity, as they are a rough estimation of individual and total dissipated energy, respectively. Further, they are more robust than other measures of storms, as duration or track lengths, due to the fact that for these latter, the value is highly influenced by the definition of when a tropical cyclone starts, whereas for the PDI and APDI, such values have little influence on the final result, as they are weighted by the smallest values of v_t^3.

What Emanuel found was a clear correlation between "unprecedented" increases in tropical SST and APDI, both for the North Atlantic (more than doubled in the last 30 years) and northwestern Pacific (75% increase), when a 1-2-1 filter was applied twice to both signals (i.e., a 3 year running average where the central point has a weight that is equal to that of both neighbors). As the upswing in SST has been generally ascribed to global warming, this author argued that the increase in APDI could be partially of anthropogenic origin.

However, *Landsea* [2005] noticed that, after applying the smoothing procedure, Emanuel failed to drop the ending points, which were not averaged and were substantially larger than the smoothed series for the North Atlantic, creating the impression of a dramatic increase in PDI values. Together with W. M. Gray (Comments on "Increasing destructiveness of tropical cyclones over the past 30 years, unpublished manuscript, available at http://arxiv.org, 2006, hereinafter referred to as Gray, unpublished manuscript, 2006), Landsea has also criticized that Emanuel reduced the values of the wind speeds before the 1970s by an excessive

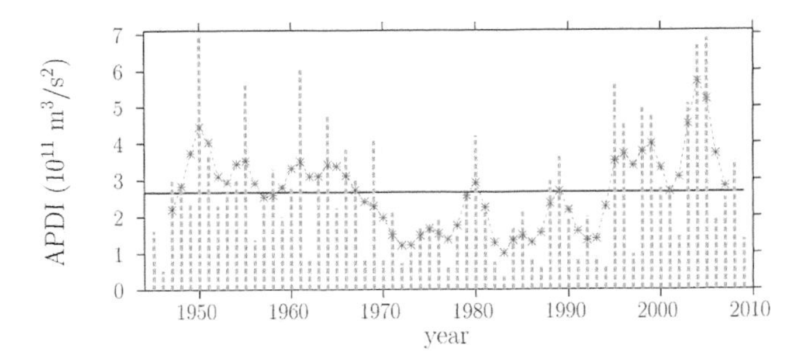

Figure 2. Annual accumulated power dissipation index (APDI) in the North Atlantic, both for the original series and for the smoothed one (by means of a 1-2-1 filter applied twice). No corrections have been applied. The mean of the original series is also shown. The year with the largest APDI is 1950, but it is followed very closely by 2005. The maximum of the smoothed series is for 2004. For the original plot, see the work of *Emanuel* [2005b].

amount (up to more than 20 knots (10 m s^{-1}) for the largest values), in order to correct an overestimation in those values. In fact, recent research has shown that surface winds in major hurricanes are stronger than previously assumed, and therefore, probably no correction at all is necessary [*Landsea*, 2005]. The North Atlantic PDI series, with no correction, is plotted in Figure 2. Other problems with the accuracy of the records, particularly for the Eastern Hemisphere are pointed out by *Landsea et al.* [2006].

Many more articles have been devoted to these complex affairs, which unfortunately cannot be abstracted in this brief overview. Before ending this section, let us just mention the work of *Chan* [2006], *Elsner et al.* [2006], *Emanuel* [2007], *Wu et al.* [2008], *Swanson* [2008], *Elsner et al.* [2008], *Aberson* [2009], and *Landsea et al.* [2010], for its relation with the issues discussed here.

3. PDI DISTRIBUTION AND POWER LAW FITS

All the studies summarized so far, counting the numbers of storms, total number of storm days, calculating NTC, ACE, or APDI, have paid attention to overall measures of annual tropical cyclone activity; that is, they were trying to answer the important question of comparing the characteristics of different seasons or longer periods. But how different are the tropical cyclones from one season to another? That is, which are the features of individual tropical cyclones in a given phase of activity or under the influence of a certain value of a climatic indicator?

The study of *Webster et al.* [2005] contains a first attempt in this direction, comparing the ratio between tropical cyclones in a given Saffir-Simpson category and tropical cyclones in all categories (1 to 5). For global aggregated data, it was found that the proportion of category 4 + 5 events increased from less than 20% in the early seventies to about 35% after 2000, with a corresponding decrease in the proportion of category 1 storms, from more than 40% to 30%. This is an indication that more major hurricanes (or typhoons) are present, in comparison, but not that the storms are becoming more intense, individually (i.e., there are more major ones in proportion, but their intensity is not necessarily record breaking). It is an open question in what part these results could be an artifact due to the incompleteness of the data for the earlier years [*Klotzbach*, 2006; *Kossin et al.*, 2007; Gray, unpublished manuscript, 2006]. In the North Atlantic, where the records are the best, these changes were much more modest, from a 20% of category 4 + 5 in 1975–1989 to 25% in 1990–2004. In fact, what Webster and coworkers were really doing was the calculation of the probability of hurricanes in a given category and how this probability distribution changes from one period to another.

However, in order to understand the changes in the behavior of individual tropical cyclones, we need to know, first of all, which are the general properties of tropical cyclones; in other words, which is their "unperturbed" nature, and only after that could we study the influences of year-to-year variability on them.

This has been attempted by one of the authors and collaborators [*Corral et al.*, 2010], analyzing the statistics of the (individual storm) PDI for long periods of time. As we have mentioned, the PDI is an approximation to dissipated energy and is therefore perhaps the most fundamental characteristic of a tropical cyclone.

In this way, the PDI for each tropical cyclone in the period and basin considered was calculated, and the resulting probability density D(PDI) was obtained, following its definition, as

$$D(\text{PDI}) = \frac{\text{Prob[value is in an interval around PDI]}}{\text{width of the interval}},$$

where there is a certain freedom in the choice of the width of the interval (or bin) [*Hergarten*, 2002] and Prob denotes probability, estimated by relative frequency of occurrence. Of course, normalization holds, $\int_0^\infty D(\text{PDI})d\text{PDI} = 1$. For more concrete details on the estimation of D(PDI), see the supplementary information of *Corral et al.* [2010].

The results for the 494 tropical cyclones of the North Atlantic during the period 1966–2009 are shown in Figure 3, in double logarithmic scale. The data are the best tracks from NOAA's National Hurricane Center [*Jarvinen et al.*, 1988; National Hurricane Center, see http://www.nhc.noaa.gov/data/hurdat/tracks1851to2010_atl_reanal.txt, 2011]; as we have seen, 1966 marks a year from which the quality and homogeneity of the records is reasonable (disregarding the possible

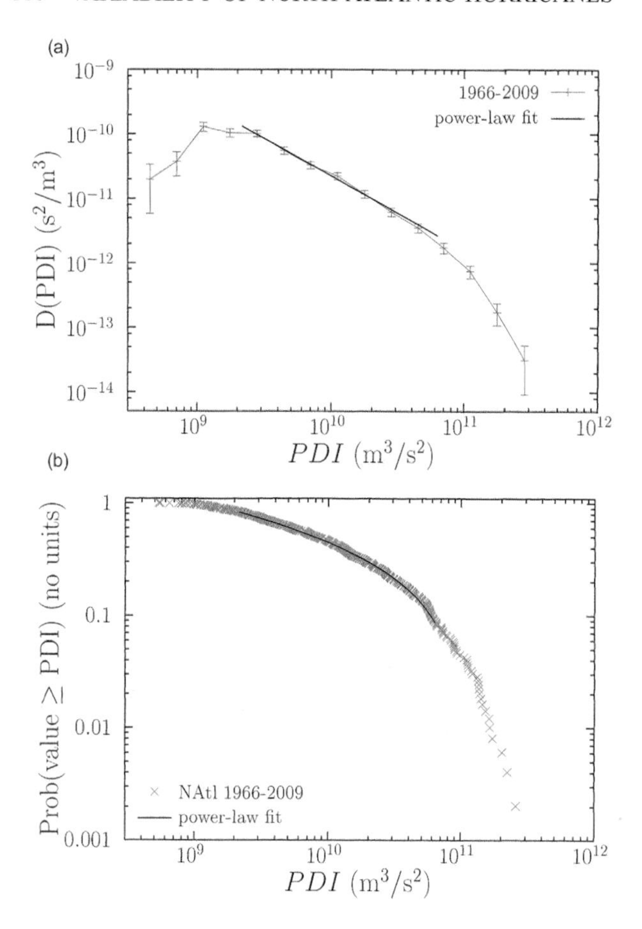

Figure 3. (a) Empirical probability density of 1966–2009 North Atlantic individual tropical cyclone PDI (494 events), together with a power law fit between 2.2×10^9 and 6.3×10^{10} m^3 s^{-2} with an exponent $\tau = 1.16$ [see *Corral et al.*, 2010]. (b) The same data and fit as in Figure 3a but for the (complement of the) cumulative distribution function. Notice that the power law distribution is no longer a straight line in a log-log plot, due to the bending effect of the upper truncation, which adds a constant to the power law [*Burroughs and Tebbens*, 2001]. In this plot, the Kolmogorov-Smirnov distance turns out to be proportional to the maximum difference between the empirical cumulative density and its fit (rather small).

overestimation of wind speeds previous to 1970). A straight line regime is apparent in Figure 3a, signaling the existence of a possible power law behavior over a certain range of PDI values,

$$D(\text{PDI}) \propto \frac{1}{\text{PDI}^{\tau}}, \text{ for } a \le \text{PDI} < b,$$

where $\propto$ indicates proportionality, τ is the exponent. and a and b are the truncation parameters.

Indeed, statistical tests support the power law hypothesis. The original reference [*Corral et al.*, 2010] used a generalization of the fitting and testing procedure proposed by *Clauset et al.* [2009]. The key point is to find the optimum values of a and b, for which the following steps are followed:

1. Fix arbitrary initial values of the truncation parameters a and b, with the only condition that $b/a > 10$, for example.

2. Find the maximum likelihood estimation of the exponent τ, by means of the maximization of

$$\ell(\tau; a, b, G) = \ln \frac{\tau - 1}{1 - (a/b)^{\tau-1}} - \tau \ln \frac{G}{a} - \ln a,$$

or, alternatively, the solution of

$$\frac{1}{\tau - 1} + \frac{a^{\tau-1}\ln(a/b)}{b^{\tau-1} - a^{\tau-1}} - \ln \frac{G}{a} = 0,$$

with G the geometric mean of the values comprised in the interval $[a, b)$.

3. Calculate the Kolmogorov-Smirnov (KS) distance between the empirical PDI distribution (defined only for the range $a \le \text{PDI} < b$) and its power law fit [*Press et al.*, 1992].

4. Look for and select the values of a and b that minimize the KS distance. Select also the corresponding τ exponent.

Once the optimal values of a and b, and from here the value of τ, are found, a p value quantifying the goodness of the fit can be calculated just by simulating synthetic data sets, as close as possible to the empirical data (with the selected power law exponent τ between a and b and resampling the empirical distribution outside this range). The previous four steps are applied then to each of these synthetic data sets (exactly in the same way as for the empirical data, in order to avoid biases), which yields a series of minimized KS distances, from which one can obtain the p value, which is the survivor probability function of the synthetic KS distances evaluated at the empirical value. To be more precise, p is defined as the probability that, for true power law distributed data, the KS distance is above the empirical KS value. In this way, very small p values are very unlikely under the null hypothesis, and one must conclude that the data are not power law distributed. For technical details about the whole procedure, see the supplementary information of *Corral et al.* [2010]. The introduction of the (arbitrary) condition $b/a > 10$ is necessary because in the case of a double truncation (from above and from below), several power law regimes can coexist in the data (in contrast to the original case of Clauset et al., $b \to \infty$), and one needs a criterion to select the most appropriate one, in our case that the range is high enough.

The results obtained from the application of this method to the North Atlantic 1966–2009 PDIs appear in Table 1, and they show that there is no reason to reject the power law hypothesis for a certain set of values of a and b. Among the

Table 1. Parameters of the Maximum Likelihood Estimation and the Kolmogorov-Smirnov Test for the Power Dissipation Index (PDI) of the 494 North Atlantic Tropical Cyclones of the Period 1966–2009, Using the Generalization of the Method of *Clauset et al.* [2009] Introduced by *Corral et al.* [2010][a]

R_{min}	a (m^3 s^{-2})	b (m^3 s^{-2})	b/a	N_{ab}	CN_{ab}/N	$\tau \pm \sigma_\tau$	d	$d\sqrt{N_{ab}}$	p Value
2	1.8×10^9	1.7×10^{10}	9.3	270	0.36	1.02 ± 0.09	2.4×10^{-2}	0.39	57 ± 5%
10	2.2×10^9	6.3×10^{10}	29.3	371	8.6	1.16 ± 0.05	2.5×10^{-2}	0.47	39 ± 5%
50	1.8×10^9	9.3×10^{10}	50.1	405	12.1	1.175 ± 0.04	3.1×10^{-2}	0.63	14 ± 3%
70	1.7×10^9	12.6×10^{10}	73.6	421	23.3	1.205 ± 0.04	4.2×10^{-2}	0.87	2 ± 1%
100	1.4×10^9	14.7×10^{10}	108.0	452	15.4	1.19 ± 0.04	5.0×10^{-2}	1.05	0%
20[b]	2.7×10^9	6.3×10^{10}	23.3	328	18.3	1.19 ± 0.06	2.4×10^{-2}	0.43	67 ± 5%

[a]The condition $b/a > R_{min}$ is imposed. N_{ab} refers to the number of tropical cyclones with PDI value between a and b; CN_{ab}/N is the constant of the power law that fits the empirical distribution between a and b when the latter is normalized from 0 to ∞, its units are $(m^3\ s^{-2})^{\tau-1}$. The uncertainty of the exponent can be obtained from its standard deviation, σ_τ [*Aban et al.*, 2006], which, in the limits of large N_{ab} and $b >> a$, is given by $\sigma_\tau \simeq (\tau - 1)/\sqrt{N_{ab}}$. Here d is the minimized KS distance and $d\sqrt{N_{ab}}$ a rescaling of that distance that should be independent on N_{ab}, for comparison purposes; a and b are determined with a resolution of 30 points per order of magnitude. The p values are calculated from 100 Monte Carlo simulations.

[b]The last row corresponds to the period 1966–2007, analyzed by *Corral et al.* [2010], where an erratum was present in the value of CN_{ab}/N.

several outcomes of the method, depending on the conditions imposed, a good choice is the one with the values of a and b around 2×10^9 and 6×10^{10} m^3 s^{-2}, with a resulting exponent $\tau = 1.16$ and a p value about 40% (see Figure 3). In general, one needs to balance the tendency to select low b/a ratios, which yield high p values, with the tendency to larger ratios and lower p.

Corral et al. [2010] checked that the behavior of the PDI distribution is very robust, showing very little variation for different time periods, or if extratropical, subtropical, wave and low stages are not taken into account, or if U.S. landfalling tropical cyclones are removed from the record, or speeds previous to 1970 are reduced by an amount of 4 m s^{-1} [*Landsea*, 1993]. In addition, the power law behavior was also found in other basins: the northeastern Pacific, the northwest Pacific, and the Southern Hemisphere as a whole (the north Indian Ocean was not analyzed due to the low number of events there and the lack of a complete satellite monitoring until recently) [*Kossin et al.*, 2007]. The resulting exponents were $\tau = 1.175$, 0.96, and 1.11, respectively, ±0.05, in the worst case.

Further, it has been pointed out that the power law PDI distribution could be a reflection of the criticality of hurricane occurrence [*Corral*, 2010], in the same way as for earthquakes, forest fires, rainfall, sandpiles, etc. [*Bak*, 1996; *Turcotte*, 1997; *Jensen*, 1998; *Sornette*, 2004; *Christensen and Moloney*, 2005]. In fact, tropical cyclone criticality could be only a part of the criticality of atmospheric convection [*Peters and Neelin*, 2006, 2009]. The parallelisms between tropical cyclone occurrence and self-organized criticality are discussed in depth in the work of *Corral* [2010]. The implications of this finding for the predictability of the phenomenon would be rather negative, in the sense that this kind of processes are characterized by an inherent unpredictability, although the differences with deterministic chaos are worth being investigated in the future.

Nevertheless, the previous statistical method was not found to be fully satisfactory when applied to a different data set [*Corral et al.*, 2011]. The problem lies in the selection criterion for a and b. First, it can provide nonoptimal values of them, which would lead to the rejection of the null hypothesis in cases in which it is true. Second, it has a strong tendency to underestimate the upper limit b (if $\tau > 1$).

Another method, with a distinct selection procedure for a and b, was introduced by *Peters et al.* [2010]. The steps are as follows:

1. Fix arbitrary initial values of the truncation parameters a and b, with no restrictions.

2. Find the maximum likelihood estimation of the exponent τ (by means of the same formula as for the previous method).

3. Calculate the KS distance between the empirical PDI distribution (defined only for the range $a \leq$ PDI $< b$) and its power law fit.

4. Calculate the p value of the fit, which we denote by q, by simulations of synthetic power law data sets, with exponent τ (the simulation of the distribution outside the interval $[a, b)$ is not necessary in this case).

5. Look for and select the values of a and b that maximize the logarithmic range, b/a, under the condition, let us say, $q >$ 20%.

In the original reference by Peters et al., the number of data in between a and b, N_{ab}, was the quantity which was maximized, rather than the log-range b/a. As in our case the power law exponent is very close to one (for which there is

Table 2. The Same as Table 1 but Using the Fitting Procedure of *Peters et al.* [2010][a]

q_{min}	a (m^3 s^{-2})	b (m^3 s^{-2})	b/a	N_{ab}	CN_{ab}/N	$\tau \pm \sigma_\tau$	D	$d\sqrt{N_{ab}}$
99%	14.7×10^9	6.8×10^{10}	5	134	4.28	1.13 ± 0.20	3.5×10^{-2}	0.41
90%	2.2×10^9	6.3×10^{10}	29	371	8.59	1.16 ± 0.05	2.5×10^{-2}	0.47
50%	2.0×10^9	9.3×10^{10}	46	395	12.1	1.175 ± 0.05	3.0×10^{-2}	0.59
20%	1.7×10^9	10.0×10^{10}	58	414	11.9	1.175 ± 0.04	3.7×10^{-2}	0.76
10%	1.7×10^9	14.7×10^{10}	86	428	30.1	1.22 ± 0.04	4.3×10^{-2}	0.90

[a]Fitting procedure maximizes the log range under the condition $q > q_{min}$, where q denotes the p value for fixed a and b.

the same number of data for each order of magnitude), there should be no relevant difference between both procedures; nevertheless, we think the present choice is more appropriate for tails with larger exponents. Of course, setting the threshold q equal to 20% is arbitrary, and we will have to check the effect of changing this value.

One has to take into account that the q value calculated in this method is not the true p value of our fitting procedure. The former is calculated for fixed, or known, a and b values, whereas the latter should take into account that both parameters are optimized. Comparing with the previous method, it seems that the change between both p values can be around a factor 2 or 3 (so caution is necessary at this point). The results of this method for the same North Atlantic data used in Table 1 are displayed now in Table 2. Indeed, a comparison between both tables show that the results are consistent, if one takes into account the difference between p and q. In summary, the existence of a power law over a range larger than one decade is well supported by the statistical tests.

4. GAMMA DISTRIBUTION OF PDI

So far, we have established that a "significant" range of the PDI probability density can be described as a power law. Deviation at small values (PDI < a) is justified through the fact that the best track records are incomplete: this is obvious, as tropical depressions are deliberately excluded from the database. Inclusion of tropical depressions in the analysis of the northwestern Pacific (which are easily available, in contrast to tropical depressions at the NHC best track records) enlarges somewhat the power law range, although the coverage of tropical depressions in that basin is far from exhaustive [*Corral et al.*, 2010].

The rapid decrease of the PDI density above the value of b has been explained as a finite size effect: tropical cyclones cannot become larger (in terms of PDI) because they are limited by the finiteness of the basin; either they reach extratropical regions (cold conditions) or they reach land, in which cases, they are dissipated as they are deprived from their warm water energy source. In this way, it has been shown how the tracks of most events with PDI > b are affected by the boundaries of the basin, and a finite-size scaling analysis showed how a reduced area (limiting it in terms of longitude) moved the cutoff to smaller values, with the cutoff defined roughly as the point in which the PDI probability density clearly departs from the power law [*Corral et al.*, 2010].

Interestingly, it has also been found in the previously mentioned reference that the season-averaged tropical SST of the North Atlantic has an influence on the PDI values similar to a finite size effect. Years with high SST lead to a larger cutoff, just the opposite of years with low SST, but keeping nearly the same value of the power law exponent (the same has been found also for the northeastern Pacific). So, the power law is a robust feature of the PDI distribution, but it is not telling us anything about the influence of other factors (finiteness of the basins and SST); rather, it is the cutoff, which is not properly defined yet, which carries this information.

In order to overcome this deficit, a function modeling the PDI distribution that includes the tail seems appropriate. Experience with finite size effects in critical phenomena (taking place in continuous phase transitions or in avalanche-evolving systems) suggests a simple exponential factor for the tail and, therefore, a gamma-like distribution for the whole domain (excepting the incompleteness for small values),

$$D(x) = \frac{1}{c\Gamma(-\beta, a/c)} \left(\frac{c}{x}\right)^{1+\beta} e^{-x/c}, \quad \text{for} \quad x \geq a,$$

where x plays the role of the PDI (only for aesthetic reasons), c represents now the cutoff, which is a scale parameter, and β is a shape parameter, with the power law exponent equal to $1 + \beta$. If $a = 0$, β has to be negative, in order to avoid the divergence of the integral of $D(x)$; however, for $a > 0$, there is no restriction on β. The (complement of the unrescaled) incomplete gamma function is

$$\Gamma(\gamma, z_o) = \int_{z_o}^{\infty} z^{\gamma-1} e^{-z} dz$$

[*Abramowitz and Stegun*, 1965], with $\gamma > 0$ for $z_o = 0$ but unrestricted for $z_o > 0$. Unfortunately, some of the numerical routines we will use are not defined for $\gamma \leq 0$, so we will need to transform

$$\frac{1}{c\Gamma(-\beta, a/c)} = \frac{\beta}{c}\left[\left(\frac{c}{a}\right)^{\beta} e^{-a/c} - \Gamma(1-\beta, a/c)\right]^{-1}$$

(integrating by parts), where, for $\beta < 1$, the incomplete gamma function on the right-hand side can be computed without any problem. An alternative approach for the evaluation of the cutoffs would have been to compute the moment ratios used by *Peters et al.* [2010].

The reason to model D(PDI) by means of a gamma distribution is due to the fact that one can heuristically understand the finiteness of a system as introducing a kind of effective correlation length. Then, it is well known that, in the simplest cases, correlations enter into probability distributions under an exponential form [*Zapperi et al.*, 1995].

In contrast to the method of fit and goodness-of-fit test explained above and in contrast also to other previous work [*Corral*, 2009], in order to avoid difficulties, we perform a simple fitting procedure here, acting directly over the estimated probability density. The value of the minimum limit a is selected from information coming from the power law fits (Tables 1 and 2) complemented by visual inspection of the plots of the empirical probability density. A reasonable assumption is $a = 2 \times 10^9$ m^3 s^{-2} (note that the empirical D(PDI) is discretized).

The fit of the shape and scale parameters, β and c, is performed using the nonlinear least-squares Marquardt-Levenberg algorithm implemented in `gnuplot`, applied to the logarithm of the dependent variable (i.e., it is the logarithm of the model density, which is fit to the logarithm of empirical density); further, the parameter introduced in the algorithm is not c but $c' = \ln c$ (i.e., we write $c = \exp(c')$). Moreover, it is

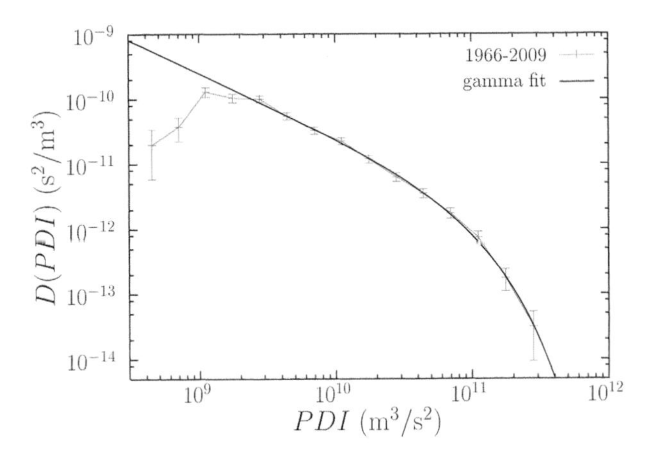

Figure 4. The same North Atlantic tropical cyclone data as in Figure 3 but with a gamma fit for PDI $\geq 2 \times 10^{10}$ m^3 s^{-2}. The exponent is $\tau = 1 + \beta = 0.98$, and the scale parameter $c = 8.1 \times 10^{10}$ m^3 s^{-2}. The fit seems reasonably good for 2 orders of magnitude.

necessary to correct that the empirical density is normalized from 0 to ∞ (i.e., not truncated, all events contribute to the normalization), whereas the model density goes from a to ∞. In order that one can fit the other, a factor N_a/N must multiply the fitting density (or, equivalently, divide the empirical one by the same factor), where N_a is the number of data points (tropical cyclones) fulfilling $x \geq a$. Due to the discretization of the empirical density, in practice, a has to coincide with the lower limit of a bin of the density, or alternatively, N_a has to be redefined accordingly.

The resulting fit for the 1966–2009 PDIs of the North Atlantic is displayed in Figure 4, and the obtained parameters are $\tau = 1 + \beta = 0.98 \pm 0.03$ and $c = (8.1 \pm 0.4) \times 10^{10}$ m^3 s^{-2}, when the empirical density is estimated with five bins per order of magnitude (and the position of those bins is the one in Figure 4). In contrast with the method used for power law fits, the results here depend on the width of the intervals of the empirical probability density (the number of bins or boxes per order of magnitude), and their position in the x axis. We have found that, in general, the influence of these factors on the parameters is more or less within the error bars given above.

Note that the exponent τ was larger than 1 for the (double truncated) power law, whereas now, $\tau = 1 + \beta$ turns out to be slightly below 1. The difference is small, but more importantly, both exponents represent different things because we are fitting different functions. If the true behavior of D(PDI) was a gamma distribution (which is, in practice, impossible to know), it is expected that a pure power law (over a certain range) would yield a larger (steepest) exponent, due to the bending effect of the exponential factor.

On the other hand, a gamma distribution with $1 + \beta \leq 1$ cannot be extrapolated to the case in which the cutoff is infinite and cannot describe vanishing finite size effects, as it would not be normalizable. If we enforce $1 + \beta \geq 1$ in the fit, we get $1 + \beta = 1.00 \pm 0.03$ and $c = (8.3 \pm 0.4) \times 10^{10}$ m^3 s^{-2}, which is not only a visually satisfactory fit but undistinguishable from the previous one. So, there is a tendency of the exponent to be smaller than one, but values slightly above one are equally acceptable.

5. SENSITIVITY OF THE TAIL OF THE PDI DISTRIBUTION TO SST

We now investigate how the splitting of the PDI distribution into two parts, one for warm SST years and another for colder years, influences the shape and scale of the resulting distributions. First, monthly SSTs with 1° spatial resolution are averaged for the tropical North Atlantic (90° to 20°W, 5° to 25°N) during the hurricane season (June to October). This yields a single SST number for each year or season [*Webster et al.*, 2005], and in this way, values above the 1966–2009

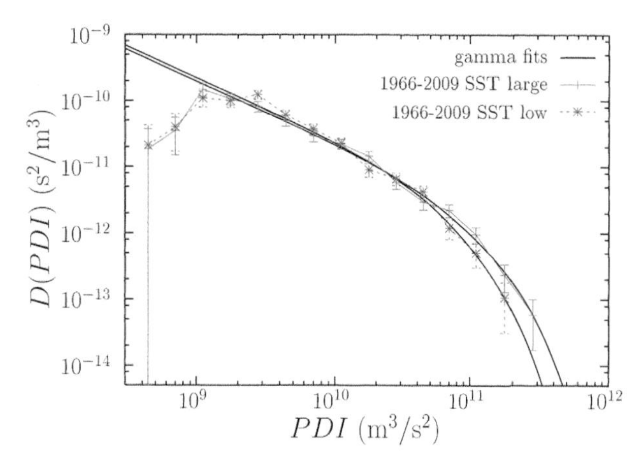

Figure 5. The same North Atlantic PDI data as for Figures 3 and 4 but separated into two distributions, one for years with sea surface temperature (SST) above the mean (large in the label) and another for below (low). The gamma fits, enforcing a constant exponent $1 + \beta = 0.93$ (see text), show an increase in the scale parameter c around 50% between cold and warm SST years.

mean are considered to denote warm years and the opposite for cold years (notice that the comparison is not done relative to a different, longer period). The average SST for warm years turns out to be 0.49°C larger than that of cold years. The SST data are those of the Hadley Centre, U.K. Meteorological Office [*Rayner et al.*, 2003; *Met Office*, 2010].

The resulting PDI distributions are displayed in Figure 5, which shows that the largest values of the PDI correspond to high SST years. If we compare the mean values of the PDI, the increase is around 40%, i.e., $\langle \mathrm{PDI} \rangle_{\mathrm{warm}} / \langle \mathrm{PDI} \rangle_{\mathrm{cold}} \simeq 1.42 \pm 0.19$, which is significantly larger than 1 (at the 99% confidence level, see Tables 3a and 3b).

The gamma fit also does a good job here, with an outcome, for high SST, $1 + \beta = 0.94 \pm 0.06$ and $c = (9.6 \pm 1) \times 10^{10}$ m^3 s^{-2}, taking $a = 2 \times 10^9$ m^3 s^{-2}. For low SST, visual inspection of the plot of the density suggests that it is more appropriate to raise the value of a to 3×10^9 m^3 s^{-2}, and then, $1 + \beta = 0.98 \pm 0.11$ and $c = (6.5 \pm 1) \times 10^{10}$ m^3 s^{-2}. This yields a $50 \pm 27\%$ increase in the value of the cutoff, which is significantly larger than zero if one assumes normality for the c parameters. Note that we do not expect differences between the ratios of the cutoffs and the ratios of the means, due to the fact that the power law exponent $1 + \beta$ turns out to be smaller than 1, and then the mean scales linearly with the cutoff (see the supplementary information of *Corral et al.* [2010]).

If, in order to make the comparison of c in the same conditions between warm and cold years, we fix $a = 3 \times 10^9$ m^3 s^{-2} and $1 + \beta = 0.93$ (this is the value obtained in this case for the whole data set, with $c = (7.8 \pm 0.4) \times 10^{10}$ m^3 s^{-2}), then we get $c = (9.6 \pm 0.5) \times 10^{10}$ m^3 s^{-2} and $(6.1 \pm 0.4) \times 10^{10}$ m^3 s^{-2}, for high and low SST years, respectively, which is not essentially different to the above case, but has smaller errors. This leads to $c_{\mathrm{warm}}/c_{\mathrm{cold}} = 1.57 \pm 0.13$, i.e., more than a 50 % increase in the largest values of the PDI for an increase in the SST equal to 0.49°C, which seems to indicate a high sensitivity of the extreme (cutoff) PDIs to the warming of the sea surface. The fits are displayed in Figure 5.

A further step is to separate the years not into two groups, warm and cold, but into three, very warm, intermediate, and very cold (in relative terms), and compare only the years with "extreme" temperatures. Taking the thresholds as the mean SST plus half its standard deviation and the mean minus half the standard deviation, we get a ratio of mean PDIs equal to 1.65 ± 0.25, for an increase in mean SST about 0.65°C (see Tables 3a and 3b). Thus, the splitting of the distributions is more pronounced for "extreme" SST.

Table 3a. Mean PDI for High and Low Values of Seasonal SST, Multivariate ENSO Index, NAO, AMO, Number n_{TC} of Tropical Cyclones, and Number of Hurricanes n_H for the North Atlantic for the Period 1966–2009[a]

	SST	MEI	NAO	AMO	n_{TC}	n_H
Variable difference	0.49°C	1.41	0.79	0.39	6.55	4.45
$\langle \mathrm{PDI} \rangle_{\mathrm{high}}$	2.55 ± 0.24	2.09 ± 0.23	2.11 ± 0.20	2.50 ± 0.24	2.39 ± 0.22	4.27 ± 0.36
$\langle \mathrm{PDI} \rangle_{\mathrm{low}}$	1.79 ± 0.17	2.29 ± 0.21	2.26 ± 0.22	1.85 ± 0.18	1.92 ± 0.21	2.83 ± 0.26
$\langle \mathrm{PDI} \rangle$ ratio	1.42 ± 0.19	0.91 ± 0.13	0.93 ± 0.13	1.35 ± 0.18	1.25 ± 0.18	1.51 ± 0.19
$\langle \mathrm{PDI} \rangle$ difference	0.76 ± 0.30	−0.20 ± 0.31[b]	−0.15 ± 0.30[b]	0.65 ± 0.30	0.48 ± 0.30[b]	1.45 ± 0.45

[a]For n_H, PDI values are calculated only for hurricanes (otherwise, we could get artificial larger ratios due to the fact that years with high number of hurricanes and less tropical storms would yield higher values of the PDI). The difference of the variables between the mean for high and low years is shown, as well as the ratio and the difference between the mean values of the PDI. Units of $\langle \mathrm{PDI} \rangle$ and its differences are 10^{10} m^3 s^{-2}. The uncertainty of the mean PDI is the standard deviation of the PDI divided by the square root of the number of data. The relative uncertainty of the ratio is the square root of the sum of square of the relative errors of each value of the means. Abbreviations are SST, sea surface temperature; MEI, multivariate El Niño–Southern Oscillation (ENSO) index; NAO, North Atlantic Oscillation; and AMO, Atlantic Multidecadal Oscillation (or mode).

[b]Nonsignificant differences (at the 95% level).

Table 3b. Mean PDI for Very High and Very Low Values of Seasonal SST, Multivariate ENSO Index, NAO, AMO, Number n_{TC} of Tropical Cyclones, and Number of Hurricanes n_H for the North Atlantic for the Period 1966–2009[a]

	SST	MEI	NAO	AMO	n_{TC}	n_H
Variable difference	0.65°C	1.92	1.15	0.52	9.47	6.32
$\langle \text{PDI} \rangle_{\text{very high}}$	2.70 ± 0.29	1.39 ± 0.17	2.43 ± 0.27	2.66 ± 0.31	2.53 ± 0.29	4.26 ± 0.41
$\langle \text{PDI} \rangle_{\text{very low}}$	1.64 ± 0.18	2.10 ± 0.24	2.22 ± 0.30	1.56 ± 0.17	1.60 ± 0.22	2.51 ± 0.31
⟨PDI⟩ ratio	1.65 ± 0.26	0.66 ± 0.11	1.09 ± 0.19	1.70 ± 0.27	1.58 ± 0.28	1.70 ± 0.26
⟨PDI⟩ difference	1.06 ± 0.34	−0.71 ± 0.30	0.21 ± 0.40[b]	1.09 ± 0.35	0.93 ± 0.36	1.75 ± 0.51

[a]For n_H, PDI values are calculated only for hurricanes (otherwise, we could get artificial larger ratios due to the fact that years with high number of hurricanes and less tropical storms would yield higher values of the PDI). The difference of the variables between the mean for high and low years is shown, as well as the ratio and the difference between the mean values of the PDI. Units of ⟨PDI⟩ and its differences are 10^{10} m^3 s^{-2}. The uncertainty of the mean PDI is the standard deviation of the PDI divided by the square root of the number of data. The relative uncertainty of the ratio is the square root of the sum of square of the relative errors of each value of the means.

[b]Nonsignificant differences (at the 95% level).

6. INFLUENCE OF EL NIÑO, THE NAO, THE AMO, AND THE NUMBER OF STORMS ON THE PDI

We can proceed in the same way for other climatic variables, or indices. Considering the El Niño–Southern Oscillation (ENSO), the multivariate ENSO index (MEI) separates warm phases of the tropical eastern Pacific (MEI > 0) from cold phases (MEI < 0), with high values of the MEI associated to El Niño and low values to La Niña [*Wolter and Timlin*, 1998]. It has been proven that El Niño phenomenon is anticorrelated with hurricane activity in the North Atlantic, in such a way that the presence of El Niño partially suppresses this activity [*Gray*, 1984], which has been detected even in the record of hurricane economic losses in the United States [*Pielke and Landsea*, 1999]. Using data from K. Wolter (MEI timeseries from Dec/Jan 1940/50 up to the present, 2010, available at http://www.esrl.noaa.gov/psd/people/klaus.wolter/MEI/table.html) and averaging for each season the months from May/June to September/October, we can distinguish between positive and negative phase seasons and, from here, obtain the corresponding PDI distributions.

We do the same for the NAO and AMO indices (averaging the indices from June to October), as well as for the annual number of tropical cyclones, n_{TC}, and the annual number of hurricanes, n_H. In the latter case, only hurricanes are accounted in the average of the PDI. High values of MEI, NAO, and AMO refer to just positive values of the indices, whereas high values of n_{TC} and n_H denote values above the 1966–2009 mean (in the same way as for the SST) and, conversely, for low values of the variables. Very high values correspond in all cases to values above the mean plus one half of the standard deviation (and conversely for very low values).

The results can be seen in Figures 6 and 7 and in Tables 3a and 3b. Comparing PDI distributions for positive and negative values of the ENSO, AMO, and NAO indices, only the ones separated by the AMO show significant differences (at the 95% level), with positive values of AMO triggering larger extreme PDI values. The lack of influence of ENSO in the PDI is in agreement with *Corral et al.* [2010].

But if the PDI distributions are compared for high enough values of the indices, both the MEI and the AMO show a clear influence on the PDIs (not the NAO). As seen in Figure 6, the presence of El Niño (very high MEI), in addition to reducing the number of North Atlantic hurricanes (which is well known), decreases the value of their PDI. Of course, just the same happens for very low values of the AMO index, which decrease both hurricane numbers [*Goldenberg et al.*, 2001] and the largest PDI values. La Niña or high values of the AMO have the opposite effect, increasing the most extreme PDIs.

The influence of the number of tropical cyclones in the values of the PDI is not significant for values of n_{TC} above or below its mean, but larger/smaller values of n_{TC} (mean ± 1/2 of standard deviation) are correlated with larger/smaller values of the PDIs, with an increase in ⟨PDI⟩ around 60% when the mean n_{TC} goes from 7.35 events to 16.8. If we consider hurricane counts, this effect is more pronounced, although care has to be taken with a kind of circular argument here: just by chance, the number of hurricanes can increase at the same time that the number of tropical storms decreases; then, it is likely that the mean PDI will be higher in this case (as the maximum sustained wind speed is correlated with PDI), but no physical effect is behind this, only statistical fluctuations. Therefore, we study the influence of the number of hurricanes on the PDI distributions of hurricanes only, finding that years with more hurricanes also have larger PDI values. This seems to indicate that the conditions for genesis and survival of the storms are correlated (compare with the conclusions in section 4.3 of *Elsner and Kara* [1999]).

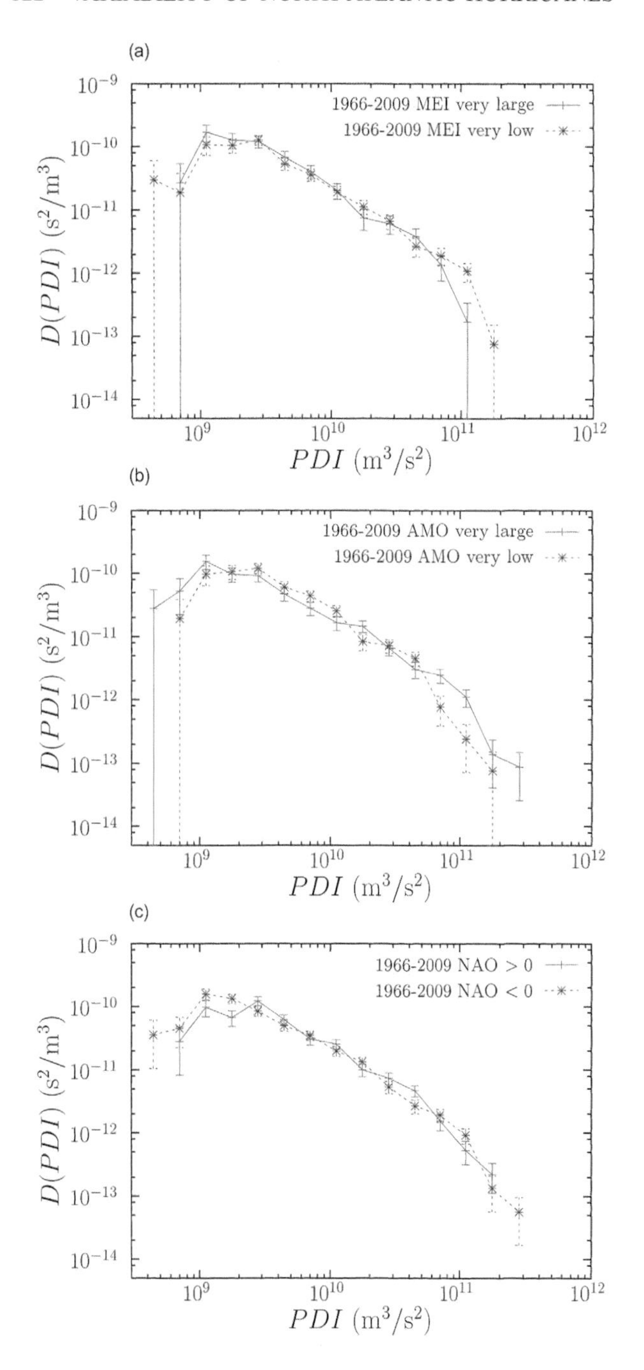

Figure 6. (a) The same North Atlantic PDI data separating this time by hurricane season multivariate El Niño–Southern Oscillation (ENSO) index above its mean plus 1/2 of the standard deviation (labeled very large) and below the mean minus 1/2 of the standard deviation (very low). (b) The same for Atlantic multidecadal mode (or oscillation). (c) The same data separating by North Atlantic oscillation (NAO) > 0 and NAO < 0. In this case, the differences are clearly nonsignificant.

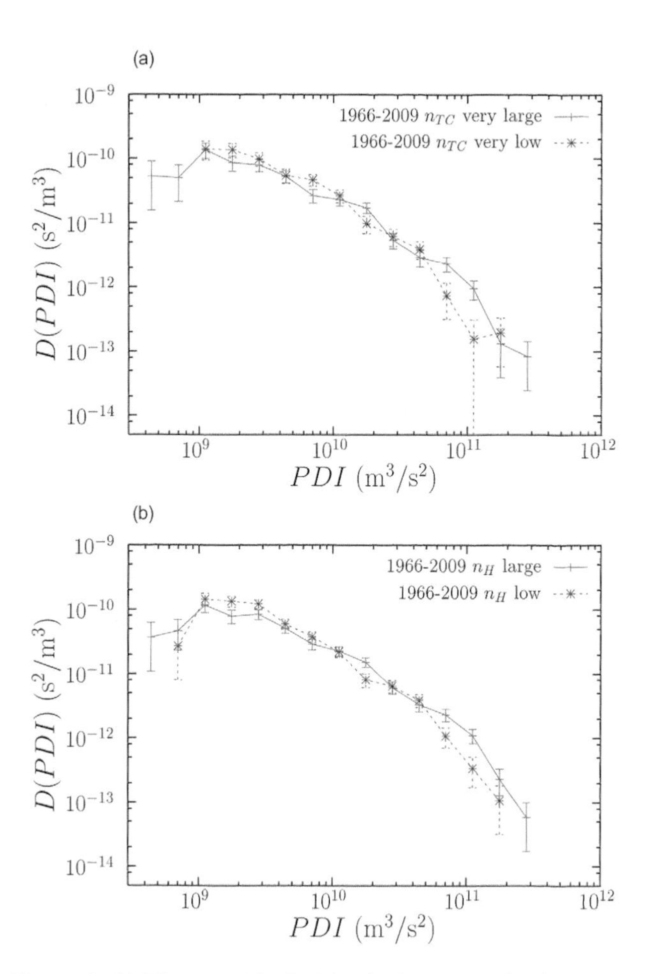

Figure 7. (a) The same North Atlantic data separating by annual number of tropical cyclones above and below its mean ± 1/2 of the standard deviation (labeled as very large and very low, respectively). (b) The same separating by number of hurricanes above or below its mean (large and low). This plot is not the one corresponding to the last column of Tables 3a and 3b, as tropical storms are not removed here. Caution has to be present in the interpretation of this case; see text.

7. SUMMARY AND CONCLUSIONS

We have illustrated how diverse metrics have been introduced by several authors in order to quantify tropical cyclone activity, for instance, major hurricane counts, total major hurricane days, major hurricane duration, NTC, etc ... In general, these metrics show a clear increase in activity in the North Atlantic since 1995, although it is difficult to associate a single cause to this phenomenon. Particularly interesting is the PDI, which was proposed by *Emanuel* [2005b] in order to estimate annual tropical cyclone kinetic energy dissipation in different basins.

In a previous work [*Corral et al.*, 2010], we demonstrated how the use of the PDI to characterize the energy dissipation of individual tropical cyclones gave coherent and robust results, leading to nearly universal power law distributions (i.e., with roughly the same exponents for different ocean basins). The outcome of this approach, in contrast to the findings obtained using other indices of activity, has important implications for the understanding of the physics of tropical cyclones, as it allows a connection with self-organized critical systems [*Corral*, 2010]. This, in turn, opens interesting questions about the limits of predictability of tropical cyclones and, in a broader context, about the compatibility between criticality and chaos in the atmosphere.

Due to the important implications that the emergence of power law distributions has, we also discuss different ways of fitting them. The main problem is not to find the power law exponent (which is the only parameter of the distribution) but to decide for which values of the variable the power law regime holds. The procedure is based on the work of *Clauset et al.* [2009], although some variations seem necessary in order to improve the performance of the method.

Whereas the power law part of the PDI distribution does not change under different climatic conditions (as it should be in a critical system), the tail of the distribution does, containing therefore precious information about external influences on the system, in particular on the finiteness of the part of the basin able to sustain tropical cyclone activity. Therefore, a probability distribution modeling not only the power law regime but also the faster decay at the largest PDI values seems very appealing. We propose the use of the gamma distribution, combining a power law with an exponential decay.

In this way, it is shown that the cutoff parameter modeling the exponential decay increases with annually averaged SST, which means that high SST has an effect on the PDI that is analogous to expanding the effective size of the part of the ocean over which tropical cyclones develop (and the opposite for low values of the annual SST). The same effect is found for the AMO index, with high AMO values leading to larger extreme PDI values. However, the El Niño phenomenon has the opposite effect, being the presence of La Niña, which triggers the largest (in dissipated energy) North Atlantic hurricanes. The number of hurricanes is also found to be positively correlated with their PDI. In contrast, for the NAO index, we do not find any significant correlation.

In conclusion, the characterization of PDI probability distribution reveals as an interesting tool not only to learn about the fundamental nature of tropical cyclones but also to evaluate the effect of different climatic indices on the energy dissipated by them.

APPENDIX A: RELATION BETWEEN PDI AND DISSIPATED ENERGY

The power dissipated at any time by a tropical cyclone can be estimated by means of the formula [*Bister and Emanuel*, 1998],

$$P(t) = \int \rho C_D v^3 \mathrm{d}^2 r,$$

where t is time, r is the spatial coordinate over the Earth's surface, ρ is the air density, C_D is the drag coefficient, and v is the modulus of the wind velocity (the latter three depending on r and t).

The formula can be simplified in practice. *Emanuel* [1998] took representative constant values for the density and the drag coefficient, $\rho = 1$ kg m^{-3}, $C_D = 0.002$, and assumed a simple velocity profile, $v(\vec{r}, t) = v_m(t) f(r/R(t))$, with $v_m(t)$ the maximum speed across the storm at time t, $R(t)$ some characteristic radius of the storm, as the radius of maximum winds, and f a scaling function that is the same for all storms. Denoting the spatial integral as $I = \int 2\pi f^3(u) u \mathrm{d}u$, then

$$P(t) \simeq \rho C_D I R^2(t) v_m^3(t) = k R^2(t) v_m^3(t).$$

Emanuel does not provide enough details about the integral I, but from the values of power, radius, and speed that he uses as an illustration, it has to be $I \simeq 2$, and then $k = 0.004$ kg m^{-3}.

One can get an estimation of the dissipated kinetic energy E just integrating $P(t)$ over time. Introducing an averaged radius over the storm, R_m, and the PDI discretization [*Emanuel*, 2005b],

$$E = \int P(t) \mathrm{d}t \simeq \rho C_D I R_m^2 \int v_m^3(t) \mathrm{d}t = k R_m^2 \mathrm{PDI}.$$

The best track records still do not provide the information required by this equation, as, in general, the storm radius is missing. However, for the northwestern Pacific, since 2001, some radii are available (Joint Typhoon Warning Center, 2011, http://www.usno.navy.mil/JTWC). Out of 11,358 6 h records for the period 2001–2010, 1671 provide nonzero values of the radius of maximum winds and the eye diameter. Averaging these values of the radii of maximum winds, we get $R_m \simeq 30$ km. If we average the square of the radius of maximum winds and take the square root, then we get as a characteristic value $R_m \simeq 35$ km, which is the value we use for the North Atlantic, then, $kR_m^2 = 4.9 \times 10^6$ kg m^{-1} and

$$E \simeq 4.9 \cdot 10^6 \quad \mathrm{PDI},$$

which yields dissipated energy in Joules if the PDI is in m^3 s^{-2}.

Under these approximations, as the North Atlantic individual storm PDI ranges between 5×10^8 and 2×10^{11} m^3 s^{-2}, the estimated dissipated energy turns out to be roughly between 3×10^{15} and 10^{18} J, i.e., between 0.6 and 200 megatons (1 megaton = 4.18×10^{15} J). Nevertheless, the real range of variation will be larger, as the variability of the radius, which we have disregarded, increases the variability of the dissipated energy.

Acknowledgments. The authors acknowledge initial support by J. E. Llebot and guidance by J. Kossin. First results of this research came through collaboration with A. Ossó. I. Bladé, A. Clauset, A. Deluca, K. Emanuel, R. D. Malmgren, and N. Moloney provided some feedback on different parts of the research. Money and a 1 year bursary for A. Ossó was obtained from Explora-Ingenio 2010 program (MICINN, Spain), grant FIS2007-29088-E. Other grants are FIS2009-09508 and 2009SGR-164. The first author is also a participant of the Consolider i-Math project.

REFERENCES

Aban, I. B., M. M. Meerschaert, and A. K. Panorska (2006), Parameter estimation for the truncated Pareto distribution, *J. Am. Stat. Assoc.*, *101*, 270–277.

Aberson, S. D. (2009). Regimes or cycles in tropical cyclone activity in the North Atlantic, *Bull. Am. Meteorol. Soc.*, *90*(1), 39–43.

Abramowitz, M., and I. A. Stegun (Eds.) (1965), *Handbook of Mathematical Functions: with Formulas, Graphs, and Mathematical Tables*, 1046 pp., Dover, New York.

Bak, P. (1996), *How Nature Works: The Science of Self-Organized Criticality*, 212 pp., Copernicus, New York.

Bell, G. D., M. S. Halpert, R. C. Schnell, R. W. Higgins, J. Lawrimore, V. E. Kousky, R. Tinker, W. Thiaw, M. Chelliah, and A. Artusa (2000), Climate assessment for 1999, *Bull. Am. Meteorol. Soc.*, *81*(6), S1–S50.

Bell, G. D., E. Blake, K. C. Mo, C. W. Landsea, R. Pasch, M. Chelliah, S. B. Goldenberg, and H. J. Diamond (2006), The record breaking 2005 Atlantic hurricane season, *Bull. Am. Meteorol. Soc.*, *87*(6), S44–S45.

Bister, M., and K. A. Emanuel (1998), Dissipative heating and hurricane intensity, *Meteorol. Atmos. Phys.*, *65*, 233–240.

Burroughs, S. M., and S. F. Tebbens (2001), Upper-truncated power laws in natural systems, *Pure Appl. Geophys.*, *158*, 741–757.

Chan, J. C. L. (2006), Comment on "Changes in tropical cyclone number, duration, and intensity in a warming environment," *Science*, *311*, 1713.

Christensen, K., and N. R. Moloney (2005), *Complexity and Criticality*, 392 pp., Imperial College Press, London.

Clauset, A., C. R. Shalizi, and M. E. J. Newman (2009), Power-law distributions in empirical data, *SIAM Rev.*, *51*, 661–703.

Corral, A. (2009), Statistical tests for scaling in the inter-event times of earthquakes in California, *Int. J. Mod. Phys. B*, *23*, 5570–5582.

Corral, A. (2010), Tropical cyclones as a critical phenomenon, in Hurricanes and Climate Change, vol. 2, edited by J. B. Elsner, pp. 81–99, Springer, Heidelberg.

Corral, A., A. Ossó, and J. E. Llebot (2010), Scaling of tropical-cyclone dissipation, *Nature Phys.*, *6*, 693–696.

Corral, A., F. Font, and J. Camacho (2011), Non-characteristic half-lives in radioactive decay, *Phys. Rev. E*, *83*, 066103, doi:10.1103/PhysRevE.83.066103.

Elsner, J. B., and A. B. Kara (1999), *Hurricanes of the North Atlantic*, 512 pp., Oxford Univ. Press, New York.

Elsner, J. B., T. Jagger, and X.-F. Niu (2000), Changes in the rates of North Atlantic major hurricane activity during the 20th century, *Geophys. Res. Lett.*, *27*(12), 1743–1746.

Elsner, J. B., A. A. Tsonis, and T. H. Jagger (2006), High-frequency variability in hurricane power dissipation and its relationship to global temperature, *Bull. Am. Meteorol. Soc.*, *87*(6), 763–768.

Elsner, J. B., J. P. Kossin, and T. H. Jagger (2008), The increasing intensity of the strongest tropical cyclones, *Nature*, *455*, 92–95.

Emanuel, K. A. (1998). The power of a hurricane: An example of reckless driving on the information superhighway, *Weather*, *54*, 107–108.

Emanuel, K. A. (2005a), *Divine Wind: The History and Science of Hurricanes*, 296 pp., Oxford Univ. Press, New York.

Emanuel, K. A. (2005b), Increasing destructiveness of tropical cyclones over the past 30 years, *Nature*, *436*, 686–688.

Emanuel, K. A. (2007), Environmental factors affecting tropical cyclone power dissipation, *J. Clim.*, *20*, 5497–5509.

Goldenberg, S. B., C. W. Landsea, A. M. Mestas-Nuñez, and W. M. Gray (2001), The recent increase in Atlantic hurricane activity: Causes and implications, *Science*, *293*, 474–479.

Gray, W. M. (1984), Atlantic seasonal hurricane frequency. Part I: El Niño and 30 mb quasi-biennial oscillation influences, *Mon. Weather Rev.*, *112*, 1649–1668.

Gray, W. M. (1990), Strong association between West African rainfall and U.S. landfall of intense hurricanes, *Science*, *249*, 1251–1256.

Gray, W. M., C. W. Landsea, P. W. Mielke Jr., and K. J. Berry (1992), Predicting Atlantic seasonal hurricane activity 6–11 months in advance, *Weather Forecasting*, *7*, 440–455.

Hergarten, S. (2002), *Self-Organized Criticality in Earth Systems*, 250 pp., Springer, Berlin.

Jarvinen, B. R., C. J. Neumann, and M. A. S. Davis (1988), A tropical cyclone data tape for the North Atlantic basin, 1886–1983: Contents, limitations, and uses, *Tech. Memo. NWS NHC 22*, NOAA, Silver Spring, Md. [Available at http://www.nhc.noaa.gov/pdf/NWS-NHC-1988-22.pdf.]

Jensen, H. J. (1998), *Self-Organized Criticality: Emergent Complex Behavior in Physical and Biological Systems*, 168 pp., Cambridge Univ. Press, Cambridge, U. K.

Kerr, R. A. (2000), A North Atlantic climate pacemaker for the centuries, *Science*, *288*, 1984–1985.

Klotzbach, P. J. (2006), Trends in global tropical cyclone activity over the past twenty years (1986–2005), *Geophys. Res. Lett.*, *33*, L10805, doi:10.1029/2006GL025881.

Knutson, T. R., C. Landsea, and K. Emanuel (2010a), Tropical cyclones and climate change: A review, in *Global Perspectives on Tropical Cyclones: From Science to Mitigation*, *Asia-Pacific Weather Climate Ser.*, vol. 4, edited by J. C. L. Chan and J. D. Kepert, pp. 243–284, World Sci., Singapore.

Knutson, T. R., J. L. McBride, J. Chan, K. Emanuel, G. Holland, C. Landsea, I. Held, J. P. Kossin, A. K. Srivastava, and M. Sugi (2010b), Tropical cyclones and climate change, *Nat. Geosci.*, *3*, 157–163.

Kossin, J. P., K. R. Knapp, D. J. Vimont, R. J. Murnane, and B. A. Harper (2007), A globally consistent reanalysis of hurricane variability and trends, *Geophys. Res. Lett.*, *34*, L04815, doi:10.1029/2006GL028836.

Landsea, C. W. (1993), A climatology of intense (or major) hurricanes, *Mon. Weather Rev.*, *121*, 1703–1713.

Landsea, C. W. (2005), Hurricanes and global warming, *Nature*, *438*, E11–E12.

Landsea, C. W. (2007), Counting Atlantic tropical cyclones back to 1900, *Eos Trans. AGU*, *88*(18), 197.

Landsea, C. W., B. A. Harper, K. Hoarau, and J. A. Knaff (2006), Can we detect trends in extreme tropical cyclones?, *Science*, *313*, 452–454.

Landsea, C. W., G. A. Vecchi, L. Bengtsson, and T. R. Knutson (2010), Impact of duration thresholds on Atlantic tropical cyclone counts, *J. Clim.*, *23*, 2508–2519.

Met Office (2010), Met Office-HadISST 1.1-Global Sea Ice Coverage and Sea Surface Temperature Data (1870–Present), http://badc.nerc.ac.uk/data/hadisst/, Br. Atmos. Data Cent., Didcot, U. K.

Neumann, C. J., B. R. Jarvinen, C. J. McAdie, and G. R. Hammer (1999), Tropical cyclones of the North Atlantic Ocean, 1871–1998, *Hist. Climatol. Ser. 6-2*, NOAA, Ashville, N. C.

Peters, O., and J. D. Neelin (2006), Critical phenomena in atmospheric precipitation, *Nature Phys.*, *2*, 393–396.

Peters, O., and J. D. Neelin (2009), Atmospheric convection as a continuous phase transition: Further evidence, *Int. J. Mod. Phys. B*, *23*, 5453–5465.

Peters, O., A. Deluca, A. Corral, J. D. Neelin, and C. E. Holloway (2010), Universality of rain event size distributions, *J. Stat. Mech.*, *2010*, P11030, doi:10.1088/1742-5468/2010/11/P11030.

Pielke, R. A., Jr., and C. N. Landsea (1999), La Niña, El Niño, and Atlantic hurricane damages in the United States, *Bull. Am. Meteorol. Soc.*, *80*(10), 2027–2033.

Press, W. H., S. A. Teukolsky, W. T. Vetterling, and B. P. Flannery (1992), *Numerical Recipes in FORTRAN: The Art of Scientific Computing*, 2nd ed., 933 pp., Cambridge Univ. Press, Cambridge, U. K.

Rayner, N. A., D. E. Parker, E. B. Horton, C. K. Folland, L. V. Alexander, D. P. Rowell, E. C. Kent, and A. Kaplan (2003), Global analyses of sea surface temperature, sea ice, and night marine air temperature since the late nineteenth century, *J. Geophys. Res.*, *108*(D14), 4407, doi:10.1029/2002JD002670.

Shepherd, J. M., and T. Knutson (2007), The current debate on the linkage between global warming and hurricanes, *Geogr. Compass*, *1*, 1–24.

Sornette, D. (2004), *Critical Phenomena in Natural Sciences: Chaos, Fractals, Selforganization and Disorder: Concepts and Tools*, 2nd ed., 528 pp., Springer, Berlin.

Swanson, K. L. (2008), Nonlocality of Atlantic tropical cyclone intensities, *Geochem. Geophys. Geosyst.*, *9*, Q04V01, doi:10.1029/2007GC001844.

Trenberth, K. (2005), Uncertainty in hurricanes and global warming, *Science*, *308*, 1753–1754.

Turcotte, D. L. (1997). *Fractals and Chaos in Geology and Geophysics*, 2nd ed., 416 pp., Cambridge Univ. Press, Cambridge.

Webster, P. J., G. J. Holland, J. A. Curry, and H.-R. Chang (2005), Changes in tropical cyclone number, duration, and intensity in a warming environment, *Science*, *309*, 1844–1846.

Wolter, K., and M. S. Timlin (1998), Measuring the strength of ENSO events: How does 1997/98 rank?, *Weather*, *53*(9), 315–324.

Wu, L., B. Wang, and S. A. Braun (2008), Implications of tropical cyclone power dissipation index, *Int. J. Climatol.*, *28*, 727–731.

Zapperi, S., K. B. Lauritsen, and H. E. Stanley (1995), Self-organized branching processes: Mean-field theory for avalanches, *Phys. Rev. Lett.*, *75*, 4071–4074.

A. Corral, Centre de Recerca Matemàtica, Edifici Cc, Campus Bellaterra, E-08193 Barcelona, Spain. (acorral@crm.cat)

A. Turiel, Institut de Ciències del Mar, CSIC, P. Marítim Barceloneta 37, E-08003 Barcelona, Spain.

Large-Scale Patterns in Hurricane-Driven Shoreline Change

Eli D. Lazarus

School of Earth and Ocean Sciences, Cardiff University, Cardiff, UK

Andrew D. Ashton

Department of Geology and Geophysics, Woods Hole Oceanographic Institution, Woods Hole, Massachusetts, USA

A. Brad Murray

Division of Earth and Ocean Sciences, Nicholas School of the Environment, Center for Nonlinear and Complex Systems
Duke University, Durham, North Carolina, USA

The effects of storm events on cross-shore beach profiles have been the subject of concerted examination by nearshore researchers for decades. Because these investigations typically span relatively short (less than a kilometer) shoreline reaches, alongshore patterns of storm-driven shoreline change at multikilometer scales remain poorly understood. Here we measure shoreline position from seven airborne lidar surveys of coastal topography, spanning 12 years (1996–2008), along a continuous ~80 km stretch of the northern North Carolina Outer Banks, United States. Two of the lidar surveys were flown in the wakes of Hurricane Bonnie (1998) and Hurricane Floyd (1999), allowing a rare window into storm-related alongshore coastline changes at large scales. In power spectra of shoreline change variance and in calculations of plan view shoreline curvature, we find evidence of transient behaviors at relatively small alongshore scales (less than a kilometer) and an interesting combination of both transient and cumulative shoreline change patterns at larger scales (1–10 km). Large-scale plan view shoreline undulations grow in amplitude during the storm intervals we examined, possibly forced by a large-scale morphodynamic instability. Long-term (decadal) shoreline adjustments, however, trend in the opposite direction, with an overall diffusion or smoothing of shoreline shape at multiple-kilometer scales, probably due to gradients in alongshore sediment transport. Although storms can significantly reshape the coastline across a wide range of scales, those changes do not necessarily accumulate to patterns of long-term change.

1. INTRODUCTION

Visit a beach after the height of a storm, and evidence of the event is obvious. Dunes that had sloping toes have been carved into flat-faced escarpments, roads are undercut, and houses can even be found in the surf zone. Coastal geomorphology texts have long emphasized the rapid movement of sand offshore as the salient effect of storms on beaches [e.g.,

Extreme Events and Natural Hazards: The Complexity Perspective
Geophysical Monograph Series 196

10.1029/2011GM001074

Davis, 1978; *Carter*, 1988; *Davis and Fitzgerald*, 2004] and with good reason: High winds and storm waves push a wedge of water onshore, allowing waves to reach more of the beach than under calm conditions. Persistent near-bed currents and breaking waves narrow the beach, transporting sediment offshore, building a sandbar. If storm surge elevates water levels high enough to submerge low-lying areas along the beach, sediment can be transported landward of the beach face and deposited behind the beach as overwash, removing that material from the immediate reach of the nearshore system [e.g., *Godfrey and Godfrey*, 1973; *Komar*, 1998].

But absent major overwashing, were you to return to the storm-battered beach in fair weather a few weeks later, you might be surprised to find those obvious signs of storm impact all but erased and the beach restored. Storm-driven formation of offshore bars, and the subsequent poststorm shoreward migration of those bars, is well documented [e.g., *Davis and Fox*, 1975; *Birkemeier*, 1979; *Egense*, 1989; *Thom and Hall*, 1991]. Sand removed from the beach during a storm event is stored offshore, typically in bars at the deep water margin of the storm surf zone, where the strength of the offshore current at the bed diminishes [*Komar*, 1998]. Then, in fair-weather conditions that tend to follow storms, long-period swell sweeps the sandbar shoreward, ultimately merging with the subaerial beach, resetting it to an effectively prestorm morphology [e.g., *Zeigler et al.*, 1959; *Sonu*, 1973; *Fucella and Dolan*, 1996]. Time-averaged images of the surf zone produce striking images of cyclical sandbar creation, migration, obliteration, and recreation with the passage of storm events [*Holman and Stanley*, 2007]. Looking along-coast, vehicle-mounted GPS shoreline surveys covering several tens of kilometers of coastline reveal that, even after large storm events, shoreline changes before and after a storm are effectively mirrored: coastal reaches that experience significant erosion see nearly equivalent magnitudes of accretion in the days and weeks following the storm event [*List and Farris*, 1999; *List et al.*, 2006].

Although homes and coastal infrastructure face short-term hazards during a storm from both flooding and temporary erosion, significant long-term hazards for coastal dwellers and development arise from accumulated changes in the shoreline itself, manifesting as cross-shore changes in shoreline position on the order of tens to hundreds of meters (or more) over decades or longer. Multidecadal records of shoreline change show zones of cumulative aggradation and erosion that vary alongshore at scales on the order of kilometers [e.g., *Schupp et al.*, 2006; North Carolina Division of Coastal Management, Oceanfront shorelines and setback: Interactive mapping, available at http://dcm2.enr.state.nc.us/Maps/shoreline_mapintro.htm, accessed January 2011]. Shoreline change might be approximately net zero over storm-event time scales, but how do the magnitudes of storm-related shoreline variability compare to magnitudes of long-term, accumulated changes measured over a range of spatial scales? Do storms affect particular spatial scales more than others? What physical insights into coastal dynamics do we gain, or what new hypotheses can we generate, from comparing fair-weather and storm-related shoreline change over a range of spatial and temporal scales?

Focusing on the well-studied sandy barrier coastline of the northern North Carolina Outer Banks, United States (Figure 1), in this chapter, we build upon existing, fair-weather analyses of shoreline change at multiple spatial and temporal scales [*Tebbens et al.*, 2002; *Lazarus and Murray*, 2007; *Lazarus et al.*, 2011] by introducing new analyses of two poststorm shorelines extracted from lidar surveys flown after Hurricane Bonnie (in 1998) and Hurricane Dennis and Hurricane Floyd (1999) (Table 1). We examine changes in shoreline position between different pairs of surveys. In some pairs, the second survey took place shortly after an extreme storm, and the change between surveys therefore reflects, in part, storm-driven changes. In other pairings, neither survey was affected by a recent extreme storm.

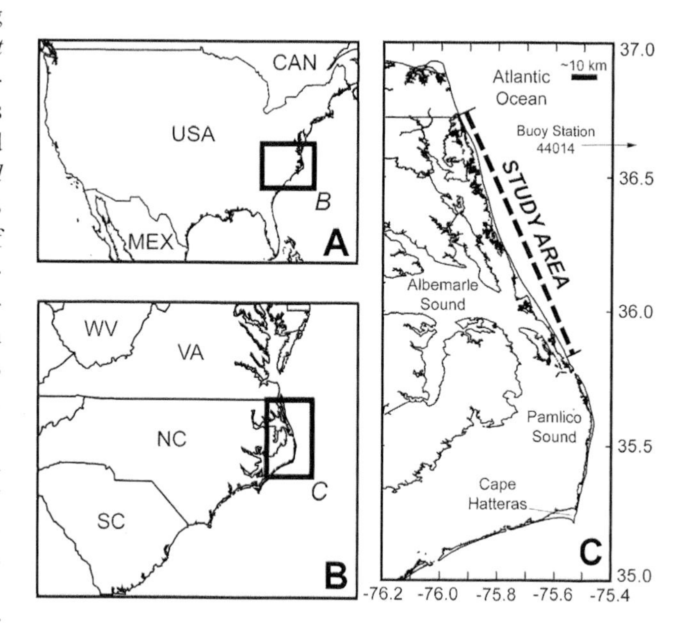

Figure 1. Our investigation focuses on approximately 80 km of uninterrupted coastline along the northern North Carolina Outer Banks, a sandy barrier island system on the U.S. Mid-Atlantic Seaboard (latitude/longitude in decimal degrees). National Buoy Data Center station 44014 (see Figure 5) is located at 36.611/-74.836, just out of the frame. Map data are courtesy of the NOAA coastline extractor, available at http://www.ngdc.noaa.gov/mgg/coast/, accessed, February 2011.)

Table 1. Lidar Surveys by Type (Fair Weather or Poststorm) and Date[a]

Survey Year	Fair Weather	Poststorm	Survey Date	Storm Date
1996	X		9–12 Oct 1996	
1997	X		15–27 Sep 1997	
1998		**X**	**1–7 Sep 1998**	**26 August**
1999		**X**	**18 Sep to 6 Oct 1999**	**4 and 16 September**
2004	X		9–13 Jul 2004	
2005	X		1 Oct to 26 Nov 2005	
2008	X		17–27 Mar 2008	

[a]Original data, with survey metadata, available through the NOAA Coastal Services Center's Digital Coast portal (http://www.csc.noaa.gov/digitalcoast/data/coastallidar/details.html, accessed January 2009). Poststorm surveys are indicated in bold.

2. SHORELINE CHANGE AND THE IMPORTANCE OF SPATIAL SCALE

2.1. *Calculating Shoreline Change*

2.1.1. Extracting a shoreline from lidar data. Although the cost of long-distance aerial beach surveys makes their collection infrequent, airborne lidar maps coastal topography over long distances with high precision [e.g., *Stockdon et al.*, 2002]. Here we extracted shorelines from the publicly available NOAA/U.S. Geological Survey/NASA lidar data sets listed in Table 1 (Digital Coast, NOAA Coastal Services Center (CSC), Coastal lidar, available at http://www.csc.noaa.gov/digitalcoast/data/coastallidar/details.html (hereinafter referred to as Digital Coast, NOAA CSC, Coastal lidar), accessed February 2007 and January 2009). To isolate a shoreline, we first converted raw (x, y, z) point clouds (0.25–8 points m^{-1}) (Digital Coast, NOAA CSC, Coastal lidar, accessed July 2011) into 5 m gridded digital elevation models, registered to the 1988 North American Vertical Datum and projected to the universal transverse Mercator North American Datum 1983 Zone 18N ellipsoid. Shorelines stretching on the order of tens of kilometers can be defined at an elevation contour; we sampled the 1 m topographic contour, which lies within 0.5 m of the mean high-water line for this region of coast [*Tebbens et al.*, 2002]. This contour retains fine-scale features in the beach while minimizing data artifacts from wave crests [*Lazarus and Murray*, 2007]. Airborne lidar data are typically vertically accurate to within ~10–40 cm [*Stockdon et al.*, 2002; Digital Coast, NOAA CSC, Coastal lidar, accessed July 2011]. With each shoreline registered and projected to the same reference frame, changes in shoreline position between two surveys can be determined by differencing the surveys.

The 1998 and 1999 lidar surveys were flown shortly after the passage of major storm events. On 26 August 1998, Hurricane Bonnie made landfall as a waning category 3 storm near Wilmington, North Carolina, south of our study area (National Hurricane Center, National Weather Service, NOAA, Archive of hurricane seasons, available at http://www.nhc.noaa.gov/pastall.shtml (hereinafter referred to as National Hurricane Center, Archive of hurricane seasons), accessed February 2011); the lidar data were acquired between 1 and 7 September, 1998. In 1999, Hurricane Dennis (a weak but lingering storm that stalled offshore for several days) and Hurricane Floyd (category 2) made landfall in North Carolina on 4 and 16 September, respectively (National Hurricane Center, Archive of hurricane seasons, accessed February 2011); the lidar data we use here was acquired 18 September 1999 [*White and Wang*, 2003; *Mitasova et al.*, 2005].

2.1.2. Examining shoreline change using wavelet analysis. We analyze the shoreline change signal using wavelets, which are scaled filter transforms that, when convolved with a signal, return coefficients constituting a spatially localized measure of signal variability at a given scale [*Hubbard*, 1996]. With the Wavelet Toolbox™ in Matlab 2009b, we applied the basic Haar wavelet in a continuous wavelet transform. The Haar wavelet is a step function of the form

$$\psi(x) = \begin{cases} 1 & 0 \le x < \frac{1}{2} \\ -1 & \frac{1}{2} < x \le 1 \\ 0 & \text{otherwise} \end{cases} \qquad (1)$$

that is translated along the signal according to

$$\psi_{jk}(x) = \psi(2^j x - k), \qquad (2)$$

where the scale exponent j is a non-negative integer and $0 \le k \le 2^j - 1$. Because of its simple shape, the Haar wavelet is less sensitive to variability at fine spatial scales but useful for identifying lower-frequency signals. To mitigate edge effects at either end of the transform, we reflected the shoreline change signal several times and used a multiple of the signal

[*Nievergelt*, 1999]. As an additional conservative step, we limited calculations to alongshore scales less than or equal to half the length of the original signal [*Lazarus et al.*, 2011].

Squaring the wavelet transform coefficients produces a measure of signal variance, which, when averaged over the length of the signal, returns the mean shoreline variance at each wavelet scale. Averaging the wavelet transforms this way is equivalent to producing a Fourier transform. The typical utility of a wavelet transform is its preservation of spatial heterogeneities within a data series; by comparison, a Fourier transform assumes that each component wavelength exists through the entire domain of the data, washing out localized information. The rationale for using wavelets in a Fourier-like fashion is that a power spectrum provides a useful summary of how transform variance depends on scale, complimenting the scale-specific, spatial details contained in the full wavelet transform [*Lazarus et al.*, 2011].

2.2. Insights From Fair-Weather Shoreline Survey Comparisons

2.2.1. Shoreline change characteristics at relatively small scales. A previous comparison of lidar-derived shoreline changes over 1 year, measured (in fair weather) for tens of kilometers alongshore, shows a log-log linear relationship between alongshore scale and the variance of shoreline change for alongshore scales spanning one order of magnitude, from approximately 100 to 1000 m [*Tebbens et al.*, 2002]. Replicating this investigation in the same study area with different fair-weather, lidar-derived surveys of shoreline change ranging from 1 to 12 years, *Lazarus et al.* [2011] confirm that mean shoreline variance at scales 20–1000 m alongshore appear log-log linear (Figure 2a), suggesting that shoreline change follows a power law over these scales, the scaling exponent being a measure of the fractal roughness of the shoreline change pattern [*Tebbens et al.*, 2002].

However, linearity in double-logarithmic plots, particularly over confined scales, is not definitive proof that a power law is the best descriptor for those data [e.g., *Sornette*, 2006]. Statistical measures of model uncertainty, such as the Akaike information criteria (AIC), help quantify how well different statistical distributions apply to data [*Akaike*, 1973; *Burnham and Anderson*, 2001, 2002]. Although the shoreline change power spectra that we investigate appear linear enough in log-log space to inspire comparison to a power law distribution, they are better described by lognormal, Weibull, and gamma functions that, with regard to AIC best fit, are statistically indistinguishable from each other [*Lazarus et al.*, 2011]. (Locally, lognormal, Weibull, and gamma functions can resemble a power law.) Statistical fitting aside, the overall slope of the spectrum in Figure 2a indicates that the larger spatial scales exhibit greater change (variance), motivating an examination of scaling relationships at these more significant large scales.

2.2.2. Extending the power spectra. Focusing on the approximately straight line section of the power spectrum calls attention away from the spectrum of variance at larger alongshore scales. When the power spectrum is extended to alongshore scales greater than a kilometer, the data exhibit local

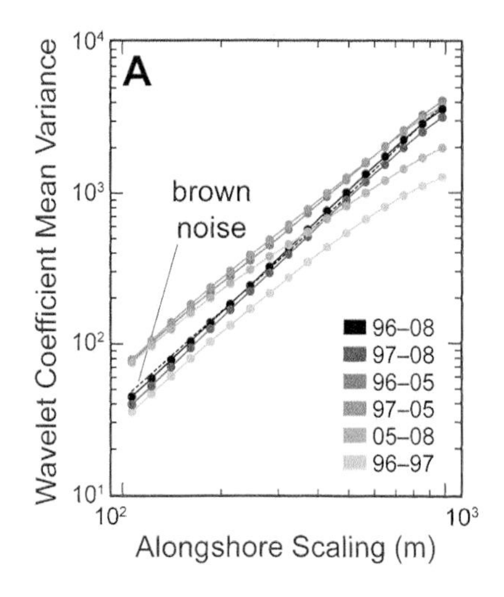

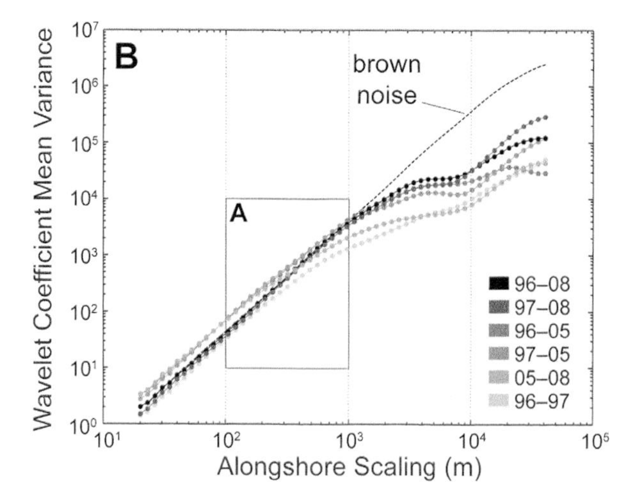

Figure 2. (a) Shoreline change power spectra, calculated for alongshore scales 10^2–10^3 m. (b) Shoreline change power spectra for a greater range of alongshore scales, with inset of Figure 2a delineated for comparison [after the work of *Lazarus et al.*, 2011]. The brown noise signal is the integral of stochastically generated white noise and is characteristically scale-invariant. The local maxima evident in shoreline change variance spectra are therefore not artifacts of the wavelet processing technique.

maxima (Figure 2b). These deviations from log-log linearity, occurring at scales between 1 and 10 km, do not arise from finite domain effects or wavelet-processing artifacts (as the brown noise spectrum in Figure 2b demonstrates), offering interesting potential clues about large-scale shoreline change [*Lazarus et al.*, 2011]. Shoreline changes measured over longer temporal intervals (>3 years) have greater variance magnitudes at large spatial scales (greater than a kilometer) than those measured over shorter time periods (1–3 years), in contrast with smaller scales (less than a kilometer), in which variance is similar across survey intervals (Figure 2b). At all scales above approximately 2 km (and below approximately 10 km), the variance tends to increase, though the rate at which variance increases varies with scale and duration between surveys.

2.3. Power Spectra for Poststorm Surveys

Adding wavelet analyses of poststorm survey pairings to the fair-weather power spectra, we find that hurricane and nonhurricane pairings exhibit similar patterns of shoreline change variance (Figure 3). At small spatial scales, all the shoreline combinations share comparable magnitudes of variance, whether the duration between surveys is short or long and whether the comparison includes a poststorm survey or not. Similar to the fair-weather comparisons, survey pairings that include poststorm data return the greatest variance at larger spatial scales. Furthermore, the largest magnitudes of variance (the greatest first maxima) still occur in the pairs that span the most time: 1996–2008 and 1999–2004 (Figure 3).

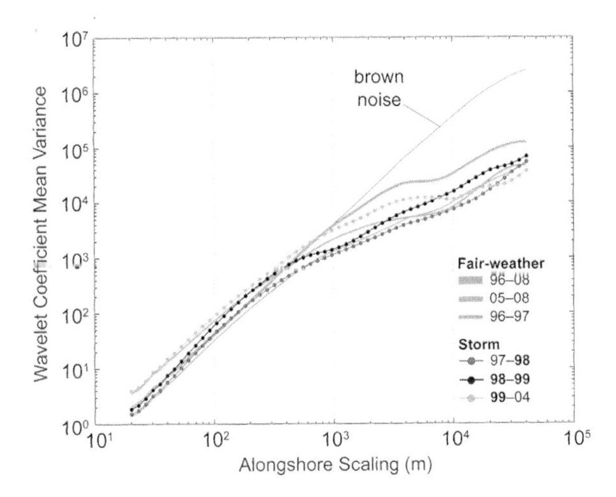

Figure 3. Shoreline change power spectra for storm years (lines with circles), compared to the 1, 3, and 12 year fair-weather spectra (gray lines, thin to thick, respectively) from Figure 2b. Storm years (1998 and 1999) are in bold.

3. DYNAMICAL MECHANISMS OF SHORELINE CHANGE

3.1. Interpreting (Apparent) Power Laws

We emphasize that these shoreline change power spectra (Figures 2 and 3) appear log-log linear but lack statistical signatures definitive enough to call them "power laws," to preempt the possible implication that power laws indicate de facto underlying scale-free dynamics.

Power laws are commonly interpreted to imply that because a quantity varies consistently between scales across a wide domain, the distribution reflects the dominance of a single dynamical process [e.g., *Bak*, 1996; *Murray*, 2007]. For many examples of self-similar patterns in nature, this interpretation of a single dominant process is likely appropriate, such as for certain dynamics of stick-slip earthquakes [e.g., *Bak et al.*, 2002], fluvial dissection and channel branching [*Pelletier*, 1999; *Jerolmack and Paola*, 2007], and granular avalanches [*Frette et al.*, 1996]. Power laws have been applied in coastal landscapes to describe fractal shoreline geometry [*Mandelbrot*, 1967], plan view crescentic patterns [*Dolan and Ferm*, 1968], and properties of cross-shore beach profiles [*Southgate and Möller*, 2000; *Barton et al.*, 2003; *Gunawardena et al.*, 2008].

But studies of shoreline change have demonstrated that distinct physical processes affect a sandy shoreline at differing spatial scales, contradicting the usual association of scale invariance with a particular dynamical interaction. Beach cusps can arise from swash zone feedback at scales of tens of meters [*Werner and Fink*, 1993; *Coco et al.*, 2003]. Larger beach changes, spanning up to a few hundred meters alongshore, can result from surf zone currents interacting with and reorganizing sandbars [*Ruessink et al.*, 2007]. Kilometer-scale shoreline changes have been associated with wave propagation over larger-scale complex nearshore bathymetric features, such as persistent shore-oblique bar fields [*McNinch*, 2004; *Schupp et al.*, 2006]. A unifying explanation for a seemingly consistent relationship between shoreline change variances across so many spatial scales therefore remains unclear, and the typical implications of a power law may not apply [e.g., *Murray*, 2007; *Solow*, 2005].

3.2. Shoreline Behavior at Large Scales

3.2.1. The role of alongshore sediment transport. Looking again at the shoreline change power spectra reminds us of three scale-related observations. First, the magnitude of the first maximum in shoreline change variance tends to increase with duration (for time scales >1 year). Second, the alongshore scale of the first maximum also increases with duration in a way that is consistent with a diffusional time/space scaling [*Lazarus et al.*, 2011]. Third, variance at smaller scales, by comparison, exhibits little dependence on duration between surveys. Collectively, these analytical results frame two inferences: (1) A diffusional scaling relationship operating at multikilometer scales suggests a diffusional process different than processes affecting more transient shoreline changes at subkilometer scales. (2) The dynamics changing coastline shape over large spatial scales are principally responsible for the greatest amounts of shoreline change [*Lazarus et al.*, 2011].

Along the northern North Carolina Outer Banks, between 1996 and 2005, at spatial scales greater than a kilometer, convex seaward promontories tended to erode landward, while concave seaward embayments tended to accrete [*Lazarus and Murray*, 2007]. One hypothesis for this relationship between large-scale, alongshore heterogeneous patterns of shoreline change and plan view shoreline shape is that, over decadal time scales and multikilometer spatial scales, cumulative shoreline change is driven by gradients in wave-forced alongshore sediment transport [*Ashton and Murray*, 2006b; *Lazarus and Murray*, 2007, 2011].

The flux of sediment alongshore, which is a function of the relative angle between the shoreline and incident waves, is maximized when the relative angle between incident deep water wave crests and the shoreline trend is approximately 45° [*Ashton et al.*, 2001; *Falqués*, 2003; *Ashton and Murray*, 2006a]. ("Deep water" in this context means seaward of the nearshore zone, or shoreface, in which seabed contours tend to approximately parallel the shoreline.) Given a plan view bump in a sandy shoreline, deep water waves with relative incident angles $<\sim 45°$ set up a divergence of alongshore sediment at the convex seaward crest of the bump causing the bump to erode (diffuse), tending to maintain a shoreline that is relatively straight in plan view. Oppositely, when the deep water incident wave climate is dominated by high-angle waves (relative angles $>\sim 45°$), convergence of alongshore sediment at the convex seaward crest of the bump (not necessarily a *directional* convergence but a *flux* convergence) causes the bump to accrete and plan view shoreline curvature to exaggerate [e.g., *Ashton et al.*, 2001; *Ashton and Murray*, 2006a]. (Shoreline promontories need to reach a minimum alongshore scale before they are large enough to trigger the shoreline instability [*Falqués and Calvete*, 2005; *List and Ashton*, 2007].)

Theoretical modeling has demonstrated an emergent consequence of this antidiffusive behavior at the landform scale: As the plan view shoreline undulations grow to finite amplitudes, the largest-amplitude bumps begin to shade out their smaller-scale neighbors from the influence of high-angle waves, causing the neighboring perturbations to diffuse away and the dominant wavelength of the plan view shoreline to increase [*Ashton et al.*, 2001; *Ashton and Murray*, 2006a; *Coco and Murray*, 2007]. Ultimately, the only parts of the shoreline that experience the net antidiffusive effect of the high-angle regional wave climate are the tips of the growing promontories. The rest of the coastline, either because of shoreline reorientation or wave shadowing from neighboring promontories, experiences a local wave climate that is dominated by low-angle waves [*Ashton and Murray*, 2006b]. Thus, even as the largest shoreline bumps advance seaward at their tips, diffusion along their flanks inhibits the growth of any new perturbations and reinforce a shoreline shape that is locally smooth in plan view [*Lazarus and Murray*, 2007].

However, no shoreline is completely straight (either as a consequence of geologic control, inheritance from previous configurations, or the development of subkilometer-scale undulations from other processes). Even if the long-term shoreline evolution trend is diffusive, occasional high-angle wave events can result in accentuation of existing shoreline curvature. If these high-angle waves occur during a storm, significant fluxes of alongshore sediment could lead to large gradients and a potentially measurable growth of shoreline instability.

3.2.2. Hypothesis for shoreline instability during storm events. A signature of these alongshore gradients would be a spatial correlation between shoreline change and plan view shoreline curvature. Here we define shoreline change as the cross-shore difference between two surveys of shoreline position and curvature as the second derivative of shoreline position. By convention, promontories are defined as having positive curvature and embayments as having negative curvature. The curvature data here are filtered with a 5 km Gaussian-type Hanning function to separate the low-frequency, multikilometer-scale shoreline shapes of interest from noisier high-frequency patterns [e.g., *Lazarus and Murray*, 2007].

Lazarus and Murray [2007] found a negative correlation between near-decadal shoreline change and shoreline curvature along a reach of the northern Outer Banks; erosion at promontories and accretion at embayments over scales greater than a kilometer is consistent with the diffusive smoothing effects of alongshore-transport gradients that result from a prevailing low-angle wave climate. The additional, post-storm surveys suggest that despite this long-term smoothing, shoreline behavior at those multikilometer scales can be locally antidiffusive during storms (Figure 4). While the general trend over the 12 years of data shown is diffusive, with gradual relaxation of curvature between 1996 and 2008 indicative of overall shoreline smoothing, the posthurricane shoreline curvatures express higher, more strongly peaked

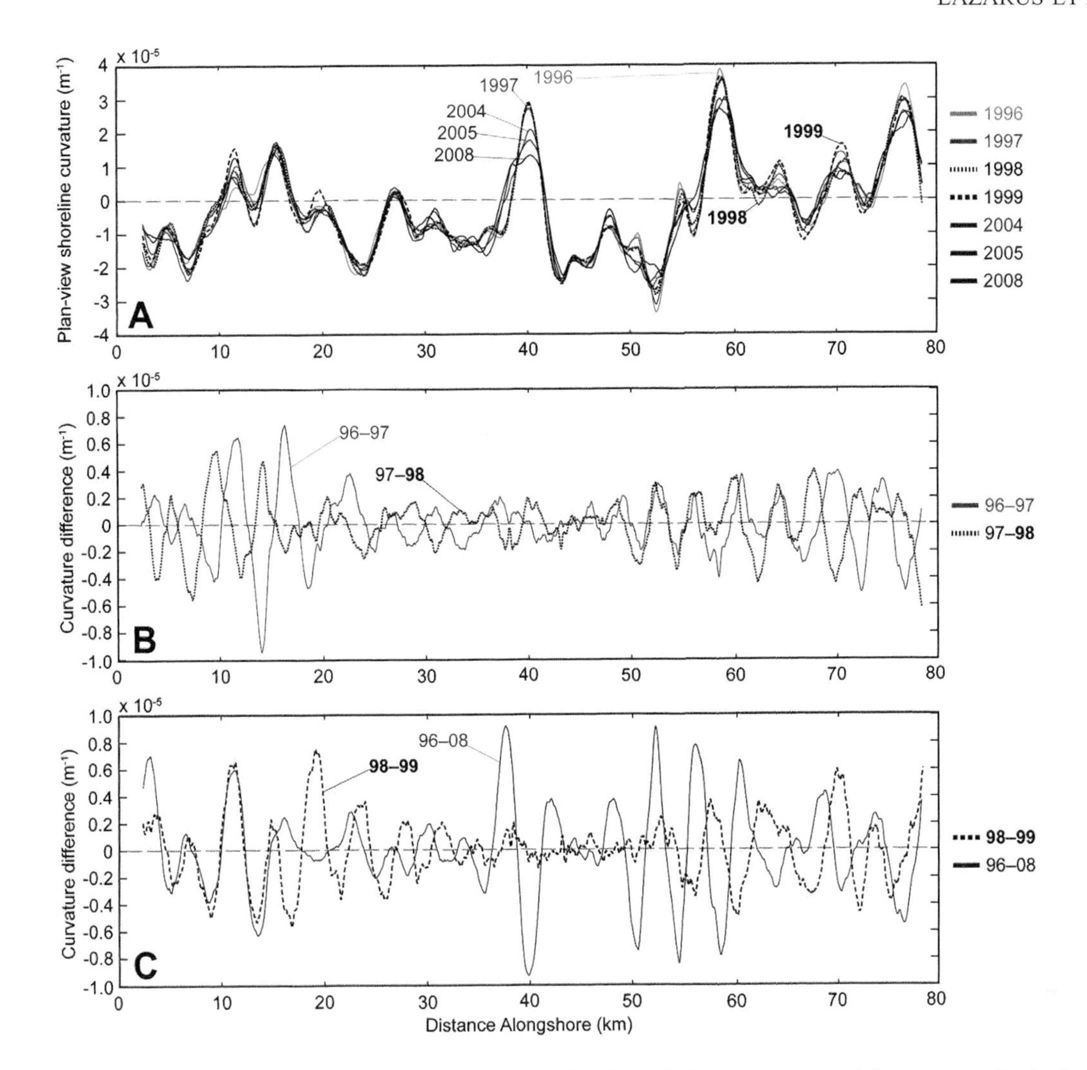

Figure 4. (a) Plan view shoreline curvature. Zones with positive (negative) curvature are subtle promontories (embayments). Hurricane events (bold) in 1998 (dotted) and 1999 (dashed) appear to have temporarily roughened the shoreline shape, which otherwise exhibits a long-term smoothing trend between 1996 and 2008. (b) Differences in curvatures between successive fair-weather and poststorm surveys and (c) poststorm surveys versus the long-term trend, both showing alongshore zones where the curvature differences are almost perfectly out of phase, consistent with temporary storm-related sharpening of plan view features that otherwise tend to smooth over time.

amplitudes at the major promontories and embayments (Figure 4).

Figure 4a shows that, at the 5 km scale, the general shape of shoreline curvature is consistent over the span of surveys plotted; the promontories and embayments do not appear to translate up and down the coastline over time. In Figures 4b and 4c, zones of curvature change (calculated by differencing the fair-weather and storm-related curvature signals) look almost perfectly out of phase in some stretches of coastline. Comparisons of fair-weather signals show certain promontories tending to blunt over time (e.g., between kilometers 10–20 and 55–65), visible as a negative change in curvature at peaks of positive curvature and positive changes in curvature where the curvature is negative (both indicative of smoothing) (Figures 4a and 4c). At those same promontories, comparisons of storm-affected shorelines exhibit the opposite trend consistent with a sharpening of some of the plan view features (Figure 4b).

One possible explanation for storm-related magnification of large-scale undulations is that high-angle, highly energetic waves could have dominated during those storms. Deep water wave data (peak direction, height, and period), recorded offshore at the NOAA buoy station 44014, northeast of the Outer Banks (Figure 1) (National Data Buoy Center,

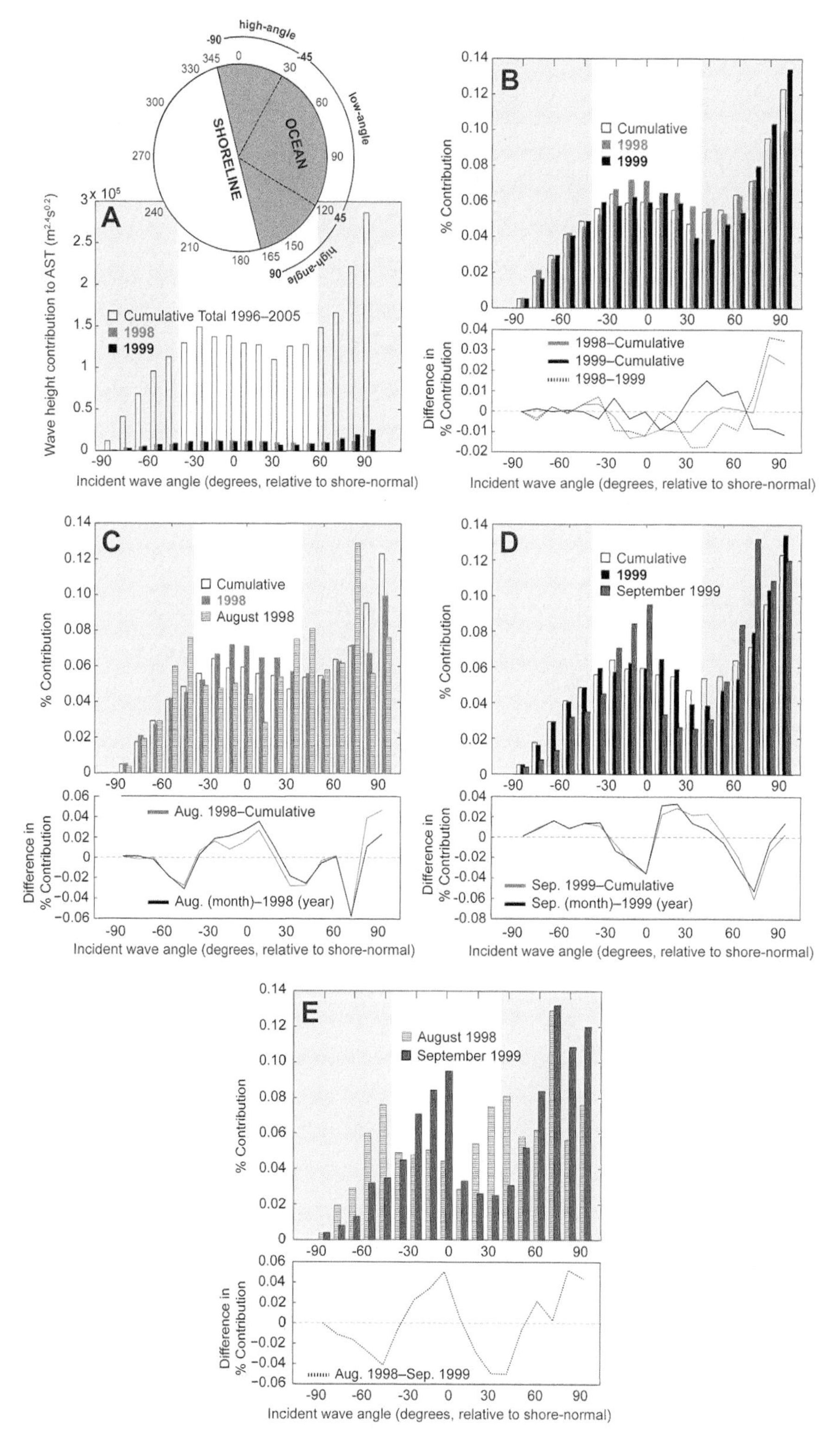

high-angle
low-angle
high-angle
SHORELINE
OCEAN
A
Wave height contribution to AST ($m^{2.4}s^{0.2}$)
Cumulative Total 1996–2005
1998
1999
B
Cumulative
1998
1999
% Contribution
Difference in % Contribution
1998–Cumulative
1999–Cumulative
1998–1999
C
Cumulative
1998
August 1998
Aug. 1998–Cumulative
Aug. (month)–1998 (year)
D
Cumulative
1999
September 1999
Sep. 1999–Cumulative
Sep. (month)–1999 (year)
E
August 1998
September 1999
Aug. 1998–Sep. 1999
Incident wave angle (degrees, relative to shore-normal)

Figure 5

Station 44014 (LLNR 550), available at http://www.ndbc.noaa.gov/station_page.php?station=44014, accessed January, 2011), can be converted into a directional spectrum of wave height contributions to alongshore sediment transport [e.g., *Ashton and Murray*, 2006b]. The near-decadal and annual (1998 and 1999) "effective" incident wave energy for this reach of shoreline have similar directional distributions, with deep water waves approaching from both nearly shore-orthogonal and high angles of incidence (Figures 5a and 5b). The distributions are skewed more to high-angle contributions, however, during the months in which hurricanes occurred (Figures 5c, 5d, and 5e). These wave data suggest that energetic influences from high angles in August, 1998, temporally associated with Hurricane Bonnie, were greater than the high-angle contributions during the 1998 year and the multiyear cumulative total. Equivalent data for September 1999 are similarly weighted toward comparatively greater contributions from high angles, likely capturing to some degree the passages of Hurricane Dennis and Hurricane Floyd. Though the relative differences between summed high-angle proportions may be small, previous modeling experiments have suggested that even slight predominance of wave energy from high angles can be enough to drive shoreline instability of the kind hypothesized here as even subtle large-scale curvature may be enough for gradients in alongshore sediment transport to be effective at changing shoreline shape [*Ashton and Murray*, 2006a, 2006b; *Slott et al.*, 2006; *Valvo et al.*, 2006, *Lazarus and Murray*, 2007].

3.3. Controls on Long-Term Shoreline Evolution

Admittedly, these incident wave distributions (Figure 5) can only serve as circumstantial evidence, limited as we are by the intervals (annual and longer) between lidar surveys; we cannot (and ought not) directly attribute all the shoreline changes reflected in the difference between the 1997 and 1998 lidar surveys, for example, to the effects of the Hurricane Bonnie event or are storm-driven waves and alongshore sediment flux the only forcing mechanisms affecting shoreline shape and evolution; interactions between incident wavefields and lithologic or bathymetric heterogeneities underlying the coastline (the coastal "geologic framework") surely influence nearshore hydrodynamics [*Bender and Dean*, 2003; *Schupp et al.*, 2006; *List et al.*, 2008; *Benedet and List*, 2008] and can also affect local supplies of beach sediment to effectively reinforce the amplitude of shoreline excursions in particular locations at particular scales [e.g., *Valvo et al.*, 2006; *Lazarus and Murray*, 2011]. The long-term diffusional trend in shoreline change is therefore even more striking, considering the potential for storm-driven amplification of shoreline excursions (Figure 4) and the presence of known heterogeneities in the underlying physical framework [e.g., *McNinch*, 2004; *Lazarus and Murray*, 2011]. Although shoreline variability could be attributable to waves interacting with complex offshore bathymetry, studies generally suggest that shorelines attain undulating, but steady state shapes in response to bathymetric irregularities [e.g., *Bender and Dean*, 2003]. To date, no comparable hypothesis from the geologic-framework perspective (i.e., broadly applicable to a variety of locales) explains why erosion and accretion patterns should reverse in such similar fashion across so many scales for two independent forcing events and still produce the demonstrated trends in long-term shoreline change. However, wavefields interacting with complex offshore bathymetry in different ways during different storms (with different wave spectra) is most likely part of the fuller explanation of the curvatures changes exhibited in Figure 4, which shows shoreline responses that vary from one portion of the coastline to another and from one storm to another (contrasting Figures 4b and 4c).

Although the increase in variance with increasing scales that our shoreline-variance spectra exhibit may be unsurprising to nonlinear dynamicists, for the broad scientific community researching coastal, shoreline, and nearshore

Figure 5. (opposite) (a) Relative-angle schematic and incident deep water wave height contributions to alongshore sediment transport (similar to "wave energy") for the general shoreline orientation (N15°W) of the northern Outer Banks. (Water depth at buoy station is 95 m.) White bars show cumulative contributions for 1996–2005, relative to the annual contributions during 1998 (gray) and 1999 (black). (b) Data in Figure 5a in terms of percentage of total energy for the periods sampled to highlight relative directional contributions from high and low angles. Wave height contributions to alongshore sediment transport, as percentages by relative deep water angle of incidence, for (c) August 1998, compared to the cumulative total and the annual distribution for 1998, and (d) September 1999, compared to the cumulative total and the annual distribution for 1998. (e) Isolated comparison of August 1998 to September 1999. High-angle regions (approximately $<-45°$ and $>45°$) are shaded in gray in Figures 5a–5e. Subplots in Figures 5b–5e show differences in contributed percentages between distributions. Energy contribution from shore-incident high-angle waves is greater during the hurricane months of August 1998 and September 1999 than during the years of 1998 and 1999 and during the near-decadal period from 1996 to 2005. A wave-energy climate dominated by high-angle waves from the southeast is consistent with subtropical hurricane paths in this region (National Hurricane Center, Archive of hurricane seasons, accessed February 2011). (Buoy data are available at http://www.ndbc.noaa.gov/station_history.php?station=44014, accessed February 2011.)

phenomena, it is unexpected to discover that changes caused by relatively well-studied processes at relatively small scales do not add up over time and that some poorly understood processes effective on larger scales apparently drive almost all of the long-term shoreline change. The coastal landscape evolves continuously, even during fair-weather conditions; however, these forcing conditions are always changing, most dramatically during major storm events. Accordingly, debate persists over whether the dominant drivers of change in these landscapes of mobile sediment are cumulative or strictly episodic, especially in a physical setting where signs of episodic change can potentially get erased by other sediment-transport events or processes [e.g., *Jerolmack and Paola*, 2010].

The combination of (1) shoreline change variance at large spatial scales increasing with survey duration [*Lazarus et al.*, 2011] and (2) the gradual relaxation (diffusion) of large-scale shoreline curvature suggests that the ongoing aggregation of shoreline position changes over longer time scales influences large-scale spatial change even more strongly than the punctuated, high-impact, antidiffusive changes wrought by strong forcing events like storms. The magnitudes of long-term (order 10 years), cumulative changes at large scales (greater than a kilometer) are still greater than the magnitudes of storm-related shoreline change at those spatial scales, at least for the hurricane events considered here (Figure 3), and simple as it may seem, the transience of small-scale changes in this system is not a trivial observation. The variance of changes at those smaller scales does not change with time; unlike the canonical variance envelope for a random walk, for example, we do not see an envelope of variability for those scales that widen with time. The lesson here is not that we expected small-scale features to aggregate into large-scale patterns and are surprised that they do not. Rather, the lesson is one familiar in complexity contexts but less so in many areas of physical science: If large-scale patterns of change are the hierarchical scale of interest, these data we analyze emphasize the importance of measuring coastal change with methods that record the dynamics that emerge at those scales because extrapolation from smaller scales may not be viable.

4. CONCLUSIONS

Our analysis of power spectra demonstrates that, along the North Carolina Outer Banks coast, the scaling dependencies of shoreline change variance induced by extreme storms do not appear to differ significantly from those of nonstorm intervals, at least at alongshore scales of a few kilometers and smaller. However, the variance of shoreline change is a squared quantity and therefore does not retain the directional sign of change (positive or negative). Examining large-scale changes in shoreline curvature reveals that the storm-related changes can be the opposite of long-term patterns, in the sense that during storms, the plan view coastline becomes rougher in some sections of the coastline, while the longer-term change data indicate an overall smoothing trend. Measuring both persistent, cumulative coastal changes and more ephemeral shoreline change effects of storms at large scales, and subsequently managing those changes, may require a different set of observational strategies [e.g., *McNinch*, 2007] than those designed to capture smaller-scale and typically ephemeral beach behaviors.

Acknowledgments. Our thanks to the National Science Foundation (grants EAR-04-44792 and EPS-0904155), the Joint Airborne Lidar Bathymetry Technical Center for Expertise (JALBTCX), the 2010 Chapman Conference on Complexity and Extreme Events in Geosciences, and the editors of this monograph.

REFERENCES

Akaike, H. (1973), Information theory as an extension of the maximum likelihood principle, in *Second International Symposium on Information Theory*, edited by B. N. Petrov and F. Csaki, pp. 267–281, Akad. Kiadó, Budapest, Hungary.

Ashton, A. D., and A. B. Murray (2006a), High-angle wave instability and emergent shoreline shapes: 1. Modeling of sand waves, flying spits, and capes, *J. Geophys. Res.*, *111*, F04011, doi:10.1029/2005JF000422.

Ashton, A. D., and A. B. Murray (2006b), High-angle wave instability and emergent shoreline shapes: 2. Wave climate analysis and comparisons to nature, *J. Geophys. Res.*, *111*, F04012, doi:10.1029/2005JF000423.

Ashton, A. D., A. B. Murray, and O. Arnoult (2001), Formation of coastline features by large-scale instabilities induced by high-angle waves, *Nature*, *414*, 296–300.

Bak, P. (1996), *How Nature Works: The Science of Self-Organized Criticality*, 226 pp., Springer, New York.

Bak, P., K. Christensen, L. Danon, and T. Scanlon (2002), Unified scaling law for earthquakes, *Phys. Rev. Lett.*, *88*(17), 178501, doi:10.1103/PhysRevLett.88.178501.

Barton, C., J. S. Dismukes, and R. A. Morton (2003), Complexity analysis of the change in shoreline position at Duck, NC, in *Coastal Sediments 2003* [CD-ROM], Am. Soc. of Civ. Eng., Reston, Va.

Bender, C. J., and R. G. Dean (2003), Wave field modification by bathymetric anomalies and resulting shoreline changes: A review with recent results, *Coastal Eng.*, *49*, 125–153.

Benedet, L., and J. H. List (2008), Evaluation of the physical process controlling beach changes adjacent to nearshore dredge pits, *Coastal Eng.*, *55*(12), 1224–1236.

Birkemeier, W. A. (1979), The effects of the 19 December 1977 coastal storm on beaches in North Carolina and New Jersey, *Shore Beach*, *47*(1), 7–15.

Burnham, K. P., and D. R. Anderson (2001), Kullback-Leibler information as a basis for strong inference in ecological studies, *Wildlife Res.*, *28*, 111–119.

Burnham, K. P., and D. R. Anderson (2002), *Model Selection and Multimodel Inference*, 496 pp., Springer, New York.

Carter, R. W. G. (1988), *Coastal Environments*, 617 pp., Academic Press, San Diego, Calif.

Coco, G., and A. B. Murray (2007), Patterns in the sand: From forcing templates to self-organization, *Geomorphology*, *91*, 271–290.

Coco, G., T. K. Burnet, B. T. Werner, and S. Elgar (2003), Test of self-organization in beach cusp formation, *J. Geophys. Res.*, *108* (C3), 3101, doi:10.1029/2002JC001496.

Davis, R. A., Jr. (Ed.) (1978), *Coastal Sedimentary Environments*, 420 pp., Springer, New York.

Davis, R. A., Jr., and D. M. Fitzgerald (2004), *Beaches and Coasts*, 419 pp., Blackwell, Malden, Mass.

Davis, R. A., Jr., and W. T. Fox (1975), Process-response mechanisms in beach and nearshore sedimentation, I, Mustang Island, Texas, *J. Sediment. Petrol.*, *45*, 852–865.

Dolan, R., and J. C. Ferm (1968), Crescentic landforms along the Atlantic Coast of the United States, *Science*, *159*, 627–629.

Egense, A. K. (1989), Southern California beach changes in response to extraordinary storm, *Shore Beach*, *57*(4), 14–17.

Falqués, A. (2003), On the diffusivity in coastline dynamics, *Geophys. Res. Lett.*, *30*(21), 2119, doi:10.1029/2003GL017760.

Falqués, A., and D. Calvete (2005), Large-scale dynamics of sandy coastlines: Diffusivity and instability, *J. Geophys. Res.*, *110*, C03007, doi:10.1029/2004JC002587.

Frette, V., K. Christensen, A. Malthe-Sørenssen, J. Feder, T. Jøssang, and P. Meakin (1996), Avalanche dynamics in a pile of rice, *Nature*, *379*, 49–52.

Fucella, J. E., and R. Dolan (1996), Magnitude of beach disturbance during northeast storms, *J. Coastal Res.*, *12*, 420–429.

Godfrey, P. J., and M. M. Godfrey (1973), Comparison of ecological and geomorphic interactions between altered and unaltered barrier island systems in North Carolina, in *Coastal Geomorphology*, edited by D. R. Coates, pp. 239–258, State Univ. of N. Y., Binghamton.

Gunawardena, Y., S. Ilic, H. N. Southgate, and H. Pinkerton (2008), Analysis of the spatio-temporal behavior of beach morphology at Duck using fractal methods, *Mar. Geol.*, *252*, 38–49.

Holman, R. A., and J. Stanley (2007), The history and technical capabilities of Argus, *Coastal Eng.*, *54*, 477–491.

Hubbard, B. B. (1996), *The World According to Wavelets: The Story of a Mathematical Technique in the Making*, 264 pp., A. K. Peters, Wellesley, Mass.

Jerolmack, D. J., and C. Paola (2007), Complexity in a cellular model of river avulsion, *Geomorphology*, *91*, 259–270.

Jerolmack, D. J., and C. Paola (2010), Shredding of environmental signals by sediment transport, *Geophys. Res. Lett.*, *37*, L19401, doi:10.1029/2010GL044638.

Komar, P. D. (1998), *Beach Processes and Sedimentation*, 2nd ed., 544 pp., Prentice-Hall, Upper Saddle River, N. J.

Lazarus, E. D., and A. B. Murray (2007), Process signatures in regional patterns of shoreline change on annual to decadal time scales, *Geophys. Res. Lett.*, *34*, L19402, doi:10.1029/2007GL031047.

Lazarus, E. D., and A. B. Murray (2011), An integrated hypothesis for regional patterns of shoreline change along the northern North Carolina Outer Banks, U.S.A., *Mar. Geol.*, *281*(1–4), 85–90, doi:10.1016/j.margeo.2011.02.002.

Lazarus, E. D., A. Ashton, A. B. Murray, S. Tebbens, and S. Burroughs (2011), Cumulative versus transient shoreline change: Dependencies on temporal and spatial scale, *J. Geophys. Res.*, *116*, F02014, doi:10.1029/2010JF001835.

List, J. H., and A. Ashton (2007), A circulation modeling approach for evaluating the conditions for shoreline instabilities, in *Coastal Sediments '07*, edited by N. C. Kraus and J. D. Rosati, pp. 327–340, Am. Soc. Civ. Eng., Reston, Va.

List, J. H., and A. S. Farris (1999), Large-scale shoreline response to storms and fair weather, in *Coastal Sediments '99*, edited by N. C. Kraus and W. G. McDougal, pp. 1324–1338, Am. Soc. of Civ. Eng., Reston, Va.

List, J. H., A. Farris, and C. Sullivan (2006), Reversing storm hotspots on sandy beaches: Spatial and temporal characteristics, *Mar. Geol.*, *226*, 261–279.

List, J. H., L. Benedet, D. M. Hanes, and P. Ruggiero (2008), Understanding differences between DELFT3D and empirical predictions of alongshore sediment transport gradients, in *Coastal Engineering 2008: Proceedings of the 31st International Conference*, edited by J. M. Smith, pp. 1864–1875, World Sci., London, U. K., doi:10.1142/9789814277426_0154.

Mandelbrot, B. (1967), How long is the coast of Britain? Statistical self-similarity and fractional dimension, *Science*, *156*, 636–638.

McNinch, J. E. (2004), Geologic control in the nearshore: Shore-oblique sandbars and shoreline erosional hotspots, Mid-Atlantic Bight, USA, *Mar. Geol.*, *211*(1–2), 121–141.

McNinch, J. E. (2007), Bar and swash imaging radar (BASIR): A mobile X-band radar designed for mapping nearshore sand bars and swash-defined shorelines over large distances, *J. Coastal Res.*, *23*(1), 59–74.

Mitasova, H., M. Overton, and R. S. Harmon (2005), Geospatial analysis of a coastal sand dune field evolution: Jockey's Ridge, North Carolina, *Geomorphology*, *72*, 204–221.

Murray, A. B. (2007), Two paradigms in landscape dynamics: Self-similar processes and emergence, in *Nonlinear Dynamics in Geophysics*, edited by A. A. Tsonis and J. B. Elsner, pp. 15–35, Springer, New York.

Nievergelt, Y. (1999), *Wavelets Made Easy*, 297 pp., Birkhäuser, Boston, Mass.

Pelletier, J. D. (1999), Self-organization and scaling relationships of evolving river networks, *J. Geophys. Res.*, *104*(B4), 7359–7375.

Ruessink, B. G., G. Coco, R. Ranasinghe, and I. L. Turner (2007), Coupled and noncoupled behavior of three-dimensional morphological patterns in a double sandbar system, *J. Geophys. Res.*, *112*, C07002, doi:10.1029/2006JC003799.

Schupp, C. A., J. E. McNinch, and J. H. List (2006), Nearshore shore-oblique bars, gravel outcrops, and their correlation to shoreline change, *Mar. Geol.*, *233*, 63–79.

Slott, J. M., A. B. Murray, A. D. Ashton, and T. J. Crowley (2006), Coastline responses to changing storm patterns, *Geophys. Res. Lett.*, *33*, L18404, doi:10.1029/2006GL027445.

Solow, A. (2005), Power laws without complexity, *Ecol. Lett.*, *8*, 361–363.

Sonu, C. J. (1973), Three-dimensional beach changes, *J. Geol.*, *81*, 42–64.

Sornette, D. (2006), *Critical Phenomena in Natural Sciences: Chaos, Fractals, Selforganization and Disorder: Concepts and Tools*, 528 pp., Springer, New York.

Southgate, H. N., and I. Möller (2000), Fractal properties of coastal profile evolution at Duck, North Carolina, *J. Geophys. Res.*, *105*(C5), 11,489–11,507.

Stockdon, H. F., A. H. Sallenger, J. H. List, and R. A. Holman (2002), Estimation of shoreline position and change using airborne topographic lidar data, *J. Coastal. Res.*, *18*, 502–513.

Tebbens, S. F., S. M. Burroughs, and E. E. Nelson (2002), Wavelet analysis of shoreline change on the Outer Banks of North Carolina: An example of complexity in the marine sciences, *Proc. Natl. Acad. Sci. U. S. A.*, *99*, suppl. 1, 2554–2560.

Thom, B. G., and W. Hall (1991), Behaviour of beach profiles during accretion and erosion dominated periods, *Earth Surf. Processes Landforms*, *16*, 113–127.

Valvo, L. M., A. B. Murray, and A. Ashton (2006), How does underlying geology affect coastline change? An initial modeling investigation, *J. Geophys. Res.*, *111*, F02025, doi:10.1029/2005JF000340.

Werner, B. T., and T. M. Fink (1993), Beach cusps as self-organized patterns, *Science*, *260*, 968–971.

White, S. A., and Y. Wang (2003), Utilizing DEMs derived from LIDAR data to analyze morphologic change in the North Carolina coastline, *Remote Sens. Environ.*, *85*, 39–47.

Zeigler, J. M., C. R. Hayes, and S. D. Tuttle (1959), Beach changes during storms on outer Cape Cod, Massachusetts, *J. Geol.*, *67*, 318–336.

A. D. Ashton, Department of Geology and Geophysics, Woods Hole Oceanographic Institution, 3609 Woods Hole Road, MS 22, Woods Hole, MA 02543, USA.

E. D. Lazarus, School of Earth and Ocean Sciences, Cardiff University, Main Building, Park Place, Cardiff CF10 3AT, UK. (LazarusED@cf.ac.uk)

A. B. Murray, Division of Earth and Ocean Sciences, Nicholas School of the Environment, Center for Nonlinear and Complex Systems, Duke University, 103 Old Chemistry Building, Box 13 90227, Durham, NC 27708, USA.

Precipitation and River Flow: Long-Term Memory and Predictability of Extreme Events

Armin Bunde, Mikhail I. Bogachev,[1] and Sabine Lennartz[2]

Institut für Theoretische Physik III, Justus-Liebig-Universität Giessen, Giessen, Germany

In this review, we discuss linear and nonlinear long-term correlations in precipitation and river flows and their influence on risk estimation. We outline the standard method for measuring linear and nonlinear correlations that can distinguish between natural fluctuations and external trends, and we use this method to show that, in general, precipitation does not show linear long-term correlation, in contrast to river flows. Both precipitation and river runoff records exhibit nonlinear long-term memory that can be detected by a multifractal detrended fluctuation analysis. Long-term memory has important consequences for the occurrence of extremes. Contrary to the intuition that extremes are uncorrelated and thus occur randomly in time, they tend to cluster in the presence of long-term memory. Here we analyze this clustering feature for the daily precipitation and river flow data. To describe the occurrence of extremes quantitatively, we determine the probability density function of the return intervals between extremes above some threshold value Q and show how it can be used to obtain the hazard function, which is crucial for a risk estimation. Using the hazard function and a Bayesian approach (receiver operator characteristic analysis), we evaluate the contributions of the linear and nonlinear memory to the predictability of precipitation and river flows.

1. INTRODUCTION

The analysis of precipitation and river flows has a long history. Already more than half a century ago, H. E. Hurst found by means of his rescaled range (R/S) analysis that annual runoff records from various rivers exhibit "long-range statistical dependencies" [*Hurst*, 1951], indicating that the fluctuations in water storage and runoff processes are self-similar over a wide range of time scales, with no single characteristic scale. Hurst's finding is now recognized as the first example for self-affine fractal behavior in empirical time series, [see, e.g., *Feder*, 1988]. The "Hurst phenomenon" was investigated on a broader empirical basis for many other natural phenomena and explained with long-term memory in the data [*Hurst et al.*, 1965; *Mandelbrot and Wallis*, 1969]. The problem is that in precipitation and runoff records with limited length, it is difficult to distinguish long-term memory from hierarchies of short-term memory and/or trends [*Mudelsee et al.*, 2003; *Montanari*, 2003; *Koscielny-Bunde et al.*, 2006; *Kantelhardt et al.*, 2006; *Ludescher et al.*, 2011].

The scaling of the fluctuations with time is reflected by the scaling of the power spectrum $E(f)$ with frequency f, $E(f) \sim f^{-\beta}$. For stationary time series, the exponent β is related to the decay of the corresponding autocorrelation function $C(s)$. For β between 0 and 1, $C(s)$ decays by a power law, $C(s) \sim s^{-\gamma}$, with $\gamma = 1 - \beta$ being restricted to the interval between 0

[1]Now at Radio System Department, Saint Petersburg Electrotechnical University, Saint Petersburg, Russia.

[2]Now at School of Engineering and School of Geosciences, University of Edinburgh, Edinburgh, UK.

Extreme Events and Natural Hazards: The Complexity Perspective
Geophysical Monograph Series 196

10.1029/2011GM001112

and 1. In this case, the mean correlation time diverges, and the system is regarded as long-term correlated. For $\beta = 0$, the data are linearly uncorrelated on long time scales and look like "white noise" in the spectrum. The exponents β and γ can also be determined from a fluctuation analysis, where the departures from the mean value are considered as increments of a random walk process. If the runoffs are uncorrelated, the fluctuation function $F_2(s)$, which is equivalent to the root-mean-square displacement of the random walk, increases as the square root of the time scale s, $F_2(s) \sim \sqrt{s}$. For long-term correlated data, the random walk becomes anomalous, and $F_2(s) \sim s^H$. The fluctuation exponent H is related to the exponents β and γ via $\beta = 1 - \gamma = 2H - 1$. For monofractal data (where only linear correlations exist), H is identical to the classical Hurst exponent. In the past, many studies using these kinds of methods have dealt with scaling properties of precipitation and river flow records and the underlying statistics [see, e.g., *Lovejoy and Schertzer*, 1991; *Turcotte and Greene*, 1993; *Gupta et al.*, 1994; *Gupta and Dawdy*, 1995; *Tessier et al.*, 1996; *Davis et al.*, 1996; *Rodriguez-Iturbe and Rinaldo*, 1997; *Pandey et al.*, 1998; *Matsoukas et al.*, 2000; *Montanari et al.*, 2000; *Peters et al.*, 2001; *Livina et al.*, 2003a, 2003b; *Mudelsee*, 2007; *Livina et al.*, 2011].

However, the conventional methods discussed above may fail when trends are present in the system. Trends are systematic deviations from the average runoff or precipitation that are caused by external processes, e.g., the construction of a water regulation device, the seasonal cycle, or a changing climate (e.g., global warming). Monotonous trends may lead to an overestimation of the existing correlations, i.e., to an overestimation of the Hurst exponent H and the power spectrum exponent β, and thus to an underestimation of γ. It is even possible that uncorrelated data, under the influence of a trend, look like long-term correlated ones when using the above analysis methods [*Bhatthacharya et al.*, 1983; *Kantelhardt et al.*, 2001; *Hu et al.*, 2001; *Ludescher et al.*, 2011]. In addition, long-term correlated data cannot simply be detrended by the common technique of moving averages, since this method destroys the correlations on long time scales (above the window size used) [see also *Schumann and Kantelhardt*, 2011]. Furthermore, it is difficult to distinguish trends from long-term correlations because stationary long-term correlated time series exhibit persistent behavior and a tendency to stay close to the momentary value. This causes positive or negative deviations from the average value for long periods of time that might look like a trend. For a discussion of these problems, see also the early work of *Klemes* [1974].

In the last years, several methods, such as wavelet techniques and detrended fluctuation analysis (DFA), have been developed that are able to determine long-term correlations in the presence of trends. For details and applications of the methods to a large number of meteorological, climatological, and biological records, we refer to *Peng et al.* [1994], *Taqqu et al.* [1995], *Bunde et al.* [2000], *Kantelhardt et al.* [2001], *Eichner et al.* [2003], *Bunde et al.* [2005], *Rybski et al.* [2006], and *Lennartz and Bunde* [2009a] for DFA applications in hydrology [see *Matsoukas et al.*, 2000; *Montanari et al.*, 2000; *Kantelhardt et al.*, 2003, 2006; *Koscielny-Bunde et al.*, 2006; *Livina et al.*, 2011].

In the recent decades, it has been realized that a single exponent H is insufficient for the full description of the scaling behavior, and thus, a multifractal description containing a set of generalized Hurst exponents $h(q)$ is required to fully quantify the fluctuations in both precipitation and runoff records [see, e.g., *Tessier et al.*, 1996; *Pandey et al.*, 1998; *Lovejoy and Schertzer*, 1991; *Bunde et al.*, 2002; *Kantelhardt et al.*, 2006; *Koscielny-Bunde et al.*, 2006, and references therein]. For each precipitation or runoff record, this multifractal description can be regarded as a "fingerprint," which, among other things, can serve as an efficient nontrivial test bed for the performance of state-of-the-art precipitation-runoff models [*Kantelhardt et al.*, 2003; *Livina et al.*, 2007]. To obtain $h(q)$, one can employ the multifractal generalization of DFA, named multifractal (MF)-DFA [*Kantelhardt et al.*, 2002] that can systematically distinguish between long-term correlations and trends [*Kantelhardt et al.*, 2006; *Koscielny-Bunde et al.*, 2006].

In the first part of our review (sections 2–4), we discuss the quantification of linear and nonlinear long-term memory in precipitation and river runoff records. Here we mainly review the results from the work of *Kantelhardt et al.* [2006] and *Koscielny-Bunde et al.* [2006]. First, in section 2, we discuss the elimination of seasonal trends that is required before further analysis is performed [see also *Livina et al.*, 2007; *Ludescher et al.*, 2011]. In section 3, we discuss linear correlations in precipitation and river flows. In section 4, we concentrate on the nonlinear memory and multifractal features of the studied records. The results show that the precipitation records are typically characterized by $h(2) \approx 1/2$, revealing a fast decay of the autocorrelation function, while the runoff records are long-term correlated above a crossover time scale of a few weeks with fluctuation exponents $h(2)$ varying in a broad range. The results seem to indicate that the persistence of the runoffs does not originate in persistence of precipitation but is rather caused by storage processes occurring in the soil and the highly intermittent spatial behavior of the rainfall. In addition, the runoff records are characterized, at large time scales, by a stronger average multifractality than the precipitation records.

In the second part of this review, we are interested in the extremes. In the presence of long-term memory, the extreme events do not appear to be independent but, instead, tend to cluster in time. Well-known examples in nature are temperature anomalies, extreme rainfalls, and floods [*Pfister*, 1998; *Glaser*, 2001; *Bunde et al.*, 2002, 2004; *Altmann and Kantz*, 2005; *Bunde et al.*, 2005]. Since the statistics of the occurrence times between extreme events in real systems is quite poor, one usually tries to extract information from events with smaller magnitudes that occur quite often and thus have enough statistics. The major issue is to find out some general scaling relations between the return intervals at low and high thresholds, which then allow extrapolation of the results to very large, extreme thresholds.

To study the occurrence of extremes, we concentrate on the return intervals r_j between single events x_i that exceed some fixed threshold Q (quantile). The process of retrieving a return interval series (r_j) with $j = 1,2,\ldots, N_Q$ from a time series (x_i) of N data points is illustrated in Figure 1. To be independent of the data distribution $P(x)$, instead of specifying the threshold Q, it is useful to specify the mean return interval, or return period R_Q, since there is a one-by-one correspondence between both quantities, $R_Q = 1/\int_Q^\infty P(x)\mathrm{d}x = N/N_Q$.

We are interested in the consequences of the long-term memory, in particular, of its nonlinear component, on the temporal occurrence of extreme events. First, in section 5, we concentrate on the probability density function (PDF) of the return intervals between single (extreme) events exceeding a certain threshold Q and briefly review the theoretical results for uncorrelated, long-term correlated, and multifractal data sets. In section 6, we discuss the related "hazard function" that determines the probability of at least one Q-exceeding event to occur within the next Δt time units given that the last event occurred t time units ago. In section 7, we discuss the predictability in precipitation and river flows due to long-term memory and evaluate the performance of a prediction algorithm that employs the hazard function. For a pure random process with statistically independent values with identical distribution, i.e., "independent and identically distributed" data (the most typical example is Gaussian white noise), also the return intervals are independent and follow a Poisson with a simple exponential PDF $P_Q(r) = (1/R_Q)\exp(-r/R_Q)$ [see, e.g., *von Storch and Zwiers*, 2002]. In contrast, long-term correlations strongly affect the statistics of return intervals between extreme events, as was shown before for both linear [see *Bunde et al.*, 2005; *Altmann and Kantz*, 2005; *Eichner et al.*, 2007; *Santhanam and Kantz*, 2008] and nonlinear memory [*Bogachev et al.*, 2007, 2008a, 2008b]. Clustering in the temporal occurrence of extremes gives rise to the prediction of anomalies that is one of the essential issues in complex systems [*Bogachev et al.*, 2007; *Bogachev and Bunde*, 2009; *Bogachev et al.*, 2009; *Bogachev and*

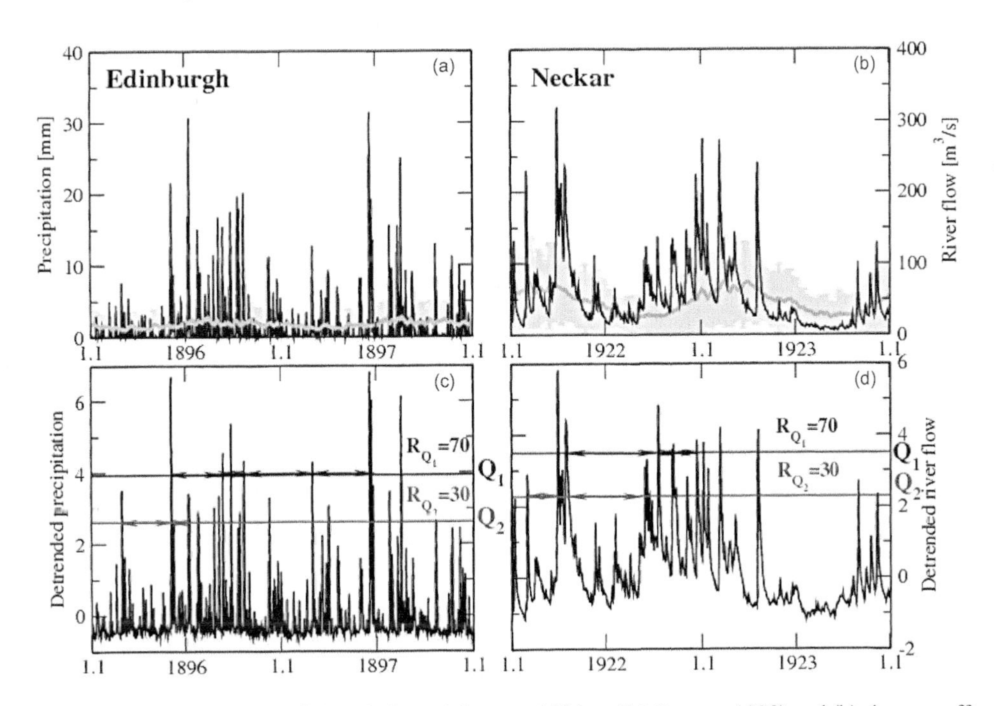

Figure 1. (a) Precipitation record of Edinburgh (from 1 January 1896 until 1 January 1898) and (b) river runoff record of the Neckar River at the station Plochingen (from 1 January 1922 until 1 January 1924). The seasonal trend (solid line) and the seasonal standard deviation $(\overline{W_i^2} - \overline{W}_i^2)^{1/2}$ (gray error bars), both smoothed over 1 week, are also shown. (c and d) Corresponding detrended data sets. The horizontal lines in the detrended records refer to two thresholds Q_1 and Q_2, with return periods R_Q of 30 and 70 days, respectively.

Bunde, 2011]. We also review here our recent results on the predictability and risk estimation in precipitation and river flows by using the return interval statistics (M. I. Bogachev et al., On long-term memory, extreme events and risk estimation in precipitation and river flows, preprint, 2011, hereinafter referred to as Bogachev et al., preprint, 2011).

2. SEASONAL DETRENDING

We consider daily river runoffs and daily precipitation records W_i measured at certain hydrological stations. The index i counts the days in the record, $i = 1,2,\ldots,N$. To eliminate the periodic seasonal trends, which may affect the analyzing methods, we concentrate on the departures $W_i - \overline{W}_i$ from the mean daily value $\overline{W}_i$. $\overline{W}_i$ is calculated for each calendar day (e.g., 1 April) by averaging over all years in the record. To eliminate also the seasonal trend in the fluctuations around $\overline{W}_i$, we divide $W_i - \overline{W}_i$ by the standard deviation $\sigma_i = (\overline{W_i^2} - \overline{W}_i^2)^{1/2}$ and obtain the seasonally detrended data [*Livina et al.*, 2007; *Ludescher et al.*, 2011]

$$x_i = (W_i - \overline{W}_i)/\sigma_i. \tag{1}$$

Figure 1 shows two standard examples: the precipitation record of Edinburgh (from 1 January 1896 to 1 January 1898) (Figure 1a) and the river runoff of the Neckar River at the station Plochingen (from 1 January 1922 to 1 January 1924) (Figure 1b). The seasonal trend is shown by a gray solid line, while the seasonal standard deviation $(\overline{W_i^2} - \overline{W}_i^2)^{1/2}$ is shown by light gray error bars. Figures 1c and 1d show the corresponding detrended data sets. The horizontal lines in the detrended records in Figure 1 refer to two thresholds, Q_1 and Q_2.

Events above Q_1 occur on average once in 70 days, i.e., the return period $R_{Q1} = 70$ days, while for the lower threshold Q_2, the return period is $R_{Q2} = 30$ days. By definition, the return period is the average overall return intervals, some of them are indicated by arrows in the figure. In the second part of this review, we are interested in the statistics of the return intervals and how we can use it for risk prediction.

3. LINEAR CORRELATIONS

Linear temporal correlations within a data set $\{x_i\}$ with zero mean can be measured via the (auto)correlation function,

$$C(s) \equiv \frac{< x_i x_{i+s} >}{< x^2 >} = \frac{1}{(N-s) < x^2 >} \sum_{i=1}^{N-s} x_i x_{i+s}, \tag{2}$$

where N is the number of days in the considered record. The x_i is uncorrelated, if $C(s)$ is zero for s positive. If linear correlations exist up to a certain time s_x, the correlation function will be positive up to s_x and vanish above s_x. For the relevant case of long-term correlations, $C(s)$ decays (for very large N) as a power law [see, e.g., *Lennartz and Bunde*, 2009b],

$$C(s) \sim (1-\gamma)s^{-\gamma}, \quad 0 < \gamma \leq 1, \tag{3}$$

such that the mean correlation time $\bar{s} = \int_0^\infty C(s)ds$ diverges.

For large time lags s, a direct calculation of $C(s)$ is hindered by possible trends in the data and by finite size effects [*Lennartz and Bunde*, 2009b]. Both problems can be overcome, to a certain extent, by using the DFA. The DFA procedure consists of four steps [*Kantelhardt et al.*, 2001]. First, we determine the cumulated sum

$$Y(i) \equiv \sum_{k=1}^{i} x_k, \quad i = 1, \ldots, N, \tag{4}$$

of the record.

Second, we divide the profile $Y(i)$ into $N_s \equiv \mathrm{int}(N/s)$ nonoverlapping segments of equal length s. Since the length N of the series is usually not a multiple of the considered time scale s, a short part at the end of the profile may remain. In order not to disregard this part of the series, the same procedure is repeated starting from the opposite end. Thereby, $2N_s$ segments are obtained altogether.

In the third step, we calculate the local trend for each of the $2N_s$ segments by fitting (least squares fit) a polynomial of order n to the data and determine the variance

$$\sigma^2(\nu, s) \equiv \frac{1}{s} \sum_{i=1}^{s} [Y((\nu-1)s + i) - p_\nu(i)]^2 \tag{5}$$

for each segment. Here $p_\nu(i)$ is the fitting polynomial representing the local trend in the segment ν. Linear, quadratic, cubic, and higher-order polynomials can be used in the fitting procedure. When linear polynomials are used, the fluctuation analysis is called DFA1, for quadratic polynomials, we have DFA2, for cubic polynomials, DFA3, etc. [see *Kantelhardt et al.*, 2001]. By definition, DFAn removes polynomial trends of the order n in the profile $Y(i)$ and of the order $n-1$ in the original series x_i.

In the fourth step, one averages the variances of all segments and takes the square root to obtain the mean fluctuation function,

$$F_2(s) \equiv \left[\frac{1}{2N_s} \sum_{\nu=1}^{2N_s} \sigma^2(\nu, s)\right]^{1/2}. \tag{6}$$

Since we are interested in how $F_2(s)$ depends on the time scale s, we have to repeat steps 2 to 4 for several time scales s. It is apparent that $F_2(s)$ will increase with increasing s. If the x_i are long-term correlated according to equation (3), $F_2(s)$ increases, for large values of s, by a power law [*Peng et al.*, 1994; *Kantelhardt et al.*, 2001],

$$F_2(s) \sim s^{h(2)} \tag{7}$$

with

$$h(2) = \begin{cases} 1 - \gamma/2, & \text{for} \quad 0 < \gamma < 1, \\ 1/2, & \text{for} \quad \gamma \geq 1 \end{cases}. \tag{8}$$

For nonstationary data, $h(2) > 1$. For example, for data generated by a random walk (Wiener process), $h(2) = 1.5$. When the autocorrelation function decreases faster than $1/s$ in time, we asymptotically have $F_2(s) \sim s^{1/2}$. For short-term correlated data described, e.g., by an exponential decay of $C(s)$, a crossover to $h(2) = 1/2$ occurs well above the correlation time $\bar{s}$.

Figure 2 shows the fluctuation functions $F_2(s)$ obtained from DFA1, DFA2, and DFA3 for three representative daily precipitation records (Figures 2a–2c) and three representative runoff records (Figures 2d–2f). In the log-log plot, the curves are approximately straight lines on large-scale s. The runoff fluctuations show a pronounced crossover at time scales of several weeks. Above the crossover, the fluctuation functions show power law behavior with exponents $h(2) \simeq 0.80$ for the Elbe, $h(2) \simeq 0.85$ for the Danube, and $h(2) \simeq 0.91$ for the Mississippi. Below the crossover, we find an effective scaling exponent $h(2) \cong 1.5$, indicating strong short-term correlations on small time scales. Approximately, the short-term correlations can be modeled by an autoregressive moving average (ARMA) process, where the correlation time is represented by the typical decay time of floods [*Kantelhardt et al.*, 2006]. This yields $h(2) = 1.5$ on short time scales in agreement with Figure 2.

For the precipitation records, there is only a very weak crossover in $F_2(s)$, and the $h(2)$ values are rather close to 0.5, indicating rapidly decaying autocorrelations. Specifically, $h(2) \simeq 0.55$ for Hamburg, $h(2) \simeq 0.50$ for Vienna, and $h(2) \simeq 0.50$ for Gothenburg, corresponding to a correlation exponent $\gamma \cong 0.9$ for Hamburg and $\gamma \cong 1$ for Vienna and Gothenburg. For the precipitation data, the higher slope at very small scales is partly a methodological artifact [*Kantelhardt et al.*, 2006].

Figure 3 shows the distributions of $h(2)$ obtained by *Kantelhardt et al.* [2006] using the DFA2 for 99 precipitation records and 42 runoff records. The figure shows that most precipitation records do not exhibit long-term correlations ($h(2) \cong 0.5$) or show only very weak long-term correlations ($h(2) \cong 0.55$); the mean value is $h(2) = 0.53 \pm 0.04$. Interestingly, for precipitation records, the largest $h(2)$ values occur

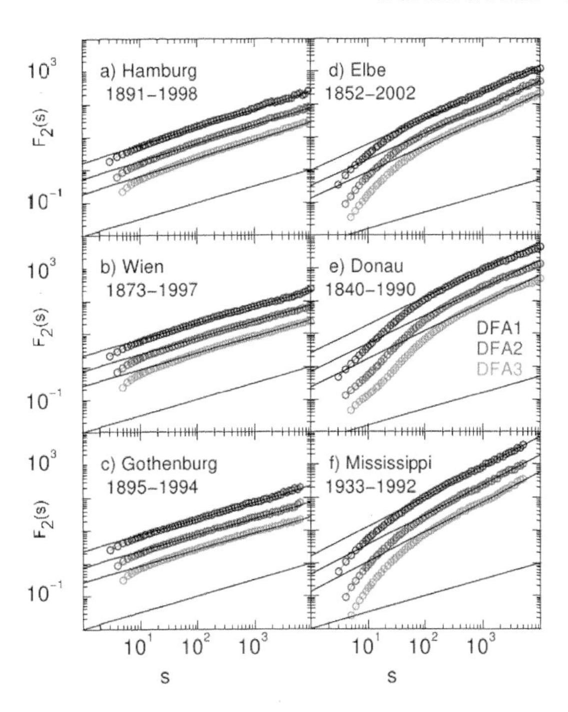

Figure 2. (top to bottom) Double logarithmic plots of the fluctuation functions $F_2(s)$ versus time scale s obtained from DFA1-DFA3 for representative precipitation and runoff records (shifted vertically for clarity). Precipitation records are for (a) Hamburg, Germany ($h(2) = 0.55 \pm 0.03$), (b) Vienna, Austria ($h(2) = 0.50 \pm 0.03$), and (c) Gothenburg, Nebraska, United States ($h(2) = 0.50 \pm 0.03$). Runoff records are for (d) Elbe River in Dresden, Germany ($h(2) = 0.80 \pm 0.03$), (e) Danube River in Orsova, Romania ($h(2) = 0.85 \pm 0.03$), and (f) Mississippi River in St. Louis, United States ($h(2) = 0.91 \pm 0.03$). The straight lines through the data have the reported slopes, and lines with slope $h(2) = 0.5$ are shown below the data for comparison with the uncorrelated case. After the work of *Kantelhardt et al.* [2006].

at the station at high altitudes. From Figures 2 and 3, one can conclude that the long-term persistence of the river flows is not due to precipitation. For extensive discussions of this point, we refer to the works of *Gupta et al.* [1994], *Gupta and Dawdy* [1995], *Kantelhardt et al.* [2006], *Koscielny-Bunde et al.* [2006], and *Mudelsee* [2007].

Figure 4 shows that the Hurst exponent $h(2)$ of the 42 runoff records from Figure 3 does not depend on the basin area A. This is in line with earlier conclusions by *Gupta et al.* [1994] for the flood peaks, where a systematic dependence on A could also not be found. Figure 4 also shows that there is no pronounced regional dependence: the rivers within a localized area (such as south Germany) tend to have nearly the same range of exponents as the international rivers. In a recent paper, *Mudelsee* [2007] studied the dependence of $h(2)$ on A for several stations on the same river. His results

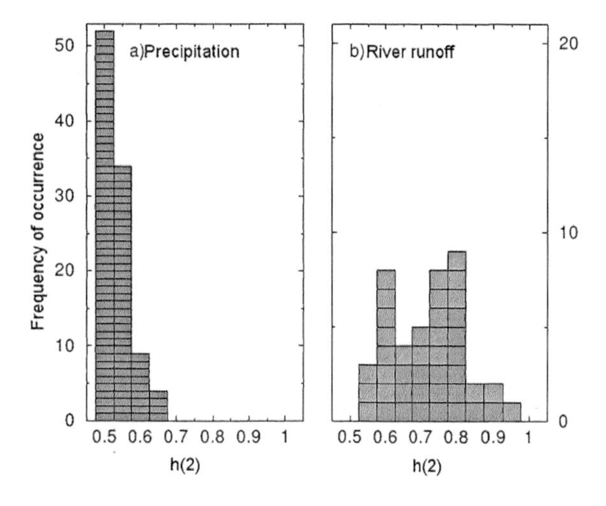

Figure 3. Histograms of the long-term fluctuation exponents $h(2)$ for (a) 99 daily precipitation records and (b) 42 daily runoff records, 18 from southern Germany and 24 from international hydrological stations. The data are specified by *Kantelhardt et al.* [2006]. The scaling exponents have been determined from power law fits of the DFA2 fluctuation function on large time scales. Each box represents the result for one hydrological or meteorological station. After the work of *Kantelhardt et al.* [2006].

indicate that for the same river, the Hurst exponent may increase down the river, when the basin size increases.

As can be seen in the figure, the exponents spread from 0.55 to 0.95. Since the correlation exponent γ is related to $h(2)$ by $\gamma = 2 - 2h(2)$, the exponent γ spreads from almost 0 to almost 1, covering the whole range from very weak to very strong correlations. It is remarkable that the three "frozen" rivers in the study (Gaula, Tana, and Dvina) are characterized by the values of $h(2)$ that are among the lowest ones. For interpreting this distinguished result, *Koscielny-Bunde et al.* [2006] noted that on permafrost ground, the lateral inflow (and hence the indirect contribution of the water storage in the catchment basin) contributes to the runoffs in a different way than on normal ground [see also *Gupta and Dawdy* 1995]. In particular, the contribution of snow melting may lead to less correlated runoffs than the contribution of rainfall, but more comprehensive studies will be needed to confirm this hypothesis.

4. NONLINEAR CORRELATIONS AND MULTIFRACTALITY

For a further characterization of precipitation and river flows, it is meaningful to extend equation (6) to a more general fluctuation function that also takes nonlinear contributions into account. To this end, following the work of *Kantelhardt et al.* [2002], we replace the variance $\sigma(\nu, s)$ in equation (6) by its q/2nd power and the square root by the qth root, where $q \neq 0$ is a real parameter,

$$F_q(s) \equiv \left[\frac{1}{2N_s}\sum_{\nu=1}^{2N_s}\sigma^{q/2}(\nu,2)\right]^{1/q}. \qquad (9)$$

In analogy to equation (7), we then define the generalized fluctuation exponent $h(q)$ by

$$F_q(s) \sim s^{h(q)}. \qquad (10)$$

For "monofractal" time series, $h(q)$ is independent of q, since the scaling behavior of the variances $\sigma(\nu,s)$ is identical for all segments ν. If, on the other hand, strong and weak fluctuations scale differently, there will be a significant dependence of $h(q)$ on q.

Figure 5 shows, for the "seasonal detrended" Edinburgh precipitation data and the Neckar runoff records of Figure 1, the generalized fluctuation function $F_q(s)$ as a function of s for $q = 1, 2, 3, 5, 8$, and 10. In the double logarithmic plots, the generalized Hurst exponents $h(q)$ are the slopes of $F_q(s)$ for large s values (straight dashed lines). One can see easily that the slopes $h(q)$ decrease monotonically with increasing q. This is an indication for nonlinear memory, which is less pronounced for the precipitation record than for the river runoff record. In Edinburgh, the precipitation record, $h(2) \cong 0.52$, and linear long-term correlations can be considered as absent. In contrast, for the Neckar River, $h(2) \cong 0.85$, and thus, pronounced linear long-term correlations occur. At small s values, deviations from the asymptotic behavior occur, which are due to the short-term correlations discussed above.

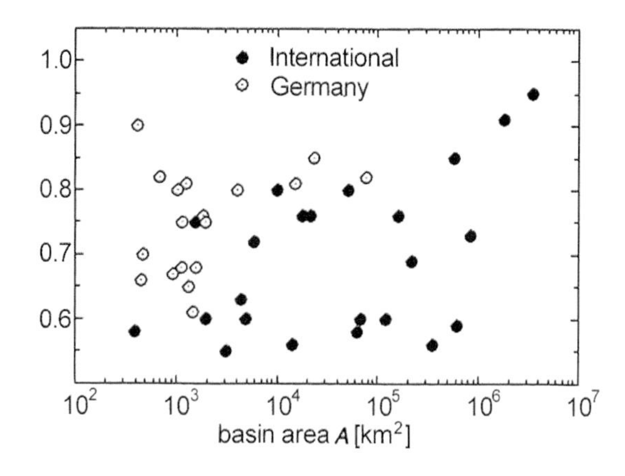

Figure 4. Long-term fluctuation exponents $h(2)$ for the same rivers as in Figure 3, as a function of the basin area A. The solid symbols refer to river flows from international stations, while the open symbols refer to records from south Germany. After the work of *Koscielny-Bunde et al.* [2006].

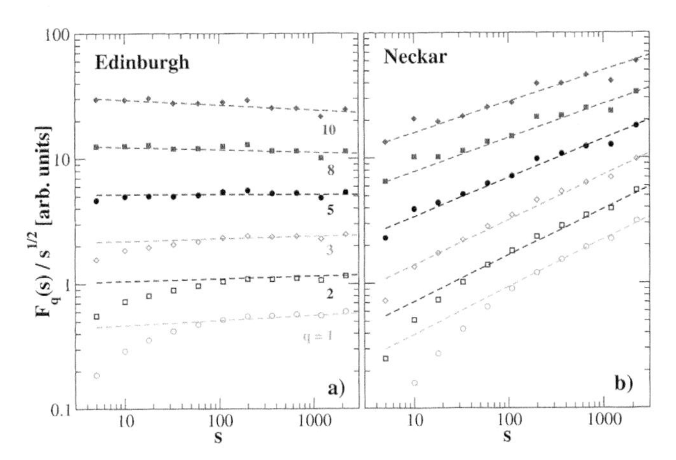

Figure 5. (top to bottom) $F_q(s)/s^{1/2}$ as a function of s for $q = 1, 2, 3, 5, 8$, and 10 for the precipitation and river runoff records of Figure 1. In the double logarithmic plots, the generalized Hurst exponents $h(q)$ are the slopes of $F_q(s)$ for large s values (straight dashed lines).

5. PROBABILITY DENSITY FUNCTION

Next, we consider the statistics of the return intervals in precipitation and river runoff records. Instead of fixing the threshold Q, we keep the return period, or the mean return interval R_Q, fixed (see Figures 1c and 1d). We are interested in the effect of linear/nonlinear memory on the PDF $P_Q(r)$ of the return intervals r. From the PDF, one can obtain the hazard function, which is an important quantity in risk estimation (see below). It is well known that for uncorrelated data sets

$$R_Q P_Q(r) = e^{-r/R_Q}, \tag{11}$$

while for long-term correlated monofractal data sets [*Bunde et al.*, 2005]

$$R_Q P_Q(r) \sim \begin{cases} \left(\frac{r}{R_Q}\right)^{\gamma-1} & , \frac{r}{R_Q} < 1 \\ e^{\left[-b\left(\frac{r}{R_Q}\right)^{\gamma}\right]} & , \frac{r}{R_Q} > 1 \end{cases} . \tag{12}$$

In a reasonable approximation (which is not exact), $P_Q(r)$ can be written as the first derivative of the Weibull distribution,

$$R_Q P_Q(r) = C(r/R_Q)^{\gamma-1} e^{-b(r/R_Q)^{\gamma}}, \tag{13}$$

where C is a normalization constant [*Eichner et al.*, 2007; *Blender et al.*, 2008; *Bogachev and Bunde*, 2009]. In both cases, for uncorrelated and linear correlated data, $R_Q P_Q(r)$ depends only on the ratio r/R_Q and not separately on Q. More recently, it has been shown that for records with strong nonlinear memory, a power law dependence of $P_Q(r)$ is observed,

$$R_Q P_Q(r) \sim (r/R_Q)^{-\delta(Q)}, \tag{14}$$

where $\delta(Q) > 1$ depends explicitly on Q [*Bogachev et al.*, 2007].

Figure 6 shows the PDFs of four representative precipitation records (left-hand side) and four representative river

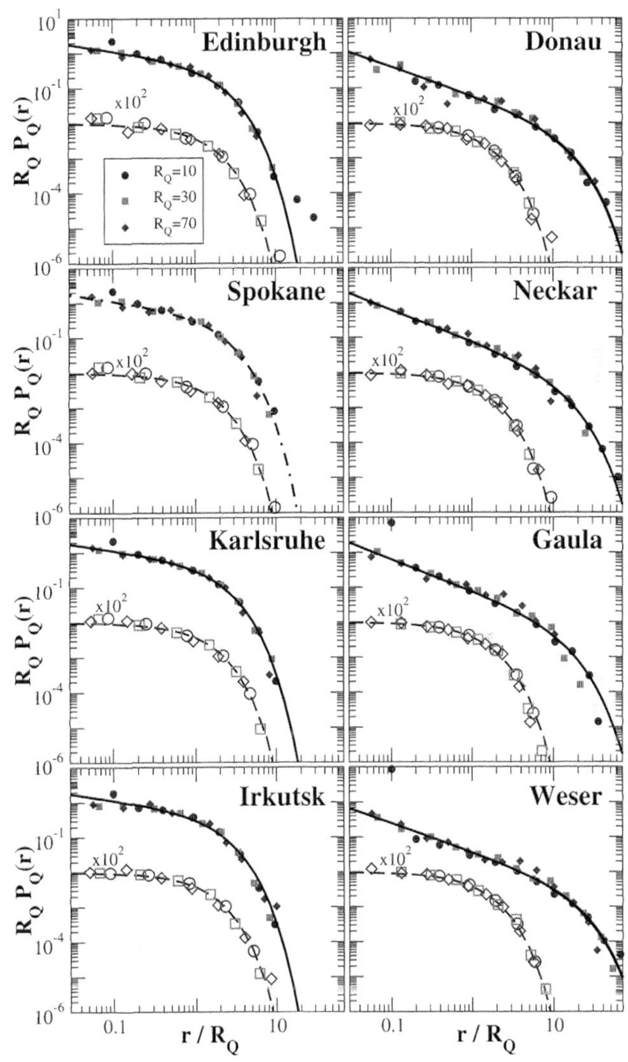

Figure 6. (left) PDFs of four representative precipitation records and (right) four representative river runoffs for $R_Q = 10$, 30, and 70. For convenience, the PDFs are plotted in rescaled form, i.e., $R_Q P_Q(r)$ as a function of r/R_Q. The dashed black line is the simple exponential e^{-x}, which holds for uncorrelated data. The solid black lines are fitted gamma distributions (see equation (15)). The PDF of the shuffled data are plotted as open symbols and have been shifted downward by a factor of 10^2 for better clarification.

runoffs (right-hand side) for R_Q = 10, 30, and 70. The precipitation records and their Hurst exponents are Edinburgh ($h(2)$ = 0.52), Spokane ($h(2)$ = 0.52), Karlsruhe ($h(2)$ = 0.56), and Irkutsk ($h(2)$ = 0.52). The river runoffs and their Hurst exponents are Danube ($h(2)$ = 0.86), Neckar ($h(2)$ = 0.85), Gaula ($h(2)$ = 0.65), and Weser ($h(2)$ = 0.80). For convenience, we have plotted the PDFs in rescaled form, i.e., $R_Q P_Q(r)$ as a function of r/R_Q. First, we consider the precipitation data. Figure 6 shows that the PDFs of the precipitation data can be well approximated by a gamma distribution (solid line),

$$f(x) \cong Cx^{\alpha-1}e^{-\lambda x}, \tag{15}$$

with $\alpha = \lambda \cong 2/3$ and $C \cong 0.56$ for all precipitation data. This curve is distinct from the expected PDF for simply linear correlated data (see equation (12)), where the initial decay is weaker. It is interesting to note that the same function, with the *same* parameters α and λ, also describes the PDFs of the interoccurrence times of quasistationary regimes in seismic catalogs [*Corral*, 2004]. Due to the multifractality, we would expect some dependence of the PDF on R_Q (see equation (14)), but since the multifractality is weak and the data are short, this feature cannot be resolved here.

The PDFs of the river runoff data (right-hand side of Figure 6) are characterized by a pronounced initial power law decay that covers more than two decades. Around r/R_Q= 10, there is a crossover toward an exponential decay. For purely linear long-term correlated data, the crossover would appear earlier (around $r/R_Q \simeq 1$), and the observed power law exponent would be smaller. Here the functional form is not described by the PDF for purely linear long-term correlated data. We consider it interesting that the PDF for the Gaula River that is characterized by a comparatively small Hurst exponent ($h(2) \cong 0.65$) has nearly the same functional form as the PDF for the Neckar River, which is characterized by a considerably larger Hurst exponent $h(2) \cong 0.85$. Accordingly, the strong deviations from the PDF of purely linear correlated data (equation (12)) are due to the nonlinear memory because strong nonlinear memory (equation (14)) would explain the observed power law with the late crossover, the large exponent, as well as the independence of the Hurst exponent.

Also, the PDFs of the river flow data are approximately gamma distributions (solid lines) (equation (15)). For Neckar and Gaula, $\alpha = \lambda \cong 0.10$ and $C \cong 0.08$, while for the Weser, $\alpha \cong 0.19$, $\lambda \cong 0.07$, and $C \cong 0.12$, and for the Donau, $\alpha \cong 0.19$, $\lambda \cong 0.10$, and $C \cong 0.13$. Like in the precipitation data, the multifractality in the river runoffs is not strong enough to generate in the PDF an initial power law regime with Q-dependent exponents larger than 1, as seen for multifractal cascade models [*Bogachev et al.*, 2007]. When we shuffle the data (open symbols), we remove both linear and nonlinear memory and obtain the result for uncorrelated data, a simple exponential (dashed black line).

We like to note that due to the linear long-term correlations, we would expect a stretched exponential decay of the PDF for large return intervals rather than an exponential decay [*Eichner et al.*, 2007]. However, due to the comparatively short length of the records, we also expect strong finite size effects to favor a simple exponential decay. We believe that the anticipated stretched exponential decay is masked by finite size effects, such that only the exponential decay can be seen, which then leads to the description of the PDF as a simple gamma distribution.

6. HAZARD FUNCTION

A central quantity in risk estimation (see Figure 7a) is the probability $W_Q(t; \Delta t)$ ("hazard function") that within the next Δt units of time, at least one event above Q occurs, if the last Q-exceeding event occurred t time units ago [*Bogachev et al.*, 2007]. This quantity is related to $P_Q(r)$ by $W_Q(t;\Delta t) = \int_t^{t+\Delta t} P_Q(r)\mathrm{d}r / \int_t^{\infty} P_Q(r)\mathrm{d}r$ and, thus, to the cumulative distribution function (CDF) $C_Q(t) = \int_0^t P_Q(r)\mathrm{d}r$ by

$$W_Q(t;\Delta t) = \frac{C_Q(t+\Delta t) - C_Q(t)}{1 - C_Q(t)}. \tag{16}$$

For uncorrelated records, $C_Q(t) = 1 - e^{-t/R_Q}$, and thus,

$$W_Q(t;\Delta t) = 1 - e^{-\Delta t/R_Q} \tag{17}$$

independent of t. For $\Delta t \ll R_Q$, $W_Q(t; \Delta t)$ reduces to

$$W_Q(t;\Delta t) \cong \frac{\Delta t}{R_Q}. \tag{18}$$

For linear long-term correlated records, we have, within the approximation of equation (13), $C_Q(t) \cong 1 - \exp(-b(t/R_Q)^\gamma)$, and thus [*Bogachev and Bunde*, 2009],

$$W_Q(t;\Delta t) \cong 1 - e^{-b\left(\frac{t}{R_Q}\right)^{\gamma}\left(\left(1+\frac{\Delta t}{t}\right)^{\gamma}-1\right)} \cong \gamma b \frac{\Delta t}{R_Q}\left(\frac{t}{R_Q}\right)^{\gamma-1} \tag{19}$$

for $\Delta t \ll t, R_Q$.

Finally, for strong nonlinear memory, equation (14) yields $C_Q(t) \sim (t/R_Q)^{1-\delta(Q)}$, and thus,

$$W_Q(t;\Delta t) \cong 1-(1+\Delta t/t)^{1-\delta(Q)} \qquad (20)$$

and

$$W_Q(t;\Delta t) \cong (\delta(Q)-1)\frac{\Delta t/R_Q}{t/R_Q} \qquad (21)$$

for $\Delta t << t$ [*Bogachev et al.*, 2009]. Here the Q (and R_Q) dependence is only in the prefactor.

For hydrological data, where the PDF can be expressed by equation (15), the situation is the following: For short and intermediate times, $t \le R_Q/\lambda$, the power law dominates the exponential decay in equation (15), and the gamma distribution can be approximated by the distribution for linear long-term correlated data (equations (12) and (13)). Thus, we expect (with $\gamma = \alpha$ and $b = \lambda$)

$$W_Q(t;\Delta t) = \frac{\int_t^{t+\Delta t} P_Q(r)\mathrm{d}r}{1-\int_0^t P_Q(r)\mathrm{d}r} \cong \alpha\lambda\frac{\Delta t}{R_Q}\left(\frac{t}{R_Q}\right)^{\alpha-1} \qquad (22)$$

for $\Delta t << t \le R_Q/\lambda$. For long times, $t >> R_Q/\lambda$, the exponential decay dominates the power law in equation (13). Thus, we expect

$$W_Q(t;\Delta t) = \frac{\int_t^{t+\Delta t} P_Q(r)\mathrm{d}r}{\int_t^{\infty} P_Q(r)\mathrm{d}r} \cong \lambda\frac{\Delta t}{R_Q} \qquad (23)$$

for $\Delta t << R_Q/\lambda << t$.

Figure 7b shows $W_Q(t;\ 1)$ for the Edinburgh precipitation data (left-hand side) and for the Neckar River flow data (right-hand side). To obtain $W_Q(t;\ 1)$, we fitted the PDF of the return intervals to equation (15) with $\alpha = \lambda \cong 2/3$ for the precipitation data and $\alpha = \lambda \cong 0.1$ for the runoff data. In both cases, the hazard function decays by a power law for small times t. For precipitation data ($\alpha \cong 2/3$), the decay is lower than for the river runoff data ($\alpha \cong 0.1$), in agreement with equation (22). For long times, t, the hazard function flattens in both cases, which is consistent with equation (23). With increasing threshold Q, R_Q increases, and the hazard function decreases as expected. In Figures 7c and 7d, the hazard function is plotted for fixed $R_Q = 70$ and $\Delta t = 5$, 10, and 20. As expected, $W_Q(t;\ \Delta t)$ increases with increasing Δt, and the decrease of the function becomes weaker. For further increasing Δt, $W_Q(t;\ \Delta t)$ becomes independent of t and approaches 1 by definition.

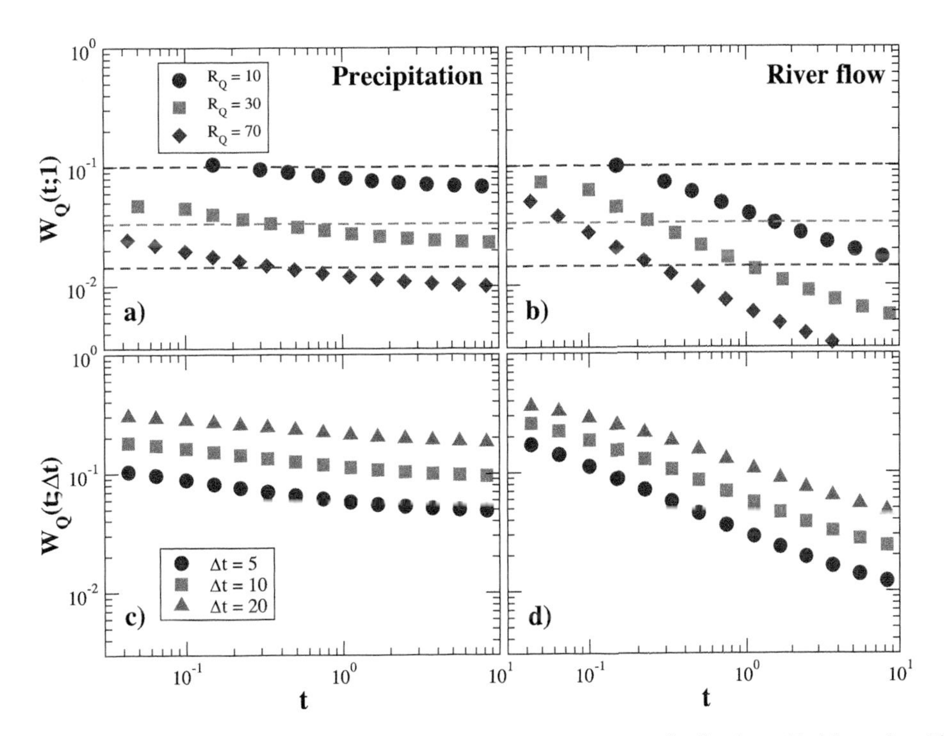

Figure 7. The hazard function $W_Q(t;1)$ determined numerically for the gamma distribution with (a) $\alpha = \lambda \cong 2/3$ and $C \cong 0.56$ that characterizes the precipitation data and (b) $\alpha = \lambda \cong 0.1$ and $C \cong 0.12$ that characterizes most of the river data for $R_Q = 10$, 30, and 70. The same function $W_Q(t;\Delta t)$ but for varying $\Delta t = 5$, 10, and 20, using the best fits for (c) precipitation and (d) river flow data, all at $R_Q = 70$.

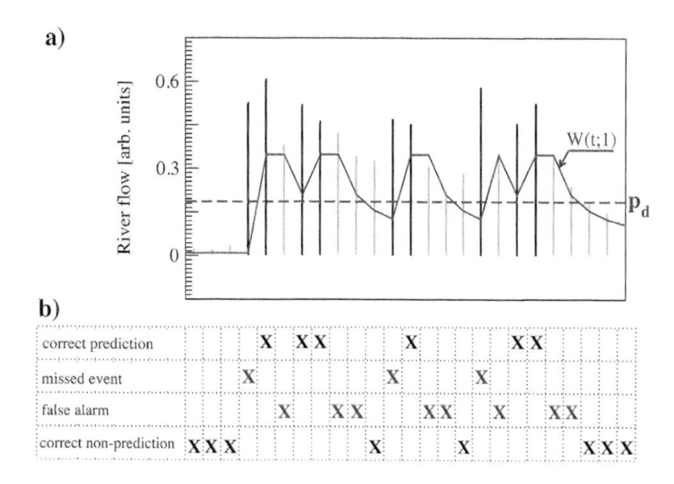

Figure 8. Extreme events above Q are shown in black, while the other events are shown in gray. (a) The solid line shows $W_Q(t;\ 1)$. (b) The actual event is above Q or below and if this event has been predicted or not is shown. The first row gives the correct predictions of extreme events, the second row the nonpredicted extremes, the third row the false alarms, and the last row the correct predicted nonextremes.

7. PREDICTION ALGORITHM AND EFFICIENCY

Next, we discuss how the hazard function $W_Q(t;\ 1)$ can be used to predict events above the threshold Q.

Figure 8a illustrates how this can be done in a decision algorithm. In the algorithm, we set up a decision threshold p_d and activate an alarm whenever $W_Q(t;\ 1)$ exceeds p_d. In Figure 8, the (extreme) events above Q are shown in black, while the other events are shown in gray. The solid line shows $W_Q(t;\ 1)$, and the dashed horizontal line is the decision threshold p_d. The top row in Figure 8b marks when an extreme event is correctly predicted, the second row when it is missed, the third row when a false alarm is given, and the last row when a nonextreme value is correctly predicted.

For a certain decision threshold p_d, the efficiency of the algorithm is generally quantified by the correct prediction rate D (number of correct predictions (first row in Figure 8b)) divided by the number of all extremes (sum of first and second rows in Figure 8b) and the false alarm rate α (number of false alarms (third row in Figure 8b)) divided by the number of all nonextremes (sum of third and fourth rows in

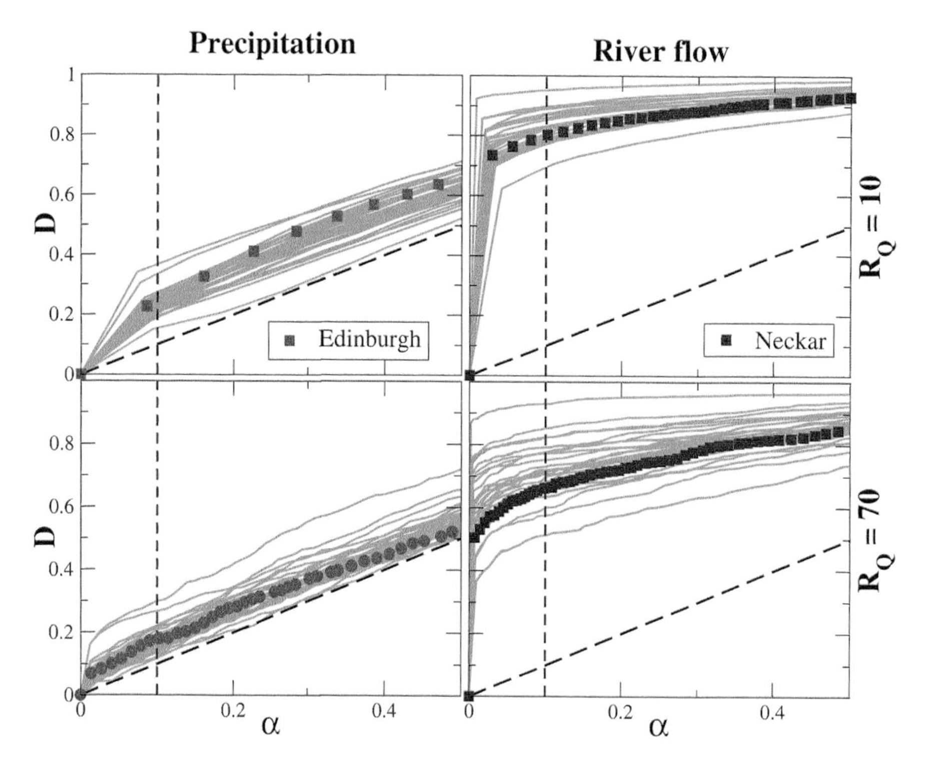

Figure 9. Receiver operator characteristic (ROC) analysis for (left) precipitation and (right) river runoff data for $R_Q = 10$ (top curves) and $R_Q = 70$ (bottom curves). The black dashed line represents the random case, for comparison. The thin vertical lines mark the 10% false alarm rate $\alpha = 0.1$.

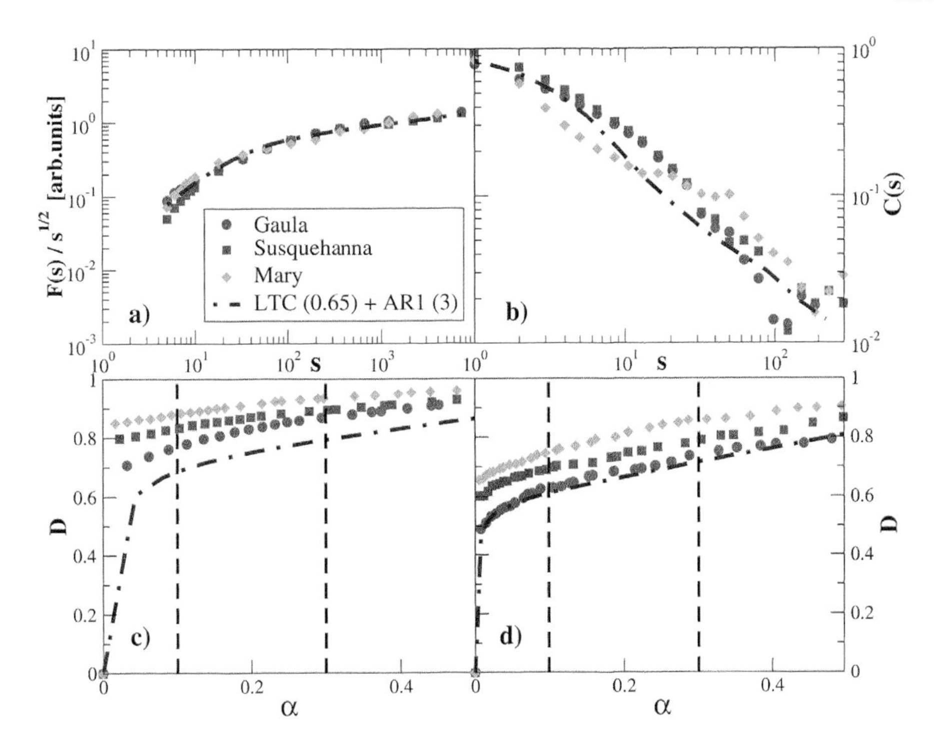

Figure 10. (a) $F_2(s)/s^{1/2}$ for three selected rivers (Gaula, Susquehanna, and Mary) characterized by low $h(2)$ values and for the fitting model (dash-dotted line). (b) Autocorrelation function for the same data records. ROC curves for predictions made in the three selected river data (symbols) and for in fitting model data (dash-dotted lines) for (c) $R_Q = 30$ and (d) 70. After the work of Bogchev et al. (preprint, 2011).

Figure 8b). The larger the correct prediction rate D is for fixed false alarm rate α, the better is the prediction provided by the algorithm. In the example of Figure 8, the prediction rate is $D = 6/9 \cong 0.67$, while the false alarm rate is $\alpha = 8/16 = 0.5$.

In general, α and D depend on p_d. The overall quantification of the prediction efficiency is usually obtained from the "receiver operator characteristic" (ROC) analysis [*Egan*, 1975; *Fawcett*, 2006; *Bogachev and Bunde*, 2011], where D is plotted versus α for *all possible* p_d values. By definition, for $p_d = 0$, $D = \alpha = 1$, while for $p_d = 1$, $D = \alpha = 0$. For $0 < p_d < 1$, the ROC curve connects the lower left corner of the panel with the upper right one (see Figure 9). If there is no memory in the data, $D = \alpha$, and the ROC curve is a straight line between both corners.

Figure 9 shows the results of the ROC analysis for all precipitation and river flow data for $R_Q = 10$ (top curves) and $R_Q = 70$ (bottom curves) for false alarm rates $\alpha < 0.5$. The results for the Edinburgh precipitation data and the runoffs of the Neckar River are highlighted. The black dashed line represents the random case, for comparison. The thin vertical lines mark the 10% false alarm rate, $\alpha = 0.1$. As expected, for the precipitation data, the prediction rate D is only slightly above the random case, due to the absence of pronounced linear and nonlinear memory. For the river flow data, in contrast, due to the interplay between linear short- and long-term correlations and nonlinear memory, the prediction rate is quite large. On the average, for $\alpha = 0.1$, the prediction rate D is 0.8 for $R_Q = 10$, while for $R_Q = 70$, the prediction rate decreases, and the average D value is around 0.7.

In general, it is not possible to separate the contribution of the nonlinear memory from the contribution of the linear memory. To a certain extent, one can estimate the contribution of the nonlinear memory when the linear short- and long-term memory is weak and the nonlinear memory is pronounced. This is the case for the Gaula, Susquehanna, and Mary Rivers.

In order to estimate the contribution of linear short- and long-term memory, we have determined fluctuation functions $F_2(s)$ and the autocorrelation functions $C(s)$ for these three records. We also determined $F_2(s)$ for synthetic data modeled by an autoregressive process of first-order (AR1), $x_{i+1} = e^{-1/\tau}x_i + \eta_i$, where η_i is long-term correlated noise with Hurst exponent $h(2) = 0.65$. We found that a correlation time $\tau = 3$ days modeled the fluctuation function best. Figures 10a and 10b show the $F_2(s)s^{1/2}$ and the autocorrelation function $C(s)$ both for the river flows and for the synthetic data.

Quite obviously, the model reproduces the linear correlations nicely, on both short and long time scales. Figures 10c and 10d show the ROC curves for the three rivers and the modeled data for $R_Q = 10$ and $R_Q = 70$, respectively. For $R_Q = 10$, the results for all three rivers are well above the linear model. The best prediction rate is for the Mary River, followed by Susquehanna and Gaula. We know from previous multifractal analysis of these rivers [*Koscielny-Bunde et al.*, 2006] that the strength of the multifractality follows the same order, with Mary River the strongest and Gaula the weakest multifractality. This indicates that the deviations of $D(\alpha)$ from the linear model are due to nonlinear memory associated with the multifractal behavior. For $R_Q = 70$ (Figure 10d), all prediction rates decreased, and the results for the Gaula River are close to the linear model, while the results of the other two rivers are still well above.

8. CONCLUSION

In this review, we have discussed the occurrence of linear and nonlinear long-term memory in precipitation and river flows and their influence on risk estimation. The important features are that both kinds of records exhibit nonlinear memory, while only river flows show also significant linear long-term correlations characterized by a Hurst exponent above 1/2. We have investigated how the linear and, in particular, the nonlinear correlations that mostly are being disregarded affect the predictability of (extreme) events above some threshold Q. By comparing the predictability in the observational records with the predictability of linear models, where only linear (short and long term) memory was present, we showed that nonlinear memory is important for the predictability in both precipitation and river flows.

Acknowledgments. We would like to thank our colleagues Eva Koscielny-Bunde, Shlomo Havlin, Jan Kantelhardt, Jan Eichner, Diego Rybski, and Peter Braun for numerous illuminating discussions on the subject. We also like to acknowledge the partial support of this work by the Deutsche Forschungsgemeinschaft (project BU 534/24-1).

REFERENCES

Altmann, E. G., and H. Kantz (2005), Recurrence time analysis, long-term correlations, and extreme events, *Phys. Rev. E*, *71*, 056106, doi:10.1103/PhysRevE.71.056106.

Bhatthacharya, R. N., V. K. Gupta, and E. C. Waymire (1983), The Hurst effect under trends, *J. Appl. Probab.*, *20*, 649–662.

Blender, R., K. Fraedrich, and K. Seinz (2008), Extreme event return times in long-term memory processes near $1/f$, *Nonlin. Processes Geophys.*, *15*, 557–565.

Bogachev, M. I., and A. Bunde (2009), On the occurrence and predictability of overloads in telecommunication networks, *Europhys. Lett.*, *86*, 66002, doi:10.1209/0295-5075/86/66002.

Bogachev, M. I., and A. Bunde (2011), On the predictability of extreme events in records with linear and nonlinear long-range memory: Efficiency and noise robustness, *Physica A*, *390*, 2240–2250.

Bogachev, M. I., J. F. Eichner, and A. Bunde (2007), Effect of nonlinear correlations on the statistics of return intervals in multifractal data sets, *Phys. Rev. Lett.*, *99*, 240601, doi:10.1103/PhysRevLett.99.240601.

Bogachev, M. I., J. F. Eichner, and A. Bunde (2008a), The effects of multifractality on the statistics of return intervals, *Eur. Phys. J. Spec. Topics*, *161*, 181–193.

Bogachev, M. I., J. F. Eichner, and A. Bunde (2008b), On the occurrence of extreme events in long-term correlated and multifractal data sets, *Pure Appl. Geophys.*, *165*, 1195–1207.

Bogachev, M. I., I. S. Kireenkov, E. M. Nifontov, and A. Bunde (2009), Statistics of return intervals between long heartbeat intervals and their usability for the online prediction of disorders, *New J. Phys.*, *11*, 063036, doi:10.1088/1367-2630/11/6/063036.

Bunde, A., S. Havlin, J. W. Kantelhardt, T. Penzel, J.-H. Peter, and K. Voigt (2000), Correlated and uncorrelated regions in heart-rate fluctuations during sleep, *Phys. Rev. Lett.*, *85*, 3736–3739.

Bunde, A., J. Kropp, and H.-J. Schellnhuber (Eds.) (2002), *Science of Disasters—Climate Disruptions, Heart Attacks, and Market Crashes*, 453 pp., Springer, Berlin.

Bunde, A., J. F. Eichner, S. Havlin, and J. W. Kantelhardt (2004), Return intervals of rare events in records with long-term persistence, *Physica A*, *342*, 308–314.

Bunde, A., J. F. Eichner, J. W. Kantelhardt, and S. Havlin (2005), Long-term memory: A natural mechanism for the clustering of extreme events and anomalous residual times in climate records, *Phys. Rev. Lett.*, *94*, 048701, doi:10.1103/PhysRevLett.94.048701.

Corral, A. (2004), Long-term clustering, scaling, and universality in the temporal occurrence of earthquakes, *Phys. Rev. Lett.*, *92*, 108501, doi:10.1103/PhysRevLett.92.108501.

Davis, A., A. Marshak, W. Wiscombe, and R. Cahalan (1996), Multifractal characterization of intermittency in nonstationary geophysical signals and fields, in *Current Topics in Nonstationary Analysis*, edited by G. Trevino et al., pp. 97–158, World Sci., Singapore.

Egan, J. P. (1975), *Signal Detection Theory and ROC-analysis: Series in Cognition and Perception*, 277 pp., Academic, New York.

Eichner, J. F., E. Koscielny-Bunde, A. Bunde, S. Havlin, and H.-J. Schellnhuber (2003), Power-law persistence and trends in the atmosphere: A detailed study of long temperature records, *Phys. Rev. E*, *68*, 046133, doi:10.1103/PhysRevE.68.046133.

Eichner, J. F., J. W. Kantelhardt, A. Bunde, and S. Havlin (2007), Statistics of return intervals in long-term correlated records, *Phys. Rev. E*, *75*, 011128.

Fawcett, T. (2006), An introduction to ROC-analysis, *Pattern Recognition Lett.*, *27*, 861–874.

Feder, J. (1988), *Fractals*, 283 pp., Plenum, New York.

Glaser, R. (2001), *Klimageschichte Mitteleuropas*, 227 pp., Wissenschaftliche Buchgesellschaft, Darmstadt, Germany.

Gupta, V. K., and D. R. Dawdy (1995), Physical interpretations of regional variations in the scaling exponents of flood quantiles, in *Scale Issues in Hydrological Modelling*, edited by J. D. Kalma, pp. 106–119, John Wiley, Chichester, U. K.

Gupta, V. K., O. J. Mesa, and D. R. Dawdy (1994), Multiscaling theory of flood peaks: Regional quantile analysis, *Water Resour. Res.*, *30*(12), 3405–3421.

Hu, K., P. C. Ivanov, Z. Chen, P. Carpena, and H. E. Stanley (2001), Effect of trends on detrended fluctuation analysis, *Phys. Rev. E*, *64*, 011114, doi:10.1103/PhysRevE.64.011114.

Hurst, H. E. (1951), Long-term storage capacity of reservoirs, *Trans. Am. Soc. Civ. Eng.*, *116*, 770–799.

Hurst, H. E., R. P. Black, and Y. M. Simaika (1965), *Long-Term Storage: An Experimental Study*, 145 pp., Constable, London, U. K.

Kantelhardt, J. W., E. Koscielny-Bunde, H. H. A. Rego, S. Havlin, and A. Bunde (2001), Detecting long-range correlations with detrended fluctuation analysis, *Physica A*, *295*, 441–454.

Kantelhardt, J. W., S. A. Zschiegner, E. Koscielny-Bunde, S. Havlin, A. Bunde, and H. E. Stanley (2002), Multifractal detrended fluctuation analysis of nonstationary time series, *Physica A*, *316*, 87–114.

Kantelhardt, J. W., D. Rybski, S. A. Zschiegner, P. Braun, E. Koscielny-Bunde, V. Livina, S. Havlin, and A. Bunde (2003), Multifractality of river runoff and precipitation: Comparison of fluctuation analysis and wavelet methods, *Physica A*, *330*, 240–245.

Kantelhardt, J. W., E. Koscielny-Bunde, D. Rybski, P. Braun, A. Bunde, and S. Havlin (2006), Long-term persistence and multifractality of precipitation and river runoff records, *J. Geophys. Res.*, *111*, D01106, doi:10.1029/2005JD005881.

Klemes, V. (1974), The Hurst phenomenon: A puzzle?, *Water Resour. Res.*, *10*(4), 675–688.

Koscielny-Bunde, E., J. W. Kantelhardt, P. Braun, A. Bunde, and S. Havlin (2006), Long-term persistence and multifractality of river runoff records: Detrended fluctuation studies, *J. Hydrol.*, *322*, 120–137.

Lennartz, S., and A. Bunde (2009a), Trend evaluation in records with long-term memory: Application to global warming, *Geophys. Res. Lett.*, *36*, L16706, doi:10.1029/2009GL039516.

Lennartz, S., and A. Bunde (2009b), Eliminating finite-size effects and detecting the amount of white noise in short records with long-term memory, *Phys. Rev. E*, *79*, 066101, doi:10.1103/PhysRevE.79.066101.

Livina, V. N., Y. Ashkenazy, P. Braun, R. Monetti, A. Bunde, and S. Havlin (2003a), Nonlinear volatility of river flux fluctuations, *Phys. Rev. E*, *67*, 042101, doi:10.1103/PhysRevE.67.042101.

Livina, V. N., Y. Ashkenazy, Z. Kizner, V. Strygin, A. Bunde, and S. Havlin (2003b), A stochastic model of river discharge fluctuations, *Physica A*, *330*, 283–290.

Livina, V. N., Z. Kizner, P. Braun, T. Molnar, A. Bunde, and S. Havlin (2007), Temporal scaling comparison of real hydrological data and model runoff records, *J. Hydrol.*, *336*, 186–198.

Livina, V. N., Y. Ashkenazy, A. Bunde, and S. Havlin (2011), Seasonality effects on nonlinear properties of hydrometeorological records, in *In Extremis: Disruptive Events and Trends in Climate and Hydrology*, edited by J. Kropp and H. J. Schellnhuber, pp. 267–284, Springer, Berlin.

Lovejoy, S., and D. Schertzer (1991), *Nonlinear Variability in Geophysics: Scaling and Fractals*, 332 pp., Springer, Dordrecht, Netherlands.

Ludescher, J., M. I. Bogachev, J. W. Kantelhardt, A. Y. Schumann, and A. Bunde (2011), On spurious and corrupted multifractality: The effects of additive noise, short-term memory and periodic trends, *Physica A*, *390*, 2480–2490.

Mandelbrot, B. B., and J. R. Wallis (1969), Some long-run properties of geophysical records, *Water Resour. Res.*, *5*(2), 321–340.

Matsoukas, C., S. Islam, and I. Rodriguez-Iturbe (2000), Detrended fluctuation analysis of rainfall and streamflow time series, *J. Geophys. Res.*, *105*(D23), 29,165–29,172.

Montanari, A. (2003), Long-range dependence in hydrology, in *Theory and Application of Long-Range Dependence*, edited by P. Doukhan, G. Oppenheim and M. S. Taqqu, pp. 461–472, Birkhäuser, Boston, Mass.

Montanari, A., R. Rosso, and M. Taqqu (2000), A seasonal fractional ARIMA model applied to the Nile River monthly flows at Aswan, *Water Resour. Res.*, *36*(5), 1249–1259.

Mudelsee, M. (2007), Long memory of rivers from spatial aggregation, *Water Resour. Res.*, *43*, W01202, doi:10.1029/2006WR005721.

Mudelsee, M., M. Börngen, G. Tetzlaff, and U. Grünwald (2003), No upward trends in the occurrence of extreme floods in central Europe, *Nature*, *425*, 166–169.

Pandey, G., S. Lovejoy, and D. Schertzer (1998), Multifractal analysis of daily river flows including extremes for basins of five to two million square kilometers, one day to 75 years, *J. Hydrol.*, *208*, 62–81.

Peng, C.-K., S. V. Buldyrev, S. Havlin, M. Simons, H. E. Stanley, and A. L. Goldberger (1994), Mosaic organization of DNA nucleotides, *Phys. Rev. E*, *49*, 1685–1689, doi:10.1103/PhysRevE.49.1685.

Peters, O., C. Hertlein, and K. Christensen (2001), A complexity view of rainfall, *Phys. Rev. Lett.*, *88*, 018701, doi:10.1103/PhysRevLett.88.018701.

Pfister, C. (1998), *Wetternachhersage: 500 Jahre Klimavariationen und Naturkatastrophen (1496–1995)*, 304 pp., Haupt Verlag Ag, Bern, Switzerland.

Rodriguez-Iturbe, I., and A. Rinaldo (1997), *Fractal River Basins: Chance and Self-Organization*, 547 pp., Cambridge Univ. Press, Cambridge, U. K.

Rybski, D., A. Bunde, S. Havlin, and H. von Storch (2006), Long-term persistence in climate and the detection problem, *Geophys. Res. Lett.*, *33*, L06718, doi:10.1029/2005GL025591.

Santhanam, M. S., and H. Kantz (2008), Return interval distribution of extreme events and long-term memory, *Phys. Rev. E*, *78*, 051113, doi:10.1103/PhysRevE.78.051113.

Schumann, A. Y., and J. W. Kantelhardt (2011), Multifractal moving average analysis and test of multifractal model with tuned correlations, *Physica A*, *390*, 2637–2654, doi:10.1016/j.physa.2011.03.002.

Taqqu, M. S., V. Teverovsky, and W. Willinger (1995), Estimators for long-range dependence: An empirical study, *Fractals*, *3*, 785–798.

Tessier, Y., S. Lovejoy, P. Hubert, D. Schertzer, and S. Pecknold (1996), Multifractal analysis and modeling of rainfall and river flows and scaling, causal transfer functions, *J. Geophys. Res.*, *101*(D21), 26,427–26,440.

Turcotte, D. L., and L. Greene (1993), A scale-invariant approach to flood-frequency analysis, *Stochastic Hydrol. Hydraul.*, *7*, 33–40.

von Storch, H., and F. W. Zwiers (2002), *Statistical Analysis in Climate Research*, 496 pp., Cambridge Univ. Press, Cambridge, U. K.

M. I. Bogachev, Radio System Department, Saint Petersburg State Electrotechnical University, St. Petersburg 97376, Russia. (Mikhail.Bogachev@mail.eltech.ru)

A. Bunde, Institut für Theoretische Physik III, Justus-Liebig-Universität Giessen, Giessen D-35390, Germany. (armin.bunde@physik.uni-giessen.de)

S. Lennartz, School of Engineering and School of Geosciences, University of Edinburgh, Edinburgh EH9 3JW, UK. (sabine.lennartz@ed.ac.uk)

Extreme Events and Trends in the Indian Summer Monsoon

V. Krishnamurthy

Center for Ocean-Land-Atmosphere Studies, Institute of Global Environment and Society, Calverton, Maryland, USA
Department of Atmospheric, Oceanic and Earth Sciences, George Mason University, Fairfax, Virginia, USA

The extreme events and trends in low-pressure systems (LPSs), rainfall, and surface air temperature over the Indian monsoon region are examined in this study using long records of daily data. A review of recent studies on the extreme events over India is presented, and results from new and extended analyses are also included. The number of days when LPSs exist (or LPS days) has an increasing trend during 1930–2003. In recent years, the LPS days of depressions have shown a decreasing trend, while the lows and cyclonic storms have shown increasing trends. The moderate rainfall events over central India (CI) and Western Ghats have decreased during 1951–2004, while heavy and very heavy rainfall events have increased during this period. This has resulted in a decreasing trend in the seasonal mean rainfall and an increasing trend in the seasonal variance of daily rainfall over the two regions. The daily mean surface air temperature and the daily maximum temperature both have increasing trends in their respective seasonal means over CI during 1969–2005. The very high temperature events of both daily mean and daily maximum have increasing trends, while the moderate events show decreasing trends.

1. INTRODUCTION

Because of the concern that climate change may be causing changes in the frequency and intensity of extreme weather events, the Intergovernmental Panel on Climate Change (IPCC) provided an assessment of the extremes based on long-term observations [*Intergovernmental Panel on Climate Change (IPCC)*, 2007]. The Fourth Assessment Report (AR4) of the IPCC examined the observational evidence for trends and changes in precipitation, temperature, tropical and extratropical cyclones, and severe local weather events. According to AR4, there has been an increase in moderate, heavy, and very heavy precipitation over Europe and the United States, although the mean annual rainfall has been decreasing. In the midlatitudes, the recent decades have seen increases in warm extremes and duration of heat waves and a reduction in cold extremes. The AR4 found evidence for increase in certain categories of tropical storms since 1970 and recent increases in the cyclone activity in the Atlantic including a record-breaking season.

The IPCC AR4 found it difficult to arrive at consistent conclusions on the changes in the extreme events, especially in precipitation, for the tropics and subtropics because of lack of sufficient data. The climate variation has tremendous socioeconomic impact in the tropical regions, especially the impact of monsoon over India and neighboring countries in South Asia. The extreme events, such as the extreme rainfall over Mumbai in July 2005 and Cyclone Nargis in May 2008, had devastating consequences on the population and economy of the region. The Indian monsoon is a seasonal phenomenon in which the winds blow from the southwest half the year and from the northeast during the other half (see review by *Krishnamurthy and Kinter* [2003]). The winds from the southwest bring copious moisture from the warm waters of the Indian Ocean over to the Indian land region. India receives most of its annual rainfall during June-July-August-September (JJAS). The southwest monsoon, also known as the Indian summer monsoon, exhibits considerable variation

Extreme Events and Natural Hazards: The Complexity Perspective
Geophysical Monograph Series 196

10.1029/2011GM001122

in rainfall on intraseasonal and interannual time scales [*Krishnamurthy and Shukla*, 2000].

A significant part of the rainfall over India is caused by synoptic-scale disturbances, which are periodically formed over the quasistationary monsoon trough during the monsoon season. A majority of the disturbances are formed over the warm waters of the Bay of Bengal and move in northwest direction. These disturbances are basically low-pressure systems (LPSs) that vary in intensity and have a life cycle of about 3–6 days and a spatial scale of 1000–2000 km [*Mooley*, 1973; *Mooley and Shukla*, 1987]. The LPSs are accompanied by intense cyclonic circulation. The India Meteorological Department (IMD) has classified the LPSs as lows (LOs), depressions (DPs), cyclonic storms (STs), severe cyclonic storms (SSs), and cyclones (CYs) based on the surface cyclonic winds with speeds in the ranges of up to 8.5, 8.5–16.5, 17–23.5, 24–31.5, and greater than 31.5 m s^{-1}, respectively [*Saha et al.*, 1981; *World Meteorological Organization*, 2010]. The CYs are also known as hurricanes or typhoons in other regions of the world.

The lack of long-term data for the tropics, as pointed out by IPCC AR4, has been addressed to some extent for the Indian region with the recent release of high-resolution daily rainfall and temperature data by the IMD [*Rajeevan et al.*, 2005, 2006; *Srivastava et al.*, 2008]. Using the rainfall data for 1951–2000, *Goswami et al.* [2006] showed that there was a significant increasing trend in the frequency and magnitude of extreme events over central India (CI), while a decreasing trend was found in moderate events. Similar increasing trend in very heavy rainfall events over CI was found on a longer time scale by *Rajeevan et al.* [2008]. In the monthly mean rainfall over different regions of India, *Pal and Al-Tabbaa* [2010] found an increasing trend in the deficit rainfall and a decreasing trend in the excess rainfall.

A data set of all the LPSs formed in the northern Indian Ocean is also available for the period 1888–2003 [*Mooley and Shukla*, 1987; *Sikka*, 2006]. The LPS data were analyzed by *Krishnamurthy and Ajayamohan* [2010] showing that these systems correspond to the most dominant daily rainfall pattern (active phase of the intraseasonal oscillation) and also contribute significantly to the seasonal mean rainfall over India. The number of days on which the LPSs were present in a season was found to have an increasing trend by *Ajayamohan et al.* [2010]. However, while the LOs showed an increasing trend, the DPs were found to have a decreasing trend.

The objective of this article is to discuss the extreme events and trends in LPSs, rainfall, and temperature over the Indian monsoon region. More details on some of the recent studies cited earlier on this topic will be discussed. Extensions of earlier studies and results of new analyses will also be presented. In the case of LPSs, the frequency of storms of different categories (based on their intensity) and their trends will be examined. The extreme events in rainfall will be analyzed for two different regions over India. A similar but new analysis of daily mean temperature and daily maximum temperature will also be presented. All the results and discussions are based on statistical analysis. Although the extreme events and trends may be related to climate change, as discussed by the reports of IPCC, it is beyond the scope of this article to discuss the possibility of such relation.

In section 2, the data sets used in this study and the method of analysis are described. The variability and trends of LPSs in the monsoon region are discussed in section 3. Section 4 discusses the trends and extreme events in rainfall over India. A similar discussion of the daily mean temperature and daily maximum temperature is provided in section 5. A summary is given in section 6.

2. DATA AND METHOD OF ANALYSIS

2.1. Data

The LPS data set used in this study contains the daily location and intensity during the life cycles of all the LPSs formed in JJAS seasons of 1888–2003 over the Bay of Bengal, the Arabian Sea, and the land points in India. The data for the period 1888–1983 come from the compilation of *Mooley and Shukla* [1987] who conducted detailed examination of daily weather reports published by the IMD. These data were prepared by using objective criteria based on the central pressure of the system. The LPSs are categorized as LOs, DPs, STs, SSs, and CYs. This categorization is similar to the IMD classification of LPSs based on the surface wind speed. The LPS data for 1984–2003 were obtained from the compilation of *Sikka* [2006], who extended the analysis using the same method and criteria of *Mooley and Shukla* [1987].

A high-resolution daily rainfall data set on 1° longitude × 1° latitude grid over the land region of India for the period 1901–2004 is used in this study. The rainfall data were prepared by the IMD using observations at 1384 gauge stations spread over India [*Rajeevan et al.*, 2005, 2006]. The IMD analysis consisted of converting the station observations into gridded data using the well-known Shepard's interpolation method. Since the IMD rainfall data do not cover the oceanic region, the satellite-based rainfall data from the Tropical Rainfall Measuring Mission (TRMM) (3B42 version) [*Huffman et al.*, 2007] are used to describe the structure of storms. The TRMM data set has a very high resolution of 0.25° longitude × 0.25° latitude but is available only for the period after 1998.

This study has also used a high-resolution daily surface air temperature data set on 1° longitude × 1° latitude grid over India for the period 1969–2005. The temperature data set has also been prepared by the IMD using daily observations at 395 stations over India and converting them into gridded data using Shepard's method [*Srivastava et al.*, 2008]. The daily mean temperature and daily maximum temperature (denoted by T_{mean} and T_{max}, respectively) from the IMD data set have been used.

For circulation data, the ERA-40 reanalysis produced by the European Centre for Medium-Range Weather Forecasts [*Uppala et al.*, 2005] are used. Daily means of zonal wind u, meridional wind v, and mean sea level pressure (MSLP) on 2.5° longitude × 2.5° latitude grid were obtained from ERA-40 for the period 1958–2001.

2.2. Method of Analysis

Most of the analyses in this study are based on simple and well-known statistical methods. However, the method used to determine the trends in time series is fairly new. In this method, known as the ensemble empirical mode decomposition (EEMD), a time series is separated into variations of different time scales [*Huang et al.*, 1998; *Huang and Wu*, 2008; *Wu and Huang*, 2009]. The EEMD decomposes a time series into empirically determined intrinsic mode functions based on the local characteristic time scale of the data. The decomposition is made robust by adding multiple noise realizations to the single observed time series, in a way mimicking multiple experimental trials with inherent uncertainties. The intrinsic mode functions are then derived from the ensemble averages. The EEMD is capable of decomposing the time series analyzed in this study into variations ranging from interannual to multidecadal time scales. However, for the purpose of this study, only the modes that represent the trends are used. The most common method used involves least-square best fit for the data and extracts linear trends, which may not be suitable representations of secular trends. The EEMD, however, provides a better representation through its nonlinear trend, which is a monotonic curve [*Huang et al.*, 1998; *Wu et al.*, 2007]. In recent years, the EEMD method has been successfully used in studying secular trends in climate data [e.g., *Wu et al.*, 2011].

3. LOW-PRESSURE SYSTEMS

Most of the LPSs are formed over the head of the Bay of Bengal, and a few are formed over the land and over the Arabian Sea [*Mooley*, 1973; *Mooley and Shukla*, 1987]. The average life cycle of the LPSs is about 3–6 days, and a new parameter called LPS day is defined as a day when an LPS is present. The LPS day is, therefore, specified by the location, date, and the intensity of the corresponding LPS. The genesis, movement, and termination of all the LPSs formed during the summer monsoon seasons (JJAS) of 1888–2003 are shown in Figure 1a. During this period, 1520 LPSs were born and existed for 6841 LPS days. On an average, during the JJAS season, the number of LPSs formed is 13, and the number of LPS days is 59. As mentioned in the previous sections, the LPSs are categorized according to their intensity. The number of LPS days for LOs, DPs, STs, SSs, and CYs are 4353, 2242, 203, 35, and 8, respectively, during JJAS of 1888-2003. The occurrences of LOs, DPs, and SSs + STs + CYs are 63.6%, 32.8%, and 3.6%, respectively. It must be noted that STs and CYs are also formed outside the monsoon season, especially during May and October. During the monsoon season, the strong easterly wind shear and the short movement over the sea restricts the LPSs from becoming CYs more often. However, the LOs and DPs are the major rain-producing systems and are relevant for this study because of their ability to contribute to extreme rainfall events.

The systems formed over the Bay of Bengal and land regions generally move northwestward, while those formed over the Arabian Sea move toward the Indian continent or to the west (Figure 1a). A quasistationary monsoon trough of thermal origin extends across India, as shown in the JJAS climatological mean of MSLP in Figure 1b, and offers a path conducive for the LPSs to move across India. The tracks of the LPSs (Figure 1a) show good correspondence with the climatological mean position of the monsoon trough (Figure 1b). The total number of LPS days for LOs and DPs on 1° longitude × 1° latitude grid for the period 1888–2003 is plotted in Figure 1c. The density is low (<25) over a large region, while higher density (>25) region is confined to part of the Gangetic Plain in CI and Bay of Bengal along the east coast. The higher density region of LOs and DPs (Figure 1c) has a good correspondence with the climatological position of the deepest part of the monsoon trough (Figure 1b). For high-intensity storms (STs, SSs, and CYs), the density is higher (>10) in a small region in the east coast and Bay of Bengal. The storms lose their intensity soon after reaching the land region.

The synoptic-scale evolution of the LPSs is illustrated with two examples in Figure 2 using TRMM rainfall and streamlines of 850 hPa horizontal wind. The LPS (A) in the first example (Figure 2a) is a LO for 2 days and becomes a DP, while the second LPS (B) (Figure 2b) starts as a DP and becomes more intense (SS and ST). The spatial scale is about 1000–2000 km, and the precipitation is generally to the south of the cyclonic circulation. It is evident that the spatial scale and the intensity of the cyclonic circulations

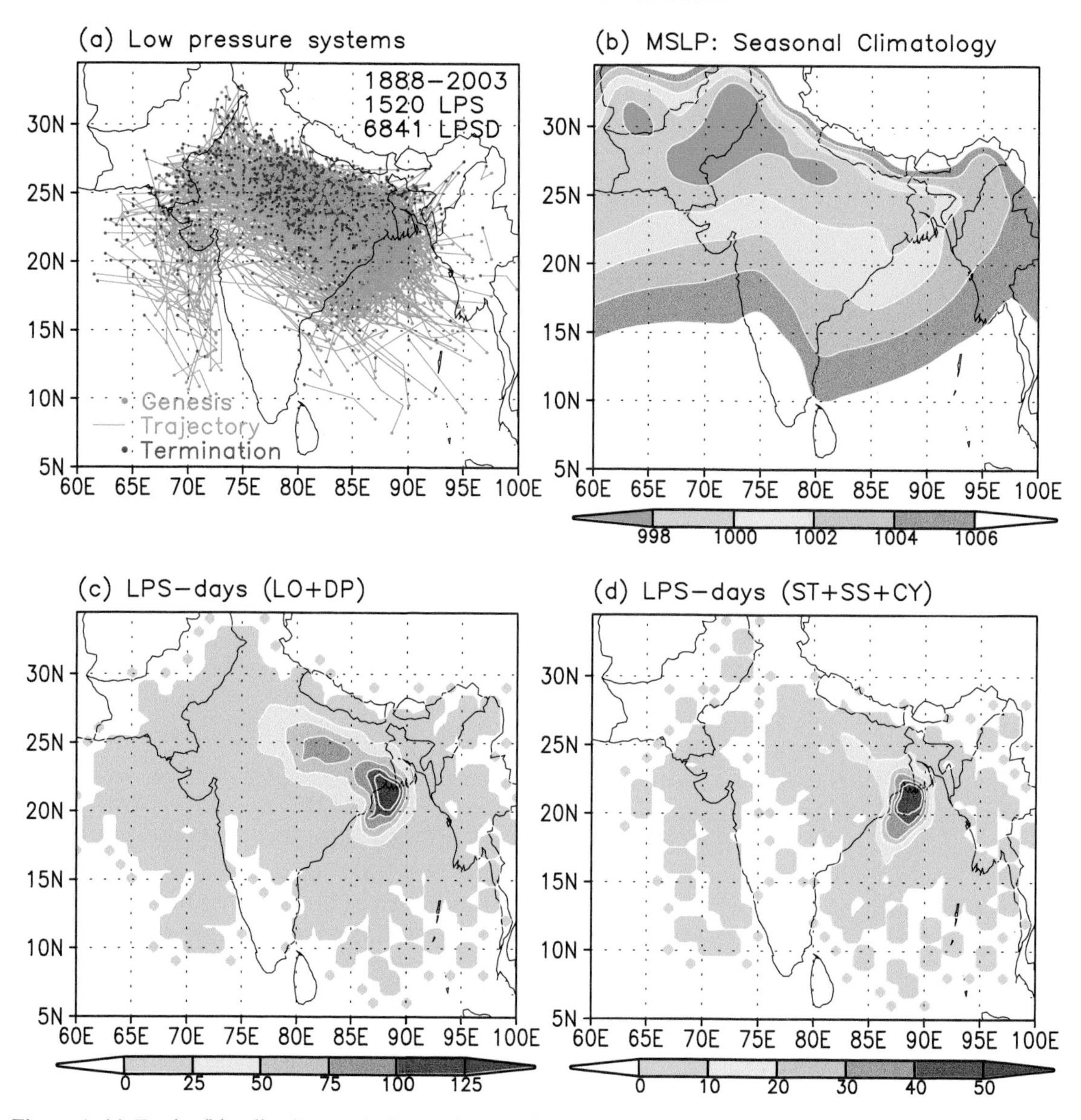

Figure 1. (a) Tracks (blue lines), genesis (green dots), and termination (red dots) of the low-pressure systems (LPSs) formed during the June-July-August-September (JJAS) monsoon seasons of 1888–2003. (b) Climatological mean (for 1958–2001) of JJAS seasonal mean of mean sea level pressure (hPa). Composite of LPS days based on (c) lows (LOs) and depressions (DPs) and (d) cyclonic storms (STs), severe cyclonic storms (SSs), and cyclones (CYs) during JJAS 1888–2003.

associated with SSs and STs are much larger than those with LOs and DPs.

Some studies have shown that there is a decreasing trend in the number of DPs since 1970 [*Rajendra Kumar and Dash*, 2001; *Sikka*, 2006]. Using the LPS data for the period 1951–2003, *Ajayamohan et al.* [2010] conducted a detailed examination of the trends in different categories of the LPS. They found significant increasing trends in the total number of LPSs and the total number of LPS days in JJAS season. The number of LPS days of LOs in JJAS also showed an increasing trend, while the number of LPS days of DPs and storms (i.e., DPs + STs + SSs + CYs) showed a decreasing trend. *Ajayamohan et al.* [2010] defined a synoptic activity index (SAI) based on the number and intensity of the LPSs geographically and related it to the extreme rainfall events, which exceed 98.3 percentile of the daily rainfall. The SAI and the extreme rainfall events averaged over CI were found to have high positive correlation as well as similar increasing trends.

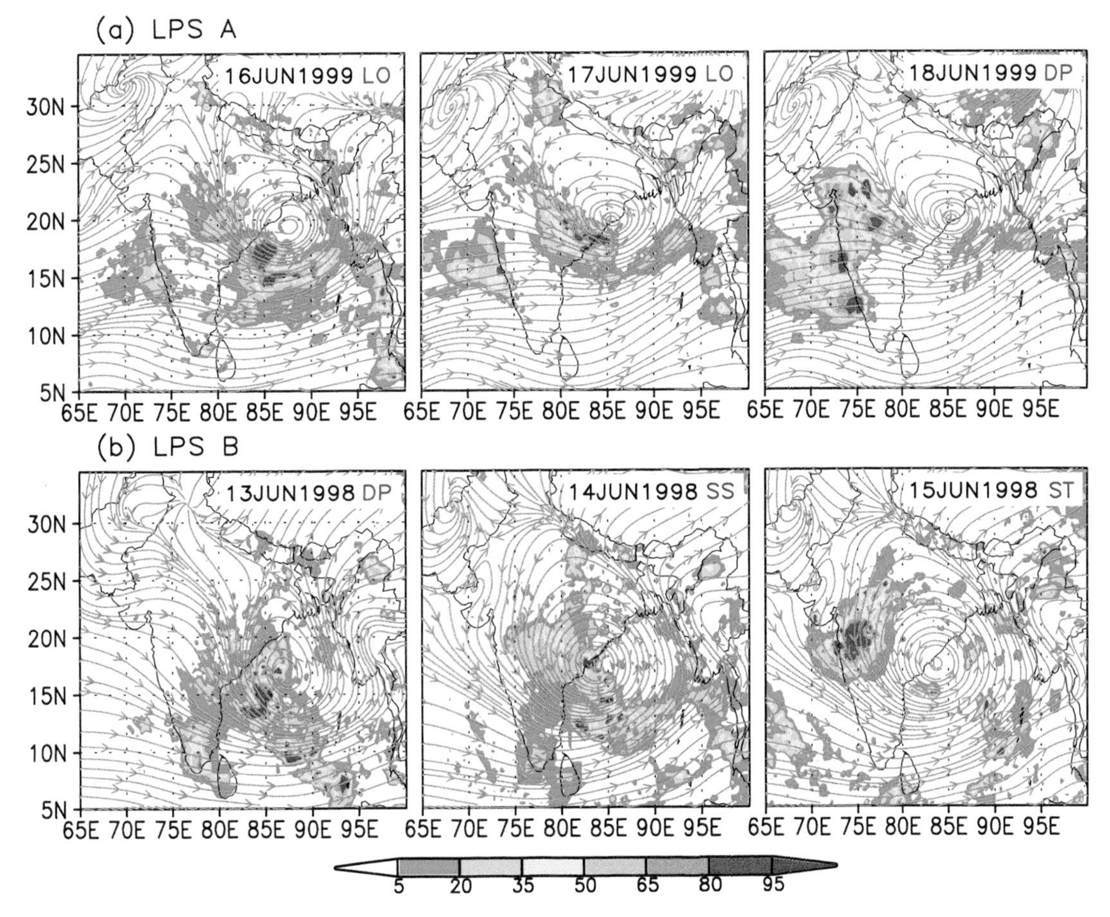

Figure 2. Evolution of two LPSs: (a) LPS A from 16 June 1999 to 18 June 1999 and (b) LPS B from 13 June 1998 to 15 June 1998. The dates and categories of the LPSs (LO, DP, ST, and SS) are indicated at the top in each plot. The rainfall (mm d^{-1}) from the Tropical Rainfall Measuring Mission data set is plotted in shaded contours, and the 850-hPa horizontal winds are plotted as streamlines.

During the years with high extreme rainfall events, the LPSs were found to be more frequent and longer lived over CI.

The variability of the LPSs is further analyzed in this study by extending the period to 1888–2003 and by considering combinations of the LPSs slightly different than those studied by *Ajayamohan et al.* [2010] and *Rajendra Kumar and Dash* [2001]. Since the focus of this study is on extreme events, the systems in higher intensity categories (STs, STs, and CYs) are combined instead of including DPs also in this category as done in other studies. The number of LPS days in JJAS of each year for LOs and DPs is shown in Figure 3a and for STs + SSs + CYs in Figure 3b. The mean LPS days for the entire period is 37.5, 19.3, and 2.1 for LOs, DPs, and STs + SSs + CYs, respectively. There is no indication of a trend in LOs until about the 1950s but shows an increasing trend in the later period (Figure 3a). The DP, however, has a decreasing trend after about 1975 and no trend before that. In the case of the storms (STs + SSs + CYs), the number of LPS days is mostly below five most of the time before 1970 but shows some increased activity after 1970 (Figure 3b).

The LPS day composites of rainfall for LOs + DPs and STs + SSs + CYs using the daily IMD rainfall data for the period 1901–2003 are shown in Figures 4a and 4b, respectively. Both the composites show heavy rainfall over the Western Ghats (WG), which may be more related to topographic effects and offshore vortices. The difference between the effects of LPSs of different intensities is seen in CI. Most of CI receives considerable rainfall during the LPS days of LOs and DPs (Figure 4a), whereas the heavy rainfall region is confined to the east coast during the LPS days of STs + SSs + CYs (Figure 4b). The composites constructed separately for LOs and DPs show similar large-scale rainfall over CI but with slightly higher magnitude in the case of DPs (figures not shown). Therefore, extreme rainfall events over

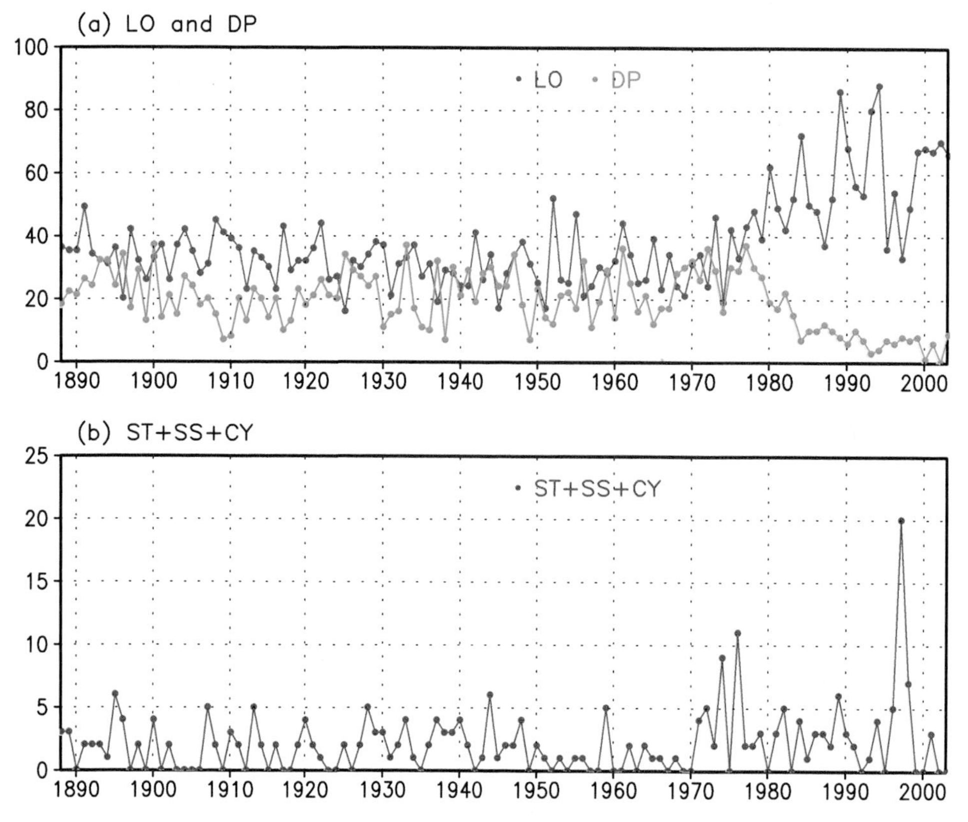

Figure 3. Number of LPS days during JJAS of each year for (a) LOs (blue) and DPs (green) separately and (b) STs, SSs, and CYs combined (red).

CI associated with the LPSs are more likely to occur during LOs and DPs. The intensity of the cyclonic circulation associated with the LPSs is shown as LPS days composites of relative vorticity for LOs + DPs and STs + SSs + CYs in Figures 4c and 4d, respectively. Both the composites have large scale and maximum at the east coast, but the magnitude is higher in the case of STs + SSs + CYs.

The EEMD method, described in section 2.2, decomposes any time series into different modes according to the time scale [*Wu and Huang*, 2009]. It is very effective in extracting nonlinear trends present in the time series. The EEMD was applied to the time series of the number of LPS days in each year's JJAS season for LOs + DPs, STs + SSs + CYs, and all LPS categories (LOs + DPs + STs + SSs + CYs) to extract their respective nonlinear trends. These trends, along with their original time series, are plotted in Figure 5. All the three cases show slightly decreasing trends from the initial time until about 1930 and then show steeper increasing trends for the rest of the period. From the trend of LOs + DPs (Figure 5a), it is seen that the average number of LPS days starts at 57.7, reaches a minimum value of 51.5 in 1932, and increases up to 71.2 in 2003. In the case of high-intensity storms (STs + SSs + CYs), the trend shows that the average number of LPS days starts at 2.3 and increases from a minimum value of 1.7 in 1929 to 4.3 in 2003 (Figure 5b). For all the LPS categories together (Figure 5c), the average number of LPS days reaches a minimum value of 53.1 in 1933 after starting from 60.3 in 1888 and increases to 74.5 days in 2003. Thus, the trends show that the JJAS LPS days increases by 1.4, 2.5, and 1.4 times the minimum values of LOs + DPs, STs + SSs + CYs, and all categories of LPSs, respectively.

4. RAINFALL

In a recent study, *Goswami et al.* [2006] investigated the trend in extreme rainfall events over India. The motivation for this study came from their observation that the seasonal mean rainfall over India does not show any trend, while the global mean surface temperature has steadily increased, and the number of severe storms has shown an increasing trend in the northern Indian Ocean. Using the daily gridded rainfall data from the IMD for the period 1951–2000, they examined the variance and extreme rainfall events over the central

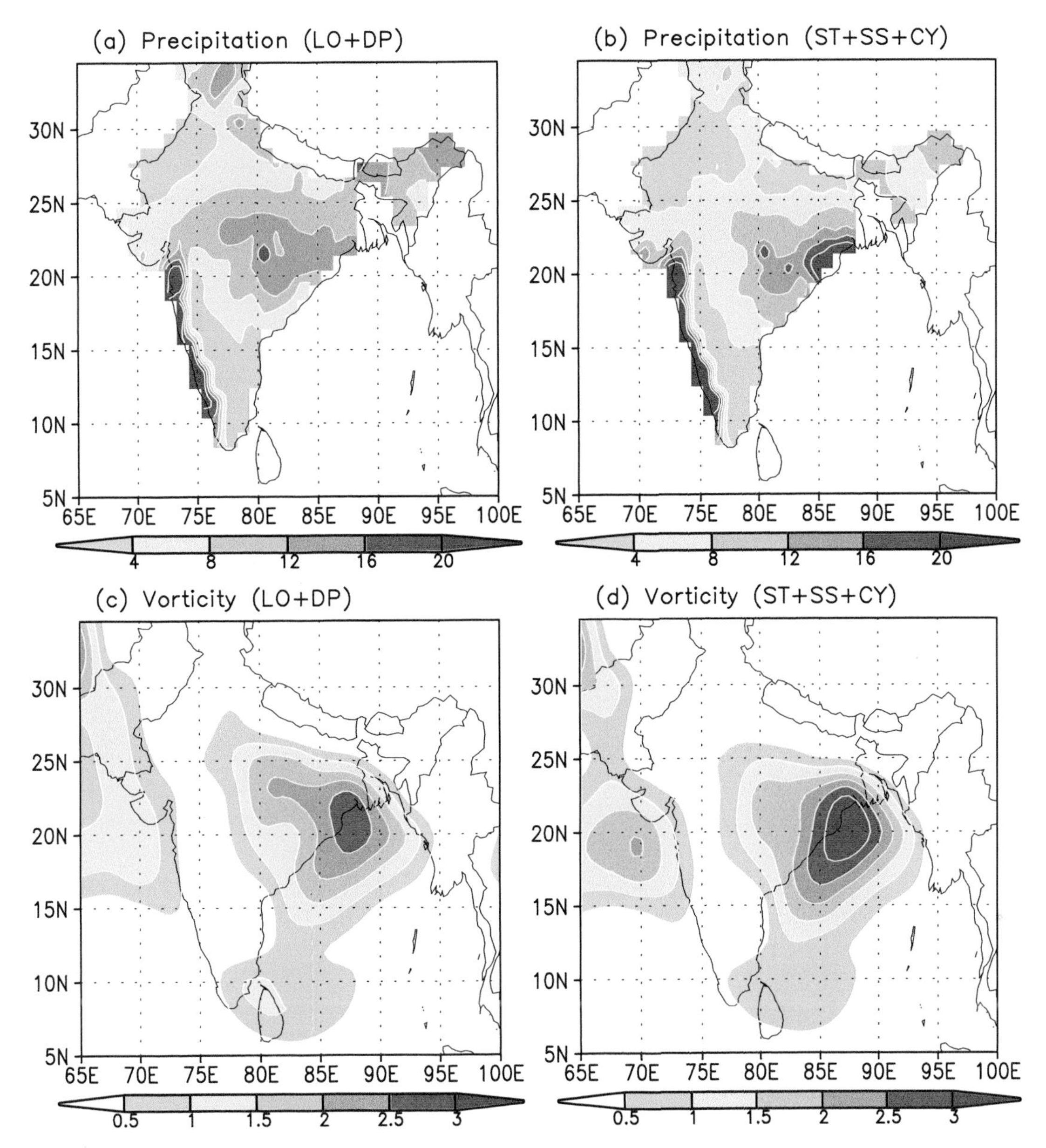

Figure 4. LPS days composites of daily rainfall (mm d^{-1}) for (a) LOs and DPs combined and (b) STs, SSs, and CYs combined for the period 1901–2003 using The India Meteorological Department rainfall data set. LPS days composites of daily 850 hPa relative vorticity (10^{-5} s^{-1}) for (c) LOs + DPs and (d) STs + SSs + CYs for the period 1958–2001 using reanalysis data.

Indian region covered by 74.5°E–86.5°E, 16.5°N–26.5°N. They showed that the interannual variability of the variance of daily rainfall anomalies in JJAS season, averaged over the central Indian region, has a significant increasing trend during 1951–2000. The number of extreme events (rainfall $\geq$ 100 mm d^{-1}) during 1981–2000 was found to be larger than that during 1951–1970. The frequency of heavy events (rainfall $\geq$ 100 mm d^{-1}) and very heavy events (rainfall $\geq$ 150 mm d^{-1}) over the central Indian region showed significant increasing trends, while the frequency of the light to moderate events (5 $\leq$ rainfall < 100 mm d^{-1}) showed significant decreasing trend. The intensity of the heavy events (99.4 to 99.9 percentile of the seasonal mean rainfall) also showed increasing trend. *Goswami et al.* [2006] suggested that the increasing trend in the heavy rainfall events is offset by the decreasing trend in the light to moderate rainfall events

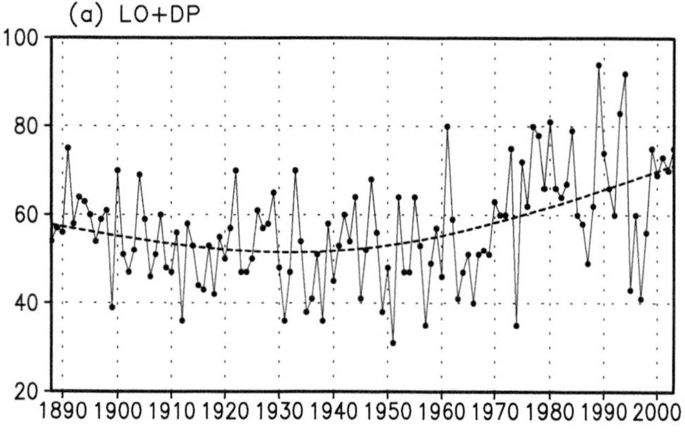

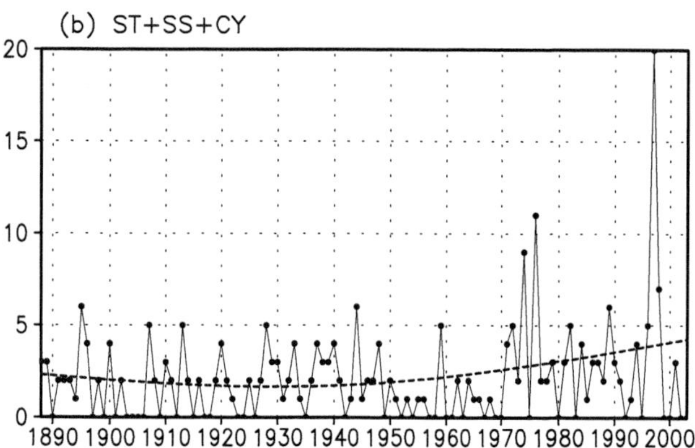

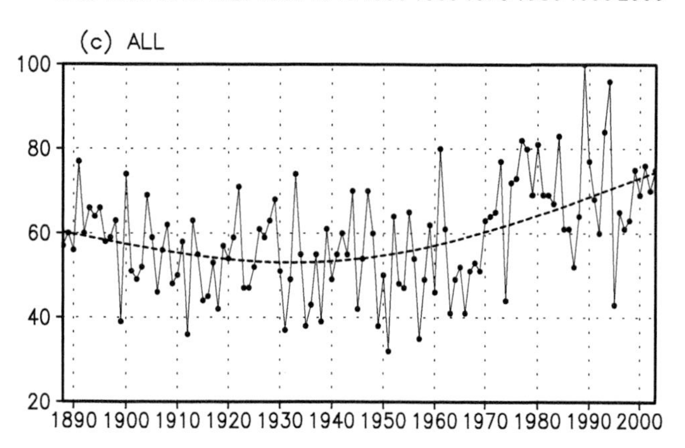

Figure 5. Number of LPS days during JJAS of each year for (a) LOs and DPs together (LOs + DPs) (solid), (b) STs, SSs, and CYs combined (STs +SSs + CYs) (solid), and (c) all categories of LPSs combined (LOs + DPs + STs + SSs + CYs) (solid). In each plot, the ensemble empirical mode decomposition (EEMD) trend of the corresponding time series of the LPS days is also plotted (dashed).

and results in no significant trend in the seasonal mean rainfall.

In this study, an analysis similar to that of *Goswami et al.* [2006] is presented for two different regions of India. The purpose of this analysis is also to illustrate the results of *Goswami et al.* [2006] and to extend it to another region of high variability. This analysis is also based on the IMD data set of daily rainfall over India for the period 1951–2004. The climatological mean of JJAS seasonal rainfall is shown in Figure 6a. The regions of maximum rainfall are in the west coast along the WG and in the northeast. Over the rest of India, CI receives the bulk of the rainfall (up to 12 mm d^{-1}), whereas less than 6 mm d^{-1} falls in the northwest and southeast regions during the monsoon season. Similar spatial structure is also seen in the standard deviation of daily rainfall for JJAS 1951–2004 (Figure 6b). The WG and the northeast region have the maximum standard deviation, and CI has considerable variability with values in the range of 12–18 mm d^{-1}. In order to examine the extreme rainfall events and trends, two regions with high seasonal mean rainfall and high variability are selected. These regions are CI covering 76°E–86°E, 19°N–26°N and the WG covering 73°E–76°E, 11°N–21°N, both of which are shown as boxes in Figure 6a. The CI region is chosen to be slightly different from that of the work of *Goswami et al.* [2006] so that it does not include the peninsular region with low rainfall and does not overlap with WG. The northeast region is not considered because of sparse station observations in that region.

The JJAS seasonal mean rainfall averaged over CI and WG are shown in Figures 6c and 6d, respectively, for the period 1951–2004. The climatological mean values of the seasonal rainfall for CI and WG are 8.2 and 12.7 mm d^{-1}, respectively. Both the time series show considerable interannual variability but do not vary in phase except for certain years such as 1961, 1972, and 1994. The difference between CI and WG in the variability of the seasonal rainfall may lie in which processes contribute to rain events. Although processes like severe thunderstorms may result in more uniform rainfall over a large spatial scale, the LOs and DPs may contribute more to the rainfall over CI, whereas topographic effects and offshore vortices may contribute more along the WG. To determine the trends, the EEMD method was applied on the time series of the JJAS seasonal mean rainfall at each grid point. The nonlinear trend modes were extracted from the EEMD at all grid points and averaged over the domains of CI and WG. The averaged EEMD trends are plotted in Figures 6c and 6d along with the original time series. Both CI and WG show decreasing trends in their seasonal mean rainfall. The average seasonal rainfall in CI decreases from 8.7 mm d^{-1} in 1951 to 7.7 mm d^{-1} in 2004, a decrease of 11.5%. Similarly, the average seasonal rainfall in WG decreases by about 8.5% going from 13.0 mm d^{-1} in 1951 to 11.9 mm d^{-1} in 2004.

The variance of the daily rainfall in the JJAS season at each grid point was computed for each year during 1951–2004. The variance was averaged over CI and WG domains, and

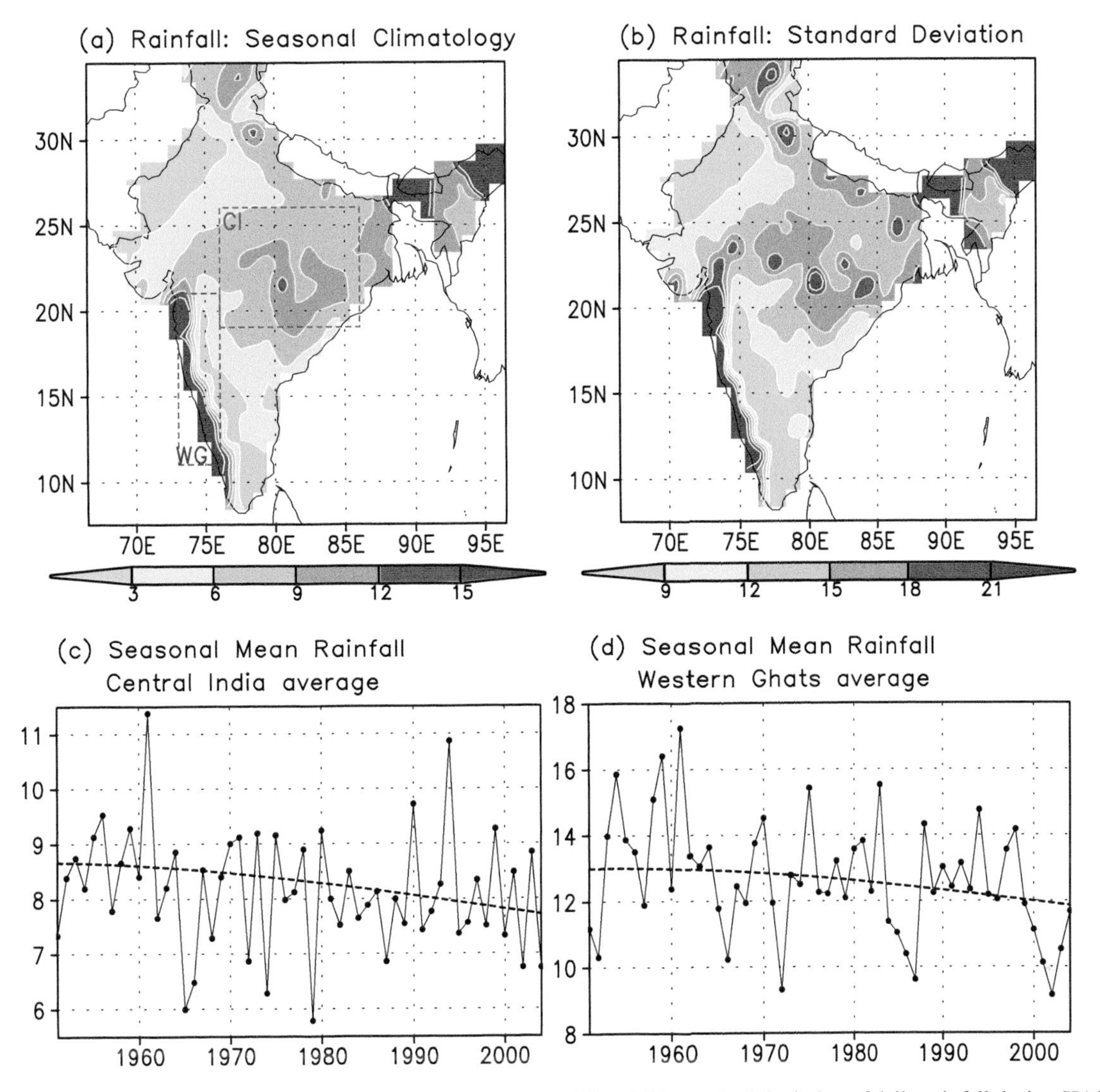

Figure 6. Climatological mean of (a) JJAS seasonal mean rainfall and (b) standard deviation of daily rainfall during JJAS monsoon season. The climatologies are averages over the period 1951–2004. JJAS seasonal mean of rainfall area averaged over (c) central India (CI) (76°–86°E, 19°–26°N) (solid) and (d) Western Ghats (WG) (73°–76°E, 11°–21°N) (solid). The domains of CI and WG used in the averaging are shown as blue and purple boxes, respectively, in Figure 6a. In Figures 6c and 6d, the corresponding area averages of the EEMD trend of the seasonal mean rainfall are also plotted (dashed). Units are in mm d^{-1}.

their time series are presented in Figure 7. The EEMD trend modes of JJAS variance averaged over CI and WG domains are plotted in Figures 7a and 7b, respectively, along with the original time series. The variance of CI (Figure 7a) agrees well with the corresponding variance (for a slightly different domain of CI) of *Goswami et al.* [2006] except that the magnitude is higher by about 50 mm^2 d^{-2}. The CI variance shows an increasing trend as seen from the EEMD mode in Figure 7a. The average variance increases from 254.0 mm^2 d^{-2} in 1951 to 265.3 mm^2 d^{-2} in 2004. This trend is slightly less steep than the corresponding trend found by *Goswami et al.* [2006]. The variance of WG (Figure 7b) is generally higher than that of CI (Figure 7a) by about 200 mm^2 d^{-2}, and their temporal variations are not in phase. The WG variance also has an increasing trend as shown by the EEMD mode (Figure 7b). The average variance of WG increases from 440.0 mm^2 d^{-2} in 1951 to 453.4 mm^2 d^{-2} in 2004.

Since the variance of daily rainfall has an increasing trend in both CI and WG, it can be argued that extreme rainfall events have increased, while moderate rainfall events have decreased in order to explain the decreasing trend in the seasonal mean rainfall in both the regions, similar to the

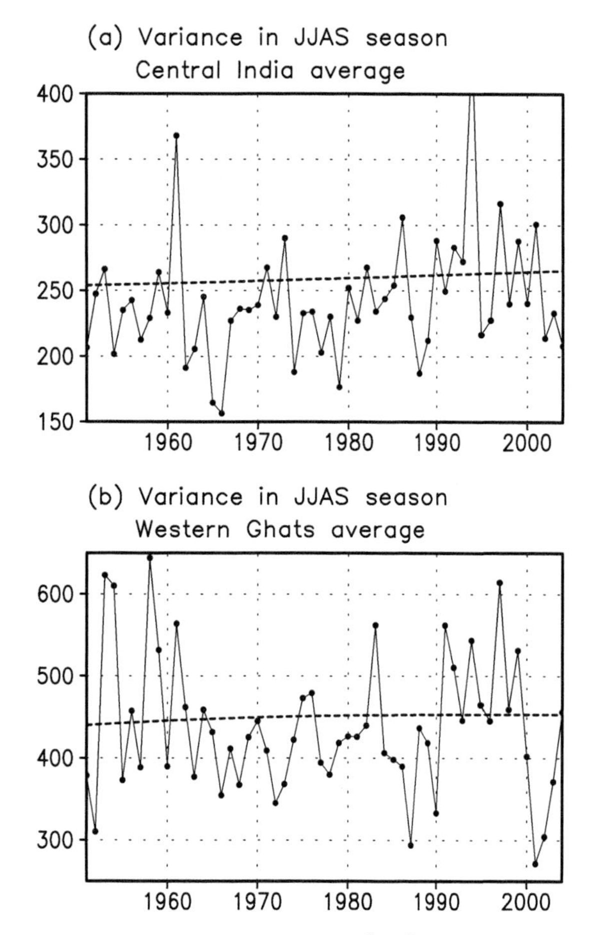

Figure 7. Variance of daily rainfall (mm^2 d^{-2}) during JJAS season of each year area averaged over (a) CI (76°–86°E, 19°–26°N) (solid) and (b) WG (73°–76°E, 11°–21°N) (solid). The corresponding area averages of the EEMD trend of the seasonal variance are also plotted (dashed).

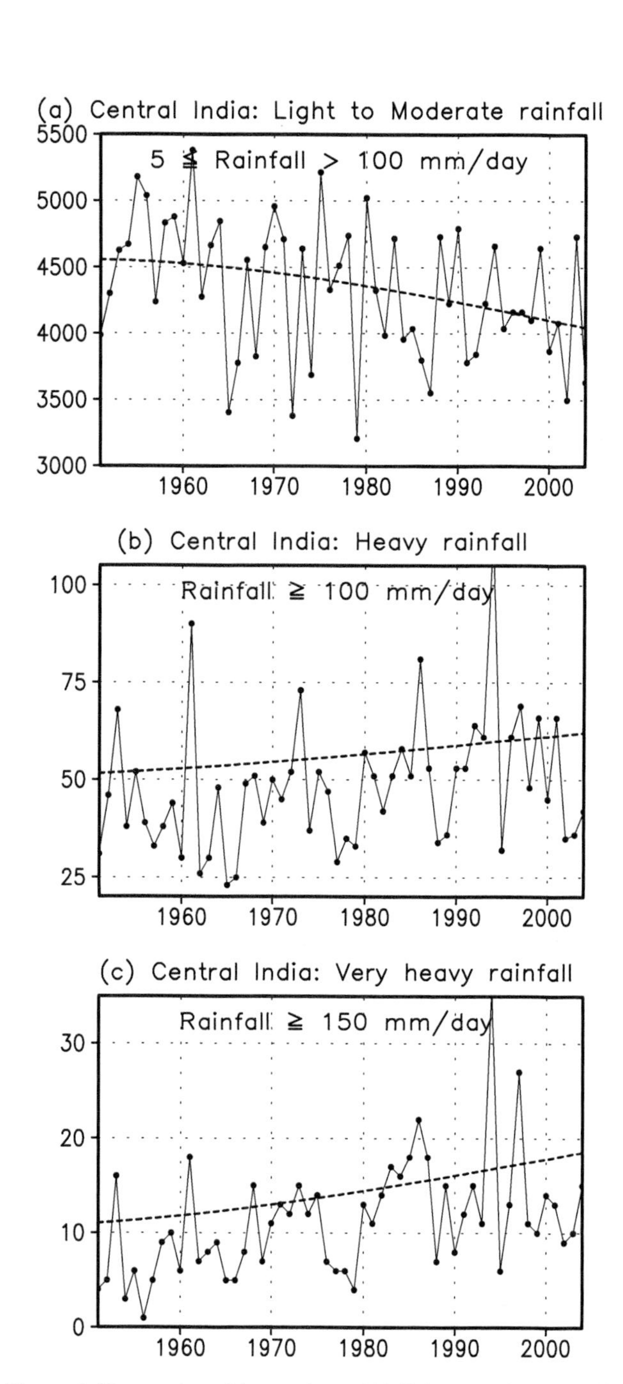

Figure 8. Time series of the number of (a) light to moderate rainfall events (5 ≤ rainfall < 100 mm d^{-1}) (solid), (b) heavy rainfall events (rainfall ≥ 100 mm d^{-1}) (solid), and (c) very heavy rainfall events (rainfall ≥ 150 mm d^{-1}) (solid) over CI (76°–86°E, 19°–26°N). The occurrence of daily rainfall of a particular category in a 1° × 1° grid is counted as one event in that category. The total number of events during JJAS of each year is plotted. The time series of the number of events in the corresponding EEMD trend are also plotted (dashed).

findings of *Goswami et al.* [2006]. To verify this, the number of events in 1° × 1° grid boxes in CI in different ranges of the magnitude of the rainfall during JJAS of each year is examined. The number of light to moderate events (5 ≤ rainfall < 100 mm d^{-1}), heavy events (rainfall ≥ 100 mm d^{-1}), and very heavy events (rainfall ≥ 150 mm d^{-1}) in CI are plotted in Figure 8, similar to the analysis of *Goswami et al.* [2006]. For each category, the EEMD was applied to the time series of the number of events at each grid box. The number of events in the EEMD trend mode of each category is also plotted in Figure 8. The light to moderate events show a decreasing trend with the average number going from 4555 in 1951 to 4047 in 2004 (Figure 8a). However, the heavy and very heavy events have increasing trends (Figures 8b and 8c). The average number of heavy events increases from 52 to 62, and the average number of very heavy events increase from 11 to 18.5 during 1951–2004. Presumably, the decreas-

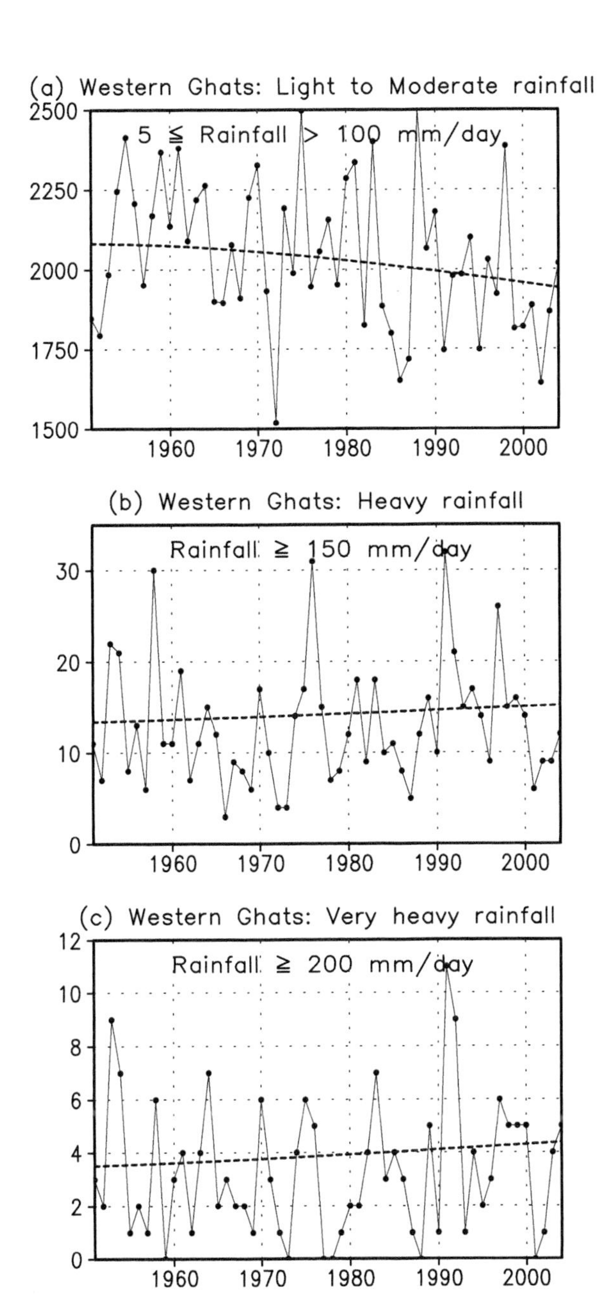

Figure 9. Time series of the number of (a) light to moderate rainfall events ($5 \leq$ rainfall < 100 mm d^{-1}) (solid), (b) heavy rainfall events (rainfall ≥ 150 mm d^{-1}) (solid), and (c) very heavy rainfall events (rainfall ≥ 200 mm d^{-1}) (solid) over WG (73°–76°E, 11°–21°N). The occurrence of daily rainfall of a particular category in a 1° × 1° grid is counted as one event in that category. The total number of events during JJAS of each year is plotted. The time series of the number of events in the corresponding EEMD trend are also plotted (dashed).

ing trend in the light to moderate events more than offsets the increasing trends of heavy and very heavy events to result in the decreasing trend of the seasonal mean rainfall shown in Figure 6c. A similar picture also emerges in the case of the rainfall events in WG. The number of light to moderate ($5 \leq$ rainfall < 100 mm d^{-1}), heavy events (rainfall ≥ 150 mm d^{-1}), and very heavy events (rainfall ≥ 200 mm d^{-1}) and their EEMD trends are shown in Figure 9. In WG also, the increasing trends in heavy and very heavy events are more than offset by the decreasing trend in light to moderate events to possibly create decreasing trend in the seasonal mean rainfall (Figure 6d).

In a recent study, *Rajeevan et al.* [2008] have examined the extreme rainfall events for a longer period covering 1901–2004. Using the same classification of rain events as *Goswami et al.* [2006], they have presented the spatial structures of the heavy and very heavy rainfall events over India. The frequency of very heavy rainfall events over a larger area of CI is shown to exhibit interdecadal variation as well as increasing trend. However, they have not identified the low-frequency variability with any of the known multidecadal oscillations.

5. SURFACE TEMPERATURE

In this section, the extreme events and trends in daily mean surface air temperature (denoted by T_{mean}) and daily maximum surface air temperature (denoted by T_{max}) over India are discussed. The IMD daily gridded data set of T_{mean} and T_{max} for the period 1969–2005 has been used in this analysis. The analysis of T_{mean} and T_{max} is similar to that of the rainfall over India presented in the previous section.

The climatological mean of JJAS seasonal mean T_{mean} for the period 1969–2005 is presented in Figure 10a. Low values of seasonal mean T_{mean} are found along the west coast and WG and in the northernmost part. The heavy rainfall received in the west coast is probably responsible for cooling the region, while the northernmost region is cooler because of its latitude and topography. The highest temperature occurs in the northwest desert region, and the entire CI and southeast region are warmer with temperatures reaching up to 31°C. The standard deviation of T_{mean} for JJAS 1969–2005 is shown in Figure 10b. The central part of India has high standard deviation with a maximum value above 2.6°C, and the southwest part of India has the lowest values. In order to examine the extreme events and trends, the region 76°–86°E, 19°–26°N in CI, which has the high variability, is selected. This is the same CI region used in the study of rainfall also (section 4). The WG region is not examined here because of its lower temperature and low variability.

The JJAS seasonal mean of T_{mean} averaged over CI is shown in Figure 10c for the period 1969–2005. The seasonal mean temperature has a climatological mean value of 28.5°C and shows considerable interannual variability. The EEMD

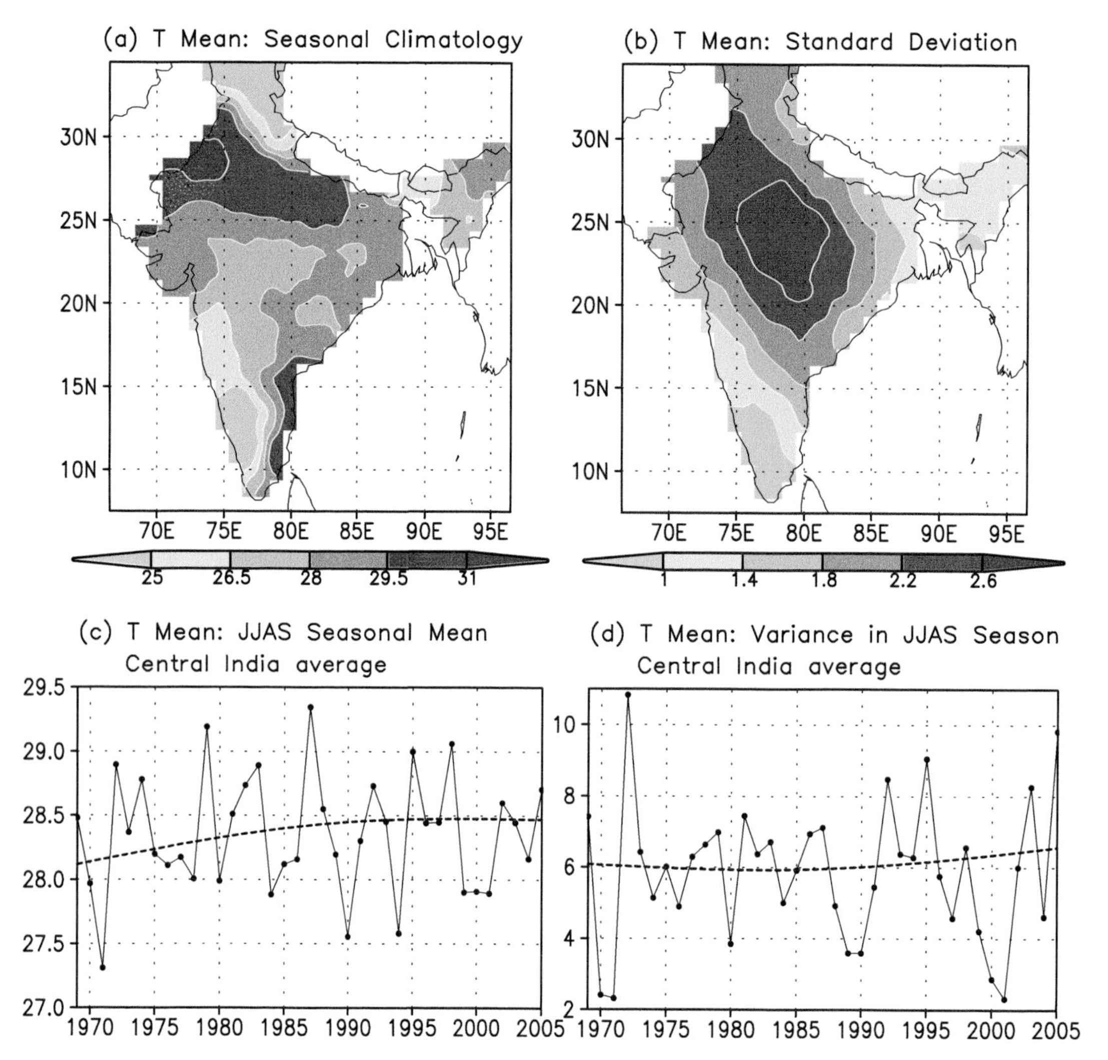

Figure 10. (a) JJAS seasonal climatological mean of daily mean surface temperature (T_{mean}) (°C) and (b) climatological mean of the standard deviation of T_{mean} (°C) during JJAS monsoon season. The climatologies are averages over the period 1969–2005. (c) JJAS seasonal mean of daily T_{mean} (°C) (solid) and (d) variance (°C^2) of daily T_{mean} during JJAS season of each year (solid) area averaged over CI (76°–86°E, 19°–26°N). In Figures 10c and 10d, the area averages of the EEMD trend of the seasonal mean of T_{mean} and the EEMD trend of the variance of T_{mean}, respectively, are also plotted (dashed).

procedure was applied on the seasonal mean time series at each point to extract the trend mode. The nonlinear trend obtained by averaging the trend mode over CI is also shown in Figure 10c. According to the trend, the average JJAS seasonal mean temperature has increased from 28.1°C to 28.5°C in CI during 1969–2005. The variance of T_{mean} during the JJAS season averaged over CI is plotted in Figure 10d. The trend obtained by the EEMD method is also averaged over CI and shown in Figure 10d. The seasonal variance also shows considerable interannual variability with a climatological mean value of 5.9°C^2. The variance has an increasing trend with the average value changing from 6.1°C^2 in 1969 to 6.6°C^2 in 2005. Thus, in the case of T_{mean}, both the seasonal mean and variance have increasing trends, although significantly during different periods, unlike in the case of rainfall.

The frequencies of moderate and extreme events in T_{mean} are examined in the same manner in which it was done with the rainfall (section 4). For this purpose, the moderate events are defined to be in the range $31 \leq T_{mean} < 34$°C, high events in $34 \leq T_{mean} < 37$°C, and very high events in $37 \leq T_{mean} < 40$°C in 1° × 1° grid boxes. For CI region, the numbers of moderate, high, and very high events during JJAS of each year are plotted in Figures 11a, 11b, and 11c,

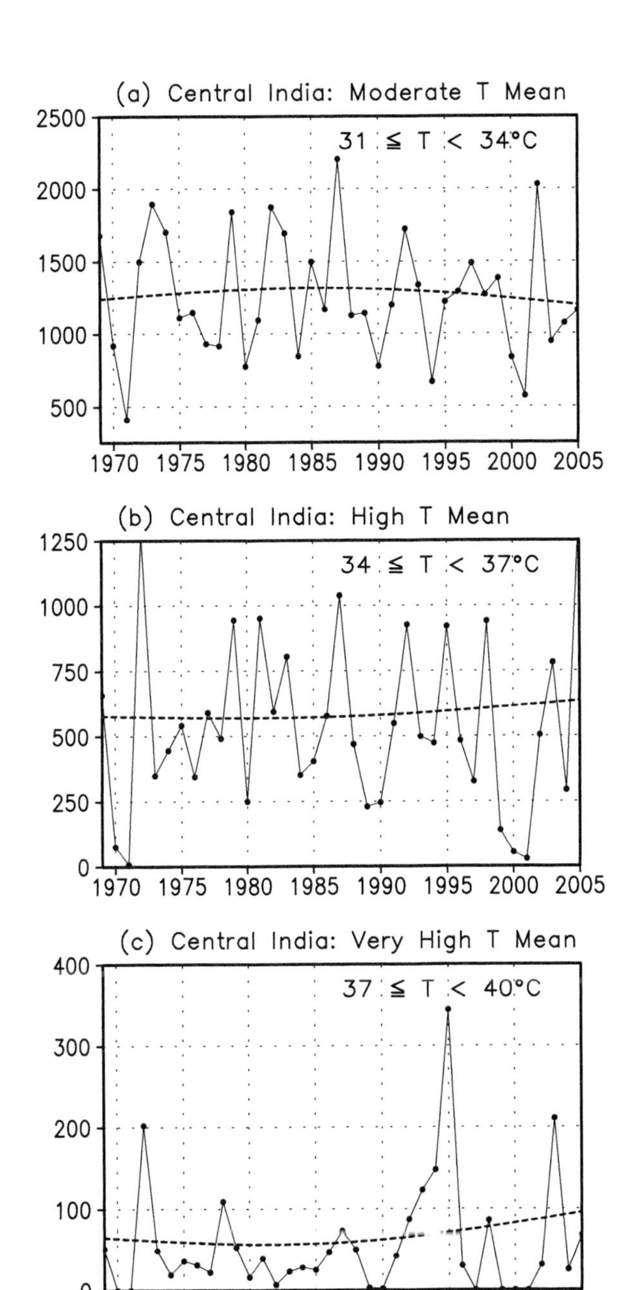

Figure 11. Time series of the number of (a) moderate temperature events (31°C ≤ T_{mean} ≥ 34°C) (solid), (b) high temperature events (34°C ≤ T_{mean} ≥ 37°C) (solid), and (c) very high temperature events (37°C ≤ T_{mean} ≥ 40°C) (solid) over CI (76°–86°E, 19°–26°N). The occurrence of daily T_{mean} of a particular category in a 1° × 1° grid is counted as one event in that category. The total number of events during JJAS of each year is plotted. The time series of the number of events in the corresponding EEMD trend are also plotted (dashed).

respectively. The corresponding trends obtained by applying the EEMD procedure on these time series are also shown. In all the three cases, there is considerable interannual variability. The moderate temperature events have a slight decreasing trend with average number changing from 1240 in 1969 to 1197 in 2005 (Figure 11a). The high and very high temperature events both show increasing trends with the average number increasing from 577 to 635 (Figure 11b) and from 65 to 95 (Figure 11c), respectively, during 1969–2005. Thus, in the case of daily mean temperature, the increasing trends in high and very high temperature events lead to increasing trends in both the seasonal mean and variance of T_{mean} (Figures 10c and 10d).

A similar analysis was also performed with the daily maximum temperature for the period 1996–2005. The climatological mean of JJAS seasonal mean T_{max} for the period 1969–2005 is presented in Figure 12a. Except for the southwest and northernmost regions, T_{max} is above 32°C over India with the highest temperature occurring in the northwest region. The spatial structure of T_{max} (Figure 12a) is similar to that of T_{mean} (Figure 10a), but the magnitude is higher by about 4°C. The standard deviation of T_{max} for the period 1969–2005, plotted in Figure 12b, shows maximum variability in the central part of India. The spatial structure is similar to that of T_{mean} (Figure 10b) but with a magnitude higher by about 0.5°C –1.0°C. The JJAS seasonal mean of T_{max} averaged over the CI region is plotted in Figure 12c along with the trend obtained by the EEMD procedure. The seasonal T_{max} has a climatological mean value of 32.6°C and an increasing trend with average T_{max} going from 32.2°C to 32.7°C during 1969–2005. The variance of T_{max} during JJAS averaged over CI and the corresponding trend from EEMD are shown in Figure 12d. The average of the variance for 1969–2005 is 12.7°C^2, which is considerably higher than the corresponding value for T_{mean}. The variance has an increasing trend, which shows that the average variance goes from 12.5°C^2 to 13.9°C^2 during 1969–2005.

For T_{max} also, the frequencies of moderate and extreme events were examined. In this case, the moderate, high, and very high events are defined as those occurring in the ranges 37 ≤ T_{max} < 40°C, 40 ≤ T_{max} < 43°C, and T_{max} ≥ 43°C, respectively, in 1° × 1° grid boxes over CI. No event was found to exceed 46°C in CI. The frequencies of moderate, high, and very high T_{max} events over CI are plotted in Figure 13, along with the corresponding EEMD trends. The moderate event has a decreasing trend in the latter part of the period, and the average number changes from 921 in 1969 to 774 in 2005 (Figure 13a). The high temperature events have almost no trend with the average number going from 585 to 570 during 1969–2005. However, the very high temperature events have an increasing trend, which shows

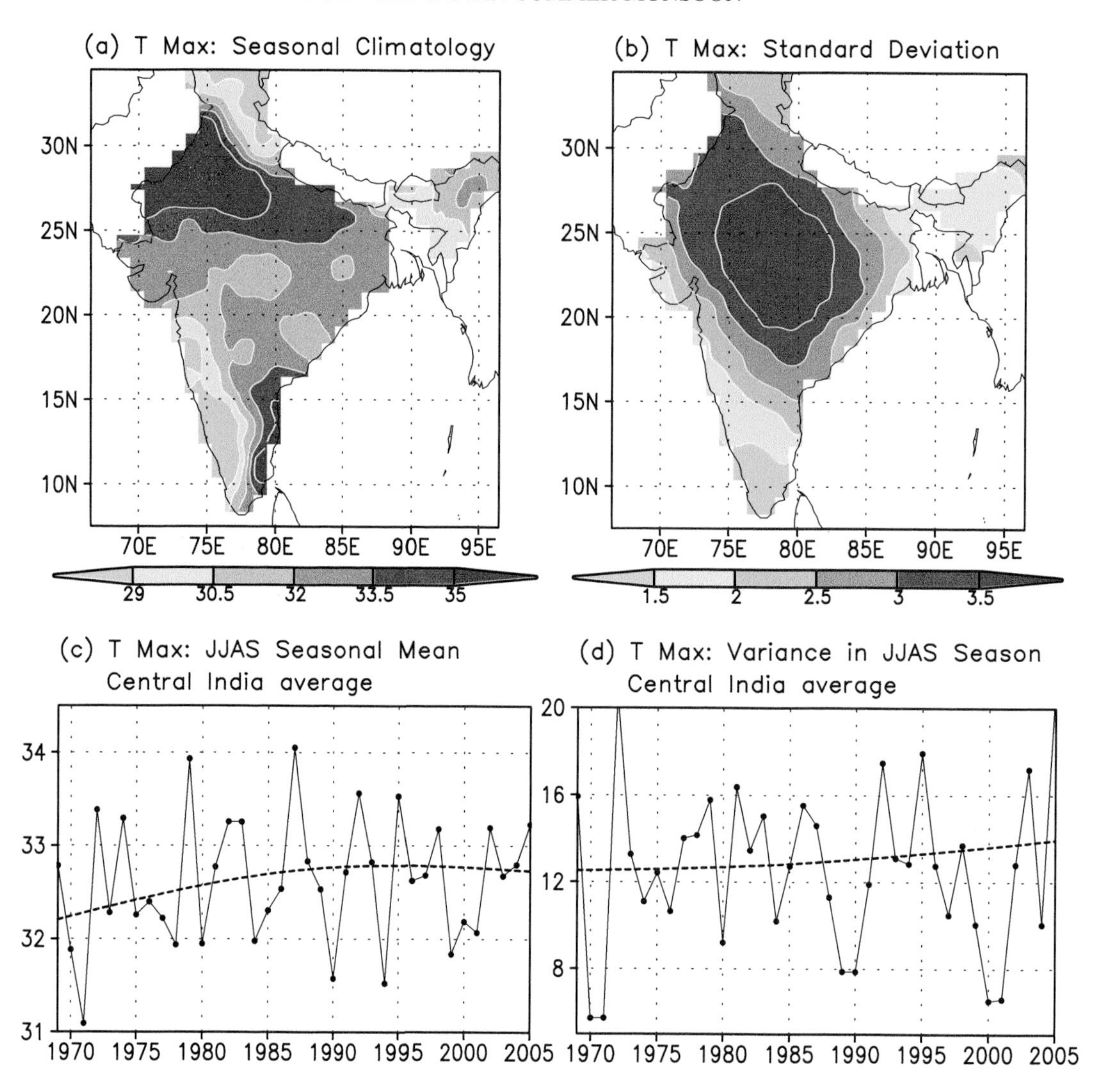

Figure 12. (a) JJAS seasonal climatological mean of daily maximum surface temperature (T_{max}) (°C) and (b) climatological mean of the standard deviation of T_{max} (°C) during JJAS monsoon season. The climatologies are averages over the period 1969–2005. (c) JJAS seasonal mean of daily T_{max} (°C) (solid) and (d) variance (°C^2) of daily T_{max} during JJAS season of each year (solid) area averaged over CI (76°–86°E, 19°–26°N). In Figures 12 c and 12d, the area averages of the EEMD trend mode of the seasonal mean of T_{max} and the EEMD trends of the variance of T_{max}, respectively, are also plotted (dashed).

that the average number changes from 156 in 1969 to 320 in 2005.

6. SUMMARY

The extreme events and trends in LPSs, rainfall, and surface air temperature in the Indian summer monsoon have been discussed in this study. A review of some recent studies was provided, and results from new analyses were presented using long records of observed data. As reported by the IPCC AR4, a better understanding of the extreme events and trends in the tropics is difficult because of lack of long observations. However, for the Indian monsoon region, high-resolution gridded data sets of rainfall and surface air temperature were made available recently by the IMD. Additionally, a very long data set of LPSs, carefully prepared by *Mooley and Shukla* [1987] and *Sikka* [2006], is also available.

For the period 1888–2003, the LPSs consist mainly of LOs and DPs, which are the main rain-producing systems. Although they are lower intensity LPSs, the LOs and DPs are relevant in this study because of their ability to cause extreme rainfall events. In recent decades, the number of LOs has

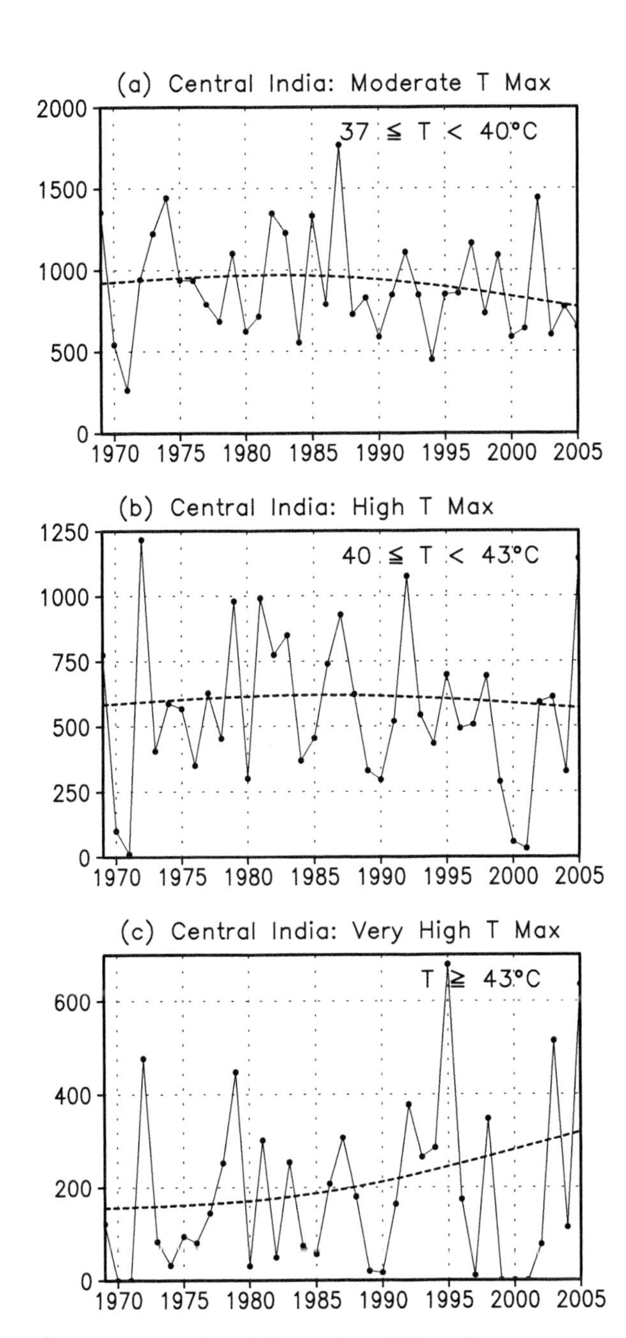

Figure 13. Time series of the number of (a) moderate temperature events (37°C ≤ T_{max} < 40°C (solid), (b) high temperature events (40°C ≤ T_{max} < 43°C) (solid), and (c) very high temperature events (T_{max} ≥ 43°C) (solid) over CI (76°–86°E, 19°–26°N). The occurrence of daily T_{max} of a particular category in a 1° × 1° grid is counted as one event in that category. The total number of events during JJAS of each year is plotted. The time series of the number of events in the corresponding EEMD trend modes are also plotted (dashed).

considerably increased, while the number of DPs has shown a marked decrease. However, the number of days of LOs and DPs in a season shows an increasing trend from 1930 to 2003. The LPS days of the STs (three higher intensity LPSs) in a season also has an increasing trend during 1929–2003.

The seasonal mean rainfall, examined over CI and the WG regions, shows a decreasing trend during 1951–2004. The seasonal rainfall has decreased by about 1–1.2 mm d^{-1} in this period. However, the variance of daily rainfall during the monsoon season has an increasing trend in both CI and WG. The light to moderate rainfall events (5 ≤ rainfall < 100 mm d^{-1}) have decreased, while heavy and very heavy rainfall events (rainfall ≥ 100 mm d^{-1}) have increased in both the regions during this period. The light to moderate rainfall events have more than offset the heavy and extreme events leading to a decreasing trend in the seasonal mean rainfall during 1951–2004. The LOs and DPs play a significant role in the rainfall over CI. It will be interesting to find out how the trends in the LPSs are related to the trends in rainfall events, especially in the extreme events. Although *Ajayamohan et al.* [2010] have examined this connection to some extent and *Krishnamurthy and Ajayamohan* [2010] have analyzed the general relation between LPSs and rainfall, specific linkage between the LPSs and extreme events and the mechanisms will be interesting topics to investigate.

The seasonal means of daily mean temperature and daily maximum temperature over CI were found to have increasing trends during 1969–2005. The variance of daily temperature in a season also showed increasing trend. While the moderate events (31 ≤ T_{mean} < 34°C) in daily mean temperature have decreasing trend, the high (34 ≤ T_{mean} < 37°C) and very high temperature (T_{mean} ≥ 37°C) events show increasing trends. In the case of daily maximum temperature, the very high temperature (T_{max} ≥ 43°C) events show increasing trends, while the moderate (37 ≤ T_{max} < 40°C) and high temperature (40 ≤ T_{max} < 43°C) events have decreasing trends.

Generally, studies that have investigated the extreme events and trends have done so within the context of increasing global temperature and climate change, similar to the concerns of IPCC in its periodic reports. For example, studies by *Goswami et al.* [2006] and *Rajeevan et al.* [2008] have shown increasing trends of sea surface temperature in the Indian Ocean, implying a relation with the increasing trends in extreme events. Although such a relation is plausible, more detailed investigation on how the extreme rainfall events over India are related to warming trend is needed. Such investigations would involve more observational analyses and model experiments such as the study by *Stowasser et al.* [2009]. Assessing the model projections under one of the climate change scenarios of IPCC AR4, *Turner and Slingo*

[2009] found that most models show increases in heavy rainfall events. Further studies are needed to understand the relation between extreme events in the Indian monsoon region and climate change.

Acknowledgments. This research was supported by grants from the National Science Foundation (ATM-0830068), National Oceanic and Atmospheric Administration (NA09OAR4310058), and National Aeronautics and Space Administration (NNX09AN50G). The author thanks Zhaohua Wu for help with EEMD and Deepthi Achuthavarier for helpful comments.

REFERENCES

Ajayamohan, R. S., W. J. Merryfield, and V. V. Kharin (2010), Increasing trend of synoptic activity and its relationship with extreme rain events over central India, *J. Clim.*, *23*, 1004–1013.

Goswami, B. N., V. Venugopal, D. Sengupta, M. S. Madhusoodanan, and P. K. Xavier (2006), Increasing trend of extreme rain events over India in a warming environment, *Science*, *314*, 1442–1445.

Huang, N. E., and Z. Wu (2008), A review on Hilbert-Huang transform: Method and its applications to geophysical studies, *Rev. Geophys.*, *46*, RG2006, doi:10.1029/2007RG000228.

Huang, N. E., Z. Shen, S. R. Long, M. C. Wu, H. H. Shih, Q. Zheng, N.-C. Yen, C. C. Tung, and H. H. Liu (1998), The empirical mode decomposition and the Hilbert spectrum for nonlinear and non-stationary time series analysis, *Proc. R. Soc. London, Ser. A*, *454*, 903–993.

Huffman, G. J., R. F. Adler, D. T. Bolvin, G. Gu, E. J. Nelkin, K. P. Bowman, Y. Hong, E. F. Stocker, and D. B. Wolff (2007), The TRMM Multi-satellite Precipitation Analysis (TMPA): Quasi-global, multi-year, combined-sensor precipitation estimates at fine scale, *J. Hydrometeorol.*, *8*, 38–55.

Intergovernmental Panel on Climate Change (IPCC) (2007), *Climate Change 2007: The physical science basis. Contribution of Working Group I to the Fourth Assessment Report of the Intergovernmental Panel on Climate Change*, edited by S. Solomon et al., 996 pp., Cambridge Univ. Press, Cambridge, U. K.

Krishnamurthy, V., and R. S. Ajayamohan (2010), Composite structure of monsoon low pressure systems and its relation to Indian rainfall, *J. Clim.*, *23*, 4285–4305.

Krishnamurthy, V., and J. L. Kinter III (2003), The Indian monsoon and its relation to global climate variability, in *Global Climate*, edited by X. Rodó and F. A. Comín, pp. 186–236, Springer, Berlin.

Krishnamurthy, V., and J. Shukla (2000), Intraseasonal and interannual variability of rainfall over India, *J. Clim.*, *13*, 4366–4377.

Mooley, D. A. (1973), Some aspects of Indian monsoon depressions and the associated rainfall, *Mon. Weather Rev.*, *101*, 271–280.

Mooley, D. A., and J. Shukla (1987), Characteristics of the westward-moving summer monsoon low pressure systems over the Indian region and their relationship with the monsoon rainfall, report, Cent. for Ocean-Land-Atmos. Interact., Univ. of Md., College Park.

Pal, I., and A. Al-Tabbaa (2010), Regional changes in extreme monsoon rainfall deficit and excess in India, *Dyn. Atmos. Oceans*, *49*, 206–214.

Rajeevan, M., J. Bhate, J. D. Kale, and B. Lal (2005), Development of a high resolution daily gridded rainfall data for the Indian region, *Met. Monogr. Climatol. 22/2005*, 26 pp., Natl. Clim. Cent., India Meteorol. Dep., Pune, India.

Rajeevan, M., J. Bhate, J. D. Kale, and B. Lal (2006), High resolution daily gridded rainfall data for the Indian region: Analysis of break and active monsoon spells, *Curr. Sci.*, *91*, 296–306.

Rajeevan, M., J. Bhate, and A. K. Jaswal (2008), Analysis of variability and trends of extreme rainfall events over India using 104 years of gridded daily rainfall data, *Geophys. Res. Lett.*, *35*, L18707, doi:10.1029/2008GL035143.

Rajendra Kumar, J., and S. K. Dash (2001), Interdecadal variations of characteristics of monsoon disturbances and their epochal relationships with rainfall and other tropical features, *Int. J. Climatol.*, *21*, 759–771.

Saha, K., F. Sanders, and J. Shukla (1981), Westward propagating predecessor of monsoon depressions, *Mon. Weather Rev.*, *109*, 330–343.

Sikka, D. R. (2006) A study on the monsoon low pressure systems over the Indian region and their relationship with drought and excess monsoon seasonal rainfall, *COLA Tech. Rep. 217*, Cent. for Ocean-Land-Atmos. Stud., Calverton, Md.

Srivastava, A. K., M. Rajeevan, and S. R. Kshirsagar (2008), Development of a high resolution daily gridded temperature data set (1969–2005) for the Indian region, *NCC Res. Rep. 8*, 18 pp., Natl. Clim. Cent., India Meteorol. Dep., Pune, India.

Stowasser, M., H. Annamalai, and J. Hafner (2009), Response of south Asian summer monsoon to global warming: Mean and synoptic systems, *J. Clim.*, *22*, 1014–1036.

Turner, A. G., and J. M. Slingo (2009), Uncertainties in future projections of extreme precipitation in the Indian monsoon region, *Atmos. Sci. Lett.*, *10*, 152–258.

Uppala, S. M., et al. (2005), The ERA-40 re-analysis, *Q. J. R. Meteorol. Soc.*, *131*, 2961–3012.

World Meteorological Organization (2010), Tropical cyclone operational plan for the Bay of Bengal and the Arabian Sea, Edition 2010, *WMO/TD 84*, Geneva, Switzerland.

Wu, Z., and N. E. Huang (2009), Ensemble empirical mode decomposition: A noise-assisted data analysis method, *Adv. Adapt. Data Anal.*, *1*, 1–41.

Wu, Z., N. E. Huang, S. R. Long, and C.-K. Peng (2007), On the trend, detrending and variability of nonlinear and non-stationary time series, *Proc. Natl. Acad. Sci. U. S. A.*, *104*, 14,889–14,894.

Wu, Z., N. E. Huang, J. M. Wallace, B. V. Smoliak, and X. Chen (2011), On the time-varying trend in global-mean surface temperature, *Clim. Dyn.*, *37*, 759–773.

V. Krishnamurthy, Center for Ocean-Land-Atmosphere Studies, IGES, 4041 Powder Mill Road, Suite 302, Calverton, MD 20705, USA. (krishna@cola.iges.org)

Empirical Orthogonal Function Spectra of Extreme Temperature Variability Decoded From Tree Rings of the Western Himalayas

R. K. Tiwari

National Geophysical Research Institute, CSIR, Hyderabad, India

R. R. Yadav

Birbal Sahni Institute of Paleobotany, Lucknow, India

K. P. C. Kaladhar Rao

National Geophysical Research Institute, CSIR, Hyderabad, India

The ability to distinguish different natural frequency modes from a complex noisy temperature record is essential for a better understanding of the climate response to internal/external forcing. Here we investigate the empirical orthogonal function and spectra of a newly reconstructed tree ring temperature variability record decoded from the western Himalayas for a period spanning 1227 A.D. to 2000 A.D., allowing frequency resolution of interdecadal and interannual oscillatory modes. The spectral analysis of first principal component (PC1) with ~61.46% variance reveals the dominance of significant solar cycles notably peaking around 81, 32, 22, and 8–14 years. Although longer solar cycles are dominant and statistically significant at more than 95% confidence level, the average 11 year solar cycle peaking at a period ranging from 8 to 14 years is less significant (not >90%) and might indicate chaotic phenomena. Similar spectral analysis of PC2 (variance 26%) and PC3 (variance 13.05%) reveals interannual oscillations peaking at a period range of 2–8 years, which are probably related to the global aspect of the El Niño–Southern Oscillation phenomena. Our present analysis in the light of the recent ocean-atmospheric model results suggests that even small variation in solar output in conjunction with the atmospheric-ocean system and other related feedback processes could cause the observed abrupt temperature variability at the time of "criticality" through the triggering mechanism.

1. INTRODUCTION

Analysis of a globally distributed set of temperature proxy records of several centuries has revealed oscillations on interdecadal (~15–35 years) and century (~50–150 years) scales [*Mann et al.*, 1995]. A matching quasicyclic pattern on the identical time scale has also been detected in historical instrumental as well as isolated climate proxies and local temperature variability records. Such quasicyclic pattern

Extreme Events and Natural Hazards: The Complexity Perspective
Geophysical Monograph Series 196

10.1029/2011GM001133

may be intrinsic to the natural climate system. Some recent modeling studies attributed these cyclic modes to the dynamical coupling of ocean and atmospheric processes and/or dynamics of thermohaline circulations. Recent studies have attributed the cause of abrupt changes in the various domains of Earth-oceans-atmosphere system to natural variability. However, the natural variability internal to the climate system alone does not explain observed nonlinear pattern of temperature variability.

There is also evidence for solar cycles in various proxies of climate records, but the possible influence of solar activity on climate has been a matter of great debate for a long time. Several researchers have reported correlations between solar variability and climate parameters. One of the earliest links between the solar variability and climate parameters has been proposed by *Eddy* [1976], who suggested that during times of few or no sunspots, e.g., during the Maunder Minimum (~1645–1715), the Sun's radiative output was reduced, leading to colder climate. Although studies as mentioned above have indicated that there is persuasive evidence for the presence of solar cycles in temperature records, the physical impact of solar contribution is minimal (~0.1%). Hence, it is somewhat difficult to explain a direct link between the solar variability and temperature variability. The possible causative forcing, therefore, may also include solar UV and galactic cosmic rays besides the postulated role of ocean circulations, radiative forcing due to greenhouse gas, aerosols, and volcanic eruption. Recently, with the addition of more years of data, a considerably good correlation has been suggested between the two phenomena, and the statistical analysis of these data also confirmed statistical significance of solar signals in temperature records [*Gray et al.*, 2010].

It is now being realized that abrupt and prevalent temperature variability with significant impacts has recurred in the past when the Earth system was forced across threshold. Evidence is emerging that the Earth's climate is a "critical phenomena" and also sensitive to small changes in solar output on centennial time scales [*Rind*, 2002; *Bond et al.*, 2001; *Shindell et al.*, 2001]. Quasicyclic manifestation of "critical forcing parameters," like solar cycles, appears in the spectra because they got imprinted at the time of major changes in the physical processes, although they do not play a vital role in driving significant changes in the underlying physical processes.

Here we investigate the temporal structure of a newly reconstructed proxy record of temperature variability of the ~past 773 years (1227–2000 A.D.), decoded from tree rings of the western Himalayas based on empirical orthogonal function (EOF) and spectral analysis. We also discuss the sensitivity of abrupt temperature variability in the light of recent modeling results of ocean-atmospheric coupling and solar variability to examine the range of temperature variability with small changes in solar outputs.

2. RECONSTRUCTED TREE RING TEMPERATURE RECORD

Global temperature variations of the Earth on multilateral time scales have been reconstructed by calibrating tree ring growth rates with instrumental records [*Mann et al.*, 1995] with considerable precision. Tree ring proxy climate indicators have been potentially used for extracting information regarding past seasonal temperature or precipitation/drought based on measurements of annual ring width. The data being analyzed here are one of the best such temperature variability reconstructions of the premonsoon season in the western Himalayas. The temperature variability records span the time period of 1227 A.D. to 2000 B.C. covering the major medieval warm and cold (Little Ice Age (LIA)) spells and most recent anthropogenic warming (Figure 1a). A detailed description of the data is presented elsewhere [*Yadav et al.*, 2004]. Figure 1 illustrates the complex time variability response of interdecadal and interannual premonsoon temperature variability for the past 775 years. Apparently, several regional and global cooling and warming episodes match well with the temperature data [*Yadav et al.*, 2004], attesting to its global significance [*Briffa et al.*, 2001; *Esper et al.*, 2002; *Cook et al.*, 2003]. The reconstructed tree ring temperature variability record shows higher variability during the recent past, since the sixteenth century, compared to the earlier past of the time series (1226–1500 A.D.). The record also exhibits unstable climate during the LIA [*Brauning*, 2001; *Zhang and Crowley*, 1989; *Yadav et al.*, 2004]. The further cooling and warming trend of the variability also matches more or less with various regional climate variability of the Asian monsoon region as reported by several workers [*Briffa et al.*, 2001; *Esper et al.*, 2002; *Cook et al.*, 2003]. This implies that the reconstructed tree ring temperature variability record analyzed here has adequately been placed both in global and regional context.

The mean temperature series is obtained from nine weather stations including both high- and low-elevation areas in the western Himalayas [*Yadav et al.*, 2004]. Temperature variability history is based on widely spread pure Himalayan cedar (*Cedrus deodara* (Roxb.) G. Don) trees, which characterizes all the sites with almost no ground vegetation and thereby minimizes individual variation in tree ring sequences induced by intertree competition [*Yadav et al.*, 2004]. The mean chronological structure is based on in total 60 radii from 45 trees, statistical features of which show that the chronology is suitable for dendroclimatic studies back to 1226 A.D. [*Yadav et al.*, 2004]. Natural and undisturbed

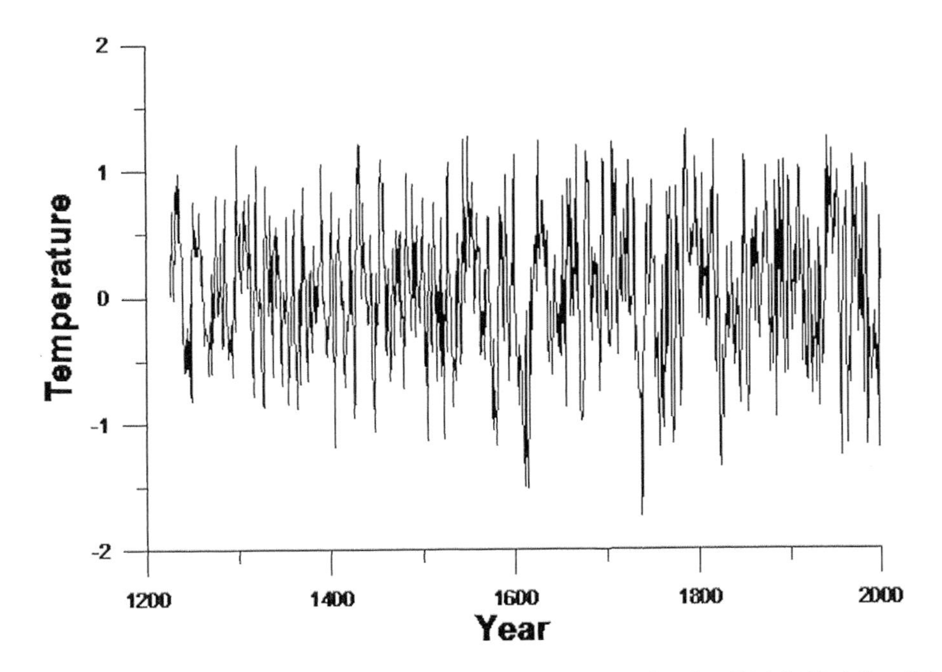

Figure 1. Normalized time series of mean premonsoon temperature anomalies for 1226–2000 A.D. of the western Himalayas. Modified from the work of *Yadav et al.* [2004].

open stands of *C. deodara* (Roxb. Ex Lambert) G. Don at different localities in Uttarkashi and Chamoli in Uttaranchal, where healthy trees were growing on rocky hill slopes, were selected for the reconstruction of the time series. Samples were collected from 16 sites in the form of increment cores with open stands of *C. deodara*. At least two cores from each tree were taken at breast height (1.4 m) of the stem [*Yadav et al.*, 2004]. The ring widths of dated samples were evaluated, and the chronologies were created by visually assigning years to rings with different widths to 0.01 mm accuracy. Visual inspection of the temperature record (Figure 1) exhibits a nonrandom pattern with some quasiperiodic oscillations. The temperature data appear to be an amalgamation of quasiperiodic, stochastic, chaotic, and random components.

3. IMPRINT OF SOLAR SIGNALS IN THE TEMPERATURE VARIABILITY RECORD

Figure 2 (top) shows 80-point moving averages of the temperature variability record of Figure 1, which apparently shows more or less the distinct intervals of lows and highs matching with the solar activities, namely, Maunder minima, Dalton minima, Sporer minima, and Wolf minima, during the past 800 years covering the fluctuations classically described for the last millennia. Figure 2 (bottom) shows the inverted scale plot of the carbon-14 record, which is indicative of solar activity for the identical periods [*Stuiver and Pearson*, 1986]. Apparently, comparison of smoothed temperature variability shows a more or less good match at least with the corresponding decreases and increases in solar activities. This provides some evidence that the phase of temperature response is consistent with solar forcing.

4. SPECTRAL ANALYSES OF EOF (PRINCIPAL COMPONENTS)

Instead of using the original data directly for the treatment of spectral analysis, we first perform principal component (PC) analysis and then utilize the PC components to perform spectral analysis, which provides a clearer picture of the trend and cyclicity in the PC variance spectra. The PCs represent independent temperature variability patterns, and the coefficient gives the pattern of variation in time. Here in the present analysis of western Himalaya temperature (WHT) temperature variability, the time series of the first three PCs (PC1 = 61.462%, PC2 = 25.485%, and PC3 = 13.05%) are constructed (Figures 3a, 3b, and 3c). The first PC explains the majority of the variance and exhibits a distinct variability pattern (Figure 3a) associated with the warming and cooling spells of underlying physical process.

The variance spectrum of the time series of the coefficients of the first PC is displayed in Figure 4a. Spectral peaks occur centered around 81, 40, 32, 22, and 10.6 years. Four peaks at around 80, 40, 30, and 22 years surpass the 95% significance level for a priori choice of period assuming a white noise null continuum. There are many spectral

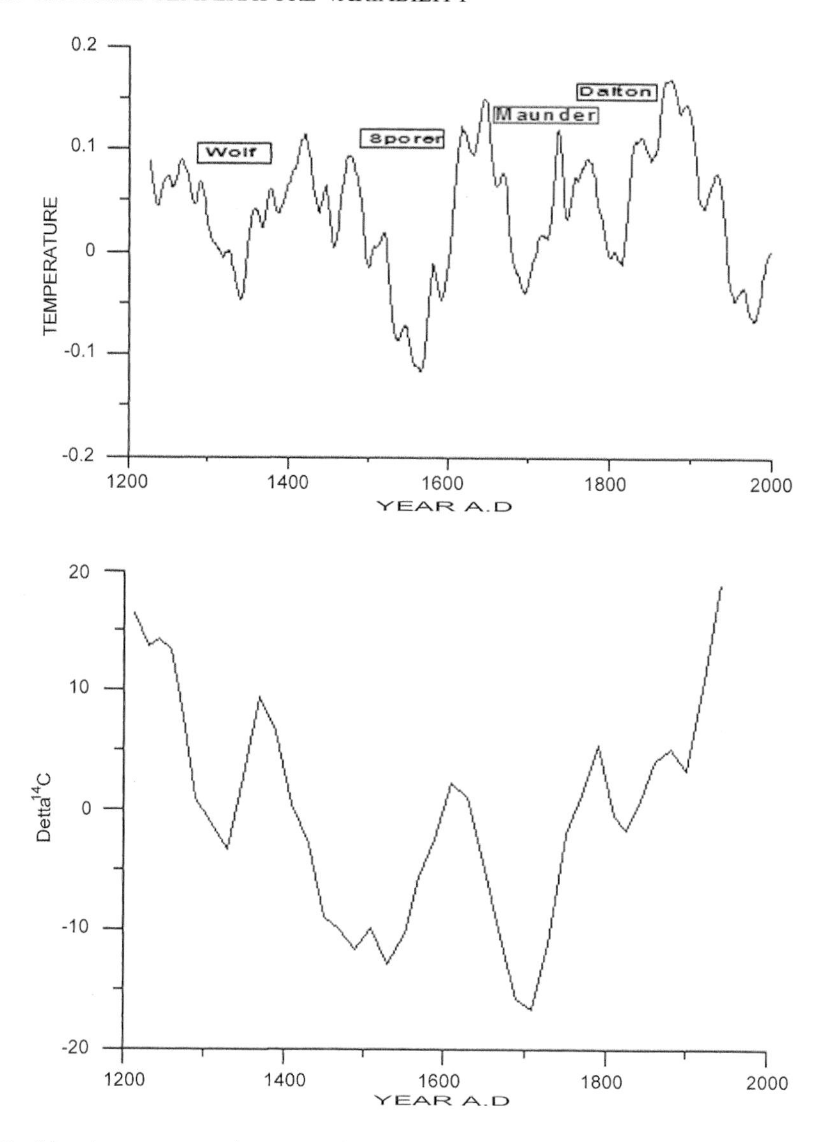

Figure 2. (top) The 80-point average running mean of the original data as shown in Figure 1. (bottom) The δ^{14} plot against the identical time period.

peaks during the period ranging from 8 to 14 years, and none is significant at more than 91% confidence level (Figure 4a). The period of any decadal signal is, therefore, seen to be poorly determined, and it is not clear that a solar cycle signal has been detected. The 11 year cycle is variable in length (varies in a range of 8–17 years), which might indicate "chaotic" phenomena. Similar spectral analysis of the PC2 time series also reveals 9.1 and 4.0 years periodicity, and analysis of the PC3 component reveals 2.0 and 1.7 years cycle (significant at 90%), which could be associated with global El Niño–Southern Oscillation (ENSO) phenomena (Figures 4b and 4c).

The wavelet spectra of PC1 components (Figure 5) show nonstationary modes in the range of 2.4, 5–8, 10–14, 20–25, 30–40, and ~80 years frequency bands above the 95% significance level. Nonstationary features apparently evident in the wavelet spectra generated using the code of *Torrence and Compo* [1998] of the western Himalayas temperature variability record might suggest a phase change/deviation in the strength of the forcing mechanism leading to fundamental reorganization in climate variability. The imprint of solar signals in the tree ring temperature record indicates external forcing response due to solar variability. Further, there is also persuasive evidence for the impact of ENSO, an outcome of coupled ocean-atmospheric processes to regional climate variability, which might signify the establishment of the contemporary climate in the region influenced by the multiple forcing functions.

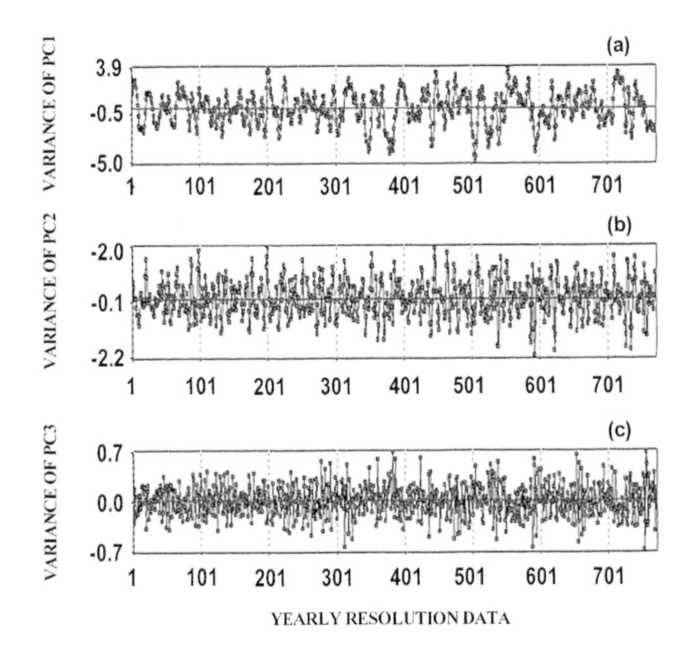

Figure 3. Principal component (a) PC1, (b) PC2, and (c) PC3 of the time series shown in Figure 1.

There is evidence for the presence of an approximately dominant mode at around 30 years. Several researchers have attributed a similar periodicity in simulation study of modern climate, which supports the role of the coupled atmosphere-ocean mode, known as the Atlantic multidecadal oscillation [*Stocker*, 1994]. This time constant is especially interesting since it also coincides with the typical time scale of anthropogenic climate modification. A similar cycle of ~30 years is also reported in the variability of southwest Indian summer monsoon precipitation record during the Bølling-Allerød [*Sinha et al.*, 2005] and tree ring $\delta^{14}C$ record [*Stuiver and Braziunas*, 1993]. Solar-induced changes in decadal oceanic circulation of large-scale climate modes, such as the Pacific decadal oscillation and the North Atlantic oscillation, could also have caused changes in atmospheric temperatures. It has been suggested that periodically driven abnormal heating anomalies affect the atmospheric temperature gradient between the land and oceans [*Ji et al.*, 2005] through the enhanced thermal contrast (warm and cool temperature fluctuations). The coupled ocean-atmosphere system appears to transport energy from the hot equatorial regions toward Himalayan territory in a quasicyclic manner. An ~30 year nonstationary cycle in wavelet spectra of temperature variability record around 1645–1715 may also be suggested as a stretching period of the 22 year solar cycle [*Miyahara et al.*, 2008].

5. DISCUSSION AND CONCLUSION

The possible connection between the solar cycles and temperature variability has been discussed in detail by several researchers in the past years [*Reid*, 1987; *Van Geel et al.*, 1999; *Newell et al.*, 1989; *Wang et al.*, 2005; *Neff et al.*, 2001; *Solanki et al.*, 2004]. However, it has been suggested that solar changes have contributed to small climate oscillations. The 11 year solar cycle is characterized by a variation 0.1% in total solar irradiance, which is too small to have significant effect on climate [*Lean*, 2000]. However, several other researchers believe that irradiance increases with number of sunspots. The effect of darkened spots on the Sun's surface is overcompensated for by the faculae, which are more brilliant zones associated with the spots [*Bard and Frank*, 2006]. Additionally, the solar UV radiation is an order of magnitude larger than the variability of total irradiance

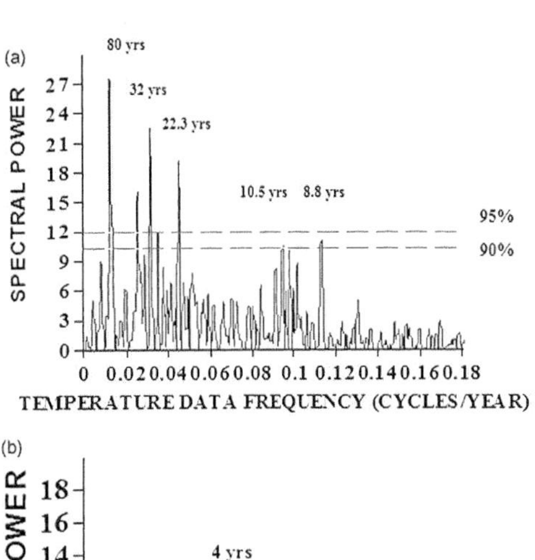

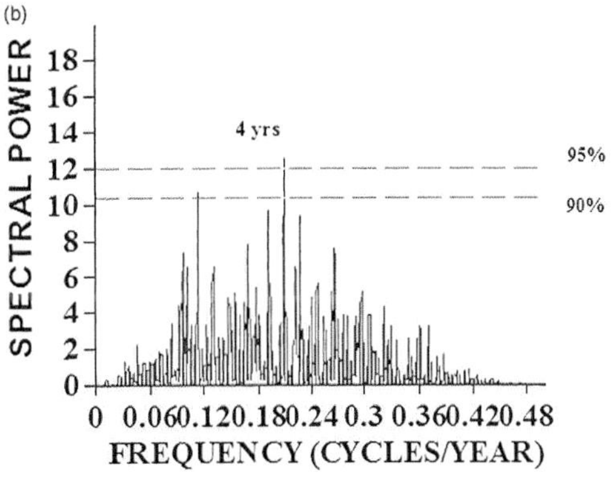

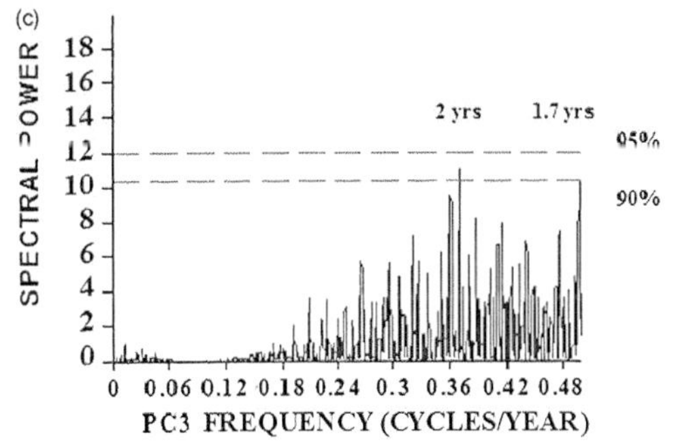

Figure 4. Fourier spectra of (a) PC1, (b) PC2, and (c) PC3. Horizontal lines indicate 95% and 90% confidence intervals of the time series shown in Figure 1.

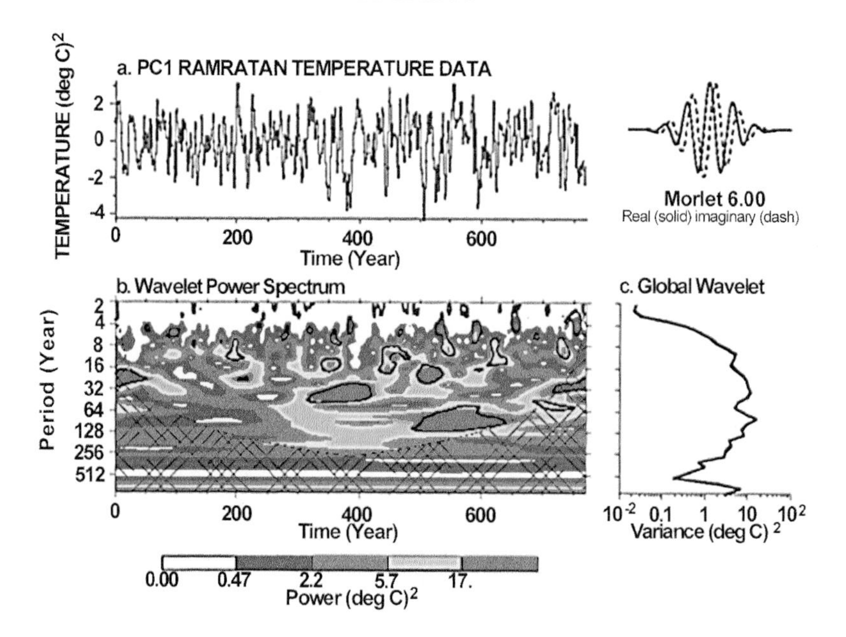

Figure 5. Wavelet spectra of PC1 temperature variability record.

[*Wang et al.*, 2005], which promotes the chemical reaction in the upper atmosphere [*Hood*, 1986; *Hood and Soukharev*, 2003]. The enhanced stratospheric ozone formation through photochemical reactions leads to further heating of the stratosphere through absorption of the excess UV radiation by ozone [*Haigh*, 1994]. This mechanism amplifies the global average warming due to the increase in irradiance by about 15%–20% [*Shindell et al.*, 2001; *Palmer et al.*, 2004]. Some recent studies indicate that solar irradiance has increased over the past 20 years [*Willson and Mordvinov*, 2003].

The present analyses reveal the mixed response of solar signal and coupled ocean-atmospheric processes (ENSO) in temperature variability data of the western Himalayas. Hence, it is prudent to discuss a coupling mechanism, which combines both the forcing of solar variability and ocean atmospheric processes. *Kodera* [2004] has shown the possible solar influence on the Indian Ocean monsoon through dynamical processes. A recent thorough review of solar influences on climate is presented by *Gray et al.* [2010]. Accordingly, variation in ocean temperatures has been found with solar cycles. Several analyses have also provided evidence of much larger responses in regional analyses, which appear to share some similarities with ENSO. Recent model studies suggest that an amplification of initially weak solar forcing of the ocean-troposphere system can occur through positive evaporation and cloud feedback [see *Meehl et al.*, 2009, and references therein]. Accordingly, the initial solar forcing can consist of the small total solar irradiance variation, which penetrates to the surface, plus the indirect radiative and dynamical effects of solar UV forcing of the stratosphere. It is uncertain, however, which is most important. The amplification occurs during a period of increasing solar radiative forcing due to positive feedback from increased evaporation in cloud-sparse subtropical regions. This causes increased moisture transport from those regions into the tropical precipitation convergence zones, increased tropical upwelling and strengthening of the Hadley and Walker circulations, increased subsidence in the subtropics leading to further reductions in cloud cover, and so on [*Meehl et al.*, 2008]. In particular, other paleoclimate proxy reconstructions over the last millennium indicate that during periods of higher solar radiative forcing (e.g., the so-called medieval warm epoch), La Niña–like conditions tended to exist in the tropical Pacific [see *Cobb et al.*, 2003; *Mann et al.*, 2009; *Marchitto et al.*, 2010, and references therein]. Conversely, during periods of low solar forcing (or during periods following major volcanic aerosol injections), E1 Niño–like conditions tended to exist. This has been studied theoretically using the simplified Zebiak-Cane model of ENSO dynamics [see *Mann et al.*, 2005, and references therein]. However, as noted, for example, by *Marchitto et al.* [2010], fully coupled global climate models do not yet adequately simulate the observed La Niña-like response to increased solar forcing. A possible modulation of the connection between the Pacific Ocean and Indian Ocean variability by solar cycle has also been provided by *Kodera et al.* [2007]. Thus, evidence for integrated quasiperiodic solar signals and ENSO in the WHT spectra provides credibility for the multiple forcing mechanisms including the role of ocean-atmospheric processes and the Sun-climate link.

It has been observed that the Sun is currently experiencing one of the most active periods/phases over the past 8000 years as estimated from the long record of carbon-14-dated tree rings. Recent correlation studies of the 11 year solar cycle and land surface temperature observations also appear to be robust but display a similar pattern in geographical distribution to those from forcing due to greenhouse gases [*Gray et al.*, 2010]. Hence, evidence of a relatively weak signal corresponding to the 11 year solar cycle in our temperature spectra may probably testify to the possible role of some other forcing mechanism in the WH temperature variability record also. The rapid global warming of the twentieth century appears to have exceeded the level that cannot be explained by solar irradiance/activity and natural variability (ENSO) alone, but the rising level of greenhouse gases such as CO_2, deforestation, land use, etc. may also have played a considerable role in a substantial proportion of this variability. However, our analyses provide evidence that relatively small solar forcing may play a significant role in century-scale WH temperature fluctuations.

On longer time scales, the relatively cold spell corresponding to the so-called Maunder Minimum (1645–1715 A.D.) and during the putative Medieval warm period (800–1200 A.D.) may have been influenced by long-term solar variations [*Gray et al.*, 2010]. As noted by *Gray et al.* [2010], there has been some controversy about whether the former was actually a global-scale cooling or was more regional. Modeling research suggests that it may have been a manifestation of a shift in the AO/NAO regime. The investigation of such a controversial mechanism remains a tantalizing issue for future researchers. Longer proxies are required both for estimates of the Sun's variations (e.g., sunspots) and for climate (e.g., tree rings) for further confirmation of such "long-term memory" processes in the temperature variability record. Such quasiperiodic modes may not act as a major forcing function alone, but it may indeed serve as a "trigger" mechanism in conjunction with the ocean-atmospheric coupling and other related feedback processes to create multiple trends in the climate system. The present study, however, would have a variety of implications for understanding the climate in the western Himalayas.

Acknowledgment. We are extremely grateful to the anonymous reviewer for his very valuable suggestion on improving the manuscript. We are also grateful to the editors for inviting and encouraging contributing to this volume. Finally, we thank the director of NGRI for his kind permission to publish this research.

REFERENCES

Bard, E., and M. Frank (2006), Climate change and solar variability: What's new under the Sun?, *Earth Planet. Sci. Lett.*, *248*, 480–493.

Bond, G., B. Kromer, J. Beer, R. Muscheler, M. N. Evans, W. Showers, S. Hoffmann, R. Lotti-Bond, I. Hajdas, and G. Bonani (2001), Persistent solar influence on North Atlantic climate during the Holocene, *Science*, *294*, 2130–2136.

Brauning, A. (2001), Climate history of the Tibetan Plateau during the last 1000 years derived from a network of Juniper chronologies, *Dendrochronologia*, *19*, 127–137.

Briffa, K. R., T. J. Osborn, F. H. Schweingruber, I. C. Harris, P. D. Jones, S. G. Shiyatov, and E. A. Vaganov (2001), Low-frequency temperature variations from a northern tree ring density network, *J. Geophys. Res.*, *106*, 2929–2941.

Cobb, K. M., C. D. Charles, H. Cheng, and R. L. Edwards (2003), El Niño/Southern Oscillation and tropical Pacific climate during the last millennium, *Nature*, *424*, 271–276.

Cook, E. R., P. J. Krusic, and P. D. Jones (2003), Dendroclimatic signals in long tree-ring chronologies from the Himalayas of Nepal, *Int. J. Climatol.*, *23*, 707–732.

Eddy, J. A. (1976), The Maunder Minimum, *Science*, *192*, 1189–1202.

Esper, J., H. Schweingruber, and M. Winiger (2002), 1300 years of climatic history for western central Asia inferred from tree rings, *Holocene*, *12*, 267–277.

Gray, L. J., et al. (2010), Solar influences on climate, *Rev. Geophys.*, *48*, RG4001, doi:10.1029/2009RG000282.

Haigh, J. D. (1994), The role of stratospheric ozone in modulating the solar radiative forcing of climate, *Nature*, *370*, 544–546.

Hood, L. L. (1986), Coupled stratospheric ozone and temperature responses to short-term changes in solar ultraviolet flux: An analysis of Nimbus 7 SBUV and SAMS data, *J. Geophys. Res.*, *91*, 5264–5276.

Hood, L. L., and B. E. Soukharev (2003), Quasi-decadal variability of the tropical lower stratosphere: The role of extratropical wave forcing, *J. Atmos. Sci.*, *60*, 2389–2403.

Ji, J. F., J. Shen, W. Balsam, J. Chen, L. Liu, and X. Q. Liu (2005), Asian monsoon oscillations in the northeastern Qinghai-Tibet Plateau since the late glacial as interpreted from visible reflectance of Qinghai Lake sediments, *Earth Planet. Sci. Lett.*, *233*, 61–70.

Kodera, K. (2004), Solar influence on the Indian Ocean Monsoon through dynamical processes, *Geophys. Res. Lett.*, *31*, L24209, doi:10.1029/2004GL020928.

Kodera, K., K. Coughlin, and O. Arakawa (2007), Possible modulation of the connection between the Pacific and Indian Ocean variability by the solar cycle, *Geophys. Res. Lett.*, *34*, L03710, doi:10.1029/2006GL027827.

Lean, J. (2000), Evolution of the Sun's spectral irradiance since the Maunder Minimum, *Geophys. Res. Lett.*, *27*, 2425–2428.

Mann, M. E., J. Park, and R. S. Bradley (1995), Global interdecadal and century-scale oscillations during the past five centuries, *Nature*, *378*, 266–270.

Mann, M. E., M. A. Cane, S. E. Zebiak, and A. Clement (2005), Volcanic and solar forcing of the tropical Pacific over the past 1000 years, *J. Clim.*, *18*, 447–456.

Mann, M. E., Z. Zhang, S. Rutherford, R. S. Bradley, M. K. Hughes, D. Shindell, C. Ammann, G. Faluvegi, and F. Ni

(2009), Global signatures and dynamical origins of the Little Ice Age and Medieval Climate Anomaly, *Science*, *326*, 1256–1259.

Marchitto, T. M., R. Muscheler, J. D. Ortiz, J. D. Carriquiry, and A. van Geen (2010), Dynamical response of the tropical Pacific Ocean to solar forcing during the early Holocene, *Science*, *330*, 1378–1381.

Meehl, G. A., J. M. Arblaster, G. Branstator, and H. van Loon (2008), A coupled air–sea response mechanism to solar forcing in the Pacific region, *J. Clim.*, *21*, 2883–2897, doi:10.1175/2007JCLI1776.1.

Meehl, G. A., J. M. Arblaster, K. Matthes, F. Sassi, and H. van Loon (2009), Amplifying the Pacific climate system response to a small 11-year solar cycle forcing, *Science*, *325*, 1114–1118.

Miyahara, H., Y. Yokoyama, and K. Masuda (2008), Possible link between multi-decadal climate cycles and periodic reversals of solar magnetic field polarity, *Earth Planet. Sci. Lett.*, *272*, 290–295.

Neff, U., S. J. Burns, A. Mangini, M. Mudelsee, D. Fleitmann, and A. Matter (2001), Strong coherence between solar variability and the monsoon in Oman between 9 and 6 kyr ago, *Nature*, *411*, 290–293.

Newell, N. E., R. E. Newell, J. Hsiung, and W. Zhongxiang (1989), Global marine temperature variation and the solar magnetic cycle, *Geophys. Res. Lett.*, *16*(4), 311–314.

Palmer, M. A., L. J. Grey, M. R. Allen, and W. A. Norton (2004), Solar forcing of climate model results, *Adv. Space Res.*, *34*, 343–348.

Reid, G. C. (1987), Influence of solar variability on global sea surface temperatures, *Nature*, *329*, 142–143.

Rind, D. (2002), The Sun's role in climate variations, *Science*, *296*, 673–677.

Shindell, D. T., G. V. Schmidt, M. E. Mann, D. Rind, and A. Waple (2001), Solar forcing of regional climate change during the Maunder Minimum, *Science*, *294*, 2149–2152.

Sinha, A., K. G. Cannariato, L. D. Stott, H.-C. Li, C.-F. You, H. Cheng, R. L. Edwards, and I. B. Singh (2005), Variability of Southwest Indian summer monsoon precipitation during the Bølling-Ållerød, *Geology*, *33*(10), 813–816.

Solanki, S. K., I. G. Usoskin, B. Kromer, M. Schussler, and J. Beer (2004), Unusual activity of the Sun during recent decades compared to the previous 11,000 years, *Nature*, *431*, 1084–1087.

Stocker, T. F. (1994), The variable ocean, *Nature*, *36*, 221–222.

Stuiver, M., and T. F. Braziunas (1993), Sun, ocean, climate and atmospheric $^{14}CO_2$: An evaluation of casual and spectral relationship, *Holocene*, *3*, 289–305.

Stuiver, M., and G. W. Pearson (1986), High-precision calibration of the radiocarbon time scale, AD 1950–500 BC, *Radiocarbon*, *28*, 805–838.

Torrence, C., and G. P. Compo (1998), A practical guide to wavelet analysis, *Bull. Am. Meteorol. Soc.*, *79*, 61–78.

Van Geel, B., O. M. Raspopov, H. Renssen, J. van der Plicht, V. A. Dergachev, and H. A. J. Meijer (1999), The role of solar forcing upon climate change, *Quat. Sci. Rev.*, *18*, 331–338.

Wang, Y., H. Cheng, R. L. Edwards, Y. He, X. Kong, Z. An, J. Wu, M. J. Kelly, C. A. Dykoski, and X. Li (2005), The Holocene Asian monsoon: Links to solar changes and North Atlantic climate, *Science*, *308*, 854–857.

Willson, R. C., and A. V. Mordvinov (2003), Secular total solar irradiance trend during solar cycles 21–23, *Geophys. Res. Lett.*, *30*(5), 1199, doi:10.1029/2002GL016038.

Yadav, R. R., W.-K. Park, J. Singh, and B. Dubey (2004), Do the western Himalayas defy global warming?, *Geophys. Res. Lett.*, *31*, L17201, doi:10.1029/2004GL020201.

Zhang, J., and T. J. Crowley (1989), Historical climate records in China and reconstruction of past climates (1470–1970), *J. Clim.*, *2*, 833–849.

K. P. C. Kaladhar Rao and R. K. Tiwari, National Geophysical Research Institute, CSIR, Uppal Road, Hyderabad-500 007, A.P., India. (rk_tiwari3@rediffmail.com)

R. R. Yadav, Birbal Sahni Institute of Paleobotany, Lucknow-226 007, U.P., India. (rryadav2000@yahoo.co.in)

On the Estimation of Natural and Anthropogenic Trends in Climate Records

S. Lennartz[1] and A. Bunde

Institut für Theoretische Physik III, Justus-Liebig-Universität Giessen, Giessen, Germany

This chapter focuses on data that exhibit long-term memory and, in addition, may be influenced by external deterministic trends. Examples for data with long-term memory are temperature and river runoff records, and deterministic trends may be caused by the increased emissions of greenhouse gases or increased urbanization, leading to global or urban warming. We show how long-term correlations can be quantified in the presence of external trends and to which extent external (approximately linear) trends can be detected in the presence of long-term memory.

1. INTRODUCTION

It is well known that the global surface air temperature as well as the global sea surface temperature has been rising in the twentieth century, with a more pronounced increase in the last 50 years. The central question is how much of this increase can be attributed to natural fluctuations, and how much is of anthropogenic origin, caused, for example, by the increasing greenhouse gas (GHG) emission or urban warming. A further question is to which extent global warming influences other climate quantities like local precipitation and river runoff data.

The detection and attribution problem [*Bloomfield and Nychka*, 1992; *Hasselmann*, 1993; *Hegerl et al.*, 1996; *Zwiers*, 1999; *Barnett et al.*, 2005; *Rybski et al.*, 2006; *Giese et al.*, 2007; *Zorita et al.*, 2008; *Lennartz and Bunde*, 2009a] plays an important role in the present climate debate. It has been recognized in the past decade that, due to processes occurring on the land surface, ocean, and cryosphere, the natural temperature fluctuations as well as river runoffs are long-term correlated [*Hurst*, 1951; *Mandelbrot and Wallis*, 1969; *Koscielny-Bunde et al.*, 1996, 1998, 2006; *Pelletier and Turcotte*, 1997, 1999; *Weber and Talkner*, 2001; *Eichner et al.*, 2003; *Fraedrich and Blender*, 2003; *Kantelhardt et al.*, 2003b, 2006; *Koutsoyannis*, 2003, 2006; *Monetti et al.*, 2003; *Rybski et al.*, 2008]. In long-term correlated data, the detection problem is quite complex, since it is difficult to distinguish a (small) external trend from natural fluctuations, as we will see in the following.

This review deals with the quantification of long-term memory in the presence of external trends and the estimation of external trends in the presence of long-term memory. In section 2, we review the occurrence of long-term memory in nature, how it is defined, and how it can be generated numerically. We show how it can be detected in the presence of additional polynomial trends. Sections 3–5 are devoted to the trend detection. In section 3, we define the quantities we need for a trend analysis like the relative trend and its exceedance probability. In section 4, which to a great extent follows *Lennartz and Bunde* [2011], we use numerical simulations and scaling theory to determine the exceedance probability and show how it can be used to estimate the anthropogenic contribution to an observed trend. The prerequisite is that the data are Gaussian distributed. In section 5, finally, we apply our results to various temperature records (which are Gaussian distributed) and, as a first approximation, also to various precipitation and river runoff data (which are not Gaussian distributed) and discuss the significance of the observed trends.

2. LONG-TERM CORRELATIONS

Apart from climate records like temperatures and river runoffs, long-term correlations are also known to occur in

[1]Now at School of Engineering and School of Geosciences, University of Edinburgh, Edinburgh, UK.

Extreme Events and Natural Hazards: The Complexity Perspective
Geophysical Monograph Series 196

10.1029/2011GM001079

physiological records like heart beat data [*Peng et al.*, 1993; *Schäfer et al.*, 1998; *Bunde et al.*, 2000; *Ashkenazy et al.*, 2001; *Kantelhardt et al.*, 2002a, 2003a] and in the volatilities of financial records [*Ding et al.*, 1983; *Liu et al.*, 1997, 1999; *Yamasaki et al.*, 2005]. In general, long-term correlated records $\{x_i\}$, $i = 1, \ldots, N$, can be characterized by the power spectral density $S(f) = |x(f)|^2$, where $\{x(f)\}, f = 0, \ldots, N/2$, is the Fourier transform of $\{x_i\}$. With increasing frequency f, $S(f)$ decays by a power law

$$S(f) \sim f^{-\beta}, \quad (1)$$

where $\beta > 0$ characterizes the long-term memory [*Malamud and Turcotte*, 1999]. For uncorrelated data $\beta = 0$.

Equation (1) can be used to generate synthetic Gaussian-distributed long-term correlated records. First, one generates uncorrelated Gaussian data. Then, one transforms them to Fourier space, multiplies the result by $f^{-\beta/2}$ (see equation (1) or, for example, *Turcotte* [1997]), and transforms them back to time space. By definition, the resulting $S(f)$ follows equation (1).

Records with $0 < \beta < 1$ are stationary and can also be characterized by an autocorrelation function $C(s) = \langle(x_i - \bar{x})(x_{i+s} - \bar{x})\rangle/\sigma^2$ that decays by a power law,

$$C(s) \sim s^{-\gamma}, \quad (2)$$

with s being the time lag. $\bar{x}$ denotes the mean value of the record, σ^2 its variance, and $\langle\ldots\rangle$ the average overall pairs (x_i, x_{i+s}). It can easily be shown that $\gamma = 1 - \beta$. Typical values of γ and β for continental temperature records are $\gamma = 0.7$ and $\beta = 0.3$, [*Koscielny-Bunde et al.*, 1998; *Pelletier and Turcotte*, 1999; *Weber and Talkner*, 2001; *Fraedrich and Blender*, 2003; *Eichner et al.*, 2003], and for local sea surface temperatures, $\gamma = 0.4$ and $\beta = 0.6$ [*Monetti et al.*, 2003].

According to equation (2), long-term correlated data have a strong persistence, such that large (small) events (i.e., temperatures) are more likely to be followed by large (small) events. As a consequence, small events tend to cluster in "valleys" and big events in "mountains," which can be clearly seen in Figure 1. If by chance a (sub-) data set starts in a valley and ends up in a mountain, the observed increase cannot be distinguished a priori from an external trend.

If the records have additional trends, then γ and β cannot be correctly determined by these methods. Due to the trend, the measured persistency is exaggerated, with β too large and γ too small. In addition both power spectrum and the autocorrelation function exhibit strong finite-size effects at small frequencies f (respectively large time lags s), which leads to an underestimation of β and an overestimation of γ [*Lennartz and Bunde*, 2009b]. The problem of determining the correlation exponent correctly, even when the data set has an overlying polynomial trend, has been solved by using the

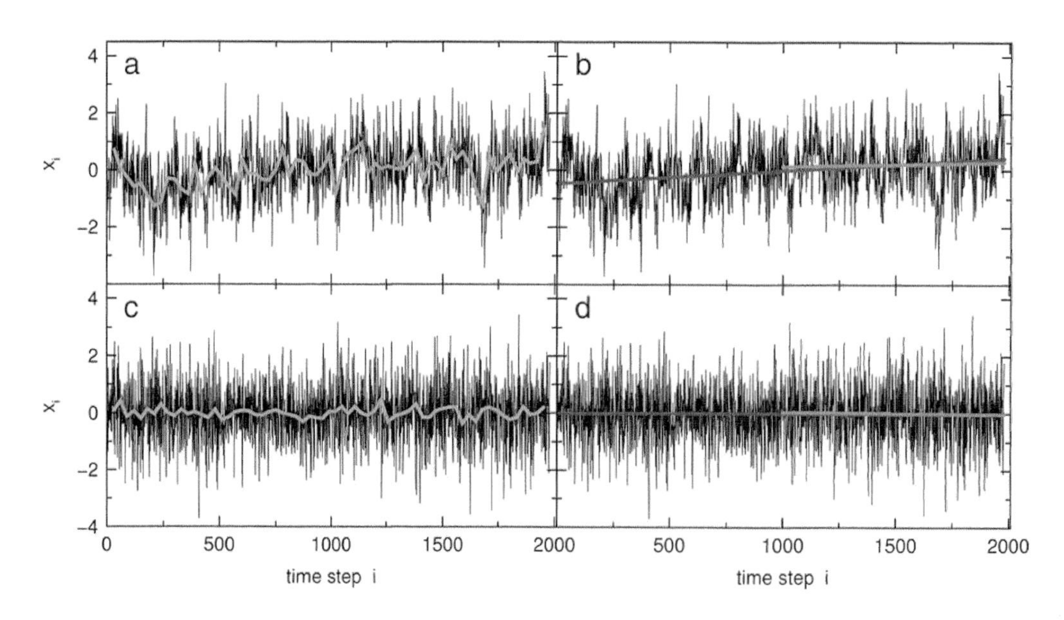

Figure 1. Mountain valley structure of long-term correlated records. (a and b) An artificial record with fluctuation exponent $\alpha = 0.85$. (c and d) An uncorrelated record with $\alpha = 0.5$. In order to emphasize the long-term behavior, the thick lines in Figures 1a and 1c represent downsampled values in windows of size 30. The straight lines in Figures 1b and 1d are linear regression fits to the records using different periods (dark gray, total; light gray, starting from 1000). One can see that, due to the mountain valley structure of long-term correlated records, steeper slopes are more likely to occur. The record in Figures 1a and 1b was created using Fourier filtering (see section 3). After *Rybski and Bunde* [2009].

detrended fluctuation analysis (DFA) and the wavelet analysis [*Peng et al.*, 1994; *Koscielny-Bunde et al.*, 1998; *Kantelhardt et al.*, 2001], where one determines the fluctuation exponent α. Since we will use DFA below, we describe it in the following. In DFA, one considers the cumulated sum (profile)

$$Y_i = \sum_{j=1}^{i} x_j, \qquad (3)$$

and divides it into k nonoverlapping windows of length s. In DFA of order n (DFAn) [*Kantelhardt et al.*, 2001], one calculates in each window ν the best polynomial fit of order n, p_ν, and subtracts it from the profile (see Figure 2 for $n = 2$)

$$Y_s(i) = Y_i - p_\nu(i). \qquad (4)$$

Next, one determines, in each time window ν of size s, the variance $F_\nu^2(s)$ of $Y_s(i)$,

$$F_\nu^2(s) = \frac{1}{s} \sum_{i(\nu)=1}^{s} Y_s^2(i), \qquad (5)$$

averages $F_\nu^2(s)$ over all k time windows, and finally obtain the fluctuation function

$$F(s) \equiv F_2(2) = \left(\frac{1}{k} \sum_{\nu=1}^{k} F_\nu^2(s) \right)^{1/2}. \qquad (6)$$

For long-term correlated records, $F(s)$ scales as

$$F(s) \sim s^\alpha. \qquad (7)$$

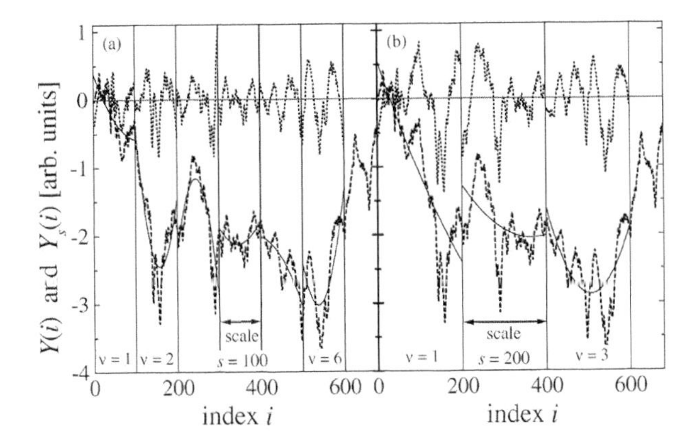

Figure 2. Illustration of the detrending procedure in the detrended fluctuation analysis. For two segment lengths (time scales) s = (a) 100 and (b) 200, the profiles Y (i) (dashed lines, defined in equation (3)), least squares quadratic fits to the profiles (solid lines), and the detrended profiles $Y_s(i)$ (dotted lines) are shown versus the index i. After *Kantelhardt et al.* [2001].

One can show that, in general, α is related to β by

$$\alpha = (1 + \beta)/2. \qquad (8)$$

The relation between α and γ ($0 < \gamma < 1$) is

$$\alpha = 1 - \gamma/2. \qquad (9)$$

For uncorrelated data, $\alpha = 1/2$, and for long-term correlated data, $\alpha > 1/2$. We like to note that for monofractal data (see below), α is identical to the Hurst exponent [*Hurst et al.*, 1965; *Feder*, 1988; *Malamud and Turcotte*, 1999; *Kantelhardt et al.*, 2006]. By definition, the exponent α is an averaged quantity. Thus, a local α value of a subrecord may differ from the global α value of the complete record. We like also to note that when in DFAn a polynomial trend of order n is subtracted from the profile, the original data set is detrended by polynomial trends of order $(n - 1)$. Thus, DFA2 eliminates the influence of linear trends on the fluctuation exponent. In the following, we will use DFA2.

The DFAn can be generalized to the multifractal-DFA [*Kantelhardt et al.*, 2002b] by considering the qth moment $F_\nu^q(s)$ instead of the second moment in equation (5), i.e.,

$$F_\nu^q(s) = \frac{1}{s} \sum_{i(\nu)=1}^{s} Y_s^q(i). \qquad (10)$$

Averaging $F_\nu^q(s)$ over all k time windows yields the generalized fluctuation function

$$F_q(s) = \left(\frac{1}{k} \sum_{\nu=1}^{k} F_\nu^q(s) \right)^{1/q}. \qquad (11)$$

In general, $F_q(s)$ scales as

$$F_q(s) \sim s^{h(q)} \qquad (12)$$

with the generalized fluctuation exponent $h(q)$. For monofractal records, $h(q)$ is independent of q and equal to the fluctuation exponent α.

Continental temperatures with a resolution above 1 week are long-term correlated and in a good approximation monofractal. Precipitation data with a resolution above 1 week are only weakly long-term correlated or uncorrelated. They show only a weak multifractality, i.e., $h(q)$ is changing only slightly with q [*Kantelhardt et al.*, 2006].

Before any correlation analysis, one has to eliminate seasonal trends in the data sets. Since monthly data are smoothened in comparison to daily data, this elimination is more reliable for monthly data than for daily data. Later, we will focus on monthly data. Also, daily temperature data are characterized by additional short-term correlations due to general weather situations (Grosswetterlagen) with correlation times between 2 days and 2 weeks [*Eichner et al.*, 2003]. Since in monthly records the short-term correlations have been averaged out and only the long-term correlations are remaining, they can be easier modeled by long-term correlated data. Another advantage is that monthly or annual data are easily available on the Internet.

To eliminate the seasonal trend in a monthly record $\{T_i\}$, and thus to standardize the data, we follow the works of *Koscielny-Bunde et al.* [2006] and *Lennartz and Bunde* [2009a]: We first calculate separately for each calendar month the mean $\langle T_i \rangle$ and subtract it from the data. Then, we determine the seasonal standard deviation of each calendar month $\langle (T_i - \langle T_i \rangle)^2 \rangle^{1/2}$ in the same way and divide the data by this value, i.e.,

$$x_i = \frac{T_i - \langle T_i \rangle}{\langle (T_i - \langle T_i \rangle)^2 \rangle^{1/2}}. \tag{13}$$

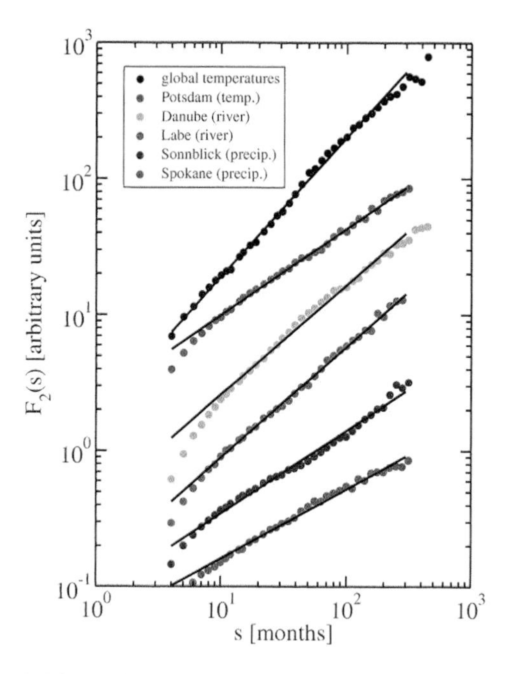

Figure 4. Fluctuation function of detrended fluctuation analysis (DFA2) for the seasonal detrended data sets shown in Figure 3 (dotted symbols). The straight lines are fitted power laws. The corresponding fluctuation exponents α are (from top to bottom) 1.02, 0.63, 0.80, 0.81, 0.61, and 0.51.

Figure 3 shows the seasonally detrended records x_i of typical temperature, river runoff, and precipitation data. Figure 4 shows the corresponding DFA2 functions, with fluctuation exponents $\alpha = 1.02$ for the global temperatures, $\alpha = 0.63$ for the temperatures in Potsdam, $\alpha = 0.80$ for the Danube River, $\alpha = 0.81$ for the Labe River, $\alpha = 0.61$ for the precipitation in Sonnblick (mountain station), and $\alpha = 0.51$ for the precipitation in Spokane; α values for further records are listed in Table 1.

In order to decide if in a record with length L and fluctuation exponent α an observed trend is of anthropogenic or natural origin, one first has to study the statistics of natural trends (i.e., its occurrence probability and its dependence on the window length L and on the fluctuation exponent α). In one of our earlier works [*Lennartz and Bunde*, 2009a], we focused on synthetic monofractal long-term correlated monthly records with $0.5 \leq \alpha \leq 1.2$ of lengths 50 and 100 years, with a Gaussian distribution of data. Here we follow *Lennartz and Bunde* [2011] and consider (in a more general way) long-term correlated (sub-) records of lengths L between 500 and 2000 with α between 0.5 and 1.5 and use scaling theory to obtain analytical expressions for all quantities of interest. The choice of α covers the vast majority of observational temperature records, which all have α exponents between 0.5 and 1.5, following approximately a Gaussian distribution and are monofractal. For annual reconstructed data, L covers the typical range of available records. For monthly observable data sets, L covers the range between roughly 40 and 160 years, which is the most frequent range for observable data. The minimum length scale $L = 500$ is motivated by the fact that, for smaller values, the error in the determination of α by DFA2 becomes too large [*Kantelhardt et al.*, 2001].

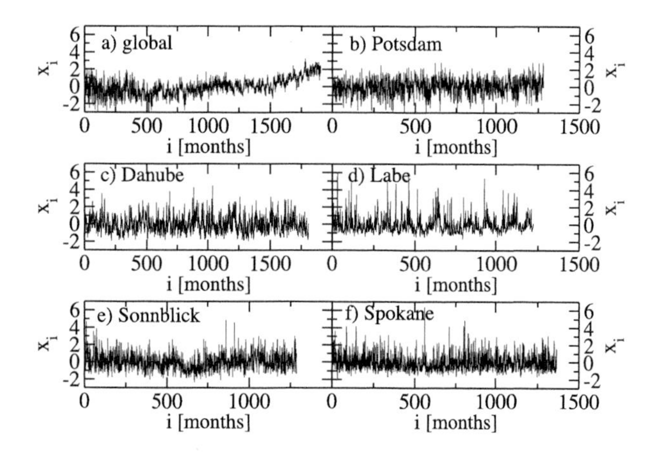

Figure 3. Seasonal detrended records of monthly averaged (a) global temperatures, (b) local temperatures in Potsdam, river runoffs of the (c) Danube and (d) Labe Rivers, and precipitation in (e) Sonnblick and (f) Spokane.

3. QUANTITIES OF INTEREST

We are interested in the probability that in a Gaussian-distributed record of length L, which is long-term correlated

Table 1. Trend Estimation in Global and Local Temperature Records[a]

	Long Records						Short Records					
Data Set	L (months)	α	Δ^{real} (°C)	x	x_Q 95%	x_Q 99%	L (months)	α	Δ^{real} (°C)	x	x_Q 95%	x_Q 99%
Global	1908	1.02	0.69	3.59	**2.40**	3.60	500	1.24	0.65	5.08	5.90	8.30
Global land air	1908	0.81	0.80	2.49	**1.00**	**1.30**	500	1.04	0.93	4.40	**3.00**	**4.30**
Global sea surface	1908	1.21	0.62	3.56	5.00	7.00	500	1.42	0.53	4.87	9.10	13.0
Arwagh (GB)	1584	0.62	0.28	0.22	0.40	0.52	500	0.64	0.13	0.16	0.70	0.99
Brno (CZ)	1296	0.60	0.73	0.45	**0.34**	0.50	500	0.67	0.49	0.39	0.80	1.25
Charleston (US)	1488	0.61	0.25	0.15	0.35	0.50	500	0.63	−0.06	−0.01	0.65	0.92
Irkutsk (RU)	1416	0.52	2.09	0.85	**0.20**	**0.28**	500	0.46	1.65	0.89	**0.20**	**0.30**
Kiev (UA)	1308	0.64	0.74	0.27	0.43	0.61	500	0.66	0.00	−0.11	0.75	1.20
Melbourne (AU)	1632	0.64	0.40	0.38	0.42	0.60	500	0.66	0.15	0.21	0.75	1.20
New York (US)	1392	0.60	2.50	1.41	0.32	0.50	500	0.57	0.32	0.70	**0.46**	**0.68**
Oxford (GB)	1740	0.64	0.85	0.61	**0.41**	**0.58**	500	0.66	0.60	0.49	0.75	1.20
Plymouth (GB)	1440	0.64	0.24	0.13	0.42	0.60	500	0.71	0.44	0.41	0.99	1.50
Potsdam (DE)	1284	0.63	0.83	0.48	**0.46**	0.60	500	0.66	0.61	0.37	0.75	1.20
Prague (CZ)	2616	0.65	0.35	0.08	0.40	0.53	500	0.64	0.95	0.60	0.70	0.99
Sonnblick (AT)	1332	0.52	1.60	0.88	**0.20**	**0.27**	500	0.54	1.02	0.66	**0.38**	**0.58**
Swerdlowsk (RU)	1260	0.61	1.68	0.46	**0.40**	0.54	500	0.66	1.37	0.46	0.75	1.20
Sydney (AU)	1404	0.61	1.21	0.59	**0.36**	0.51	500	0.56	0.59	0.61	**0.42**	0.62
Tomsk (RU)	1427	0.57	1.34	0.40	**0.27**	**0.40**	500	0.56	1.33	0.53	**0.42**	0.62
Vancouver (US)	1116	0.66	−0.70	−0.37	0.52	0.70	500	0.64	−0.64	−0.37	0.70	0.99
Wien Hohewarte (AT)	1500	0.55	1.69	0.96	**0.22**	**0.30**	500	0.64	0.91	0.66	0.70	0.99

[a]Here x_Q is highlighted in boldface if the observed relative trend x is outside the corresponding confidence interval $[-x_Q, x_Q]$ and is thus significant.

and characterized by a certain fluctuation exponent α, an increase of size Δ occurs. The determination of Δ is not unique. One can either determine the mean value of m successive data points at the beginning and at the end of the record and then obtain Δ from the difference between them (this treatment has been followed, for example, by *Rybski et al.* [2006]), or one can use the linear regression method (i.e., fitting a straight line with least mean square displacement to the data) where Δ is obtained as the slope of the regression line times L. Here we follow the second path. From the regression analysis (see Figure 5), we also obtain the standard deviation σ_t around the regression line, which is a measure for the fluctuations in the considered time interval. It is important to note that, in contrast to the normal standard deviation σ around the mean value, σ_t is not affected by an external trend and thus represents the natural fluctuations. In the following, we are interested in the dimensionless relative trend $x \equiv \Delta/\sigma_t$. For determining the fluctuation exponent α of the record, we use DFA2, since it allows to eliminate the influence of linear trends in the data. Accordingly, a given record is characterized by (1) its length L, (2) its relative trend $x = \Delta/\sigma_t$, and (3) its fluctuation exponent α obtained by DFA2.

To estimate the contribution of an external linear trend to an observed relative trend, we need to know the probability $P(x, \alpha; L)dx$ that in a given purely long-term correlated record of length L characterized by α, a relative trend between x and $x + dx$ occurs. By definition, the probability density function (PDF) $P(x, \alpha; L)$ is symmetric in x [i.e.,

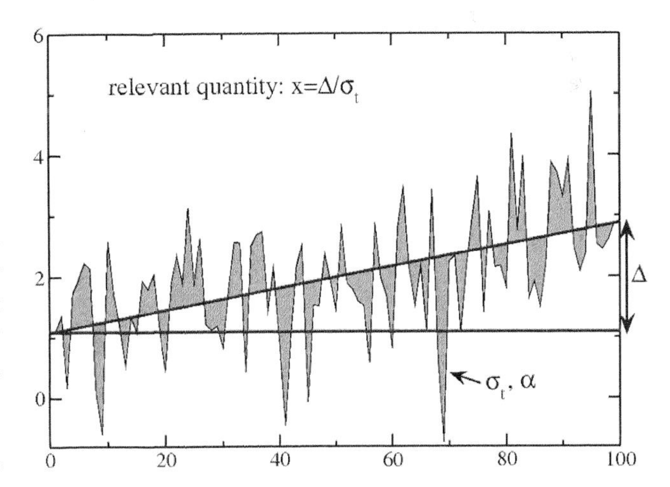

Figure 5. Visualization of the considered quantities Δ, σ_t, and α. Δ is the total increase in the time window and is measured by a linear regression line. σ_t is the standard deviation around this regression line. The fluctuation exponent α characterizes the memory of the data in this time window. The relevant quantity in which we are interested is the dimensionless fraction Δ/σ_t when a certain value for α is measured.

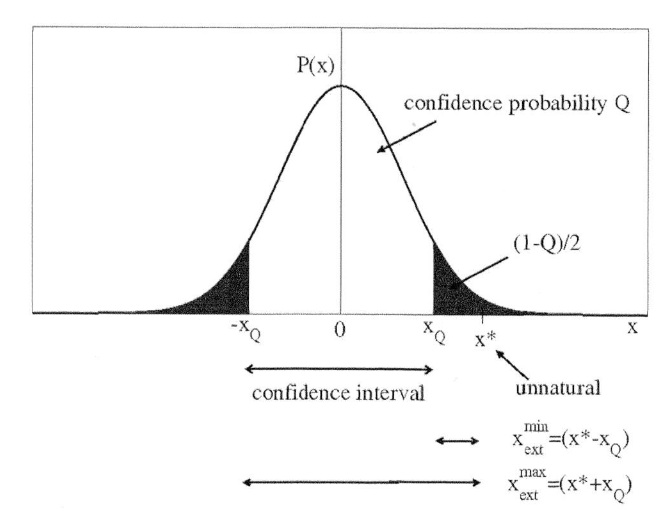

Figure 6. Sketch of a probability density function (PDF) $P(x)$. The integral over the white area defines the confidence probability Q. The lower and upper bounds $-x_Q$ and x_Q define the confidence interval $[-x_Q, x_Q]$. If an event x^* is outside this interval, it is considered unnatural. In this case, the minimum external relative trend is $x^* - x_Q$.

$P(x, \alpha; L) = P(-x, \alpha; L)]$. From $P(x, \alpha; L)$, we then obtain the exceedance probability

$$W(x, \alpha; L) = \int_x^{\infty} P(x', \alpha; L)\mathrm{d}x' \tag{14}$$

that an observed relative trend is larger than x.

An important quantity in the further analysis is the confidence interval $[-x_Q, x_Q]$ (see Figure 6). By definition, relative trends within this interval occur with confidence probability Q; that is,

$$Q = \int_{-x_Q}^{x_Q} P(x', \alpha; L)\mathrm{d}x' = 1 - 2W(x_Q, \alpha; L). \tag{15}$$

Accordingly, x_Q can be obtained from the inverse function of W; that is,

$$x_Q = W^{-1}((1-Q)/2, \alpha; L). \tag{16}$$

For $Q = 0.95$, for example, $[-x_Q, x_Q]$ defines the 95% confidence interval, which we will use later as a criterion to distinguish between natural and unnatural trends (see Figure 6). In the presence of an external linear trend, the total trend is the sum of the natural trend and the external trend. If an observed relative trend x is larger than x_Q, it is considered unnatural. In this case, the Q-dependent minimum and maximum external trends are given by

$$\Delta_{\text{ext}}^{\min} = \sigma_t(x - x_Q) \tag{17}$$

and

$$\Delta_{\text{ext}}^{\max} = \sigma_t(x + x_Q). \tag{18}$$

By definition, the uncertainty in the external trend (error bar) $(\Delta_{\text{ext}}^{\max} - \Delta_{\text{ext}}^{\min})/2$ is equal to the confidence interval times σ_t and

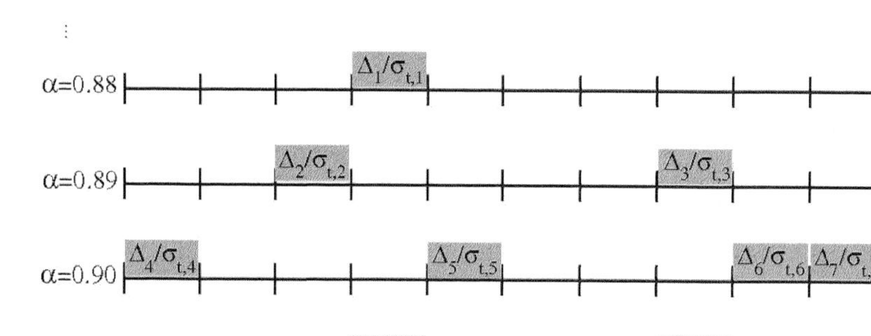

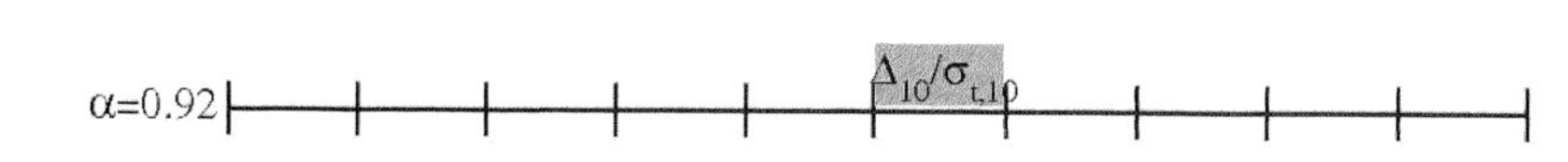

Figure 7. Visualization of the data generation procedure. For each global α value, we generated several long data sets (long lines), which we divided into smaller subsequences of size L. In each of these subsequences, we measured the local α value by DFA2. Next we consider only those subsequences where the measured α value is in a certain range; for example, $\alpha = 0.90 \pm 0.01$ (marked subsequences). In all these sequences, we measure the relative trend $x_i = \Delta_i/\sigma_{t,i}$, as described in Figure 5, to obtain the histogram $h(x)$, which, after normalization, yields the desired PDF $P(x,\alpha;L)$.

thus depends only on the confidence probability Q and not on the observed trend Δ.

4. NUMERICAL ESTIMATION OF THE EXCEEDANCE PROBABILITY

In the following, we describe the numerical estimation of the exceedance probability [*Lennartz and Bunde*, 2011]. First, for 160 α values ranging from 0.41 to 2.00, 100 synthetic records of length 2^{21} were generated, respectively, as described in section 2, with $\beta = 2\alpha - 1$. With this technique, Gaussian-distributed data are generated. Here we focus on Gaussian-distributed data but would like to note that, in a considerably more elaborate iterative procedure suggested by *Schreiber and Schmitz* [1996], (monofractal) long-term correlated records with any other distribution can also be generated. Since we are interested in the analysis of trends in smaller records, which typically vary between 500 and 2000

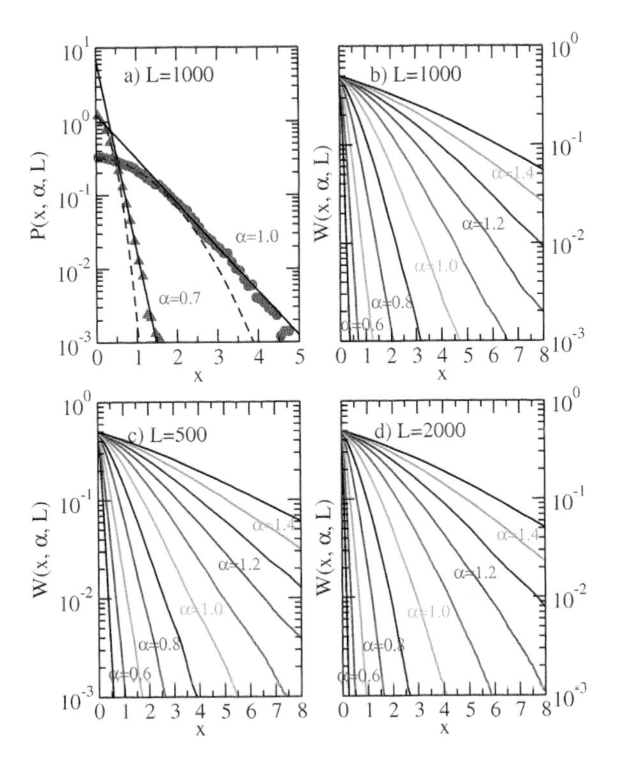

Figure 8. (a) PDF $P(x, \alpha; L)$ of the relative temperature increase $x = \Delta/\sigma_t$ of synthetic long-term correlated records in a time window $L = 1000$. The circles represent the PDF for $\alpha = 1.0$, and the triangles are for $\alpha = 0.7$. The dashed lines are Gaussians, which fit the data best for small arguments, while the straight lines are exponentials, which fit the data best for large arguments. Cumulative probability $W(x, \alpha; L)$ for (b) $L = 1000$, (c) $L = 500$, and (d) $L = 2000$ for fluctuation exponents α between 0.5 and 1.5 (from left to right). After *Lennartz and Bunde* [2011].

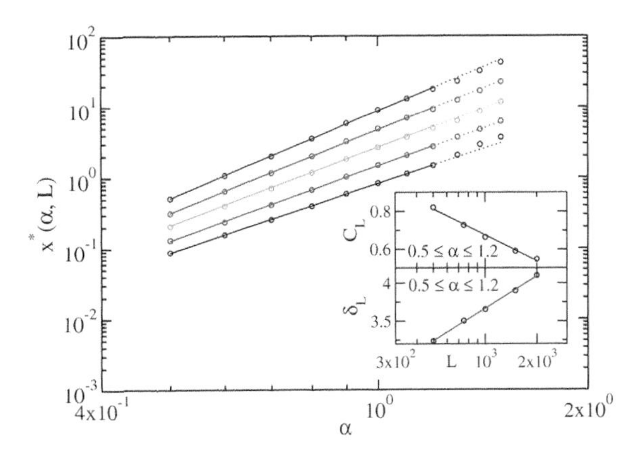

Figure 9. (bottom to top) Inverse slopes $x^*(\alpha, L)$ of the exponential tails of $W(x, \alpha; L)$ as a function of α for $L = 500$, 750, 1000, 1500, and 2000 (circles, shifted upward for clarity by a factor of 1, 2, 4, 8, and 16, respectively). The values can be approximated by power laws (straight lines), where the L dependence of the prefactor C_L and of the exponent δ_L are shown in the insets (circles). Both parameters can be approximated by logarithmic functions (straight lines).

data points, we extracted from the long records, for each α value, 25,000 subsequences of lengths $L = 500$, 750, 1000, 1500, and 2000. In each subrecord of length L, we determined (1) the local α values by DFA2 as the slope of the logarithmic fluctuation function between $s = 10$ and $s = 100$ and (2) the relative trends $x = \Delta/\sigma_t$. It is essential to determine α in each subrecord, since the local α values of the subrecords usually differ from the global one [see, e.g., *Rybski and Bunde*, 2009].

To estimate $P(x, \alpha; L)$ for fixed L, we proceed as described in Figure 7. We divide the local α values into windows of size 0.02 and focus on the most interesting regime between 0.5 and 1.5. Most natural records, where long-term memory is important, are characterized by α values in this regime. In each α window of length L, we determined the histogram, which after normalization yielded the desired PDF, $P(x, \alpha; L)$.

Figure 8a shows, for $L = 1000$, the resulting $P(x, \alpha; L)$ in a semilogarithmical plot for $\alpha = 0.7 \pm 0.01$ (triangles) and $\alpha = 1.0 \pm 0.01$ (circles). Since $P(x, \alpha; L)$ is symmetric in x, only positive values of x are shown. One can see clearly that for small relative trends x, the curves are Gaussian (dashed lines), while for large x, the curves follow a simple exponential (solid lines). The figure also shows that the PDFs broaden with increasing Hurst exponent α, when the mountain-valley structure of the long-term correlated record becomes more pronounced (i.e., large natural trends become more likely with increasing α). From $P(x, \alpha; L)$, we obtained, by direct summation, the desired exceedance probability $W(x, \alpha; L)$ (see equation (14)).

Figures 8b–8d show $W(x, \alpha; L)$ for $L = 500$, 1000, and 2000. By definition, $W(0, \alpha; L) = 1/2$ (i.e., all curves intersect

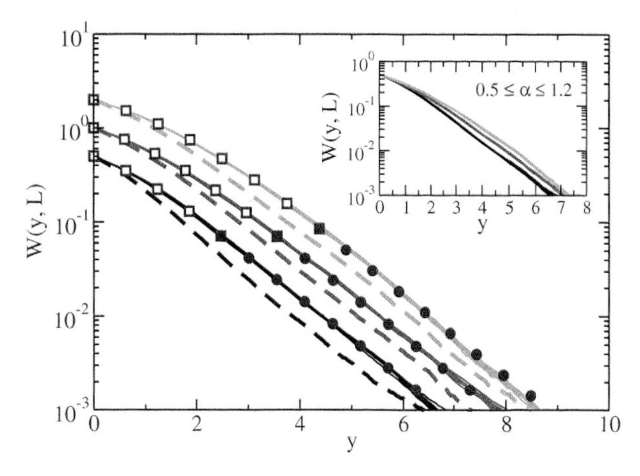

Figure 10. Cumulative probability $W(y, L)$ of the rescaled parameter $y = x/x^*(\alpha, L)$ for $L = 500$ (black), $L = 1000$ (dark gray), and $L = 2000$ (light gray) for α between 0.5 and 1.2 (solid lines). The curves for $\alpha = 1.5$ are shown as dashed lines. The curves for $L = 1000$ and 2000 have been shifted upward by a factor 2 and 4, respectively, for clarity. It can be seen clearly that the curves scale only for $\alpha \leq 1.2$. In the inset is shown that $W(y, L)$ still has an L dependence. $W(y, L)$ from equation (26), where w_L has been obtained numerically, is shown as open squares and solid circles. The agreement between the solid lines and the symbols is very good.

at $x = 0$). As also seen in the PDF, large relative trends become more likely with an increasing fluctuation exponent α. The figure also shows that the exceedance probability exhibits a remarkable L dependence such that, for fixed α, large relative trends become more likely for small (sub-) records than for large ones. This is also reasonable: Due to the mountain-valley structure of long-term correlated data, a large trend can be seen if the subrecord starts in a valley and ends up in a mountain. In a short subrecord, it is very likely that the mountain is just the next one, and the standard deviation around the regression line will be small, resulting in a very large relative trend. Conversely, in a long subrecord, it is very likely that the mountain is not the first one; that is, between the start and the end, there are several mountains and valleys in between. Thus, the standard deviation around the regression line will be large. This results in a smaller relative trend in comparison with a short subrecord.

From Figures 8b–8d, one can obtain graphically the boundary x_Q of the confidence interval. For example, for $Q = 0.99$, x_Q is obtained from the intersection of $W(x, \alpha; L)$ and $(1 - Q)/2$ (see equation (16)). For $L = 2000$ and $\alpha = 1.0$, the intersection is at $x_Q \approx 3.3$.

According to the behavior of the PDF, the exceedance probability for small arguments x can be expressed by the error function, while for large arguments x, $W(x, \alpha; L)$ decays exponentially. As a consequence, the inverse of the asymptotic slopes in the semilogarithmic plots of $W(x, \alpha; L)$ can be regarded as the characteristic scales $x^*(\alpha, L)$ of the relative trends. Figure 9 shows that

$$x^*(\alpha, L) = C_L \alpha^{\delta_L}, \tag{19}$$

where both C_L and δ_L depend logarithmically on L:

$$C_L \approx C^{(0)} + C^{(1)} ln(L) \tag{20}$$

$$\delta_L \approx \delta^{(0)} + \delta^{(1)} ln(L) \tag{21}$$

with $C^{(0)} \approx 2.04$, $C^{(1)} \approx -0.20$, $\delta^{(0)} \approx -0.57$, and $\delta^{(1)} \approx 0.61$ as shown in the insets of Figure 9.

To see if scaling exists, we plotted the exceedance probabilities (for fixed L) versus $y = x/x^*(\alpha, L)$. Figure 10 shows that, for $0.5 \leq \alpha \leq 1.2$ and $L = 500$, 1000, and 2000, the data collapse to a single scaling function $W(y, L)$ that depends explicitly on L (as seen in the inset). For larger α values, there is no data collapse, as is also shown in Figure 10, where the dashed lines refer to $\alpha = 1.5$. In the following analytical discussion, we focus on the scaling regime $0.5 \leq \alpha \leq 1.2$.

To find the y and L dependence of $W(y, L)$, it is convenient to go back to the corresponding rescaled PDF $P(y, L)$. Figures 8a and 10 suggest that $P(y, L)$ has the form

$$P(y, L) = \begin{cases} \dfrac{n_L}{\sqrt{2\pi} w_L} \exp(-y^2/2w_L^2) & y \leq y_c \\ m_L \exp(-y) & y \geq y_c \end{cases}. \tag{22}$$

In equation (22), we assumed for simplicity that there is no transition regime between the Gaussian and the exponential behavior such that, at $y = y_c$, the Gaussian behavior ends, and

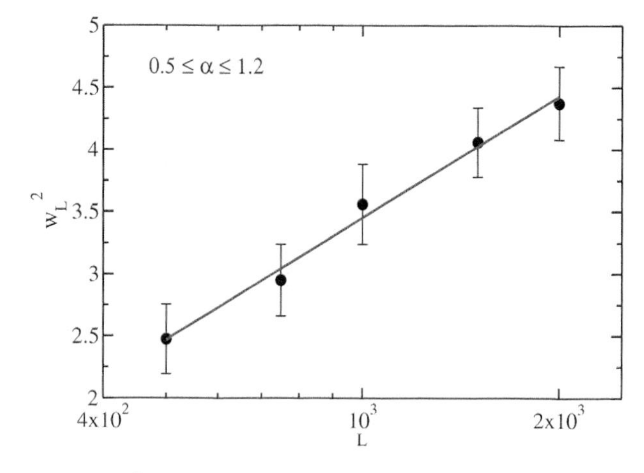

Figure 11. w_L^2 as a function of L. The error bars are estimated by the error of m_L together with equation (24). Within the error bars, w_L^2 can be approximated by $w_L^2 = -6.32 + 1.41\ln(L)$. After *Lennartz and Bunde* [2011].

the exponential behavior starts. We then used this assumption, which is satisfied to a great extent by our numerical results, to relate the parameters n_L, m_L, w_L, and y_c to each other. The normalization condition $\int_0^\infty P(y,L)\mathrm{d}y = 1/2$, together with the continuity of $P(y, L)$ and its first derivative $P'(y, L)$ at $y = y_c$, yields

$$y_c = w_L^2 \tag{23}$$

and

$$m_L = m(w_L) = n_L \frac{e^{w_L^2/2}}{\sqrt{2\pi}w_L} = \frac{e^{w_L^2}}{2+\sqrt{2\pi}w_L e^{w_L^2/2}\mathrm{erf}(w_L/\sqrt{2})}, \tag{24}$$

where erf(z) is the error function. Accordingly, all parameters can be expressed by w_L. To determine w_L, we first determined m_L from the asymptotic exponential part of $P(y, L)$ and then used equation (24) to obtain w_L numerically.

Figure 11 shows our numerical result for w_L^2. Within the error bars, to a good approximation, w_L^2 increases logarithmically with L:

$$w_L^2 \approx w^{(0)} + w^{(1)}\ln(L), \tag{25}$$

with $w^{(0)} \approx -6.32$ and $w^{(1)} \approx 1.41$.

From equations (22)–(24), the rescaled cumulative probability $W(y,L) = \int_y^\infty P(y',L)\mathrm{d}y'$ can be easily obtained:

$$W(y,L) = \begin{cases} \frac{1}{2} - \frac{n(w_L)}{2}\mathrm{erf}(y/(\sqrt{2}w_L)) & y \le w_L^2 \\ m(w_L)e^{-y} & y \ge w_L^2 \end{cases}. \tag{26}$$

To show that equation (26) together with equation (25) gives an excellent description of the rescaled exceedance probability $W(y, L)$, we have compared this result (symbols in Figure 10) with our simulations. The figure shows that the agreement is excellent in the scaling regime (i.e., for α between 0.5 and 1.2). Thus, we can use equation (26) with equations (24)–(25) to obtain the bounds $\pm y_Q(L)$ of the rescaled confidence interval (similar to equation (16)). If the confidence probability Q is large (close to 1), then only the exponential part of $W(y, L)$ has to be considered, and $y_Q(L)$ can be obtained analytically:

$$y_Q(L) = W^{-1}((1-Q)/2, L) = \ln\left(\frac{2m(w_L)}{1-Q}\right). \tag{27}$$

This yields, for the bounds $x_Q(\alpha, L) = y_Q(L)x_Q^*(\alpha, L)$ of the confidence interval,

$$x_Q(\alpha,L) = C_L\alpha^{\delta_L} \left\{ w_L^2 + \ln\left(\frac{2}{1-Q}\right) - \ln\left(2+\sqrt{2\pi}w_L e^{w_L^2/2}\mathrm{erf}(w_L/\sqrt{2})\right)\right\}, \tag{28}$$

with C_L from equation (20), δ_L from equation (21), and w_L from equation (25).

Figure 12 compares the α and L dependence of the analytical expression (28) (solid lines) with the numerical results (circles) for the confidence probabilities $Q = 95\%$ (left-hand side) and $Q = 99\%$ (right-hand side). The figure shows that there is a very good agreement between the numerical and the analytical results, as long as α is between 0.5 and 1.2. This is expected, since equation (28) is based on the scaling hypothesis, which is only valid for α between 0.5 and 1.2.

5. APPLICATION TO CLIMATE DATA

From Figure 12, one can immediately obtain, in a data set of length L characterized by the DFA2 fluctuation exponent α, the boundary of the confidence intervals of interest, and by comparing x_Q with a measured relative trend $x = \Delta/\sigma_t$, one can decide if the measured trend is significant or not.

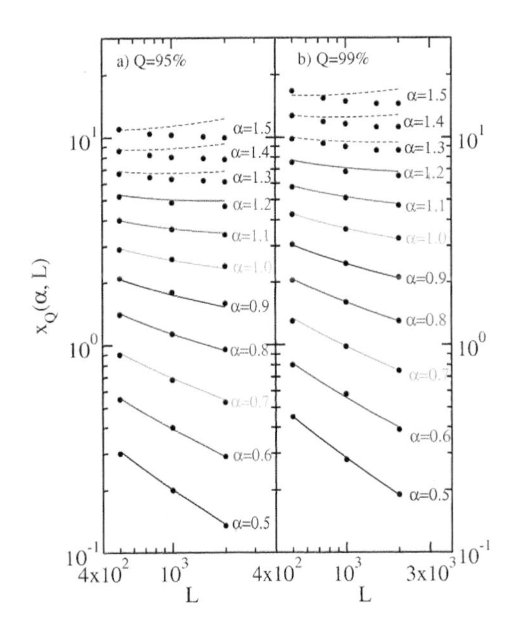

Figure 12. (bottom to top) Boundaries $x_Q(\alpha, L)$ from equation (28) of the confidence interval as a function of L for (a) $Q = 95\%$ and (b) $Q = 99\%$ for α between 0.5 and 1.2 (solid lines) and between 1.3 and 1.5 (dashed lines). For comparison, the numerically estimated values for $x_Q(\alpha, L)$ are also shown (solid circles). One can see clearly that equation (28) is only a very good approximation for $\alpha \le 1.2$ or $L \approx 500$. For larger α values and $L > 500$, one has to interpolate the numerical values. After *Lennartz and Bunde* [2011].

Table 2. Trend Estimation in River Runoff Records[a]

	Long Records						Short Records					
Data Set	L (months)	α	Δ^{real} ($m^3 s^{-1}$)	x	x_Q 95%	x_Q 99%	L (months)	α	Δ^{real} ($m^3 s^{-1}$)	x	x_Q 95%	x_Q 99%
Barron River (Myola, AU)	948	0.96	−2.9	0.18	2.30	3.20	500	0.92	−9.7	−0.32	2.30	3.30
Danube (Orsova, RO)	1812	0.80	−1.5	−0.02	1.00	1.20	500	0.83	−196	−0.09	1.60	2.20
Gaula (Haga Bru, NO)	1080	0.66	8.5	0.14	0.55	0.70	500	0.64	−4.4	−0.18	0.70	0.99
Labe (Decin, CZ)	1224	0.81	−2.5	−0.03	1.10	1.50	500	0.79	113	0.66	1.20	1.90
Maas (Borgharen, NL)	960	0.85	−40	−0.27	1.46	2.00	500	0.83	19	0.13	1.60	2.20
Niger (Koulikoro, ML)	948	0.87	−198	−0.28	1.60	2.20	500	0.87	−565	−1.15	1.50	2.60
Susquehanna (Harrisb., US)	1152	0.63	67	0.12	0.42	0.60	500	0.60	25	0.11	0.55	0.80
Thames (Kingston, GB)	1356	0.93	14	0.46	1.89	2.70	500	0.87	−0.1	−0.06	1.80	2.60
Weser (Vlotho, DE)	2052	0.81	−37	−0.35	1.00	1.30	500	0.96	4.3	0.09	2.50	3.70
Zaire (Kinshasa, CD)	972	1.07	6025	1.21	3.50	4.80	500	1.06	2986	0.47	3.50	5.00

[a]All relative trends x are within the 95% and 99% confidence interval $[-x_Q, x_Q]$ and thus are not significant.

We apply our analysis to (1) global temperature records provided by the Hadley Centre (http://www.cru.uea.ac.uk/cru/data/temperature/) [*Brohan et al.*, 2006] and (2) representative local temperature records provided by the Potsdam Institute for Climate Impact Research (PIK). We also consider (3) representative precipitation records, also provided by the PIK, and (4) representative river runoff records provided by the Water Management Authorities of Bavaria and Baden-Württemberg and the Global Runoff Data Centre in Koblenz (Germany). The data are listed in Table 1. These data were used in the works of *Eichner et al.* [2003], *Kantelhardt et al.* [2006], and *Koscielny-Bunde et al.* [2006], where they are also described.

We would like to note again that, in contrast to the temperature data, precipitation and river flows show (on the time scales considered here) weak multifractal behavior with nonlinear memory [*Kantelhardt et al.*, 2006; *Koscielny-Bunde et al.*, 2006; *Schertzer and Lovejoy*, 1991; *Schertzer et al.*, 1997], and their data distribution differs from Gaussians. Accordingly, our assumption of monofractal behavior with a Gaussian distribution of the data is not a priori satisfied. However, we can argue that we can still apply our methodology, to a reasonable approximation, to these data. The reason for this is that the characteristic mountain-valley behavior in long-term correlated records is mainly due to the linear persistence being described by the fluctuation exponent α. The nonlinear correlations solely add some "bursty" behavior to it that does not show up in α but in higher moments of the fluctuation function and does not affect the mountain-valley structure. The same is true for the deviations

Table 3. Trend Estimation in Precipitation Records[a]

	Long Records						Short Records					
Data Set	L (months)	α	Δ^{real} (mm)	x	x_Q 95%	x_Q 99%	L (months)	α	Δ^{real} (mm)	x	x_Q 95%	x_Q 99%
Albany (US)	1452	0.48	0.06	0.06	0.14	0.18	500	0.49	0.41	0.34	**0.28**	0.41
Charleston (US)	1487	0.53	−0.36	−0.19	0.21	0.28	500	0.51	0.05	0.02	0.31	0.49
Hamburg (DE)	1296	0.59	0.20	0.18	0.31	0.42	500	0.66	0.22	0.20	0.75	1.20
Irkutsk (RU)	1356	0.52	0.43	0.64	**0.20**	**0.27**	500	0.53	0.18	0.19	0.36	0.53
München (DE)	768	0.69	−0.24	−0.24	0.75	1.01	500	0.69	−0.37	−0.29	0.85	1.30
Norfolks Ils. (AU)	744	0.68	−1.42	−0.61	0.73	1.00	500	0.74	−1.96	−0.83	1.10	1.60
Perm (RU)	1356	0.51	0.23	0.24	**0.18**	0.25	500	0.51	0.38	0.39	**0.32**	0.50
Seoul (KR)	1056	0.59	0.63	0.27	0.39	0.52	500	0.58	0.08	0.13	0.49	0.71
Sonnblick (AT)	1289	0.61	−0.11	−0.02	0.39	0.55	500	0.53	0.26	0.17	0.36	0.53
Spokane (US)	1367	0.51	−0.08	−0.08	0.19	0.28	500	0.54	−0.13	−0.12	0.38	0.58

[a]Here x_Q is highlighted in boldface if the observed relative trend x is outside the corresponding confidence interval $[-x_Q, x_Q]$ and is thus significant.

of the distribution function from a Gaussian, as long as they are not characterized by "fat" tails. This is the case for both precipitation and river runoff data.

Tables 1, 2, and 3 also show, for each data set, the length L of the considered time window (full length and the last 500 months of the record), the fluctuation exponent α, and the relative trend x in this time window. When x, α, and L are known, we can easily read the desired bound of the confidence interval from Figure 12. We show the bounds in Tables 1–3 and highlight them in boldface if the observed relative trend is outside this interval and thus significant. In addition to the relative trend x, we also specify the linear trend Δ^{real} obtained by linear regression of the original monthly data set (without seasonal detrending) in the original units, for example, a temperature increase in °C. To obtain the upper and lower bound Δ^{real}_{min} and Δ^{real}_{max}, respectively, of the anthropogenic contribution to the observed trend Δ^{real}, we note that the fraction between the minimal (maximal) trend and the observed trend is (for large $L >> 12$) approximately independent of the seasonal detrending; that is,

$$\frac{\Delta^{real}_{min}}{\Delta^{real}} = \frac{\Delta^{ext}_{min}}{\Delta} = \frac{x - x_Q}{x} \tag{29}$$

$$\frac{\Delta^{real}_{max}}{\Delta^{real}} = \frac{\Delta^{ext}_{max}}{\Delta} = \frac{x + x_Q}{x}, \tag{30}$$

yielding

$$\Delta^{real}_{min} = \Delta^{real}(1 - x_Q/x) \tag{31}$$

$$\Delta^{real}_{max} = \Delta^{real}(1 + x_Q/x). \tag{32}$$

Most local temperatures available have been obtained in or near big cities that have grown considerably in the twentieth century. In many cases, the side effect of urban growth is a temperature increase (urban warming), which is a local effect and must not be confused with global warming due to an enhanced concentration of GHGs [*Jones et al.*, 2008; *Hansen et al.*, 2010; *Parker*, 2004; *Peterson*, 2003]. In large cities, both effects are superimposed and may lead to a larger temperature increase in the cities than in the countryside. We consider it as a surprise that only 5 of 17 stations considered show a significant trend on the 500 month scale. It can be verified immediately from Figure 12 that even the temperature increase in Prague of $\Delta = 0.95$°C corresponding to $x = 0.60$ is not significant. On larger scales of about 150 years, we see a considerable significant warming in 9 of 17 stations.

For the global data, due to the averaging procedure, the standard deviation σ_t is considerably smaller than for the local records. For the land air temperatures, this leads to significant trends in both the 95% and 99% confidence interval. In contrast, for the global sea surface temperatures, the temperature increase is not large enough to be significant, both for the last 500 months and the whole period of 1908 months. As a consequence, the global temperature, over 1908 months, shows a significant temperature increase just short of the 99% confidence interval, while for the last 500 months, the observed trend (even though larger than the trend over the longer period of 1908 months) is not significant.

There are two reasons why the increase in the sea surface temperatures is less significant. (1) The increase in the sea surface temperatures (0.62°C) is considerably smaller than in the land air temperatures (0.80°C). (2) For the sea surface temperatures, the fluctuation exponent α is quite large (greater than 1). As a result, the confidence interval is enhanced, as can be read from Figure 12.

Table 2 indicates that none of the river runoff data show a significant trend, neither in the whole period of time nor in the last 500 months. In contrast, the precipitation data in Table 3 show a mixed picture. In the last 500 months, only the precipitation of Albany and Perm show a significant trend. But over the whole period of time ($\approx$ 1400 months), there is no significant trend in Albany; only Irkutsk and Perm show a significant trend.

6. CONCLUSION

We have summarized how long-term correlations are defined, how they can be measured in the presence of additional trends by using the DFA, and how artificial long-term correlated Gaussian-distributed records can be generated. Following *Lennartz and Bunde* [2011], we have considered synthetic long-term correlated (monofractal) Gaussian-distributed records of length L between 500 and 2000 that are characterized by a fluctuation exponent α between 0.5 and 1.5 and estimated the exceedance probability $W(\Delta/\sigma_t, \alpha; L)$ to observe an increase Δ, obtained by linear regression analysis, in units of the standard deviation σ_t around the regression line. The result is that the exceedance probability scales for α between 0.5 and 1.2 and an analytical formula for the scaling function has been presented.

The analytical description of $W(\Delta/\sigma_t, \alpha; L)$ allows, for any confidence probability Q, to estimate the confidence interval of natural trends for arbitrary α between 0.5 and 1.2 and arbitrary lengths between 500 and 2000. For larger α values, where the exceedance probability does not scale, we also obtained, for confidence probabilities $Q = 0.95$ and $Q = 0.99$, numerical values of the confidence interval. If an observed trend is outside the confidence interval, it can be considered significant and unnatural, but the underlying probability Q is a certain matter of taste. For a conservative estimation, one may be inclined to consider a relatively large confidence probability $Q = 0.99$, while in more relaxed estimations, confidence probabilities around $Q = 0.95$ are sufficient. The advantage of the method discussed here is that one can obtain the confidence intervals for any Q in a simple (and quasi-analytical) way. The confidence interval allows an estimation of the upper and lower bound of the anthropogenic contribution to the observed trend.

We have used these results to estimate the significance of trends in temperature data, which show more significant upward trends in the longer time periods than in the last 500 months. For river runoff and precipitation data, our results are not exact but can be used as a first-order approximation. For river runoffs, they do not indicate any significant trend, while in the precipitation data, only very few significant trends are observed.

Finally, we would like to emphasize that, for estimating the size of a possible external trend, it is not sufficient to look only at the lower bound of the confidence interval, but the upper bound must also be considered. With increasing strength of the correlations (i.e., with increasing fluctuation exponent α), the confidence interval increases and, thus, so does the uncertainty about the trend.

REFERENCES

Ashkenazy, Y., P. C. Ivanov, S. Havlin, C.-K. Peng, A. L. Goldberger, and H. E. Stanley (2001), Magnitude and sign correlations in heartbeat fluctuations, *Phys. Rev. Lett.*, *86*, 1900–1903.

Barnett, T., et al. (2005), Detecting and attributing external influences on the climate system: A review of recent advances, *J. Clim.*, *18*(9), 1291–1314.

Bloomfield, P., and D. Nychka (1992), Climate spectra and detecting climate change, *Clim. Change*, *21*, 275–287.

Brohan, P., J. J. Kennedy, I. Harris, S. F. B. Tett, and P. D. Jones (2006), Uncertainty estimates in regional and global observed temperature changes: A new data set from 1850, *J. Geophys. Res.*, *111*, D12106, doi:10.1029/2005JD006548.

Bunde, A., S. Havlin, J. W. Kantelhardt, T. Penzel, J.-H. Peter, and K. Voigt (2000), Correlated and uncorrelated regions in heart-rate fluctuations during sleep, *Phys. Rev. Lett.*, *85*, 3736–3739.

Ding, Z., C. W. J. Granger, and R. F. Engle (1983), A long memory property of stock market returns and a new model, *J. Empirical Finan.*, *1*, 83–106.

Eichner, J. F., E. Koscielny-Bunde, A. Bunde, S. Havlin, and H.-J. Schellnhuber (2003), Power-law persistence and trends in the atmosphere: A detailed study of long temperature records, *Phys. Rev. E*, *68*, 046133, doi:10.1103/PhysRevE.68.046133.

Feder, J. (1988), *Fractals*, 310 pp., Plenum, New York.

Fraedrich, K., and R. Blender (2003), Scaling of atmosphere and ocean temperature correlations in observations and climate models, *Phys. Rev. Lett.*, *90*(10)108501, doi:10.1103/PhysRevLett.90.108501.

Giese, E., I. Mossig, D. Rybski, and A. Bunde (2007), Long-term analysis of air temperature trends in Central Asia, *Erdkunde*, *61* (2), 186–202.

Hansen, J., R. Ruedy, M. Sato, and K. Lo (2010), Global surface temperature change, *Rev. Geophys.*, *48*, RG4004, doi:10.1029/2010RG000345.

Hasselmann, K. (1993), Optimal fingerprints for the detection of time-dependent climate change, *J. Clim.*, *6*(10), 1957–1971.

Hegerl, G. C., H. von Storch, K. Hasselmann, B. D. Santer, U. Cubasch, and P. D. Jones (1996), Detecting greenhouse-gas-induced climate change with an optimal fingerprint method, *J. Clim.*, *9*(10), 2281–2306.

Hurst, H. E. (1951), Long-term storage capacity of reservoirs, *Trans. Am. Soc. Civ. Eng.*, *116*, 770–799.

Hurst, H. E., R. Black, and Y. M. Sinaika (1965), *Long-Term Storage in Reservoirs: An Experimental Study*, 145 pp., Constable, London, U. K.

Jones, P. D., D. H. Lister, and Q. Li (2008), Urbanization effects in large-scale temperature records, with an emphasis on China, *J. Geophys. Res.*, *113*, D16122, doi:10.1029/2008JD009916.

Kantelhardt, J. W., E. Koscielny-Bunde, H. H. A. Rego, S. Havlin, and A. Bunde (2001), Detecting long-range correlations with detrended fluctuation analysis, *Physica A*, *295*(3–4), 441–454.

Kantelhardt, J. W., Y. Ashkenazy, P. C. Ivanov, A. Bunde, S. Havlin, T. Penzel, H. J. Peter, and H. E. Stanley (2002a), Characterization of sleep stages by correlations in the magnitude and sign of heartbeat increments, *Phys. Rev. E*, *65*, 051908, doi:10.1103/PhysRevE.65.051908.

Kantelhardt, J. W., S. A. Zschiegner, E. Koscielny-Bunde, S. Havlin, A. Bunde, and H. E. Stanley (2002b), Multifractal detrended fluctuation analysis of nonstationary time series, *Physica A*, *316*, 87–114.

Kantelhardt, J. W., T. Penzel, S. Rostig, H. F. Becker, S. Havlin, and A. Bunde (2003a), Breathing during REM and non-REM sleep: Correlated versus uncorrelated behaviour, *Physica A*, *319*, 447–457.

Kantelhardt, J. W., D. Rybski, S. A. Zschiegner, P. Braun, E. Koscielny-Bunde, V. Livina, S. Havlin, and A. Bunde (2003b), Multifractality of river runoff and precipitation: Comparison of fluctuation analysis and wavelet methods, *Physica A*, *330*, 240–245.

Kantelhardt, J. W., E. Koscielny-Bunde, D. Rybski, P. Braun, A. Bunde, and S. Havlin (2006), Long-term persistence and multifractality of precipitation and river runoff records, *J. Geophys. Res.*, *111*, D01106, doi:10.1029/2005JD005881.

Koscielny-Bunde, E., A. Bunde, S. Havlin, and Y. Goldreich (1996), Analysis of daily temperature fluctuations, *Physica A*, *231*, 393–396.

Koscielny-Bunde, E., A. Bunde, S. Havlin, H. E. Roman, Y. Goldreich, and H. J. Schellnhuber (1998), Indication of a universal persistence law governing atmospheric variability, *Phys. Rev. Lett.*, *81*(3), 729–732.

Koscielny-Bunde, E., J. W. Kantelhardt, P. Braun, A. Bunde, and S. Havlin (2006), Long-term persistence and multifractality of river runoff records: Detrended fluctuation studies, *J. Hydrol.*, *322*, 120–137.

Koutsoyannis, D. (2003), Climate change, the Hurst phenomenon, and hydrological statistics, *Hydrol. Sci. J.*, *48*(1), 3–24.

Koutsoyannis, D. (2006), A toy model of climate variability with scaling behaviour, *J. Hydrol.*, *322*, 25–48.

Lennartz, S., and A. Bunde (2009a), Trend evaluation in records with long-term memory: Application to global warming, *Geophys. Res. Lett.*, *36*, L16706, doi:10.1029/2009GL039516.

Lennartz, S., and A. Bunde (2009b), Eliminating finite-size effects and detecting the amount of white noise in short records with long-term memory, *Phys. Rev. E*, *79*, 066101, doi:10.1103/PhysRevE.79.066101.

Lennartz, S., and A. Bunde (2011), Distribution of natural trends in long-term correlated records: A scaling approach, *Phys. Rev. E*, *84*, 021129, doi:10.1103/PhysRevE.84.021129.

Liu, Y. H., P. Cizeau, M. Meyer, C.-K. Peng, and H. E. Stanley (1997), Correlations in economic time series, *Physica A*, *245*, 437–440.

Liu, Y. H., P. Gopikrishnan, P. Cizeau, M. Meyer, C.-K. Peng, and H. E. Stanley (1999), Statistical properties of the volatility of price fluctuations, *Phys. Rev. E*, *60*, 1390–1400.

Malamud, B. D., and D. L. Turcotte (1999), Self-affine time series: I. Generation and analyses, *Adv. Geophys.*, *40*, 1–90.

Mandelbrot, B. B., and J. R. Wallis (1969), Some long-run properties of geophysical records, *Water Resour. Res.*, *5*, 321–340.

Monetti, R. A., S. Havlin, and A. Bunde (2003), Long term persistence in the sea surface temperature fluctuations, *Physica A*, *320*, 581–590.

Parker, D. E. (2004), Climate: Large-scale warming is not urban, *Nature*, *432*, 290, doi:10.1038/432290a.

Pelletier, J. D., and D. L. Turcotte (1997), Long-range persistence in climatological and hydrological time series: Analysis, modeling and application to drought hazard assessment, *J. Hydrol.*, *203*, 198–208.

Pelletier, J. D., and D. L. Turcotte (1999), Self-affine time series: II. Applications and models, *Adv. Geophys.*, *40*, 91–166.

Peng, C.-K., J. Mietus, M. Hausdorff, S. Havlin, H. E. Stanley, and A. L. Goldberger (1993), Long-range anticorrelations and non-Gaussian behavior of the heartbeat, *Phys. Rev. Lett.*, *70*, 1343–1346.

Peng, C. K., S. V. Buldyref, S. Havlin, S. Simons, H. E. Stanley, and A. L. Goldberger (1994), Mosaic organisation of DNA nucleotides, *Phys. Rev. E*, *49*, 1685–1689.

Peterson, T. C. (2003), Assessment of urban versus rural in situ surface temperatures in the contiguous United States: No difference found, *J. Clim.*, *16*, 2941–2959.

Rybski, D., and A. Bunde (2009), On the detection of trends in long-term correlated records, *Physica A*, *388*, 1687–1695, doi:10.1016/j.physa.2008.12.026.

Rybski, D., A. Bunde, S. Havlin, and H. von Storch (2006), Long-term persistence in climate and the detection problem, *Geophys. Res. Lett.*, *33*, L06718, doi:10.1029/2005GL025591.

Rybski, D., A. Bunde, and H. von Storch (2008), Long-term memory in 1000-year simulated temperature records, *J. Geophys. Res.*, *113*, D02106, doi:10.1029/2007JD008568.

Schäfer, C., M. G. Rosenblum, J. Kurths, and H. H. Abel (1998), Heartbeat synchronized with ventilation, *Nature*, *392*, 239–240.

Schertzer, D., and S. Lovejoy (1991), *Nonlinear Variability in Geophysics: Scaling and Fractals*, 332 pp., Springer, New York.

Schertzer, D., S. Lovejoy, F. Schmitt, Y. Chigirinskaya, and D. Marsan (1997), Multifractal cascade dynamics and turbulent intermittency, *Fractals*, *5*(3), 427–471.

Schreiber, T., and A. Schmitz (1996), Improved surrogate data for nonlinearity tests, *Phys. Rev. Lett.*, *77*, 635–638.

Turcotte, D. L. (1997), *Fractals and Chaos in Geology and Geophysics*, 2nd ed., 416 pp., Cambridge Univ. Press, Cambridge, U. K.

Weber, R. O., and P. Talkner (2001), Spectra and correlations of climate data from days to decades, *J. Geophys. Res.*, *106*, 20,131–20,144.

Yamasaki, K., L. Muchnik, S. Havlin, A. Bunde, and H. E. Stanley (2005), Scaling and memory in volatility return intervals in financial markets, *Proc. Natl. Acad. Sci. U. S. A.*, *102*, 9424–9428.

Zorita, E., T. F. Stocker, and H. von Storch (2008), How unusual is the recent series of warm years?, *Geophys. Res. Lett.*, *35*, L24706, doi:10.1029/2008GL036228.

Zwiers, F. W. (1999), The detection of climate change, in *Anthropogenic Climate Change*, edited by H. von Storch and G. Flöser, pp. 163–209, Springer, New York.

A. Bunde, Institut für Theoretische Physik III, Justus-Liebig-Universität Giessen, D-35392 Giessen, Germany. (armin.bunde@uni-giessen.de)

S. Lennartz, School of Engineering and School of Geosciences, University of Edinburgh, Edinburgh EH9 3JL, UK. (sabine.lennartz@ed.ac.uk)

Climate Subsystems: Pacemakers of Decadal Climate Variability

Anastasios A. Tsonis

Atmospheric Sciences Group, Department of Mathematical Sciences, University of Wisconsin-Milwaukee, Milwaukee, Wisconsin, USA

This paper is a synthesis of work spanning the last 25 years. It is largely based on the use of climate networks to identify climate subsystems and to subsequently study how their collective behavior explains decadal variability. The central point is that a network of coupled nonlinear subsystems may at times synchronize. If during synchronization the coupling between the subsystems increases, the synchronous state may be destroyed, shifting climate to a new regime. This climate shift manifests itself as a change in global temperature trend. This mechanism, which is consistent with the theory of synchronized chaos, appears to be a very robust mechanism of the climate system. It is found in the instrumental records and in forced and unforced climate simulations, as well as in proxy records spanning several centuries.

1. INTRODUCTION

The story in this chapter starts in the mid-1980s when new and exciting approaches to nonlinearly analyze time series "hit the market." At that time, very few in the atmospheric sciences community had heard terminology such as "fractals," "chaos theory," "strange attractors," and the like. A few innovative scientists, however, were experimenting with these new ideas and soon reports of "fractality" and "low dimensionality" in climate records, and other geophysical data begun to surface. These climate records represented dynamics over different time scales ranging from very long (thousands of years) [*Nicolis and Nicolis*, 1984] to very short (hours) [*Tsonis and Elsner*, 1988]. Virtually every report suggested underlying attractors of dimensions between 3 and 8. These early results suggested that climate variability may indeed be described by only a few equations. This resulted in both enthusiasm and hope that climate variability may be tamed after all, and in fierce opposition. Fortunately, this "tug of war" did not eliminate interest in this new theory; rather, it led to a deeper understanding of the nonlinear character of nature and to new insights about the properties of the climate system. This chapter is a small part of what we have learned so far, and it largely draws from my work over the years.

Extreme Events and Natural Hazards: The Complexity Perspective
Geophysical Monograph Series 196

10.1029/2011GM001053

The initial opposition to those dimension estimates seemed to be that in all these studies, the sample size was simply too small. While this issue has been debated extensively [*Smith*, 1988; *Nerenberg and Essex*, 1990; *Tsonis*, 1992, *Tsonis et al.*, 1994], it has not been settled beyond doubts. In a sense, it is naïve to imagine that our climate system (a spatially extended system of infinite dimensional state space) is described by a grand attractor let alone a low-dimensional attractor. If that were the case, then all observables representing different processes should have the same dimension, which is not suggested from the myriad of reported dimensions. *Tsonis and Elsner* [1989] suggested that if low-dimensional attractors exist, they are associated with subsystems each operating at different space and/or time scales. In his study on dimension estimates, *Lorenz* [1991] concurs with the suggestion of *Tsonis and Elsner* [1989]. These subsystems may be nonlinear and exhibit a variety of complex behaviors. All subsystems are connected with each other, as in a web, with various degrees of connectivity. Accordingly, any subsystem may transmit "information" to another subsystem thereby perturbing its behavior. This "information" plays the role of an ever-present external noise, which perturbs the subsystem, and depending on the connectivity of a subsystem to another subsystem, the effect can be dramatic or negligible. Subsystems with weak connectivities will be approximately "independent," and as such, they may exhibit low-dimensional chaos. It is also possible that the connectivity between subsystems may vary in time, and this effect may dictate the variability of the climate system.

Given the above, the question arises. If subsystems exist in the climate system, what are they and what physics can we infer from them?

2. SEARCHING FOR SUBSYSTEMS

2.1. Methods and Results From Observations

Answers on the nature, geographical basis, and physical mechanisms underlying these subsystems are provided by recent developments in graph theory and networks. Networks relate to the underlying topology of complex systems with many interacting parts. They have found many applications in many fields of sciences. In the interest of completeness, a short introduction to networks is offered next.

A network is a system of interacting agents. In the literature, an agent is called a node. The nodes in a network can be anything. For example, in the network of actors, the nodes are actors that are connected to other actors if they have appeared together in a movie. In a network of species, the nodes are species that are connected to other species they interact with. In the network of scientists, the nodes are scientists that are connected to other scientists if they have collaborated. In the grand network of humans, each node is an individual, which is connected to people he or she knows. There are four basic types of networks.

2.1.1. Regular (ordered) networks. These networks are networks with a fixed number of nodes, each node having the same number of links connecting it in a specific way to a number of neighboring nodes (Figure 1, left). If each node is linked to all other nodes in the network, then the network is a fully connected network. When the number of links per node is high, regular networks have a high (local) clustering coefficient. In this case, loss of a number of links does not break the network into noncommunicating parts. In this case, the network is stable, which may not be the case for regular networks with small local clustering. Also, unless networks are fully connected, they have a large diameter. The diameter of a network is defined as the maximum shortest path between any pair of its nodes. It relates to the characteristic path length, which is the average number of links in the shortest path between two nodes. The smaller the diameter, the easier is the communication in the network.

2.1.2. Classical random networks. In these networks, the nodes are connected at random (Figure 1, right). In this case, the degree distribution is a Poisson distribution (the degree distribution, p_k, gives the probability that a node in the network is connected to k other nodes). The problem with these networks is that they have very small clustering coefficient and, thus, are not very stable. Removal of a number of nodes at random may fracture the network to noncommunicating parts. On the other hand, they are characterized by a small diameter. Far away nodes can be connected as easily as nearby nodes. In this case, information may be transported all over the network much more efficiently than in ordered networks. Thus, random networks exhibit efficient information transfer, but they are not stable.

2.1.3. Small-world networks. In nature, we should not expect to find either very regular or completely random networks. Rather, we should find networks that are efficient in processing information and at the same time are stable. Work in this direction led to a new type of network, which was proposed 12 years ago by the American mathematicians Duncan Watts and Steven Strogatz and is called *small-world* network [*Watts and Strogatz*, 1998]. A "small-world" network is a superposition of regular and classical random graphs. Such networks exhibit a high degree of local clustering, but a small number of long-range connections make

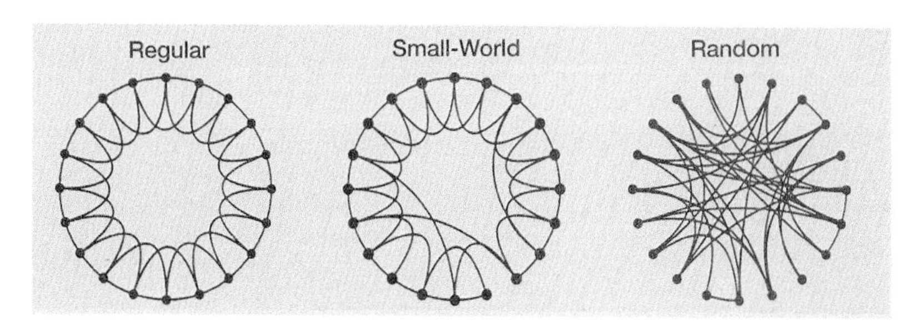

Figure 1. Illustration of a regular, a small-world, and a random network [after *Watts and Strogatz*, 1998]. Reprinted by permission from Macmillan Publishers Ltd: *Nature*, copyright 1998.

them as efficient in transferring information as random networks. Those long-range connections do not have to be designed. A few long-range connections added at random will do the trick (Figure 1, middle). The degree distribution of small-world networks is also a Poisson distribution.

2.1.4. Networks with a given degree distribution. The "small-world" architecture can explain phenomena such as the 6° of separation (most people are friends with their immediate neighbors, but we all have one or two friends a long way away), but it really is not a model found often in the real world. In the real world, the architecture of a network is neither random nor small-world, but it comes in a variety of distributions such as truncated power law distributions, Gaussian distributions, power law distributions, and distributions consisting of two power laws separated by a cutoff value (for a review, see *Strogatz* [2001]). The most interesting and common of such networks are the so-called *scale-free* networks. Consider a map showing an airline's routes. This map has a few hubs connecting with many other points (supernodes) and many points connected to only a few other points, a property associated with power law distributions. Such a map is highly clustered, yet it allows motion from a point to another far away point with just a few connections. As such, this network has the *property* of small-world networks, but this property is not achieved by local clustering and a few random connections. It is achieved by having a few elements with large number of links and many elements having very few links. Thus, even though they share the same property, the architecture of scale-free networks is different than that of "small-world" networks. Such inhomogeneous networks have been found to pervade biological, social, ecological, and economic systems, the Internet, and other systems [*Albert et al.*, 1999; *Liljeros et al.*, 2001; *Jeong et al.*, 2001; *Pastor-Satorras and Vespignani*, 2001; *Bouchaud and Mezard*, 2000; *Barabasi and Bonabeau*, 2003]. These networks are referred to as scale-free because they show a power law distribution of the number of links per node. Lately, it was also shown that, in addition to the power law degree distribution, many real scale-free networks consist of self-repeating patterns on all length scales [*Song et al.*, 2005]. These properties are very important because they imply some kind of self-organization within the network. Scale-free networks are not only efficient in transferring information, but due to the high degree of local clustering, they are also very stable [*Barabasi and Bonabeau*, 2003]. Because there are only a few super nodes, chances are that accidental removal of some nodes will not include the super nodes. In this case, the network would not become disconnected. This is not the case with weakly connected regular or random networks (and to a lesser degree with small-world networks), where accidental removal of the same percentage of nodes makes them more prone to failure [*Barabasi and Bonabeau*, 2003].

The topology of the network can reveal important and novel features of the system it represents [*Albert and Barabasi*, 2002; *Strogatz*, 2001; *Costa et al.*, 2007]. One such feature is communities [*Newman and Girvan*, 2004]. Communities represent groups of densely connected nodes with only a few connections between groups. It has been conjectured that each community represents a subsystem, which operates relatively independent of the other communities [*Arenas et al.*, 2006]. Thus, identification of these communities can offer useful insights about dynamics. In addition, communities can be associated to network functions such as in metabolic networks where certain groups of genes have been identified that perform specific functions [*Holme et al.*, 2003; *Guimera and Amaral*, 2005]. Recently, concepts from network theory have been applied to climate data organized as networks with impressive results [*Tsonis et al.*, 2006, 2007, 2008; *Tsonis and Swanson*, 2008; *Yamasaki et al.*, 2008; *Gozolchiani et al.*, 2008; *Swanson and Tsonis*, 2009; *Elsner et al.*, 2009; *Tsonis et al.*, 2010].

Figure 2 is an example of a climate network showing the area weighted connectivity (number of edges) at each geographic location for the 500 hPa height field [*Tsonis et al.*, 2006]. More accurately, it shows the fraction of the total global area that a point is connected to. This is a more appropriate way to show the architecture of the network because the network is a continuous network defined on a sphere. These data are derived from the global National Centers for Environmental Prediction/National Center for Atmospheric Research (NCEP/NCAR) atmospheric reanalysis data set [*Kistler et al.*, 2001]. The data are arranged on a grid with a resolution of 10° latitude × 10° longitude. This results in 36 points in the east-west direction and 19 points in the north-south direction for a total of $n = 684$ points. These 684 points are assumed to be the nodes of the network. For each grid point, monthly values from 1950 to 2005 are available. From the monthly values, anomaly values (actual value minus the climatological average for each month) are then produced. Even though the leading order effect of the annual cycle is removed by producing anomaly values, some of it is still present as the amplitude of the anomalies is greater in the winter than in the summer. For this reason, in order to avoid spurious high values of correlations, only the values for December–February in each year were considered. Thus, for each grid point, we have a time series of 168 anomaly values. In order to define the links between the nodes for either network, the correlation coefficient at lag zero (r) between the time series of all possible pairs of nodes [$n(n-1)/2 = 232,903$ pairs] is estimated. Note that since the

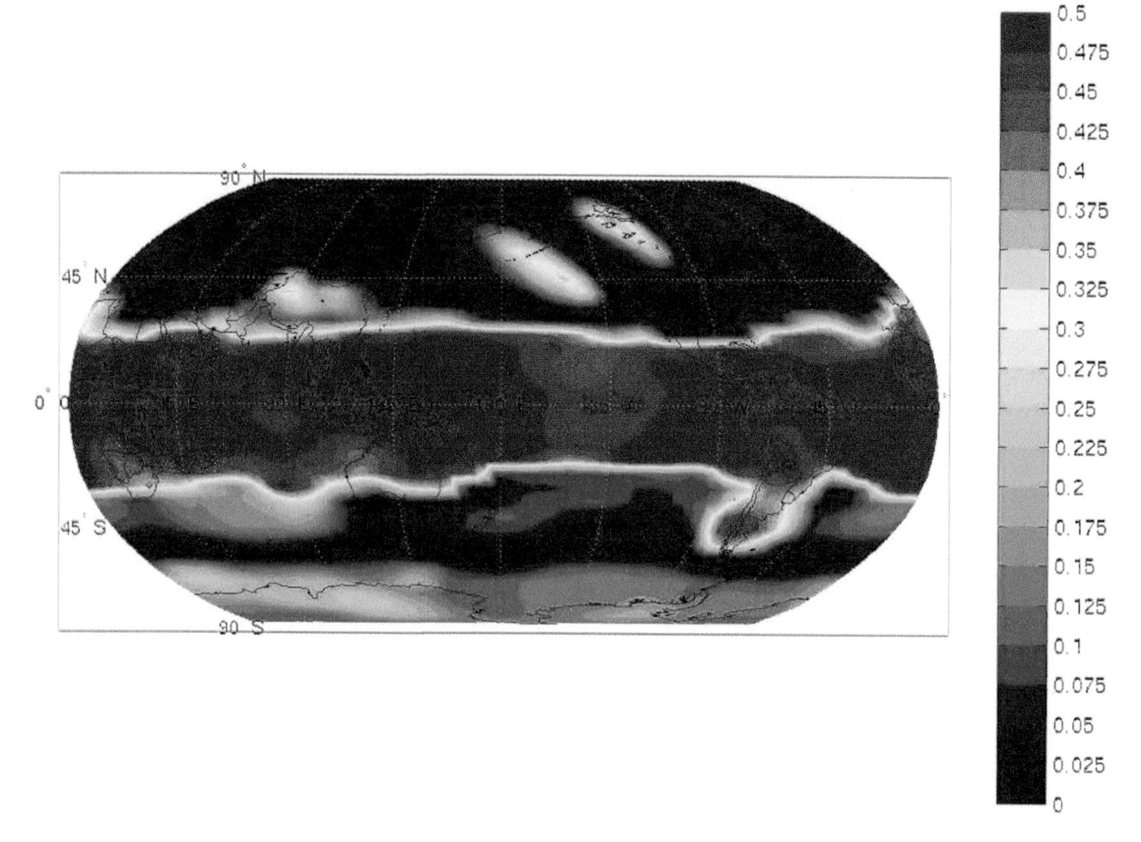

Figure 2. Total number of links (connections) at each geographic location. More accurately, it shows the fraction of the total global area that a point is connected to. This is a more appropriate way to show the architecture of the network because the network is a continuous network defined on a sphere. The uniformity observed in the tropics indicates that each node possesses the same number of connections. This is not the case in the extratropics where certain nodes possess more links than the rest.

values are monthly anomalies, there is very little autocorrelation in the time series.

A pair is considered as connected if the absolute value of their cross-correlation $|r| \geq 0.5$. This criterion is based on parametric and nonparametric significance tests. According to the t test with $N = 168$, a value of $r = 0.5$ is statistically significant above the 99% level. In addition, randomization experiments where the values of the time series of one node in a pair are scrambled and then are correlated to the unscrambled values of the time series of the other node indicate that a value of $r = 0.5$ will not arise by chance. The choice of $r = 0.5$, while it guarantees statistical significance, is somewhat arbitrary. We find that while other values might affect the connectivity structure of the network, the effect of different correlation thresholds (between 0.4 and 0.6) does not affect the conclusions. Obviously, as $|r| \rightarrow 1$, we end up with a random network, and as $r \rightarrow 0$, we remain with just one fully connected community. The use of the correlation coefficient to define links in networks is not new. Correlation coefficients have been used to successfully derive the topology of gene expression networks [*Farkas et al.*, 2003] and to study financial markets [*Mantegna*, 1999].

Returning to Figure 2, we observe two very interesting features. In the tropics, it appears that all nodes posses more or less the same (and high) number of connections, which is a characteristic of fully connected networks. In the extratropics, it appears that certain nodes posses more connections than the rest, which is a characteristic of scale-free networks. In the Northern Hemisphere, we clearly see the presence of regions where such supernodes exist in China, North America, and the northeast Pacific Ocean. Similarly, several supernodes are visible in the Southern Hemisphere. These differences between tropics and extratropics have been delineated in the corresponding degree distributions, which suggest that indeed the extratropical network is a scale-free network characterized by a power law degree distribution [*Tsonis et al.*, 2006]. As is the case with all scale-free networks, the extratropical network is also a small-world network [*Tsonis et al.*, 2006].

An interesting observation in Figure 2 is that supernodes may be associated with major teleconnection patterns. For example, the supernodes in North America and northeast Pacific Ocean are located where the well-known Pacific North America (PNA) pattern [*Wallace and Gutzler*, 1981] is found. In the Southern Hemisphere, we also see supernodes over the southern tip of South America, Antarctica, and South Indian Ocean that are consistent with some of the features of the Pacific South America pattern [*Mo and Higgins*, 1998]. Interestingly, no such super nodes are evident where the other major pattern, the North Atlantic Oscillation (NAO) [*Thompson and Wallace*, 1998; *Pozo-Vazquez et al.*, 2001; *Huang et al.*, 1998] is found. This, however, does not indicate that NAO is an insignificant feature of the climate system. Since NAO is not strongly connected to the tropics, the high connectivity of the tropics with other regions is masking NAO out [*Tsonis et al.*, 2008].

Once the edges in a network have been defined, one can proceed with identifying the communities. Several methods of identifying communities were used. The first is based on the notion of node betweenness [*Girvan and Newman*, 2002]. For any node i, node betweenness is defined as the number of shortest paths between pairs of other nodes that run through i. The algorithm extends this definition to the case of edges, defining the "edge betweenness" of an edge as the number of shortest paths between pairs of nodes that run along this edge. If a network contains communities or groups that are only loosely connected by a few intergroup edges (think of bridges connecting different sections of New York City, for example), then all shortest paths between two nodes belonging to different communities must go along one of these few edges. Thus, the edges connecting communities will have high edge betweenness. By removing these edges, the groups are separated from one another thereby revealing the underlying community structure of the network.

Figure 3 illustrates the basics behind this algorithm. The setup is adapted from the work of *Newman and Girwan* [2004]. We start with a "source" node s, which is connected to six other nodes according to the simple network shown in Figure 3. This node is assigned a distance $d_s = 0$ and a weight $w_s = 1$. Then, each node i adjacent to s (i.e., nodes 1 and 3) is given a distance $d_i = d_s + 1$ and a weight $w_i = w_s$ (1,1 and 1,1, respectively). Then, each node j adjacent to nodes i is given a distance $d_j = d_i + 1$ and weight $w_j = w_i$ and so on. This procedure results in the pairs of values shown for each node. Once distances and weights have being assigned, one finds those nodes such that no shortest paths between

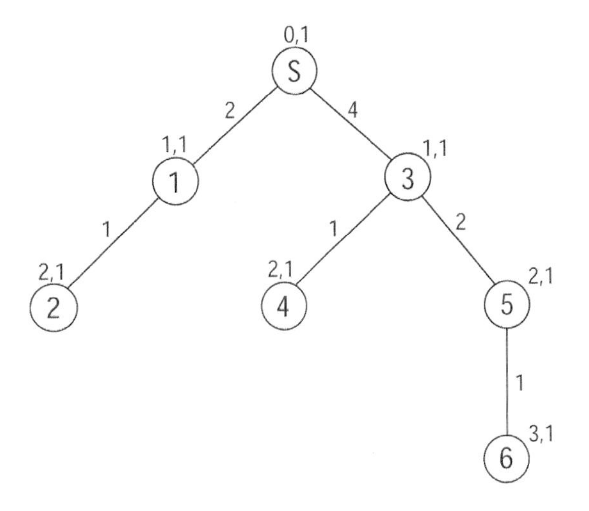

Figure 3. An illustration of the calculation of the shortest-path betweenness. See text for details.

any other node and s pass through them. Such nodes are nodes 2, 4, and 6. We call these nodes "leaves" and denote them as l. For each node i neighboring l, we assign a score to their edge of w_i/w_l. Accordingly, the score of the edges connecting pairs 1 and 2, 3 and 4, and 5 and 6 is equal to 1. Then, we start from the edge that is farther from s (that will be the edge connecting nodes 3 and 5) and assign a score that is 1 plus the sum of the scores on the neighboring edges immediately below it, multiplied by w_i/w_j where node j is farther from s than node i. We then move upward repeating the process until node s is reached. This procedure will identify the edge connecting nodes s and 3 as having the highest edge betweenness. By removing it, we remain with two communities, one consisting of nodes s, 1, and 2 and another consisting of nodes 3, 4, 5, and 6. We can then continue splitting the network by removing the edge with the second highest betweenness and so on. In applications with real data, however, this approach may not work well. The reason is that it may be that there is more than one edge connecting communities. In this case, there is no guarantee that all those edges will have high betweenness. We can only be sure that at least one will have high betweenness. For this reason, the algorithm repeats the whole process for all nodes in the network (i.e., considering as the source node each node in the network at a time) and sums all the scores. This gives the total betweenness for all edges in the network. We then have to repeat this calculation for each edge removed from the network. The network is then divided into two communities by removing the edge whose removal resulted in the lowest sum, and so on.

In order to quantify the strength of a community structure, we use a measure called the *modularity*. For a particular division into n communities, we define a $n \times n$ symmetric matrix **e** whose element e_{ij} is the fraction of all connections in the network that link nodes in community i to community j. The modularity [*Newman and Girvan*, 2004; *Girvan and Newman*, 2002; *Newman*, 2006] is defined as

$$Q = Tr\mathbf{e} - ||\mathbf{e}^2||,$$

where Tr is the trace of matrix **e** (the sum of the elements along the main diagonal), and $||x||$ indicates the sum of the elements of matrix $\boldsymbol{x}$. The modularity measures the fraction of the edges in the network that connect nodes of the same type (within-community nodes) minus the expected value of the same quantity in a network with the same community divisions but random connections between the nodes. Modularity ranges from zero (for random networks) to one (for very strong community structure). The optimum community division is found by estimating at how many communities (at what n) Q is maximum.

Figure 4 (left) shows the division into communities of three observed climate fields (the 500 hPa height, the global sea level pressure, and the global surface temperature fields [*Tsonis et al.*, 2010]. All fields are derived from the NCEP/NCAR atmospheric reanalysis data set [*Kistler et al.*, 2001]. Figure 4 (top) shows the community structure of the 500 hPa height network (shown in Figure 2), Figure 4 (middle) shows that of the sea level pressure network, and Figure 4 (bottom) shows that of the surface temperature network. The total number of communities is 47, 15, and 58, respectively. Many of these communities, however, consist of very few points in the boundaries between a small number of dominant communities (think of a country whose population is dominated by two races but also includes small groups of other races). As is evident in Figure 4, the effective number of communities is, arguably, four in all three networks (delineated as purple, blue, green, and yellow-red areas). The modularity of the networks is 0.49, 0.56, and 0.59, respectively, indicating networks that lie between completely random and strongly communal.

Importantly, this purely mathematical approach results in divisions that have connections with actual physics and dynamics. For example, in Figure 4 (top left) (500 hPa height network), we see that three of the effective four communities correspond to a latitudinal division 90°S–30°S, 30°S–30°N, and 30°N–90°N. This three-zone separation is not a trivial separation into Northern Hemisphere winter, Southern Hemisphere summer, and the rest of the world because when we repeat the analysis with yearly averages rather than seasonal values, we also see evidence of this three-zone separation. This separation is consistent with the transition from a barotropic atmosphere (where pressure depends on density only; appropriate for the tropics-subtropics) to a baroclinic atmosphere (where pressure depends on both density and temperature; appropriate for higher latitudes). Another possibility is that it reflects the well-known three-zone distribution of variance of the surface pressure field. Within the third community (green area) another community (yellow-red) is embedded. This community is consistent with the presence of major atmospheric teleconnection patterns such as the PNA pattern and the NAO [*Wallace and Gutzler*, 1981; *Barnston and Livezey*, 1987]. We note here that NAO (which has been lately suggested of being a three-pole pattern rather than a dipole) [*Tsonis et al.*, 2008] and Arctic Oscillation (AO) [*Thompson and Wallace*, 1998] are often interpreted as manifestations of the same dynamical mode, even though in some cases more physical meaning is given to NAO [*Ambaum et al.*, 2001]. In any case, here we do not make a distinction between NAO and AO.

In the sea level pressure network, we see again the latitudinal division into three communities. Here the purple area extends into the eastern Pacific as far as 30°N. This feature is consistent with the interhemispheric propagation of Rossby waves via the well-documented eastern Pacific corridor [*Webster and Holton*, 1982; *Tsonis and Elsner*, 1996]. The fourth community (yellow-red) embedded within the third community (green) is found over areas in the Northern Hemisphere where cyclogenesis is more frequently found. Note that PNA relates to anomalies in the forcing of extratropical quasistationary waves, the NAO arises from wave-mean flow interaction, and El Niño–Southern Oscillation (ENSO) is known to affect extratropical cyclone variability [*Eichler and Higgins*, 2006; *Wang and Fu*, 2000; *Held et al.*, 2002; *Favre and Gershunov*, 2006, 2009].

Figure 4. (opposite) Community structure in three climate networks. The networks are constructed from (top) the 500 hPa height field, (middle) the surface pressure field, and (bottom) the surface temperature field. Networks constructed (left) from observations and (right) from model simulations. The numbers below the shading key indicate the total number of communities. Because the total number of communities is not necessarily the same in each network, the color scheme used to show the spatial delineation of the communities is not the same in each plot. This means that the same community may be represented by a different color in the observations and in the model. The reverse may also be possible: the same color may not represent the same community. What we should compare in Figure 4 is the spatial distribution and structure of communities in observations and model (see text for more details).

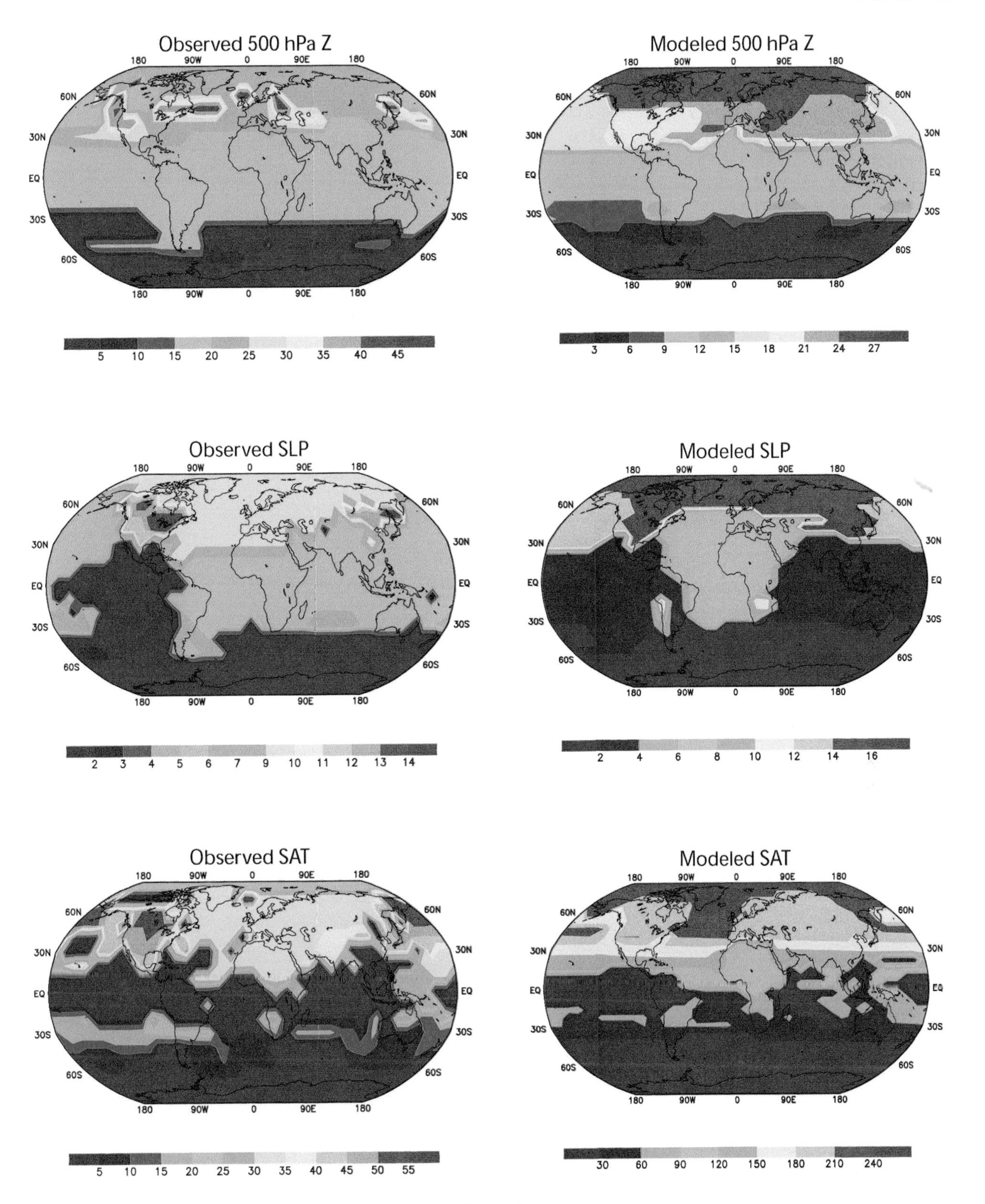

Observed 500 hPa Z
Modeled 500 hPa Z
5 10 15 20 25 30 35 40 45
3 6 9 12 15 18 21 24 27
Observed SLP
Modeled SLP
2 3 4 5 6 7 9 10 11 12 13 14
2 4 6 8 10 12 14 16
Observed SAT
Modeled SAT
5 10 15 20 25 30 35 40 45 50 55
30 60 90 120 150 180 210 240

Figure 4

In the temperature network, we see a major subdivision into two communities, one covering the Southern Hemisphere (purple) and the other the Northern Hemisphere (green). This appears to be a result of the fact that the Southern Hemisphere is covered mostly by water (which moderates surface temperatures), whereas the Northern Hemisphere is mostly land (where temperature variability is greater). In the Northern Hemisphere, we observe that embedded in the green community, there is a separate community (yellow-red) over North America and northeast Asia where cold winter air outbreaks normally occur. This could be an influence of the Pacific Decadal Oscillation (PDO), which is known to affect the North American temperature anomaly structure [*Mantua et al.*, 1997]. The green community, which includes the North Atlantic and Europe, may reflect the moderation of temperature fluctuations in the North Atlantic and Europe due to the Gulf Stream as a different community. In the Southern Hemisphere, embedded within the purple community, we see a separate community (blue) along the 30°S latitude. This community probably arises from the corresponding Southern Hemisphere storm tracks, which modify temperature fluctuations. Alternatively, it may be due to the presence of the subtropical belt. We note here that the location of some of the communities and their attributed teleconnection patterns may not be exactly the location of these patterns as delineated by empirical orthogonal function (EOF) analysis because they are different methods.

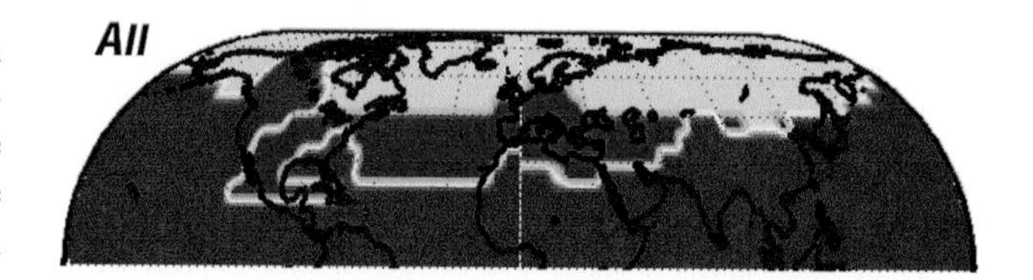

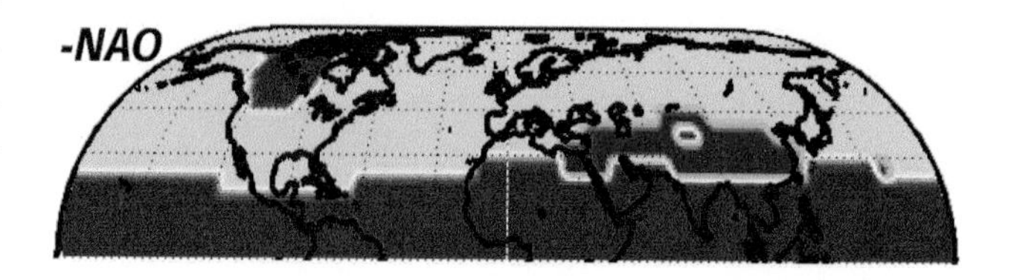

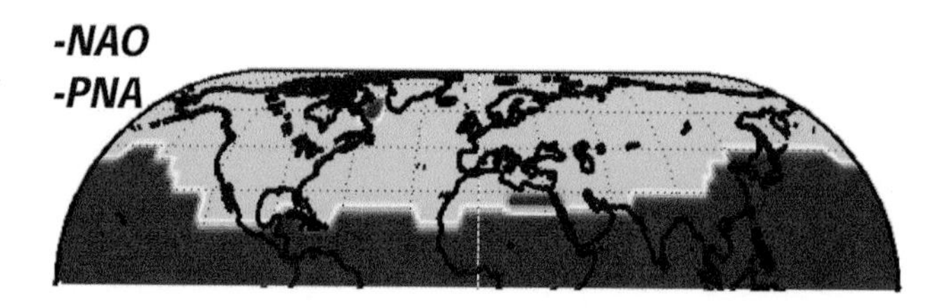

Figure 5. Community structure in the network for the observed 500 hPa height field in (top) the Northern Hemisphere as well as in the networks of synthetic 500 hPa height fields (middle) without the North Atlantic Oscillation (NAO) and (bottom) without the NAO and Pacific North America (PNA) (see text for details).

2.2. Sensitivity Analysis

The above analysis was repeated using detrended data as well as resampled data to obtain grids of equal area. The results were very similar to those in Figure 4. The division into communities for all three networks appears consistent with known dynamics and highlights the aspects of those dynamics that dominate interannual to decadal climate variability. In order to further strengthen the connection between communities and teleconnection patterns, the following experiment was performed: The observed 500 hPa field height in the Northern Hemisphere was considered, and as before, the corresponding network and its community structure were produced. This is the "null model," and its community structure is shown in Figure 5 (top) where three major communities (brown, green, and blue) are observed. Then, following the technique reported by *Tsonis et al.* [2008] and considering that EOF2 of the 500 hPa field is associated with NAO, a 500 hPa height field without the NAO was produced (by projecting the data to all but EOF2), the corresponding network was constructed, and its community structure was found (Figure 5, middle). Similarly, by considering that EOF1 of the 500 hPa filed is associated with the PNA, the PNA is removed as well (by projecting the data to all but EOF1 and EOF2) thereby producing a field without the PNA and without NAO. Then, the network is constructed, its community structure is found (Figure 5, bottom). It is observed that when NAO is removed, the blue community centered in the middle North Atlantic (where NAO occurs) has largely disappeared. A community identified by the blue color is still present but has shifted eastward and away from the area where NAO occurs. The brown community includes the tropics and a spot in North America located where the PNA pattern is found. This spot remains over the same area when the NAO is removed. That this spot is associated with the community, which includes the tropics, is consistent with the fact that the PNA is a linear response to tropical forcing. When the PNA is removed as well, the brown spot disappears. In Figure 5 (bottom), there is no sign of any kind that will be consistent with the PNA or NAO. This indicates that an "alternative model" that does not contain the kind of structure in which we are interested gives results outside the range of behavior produced by the "null model." Thus, attributing communities to teleconnection patterns is justified.

A number of additional network community sensitivity experiments were carried out. These experiments examined the network structure for spatiotemporal variability that is white with respect to time and spatially correlated with a decorrelation length of 3000 km on the sphere. A typical example of the character of the communities resulting from analyzing this type of noise is shown in Figure 6. In general, the following can be observed: (1) The number of communities is roughly the area of the sphere divided by the area of a circle of radius equal to the decorrelation length, i.e., for 3000 km, there are consistently 17–20 communities. (2) The community structure is nonrobust, as different realizations of the noise result in the same qualitative structure of communities (i.e., their number), but their position on the sphere is shifted. Even adding or subtracting 1 "year" from the synthetic data causes a leading order rearrangement in the position of the communities. Neither of these properties is consistent with what we observe for communities in the sea air temperature, sea level pressure, or 500 hPa height fields. In those observed fields, the number of communities is much smaller (about 5), and the structure of the communities is robust to adding or subtracting years from the analysis. Hence, it appears as if the observed community structure is not consistent with the structure arising from spatially correlated Gaussian white noise, and specifically, a null hypothesis that the observed structure can be reproduced using spatially correlated Gaussian white noise can safely be rejected.

2.3. Model Results

Returning to Figure 4, the right plots show the division into communities of networks constructed from model simulation of these three fields. The model used is the Geophysical Fluid Dynamics Laboratory (GFDL) CM2.1 coupled ocean/atmosphere model [*Delworth et al.*, 2006; *Gnanadesikan et al.*, 2006; *Wittenberg et al.*, 2006; *Stouffer et al.*, 2006]. The 1860 preindustrial conditions control run in the years 1950–2005 was used here. The resolution of these fields is also $10° \times 10°$. For the simulated fields, the number of communities is 27, 16, and 249, respectively. However, as with the observations, many of these communities include very few points in the boundaries of what appears to be about five dominant communities. The modularities of the networks are 0.55, 0.49, and 0.80, respectively, indicating again large differences from random networks. Note that because the number of communities is not necessarily the same in the model and in the observations, the color scheme used to show the spatial delineation of the communities is not the

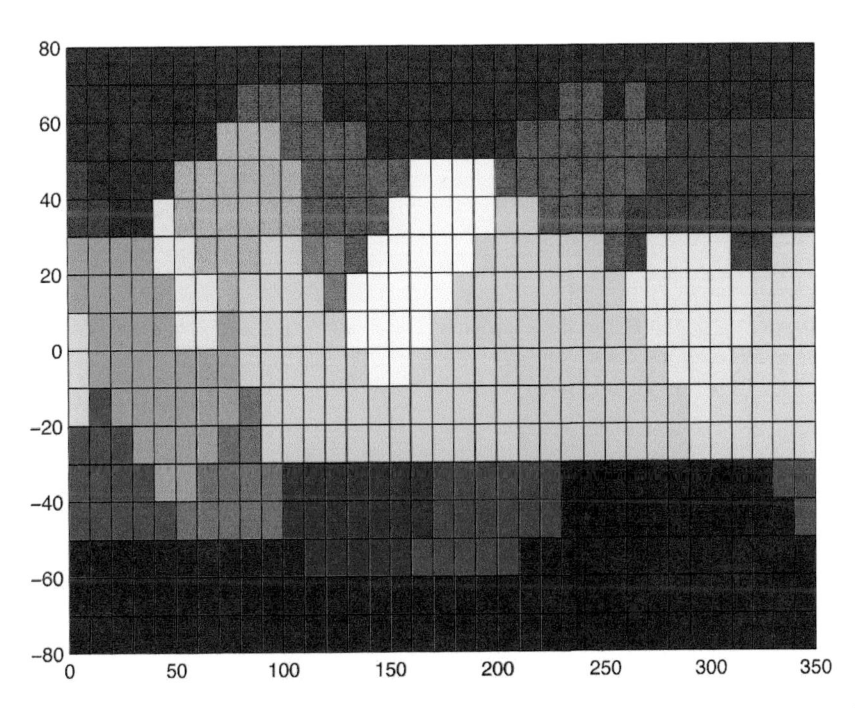

Figure 6. Community structure for a spatiotemporal variability that is white with respect to time and spatially correlated with a decorrelation length of 3000 km on the sphere. This structure is not consistent with the community structures shown in Figure 4. Thus, a null hypothesis that the observed structure in Figure 4 can be reproduced using spatially correlated Gaussian white noise can safely be rejected.

same in both observations and model simulations. This means that the same community may be represented by different colors in the observations and in the model. The reverse may also be possible: the same color may not represent the same community. What is relevant in comparing model simulations with observations in Figure 4 is the spatial distribution and structure of the most highly populated communities regardless of their assigned color.

The observed general subdivision into the three latitudinal zones in the 500 hPa height and sea level pressure networks, as well as the North Hemisphere-South Hemisphere division in the surface temperature network is well captured by the model. However, the detailed structure of the communities, especially in the Northern Hemisphere, is not well captured by the model simulations. This indicates limitations in the models' ability to adequately describe certain dynamical aspects. For example, in the temperature network (Figure 4, bottom right), the model does not reproduce a common community for the North Atlantic and Europe. This is consistent with insufficient resolution of oceanic sea surface temperatures (SSTs) and air-sea fluxes in the model to accurately delineate the moderation over Europe due to the heat transport of the Gulf Stream. Similarly, in the 500 hPa height network (Figure 4, top right), the model has difficulty reproducing the community structure associated with teleconnection patterns, indicating model deficiencies at simulating interannual atmospheric variability. In the sea level pressure network (Figure 4, middle right), it misses the Rossby waves emanation from the Southern Hemisphere to the Northern Hemisphere, marking model deficiencies in simulating the upper atmospheric flow over the tropical Pacific, a well-known problem in climate simulation [*Hack et al.*, 1998]. In spite of these structural differences, both model and observations agree that the number of communities is rather small.

This conclusion was verified by repeating the analysis using two additional different algorithms to divide the networks into communities. The first of these two approaches is based on finding the eigenvalues and eigenvectors of the modularity matrix [*Newman*, 2006]. The modularity matrix of the complete network is a real symmetric matrix B with elements

$$b_{ij} = a_{ij} - \frac{k_i k_j}{2M},$$

where a_{ij} are the elements of the adjacency matrix (equal to one if i and j are connected and zero otherwise), k_i is the number of connections of node i, and M is the total number of connections in the network. The method to split the network begins by estimating the eigenvalues and eigenvectors of the modularity matrix of the complete network. If all the elements in the eigenvector are of the same sign, there is only one community. Otherwise, according to the sign of the elements of this vector, the network is divided into two parts (communities). Next, this process is repeated recursively for each community (using now the community modularity matrix), and so on. The other approach is based on a Bayesian approach to network modularity [*Hofman and Wiggins*, 2008]. The results from these two approaches (not shown) are not significantly different from those in Figure 4. All methods delineate similar community structure with four to five effective communities. Note that due to spatial correlations, community structures do not change significantly when we consider the fields at a higher spatial resolution.

The above analysis brings up the more general question as to whether or not EOF analysis (which is based on variance explained) is indeed the appropriate method to study climate signals or oscillations. If variance is more important than how the system works (i.e., underlying topology), then EOF analysis may be a better approach. Otherwise, approaches like the network approach may be more appropriate. If there is an advantage of using networks, it is that the delineation of the major components is done "holistically" (meaning not in a set of EOFs), and it does not depend on the methodology and the assumptions used in estimating EOFs. The differences between the two methodologies may also account for possible mismatches in space between teleconnection patterns derived using network and EOF analysis. In any case, this is an important area of research, and it should be further pursued.

It thus appears that the full complexity of the climate system is, over the time scales used here, suppressed into a small number of relevant communities (subsystems). These communities involve major teleconnection patterns and climate modes such as the PNA and NAO, communication between the Southern Hemisphere and Northern Hemisphere, storm track dynamics, and the barotropic and baroclinic property of lower and higher latitudes, respectively. Having established the existence of subsystems, we then ask the question: what is their role and can their interaction explain decadal climate variability?

3. INTERACTION BETWEEN SUBSYSTEMS

One of the most important events in recent climate history is the climate shift in the mid-1970s [*Graham*, 1994]. In the Northern Hemisphere 500 hPa atmospheric flow, the shift manifested itself as a collapse of a persistent wave-3 anomaly pattern and the emergence of a strong wave-2 pattern. The shift was accompanied by SST cooling in the central Pacific and warming off the coast of western North America [*Miller et al.*, 1994]. The shift brought sweeping long-range changes in the climate of Northern Hemisphere. Incidentally, after "the dust settled," a new long era of frequent El Niño events superimposed on a sharp global temperature increase begun. While several possible triggers for the shift have been suggested and investigated [*Graham*, 1994; *Miller et al.*, 1994; *Graham et al.*, 1994], the actual physical mechanism that led to this shift is not known. Understanding the dynamics of such phenomena is essential for our ability to make useful prediction of climate change. A major obstacle to this understanding is the extreme complexity of the climate system, which makes it difficult to disentangle causal connections leading to the observed climate behavior. Next, a novel approach is presented, which reveals an important new mechanism in climate dynamics and explains several aspects of the observed climate variability in the late twentieth century.

First, a network from four major climate indices was constructed. The indices represent the PDO, the NAO, the ENSO, and the North Pacific Index (NPI) [*Barnston and Livezey*, 1987; *Hurell*, 1995; *Mantua et al.*, 1997; *Trenberth and Hurrell*, 1994]. These indices represent regional but dominant modes of climate variability, with time scales ranging from months to decades. NAO and NPI are the leading modes of surface pressure variability in northern Atlantic and Pacific Oceans, respectively, the PDO is the leading mode of SST variability in the northern Pacific, and ENSO is a major signal in the tropics. Together, these four modes capture the essence of climate variability in the Northern Hemisphere. Each of these modes is assumed to represent a subsystem involving different mechanisms over different geographical regions. Indeed, some of their dynamics have been adequately explored and explained by simplified models, which represent subsets of the complete climate system and which are governed by their own dynamics [*Elsner and Tsonis*, 1993; *Schneider et al.*, 2002; *Marshall et al.*, 2001; *Suarez and Schopf*, 1998]. For example, ENSO has been modeled by a simplified delayed oscillator in which the slower adjustment timescales of the ocean supply the system with the memory essential to oscillation. Monthly mean values in the interval 1900–2000 are available for all indices (see http://climatedataguide.ucar.edu/guidance/north-pacific-index-npi-trenberth-and-hurrell-monthly-and-winter).

These four climate indices are assumed to form a network of interacting nodes [*Tsonis et al.*, 2007]. A commonly used measure to describe variations in the network's topology is the mean distance $d(t)$ [*Onnela et al.*, 2005]:

$$d(t) = \frac{2}{N(N-1)} \sum_{d^t_{ij} \in D^t} d^t_{ij}. \tag{1}$$

Here t denotes the time in the middle of a sliding window of width Δt, $N = 4$; $i, j = 1, \ldots, N$, and $d^t_{ij} = \sqrt{2\left(1 - |\rho^t_{ij}|\right)}$, where ρ^t_{ij} is the cross-correlation coefficient between nodes i and j in the interval $[t - (\Delta t - 1)/2, t + (\Delta t - 1)/2]$, and D^t is the $N \times N$ distance matrix. The sum is taken over the upper triangular part (or the distinct elements of D^t). The above formula uses the absolute value of the correlation coefficient because the choice of sign of indices is arbitrary. The distance can be thought of as the average correlation between all possible pairs of nodes and is interpreted as a measure of the synchronization of the network's components. Synchronization between nonlinear (chaotic) oscillators occurs when their corresponding signals converge to a common, albeit irregular, signal. In this case, the signals are identical, and their cross-correlation is 1. Thus, a distance of zero corresponds to a complete synchronization, and a distance of $\sqrt{2}$ signifies a set of uncorrelated nodes.

Figure 7a shows the distance as a function of time for a window length of $\Delta t = 11$ years, with tick marks corresponding to the year in the middle of the window. The correlations (and thus distance values for each year) were computed based on the annual mean indices constructed by averaging the monthly indices over the period of November–March. The dashed line parallel to the time axis in Figure 7a represents the 95% significance level associated with the null hypothesis that the observed indices are sampled from a population of a 4-D autoregressive-1 (AR-1) process driven by a spatially (cross-index) correlated Gaussian noise; the parameters of the AR-1 model and the covariance matrix of the noise are derived from the full time series of the observed indices. This test assumes that the variations of the distance with time seen in Figure 7a are due to sampling associated with a finite-length (11 years) sliding window used to compute the local distance values. Retaining overall cross-correlations in constructing the surrogates makes this test very stringent. Nevertheless, we still find that at five times (1910s, 1920s, 1930s, 1950s, and 1970s), distance variations fall below the 95% significance level. We therefore conclude that these features are not likely to be due to sampling limitations, but they represent statistically significant synchronization events. Note that the window length used in Figure 7a is a compromise between being long enough to estimate correlations but not too long to "dilute" transitions. Nevertheless, the observed synchronizations are insensitive to the window size in a wide range of 7 years $\leq \Delta t \leq$ 15 years.

An important aspect in the theory of synchronization between coupled nonlinear oscillators is coupling strength. It is vital to note that synchronization and coupling are not interchangeable; for example, it is trivial to construct a pair of coupled simple harmonic oscillators whose displacements

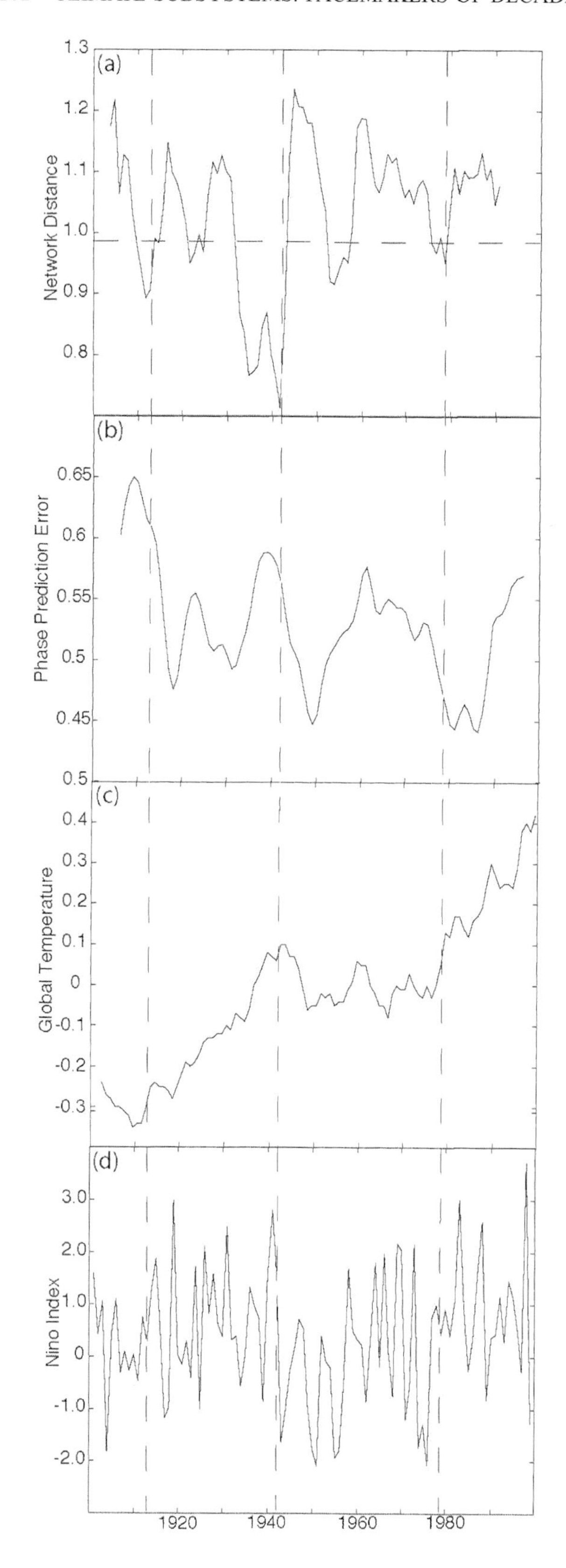

are in quadrature (and, hence, perfectly uncorrelated), but whose phases are strongly coupled [*Vanassche et al.*, 2003]. As such, coupling is best measured by how strongly the phases of different modes of variability are linked. The theory of synchronized chaos predicts that in many cases when such systems synchronize, an increase in coupling between the oscillators may destroy the synchronous state and alter the system's behavior [*Heagy et al.*, 1995; *Pecora et al.*, 1997]. In view of the results above, the question thus arises as to how the synchronization events in Figure 7a relate to coupling strength between the nodes. It should be noted that in this study, the focus is in the complete synchronization among the nodes, rather than weaker types of synchronization, such as phase synchronization [*Boccaletti et al.*, 2002; *Maraun and Kurths*, 2005] or clustered synchronization [*Zhou and Kurths*, 2006], which are also important in climate interactions.

For our purposes here, if future changes in the phase between pairs of climate modes can be readily predicted using only information about the current phase, those modes may be considered strongly coupled [*Smirnov and Bezruchko*, 2003]. Here in order to study coupling, symbolic dynamics are used. For any given time series point, we can define a symbolic phase by examining the relationship between that point and its nearest two neighbors in time. As shown in Figure 8, if the three points are sequentially increasing, we can assign to the middle point a phase of 0 while, if they are sequentially decreasing, a phase of π. Intermediate values then follow. Notice that this procedure is totally nonparametric, as it does not compare the actual values of the points aside from whether a point is larger or smaller than its neighbors. The advantage of this approach is that it is blind to ultralow-frequency variability, i.e., decadal scale and longer. Use of symbolic dynamics is appropriate in this case, as we are primarily interested in changes in the synchronicity and coupling of climate modes over decadal time scales. The symbolic phase ϕ_n^j is constructed separately for the four climate indices, where j denotes the index, and n denotes the

Figure 7. (opposite) (a) The distance (see definition in text) of a network consisting of four observed major climate modes as a function of time. This distance is an indication of synchronization between the modes with smaller distance implying stronger synchronization. The parallel dashed line represents the 95% significance level associated with a null hypothesis of spatially correlated red noise. (b) Coupling strength between the four modes as a function of time. (c) The global surface temperature record. (d) Global sea surface temperature El Niño–Southern Oscillation index. The vertical lines indicate the time when the network goes out of synchronization for those cases where synchronization is followed by a coupling strength increase.

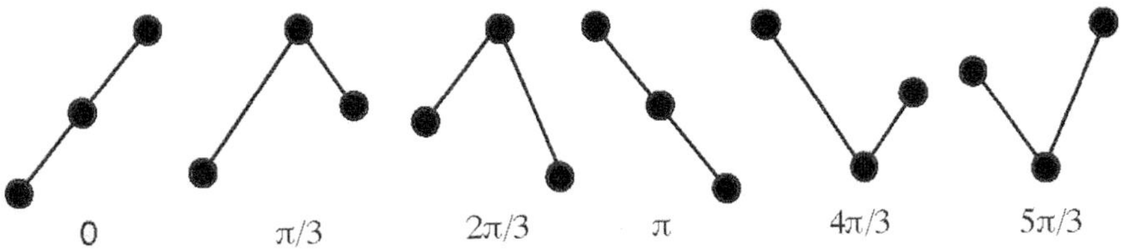

Figure 8. The six states for the symbolic phase construction. The points in each triplet correspond to three consecutive points in a time series, and their relative vertical positions to each other indicate their respective values.

year. The phases for a given year n are represented by the complex phase vector $\overrightarrow{Z_n}$ with elements $Z_n^j = \exp(i\phi_n^j)$. The predictability of this phase vector from year to year provides a measure of the coupling and is determined using the least squares estimator

$$\overrightarrow{Z_{n+1}^{est}} = \mathbf{M}\overrightarrow{Z_n}, \qquad (2)$$

where $\mathbf{M} = [\mathbf{Z}_{+}\mathbf{Z}^{\mathrm{T}}]\,[\mathbf{Z}\mathbf{Z}^{\mathrm{T}}]^{-1}$ is the least squares predictor. Here $\mathbf{Z}$ and $\mathbf{Z}_{+}$ are the matrices whose columns are the vectors $\overrightarrow{Z_n}$ and $\overrightarrow{Z_{n+1}}$, respectively, constructed using allyears. A measure of the coupling then is simply $||\overrightarrow{Z_{n+1}^{est}} - \overrightarrow{Z_{n+1}}||^2$, where strong coupling is associated with small values of this quantity, i.e., good phase prediction. Note that only three values are used to define phases rather than four or five or any other number. The reason is that the possible number of permutations of m values is $m!$ Thus, if $m > 3$, there is at least 24 possible permutations, which will not result in large sample sizes to evaluate the predictability of the phase vector.

This quantity is plotted in Figure 7b. Figures 7c and 7d show the global surface temperature (http://data.giss.nasa.gov/gistemp/) and El Niño index in our period. Figure 7 tells a remarkable story. First, let us consider the event in the 1910s. The network synchronizes at about 1910. At that time, the coupling strength begins to increase. Eventually, the network comes out of the synchronous state sometime in late 1912, early 1913 (marked by the left vertical line). The destruction of the synchronous state coincides with the beginning of a sharp global temperature increase and a tendency for more frequent and strong El Niño events. The network enters a new synchronization state in the early 1920s, but this is not followed by an increase in coupling strength. In this case, no major shifts are observed in the behavior of global temperature and ENSO. Then, the system enters a new synchronization state in the early 1930s. Initially, this state was followed by a decrease in coupling strength, and again, no major shifts are observed. However, in the early 1940s, the still present synchronous state is subjected to an increase in coupling strength, which soon destroys it (at the time indicated by the middle vertical line). As the synchronous state is destroyed, a new shift in both temperature trend and ENSO variability is observed. The global temperature enters a cooling regime, and El Niño events become much less frequent and weaker. The network synchronizes again in the early 1950s. This state is followed by a decrease in coupling strength, and as was the case in 1920s, no major shifts occur. Finally, the network synchronizes again in the mid-1970s. This state is followed by an increase in coupling strength, and incredibly, as in the cases of 1910s and 1940s, synchronization is destroyed (at the time marked by the right vertical line), and then, climate shifts again. The global temperature enters a warming regime, and El Niño events become frequent and strong. The fact that around 1910, 1940, and in the late 1970s climate shifted to a completely new state indicates that synchronization followed by an increase in coupling between the modes leads to the destruction of the synchronous state and the emergence of a new state.

The above mechanism was also found in three climate simulations. The first two are from the GFDL CM2.1 coupled ocean/atmosphere model [*Delworth et al.*, 2006; *Gnanadesikan et al.*, 2006; *Wittenberg et al.*, 2006; *Stouffer et al.*, 2006]. The first simulation is an 1860 preindustrial condition, 500 year control run, and the second is the SRESA1B, which is a "business as usual" scenario with CO_2 levels stabilizing at 720 ppmv at the close of the twenty-first century [*Intergovernmental Panel on Climate Change*, 2001]. The third simulation is a control run from the ECHAM5 model [*Tsonis et al.*, 2007; *Wang et al.*, 2009]. From these model outputs, we constructed the same indices (in the periods of 100–200 years, twenty-first century, and years 240–340, respectively) and repeated the above procedure to study synchronization and coupling in the corresponding networks. In all, we found seven synchronization events in the model simulations. As with the five synchronization events in the observations in the twentieth century shown in Figure 7, here as well, without an exception in all cases when the major modes of variability in the Northern Hemisphere are synchronized, an increase in the coupling strength destroys the synchronous state and causes climate to shift to a new state. Importantly, the mechanism is found in both forced and unforced simulations indicating that the mechanism is a result of the natural variability of the climate system. Lately, *Swanson and Tsonis* [2009] extended the analysis to the updated observations in the twenty-first century and discovered yet another consistent event signaling a new climate shift in the beginning of the twenty-first century.

The shifts described above are based on careful visual examination of the results. Once shifts have been visually identified, one can statistically test their significance. From the above results, it was observed [*Tsonis et al.*, 2007; *Wang et al.*, 2009] that most often a shift in global temperature can manifest itself as a trend change, but in a couple of cases, it shows as a jump. Changes in ENSO variability, on the other hand, can come in more ways. In this case, the possible regimes are five. A regime of more frequent El Niño events, a regime of more frequent La Niña events, a regime of alternating strong El Niño and La Niña events, a regime of no activity or alternating weak El Niño and La Niña events, and a regime where the spacing between El Niño and La Niña events is irregular. In all those regimes, the distribution of ENSO index is different, and as such, the Mann-Whitney rank sum test can be used to test for differences before and during a shift or between shifts. The same test can be used to test differences in global temperature tendency before and after a jump. In cases when a temperature tendency shift manifests itself as a trend change, the t test can be used. In all, 12 synchronization events and eight shifts occurred in observations and model simulations (not including the suspected shift in the twenty-first century observations). For all shifts (three in observations and five in the models), it was found that the change in ENSO variability is significant at the 90% or higher confidence level, whereas the change in temperature tendency is significant at the 95% or higher confidence level (see supplementary material of *Wang et al.* [2009]).

The above results refer to the collective behavior of the four major modes used in the network. As such, they do not offer insights on the specific details of the mechanism. For example, do small distance values (strong synchronization) result from all modes synchronizing or from a subset of them? When the network is synchronized, does the coupling increase require that all modes must become coupled with each other? To answer these questions, *Wang et al.* [2009] split the network of four modes into its six pair components and investigated the contribution of each pair in each synchronization event and in the overall coupling of the network. It was found that one mode is behind all climate shifts. This mode is the NAO. This North Atlantic mode is without exception the common ingredient in all shifts, and when it is not coupled with any of the Pacific modes, no shift ensues. In addition, in all cases where a shift occurs, NAO is necessarily coupled to North Pacific. In some cases, it may also be coupled to the tropical Pacific (ENSO) as well, but in none of the cases, NAO is only coupled to ENSO. Thus, results indicate that not only NAO is the instigator of climate shifts but also that the likely evolution of a shift has a path where the North Atlantic couples to the North Pacific, which, in turn, couples to the tropics. Solid dynamical arguments and past work offer a concrete picture of how the physics may play out. NAO with its huge mass rearrangement in North Atlantic affects the strength of the westerly flow across midlatitudes. At the same time through its "twin," the AO, it impacts sea level pressure patterns in the northern Pacific. This process is part of the so-called intrinsic midlatitude Northern Hemisphere variability [*Vimont et al.*, 2001, 2003]. Then, this intrinsic variability through the seasonal footprinting mechanism [*Vimont et al.*, 2001, 2003] couples with equatorial wind stress anomalies, thereby acting as a stochastic forcing of ENSO. This view is also consistent with a recent study showing that PDO modulates ENSO [*Gershunov and Barnett*, 1998; *Verdon and Franks*, 2006]. Another possibility of how NAO couples to the North Pacific may be through the five-lobe circumglobal waveguide pattern [*Branstator*, 2002]. It has been shown that this waveguide pattern projects onto NAO indices, and its features contribute to variability at locations throughout the Northern Hemisphere. Finally, the North Atlantic variations have been linked to the Northern Hemisphere mean surface temperature multidecadal variability through redistribution of heat within the northern Atlantic with the other oceans left free to adjust to these Atlantic variations [*Zhang et al.*, 2007]. Thus, NAO, being the major mode of variability in the northern Atlantic, impacts both ENSO variability and global temperature variability. Recently, a study has shown how ENSO with its effects on PNA can, through vertical propagation of Rossby waves, influence the lower stratosphere and how in turn the stratosphere can influence NAO through downward progression of Rossby wave [*Ineson and Scaife*, 2009]. These results, coupled with our results, suggest the following 3-D superloop: NAO → PDO → ENSO → PNA → stratosphere → NAO, which captures the essence of decadal variability in the Northern Hemisphere and possibly the globe.

It is interesting to compare Figure 2 and the top left plot of Figure 4. Apparently, there are similarities (the three-zone separation, for example), but the community algorithm identifies NAO clearly, whereas in Figure 2, as we mentioned earlier, NAO is masked. Owing to barotropic conditions in the tropical areas, communication via gravity waves is fast, and as result, the information flows very efficiently resulting in a fully connected network in the tropics. In the extratropics, supernodes are found in locations where major teleconnection patterns are found, which, in turn, define distinct communities in the network. It may be that in spatially extended systems with spatial correlations extending over a

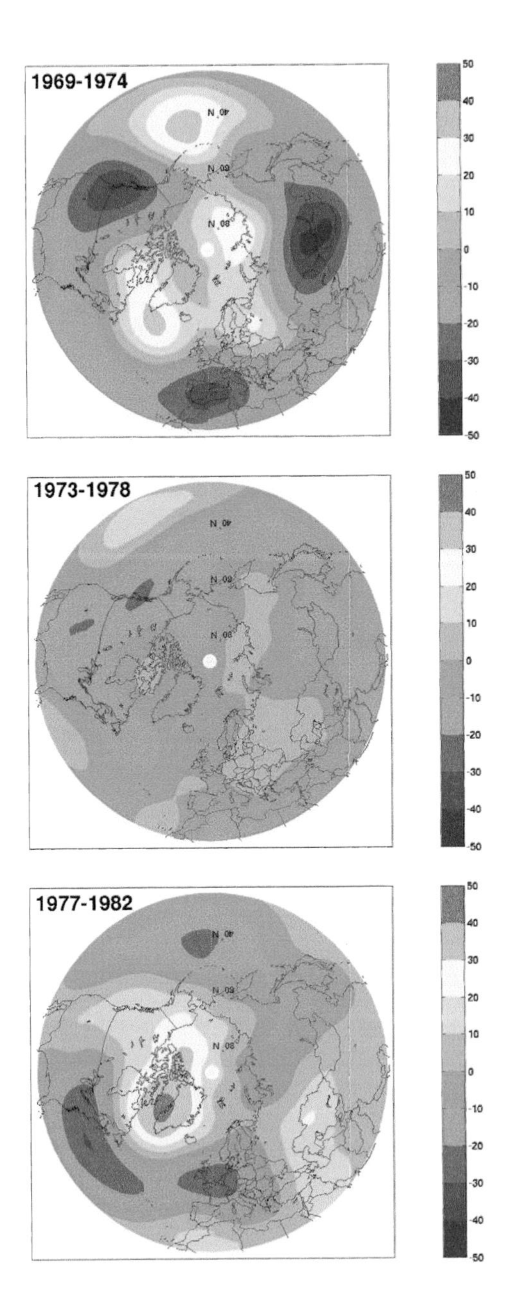

Figure 9. The 500 hPa anomaly field composites for the periods (top) 1969–1974, (middle) 1973–1978, and (bottom) 1977–1982. A wave-3 pattern is visible in the top plot with PNA and NAO in its negative phase being present. In the middle plot, both NAO and PNA have, for all practical purposes, disappeared. In the bottom plot, the field emerges as a wave-2 pattern with NAO in its positive phase. As we explain in the text, this transition (known as the climate shift of the 1970s) is consistent with our conjecture that removal of supernodes makes the (climate) network unstable and more prone to failure (breakdown of a regime and emergence of another regime).

characteristic scale, the connectivity pattern is related to community structure. In any case, since the presence of super nodes makes the network stable and efficient in transferring information, it was speculated and shown that, indeed, teleconnection patterns act as climate stabilizers. *Tsonis et al.* [2008] have shown that removal of teleconnection patterns from the climate system results in less stable networks, which make an existing climate regime unstable and more likely to shift to a new regime. Indeed, they showed that this process may be behind the climate shift of the 1970s and related to the dynamical mechanism for major climate shifts discussed above. Figure 9 shows 500 hPa anomaly composites for three 5 year periods in the 1970s and early 1980s. In the early 1970s (Figure 9, top), the 500 hPa anomaly field is dominated by the presence of a wave-3 pattern with both the PNA and NAO (in its negative phase) being very pronounced. In the mid-1970s (Figure 9, middle), this field is very weak, and both NAO and PNA have, for all practical purposes, disappeared. After that (Figure 9, bottom), the field becomes strong again, but a new wave-2 pattern with a very pronounced positive NAO has emerged. This shift is known as the climate shift of the 1970s. According to the *Tsonis et al.* [2007] mechanism for major climate shifts, climate modes may synchronize. Once in place, the synchronized state may become unstable and shift to a new state. The results of *Tsonis et al.* [2008] and those in Figure 9 suggest a connection between stability, synchronization, coupling of major climate modes, and climate shifts. This point is the subject of our continuing work in this area, and more results will be forthcoming in the future.

As a final note before concluding, it should be mentioned that lately, *Tsonis and Swanson* [2011] extended their approach to consider proxy data for climate modes going back several centuries. While noise in the proxy data, in some cases, masks the mechanism, it was found that significant coherence between both synchronization and coupling and global temperature exists. These results provide further support that the mechanism for climate shifts discussed here is a robust feature of the climate system.

4. CONCLUSIONS

The above synthesis describes some new approaches that have been applied lately to climate data. The findings presented here may settle the issue of dimensionality of climate variability over decadal scales, as they support the view that over these scales, climate collapses into distinct subsystems whose interplay dictates decadal variability. At the same time, these results provide clues as to what these subsystems might be. As such, while "weather" may be complicated, "climate" may be complex but not complicated. Moreover, it appears that the interaction between these subsystems may be largely responsible for the observed decadal climate variability. A direct consequence of these results is that a dynamical reconstruction directly from a small number of climate modes/subsystems may be attempted to extract differential equations, which model the network of major modes. Such an approach may provide an alternative and direct window to study low-frequency variability in climate. Work in this area is in progress and will be reported in the future elsewhere.

Acknowledgment. This work was supported by NSF grant AGS-0902564.

REFERENCES

Albert, R., and A.-L. Barabasi (2002), Statistical mechanics of complex networks, *Rev. Mod. Phys.*, *74*, 47–101.

Albert, R., H. Jeong, and A.-L. Barabasi (1999), Diameter of the World Wide Web, *Nature*, *401*, 130–131.

Ambaum, M. H. P., B. J. Hoskins, and D. B. Stephenson (2001), Arctic Oscillation or North Atlantic Oscillation?, *J. Clim.*, *14*, 3495–3507.

Arenas, A., A. Diaz-Guilera, and C. J. Perez-Vicente (2006), Synchronization reveals topological scales in complex networks, *Phys. Rev. Lett.*, *96*, 114102, doi:10.1103/PhysRevLett.96.114102.

Barabasi, A.-L., and E. Bonabeau (2003), Scale-free networks, *Sci. Am.*, *288*, 60–69.

Barnston, A. G., and R. E. Livezey (1987), Classification, seasonality, and persistence of low-frequency atmospheric circulation patterns, *Mon. Weather Rev.*, *115*, 1083–1126.

Branstator, G. (2002), Circumglobal teleconnections, the jet stream waveguide, and the North Atlantic Oscillation, *J. Clim.*, *15*, 1893–1910.

Boccaletti, S., J. Kurths, G. Osipov, D. J. Valladares, and C. S. Zhou (2002), The synchronization of chaotic systems, *Phys. Rep.*, *366*, 1–101.

Bouchaud, J.-P., and M. Mezard (2000), Wealth condensation in a simple model of economy, *Physica A*, *282*, 536–540.

Costa, L. D. F., F. A. Rodrigues, G. Travieso, and P. R. Villas Boas (2007), Characterization of complex networks: A survey of measurements, *Adv. Phys.*, *56*, 167–242.

Delworth, T. L., et al. (2006), GFDL's CM2 global coupled climate models. Part I: Formulation and simulation characteristics, *J. Clim.*, *19*, 643–674.

Eichler, T., and R. W. Higgins (2006), Climatology and ENSO-related variability of North American extratropical cyclone activity, *J. Clim.*, *19*, 2076–2093.

Elsner, J. B., and A. A. Tsonis (1993), Nonlinear dynamics established in the ENSO, *Geophys. Res. Lett.*, *20*(3), 213–216.

Elsner, J. B., T. H. Jagger, and E. A. Fogarty (2009), Visibility network of United States hurricanes, *Geophys. Res. Lett.*, *36*, L16702, doi:10.1029/2009GL039129.

Farkas, I. J., H. Jeong, T. Vicsek, A.-L. Barabási, and Z. N. Oltvai (2003), The topology of the transcription regulatory network in the yeast *Saccharomyces cerevisiae*, *Physica A*, *318*, 601–612.

Favre, A., and A. Gershunov (2006), Extra-tropical cyclonic/anticyclonic activity in north-eastern Pacific and air temperature extremes in western North America, *Clim. Dyn.*, *26*, 617–629.

Favre, A., and A. Gershunov (2009), North Pacific cyclonic and anticyclonic transients in a global warming context: Possible consequences for western North American daily precipitation and temperature extremes, *Clim. Dyn.*, *32*, 969–987.

Gershunov, A., and T. P. Barnett (1998), Interdecadal modulation of ENSO teleconnections, *Bull. Am. Meteorol. Soc.*, *79*, 2715–2725.

Girvan, M., and M. E. J. Newman (2002), Community structure in social and biological networks, *Proc. Natl. Acad. Sci. U. S. A.*, *99*, 7821–7826.

Gnanadesikan, A., et al. (2006), GFDL's CM2 global coupled climate models. Part II: The baseline ocean simulation, *J. Clim.*, *19*, 675–697.

Gozolchiani, A., K. Yamasaki, O. Gazit, and S. Havlin (2008), Pattern of climate network blinking links follow El Niño events, *Europhys. Lett.*, *83*, 28005, doi:10.1209/0295-5075/83/28005.

Graham, N. E. (1994), Decadal scale variability in the tropical and North Pacific during the 1970s and 1980s: Observations and model results, *Clim. Dyn.*, *10*, 135–162.

Graham, N. E., T. P. Barnett, R. Wilde, M. Ponater, and S. Schubert (1994), On the roles of tropical and mid-latitude SSTs in forcing interannual to interdecadal variability in the winter Northern Hemisphere circulation, *J. Clim.*, *7*, 1500–1515.

Guimera, R., and L. A. N. Amaral (2005), Functional cartography of complex metabolic network, *Nature*, *433*, 895–900.

Hack, J. J., J. T. Kiehl, and J. W. Hurrell (1998), The hydrologic and thermodynamic characteristics of NCAR CCM3, *J. Clim.*, *11*, 1179–1206.

Heagy, J. F., L. M. Pecora, and T. L. Carroll (1995), Short wavelength bifurcations and size instabilities in coupled oscillator systems, *Phys. Rev. Lett.*, *74*, 4185–4188.

Held, I. M., M. Ting, and H. Wang (2002), Northern winter stationary waves: Theory and modeling, *J. Clim.*, *15*, 2125–2144.

Hofman, J. M., and C. H. Wiggins (2008), A Bayesian approach to network modularity, *Phys. Rev. Lett.*, *100*, 258701, doi:10.1103/PhysRevLett.100.258701.

Holme, P., M. Huss, and H. Jeong (2003), Subnetwork hierarchies of biochemical pathways, *Bioinformatics*, *19*, 532–543.

Huang, J., K. Higuchi, and A. Shabbar (1998), The relationship between the North Atlantic Oscillation and El Niño-Southern Oscillation, *Geophys. Res. Lett.*, *25*(14), 2707–2710.

Hurrell, J. W. (1995), Decadal trends in the North Atlantic oscillation regional temperature and precipitation, *Science*, *269*, 676–679.

Ineson, S., and A. A. Scaife (2009), The role of the stratosphere in the European climate response to El Niño, *Nature Geosci.*, *2*, 32–36.

Intergovernmental Panel on Climate Change (2001), *Climate Change 2001: The Scientific Basis: Contribution of Working Group I to the Third Assessment Report of the Intergovernmental*

Panel on Climate Change, edited by J. T. Houghton et al., 881 pp., Cambridge Univ. Press, New York.

Jeong, H., S. Mason, A.-L. Barabasi, and Z. N. Oltvai (2001), Lethability and centrality in protein networks, *Nature*, *411*, 41–42.

Kistler, R., et al. (2001), The NCEP/NCAR 50-year reanalysis: Monthly means, CD-ROM and documentation, *Bull. Am. Meteorol. Soc.*, *82*, 247–267.

Liljeros, F., C. Edling, L. N. Amaral, H. E. Stanley, and Y. Aberg (2001), The web of human sexual contacts, *Nature*, *411*, 907–908.

Lorenz, E. N. (1991), Dimension of weather and climate attractors, *Nature*, *353*, 241–244.

Mantegna, R. N. (1999), Hierarchical structure in financial markets, *Eur. Phys. J. B*, *11*, 193–197.

Mantua, N. J., S. R. Hare, Y. Zhang, J. M. Wallace, and R. C. Francis (1997), A Pacific interdecadal climate oscillation with impacts on salmon production, *Bull. Am. Meteorol. Soc.*, *78*, 1069–1079.

Maraun, D., and J. Kurths (2005), Epochs of phase coherence between El Niño/Southern Oscillation and Indian monsoon, *Geophys. Res. Lett.*, *32*, L15709, doi:10.1029/2005GL023225.

Marshall, J., et al. (2001), North Atlantic climate variability: Phenomena, impacts and mechanisms, *Int. J. Climatol.*, *21*, 1863–1898.

Miller, A. J., D. R. Cayan, T. P. Barnett, N. E. Craham, and J. M. Oberhuber (1994), The 1976–77 climate shift of the Pacific Ocean, *Oceanography*, *7*, 21–26.

Mo, K. C., and R. W. Higgins (1998), The Pacific-South America modes and tropical convection during the Southern Hemisphere winter, *Mon. Weather Rev.*, *126*, 1581–1596.

Nerenberg, M. A. H., and C. Essex (1990), Correlation dimension and systematic geometric effects, *Phys. Rev. A*, *42*, 7065–7074.

Newman, M. E. J. (2006), Modularity and community structure in networks, *Proc. Natl. Acad. Sci. U. S. A.*, *103*, 8577–8582.

Newman, M. E. J., and M. Girvan (2004), Finding and evaluating community structure in networks, *Phys. Rev. E*, *69*, 026113, doi:10.1103/PhysRevE.69.026113.

Nicolis, C., and G. Nicolis (1984), Is there a climatic attractor?, *Nature*, *311*, 529–532.

Onnela, J.-P., J. Saramaki, J. Kertesz, and K. Kaski (2005), Intensity and coherence of motifs in weighted complex networks, *Phys. Rev. E*, *71*, 065103, doi:10.1103/PhysRevE.71.065103.

Pastor-Satorras, R., and A. Vespignani (2001), Epidemic spreading in scale-free networks, *Phys. Rev. Lett.*, *86*, 3200–3203.

Pecora, L. M., T. L. Carroll, G. A. Johnson, and D. J. Mar (1997), Fundamentals of synchronization in chaotic systems, concepts, and applications, *Chaos*, *7*, 520–543.

Pozo-Vazquez, D., M. J. Esteban-Parra, F. S. Rodrigo, and Y. Castro-Diez (2001), The association between ENSO and winter atmospheric circulation and temperature in the North Atlantic region, *J. Clim.*, *14*, 3408–3420.

Schneider, N., A. J. Miller, and D. W. Pierce (2002), Anatomy of North Pacific decadal variability, *J. Clim.*, *15*, 586–605.

Smirnov, D. A., and B. P. Bezruchko (2003), Estimation of interaction strength and direction from short and noisy time series, *Phys. Rev.*, *68*, 046209, doi:10.1103/PhysRevE.68.046209.

Smith, L. A. (1988), Intrinsic limits on dimension calculations, *Phys. Lett. A*, *133*, 283–288.

Song, C., S. Havlin, and H. A. Makse (2005), Self-similarity of complex networks, *Nature*, *433*, 392–395.

Stouffer, R. J., et al. (2006), GFDL's CM2 global coupled climate models. Part IV: Idealized climate response, *J. Clim.*, *19*, 723–740.

Strogatz, S. H. (2001), Exploring complex networks, *Nature*, *410*, 268–276.

Suarez, M. J., and P. S. Schopf (1998), A delayed action oscillator for ENSO, *J. Atmos. Sci.*, *45*, 549–566.

Swanson, K. L., and A. A. Tsonis (2009), Has the climate recently shifted?, *Geophys. Res. Lett.*, *36*, L06711, doi:10.1029/2008GL037022.

Thompson, D. W. J., and J. M. Wallace (1998), The Arctic Oscillation signature in the wintertime geopotential height and temperature fields, *Geophys. Res. Lett.*, *25*(9), 1297–1300.

Trenberth, K. E., and J. W. Hurrell (1994), Decadal atmospheric-ocean variations in the Pacific, *Clim. Dyn.*, *9*, 303–319.

Tsonis, A. A. (1992), *Chaos: From Theory to Applications*, 274 pp., Plenum, New York.

Tsonis, A. A., and J. B. Elsner (1988), The weather attractor over very short timescales, *Nature*, *33*, 545–547.

Tsonis, A. A., and J. B. Elsner (1989), Chaos, strange attractors and weather, *Bull. Am. Meteorol. Soc.*, *70*, 16–23.

Tsonis, A. A., and J. B. Elsner (1996), Mapping the channels of communication between the tropics and higher latitudes in the atmosphere, *Physica D*, *92*, 237–244.

Tsonis, A. A., and K. L. Swanson (2008), Topology and predictability of El Niño and La Niña networks, *Phys. Rev. Lett.*, *100*, 228502, doi:10.1103/PhysRevLett.100.228502.

Tsonis, A. A., and K. L. Swanson (2011), Climate mode co-variability and climate shifts, *Int. J. Bifurcat. Chaos*, *21*, 3549–3556, doi:10.1142/S0218127411030714.

Tsonis, A. A., G. N. Triantafyllou, and J. B. Elsner (1994), Searching for determinism in observed data: A review of the issues involved, *Nonlinear Processes Geophys.*, *1*, 12–25.

Tsonis, A. A., K. L. Swanson, and P. J. Roebber (2006), What do networks have to do with climate?, *Bull. Am. Meteorol. Soc.*, *87*, 585–595, doi:10.1175/BAMS-87-5-585.

Tsonis, A. A., K. Swanson, and S. Kravtsov (2007), A new dynamical mechanism for major climate shifts, *Geophys. Res. Lett.*, *34*, L13705, doi:10.1029/2007GL030288.

Tsonis, A. A., K. L. Swanson, and G. Wang (2008), On the role of atmospheric teleconnection in climate, *J. Clim.*, *21*, 2990–3001.

Tsonis, A. A., G. Wang, K. L. Swanson, F. A. Rodrigues, and L. D. F. Costa (2010), Community structure and dynamics in climate networks, *Clim. Dyn.*, *37*, 933–940, doi:10.1007/s00382-010-0874-3.

Vanassche, P., G. G. E. Gielen, and W. Sansen (2003), Behavioral modeling of (coupled) harmonic oscillators, *IEEE Trans. Comput.-Aided Design Integrated Circuits Syst.*, *22*, 1017–1027.

Verdon, D. C., and S. W. Franks (2006), Long-term behaviour of ENSO: Interactions with the PDO over the past 400 years inferred from paleoclimate records, *Geophys. Res. Lett.*, *33*, L06712, doi:10.1029/2005GL025052.

Vimont, D. J., D. S. Battisti, and A. C. Hirst (2001), Footprinting: A seasonal connection between the tropics and mid-latitudes, *Geophys. Res. Lett.*, *28*, 3923–3926.

Vimont, D. J., J. M. Wallace, and D. S. Battisti (2003), The seasonal footprinting mechanism in the Pacific: Implications for ENSO, *J. Clim.*, *16*, 2668–2675.

Wallace, J. M., and D. S. Gutzler (1981), Teleconnections in the geopotential height field during the Northern Hemisphere winter, *Mon. Weather Rev.*, *109*, 784–812.

Wang, G., K. L. Swanson, and A. A. Tsonis (2009), The pacemaker of major climate shifts, *Geophys. Res. Lett.*, *36*, L07708, doi:10.1029/2008GL036874.

Wang, H., and R. Fu (2000), Influence of ENSO SST anomalies and water storm-tracks on the interannual variability of the upper tropospheric water vapor over the Northern Hemisphere extratropics, *J. Clim.*, *13*, 59–73.

Watts, D. J., and S. H. Strogatz (1998), Collective dynamics of 'small-world' networks, *Nature*, *393*, 440–442.

Webster, P. J., and J. R. Holton (1982), Wave propagation through a zonally varying basic flow: The influences of mid-latitude forcing in the equatorial regions, *J. Atmos. Sci.*, *39*, 722–733.

Wittenberg, A. T., A. Rosati, N.-C. Lau, and J. J. Ploshay (2006), GFDL's CM2 global coupled climate models. Part III: Tropical Pacific climate and ENSO, *J. Clim.*, *19*, 698–722.

Yamasaki, K., A. Gozolchiani, and S. Havlin (2008), Climate networks around the globe are significantly affected by El Niño, *Phys. Rev. Lett.*, *100*, 228501, doi:10.1103/PhysRevLett.100.228501.

Zhang, R., T. L. Delworth, and I. M. Held (2007), Can the Atlantic Ocean drive the observed multidecadal variability in Northern Hemisphere mean temperature?, *Geophys. Res. Lett.*, *34*, L02709, doi:10.1029/2006GL028683.

Zhou, C. S., and J. Kurths (2006), Dynamical weights and enhanced synchronization in adaptive complex networks, *Phys. Rev. Lett.*, *96*, 164102, doi:10.1103/PhysRevLett.96.164102.

A. A. Tsonis, Atmospheric Sciences Group, Department of Mathematical Sciences, University of Wisconsin-Milwaukee, Milwaukee, WI 53201-0413, USA. (aatsonis@uwm.edu)

Dynamical System Exploration of the Hurst Phenomenon in Simple Climate Models

O. J. Mesa

Escuela de Geociencias y Medio Ambiente, Universidad Nacional de Colombia, Medellín, Colombia

V. K. Gupta

Department of Civil, Environmental and Architectural Engineering and Cooperative Institute for Research in Environmental Sciences University of Colorado, Boulder, Colorado, USA

P. E. O'Connell

School of Civil Engineering and Geosciences, Newcastle University, Newcastle, UK

The Hurst phenomenon, which reflects long-term fluctuations in geophysical time series, has attracted the attention of the research community for nearly 60 years due to its practical and theoretical importance. Yet a geophysical understanding of the Hurst phenomenon has remained elusive despite mutually conflicting hypotheses of long-term memory and nonstationarity proposed in the hydrology literature. The hypothesis of long-term memory in the climate system has recently gathered support from climate scientists through the publication of spectra that are consistent with the Hurst phenomenon for decadal to centennial time scales. A possible geophysical basis for this statistical feature has been explored elsewhere using complex global climate models. Here we explore a complementary pathway by means of the dynamics of nonuniformly mixing systems. As these systems pass bifurcations, they can exhibit the characteristics of tipping points and critical transitions that are observed in many complex dynamical systems, ranging from ecosystems to financial markets and the climate. The specific case of a Hopf bifurcation, which marks the transition from a stable system to a cyclic system, is investigated using simple dynamical systems that have been used as climate models involving energy balance and negative feedback with and without the hydrological cycle. Analyses of the model simulations reveal complex nonstationary behavior that inhibits consistent interpretation of the Hurst exponent using three well-known estimation methods. Nonetheless, the results demonstrate that dynamical systems methods can play an important role in further work on understanding the links between climate dynamics, long-term memory, nonstationarity, and the Hurst phenomenon.

1. INTRODUCTION

Harold E. Hurst [*Hurst*, 1951] discovered an anomalous behavior in his statistical analysis of paleohydrologic and climatic time series, while researching the design of long-term

Extreme Events and Natural Hazards: The Complexity Perspective
Geophysical Monograph Series 196

10.1029/2011GM001081

storage on the River Nile. It came to be known as the "Hurst phenomenon." Briefly, the reservoir size necessary to regulate the average flow depends upon the range, defined as the difference between the maximum and the minimum of the residual partial sums of the inflows. The range depends on record length. Hurst's analysis of various geophysical time series indicated that the range rescaled by the sample standard deviation, denoted by R_n^*, grows with the record length n to a power H greater than 0.5, typically close to 0.7. This finding contradicted his own deduction that, for an independent random process, and for large n, H should be 0.5. Subsequently, rigorous mathematical results [*Feller*, 1951; *Bhattacharya et al.*, 1983] confirmed this result. Furthermore, subsequent research led to the implication that a time series with $H \neq 0.5$ is a realization of a stochastic process that could be either nonstationary or stationary with a nonsummable correlation function that results in long-range dependence. Careful analysis of the spectra of climate records confirms the presence of the Hurst phenomenon [*Huth et al.*, 2001; *Syroka and Toumi*, 2001]. But despite its practical and theoretical importance, a geophysical understanding of the Hurst phenomenon has remained elusive [*Klemeš*, 1974; *Samorodnitsky*, 2007].

How well do climate model simulations represent the low-frequency behavior of the observed variables, say precipitation, in historical records? This issue is very important, as extended and persistent droughts, for instance, can be manifestations of low-frequency climate behavior. The answer to that question would imply whether climate models are able to reproduce droughts with severity and length consistent with historic observations and, therefore, their ability to predict future drought statistics [*Lettenmaier and Wood*, 2009]. In fact, it is well established that the reliability of water supply reservoirs is closely related to persistence in below average inflows [*Jain and Eischeid*, 2008]. *Fraedrich et al.* [2009] report that H in a 1000 year simulation of the present-day climate with a complex coupled atmosphere-ocean general circulation model in a present-day constant greenhouse gas environment agrees closely with H estimated from the observations (their Figure 3). M. Rutten et al. (Six fat years, six lean years: Low persistence in general circulation model rainfall projection, submitted to *Nature Geosciences*, 2011) report unpublished evidence in the opposite direction. Their analysis of global gridded observed precipitation records from the twentieth century and twentieth century runs from four global climate models indicates that the observations had H values that were considerably larger than the values computed from global climate model (GCM) precipitation sequences and that this was true for all of the GCMs tested. It is possible that the opposite behavior in these two reports can be explained by the use of a coupled model versus a simple atmospheric model. This was suggested in another context by *Syroka and Toumi* [2001]. The paper by *Koutsoyiannis et al.* [2008] concludes that GCMs do not reproduce natural interannual fluctuations and, generally, underestimate the variance and the Hurst coefficient of the observed series. This issue is a topic of current debate.

One line of enquiry that needs much greater attention than has been given until now is the role of dynamical system theory in understanding the link between climate dynamics and the Hurst phenomenon. The mathematical theory of dynamical systems has made remarkable progress in understanding how deterministic nonlinear equations behave in a random-like way. It is well known that the topological and geometric structure of the orbits, even for simple dynamical systems, can be extremely complicated. However, it is possible to obtain significant results by focusing on the statistical properties of these orbits rather than on their precise deterministic structure. In fact, chaos theory shows that sensitive dependence on initial conditions can give rise to mixing of the orbits in the phase space, which explains the random-like behavior. A possible explanation for the widespread observations of the Hurst phenomenon could be that the implied long-range dependence is a consequence of the deterministic component in the natural climate system. For instance, one can read sentences like: "It's important to realize that there is nothing magic about processes with long term persistence. This is simply a property that complex systems—like the climate—will exhibit in certain circumstances" (G. Schmidt, Hypothesis testing and long range memory, available at http://www.realclimate.org/index.php/archives/2008/08/hypothesis-testing-and-long-term-memory/, 2008). But what those certain circumstances are has not been clear, as the correlation structure of most of the chaotic systems studied so far decays exponentially fast, implying short-range dependence [*Viana*, 1997; *Bonatti et al.*, 2005]. The main idea that we will explore here is the existence of long-range fluctuations, possibly deriving from either nonstationarity or long-range dependence, which is consistent with the Hurst phenomenon for a particular class of "nonuniformly mixing systems." Long-range fluctuations arise when orbits spend long times near neutral fixed points in an overall mixing system. This results in a "slowing down of the mixing," which gives rise to either slower than exponential rates of decay of correlation or to nonstationarity; either one or both provide a dynamical basis for understanding the Hurst phenomenon. The work of *Scheffer et al.* [2009] is directly related to this approach, both through the dynamical systems ideas of bifurcations and tipping points and through the use of time series analysis methods to detect the critical slowing down near those transitions. They reviewed important implications of critical slowing down as early warning signals for critical transitions in many fields besides the climate system ranging from ecosystems to financial markets.

The paper is organized as follows: in section 2, we present a brief review of previous work on the Hurst phenomenon from the related hydrology literature and from the analysis of climate records that indicate either long-range dependence or nonstationarity. A brief review of the related climate-science literature is given in section 3. We explore a geophysical understanding of the Hurst phenomenon by means of the dynamics of nonuniformly mixing systems. A brief review of the related dynamical systems literature is given in section 4. In section 5, we consider the simple nonlinear, Daisyworld model that *Watson and Lovelock* [1983] first introduced. *Nevison et al.* [1999] relaxed the assumption of exact balance between incoming and outgoing solar radiation, which showed that it exhibits self-sustained oscillations. We force it with the variable solar insolation in the past 100 ky. Our numerical experiment is designed to understand how the slow decay of the autocorrelation function or a nonstationary behavior can come from the proximity to a neutral fixed point of the dynamical system. Next, in section 6, we consider a generalization of the Daisyworld model that includes the hydrologic cycle. Finally, in section 7, noise is introduced in the Daisyworld model, which represents the effect of all those spatial physical processes that are not explicitly represented in the dynamical equations. These three sets of investigations illustrate our steps toward understanding the Hurst phenomenon based on the dynamics of simple nonlinear climate models. Section 8 ends with concluding remarks.

2. THE HURST PHENOMENON: REVIEW OF THEORETICAL AND ESTIMATION CONCEPTS

Let q_t represent the average water inflow discharge over the period $(t - \Delta t, t)$, into an ideal reservoir, which is neither empty nor full. Let the record length be n, $\Delta t = 1$ and suppose that in each time period a constant amount of water is extracted from the reservoir, and that it is equal to the sample mean $\overline{q}$. Therefore, the continuity equation gives the volume of water stored in the reservoir at time t as $S_t = S_{t-1} + (q_t - \overline{q})$. Let us further assume $S_0 = 0$. For these conditions, define the range of S_n as,

$$R_n = \max_{\{0 \le t \le n\}} S_t - \min_{\{0 \le t \le n\}} S_t. \quad (1)$$

Hurst [1951] found that the rescaled range $R_n^* = R_n/S_n$, where S_n is the sample standard deviation, grows with the record length n to a power H greater than 0.5, typically close to 0.7. By contrast, according to the functional central limit theorem (FCLT) of probability theory, for a very general class of finite variance stochastic processes,

$$E[R_n^*] \sim \left(\frac{1}{2}\pi\theta n\right)^{0.5}, \quad \text{with} \quad \theta = \sum_{k=-\infty}^{\infty} \rho_k. \quad (2)$$

The symbol $\sim$ here means that the ratio of the two sides converges to 1 as $n \to \infty$, and ρ_k denotes the sequence of lag-k correlation coefficients [*Bhattacharya et al.*, 1983]. The integral of the correlation function θ is also known as the "scale of fluctuation of the process" in turbulence [*Taylor*, 1922]. The deviation of the empirically observed values of H from 0.5 came to be known as the "Hurst phenomenon." Clearly, because of the FCLT, for the Hurst phenomenon to hold, it would be necessary to violate at least one of the two hypotheses of the theorem, namely, stationarity or the existence of finite θ. A stationary process with finite θ is known as a "short-memory" process. Existence of θ depends on the rate of convergence of ρ_k to zero. For example, if ρ_k decays at a slow rate, say as a power law, θ does not exist. Such stochastic processes have "long-range dependence" or "infinite memory". It is important to note here that the rate of decay of ρ_k to zero will play a similar role in the theory of deterministic dynamical systems that is discussed later. These definitions are based on second-order properties (correlations). The infinite variance case will be dealt with later. One can define stronger notions based on the whole dependence structure of stationary stochastic processes, but the weaker notions are enough for our purposes [*Samorodnitsky*, 2007].

From an applied viewpoint, a hydrologist or a climatologist is primarily interested in decadal to centennial time scales in relation to droughts, or the impact of all human forcings, and not only greenhouse gases, on climate change [*Pielke et al.*, 2009], and the persistence/clustering of these events that can arise due to long-term fluctuations in the climate system. Typically, this is the range where H values corresponding to stationary processes are observed, which allows us to make statistical inferences about practical issues such as risks. As the above definition of "infinite memory" is a theoretical notion defined in terms of infinite limits, applications to real-world problems instigated considerations of preasymptotic issues in the hydrology literature. The preasymptotic behavior of some stationary processes for which H is asymptotically 0.5 can exhibit $H > 0.5$ [*Boes and Salas*, 1978; *O'Connell*, 1971]. However, since our focus here is on a geophysical understanding of the Hurst phenomenon, we will only consider concepts, both statistical and dynamical, that apply in an asymptotic sense rather than preasymptotically.

The concepts of similarity and self-similarity are very important for many branches of science and engineering. For a self-similar process $Z(t)$, $t \geq 0$, if the time scale is changed, the resulting process has the same probabilistic structure as a scaled version of the original. Specifically, for $a > 0$, there exists a positive constant b such that

$$Z(at) \stackrel{d}{=} bZ(t). \qquad (3)$$

For a nontrivial, self-similar, stationary stochastic process, there exists a unique positive exponent H such that $b = a^H$. Self-similar processes do not have a characteristic scale in the sense that both the local and the large time properties are similar. The geometrical structure of trajectories can be characterized by the fractal dimension. But if the local and long-range properties of a stochastic process are independent, then self-similarity is violated in those cases [*Gneiting and Schlather*, 2004].

For the finite variance case, let $\rho(s)$ be the correlation at lag s of a continuous stationary stochastic process. The asymptotic behavior of $\rho(s)$ as $s \to \infty$ determines the presence or absence of long-range dependence. If

$$\lim_{s\to\infty} \rho(s) \sim \frac{1}{|s|^{1-\beta}},\ \beta \in (0,1), \text{ then } H = (1+\beta)/2. \qquad (4)$$

Long-range dependence, or persistence, corresponds to the case $H \in \left(\frac{1}{2}, 1\right)$. If $H < \frac{1}{2}$, one has antipersistence, and short-frequency fluctuations dominate. In the 1-D case, this can be expressed in terms of the spectral density

$$P(f) = \int_{-\infty}^{\infty} e^{-2\pi i f s} \rho(s)\ ds. \qquad (5)$$

For the long-range dependence case,

$$P(f) \sim |f|^{-\beta} \quad \text{as} \quad |f| \to 0. \qquad (6)$$

Now the local properties of realizations depend on $\rho(s)$ near $s = 0$. If

$$1 - \rho(s) \sim |s|^{-\kappa} \quad \text{as} \quad s \to 0, \quad \text{for some} \quad \kappa \in (0,2), \qquad (7)$$

then the fractal dimension of the trajectories in an m dimensional space is [*Gneiting and Schlather*, 2004]

$$D = m + 1 - \kappa/2. \qquad (8)$$

A well-known example of a self-similar, finite variance process is the fractional Brownian motion (FBM), a zero mean Gaussian process ($B_H(t)$, $t \geq 0$), with $B_h(0) = 0$ and $E[B_H(t) - B_H(s)]^2 = \sigma^2 |t - s|^{2H}$, for some $\sigma > 0$ and $H \in (0,1)$. This process has stationary increments. In case $H = 1/2$, it becomes the usual Brownian motion. Clearly, it has the self-similarity property $B_H(ct) \stackrel{d}{=} c^H B_H(t)$, for any $c > 0$, $t \geq 0$. Fractional Gaussian noise (FGN) is a discrete time process defined by the FBM increments $X_n = B_H(n) - B_H(n-1)$, for $n = 1,2,\ldots$. This process is stationary because of the stationarity of the increments of $B_H(t)$. Its covariance is given by $\mathrm{Cov}(X_{j+n}, X_j) = \frac{1}{2}\sigma^2[(n+1)^{2H} + |n-1|^{2H} - 2n^{2H}]$. Using the self-similarity property, one can show for this process that R_n^* grows at the rate n^H [*Samorodnitsky*, 2007].

A different but important approach to statistically understanding the Hurst phenomenon in a preasymptotic sense is the so called nonstationarity of the mean. Following the works of *Hurst* [1957], *Klemeš* [1974], and *Boes and Salas* [1978], if one constructs a process where the mean of a short-memory stationary process is fluctuating randomly over time and across the realizations, a realization may appear to be nonstationary, but the underlying stochastic process is stationary. It will exhibit dependence/autocorrelation, but there is no memory. For a particular case of a shifting mean model, *Boes and Salas* [1978] showed that it has the same autocorrelation function as an autoregressive moving average ARMA(1,1) model, realizations of which can, for selected parameter values, exhibit the Hurst phenomenon over "long" time scales [*O'Connell*, 1971]. This preasymptotic behavior mimics FGN over a finite span of dependence, as exhibited by rescaled range plots, which impacts low-frequency variability and drought persistence. These are issues of practical concern. But as mentioned before, we have chosen to focus on asymptotic behavior rather than preasymptotic for a geophysical understanding of the Hurst phenomenon.

Bhattacharya et al. [1983] proved that, for certain class of trends, the Hurst phenomenon also arises in models with finite memory or short-range dependence. In particular, let X_n be a short-memory stationary stochastic process, for instance, a white noise, and $f(n) = c(n + n_0)^\mu + n_1$, is a deterministic trend for some positive parameters c and n_0, and arbitrary constants μ and n_1. The process $Y_n = X_n + f(n)$ exhibits the Hurst phenomenon. For this example, the trend f may be very small, even difficult to detect. Nevertheless, the Hurst exponent depends on the exponent of the trend μ as is shown in Figure 1. It is worthwhile noticing the discontinuity at $\mu = 0$. Whether most of the paleoclimatic and paleohydrologic natural processes exhibit long-term or short-term persistence has been debated for more than 50 years. Likewise, trend tests depend critically upon the long-range dependence structure of stochastic processes that are assumed to describe the records [*Cohn and Lins*, 2005]. The importance of this issue to the study of climate

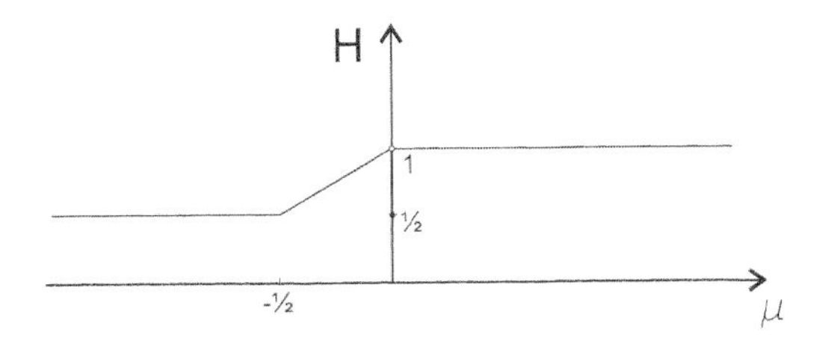

Figure 1. The relationship between the Hurst exponent H and the trend exponent μ. From the work of *Bhattacharya et al.* [1983], reprinted by permission of The Applied Probability Trust. Copyright © Applied Probability Trust 1983.

change is evident, but we will not pursue it any further in our review here.

The infinite variance case needs separate consideration because there exists the possibility that self-similarity comes from non-Gaussian stable distributions. In this case, the scaling exponent can have contributions from the long-range dependence and/or from the stable exponent of the distribution. *Mandelbrot and Wallis* [1968] referred to the first source as the Joseph effect and to the second as Noah effect using biblical references. A system that exhibits sharp discontinuities, due to large and abrupt changes, can result in infinite variance. Both the sources can occur simultaneously in natural systems. An adequate representation of both properties is very important in applications. For instance, the estimation of the probability of occurrence of extreme events is very sensitive to the existence of the second moment.

The general case of self-similar processes with infinite variance is complex [*Samorodnitsky*, 2007]. To make the main points clear, let us take a typical model, the so-called linear fractional stable motion (LFSM) that simultaneously has long-range dependence and infinite variance. It is constructed following the model corresponding to the FBM case but with non-Gaussian stable increments. Recall that a family of independent random variables, $X_1, X_2, \ldots,$ has a stable distribution with characteristic exponent α, if $X_1 + X_2 + \ldots + X_n \overset{d}{=} n^{1/\alpha} X_1$, for $0 < \alpha \leq 2$. In the symmetric case, the complementary cumulative distribution function decays with a power law, $P(X > x) \sim x^{\alpha}$.

From the LFSM, one constructs the so-called linear fractional stable noise, the increments of which are stationary with self-similar scaling exponent given by C. L. E. Franzke et al. (Robustness of estimators of long-range dependence and self-similarity under non-Gaussianity, Arxiv preprint arXiv:1101.5018, hereinafter referred to as Franzke et al., preprint, 2011)

$$\vartheta = H - \frac{1}{2} + \frac{1}{\alpha}, \quad 0 < \vartheta < 1. \tag{9}$$

Notice that we use a different notation than given by Franzke et al. (preprint, 2011) or *Mandelbrot* [2002]. For us, $H - \frac{1}{2}$ represents the contribution to the scaling exponent coming from the long-range persistence, and $1/\alpha$ represent the contribution to the scaling coming from the non-Gaussian stable distribution; their sum, ϑ is the total scaling exponent. The distinction between H and ϑ is important. Not all observed time series have Gaussian fluctuations, and some of the statistical methods used to analyze time series may be insensitive to heavy tails, detecting H and not ϑ or the other way. It would be useful to be able to estimate both. Notice that the extreme case $\alpha = 2$ corresponds to the Gaussian distribution, and $\vartheta = H$, as it should be.

Clearly, infinite variance is another possible cause for the FCLT to break down in addition to nonstationarity and long-range dependence. In this case, even for a stationary short-memory process, the convergence is not to a Gaussian process, and the theoretical result is called a noncentral limit theorem [*Samorodnitsky*, 2007, p. 202]. Nevertheless, as *Mandelbrot and Taqqu* [1979] showed, infinite variance alone cannot explain the Hurst phenomenon as was suggested in the early stages by *Moran* [1964]. This is because of the normalizing feature of the rescaled range (R/S) statistic.

Estimation of H, or ϑ, or both from sample records is a delicate issue because record lengths are generally limited. In the early stages, most of the estimators of H were defined in terms of the slope of the linear regression of R_n^* versus n in log-log space. Later spectral estimators using peridograms were introduced. The list was later enriched with fluctuation analysis (FA), wavelets, variable bandwidth, and other methods. Of these, the last two estimate the self-similar scaling exponent ϑ, whereas the rest are H estimation methods. There is a long list of applications of such estimators to a wide variety of records that are not reviewed here [see, e.g., *Beran*, 1994; Franzke et al., preprint, 2011].

Important practical issues arise in water resources management that are related to an adequate interpretation of climate variability. For instance, understanding the fluctuations of the past multidecadal streamflow regimes is critical to assessing the severity of recent droughts in the western North America [*Jain et al.*, 2002]. Recently, *Koscielny-Bunde et al.* [2006] studied temporal correlations and multifractal properties of long river discharge records from 41 hydrological stations around the globe. They used detrended fluctuation analysis (DFA) explained below and inferred long-range correlations in streamflows but not in corresponding precipitation time series. A geophysical understanding of this observation remains open.

A brief discussion of some methods that have been used to estimate the Hurst exponent H follows. Given the time series, q_t, not necessarily discharge, of record length N, first divide it into $N_n = N/n$ nonoverlapping segments of size n. The R/S diagram is a log-log plot of the rescaled adjusted range R_n^* (equation (1)) for each segment against the size of the segment n. The slope of this diagram has been used as an estimator of the Hurst exponent H. Underlying this estimation procedure is the assumption of a self-similar stationary process. More general methods have been developed that are based on estimating fluctuations around trends for similarly defined segments of length n and thereby inferring the existence of long-term memory in such fluctuations. The starting point for defining such methods is the quantity S_t (equation (1)), which, in the probability literature, corresponds to the position of a random walker at time t. He moves a step of length $q_t - \overline{q}$ from his previous position at S_{t-1}. If the steps are independent random variables, the mean square displacement grows proportional to t, but long-range dependence can make the rate of growth superlinear. The estimation of the rate of growth of the root mean square displacement is the basis of the FA method [*Koscielny-Bunde et al.*, 2006]. For each of the segments, one estimates the fluctuation by means of $F^2(k, n) = (S_{kn} - S_{(k-1)\,n})^2$ and the mean fluctuation as

$$F(n) = \sqrt{\frac{1}{N_n}\sum_k F^2(k,n)}. \tag{10}$$

Again the slope in a log-log plot of $F(n)$ versus n is used as an estimator of the Hurst exponent. This is known as the FA diagram.

However, the R/S and FA methods may fail when trends are present in the record, leading to an overestimation of the Hurst exponent. At the same time, stationary long-term correlated time series exhibit a tendency for positive or negative deviations from the average value to occur for long periods of time that might look like a trend. For this reason, the FA diagrams have been generalized into DFA. To eliminate the trends, a polynomial y_n of order p is fitted to the data in each segment k of length n by least squares. The variance of the residual mass curve from the fitted polynomial is estimated as

$$F_p^2(k,s) = \frac{1}{n}\sum_{t=1}^{n}(S_{(k-1)n+t} - y_n(t))^2. \tag{11}$$

Using a similar averaging as in equation (10), one estimates the mean detrended fluctuation of order p, $F_p(n)$, and the corresponding slope of the log-log diagram of $F_p(n)$ versus n is used to estimate the Hurst exponent for stationary processes, the so-called DFA$_p$. Usually, the linear and quadratic polynomial fits are enough.

Another common method for estimating the Hurst exponent from a time series is by means of its power spectral density. As shown in equation (5), the slope $-\beta$ of the log-log plot of $P(f)$ versus f, when $f \to 0$, is related to the Hurst exponent through $H = (1 + \beta)/2$. For the time series considered in this paper, we estimated the power spectral density by means of the multitaper method [*Percival and Walden*, 1993]. The multitaper method uses discrete prolate spheroidal sequences as tapers. These tapers are chosen both to minimize frequency leaking due to discrete sampling and to produce approximately independent spectral estimates that are then averaged to achieve maximum sample variance reduction. *Heneghan and McDarby* [2000] have established the relationship between DFA analysis (notice that they work with the variance instead of the standard deviation) and power spectral density for self-similar FGN and FBM. For nonstationary processes, the correlation function is not well defined because it is a function not only of the lag separation but also of the particular time itself. The same happens with its Fourier transform, the power spectral density. Nevertheless, there is a generalization using the ideas of the Wigner-Ville spectrum to obtain a limiting power law relation between the exponents similar to the ones we reported above for the stationary case (equation (4)) [*Heneghan and McDarby*, 2000]. In this case, the relation is

$$H = (\beta - 1)/2. \tag{12}$$

Therefore, for FBM, a nonstationary process with self-similar increments, the exponent of the DFA diagram, $H + 1$, is in the range [1, 2], β, the spectral exponent, is in the range of [1, 3], and one applies equation (12). Whereas, for the stationary FGN case, the exponent of the DFA diagram, H, is in the range [0, 1], β is in the range [−1, 1], and one applies equation (4). It should be remembered that these relations are only valid under the hypothesis of self-similarity.

Figure 2 presents the application of the above methods to one of the longest recorded temperature time series corresponding to the historical reconstruction of the monthly central England temperature record [*Parker et al.*, 1992]. The slope of the DFA diagram is 0.71, and the spectral exponent β is 0.4. Applying equation (4) for a stationary process gives $H = 0.7$. The R/S diagram gives a slope of 0.81, which is somewhat larger than 0.7. Nevertheless, DFA methods are now considered more appropriate than other methods for nonstationary time series. However, the time series in Figure 2 does not seem to show any evidence of nonstationarity.

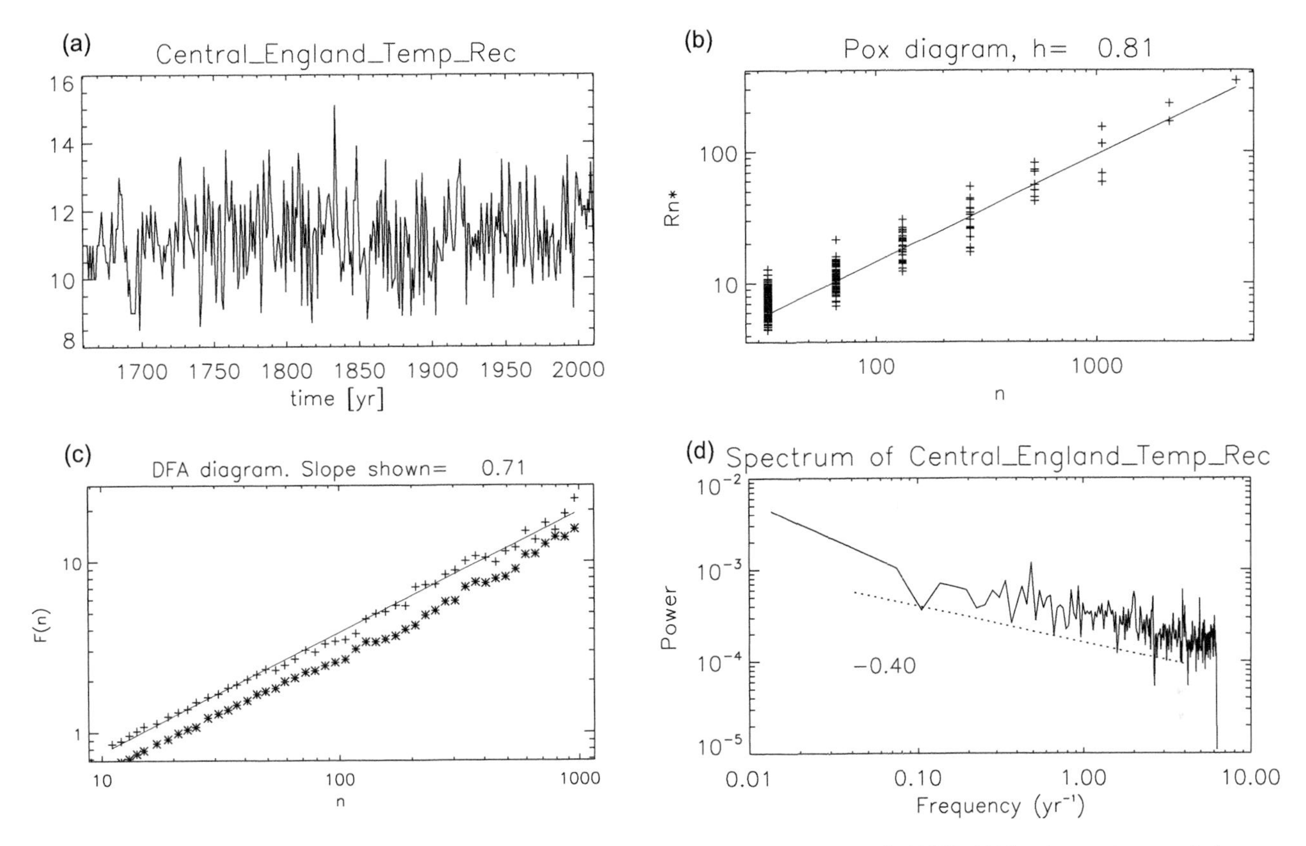

Figure 2. Central England temperature reconstruction in degrees Celsius [*Parker et al.*, 1992]. (a) The time series sampled annually in April. Other computations are with monthly series of anomalies with respect to the long-term monthly means. (b) Corresponding R/S diagram. (c) Detrended fluctuation analysis (DFA) diagram, plus signs for DFA1 and asterisks for DAF2. (d) Corresponding spectrum.

3. STOCHASTIC CLIMATE MODELING AND LONG-TERM MEMORY

Spectra of the Earth's surface temperature indicate that climatic processes exhibit a range of scales of variability [*Huybers and Curry*, 2006]. It is possible that there is coupling in the climate system that links these scales of variability, so that understanding of variability at any one time scale would require some understanding of variability as a whole. This feature is reflected, for instance, in the spectra of the Earth's surface temperature that has a continuum of energy at all time scales, whose origins are not completely understood. *Pelletier* [1998, 2002, Figure 5] shows that for frequencies approximately in the range $f = 1/10^0$ to $1/10^3$, the power spectrum $P(f)$ is proportional to $f^{-\beta}$, where $\beta = 0.5$, which corresponds to $H = 0.75 = (1 + \beta)/2$. *Fraedrich et al.* [2009, Figure 1] present a spectrum of global climatic variability where, over a similar range of frequency, $P(f) \sim f^{-\beta}$ with $\beta = 0.3$, which gives $H = 0.65$. This range of frequencies is of particular interest when estimating the frequency and persistence of droughts, and it suggests the existence of long-term memory over it. At longer time scales, paleoclimate records indicate β is closer to two, a nonstationary Brownian random walk [*Pelletier*, 1998, 2002; *Fraedrich et al.*, 2009].

Huybers and Curry [2006] have raised the very important point that the continuum in a record is as important as the peaks and deserves a physical explanation. *Huybers and Curry* [2006] show that the spectral power $P(f) \sim f^{-\beta}$, follows a power law. The exponent β is inversely related to the amplitude of the annual cycle and is different depending on the time scale. It is larger for longer time scales. It exhibits land-sea contrast in the instrumental record, generally one over the oceans and zero over the continents. There is also latitudinal contrast, with β larger near the equator. The existence of two separate scaling regimes at high and low frequencies, from 1 to 100 years and from 100 to 10^6 years, respectively, suggests distinct modes of climate variability. The estimated exponents are 0.37 and 1.64 for high-latitude continental records and short and long time scales, respectively, and similarly 0.56 and 1.29 for tropical sea surface temperature records. These observations have a direct bearing on the Hurst phenomenon. The Hurst coefficients, which are of the order of 0.68 for high-latitude continental records and 0.78 for tropical sea surface temperatures, both for shorter than centennial time scales, suggest that the climate system has long memory. For longer time scales, corresponding Hurst coefficients are larger than 1, suggesting nonstationarity. These observations need dynamical explanations. For shorter than centennial time scales, high-frequency variability seems to accumulate into progressively larger and longer period variations. For longer than centennial time scales, one might conjecture [*Huybers and Curry*, 2006] that low-frequency forcing (Milankovitch) cascades toward higher frequencies, which is analogous to turbulence [*Kolmogorov*, 1962].

To illustrate the above observations, Figures 3 and 4 show the time series, rescaled range R/S diagram, DFA diagram, and spectrum for two classical paleoclimatic time series: Oxygen isotopes ($\delta^{18}O$ in ‰) for the last 10,000 years derived from the Greenland Ice Core Project ice record, Greenland [*Johnsen et al.*, 1997], and reconstructed temperature differences with respect to the recent mean value in Antarctic (the Vostok record) [*Petit et al.*, 2001]. For the Greenland oxygen isotopes record, all the methods give approximately similar Hurst exponents. The R/S diagram gives 0.75, the DFA diagram 0.69, and the spectral exponent $\beta = 0.28$ gives $H = 0.64$ based on equation (4) for a stationary process. This accords with the analysis of *Blender et al.* [2006]. For the Vostok ice record in Figure 4, the same three methods of estimation give different values. Whereas the R/S diagram gives $H = 0.89$, the DFA diagram has a slope of 1.45, and the spectral exponent gives $\beta \approx 1.5$. This value of $\beta \approx 1.5$ is similar to the one reported in the work of *Huybers and Curry* [2006] for ice core records. It is interesting that the nonstationary relation (12) gives $H \approx 0.25$ from the spectrum and $H \approx 0.4$ from the DFA slope, which are not mutually consistent. These results suggest that the FBM theoretical model is not appropriate for this record, which appears to exhibit some quasiperiodic, nonstationary behavior. The Vostok time series suggests a skewed marginal distribution and also appears to exhibit time irreversibility. The pattern that emerges from the three cases presented above is that the

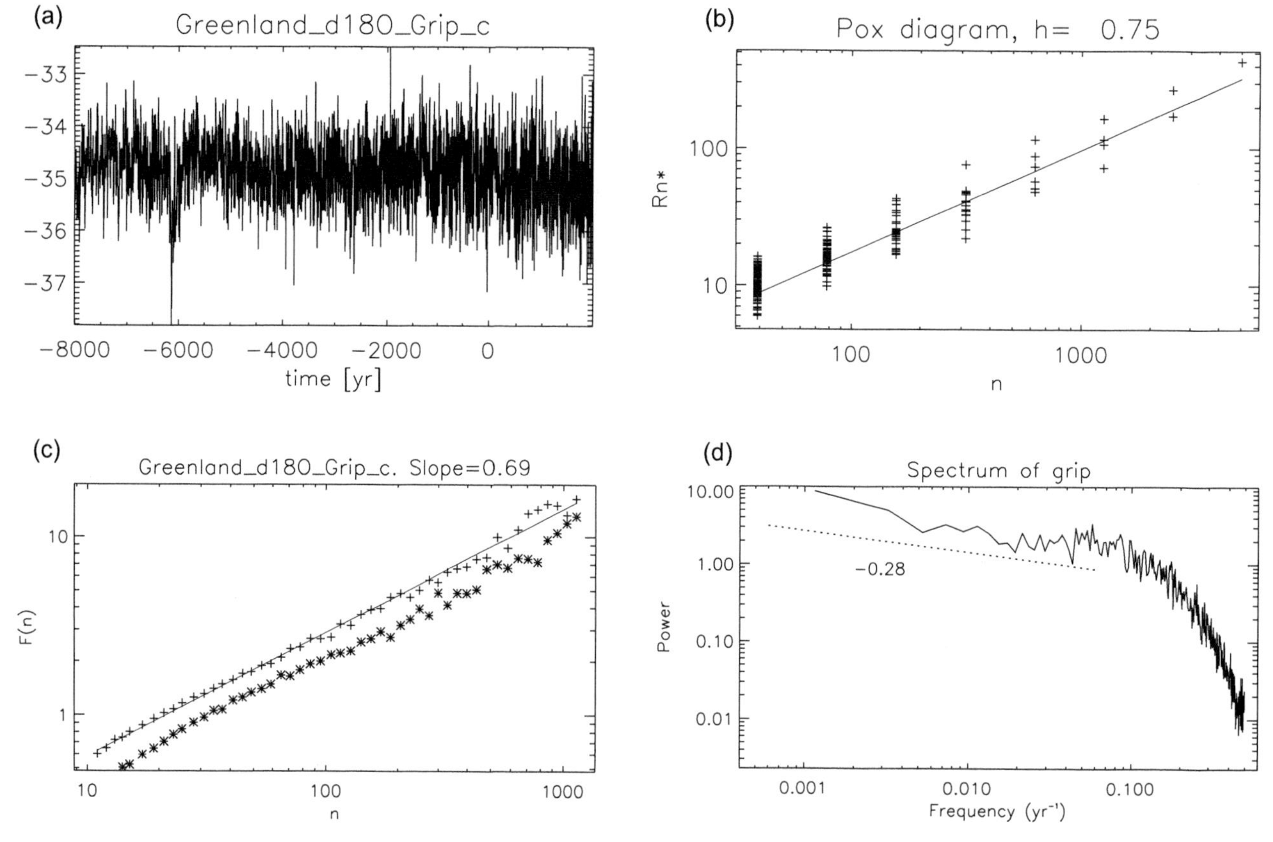

Figure 3. Oxygen isotopes ($\delta^{18}O$ in %) derived from the Greenland Ice Core Project ice record, Greenland [*Johnsen et al.*, 1997]. (a) The time series. (b) Corresponding R/S diagram. (c) DFA analysis diagrams, plus signs for DFA1 and asterisks for DAF2. (d) Estimation of the corresponding spectrum. For comparison purposes, a line with a given slope is shown.

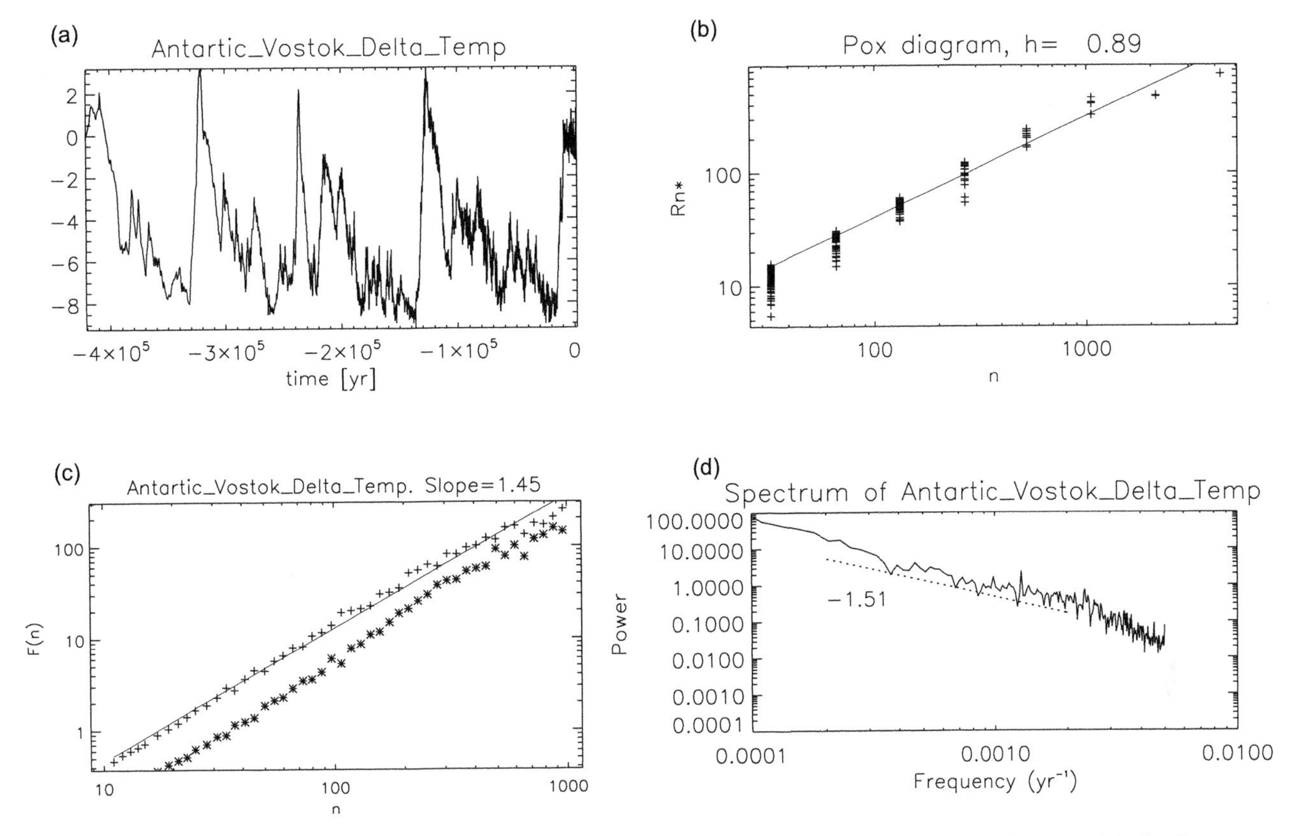

Figure 4. Antarctic temperature changes derived from the Vostok ice record, Antarctica [*Petit et al.*, 2001]. (a) The time series. (b) Corresponding R/S diagram. (c) DFA analysis diagrams, plus signs for DFA1and asterisks for DAF2. The line is fitted to values of DFA1. (d) Estimation of the corresponding spectrum. For comparison purposes, a line with a given slope is shown.

methods of estimation are adequate for stationary records or records with a weak trend. For those cases, the R/S diagram gives a somewhat larger value of the Hurst exponent than the DFA and spectral methods. But for more complex, and possibly nonstationary time series, like the Vostok record, there is no consistency among the results from the three estimation methods. This observation plays an important role in gaining a dynamical system understanding of the Hurst exponent that is given later in this paper.

Fraedrich et al. [2009] advanced a possible explanation of the spectrum of the Earth's near-surface temperature. They concluded that this low-frequency variability can be modeled by complex state-of-the-art climate models. Besides, the paper contains a conceptual model for the low-frequency sea surface temperature $1/f$ scaling that consists of a stochastically forced two-layer heat diffusion model of the global ocean, with extremely different diffusivities, representing a mixed layer on top of a deep abyssal ocean; the atmospheric forcing needs to be white. It is worthwhile to report that the scaling exponents they found vary with the type of variable, the location, and the range of time scales of the analysis. For instance, in the atmosphere, they reported that convective available potential energy showed a $1/f$ spectrum within 1 to 30 days, while temperature, wind speed, and moisture show this spectrum within periods of 1 h to 10 days and that the observed sea surface temperature in the North Atlantic showed a $1/f$ spectrum on intra-annual time scales. In the inner continents, they found that memory is absent (white noise) and that, in coastal regions and areas under maritime influence, a transition between white and $1/f$ noise was observed.

The understanding of climate variability and change has improved significantly from the analysis of paleoclimate records. At millennial and longer time scales, which extend beyond the time scales for which Hurst phenomenon has been analyzed, theories and models have been advanced to interpret the observed fluctuations as a result of insolation variations due to changes in the Earth's orbit eccentricity, tilt, and precession [*Paillard*, 2001; *Ghil and Childress*, 1987]. These models incorporate, in one form or another, nonlinear feedback, thresholds, rectification, or modulation of signals. Others propose that the observed fluctuations could be explained by stochastic resonance or chaotic mechanisms, with or without insolation forcing. Some of these models may be able to simulate key features of the climate records. However, important uncertainties about the processes and the role of feedback, parameters, and boundary conditions in the models remain, and we still lack a sufficient theory for a comprehensive interpretation of the paleoclimate records and for better prediction of future global climate variability and change [*Rial et al.*, 2005].

Among the simpler models that have been proposed to explain climate variability, one can mention the simplest ones [*Calder*, 1974; *Imbrie and Imbrie*, 1980] that have a very strong dependence on the values of their parameters and some success in reproducing features of the paleorecords. Also, there are threshold models [*Paillard*, 1998, 2001], stochastic resonance models [*Hasselmann*, 1976; *Pelletier*, 2003], coupled nonlinear differential and delayed equations [*Rial et al.*, 2004, 2005]; and deterministic chaotic models [*Huybers*, 2004]. At interannual time scales, the El Niño–Southern Oscillation signal has been modeled as an output of simple models [*Wang*, 2001], including the linear inverse model of *Penland* [1989, 2003], *Penland and Sardeshmukh* [1995], and *Sura et al.* [2005].

Important theoretical considerations about the nature of climate models and the importance of stochastic components have been made for over three decades [*Hasselmann*, 1976, 1977, 1999; *Lemke*, 1977; *Arnold*, 2001; *Imkeller and von Storh*, 2001; *Kleeman*, 2008; *Palmer and Williams*, 2008; *Majda et al.*, 2008; *Penland and Ewald*, 2008; *Dijkstra et al.*, 2008; *Neelin et al.*, 2008]. However, we do not review these important contributions here because they are not directly relevant to the objectives of our paper.

4. REVIEW OF DYNAMICAL SYSTEMS THEORY

The time evolution of a natural process, say the climate system, can be described by a transformation f of some set M in the phase space into itself. Observable quantities, physical quantities, correspond to real functions, ϕ, defined on the set M. In probability terminology, observable quantities correspond to random variables. A time series corresponds to a sequence $\{\phi(f^t(z)), t = 0,1,\ldots\}$, where $z \in M$ is the initial condition of the system, and f^t represents the t-th iterate of f. Implicit in this representation is an interval of time Δt, and a discrete time t, but a continuous time representation of the system is also possible by means of differential equations. From a practical viewpoint, one is tempted to identify the latter with the former case, given the discrete integration in numerical integration schemes. Nevertheless, there are important theoretical differences that we will remark upon later. The discrete case is called a map iteration dynamics and, the continuous case, a flow. The following review, adapted from the works of *Viana* [1997] and *Bonatti et al.* [2005], is expository in nature.

We are interested in the case where the dynamics of f is complicated or erratic. As the *Lorenz* [1963] example illustrates, even in the continuous case, with solutions that depend continuously on the initial conditions, there is sensitive dependence on initial condition. This means that trajectories corresponding to different but close initial conditions separate with time and apparently become uncorrelated. The initial condition is somehow forgotten, and the dynamics are difficult to describe in deterministic terms; therefore, for large time scales, statistical characterization is necessary.

A first basic problem is the existence of asymptotic time averages

$$E_z(\phi) = \lim_{n\to\infty} \frac{1}{n} \sum_{t=0}^{n-1} \phi(f^t(z)). \qquad (13)$$

This is known as an ergodic problem. It corresponds to the law of large numbers in probability theory. For practical reasons, it is important that an asymptotic average exists for many initial conditions z and be independent of them. In fact, this is usually assumed implicitly in real-world applications. The affirmative answer to this problem depends upon the notion of f invariant probability measures, v for which $v(f^{-1}(A)) = v(A)$ for every measurable set $A \subset M$. By a general result of ergodic theory, there exists at least one invariant measure under some mild conditions on the map f. Then, the Birkhoff ergodic theorem states that the limit in equation (13) exists for almost every initial point z, with respect to any f invariant measure. A map f may have many invariant measures, but often, there is only one that is physically relevant.

Suppose that, for every continuous real function ϕ defined on M, the average $E_z(\phi)$ exists and is independent of z, for any z in some set $B \subseteq M$ of positive probability. This defines the nonnegative linear operator $\phi \mapsto E(\phi) = E_z(\phi)$, on the space of all real continuous functions on M. By the representation theorem of functional analysis, there exists a measure μ corresponding to this operator such that

$$\int \phi d\mu = E(\phi) = \lim_{n\to\infty} \frac{1}{n} \sum_{t=0}^{n-1} \phi(f^t(z)), \text{for any } z \in B. \qquad (14)$$

The measure μ can be physically observed by sampling and time averaging for random choices of z. This is the basis for the very important definition of a *physical measure* or Sinai-Ruelle-Bowen (SRB) measure for f. An f invariant probability measure μ is a *physical measure* for f, if there exists a set of initial states of positive volume B_E, such that, for any $z \in B_E$

$$\int \phi d\mu = E(\phi) = \lim_{n\to\infty} \frac{1}{n} \sum_{t=0}^{n-1} \phi(f^t(z)), \qquad (15)$$

for every continuous function ϕ. The set B_E is the ergodic basin of μ. Notice that SRB measures are physically relevant.

Suppose that f admits a unique SRB measure μ; we represent that dynamical system by means of the pair (f, μ). The next question, and one very much related to our long-range dependence problem, is whether, and how fast, memory of the initial states is lost by the system (f, μ) as it evolves. In mathematical terms, one defines this by means of the *correlation function*

$$c_{\varphi,\psi}(n) = \int (f^n\phi)\psi d\mu - \int \phi d\mu \cdot \int \psi d\mu. \qquad (16)$$

Compare this definition with the probabilistic definition. They are essentially equivalent because dynamical systems with a unique invariant SRB measure are stationary stochastic processes.

A dynamical system (f, μ) is *mixing* if $c_{\phi,\psi}(n) \to 0$ as $n \to \infty$ for every pair of observable quantities ϕ, ψ. If the rate is exponential, the system is called *exponentially mixing*. Clearly, exponentially mixing corresponds to the statistical concept of short memory.

It is necessary to characterize the deviations of finite time averages around their expected value. In probability theory, this is done by means of the central limit theorem (CLT) and the law of large deviations (LLD). There are corresponding results for dynamical systems with an invariant measure (f, μ) that depends on the rate of correlation decay.

An observable quantity ϕ satisfies the CLT for (f, μ) if there exists $\sigma > 0$ such that for any open interval, $I \subset \Re$,

$$\mu\left\{z \in B_E; \frac{1}{\sqrt{n}} \sum_{t=0}^{n-1} (\phi(f^t(z)) - \int \phi d\mu) \in I\right\} \to \frac{1}{\sqrt{2\pi}\sigma} \int_I e^{-\frac{y^2}{2\sigma^2}} dy, \qquad (17)$$

as n goes to ∞. Similarly, ϕ satisfies the LLD if, given any $\epsilon > 0$, there is $q > 0$, depending on ϵ, such that

$$\mu\left\{ z \in B_E; \left|\frac{1}{n} \sum_{t=0}^{n-1} (\phi(f^t(z)) - \int \phi d\mu)\right| > \epsilon\right\} \leq e^{-nq}, \qquad (18)$$

for every large n. There are general results that establish conditions for these two important theorems. Before proceeding to consider some special cases, there is still a very important general problem relating to the structural stability of the systems.

In physical systems, one needs to resort to approximations, and the system f represents only the "important" part of the real system. In that case, it is very pertinent to ask about the consequences of perturbing f with "close" functions g. If the long-term dynamics of the perturbed system remain close to the original one, there is structural stability. In some cases, if the "rest" of the real system that is not taken into account in f is not known, or is too complex, one can represent those influences as random noise, and the stability property of the system that one wants is stochastic stability. This can be expressed more precisely as follows. For each $\epsilon > 0$, and for $j \geq 0$, consider functions f_j, chosen randomly and independently in the ε neighborhood of f. Then, under some general conditions, there exist probability measures μ_ϵ such that the averages of iterates $z_1 = f_1(z)$, $z_j = f_j(z_{j-1})$, $j > 2$, converge to

$$\lim_{n\to\infty} \frac{1}{n} \sum_{j=0}^{n-1} \phi(z_j) = \int \phi d\mu_\epsilon \qquad (19)$$

for many continuous functions ϕ and initial conditions $z_0 = z$. Then the system is *stochastically stable* if μ_ϵ converges to μ when $\epsilon \to 0$, in the sense that

$$\int \phi d\mu_\epsilon \to \int \phi d\mu, \text{ as } \epsilon \to 0 \qquad (20)$$

for all continuous functions ϕ. Systems with exponential decay of correlations tend to have stochastic stability [*Viana*, 1997].

It is known that a wide range of rates of decay of correlations can occur in different systems, from exponential to stretched exponential, to logarithmic and to polynomial ($c_{\phi,\psi}(n) \sim n^{1-1/\kappa}$, for some $\kappa \in (0,1)$). Notice that a polynomial decay rate corresponds to a power law.

For discrete dynamical systems, an important mathematical result is that a "uniformly expanding" system implies the existence of a unique invariant absolutely continuous measure, the CLT, and stochastic stability [*Viana*, 1997]. Therefore, slower rates of decay of correlation can only come from nonuniformly expanding maps. A map is expanding, but not necessarily uniformly expanding, if there exists $\lambda > 0$ such that

$$\liminf_{n\to\infty} \frac{1}{n} \sum_{i=0}^{n-1} \log \|Df^{-1}_{f^i(x)}\|^{-1} > \lambda \qquad (21)$$

for almost every $x \in M$.

A simple way to weaken the uniformly expanding condition is to consider 1-D maps, expanding at every point but a neutral fixed point p; this means that $f(p) = p$ and that $|f'(x)| > 1$ for every $x \neq p$ and $f'(p) = 1$. The fixed point p is repelling, but nearby points remain close to p much longer than in the uniformly expanding case. This is characteristic of intermittency. Along these lines, there are known examples, both for the discrete and continuous cases [*Alves and Viana*, 2002; *Luzzatto*, 2006; *Bruin et al.*, 2003], which are not uniformly expanding and exhibit slow rates of decay of correlations. There are even cases with such a slow rate of decay that they do not admit absolutely continuous invariant probability measures.

Such intermittent interval maps provide the simplest examples of "chaotic" dynamical systems with slow speed of mixing, and in the framework of thermodynamic formalism, they may exhibit phase transitions and dynamical ζ functions with nonpolar singularities. Motivations to investigate such maps come from hydrodynamics [*Luzzatto*, 2006; *Liverani et al.*, 1999; *Isola*, 1999; *Pomeau and Manneville*, 1980]. Let f be a map from [0,1] to itself given by $f(x) \approx x + x^2\varphi(x)$, for $x \in [0, c)$ and $f(x) = x/c - 1$, for $x \in [c, 1]$, where $\approx$ means equality in the limit as $x \to 0$ of the two sides as well as equality of the derivatives of order one and two, and c is some constant in (0,1), say 1/2. Depending on the conditions on the function φ, one gets a different behavior. In this case, 0 is a fixed neutral point ($f'(0) = 1$), which means that, at zero, there is no expansion nor contraction. Properties of φ will determine if the map is expanding in its vicinity. For example, if f is C^2 continuous, it does not admit any absolutely continuous invariant probability measure. This occurs, for example, for the case $\varphi = 1$. The orbits in this case remain most of the time near 0. As a consequence, the invariant measure is infinite for any neighborhood of 0. A different case results if $\varphi = x^{-\kappa}$ for some $\kappa \in (0,1)$. In this case, f admits an ergodic invariant absolutely continuous measure with a power law rate of mixing.

The main mechanism for this behavior has been explained as coming from a saddle node bifurcation, this is the so-called *Pomeau and Manneville* [1980] type I intermittency. Type II intermittency occurs near a Hopf bifurcation. The Daisyworld model [*Nevison et al.*, 1999], analyzed in this article, is a simple energy balance climate model that exhibits Hopf bifurcation. Therefore, we explore its possible role in understanding the Hurst phenomenon. Bifurcations play a central role in the low-frequency variability of climate dynamics [*Dijkstra and Ghil*, 2005].

Scheffer et al. [2009, p. 53], in their review article, referred to critical transitions as the fundamental shifts that occur in systems when they pass bifurcations. They further state that "It is notably hard to predict such critical transitions, because the state of the system may show little change before the tipping point is reached. Also, models of complex systems are usually not accurate enough to predict reliably where critical thresholds may occur. Critical transitions in cyclic and chaotic systems are associated with different classes of bifurcation." The specific case of a Hopf bifurcation, which marks the transition from a stable system to a cyclic system, "is signalled by critical slowing down." We investigate Hopf bifurcation in the next three sections in this article.

We wish to underscore that some extremely important systems, such as the climate or ocean circulation, are singular and afford us limited opportunity to learn by studying many similar transitions. Moreover, developing accurate models to predict thresholds in complex systems, such as climate, is a daunting task because we simply do not understand all the relevant mechanisms and feedback sufficiently well [*Pielke et al.*, 2009]. *Scheffer et al.* [2009] illustrated the generic character of the early warning signals in a wide variety of complex systems based on the concept of "critical slowing down" investigated here, as they seem to occur largely independently of the precise mechanism involved. In conclusion, our perspective is that, by understanding the possibility of a critical transition in the climate system using simple climate models and a wide variety of data sets, we are likely to produce a more robust understanding than developing and interpreting the mean state of complicated climate models.

5. LONG-TERM STATISTICAL DEPENDENCE IN A SIMPLE CLIMATE MODEL WITHOUT NOISE

Planetary climate is a highly complex nonlinear system. Positive and negative feedback are a key component of the climate system [*National Research Council*, 2003]. We used the presence of feedback as the primary criterion in selecting a simple model for the climate system. *Watson and Lovelock* [1983], in defense of the Gaia hypothesis, originally developed the Daisyworld model as a parable. It provided a mathematical illustration of how the energy balance coupled with negative feedback from the biosphere self-regulate the climate of Daisyworld as an emergent property. The Daisyworld model has attracted a great deal of interest from the scientific community over a quarter of a century. *Wood et al.* [2008] have published a comprehensive review of the extensive literature surrounding Daisyworld.

Briefly, the Daisyworld model, as a dynamical system, is a set of coupled nonlinear differential equations. It has no spatial dimensions, so it is called a 0-D model following the work of *Ghil and Childress* [1987]. The differential equations and parameters are taken from the work of *Nevison et al.* [1999, hereinafter NGK] (NGK Daisyworld),

$$\frac{dT_e}{dt} = \frac{S_0 L}{c_p}(1 - A) - \frac{\varsigma}{c_p} T_e^4, \tag{22}$$

$$\frac{da_b}{dt} = a_b(x\beta_b - \gamma), \tag{23}$$

$$\frac{da_w}{dt} = a_w(x\beta_w - \gamma). \tag{24}$$

The dependent variables are the effective temperature T_e [K] and the fractions of the planet covered by black and white daisies, a_b and a_w, respectively. The independent variable is time t [year]. Auxiliary variables that are used to facilitate writing the differential equations include the planetary albedo, $A = (1 - a_b - a_w)A_s + a_b A_b + a_w A_w$; $A_s = \frac{1}{2}$, $A_w = \frac{3}{4}$, and $A_b = \frac{1}{4}$; the proportion of fertile area that is not covered by daisies, $x = 1 - a_b - a_w$; and the growth rate of each species ($y = b$ or $y = w$) of daisy, $\beta_y = \max\{0, 1 - ((T_{op} - T_y)/17.5)^2\}$ that depends on the local temperature of each species $T_y = q(A - A_y) + T_e$, $c_p = [3 \times 10^{10}]\text{Jm}^{-2}\text{K}^{-1}$, $q = [20]\text{K}$. Other parameters are the Stefan-Boltzmann constant $\varsigma = 1.789\ \text{Jyr}^{-1}\ \text{m}^{-2}\ \text{K}^{-4}$, the present-day constant flux of solar radiation reaching the planet $S_0 = 2.89 \times 10^{10}\ \text{Jm}^{-2}\ \text{yr}^{-1}$, $\gamma = 0.3\ \text{year}^{-1}$.

Wood et al. [2008], and references therein, reviewed the dynamics of the unforced Dasiyworld model (L is a constant parameter, say L_0). We give brief highlights of the known results that are pertinent here. Depending on the values of the solar luminosity parameter L_0, there are different fixed points. One possibility, E_0, when $a_b = a_w = 0$, corresponds to a no-life situation with equilibrium temperature obtained from taking equation (22) to be equal to zero. Two other possibilities, E_w and E_b, are for solutions with only one species present. Another possibility, E_{bw}, is with the two species present (see *Weber* [2001, Figure 1] for details). The equilibrium temperature is, respectively, higher (lower) than the one corresponding to the no-life situation when the black (white) daisies dominate. In the range where both species are present, the equilibrium temperature decreases with increasing solar luminosity.

It is important to remark that the heat transport parameterization in Daisyworld contains assumptions that restrict its temperature regulating properties to adjustments of the biota alone, as *Weber* [2001] pointed out. That the regulation of local temperatures, or "homeostasis," is proscribed by the biota independently of solar forcing may be regarded as a conceptual setback for the Daisyworld parable. Additionally, the strength of the model's homeostatic behavior, its main result, is forced to depend on an arbitrary parameter associated with the heat transport.

Nordstrom et al. [2004] constructed a new version of the Daisyworld model based on local heat balances. They used a simple heat transport parameterization [*Budyko*, 1969]. Their dynamic area fraction model (DAFM) formulation removed the assumption of perfect local homeostasis through the albedo-dependent local heat transfer equation. From DAFM, they interpreted the Daisyworld heat transport parameter physically, thus removing the artificiality in its parameter set. DAFM representation results in global temperature regulation, despite the removal of the assumption of perfect local homeostasis. These results are surprising and lend weight to the Daisyworld parable. But DAFM is a more complex mathematical model than the Daisyworld. Therefore, for the present purposes, we stay with the original Daisyworld model in a slightly modified form [*Nevison et al.*, 1999].

The solutions corresponding to the equilibrium E_{bw} are periodic, and their eigenvalues are complex. There is a Hopf bifurcation for a value of $L_0 = L_{0b} \simeq 0.739$ in the parameter space. At this point, the eigenvalue is purely imaginary. As L_0 grows from values less than L_{0b}, the system makes a continuous transition from a stable fixed point to a stable periodic orbit. For larger $L_0 \simeq 1.36$, there is a similar bifurcation, but in our application, we concentrate on L_{0b}. Note that, at an equilibrium point, solutions of the ordinary differential equation are of the form $\sum c_i \exp(\rho_1 t)$, i.e., a linear combination of exponentials of eigenvalues of the linear operator multiplied by time. The inverse of the real part of an eigenvalue is a relaxation time. Hence, a critical point is characterized by an infinite relaxation time. The vicinity of a limit point is characterized by critical slowing down. This is the reason we are looking for long-range dependence near bifurcation points as mentioned earlier.

We apply recent developments from the theory of dynamical systems to explore the existence and a possible explanation of the Hurst phenomenon in Daisyworld model outputs. Therefore, exploring the existence of the Hurst phenomenon in this model provides a potential link between three disconnected areas of research, physical climate models, dynamical systems theory, and statistical hydrology. We consider a time-dependent dimensionless measure of the solar luminosity relative to present day $L = L_0 I(t)$ that includes both the annual cycle and the long time changes in orbital parameters. The deterministic quasiperiodic part $I(t)$ is computed using standard equations [*Hartmann*, 1994],

$$I(t) = \frac{c_\phi}{\pi}\left(\frac{\bar{d}}{r(t)}\right)^2 [h_0(t)\sin\phi\sin\delta(t) + \cos\phi\cos\delta(t)\sin h_0(t)], \tag{25}$$

where r and $\bar{d}$ are, respectively, the distance and the mean distance between the Earth and the Sun, h_0 is the hour angle at sunset, ϕ is the latitude, δ is the solar declination, which depends on the latitude, the obliquity Φ, and the true anomaly θ (the position of the Earth in the orbit expressed by means of the angle from the perihelion, $\sin\delta(t) = \sin(\theta(t) + \Lambda(t))\sin\Phi(t)$, where Λ is the longitude of the perihelion). Both the distance r and the true anomaly θ are functions of time and orbit eccentricity $e(t)$. In our simulations, we use a constant $c_\phi = 1.58$ equal to the ratio of the average annual planet insolation and that at the latitude used ($\phi = 65°$). The orbital parameters e, Φ, and Λ are not constant at millennium time scales due to small-order effects coming from other planets, the moon, and the nonspherical shape of the Earth. In order to compute the solar forcing in equation (25), we use a series of Φ, Λ, and e that are available from the work of *Laskar et al.* [2004]. Without the solar forcing, the dynamics is quite dull. Numerical experiments were run using a Runge-Kutta integration scheme, with $\Delta t = 0.25$ year and going 125,000 years into the future. Figure 5 presents the spectrum of the solar forcing showing the peaks associated

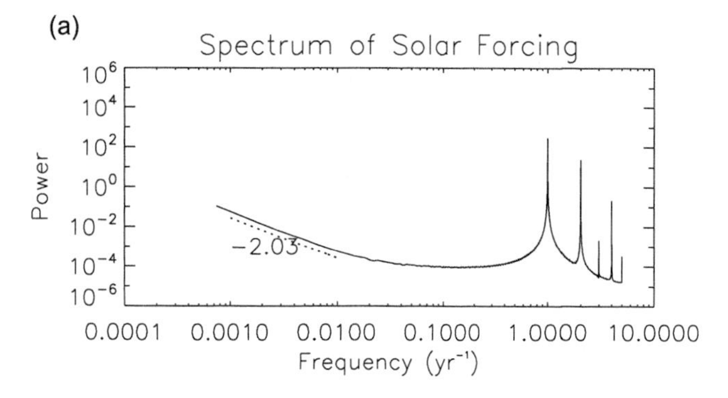

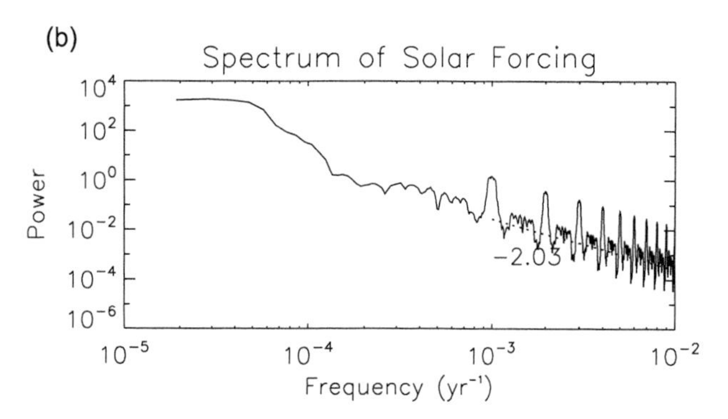

Figure 5. Spectral power density of the solar forcing used in the models. Estimation uses multitaper methods. (a) A smooth version that averages out the peaks corresponding to low frequencies but useful to estimate the slope in the log-log plot corresponding to the exponent β. (b) Those peaks in a less smoothed estimation.

with the annual cycle and its harmonics, smaller peaks at Milankovitch frequencies, and an exponent $\beta = 2$ coming from the quasiperiodicity.

In Figures 6 and 7, we present numerical results for two cases with different parameters L_0, approaching the bifurcation point from above. When L_0 is far from a bifurcation point, the output of the runs is just periodic. That is why we will consider the parameters that are close to a bifurcation point. It is clear from the time series that the trajectories spend a long time near the neutral fixed point. As the fixed point is not periodic, the trajectories look like the laminar phase described by *Pomeau and Manneville* [1980], whereas the periodic behavior corresponds to the burst phase. The forcing of this nonlinear limit cycle dynamics with values near the bifurcation point produces this complex picture. For $L_0 = 0.759$, closer to the bifurcation point, the time series shows longer periods near the neutral fixed point; the corresponding time series for the variable a_w, fraction of the planet cover with white daises, remains close to zero at those epochs. The simulations suggest nonstationary behavior in the variance and covariance; this is reflected in the nonlinear S-shape of the DFA diagrams, which are difficult to interpret in terms of a unique exponent. The slope of the DAF1 in the middle of the range is about 1.9, and for DAF2 it is 2.7, but both saturate to near zero slope for larger n. It has been recognized that, although the DFA method works well for certain types of nonstationary time series (especially slowly varying trends), it is not designed to handle all possible nonstationarities in real-world data [*Chen et al.*, 2002; *Hu*, 2004]. Apparently, the quasiperiodic behavior in the Daisyworld outputs is responsible for the S-shape of the DFA diagram in Figure 6, and a double S-shaped DFA diagram in Figure 7. The power spectral density exponent of −2 is in

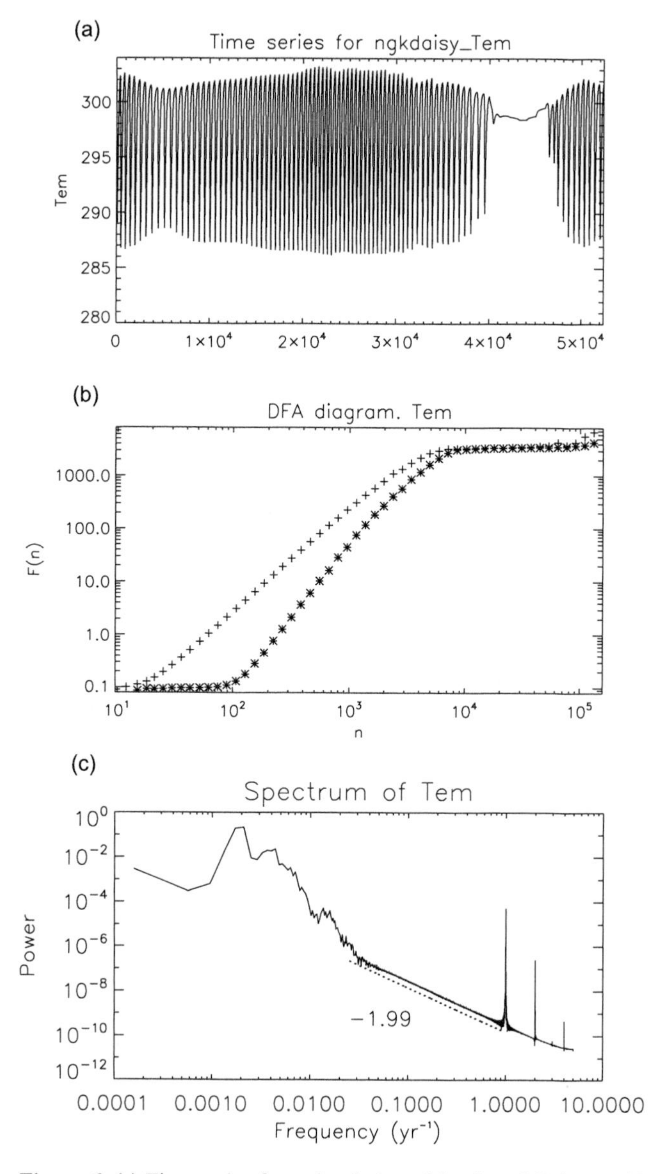

Figure 6. (a) Time series for a simulation of the forced Daisyworld model, (b) DFA diagram, and (c) power spectral density for an amplitude of the forcing $L_0 = 0.80$ (without noise).

(a)

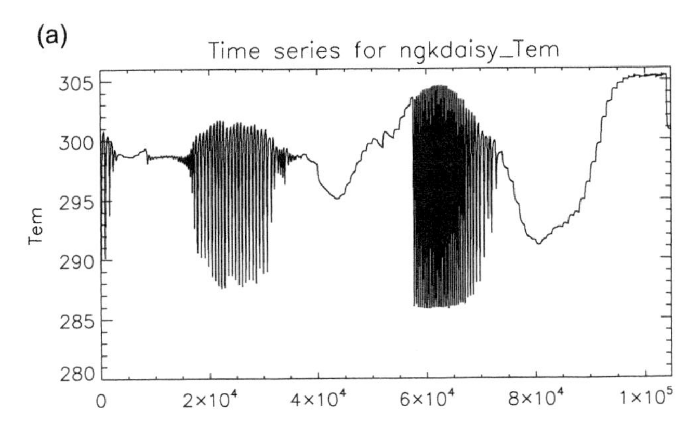

(b)

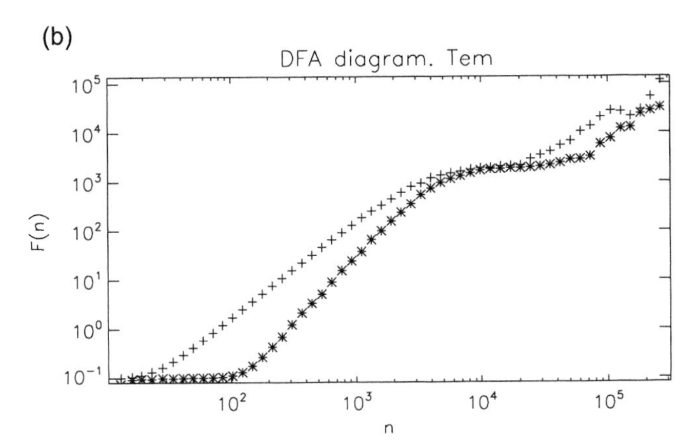

Figure 7. Similar to Figure 6 for (a) an amplitude of the forcing $L_0 = 0.759$, closer to the bifurcation point (without noise). The power spectral density is not included because it is similar to the previous case with the same exponent of −2.0. (b) DFA diagram showing a double S-shaped curve.

the range of nonstationarity, again probably due to the quasiperiodic behavior of the solar forcing. Therefore, these DFA and spectra diagrams cannot be interpreted in terms of a single value of H. There is a great influence of the quasiperiodic behavior of the forcing function. Also, the lack of linearity in the DFA diagrams, and the resulting inconsistent estimates of H from DFA and the spectra indicate that the outputs do not conform to self-similar assumptions for the increments. The intermittent behavior we are observing is nonstationary, and it is taxing all the analysis methods. This identifies an important topic for future research.

6. LONG-TERM STATISTICAL DEPENDENCE IN A SIMPLE HYDROCLIMATIC DAISYWORLD MODEL WITHOUT NOISE

The water cycle is among the most significant components of the climate system and involves, for example, cloud radiation, ice albedo, and land use feedback [*NRC*, 2003]. We consider cloud radiation feedback within the Daisyworld framework using the model of *Salazar and Poveda* [2009]. It is called Daisyworld with hydrological cycle and clouds (DAWHYC). Just like the modified original Daisyworld model analyzed above in section 5, the model of *Salazar and Poveda* [2009] is a 0-D model. The *Salazar and Poveda* [2009] approach takes the principal concepts of the role of the hydrologic cycle from the work of *Nordstrom et al.* [2005] without the added dynamical complications that their DAFM contains. *Nordstrom et al.* [2005] generalized their DAFM [*Nordstrom et al.*, 2004] to include the hydrologic cycle and clouds.

The main idea of DAWHYC is to take into account the cloud cover area fraction a_c, with different albedo A_c, as a new variable. The radiative energy conservation equation (22) is therefore modified to

$$\frac{dT_e}{dt} = \frac{S_0 L}{c_p}(1 - a_c A_c)(1 - A) - \frac{\varsigma}{c_p} T_e^4 + \frac{a_c \varsigma}{c_p} T_c^4. \qquad (26)$$

Clouds emit at a temperature T_c that is computed from the surface temperature T_e according to $T_c = T_e - \Gamma z_c$, a linear equation with lapse rate Γ and cloud base height z_c. Both Γ and z_c are fixed parameters within a simulation.

The new variable a_c comes with a new equation that represents the water balance in the atmosphere

$$\frac{da_c}{dt} = (1 - a_c)E_p - a_c P_p, \qquad (27)$$

with E_p a fraction representing evaporation and P_p precipitation. Both are defined by means of plausible empirical relations, E_p depends strongly on temperature

$$P_p = \frac{1}{P_{\max}} \left(\frac{a_c}{m}\right)^{1/n} \qquad (28)$$

$$E_p(T) = 192 \left(\frac{10T}{I(T)}\right)^{k(T)}, \qquad (29)$$

where $n = 0.1$, $m = 0.35$, $T = T_e - 273$, $I(T) = 12(T/5)^{1.5}$, and $k(T) = 0.49 + 0.018I(T) - 7.7 \times 10^{-5} I^2(T) + 6.7 \times 10^{-7} I^3(T)$.

Salazar and Poveda [2009] provide a more complete description of the model and a thorough discussion of its physical basis and an analysis of results for the whole range of parameters. In many aspects, the dynamics of DAWHYC is similar to those of the NGK Daisyworld model. For instance, *Weber* [2001, Figure 1] shows that the fixed points of the system under varying solar forcings are similar to those in the work of *Salazar and Poveda* [2009, Figure 7]. For our purpose, this is important as changes in the solar forcing parameter L_0 produce Hopf bifurcations that are exactly what we are exploring. In fact, Figures 8 and 9 represent, for the DAWHYC model, the same behavior in the DFA diagrams and spectra that was represented in Figures 6 and 7 for the NGK Daisyworld model.

Salazar and Poveda [2009] found that the DAWHYC's dynamics is not only governed by the external forcing and heat capacity (as in NGK) but also by the role of the hydrological cycle and clouds. In particular, the cloud parameters control the frequency and amplitude of the self-sustaining oscillations. For instance, in DAWHYC, for values of the solar forcing parameter that do produce self-sustaining oscillations in the NGK model and for some values of the cloud parameter, there are no oscillations. This is evident in the work of *Salazar and Poveda* [2009, Figure 6]. Therefore, changes in cloud albedo and height within DAWHYC can

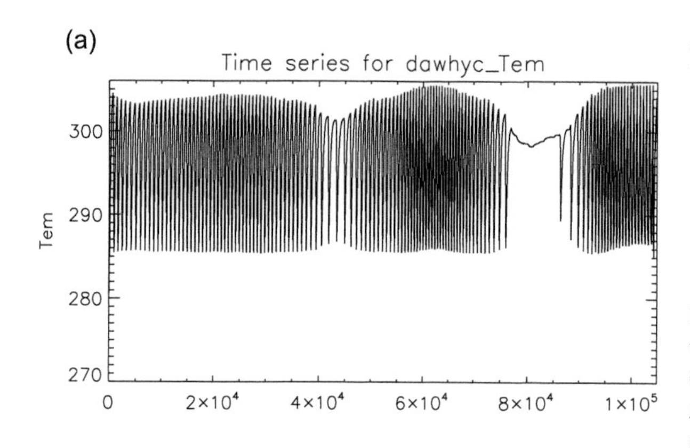

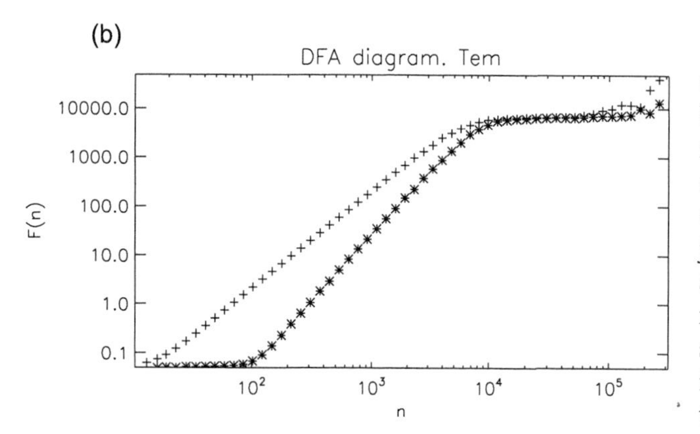

Figure 8. Similar to Figure 7 for (a) the forced DAWHYC model with $A_c = 0.6$ and z_c= 4 km and amplitude of the forcing $L_0 = 0.75$. (b) DFA diagram showing an S-shaped curve, but for larger n, the slope becomes positive. The power spectral density is similar to the previous case with the same exponent of −2.0.

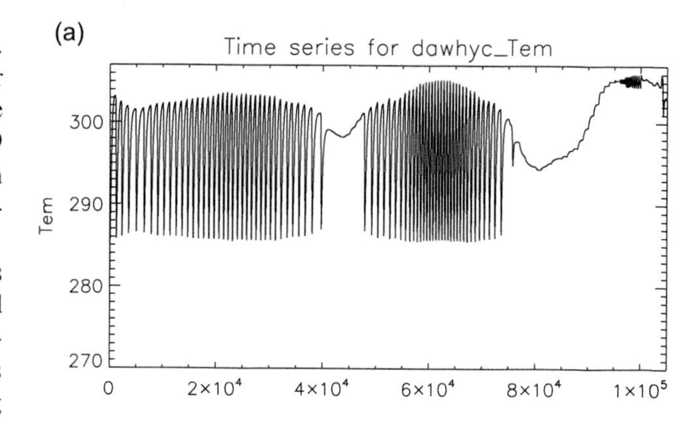

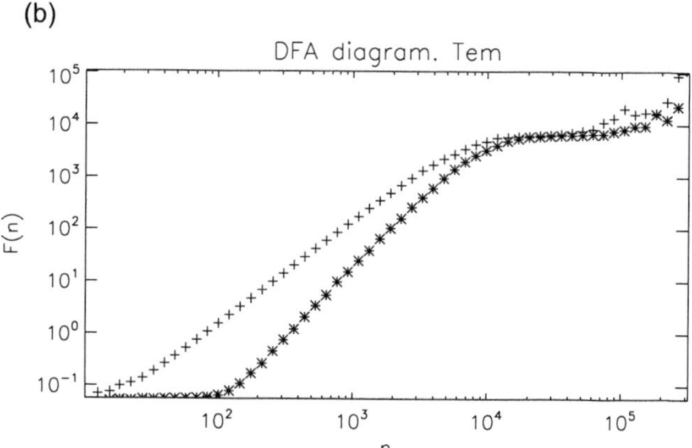

Figure 9. Similar to Figure 6 for (a) the forced DAWHYC model with A_c = 0.6 and z_c= 4 km and amplitude of the forcing L_0 = 0.7125. (b) DFA diagram showing an S-shaped curve, but for larger n, the slope becomes positive. The power spectral density (not shown) is similar to the previous case with the same exponent of −2.0.

significantly modify the dynamics, even to produce Hopf bifurcations. In Figure 10, we present the application of the methods to study long-term memory for the output of this system near a new bifurcation point that comes from the role of the hydrologic cycle. Again, this case confirms our conjecture that a nonlinear system of differential equations near a neutral fixed point develops a critical slowing down, intermittency, and low-frequency oscillations. Again, the time series, and the corresponding spectra, indicate nonstationarity. In fact, the frequency exponents in the spectra are of the order of −2, values that correspond to nonstationary FBM processes with self-similar increments. The DFA diagram shows an S-shaped curve, but for larger n, the slope becomes positive, which suggests nonconformance with the assumption of self-similar increments. It is therefore unclear how to interpret it in terms of memory of the process under

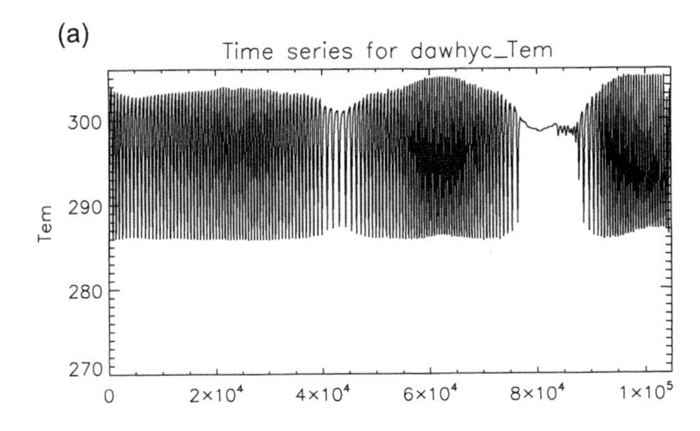

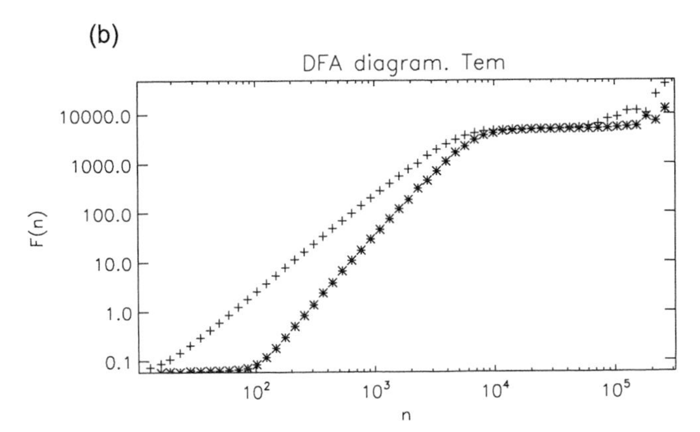

Figure 10. Similar to Figure 6 for (a) the forced DAWHYC model with $A_c = 0.7$ and $z_c = 6.3$ km and amplitude of the forcing $L_0 = 1.0$. (b) DFA diagram pattern similar to Figures 8 and 9. The power spectral density (not shown) is similar to the previous case with the same exponent of −2.0.

consideration. As mentioned earlier, it identifies an important topic of future research.

7. LONG-TERM STATISTICAL DEPENDENCE IN A SIMPLE CLIMATE MODEL WITH NOISE

There are at least two issues to consider in the analysis of the previous simulations. First, Daisyworld simulations must be interpreted as globally averaged, whereas the Hurst phenomenon has been observed at the point scale from observational records. Comparing these two must take account of the fact that, for instance, observational temperature records display much greater year to year variability than the globally averaged temperature records that have been produced to make the case for global warming [*Jones and Moberg*, 2003]. The simulations corresponding to the no-noise case can be viewed as the (smooth) globally averaged temperature on Daisyworld. This reflects the periodic nature of the output from the *Nevison et al.* [1999] version of Daisyworld and *Salazar and Poveda* [2009] from the DAWHYC, modulated by the forcing function. From what we observe for the Earth's climate system, a large amount of variability/noise is observed in locally observed climate because of the huge spatial variability within the climate system. Therefore, we argue that, to enable us to compare the outputs of Daisyworld with observational records and assess the Hurst phenomenon, we need to add a substantial component of noise into the input forcing function to reproduce this variability. This provides a rationale for adding noise into the input function to go from smooth globally averaged temperature on Daisyworld to something resembling what we would see at local scale on Daisyworld, without resorting to the complexity of introducing space explicitly. For that reason, in this section, we explore a case in which we add random noise to the solar forcing. It represents the effect of all those physical spatial processes that are not explicitly represented in the dynamical equations.

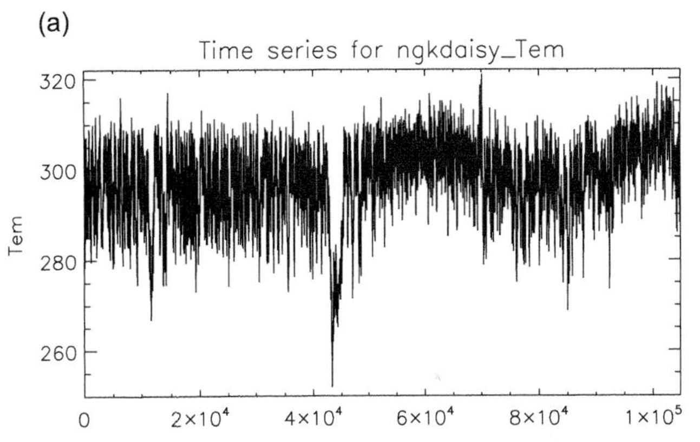

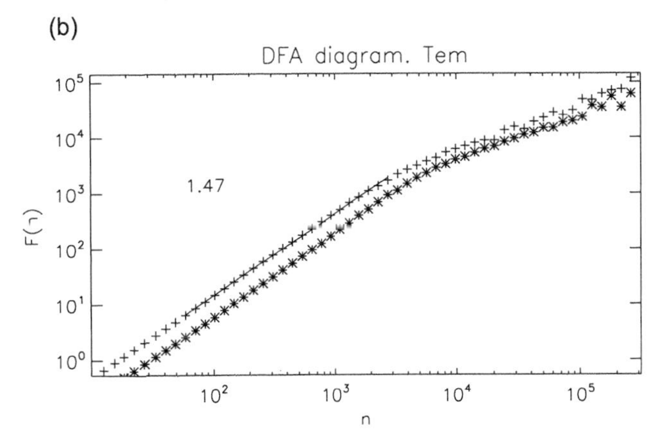

Figure 11. Similar to Figure 6 for (a) the forced NGK Daisyworld model with random noise forcing with standard deviation, $\sigma = 5.0$, and amplitude of the forcing $L_0 = 0.819$. (b) DFA diagram is more linear. The slope of the power spectral density is −2.0 (not shown).

The second issue is that Daisyworld is a simple model lacking many components of the climate system. There is a tradition in climate science of considering energy balance climate models [*Ghil*, 2002] as a fundamental tool to understand the nonlinear interplay of feedback. Daisyworld does not include, for instance, ice albedo feedback effects and many others. For that reason, comparisons of Daisyworld outputs with real climate time series is limited.

In this section, we add to the solar forcing a white random noise to the previously considered forcing, this is $L = L_0 I(t) + \sigma \dot{W}_t$. Therefore, the forcing includes both the annual cycle and the long time changes in orbital parameters, plus a random white noise $\dot{W}_t$. At this point, no attempt is made to quantify the strength σ of this random forcing, for the objective of the experiments is to investigate the structure of the low-frequency variability of the resulting time series.

We performed several runs, but we show one case for the NGK Daisyworld model with $L_0 = 0.819$, which is relatively far from the bifurcation point $L_0^b = 0.739$, with the standard deviation of the noise $\sigma = 5$ (Figure 11). One can see that the S-shaped DFA curve is more linear, suggesting better conformance with the assumption of self-similar increments. The other case shown (Figure 12) is for the DAWHYC model with $L = 1$ (away from the bifurcation point) and cloud parameters $A_c = 0.7$ and $z_c = 6.3$ km. In both cases, the simulated time series look more similar to real geophysical time series than in the previous runs, and the estimated Hurst exponents are significantly larger than 0.5. The spectral exponent of -2 indicates nonstationarity, but an interpretation of memory from the DFA diagrams remains unclear as mentioned before. How to get consistent estimates of H from different estimating methods for nonstationary time series identifies an important topic for further research.

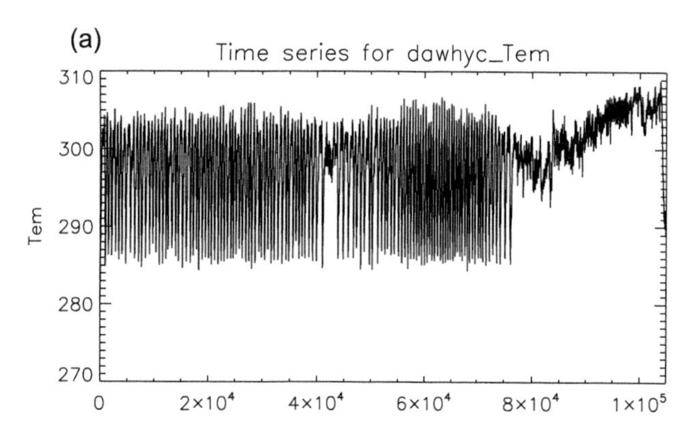

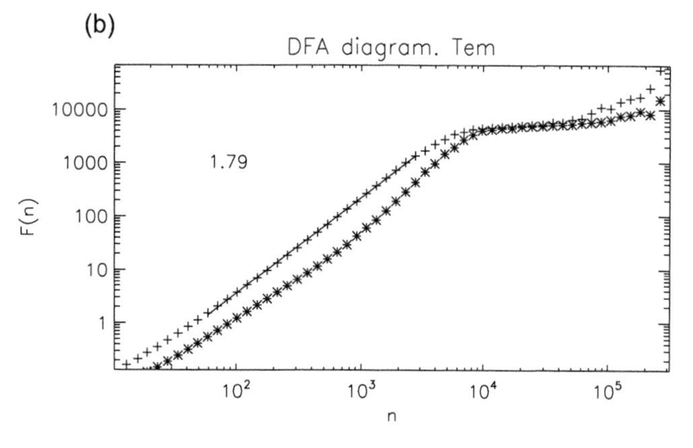

Figure 12. Similar to Figure 6 for (a) the DAWHYC Daisyworld model with hydrologic cycle with random noise forcing with standard deviation $\sigma = 2.0$. The cloud parameters are $A_c = 0.7$ and $z_c =$ 6.3 km, and the amplitude of the forcing is $L_0 = 1.0$. (b) Corresponding DFA diagram showing an S-shaped curve.

8. CONCLUDING REMARKS

We have achieved the following objectives in this paper. First, we have reviewed pertinent literature from the field of statistical hydrology. There has been a debate for several decades in the statistical hydrology literature, which has called for a physical understanding of the Hurst phenomenon. During this time period, paleoclimate science dealing with low-frequency climate variability has made major progress, but communication between these two communities has been minimal. The existence of the Hurst phenomenon from statistical hydrology provides a common format for this purpose because it represents either long-term memory or nonstationarity or a combination of the two. Second, simple and complex physical climate models have been developed in the last 40 years, but connection of these models with the Hurst phenomenon needs more emphasis than has been given so far. Third, to contribute to a potential link between these three fields, we reviewed and applied recent developments from the mathematical theory of dynamical systems in understanding how deterministic nonlinear equations behave in a random-like manner. In particular, we investigated the solutions of a simple (0-D) nonlinear deterministic climate model that includes energy balance and negative feedback and produces self-regulation of climate. We chose the Daisyworld model without and with the hydrological cycle in it. Our numerical experiments confirmed that the dynamics of a simple nonlinear climate model with periodic solutions can produce complex intermittent behavior near a neutral fixed point when forced by a quasiperiodic solar forcing. In our case, the neutral fixed point is at a Hopf bifurcation in the parameter space. Our original conjecture for this possibility came from the theoretical results of *Liverani et al.* [1999] and *Bruin et al.* [2003] for the so-called intermittency maps.

In order to analyze statistical memory in the presence of certain kinds of nonstationarity, we selected three statistical methods to estimate the Hurst exponent: (1) the R/S diagram, which has been widely used in statistical hydrology, (2) spectral analysis, which has been preferred in the climate and paleoclimate sciences, and (3) DFA, which has been used in all these fields. These methods are well known to give consistent results for stationary time series obeying self-similarity or nonstationary time series obeying self-similar FBM behavior. Unexpectedly, these methods did not give a clear interpretation when applied to the outputs of the Daisyworld model, probably due to nonstationarity and/or violation of self-similar hypotheses. It presents an important topic for further research (see, e.g., Franzke et al., preprint, 2011).

The Hurst effect has been observed for records in the range 100–1000 years, where it is plausible that the realizations come from a stationary stochastic process. The model realizations shown in the paper, when viewed as they are, or blown up, are not physically plausible in this regard. However, they have relevance in the context of climate change on geological time scales, as indicated by the climate spectra where nonstationarity is present. The results presented here should be interpreted for what they are: simulations from very simple climate models that exhibit interesting behavior from a dynamical system perspective due to critical slowing down at a Hopf bifurcation. In this connection, *Scheffer et al.* [2009, p. 54] stated that "In summary, the phenomenon of critical slowing down leads to three possible early-warning signals in the dynamics of a system approaching a bifurcation: slower recovery from perturbations, increased autocorrelation and increased variance."

Because droughts and, in general, other low-frequency climate behavior have enormous economical and societal consequences, climate model simulations need to be improved in this regard. The issues of nonstationarity and infinite memory are also very relevant, both from practical and theoretical viewpoints, particularly in relation to climate change scenarios. In summary, it appears that exploring the notion of long-range dependence for physically pertinent dynamical systems is promising because it allows us to concentrate on the behavior of really important characteristics of a complex system from the physical point of view and to understand human-climate (Earth) system interactions within the framework of dynamical systems theory.

Acknowledgments. OM acknowledges CIRES funding for a visiting fellowship that helped us initiate this work. VG acknowledges partial support from NSF grant EAR 1005311 for this research. VG also acknowledges support from EPSRC Platform Grant: "Earth Systems Engineering: Systems engineering for sustainable adaptation to global change" for a visit to the Centre for Earth Systems Engineering Research (CESER) at Newcastle University, UK. The comments of an anonymous reviewer were very helpful in clarifying key issues and improving this paper and are much appreciated. Finally, we are indebted to the late Vit Klemeš for laying down the challenge to the scientific community many years ago of finding a physical explanation for the Hurst phenomenon. We hope that he would have appreciated our modest step in this direction.

REFERENCES

Alves, J., and M. Viana (2002), Statistical stability for robust classes of maps with non-uniform expansion, *Ergodic Theory Dyn. Syst.*, *22*(1), 1–32.

Arnold, L. (2001), Hasselman's program revisited: The analysis of stochasticity in deterministic climate models, in *Stochastic Climate Models*, edited by P. Imkeller and J. von Storh, pp. 141–157, Birkhäuser, Basel, Switzerland.

Beran, J. (1994), *Statistics for Long-Memory Processes*, 315 pp., Chapman and Hall, New York.

Bhattacharya, R. N., V. K. Gupta, and E. Waymire (1983), The Hurst effect under trends, *J. Appl. Probab.*, *20*(3), 649–662.

Blender, R., K. Fraedrich, and B. Hunt (2006), Millennial climate variability: GCM-simulation and Greenland ice cores, *Geophys. Res. Lett.*, *33*, L04710, doi:10.1029/2005GL024919.

Boes, D. C., and J. D. Salas (1978), Nonstationarity of the mean and the Hurst phenomenon, *Water Resour. Res.*, *14*(1), 135–143.

Bonatti, C., L. Díaz, and M. Viana (2005), Encyclopaedia of Mathematical Sciences, vol. 102, *Dynamics Beyond Uniform Hyperbolicity: A Global Geometric and Probabilistic Perspective*, edited by J. Frõhlich, S. Novikov, and D. Ruelle, 402 pp., Springer, Berlin.

Bruin, H., S. Luzzatto, and S. VanStrien (2003), Decay of correlations in one-dimensional dynamics, *Ann. Sci. Ecole Norm. Super.*, *36*, 621–646.

Budyko, M. (1969), The effect of solar radiation variations on the climate of the Earth, *Tellus*, *21*(5), 611–619.

Calder, N. (1974), Arithmetic of ice ages, *Nature*, *252*, 216–218.

Chen, Z., P. C. Ivanov, K. Hu, and H. E. Stanley (2002), Effect of nonstationarities on detrended fluctuation analysis, *Phys. Rev. E*, *65*(4), 041107, doi:10.1103/PhysRevE.65.041107.

Cohn, T. A., and H. F. Lins (2005), Nature's style: Naturally trendy, *Geophys. Res. Lett.*, *32*, L23402, doi:10.1029/2005GL024476.

Dijkstra, H. A., and M. Ghil (2005), Low-frequency variability of the large scale ocean circulation: A dynamical systems approach, *Rev. Geophys.*, *43*, RG3002, doi:10.1029/2002RG000122.

Dijkstra, H. A., L. M. Frankcombe, and A. S. von der Heydt (2008), A stochastic dynamical systems view of the Atlantic Multidecadal Oscillation, *Philos. Trans. R. Soc. A*, *366*(1875), 2543–2558, doi:10.1098/rsta.2008.0031.

Feller, W. (1951), The asymptotic distribution of the range of sums of independent random variables, *Ann. Math. Stat.*, *22*, 427–432.

Fraedrich, K., R. Blender, and X. Zhu (2009), Continuum climate variability: Long-term memory scaling, and 1/f-noise, *Int. J. Mod. Phys. B*, *23*(28–29), 5403–5416.

Ghil, M. (2002), Natural climate variability, in *Encyclopedia of Global Environmental Change*, vol. 1, edited by T. Munn, pp. 544–549, John Wiley, Chichester, U. K.

Ghil, M., and S. Childress (1987), *Topics in Geophysical Fluid Dynamics: Atmospheric Dynamics, Dynamo Theory, and Climate Dynamics*, 527 pp., Springer, New York.

Gneiting, T., and M. Schlather (2004), Stochastic models that separate fractal dimension and the Hurst effect, *SIAM Rev.*, *46*(2), 269–282.

Hartmann, D. L. (1994), *Global Physical Climatology*, *Int. Geophys. Ser.*, vol. 56, 411 pp., Academic Press, San Diego, Calif.

Hasselmann, K. (1976), Stochastic climate models: Part I. Theory, *Tellus*, *28*, 473–485.

Hasselmann, K. (1977), Stochastic climate models: Part II. Applications to sea-surface temperature anomalies and thermocline variability, *Tellus*, *29*, 289–305.

Hasselmann, K. (1999), Climate change: Linear and nonlinear signature, *Nature*, *398*, 755–756.

Heneghan, C., and G. McDarby (2000), Establishing the relation between detrended fluctuation analysis and power spectral density analysis for stochastic processes, *Phys. Rev. E*, *62*(5), 6103–6110, doi:10.1103/PhysRevE.62.6103.

Hu, H. (2004), Decay of correlations for piecewise smooth maps with indifferent fixed points, *Ergodic Theory Dyn. Syst.*, *24*(2), 495–524.

Hurst, H. (1951), Long term storage capacity of reservoirs, *Trans. Am. Soc. Civ. Eng.*, *116*, 776–808.

Hurst, H. (1957), A suggested statistical model of some time series which occur in nature, *Nature*, *180*, 494.

Huth, R., J. Kyselý, and M. Dubrovský (2001), Time structure of observed, GCM-simulated, downscaled, and stochastically generated daily temperature series, *J. Clim.*, *14*(20), 4047–4061, doi:10.1175/1520-0442(2001)014<4047:TSOOGS>2.0.CO;2.

Huybers, P. (2004), On the origins of the ice ages: Insolation forcing, age models, and nonlinear climate change, Ph.D. thesis, 245 pp., Mass. Inst. of Technol., Cambridge.

Huybers, P., and W. Curry (2006), Links between the annual, Milankovitch, and continuum of temperature variability, *Nature*, *441*, 329–332.

Imbrie, J., and J. Z. Imbrie (1980), Modelling the climatic response to orbital variations, *Science*, *207*, 943–953.

Imkeller, P., and J. von Storh (2001), *Stochastic Climate Models*, Birkhäuser, Basel, Switzerland.

Isola, S. (1999), Renewal sequences and intermittency, *J. Stat. Phys.*, *97*, 263–280.

Jain, S., and J. K. Eischeid (2008), What a difference a century makes: Understanding the changing hydrologic regime and storage requirements in the Upper Colorado River basin, *Geophys. Res. Lett.*, *35*, L16401, doi:10.1029/2008GL034715.

Jain, S., C. A. Woodhouse, and M. P. Hoerling (2002), Multidecadal streamflow regimes in the interior western United States: Implications for the vulnerability of water resources, *Geophys. Res. Lett.*, *29*(21), 2036, doi:10.1029/2001GL014278.

Johnsen, S. J., et al. (1997), The $\delta^{18}O$ record along the Greenland Ice Core Project deep ice core and the problem of possible Eemian climatic instability, *J. Geophys. Res.*, *102*(C12), 26,397–26,410.

Jones, P., and A. Moberg (2003), Hemispheric and large-scale surface air temperature variations: An extensive revision and an update to 2001, *J. Clim.*, *16*(2), 206–223.

Kleeman, R. (2008), Stochastic theories for the irregularity of ENSO, *Philos. Trans. R. Soc. A*, *366*(1875), 2509–2524, doi:10.1098/rsta.2008.0048.

Klemeš, V. (1974), The Hurst phenomenon: A puzzle?, *Water Resour. Res.*, *10*(4), 675–688.

Kolmogorov, A. N. (1962), A refinement of previous hypotheses concerning the local structure of turbulence in a viscous incompressible fluid at high Reynolds number, *J. Fluid Mech.*, *13*, 82–85.

Koscielny-Bunde, E., J. W. Kantelhardt, P. Braun, A. Bunde, and S. Havlin (2006), Long-term persistence and multifractality of river runoff records: Detrended fluctuation studies, *J. Hydrol.*, *322*, 120–137.

Koutsoyiannis, D., A. Efstratiadis, N. Mamassis, and A. Christofides (2008), On the credibility of climate predictions, *Hydrol. Sci. J.*, *53*(4), 671–684.

Laskar, J., P. Robutel, F. Joutel, M. Gastineau, A. C. M. Correia, and B. Levrard (2004), A long-term numerical solution for the insolation quantities of the Earth, *Astron. Astrophys.*, *428*, 261–285.

Lemke, P. (1977), Stochastic climate models: Part 3. Application to zonally averaged energy models, *Tellus*, *29*(5), 385–392.

Lettenmaier, D. P., and E. F. Wood (2009), Water in a changing climate: Challenges for GEWEX, *GEWEX News*, *19*(2), 3–5.

Liverani, C., B. Saussol, and S. Vaienti (1999), A probabilistic approach to intermittency, *Ergodic Theory Dyn. Syst.*, *19*, 671–685.

Lorenz, E. (1963), Deterministic nonperiodic flow, *J. Atmos. Sci.*, *20*, 130–141.

Luzzatto, S. (2006), Stochastic-like behaviour in nonuniformly expanding maps, in *Handbook of Dynamical Systems*, vol. 1, edited by B. Hasselblatt and A. Katok, pp. 265–326, Elsevier, Amsterdam.

Majda, A. J., C. Franzke, and B. Khouider (2008), An applied mathematics perspective on stochastic modelling for climate, *Philos. Trans. R. Soc. A*, *366*(1875), 2427–2453, doi:10.1098/rsta.2008.0012.

Mandelbrot, B. (2002), *Gaussian Self-Affinity and Fractals*, 654 pp., Springer, New York.

Mandelbrot, B., and M. Taqqu (1979), Robust R/S analysis of long run serial correlation, in *Proceedings of 42nd Session ISI*, pp. 69–99, Int. Stat. Inst., The Hague, Netherlands.

Mandelbrot, B. B., and J. R. Wallis (1968), Noah, Joseph, and operational hydrology, *Water Resour. Res.*, *4*(5), 909–918.

Moran, P. (1964), On the range of cumulative sums, *Ann. Inst. Stat. Math.*, *16*(1), 109–112.

National Research Council (2003), *Understanding Climate Change Feedbacks*, 166 pp., Natl. Acad. Press, Washington, D. C.

Neelin, J. D., O. Peters, J. W.-B. Lin, K. Hales, and C. E. Holloway (2008), Rethinking convective quasi-equilibrium: Observational constraints for stochastic convective schemes in climate models, *Philos. Trans. R. Soc. A*, *366*(1875), 2579–2602, doi:10.1098/rsta.2008.0056.

Nevison, C., V. Gupta, and L. Klinger (1999), Self-sustained temperature oscillations on Daisyworld, *Tellus, Ser. B*, *51*, 806–814.

Nordstrom, K., V. Gupta, and T. Chase (2004), Salvaging the Daisyworld parable under the dynamic area fraction framework, in *Scientists Debate Gaia: The Next Century*, edited by S. H. Schneider et al., pp. 241–253, MIT Press, Cambridge, Mass.

Nordstrom, K., V. Gupta, and T. Chase (2005), Role of the hydrological cycle in regulating the planetary climate system of a simple nonlinear dynamical model, *Nonlinear Processes Geophys.*, *12*(5), 741–753.

O'Connell, P. (1971), A simple stochastic modelling of Hurst's law, in *International Symposium on Mathematical Models in Hydrology, Int. Ass. Sci. Hydol.*, *1*, 169–187.

Paillard, D. (1998), The timing of Pleistocene glaciations from a simple multiple state climate model, *Nature*, *391*, 378–381.

Paillard, D. (2001), Glacial cycles: Toward a new paradigm, *Rev. Geophys.*, *39*(3), 325–346.

Palmer, T., and P. Williams (2008), Introduction. Stochastic physics and climate modelling, *Philos. Trans. R. Soc. A*, *366*(1875), 2419–2425, doi:10.1098/rsta.2008.0059.

Parker, D., T. Legg, and C. Folland (1992), A new daily central England temperature series, 1772–1991, *Int. J. Climatol.*, *12*(4), 317–342.

Pelletier, J. (1998), The power-spectral density of atmospheric temperature from time scales of 10^{-2} to 10^{6} yr, *Earth Planet. Sci. Lett.*, *158*, 157–164.

Pelletier, J. (2002), Natural variability of atmospheric temperatures and geomagnetic intensity over a wide range of time scales, *Proc. Natl. Acad. Sci. U. S. A.*, *99*, suppl. 1, 2546–2553.

Pelletier, J. D. (2003), Coherence resonance and ice ages, *J. Geophys. Res.*, *108*(D20), 4645, doi:10.1029/2002JD003120.

Penland, C. (1989), Random forcing and forecasting using principal oscillation pattern analysis, *Mon. Weather Rev.*, *117*(10), 2165–2185.

Penland, C. (2003), A stochastic approach to nonlinear dynamics: A review, *Bull. Am. Meteorol. Soc.*, *84*(7), ES43–ES52.

Penland, C., and B. D. Ewald (2008), On modelling physical systems with stochastic models: Diffusion versus Lévy processes, *Philos. Trans. R. Soc. A*, *366*(1875), 2455–2474, doi:10.1098/rsta.2008.0051.

Penland, C., and P. D. Sardeshmukh (1995), The optimal growth of tropical sea surface temperature anomalies, *J. Clim.*, *8*(8), 1999–2024.

Percival, D., and A. Walden (1993), *Spectral Analysis for Physical Applications: Multitaper and Conventional Univariate Techniques*, 612 pp., Cambridge Univ. Press, Cambridge, U. K.

Petit, J., et al. (2001), Vostok ice core data for 420,000 years, IGBP PAGES/World Data Center for Paleoclimatology Data Contribution Series 2001-076, http://www.ncdc.noaa.gov/paleo/icecore/antarctica/vostok/vostok.html, NOAA Paleoclimatol. Program, Boulder, Colo.

Pielke, R., Sr., et al. (2009), Climate change: The need to consider human forcings besides greenhouse gases, *Eos Trans. AGU*, *90*(45), 413, doi:10.1029/2009EO450008.

Pomeau, Y., and P. Manneville (1980), Intermittent transition to turbulence in dissipative dynamical systems, *Commun. Math. Phys.*, *74*(2), 189–197.

Rial, J., et al. (2004), Nonlinearities, feedbacks and critical thresholds within the Earth's climate system, *Clim. Change*, *65*, 11–38.

Rial, J., D. Noone, K. Nordstrom, T. Chase, and V. Gupta (2005), Toward a theory for millennial-scale climate variability through application of MEP in a simple dynamical model, *Eos Trans. AGU*, *86*(52), Fall Meet. Suppl., Abstract NG23A-0091.

Salazar, F., and G. Poveda (2009), Role of a simplified hydrological cycle and clouds in regulating the climate–biota system of Daisyworld, *Tellus, Ser. B*, *61*(2), 483–497.

Samorodnitsky, G. (2007), Long range dependence, *Found. Trends Stochastic Syst.*, *1*, 163–257.

Scheffer, M., J. Bascompte, W. Brock, V. Brovkin, S. Carpenter, V. Dakos, H. Held, E. Van Nes, M. Rietkerk, and G. Sugihara (2009), Early-warning signals for critical transitions, *Nature*, *461*(7260), 53–59.

Sura, P., M. Newman, C. Penland, and P. Sardeshmukh (2005), Multiplicative noise and non-Gaussianity: A paradigm for atmospheric regimes?, *J. Atmos. Sci.*, *62*, 1391–1409.

Syroka, J., and R. Toumi (2001), Scaling and persistence in observed and modeled surface temperature, *Geophys. Res. Lett.*, *28* (17), 3255–3258, doi:10.1029/2000GL012273.

Taylor, G. I. (1922), Diffusion by continuous movements, *Proc. London Math. Soc.*, *S2–20*, 196–212.

Viana, M. (1997), Stochastic dynamics of deterministic systems, *Lect. Notes 21*, Braz. Math. Colloq., Inst. Nac.de Mat. Pura e Apl., Rio de Janeiro, Brazil.

Wang, C. (2001), A unified oscillator model for the El Niño-Southern Oscillation, *J. Clim.*, *14*, 98–115.

Watson, A., and J. Lovelock (1983), Biological homeostasis of the global environment: The parable of Daisyworld, *Tellus, Ser. B*, *35* (4), 284–289.

Weber, S. L. (2001), On homeostasis in Daisyworld, *Clim. Change*, *48*, 465–485.

Wood, A. J., G. J. Ackland, J. G. Dyke, H. T. P. Williams, and T. M. Lenton (2008), Daisyworld: A review, *Rev. Geophys.*, *46*, RG1001, doi:10.1029/2006RG000217.

V. K. Gupta, Department of Civil, Environmental and Architectural Engineering and Cooperative Institute for Research in Environmental Sciences, University of Colorado, Boulder, CO 80309, USA.

O. J. Mesa, Escuela de Geociencias y Medio Ambiente, Facultad de Minas, Sede Medellín, Universidad Nacional de Colombia, Carrera 80 No 65-223, Bloque M2-316, Medellín 050041, Colombia. (ojmesa@unal.edu.co)

P. E. O'Connell, School of Civil Engineering and Geosciences, Newcastle University, Newcastle upon Tyne NE1 7RU, UK.

Low-Frequency Weather and the Emergence of the Climate

S. Lovejoy

Department of Physics, McGill University, Montreal, Quebec, Canada

D. Schertzer

LEESU, Ecole des Ponts ParisTech, Université Paris Est, Paris, France
Météo France, Paris, France

We survey atmospheric variability from weather scales up to several hundred kiloyears. We focus on scales longer than the critical $\tau_w \approx 5$–20 day scale corresponding to a drastic transition from spectra with high to low spectral exponents. Using anisotropic, intermittent extensions of classical turbulence theory, we argue that τ_w is the lifetime of planetary-sized structures. At τ_w, there is a dimensional transition; at longer times the spatial degrees of freedom are rapidly quenched, leading to a scaling "low-frequency weather" regime extending out to $\tau_c \approx 10$–100 years. The statistical behavior of both the weather and low-frequency weather regime is well reproduced by turbulence-based stochastic models and by control runs of traditional global climate models, i.e., without the introduction of new internal mechanisms or new external forcings; hence, it is still fundamentally "weather." Whereas the usual (high frequency) weather has a fluctuation exponent $H > 0$, implying that fluctuations *increase* with scale, in contrast, a key characteristic of low-frequency weather is that $H < 0$ so that fluctuations *decrease* instead. Therefore, it appears "stable," and averages over this regime (i.e., up to τ_c) define climate *states*. However, at scales beyond τ_c, whatever the exact causes, we find a new scaling regime with $H > 0$; that is, where fluctuations again *increase* with scale, climate states thus appear unstable; this regime is thus associated with our notion of *climate change*. We use spectral and difference and Haar structure function analyses of reanalyses, multiproxies, and paleotemperatures.

1. INTRODUCTION

1.1. What Is the Climate?

Notwithstanding the explosive growth of climate science over the last 20 years, there is still no clear universally accepted definition of what the climate *is* or what is almost the same thing, what distinguishes the weather from the climate. The core idea shared by most climate definitions is famously encapsulated in the dictum: "The climate is what you expect, the weather is what you get" (see *Lorenz* [1995] for a discussion). In more scientific language, "Climate in a narrow sense is usually defined as the average weather, or more rigorously, as the statistical description in terms of the mean and variability of relevant quantities over a period of time ranging from months to thousands or millions of years" [*Intergovernmental Panel on Climate Change*, 2007, p. 942].

Extreme Events and Natural Hazards: The Complexity Perspective
Geophysical Monograph Series 196

10.1029/2011GM001087

An immediate problem with these definitions is that they fundamentally depend on subjectively defined averaging scales. While the World Meteorological Organization defines climate as 30 years or longer variability, a period of 2 weeks to a month is commonly used to distinguish weather from climate so that even with these essentially arbitrary periods, there is still a range of about a factor 1000 in scale (2 weeks to 30 years) that is up in the air. This fuzzy distinction is also reflected in numerical climate modeling since global climate models are fundamentally the same as weather models but at lower resolutions, with a different assortment of subgrid parametrizations, and they are coupled to ocean models and, increasingly, to cryosphere, carbon cycle, and land use models. Consequently, whether we define the climate as the long-term weather statistics, or in terms of the long-term interactions of components of the "climate system," we still need an objective way to distinguish it from the weather. These problems are compounded when we attempt to objectively define climate *change*.

However, there is yet another problem with this and allied climate definitions: they imply that climate dynamics are nothing new, that they are simply weather dynamics at long time scales. This seems naïve since we know from physics that when processes repeat over wide-enough ranges of space or time scales, qualitatively new low-frequency laws should emerge. These "emergent" laws could simply be the consequences of long-range statistical correlations in the weather physics in conjunction with qualitatively new climate processes, due to either internal dynamics or to external forcings, their nonlinear synergy giving rise to emergent laws of climate dynamics.

1.2. Using the Type of Scaling Variability to Determine the Dynamical Regime

The atmosphere is a nonlinear dynamical system with interactions and variability occurring over huge ranges of space and time scales (millimeters to planet scales, milliseconds to billions of years, ratios $\approx 10^{10}$ and $\approx 10^{20}$, respectively), so that the natural approach is to consider it as a hierarchy of processes each with wide-range scaling, i.e., each with nonlinear mechanisms that repeat scale after scale over potentially wide ranges. Following the works of *Lovejoy and Schertzer* [1986], *Schmitt et al.* [1995], *Pelletier* [1998], *Koscielny-Bunde et al.* [1998], *Talkner and Weber* [2000], *Blender and Fraedrich* [2003], *Ashkenazy et al.* [2003], *Huybers and Curry* [2006], and *Rybski et al.* [2008], this approach is increasingly superseding earlier approaches that postulated more or less white noise backgrounds with a large number of spectral "spikes" corresponding to many different quasiperiodic processes. This includes the slightly more sophisticated variants [e.g., *Mitchell*, 1976], which retain the spikes but replace the white noise with a hierarchy of Ornstein-Uhlenbeck processes (white noises and their integrals). In the spectrum, these appear as "spikes" and "shelves" (see also *Fraedrich et al.* [2009] for a hybrid, which includes a single (short) scaling regime).

Over the past 25 years, scaling approaches have also been frequently applied to the atmosphere, mostly at small or regional scales but, in the last 5 years, increasingly to global scales. This has given rise to a new scaling synthesis covering the entire gamut of meteorological scales from milliseconds to beyond the $\approx$10 day period, which is the typical lifetime of planetary structures, i.e., the weather regime. In the *Lovejoy and Schertzer* [2010] review, it was concluded that the theory and data were consistent with wide range but anisotropic spatial scaling and that the lifetime of planetary-sized structures provides the natural scale at which to distinguish weather and a qualitatively different lower-frequency regime. Figure 1a shows a recent composite indicating the three basic regimes covering the range of time scales from $\approx$100 kyr down to weather scales.

The label "weather" for the high-frequency regime seems clearly justified and requires no further comment. Similarly the lowest frequencies correspond to our usual ideas of

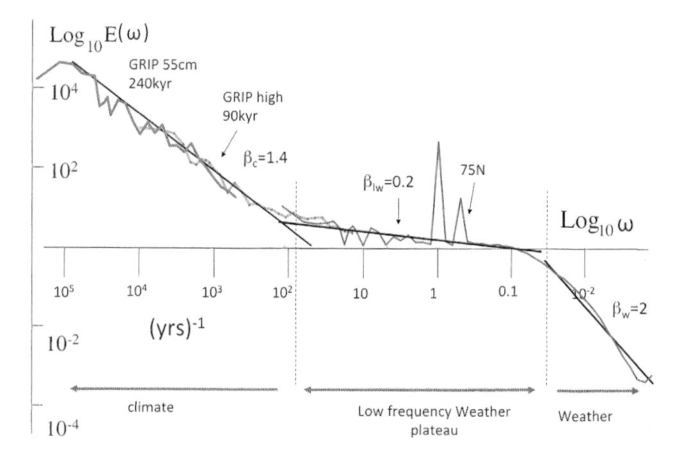

Figure 1a. A modern composite based only on two sources: the Greenland Ice Core Project (GRIP) core (paleotemperatures from Summit Greenland) and the Twentieth Century Reanalyses (20CR) at the same latitude (75°N, thin line on the right). All spectra have been averaged over logarithmically spaced bins, 10 per order of magnitude, and the 20CR spectra have been averaged over all 180° longitude, 2° × 2° elements; frequency units are (years)$^{-1}$. The thin gray line with points is the mean of the GRIP 5.2 resolution data for the last 90 kyr, and the (lowest) frequency line is from the lower-resolution (55 cm) GRIP core interpolated to 200 year resolution and going back 240 kyr. The black reference lines have absolute slopes $\beta_{lw} = 0.2$, $\beta_c = 1.4$, and $\beta_w = 2$ as indicated. The arrows at the bottom indicate the basic qualitatively different scaling regimes.

multidecadal, multicentennial, multimillennial variability as "climate." However, labeling the intermediate region "low-frequency weather" (rather than say "high-frequency climate") needs some justification. The point is perhaps made more clearly with the help of Figure 1b, which shows a blowup of Figure 1a with both globally and locally averaged instrumentally based spectra as well as the spectrum of the output of the stochastic fractionally integrated flux (FIF) model [*Schertzer and Lovejoy*, 1987] and the spectrum of the output of a standard global climate model (GCM) "control run," i.e., without special anthropogenic, solar, volcanic, orbital, or other climate forcings. This regime is, therefore, no more than "low-frequency weather"; it contains no new internal dynamical elements or any new forcing mechanism. As we discuss below, whereas the spectra from data (especially when globally averaged) begin to rise for frequencies below $\approx$(10 years)$^{-1}$, both the FIF and GCM control runs maintain their gently sloping "plateau-like" behaviors out to at least (500 years)$^{-1}$ (note that we shall see that the "plateau" is not perfectly flat, but its logarithmic slope is small, typically in the range -0.2 to -0.6). Similar conclusions for the control runs of other GCMs at even lower frequencies were found by *Blender et al.* [2006] and *Rybski et al.* [2008] so that it seems that in the absence of external climate forcings, the GCMs reproduce the low-frequency weather regime but not the lower-frequency strongly spectrally rising regime that requires some new *climate* ingredient. The aim of this chapter is to understand the natural variability so that the important question of whether or not GCMs with realistic forcings might be able to reproduce the low-frequency climate regime is outside our present scope. Certainly, existing studies of the scaling of forced GCMs [*Vyushin et al.*, 2004; *Blender et al.*, 2006; *Rybski et al.*, 2008] have consistently reported unique low-frequency weather, but not climate, exponents, and this even at the lowest-simulated frequencies.

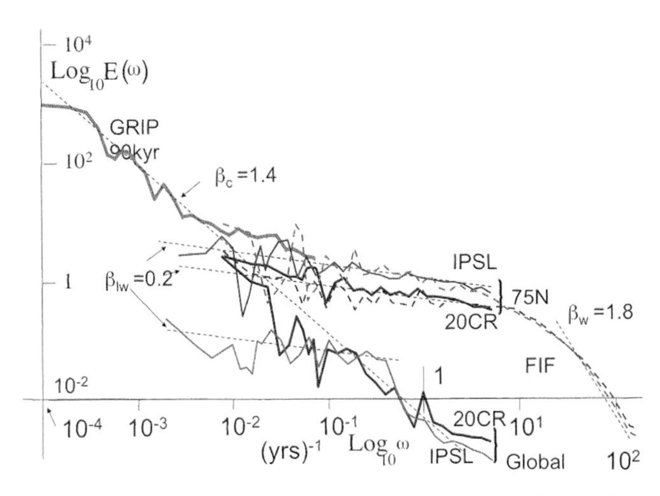

Figure 1b. A comparison of the spectra of temperature fluctuations from the GRIP Greenland Summit core, last 90 kyr, 5.2 year resolution (thick gray, upper left), monthly 20CR reanalysis (black), global (bottom) and 2° resolution (top at 75°N), and a 500 year control run of the monthly Institut Pierre Simon Laplace (IPSL) GCM (gray) used in the 4th IPCC report, at the corresponding resolutions. Frequency units are (years)$^{-1}$. The dashed lines are the detrended daily, grid-scale data (75°N 20CR, dashed dark) and the cascade-based fractionally integrated flux (FIF) simulation (dashed light), both adjusted vertically to coincide with the analyses of the other, monthly scale data. Reference lines with slopes $\beta_c = 1.4$, $\beta_{lw} = 0.2$, $\beta_w = 1.8$ are shown. Notice that the IPSL control run, which lacks external climate forcing and is therefore simply low-frequency weather as well as the low-frequency extension of the cascade-based FIF model, continues to have shallow spectral slopes out to their low-frequency limits, whereas the globally averaged 20CR and paleospectra follow $\beta_c \approx 1.4$ to roughly follow their low-frequency limits. Hence, the plateau is best considered "low-frequency weather": the true climate regime has a much steeper spectrum determined either by new low-frequency internal interactions or by the low-frequency climate "forcing" (solar, orbital, volcanic, anthropogenic, or other); Figure 1a shows that the $\beta_c \approx 1.4$ regime continues to $\approx$(100 kyr)$^{-1}$.

2. TEMPORAL SCALING, WEATHER, LOW-FREQUENCY WEATHER, AND THE CLIMATE

2.1. Discussion

Although spatial scaling is fundamental for weather processes, time scales much greater than $\tau_w \approx 10$ days the spatial degrees of freedom essentially collapse (via a "dimensional transition," section 2.5), so that we focus only on temporal variability. Indeed, *Lovejoy and Schertzer* [2012] argue that the spatial variability in the low-frequency weather regime is very large and is associated with different climatic zones. However, it is primarily due to even lower frequency climate-scale processes, so we do not pursue it here.

In order to simplify things as much as possible in section 2, we will only use spectra. Consider a random field $f(t)$ where t is time. Its "spectral density" $E(\omega)$ is the average total contribution to the variance of the process due to structures with frequency between ω and $\omega + d\omega$ (i.e., due to structures of duration $\tau = 2\pi/\omega$, where τ is the corresponding time scale). $E(\omega)$ is thus defined as

$$E(\omega) = \langle |\widetilde{f(\omega)}|^2 \rangle, \tag{1}$$

where $\widetilde{f(\omega)}$ is the Fourier transform of $f(t)$, and the angular brackets indicate statistical averaging. Here $\langle f(t)^2 \rangle$ is thus the total variance (assumed to be independent of time), so that the spectral density thus satisfies $\langle f(t)^2 \rangle = \int_0^\infty E(\omega) d\omega$.

In a scaling regime, we have power law spectra:

$$E(\omega) \approx \omega^{-\beta}. \quad (2)$$

If the time scales are reduced in scale by factor λ, we obtain $t \rightarrow \lambda^{-1}t$; this corresponds to a "blow up" in frequencies: $\omega \rightarrow \lambda\omega$; and the power law $E(\omega)$ (equation (2)) maintains its form: $E \rightarrow \lambda^{-\beta}E$ so that E is "scaling," and the spectral exponent β is "scale invariant." Thus, if empirically we find E to be of the form equation (2), we take this as evidence for the scaling of the field f. Note that numerical spectra have well-known finite size effects; the low-frequency effects have been dealt with below using standard "windowing" techniques (here a Hann window was used to reduce spectral leakage).

2.2. Temporal Spectral Scaling in the Weather Regime

One of the earliest atmospheric spectral analyses was that of the work of *Van der Hoven* [1957], whose graph is at the origin of the legendary "mesoscale gap," the supposedly energy-poor spectral region between roughly 10–20 min and ≈4 days (ignoring the diurnal spike). Even until fairly recently, textbooks regularly reproduced the spectrum (often redrawing it on different axes or introducing other adaptations), citing it as convincing empirical justification for the neat separation between low-frequency isotropic 2-D turbulence, identified with the weather, and high-frequency isotropic 3-D "turbulence." This picture was seductive since if the gap had been real, the 3-D turbulence would be no more than an annoying source of perturbation to the (2-D) weather processes.

However, it was quickly and strongly criticized (e.g., by *Goldman* [1968], *Pinus* [1968], *Vinnichenko* [1969], *Vinnichenko and Dutton* [1969], *Robinson* [1971], and indirectly by *Hwang* [1970]). For instance, on the basis of much more extensive measurements, *Vinnichenko* [1969] commented that even if the mesoscale gap really existed, it could only be for less than 5% of the time; he then went on to note that Van der Hoven's spectrum was actually the superposition of four spectra and that the extreme set of high-frequency measurements were taken during a single 1 h long period during an episode of "near-hurricane" conditions, and these were entirely responsible for the high-frequency "bump."

More modern temporal spectra are compatible with scaling from dissipation scales to ≈5–20 days. Numerous wind and temperature spectra now exist from milliseconds to hours and days showing, for example, that $\beta \approx 1.6$ and 1.8 for v and T, respectively; some of this evidence is reviewed by *Lovejoy and Schertzer* [2010, 2012]. Figure 2 shows an example of the hourly temperature spectrum for frequencies down to (4 years)$^{-1}$. According to Figure 2, it is plausible that the

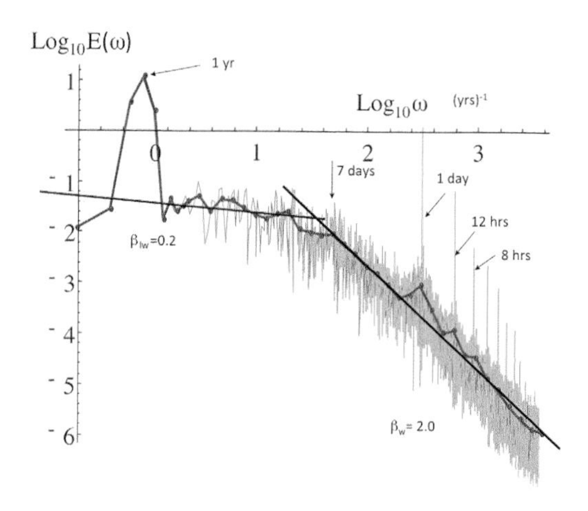

Figure 2. Scaling of hourly surface temperatures from four stations in the northwest United States for 4 years (2005–2008), taken from the U.S. Climate Reference Network. One can see that in spite of the strong diurnal cycle (and harmonics), the basic scaling extends to about 7 days. The reference lines (with absolute slopes 0.2, 2) are theoretically motivated for low-frequency weather and weather scales, respectively. The spectra of hourly surface temperature data are from four nearly colinear stations running northwest-southeast in the United States (Lander, Wyoming; Harrison, Nebraska; Whitman, Nebraska; and Lincoln, Nebraska. The gray line is the raw spectrum; the thick line is the spectrum of the periodically detrended spectrum, averaged over logarithmically spaced bins, 10 per order of magnitude.

scaling in the temperature holds from small scales out to scales of ≈5–10 days, where we see a transition. This transition is essentially the same as the low-frequency "bump" observed by Van der Hoven; its appearance only differs because he used a $\omega E(\omega)$ rather than $\log E(\omega)$ plot.

2.3. Temporal Spectral Scaling in the Low-Frequency Weather-Climate Regime

Except for the annual cycle, the roughly flat low-frequency spectral transitions in Figure 2 (and Figure 1a between ≈10 days and 10 years) are qualitatively reproduced in all the standard meteorological fields, and the transition scale τ_w is relatively constant. Figure 3 shows estimates of τ_w, estimated using reanalyses taken from the Twentieth Century Reanalysis project (20CR) [*Compo et al.*, 2011] on 2° × 2° grid boxes. Also shown are estimates of τ_c, the scale where the latter ends and the climate regime begins. Between τ_w and τ_c is the low-frequency weather regime; it covers a range of a factor ≈1000 in scale. Also shown are estimates of the planetary-scale eddy turnover time discussed below.

Figure 4 shows the surface air temperature analysis out to lower frequencies and compares this with the corresponding spectrum for sea surface temperatures (SSTs) (section 2.7). We see that the ocean behavior is qualitatively similar except

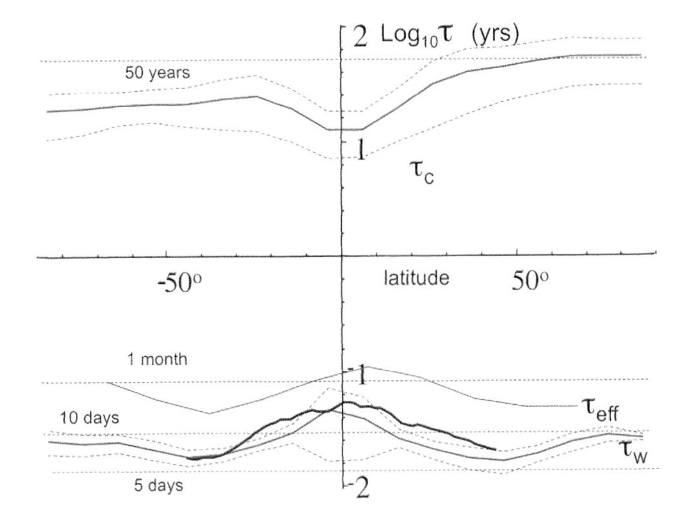

Figure 3. Variation of (bottom) τ_w and (top) τ_c as a function of latitude as estimated from the 138 year long 20CR reanalyses, 700 mb temperature field, compared with (bottom) the theoretically predicted planetary-scale eddy turnover time (τ_{eddy}, black) and the effective external scale (τ_{eff}) of the temperature cascade estimated from the European Centre for Medium-Range Weather Forecasts interim reanalysis for 2006 (thin gray). The τ_w estimates were made by performing bilinear log-log regressions on spectra from 180 day long segments averaged over 280 segments per grid point. The thick gray curves show the mean over all the longitudes; the dashed lines are the corresponding longitude to longitude 1 standard deviation spreads. The τ_c were estimated by bilinear log-log fits on the Haar structure functions applied to the same data but averaged monthly.

that the transition time scale τ_o is ≈ 1 year, and the "low-frequency ocean" exponent $\beta_{lo} \approx 0.6$ is a bit larger than the corresponding $\beta_{lw} \approx 0.2$ for the air temperatures ("*lw*," "*lo*" for "low" frequency "weather" and "ocean," respectively). For comparison, we also show the best fitting Orenstein-Uhlenbeck processes (essentially integrals of white noises, i.e., with $\beta_w = 2$, $\beta_{lw} = 0$); these are the basis of stochastic linear modeling approaches [e.g., *Penland*, 1996]. We see that they are only rough approximations to the true spectra.

To underline the ubiquity of the low-frequency weather regime, its low β character, and to distinguish it from the higher-frequency weather regime, this regime was called the "spectral plateau" [*Lovejoy and Schertzer*, 1986], although it is somewhat of a misnomer since it is clear that the regime has a small but nonzero logarithmic slope whose negative value we indicate by β_{lw}. The transition scale τ_w was also identified as a weather scale by *Koscielny-Bunde et al.* [1998].

2.4. The Weather Regime, Space-Time Scaling, and Some Turbulence Theory

In order to understand the weather and low-frequency weather scaling, we briefly recall some turbulence theory using the example of the horizontal wind *v*. In stratified scaling turbulence (the 23/9D model) [*Schertzer and Lovejoy*, 1985a; *Schertzer et al.*, 2012], the energy flux ε dominates the horizontal, and the buoyancy variance flux ϕ dominates the vertical so that horizontal wind fluctuations Δ*v* (e.g., differences, see section 3) follow

$$\Delta v(\Delta x) = \varepsilon^{1/3}\Delta x^{H_h}, \quad H_h = 1/3, \tag{3a}$$

$$\Delta v(\Delta y) = \varepsilon^{1/3}\Delta y^{H_h}, \quad H_h = 1/3, \tag{3b}$$

$$\Delta v(\Delta z) = \phi^{1/5}\Delta z^{H_v}, \quad H_v = 3/5, \tag{3c}$$

$$\Delta v(\Delta t) = \varepsilon^{1/2}\Delta t^{H_\tau}, \quad H_\tau = 1/2, \tag{3d}$$

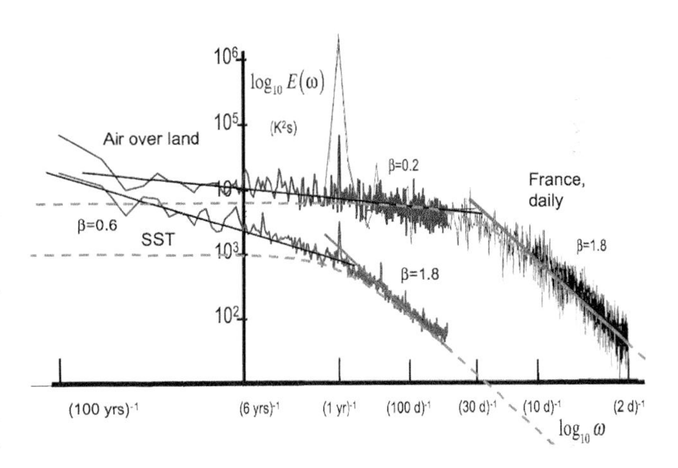

Figure 4. Superposition of the ocean and atmospheric plateaus showing their great qualitative similarity. (bottom left) A comparison of the monthly SST spectrum) and (top) monthly atmospheric temperatures over land for monthly temperature series from 1911 to 2010 on a 5° × 5° grid (the NOAA National Climatic Data Center data). Only those near complete series (missing less than 20 months out of 1200) were considered; there were 465 for the SST and 319 for the land series; the missing data were filled using interpolation. The reference slopes correspond to (top) β = 0.2, (bottom left) 0.6, and (bottom right) 1.8. A transition at 1 year corresponds to a mean ocean $\varepsilon_o \approx 1 \times 10^{-8}$ m^2 s^{-3}. The dashed lines are Orenstein-Uhlenbeck processes (of the form $E(\omega) = \sigma^2/(\omega^2 + a^2)$, where σ and *a* are constants) used as the basis for stochastic linear forcing models. (right) The average of five spectra from 6 year long sections of a 30 year series of daily temperatures at a station in France (black) (taken from the work of *Lovejoy and Schertzer* [1986]). The gray reference line has a slope 1.8. The relative up-down placement of this daily spectrum with the monthly spectra (corresponding to a constant factor) was determined by aligning the atmospheric spectral plateaus (i.e., the black and gray spectra). The raw spectra are shown (no averaging over logarithmically spaced bins).

where Δx, Δy, Δz, Δt are the increments in horizontal coordinates, vertical coordinate, and time, respectively, and the exponents H_h, H_v, H_τ are "fluctuation" or "nonconservation" exponents in the horizontal, vertical, and time, respectively. Since the mean fluxes are independent of scale (i.e., $\langle\varepsilon\rangle$, $\langle\phi\rangle$ are constant, angular brackets indicate ensemble averaging), these exponents express how the mean fluctuations Δv increase ($H > 0$) or decrease ($H < 0$) with the scale (e.g., the increments, when $H > 0$, see below). Equations (3a) and (3b) describe the real space horizontal [*Kolmogorov*, 1941] scaling, and equation (3c) describes the vertical Bolgiano-Obukhov [*Bolgiano*, 1959; *Obukhov*, 1959] scaling for the velocity. The anisotropic Corrsin-Obukov law for passive scalar advection is obtained by the replacements $v \rightarrow \rho$; $\varepsilon \rightarrow \chi^{3/2}\varepsilon^{-1/2}$, where ρ is the passive scalar density, χ is the passive scalar variance flux [*Corrsin*, 1951; *Obukhov*, 1949].

Before proceeding, here are a few technical comments. In equation (3), the equality signs should be understood in the sense that each side of the equation has the same scaling properties: the FIF model (Figure 1b) is essentially a more precise interpretation of the equations in terms of fractional integrals of order H. Ignoring intermittency (associated with multifractal fluxes, which we discuss only briefly below), the spectral exponents are related to H as $\beta = 1 + 2H$ so that $\beta_h = 5/3$, $\beta_v = 11/5$, $\beta_\tau = 2$. Finally, although the notation "H" is used in honor of E. Hurst, in multifractal processes, it is generally *not* identical to the Hurst (i.e., rescaled range, "R/S") exponent, the relationship between the two is nontrivial.

Although these equations originated in classical turbulence theory, the latter were all spatially statistically isotropic so that the simultaneous combination of the horizontal laws (3a) and (3b) with the vertical law (3c) is nonclassical. Since the isotropy assumption is very demanding, the pioneers originally believed that the classical laws would hold over only scales of scales of hundreds of meters. The anisotropic extension implied by equations (3a)–(3d) is itself based on a generalization of the notion of scale invariance and thus has the effect of radically changing the potential range of validity of the laws. For example, even the finite thickness of the troposphere, which in isotropic turbulence would imply a scale break at around 10 km, no longer implies a break in the scaling. Beginning with the work of *Schertzer and Lovejoy* [1985b], it has been argued that atmospheric variables including the wind do indeed have wide range (anisotropic) scaling statistics (see the review by *Lovejoy and Schertzer* [2010]).

In addition, the classical turbulence theories were for spatially uniform ("homogeneous") turbulence in which the fluxes were quasiconstant (e.g., with Gaussian statistics). In order for these laws to apply up to planetary scales, starting in the 1980s [*Parisi and Frisch*, 1985; *Schertzer and Lovejoy*, 1985b], they were generalized to strongly variable (intermittent) multiplicative cascade processes yielding multifractal fluxes so that, although the mean flux statistics $\langle\varepsilon_{\Delta x}\rangle$ remain independent of scale Δx, the statistical moments:

$$\langle\varepsilon_{\Delta x}^{q}\rangle \approx \Delta x^{-K(q)}, \tag{4}$$

where $\varepsilon_{\Delta x}$ is the flux averaged over scales Δx, and $K(q)$ is the (convex) moment scaling exponent. Although $K(1) = 0$, for $q \neq 1$, $K(1) \neq 0$ so that $\langle\varepsilon_{\Delta x}{}^{q}\rangle$ is strongly scale *dependent*, the fluxes are thus the densities of singular multifractal measures.

Along with the spatial laws (equations (3a)–(3c)), we have included equation (3d), which is the result for the pure time evolution in the absence of an overall advection velocity; this is the classical Lagrangian version of the Kolmogorov law [*Inoue*, 1951; *Landau and Lifschitz*, 1959]; it is essentially the result of dimensional analysis using ε and Δt rather than ε and Δx. Although Lagrangian statistics are notoriously difficult to obtain empirically (see, however, *Seuront et al.* [1996]), they are roughly known from experience and are used as the basis for the space-time or "Stommel" diagrams that adorn introductory meteorology textbooks (see *Schertzer et al.* [1997] and *Lovejoy et al.* [2000] for scaling adaptations).

Due to the fact that the wind is responsible for advection, the spatial scaling of the horizontal wind leads to the temporal scaling of all the fields. Unfortunately, space-time scaling is somewhat more complicated than pure spatial scaling. This is because, at meteorological time scales, we must take into account the mean advection of structures and the Galilean invariance of the dynamics. The effect of the Galilean invariance/advection of structures is that the temporal exponents in the Eulerian (fixed Earth) frame become the same as in the horizontal direction (i.e., in (x,y,t) space, we have "trivial anisotropy," i.e., with "effective" temporal exponent $H_{\tau\mathrm{eff}} = H_h$) [*Lovejoy et al.*, 2008; *Radkevitch et al.*, 2008]. At the longer time scales of low-frequency weather, the scaling is broken because the finite size of the Earth implies a characteristic lifetime ("eddy turnover time") of planetary-scale structures. Using equations (3a) and (3b) with $\Delta x = L_e$, we obtain: $\tau_{\mathrm{eddy}} = L_e/\Delta v(L_e) = \varepsilon_w{}^{-1/3}L_e{}^{2/3}$, where $L_e = 20{,}000$ km, the size of the planet, and ε_w is the globally averaged energy flux density.

Considering time scales longer than τ_{eddy}, the effect of the finite planetary size implies that the spatial degrees of freedom become ineffective (there is a "dimensional transition") so that instead of interactions in (x,y,z,t) space for long times, the interactions are effectively only in t space, and this implies a drastic change in the statistics, summarized in the next section.

2.5. Low-Frequency Weather and the Dimensional Transition

To obtain theoretical predictions for the statistics of atmospheric variability at time scales $\tau > \tau_w$ (i.e., in the low-frequency weather regime), we can take the FIF [*Schertzer and Lovejoy*, 1987] model that produces multifractal fields respecting equation (3) and extend it to processes with outer time scales $\tau_c >> \tau_w$. The theoretical details are given by *Lovejoy and Schertzer* [2010, 2012], but the upshot of this is that we expect the energy flux density ε to factor into a statistically independent space-time weather process $\varepsilon_w(\underline{r},t)$ and a low-frequency weather process $\varepsilon_{lw}(t)$, which is only dependent on time:

$$\varepsilon(\underline{r},t) = \varepsilon_w(\underline{r},t)\varepsilon_{lw}(t). \tag{5}$$

In this way, the low-frequency energy flux $\varepsilon_{lw}(t)$, which physically is the result of nonlinear radiation/cloud interactions, multiplicatively modulates the high-frequency space-time weather processes. *Lovejoy and Schertzer* [2012] discuss extensions of this model to include even lower frequency space-time climate processes; it is sufficient to include a further climate flux factor in equation (5).

The theoretical statistical behavior of $\varepsilon_{lw}(t)$ is quite complex to analyze and has some surprising properties. Some important characteristics are the following: (1) At large temporal lags Δt, the autocorrelation $\langle\varepsilon_{lw}(t)\ \varepsilon_{lw}(t-\Delta t)\rangle$ ultimately decays as Δt^{-1}, although very large ranges of scale may be necessary to observe it. (2) Since the spectrum is the Fourier transform of the autocorrelation, and the transform of a pure Δt^{-1} function has a low- (and high) frequency divergence, the actual spectrum of the low-frequency weather regime depends on its overall range of scales $\Lambda_c = \tau_c/\tau_w$. From Figure 3, we find that to within a factor of ≈ 2, the mean Λ_c over the latitudes is ≈ 1100. (3) Over surprisingly wide ranges (factors of 100–1000 in frequency for values of Λ_c in the range 2^{10}–2^{16}), one finds "pseudoscaling" with nearly constant spectral exponents β_{lw}, which are typically in the range 0.2–0.4. (4) The statistics are independent of H and only weakly dependent on $K(q)$.

In summary, we therefore find for the overall weather/low frequency weather (FIF) model

$$\begin{aligned} E(k) &\approx k^{-\beta_w}; & k &> L_e^{-1} \\ E(\omega) &\approx \omega^{-\beta_w}; & \omega &> \tau_w^{-1}, \\ E(\omega) &\approx \omega^{-\beta_{lw}}; & \tau_c^{-1} &< \omega < \tau_w^{-1}, \end{aligned} \tag{6}$$

where k is the modulus of the horizontal wave vector, τ_c is the long external climate scale where the low-frequency weather regime ends (see discussion below), and β_l, β_{lw} are

$$\begin{aligned} \beta_w &= 1 + 2H_w - K(2) \\ 0.2 &< \beta_{lw} < 0.6. \end{aligned} \tag{7}$$

In the low-frequency weather regime, the intermittency (characterized by $K(q)$) decreases as we average the process over scales $>\tau_w$ so that the low-frequency weather regime has an effective fluctuation exponent H_{lw}:

$$H_{lw} \approx -(1 - \beta_{lw})/2. \tag{8}$$

The high-frequency weather spectral exponents β_w, H_w are the usual ones, but the low-frequency weather exponents β_{lw}, H_{lw} are new. Since $\beta_{lw} < 1$, we have $H_{lw} < 0$, and using $0.2 < \beta_{lw} < 0.6$ corresponding to $-0.4 < H_{lw} < -0.2$, this result already explains the preponderance of spectral plateau β around that value already noted. However, as we saw in Figure 4, the low-frequency ocean regime has a somewhat high value $\beta_{oc} \approx 0.6$, $H_{oc} \approx -0.2$; *Lovejoy and Schertzer* [2012] use a simple coupled ocean-atmosphere model to show how this could arise as a consequence of double (atmosphere and ocean) dimensional transitions. The fact that $H_{lw} < 0$ indicates that the mean fluctuations decrease with scale so that the low-frequency weather and ocean regimes are "stable."

2.6. The Transition Time Scale From Weather to Low-Frequency Weather Using "First Principles"

Figures 1–4 show evidence that temporal scaling holds from small scales to a transition scale τ_w of around 5–20 days, which we mentioned was the eddy turnover time (lifetime) of planetary-scale structures. Let us now consider the physical origin of this scale in more detail. In the famous [*Van der Hoven*, 1957] $\omega E(\omega)$ versus $\log\omega$ plot, its origin was argued to be due to "migratory pressure systems of synoptic weather-map scale." The corresponding features at around 4–20 days notably for temperature and pressure spectra were termed "synoptic maxima" by *Kolesnikov and Monin* [1965] and *Panofsky* [1969] in reference to the similar idea that it was associated with synoptic-scale weather dynamics (see *Monin and Yaglom* [1975] for some other early references).

More recently, *Vallis* [2010] suggested that τ_w is the basic lifetime of baroclinic instabilities, which he estimated using the inverse Eady growth rate: $\tau_{\text{Eady}} \approx L_d/U$, where the deformation rate is $L_d = NH/f_0$, f_0 is the Coriolis parameter, and H is the thickness of the troposphere, where N is the mean Brunt-Väisällä frequency, and U is the typical wind. The Eady growth rate is obtained by linearizing the equations about a hypothetical state with both uniform shear and stratification across the troposphere. By taking $H \approx 10^4$ m, $f_0 \approx 10^{-4}$ s^{-1}, $N = 0.01$ Hz, and $U \approx 10$ m s^{-1}, Vallis obtained the estimate $L_d \approx 1000$ km. Using the maximum Eady growth rate theoretically introduces a numerical factor 3.3 so that the actual predicted inverse growth rate is: $3.3\tau_{\mathrm{Eady}} \approx 4$ days. Vallis similarly argued that this also applies to the oceans but with $U \approx 10$ cm s^{-1} and $L_d \approx$ 100 km yielding $3.3\tau_{\mathrm{Eady}} \approx 40$ days. The obvious theoretical problem with using τ_{Eady} to estimate τ_w is that the former is expected to be valid in homogeneous, quasilinear systems, whereas the atmosphere is highly heterogeneous with vertical and horizontal structures (including strongly nonlinear cascade structures) extending throughout the troposphere to scales substantially larger than L_d. Another difficulty is that although the observed transition scale τ_w is well behaved at the equator (Figure 3), f_0 vanishes implying that L_d and τ_{Eady} diverge: using τ_{Eady} as an estimate of τ_w is at best a midlatitude approximation. Finally, there is no evidence for any special behavior at length scales near L_d ≈ 1000 km.

If there is (at least statistically) a well-defined relation between spatial scales and lifetimes (the "eddy turnover time"), then the lifetime of planetary-scale structures τ_{eddy} is of fundamental importance. The shorter period ($\tau < \tau_{\mathrm{eddy}}$) statistics are dominated by structures smaller than planetary size, whereas for $\tau > \tau_{\mathrm{eddy}}$, they are dominated by the statistics of many lifetimes of planetary-scale structures. It is therefore natural to take $\tau_w \approx \tau_{\mathrm{eddy}} = \varepsilon_w^{-1/3} L_e^{2/3}$.

In order to estimate τ_w, we therefore need an estimate of the globally averaged flux energy density ε_w. We can estimate ε_w by using the fact that the mean solar flux absorbed by the Earth is $\approx$200 W m^{-2} [e.g., *Monin*, 1972]. If we distribute this over the troposphere (thickness $\approx 10^4$ m), with mean air density ≈ 0.75 kg m^{-3}, and we assume a 2% conversion of energy into kinetic energy [*Palmén*, 1959; *Monin*, 1972], then we obtain a value $\varepsilon_w \approx 5 \times 10^{-4}$ m^2 s^{-3}, which is indeed typical of the values measured in small-scale turbulence [*Brunt*, 1939; *Monin*, 1972]. Using the European Centre for Medium-Range Weather Forecasts (ECMWF) interim reanalysis to obtain a modern estimate of ε_w, *Lovejoy and Schertzer* [2010] showed that although ε is larger in midlatitudes than at the equator and that at 300 mb it reaches a maximum, the global tropospheric average is $\approx 10^{-3}$ m^2 s^{-3}. They also showed that the latitudinally varying ε explains to better than ±20% the latitudinal variation of the hemispheric antipodes velocity differences (using $\Delta v = \varepsilon^{1/3} L_e^{1/3}$). They concluded that the solar energy flux does a good job of explaining the horizontal wind fluctuations up to planetary scales. In addition, we can now point to Figure 3, which shows that the latitudinally varying ECMWF estimates of $\varepsilon_w(\theta)$ do indeed lead to $\tau_{\mathrm{eddy}}(\theta)$ very close to the direct 20CR $\tau_w(\theta)$ estimates for the 700 mb temperature field.

2.7. Ocean "Weather" and "Low-Frequency Ocean Weather"

It is well known that for months and longer time scales, ocean variability is important for atmospheric dynamics; before explicitly attempting to extend this model of weather variability beyond τ_w, we must therefore consider the role of the ocean. The ocean and the atmosphere have many similarities; from the preceding discussion, we may expect analogous regimes of "ocean weather" to be followed by an ocean spectral plateau both of which will influence the atmosphere. To make this more plausible, recall that both the atmosphere and ocean are large Reynolds' number turbulent systems and both are highly stratified, albeit due to somewhat different mechanisms. In particular, there is no question that at least over some range, horizontal ocean current spectra are dominated by the ocean energy flux ε_o. It roughly follows that $E(\omega) \approx \omega^{-5/3}$ and presumably in the horizontal: $E(k) \approx k^{-5/3}$ (i.e., $\beta_o = 5/3$, $H_o = 1/3$) [see, e.g., *Grant et al.*, 1962; *Nakajima and Hayakawa*, 1982]. Although surprisingly few current spectra have been published, the recent use of satellite altimeter data to estimate sea surface height (a pressure proxy) has provided relevant empirical evidence that $k^{-5/3}$ continues out to scales of at least hundreds of kilometers, refueling the debate about the spectral exponent and the scaling of the current [see *Le Traon et al.*, 2008].

Although empirically the current spectra (or their proxies) at scales larger than several hundred kilometers are not well known, other spectra, especially those of SSTs, are known to be scaling over wide ranges, and due to their strong nonlinear coupling with the current, they are relevant. Using mostly remotely sensed IR radiances, and starting in the early 1970s, there is much evidence for SST scaling up to thousands of kilometers with $\beta \approx 1.8$–2, i.e., nearly the same as for the atmospheric temperature (see, e.g., *McLeish* [1970], *Saunders* [1972], *Deschamps et al.* [1981, 1984], *Burgert and Hsieh* [1989], *Seuront et al.* [1996], and *Lovejoy et al.* [2000] and a review by *Lovejoy and Schertzer* [2012]).

If, as in the atmosphere, the energy flux dominates the horizontal ocean dynamics, then we can use the same methodology as in the previous subsection (basic turbulence theory (the Kolmogorov law) combined with the mean ocean energy flux ε_o) to predict ocean eddy turnover time and hence the outer scale τ_o of the ocean regime. Thus, for ocean gyres and eddies of size l, we expect there to be a characteristic eddy turnover time (lifetime) $\tau = \varepsilon^{-1/3} l^{2/3}$ with a critical "ocean weather"-"ocean climate" transition time scale obtained when $l = L_e$: $\tau_o = \varepsilon_o^{-1/3} L_e^{2/3}$. Again, we expect a fundamental difference in the statistics for fluctuations of duration $\tau < \tau_o$, the ocean equivalent of "weather" with a turbulent spectrum with roughly $\beta_o \approx 5/3$ (at least for the current), and for durations $\tau > \tau_o$, the ocean "climate" with a shallow ocean spectral regime with $\beta \approx <1$. Since the spatial β for temperature in the atmosphere and ocean are very close, if the β for the current and wind are also close, then so will the β for the temporal temperature spectra.

In order to test this idea, we need the globally averaged ocean current energy flux, ε_o. As expected, ε_o is highly intermittent [see *Robert*, 1976; *Clayson and Kantha*, 1999; *Moum et al.*, 1995; *Lien and D'Asaro*, 2006; *Matsuno et al.*, 2006], and as far as we know, the only attempt to estimate its global average was *Lovejoy and Schertzer* [2012], who used ocean drifter maps of eddy kinetic energy. They found that $\varepsilon_o \approx 10^{-8}$ m^2 s^{-3} is a reasonable global estimate for the surface layer (it decreases quite rapidly with depth). Using the formula $\tau_o = \varepsilon_o^{-1/3} L_e^{2/3}$ and ε_o in the range 1×10^{-8} to 8×10^{-8} m^2 s^{-3}, we find $\tau_o \approx 1$–2 years; compare this with the values for the atmosphere: $\varepsilon_w \approx 10^{-3}$ m^2 s^{-3}, $\tau_w \approx$ 10 days.

This provides us with a prediction for the SST spectrum: $E(\omega) \approx \omega^{-1.8}$ for ω (1 year)$^{-1}$ followed by a transition to a much flatter plateau (here $\approx \omega^{-0.6}$) for the lower frequencies (see Figure 4, which compares the ocean and air over land spectra). While the latter spectrum is, as expected, essentially a pure spectral plateau (with $\beta_{lw} \approx 0.2$, the value cited earlier), we see that the SST spectrum is essentially the same ($\beta_o \approx \beta_w \approx 1.8$) except that $\beta_{lo} \approx 0.6$ and $\tau_o \approx 1$ year. This basic "crossover" to an exponent $\beta_{lo} \approx 0.6$ was already noted by *Monetti et al.* [2003], who estimated it as 300 days. Note also the rough convergence of the spectra at about 100 years, which implies that the land and ocean variability become equal and also the hint that there is a low-frequency rise in the land spectrum for periods $>\approx 30$ years.

2.8. Other Evidence for the Spectral Plateau

Various published scaling composites such as Figures 1a and 1b give estimates for the low-frequency weather exponent β_{lw}, the climate exponent β_c, and the transition scale τ_c; they agree on the basic picture while proposing somewhat different parameter values and transition scales τ_c. For example, *Huybers and Curry* [2006] studied many paleoclimate series as well as the 60 year long National Centers for Environmental Prediction (NCEP) reanalyses concluded that for periods of months up to about 50 years, the spectra are scaling with midlatitude β_{lw} larger than the tropical β_{lw} (their values are 0.37 ± 0.05, 0.56 ± 0.08). Many analyses in the spectral plateau regime have been carried out using in situ data with the detrended fluctuation analysis (DFA) method [*Fraedrich and Blender*, 2003; *Koscielny-Bunde et al.*, 1998; *Bunde et al.*, 2004], SSTs [*Monetti et al.*, 2003], and $\approx$1000 year long Northern Hemisphere reconstructions [*Rybski et al.*, 2006]; see also the works of *Lennartz and Bunde* [2009] and *Lanfredi et al.* [2009] and see the work of *Eichner et al.* [2003], for a review of many scaling analyses and their implications for long-term persistence/memory issue. From the station analyses, the basic conclusions of *Fraedrich and Blender* [2003] were that over land, $\beta \approx 0$–0.1, whereas over the ocean, $\beta \approx$ 0.3; *Eichner et al.* [2003] found $\beta \approx 0.3$ over land and using NCEP reanalyses; *Huybers and Curry* [2006] found slightly higher values and noted an additional latitudinal effect (β is higher at the equator). At longer scales, *Blender et al.* [2006] analyzed the anomalous Holocene Greenland paleotemperatures finding $\beta \approx 0.5$ (see section 4.2). Other pertinent analyses are of global climate model outputs and historical reconstructions of the Northern Hemisphere temperatures; these are discussed in detail in section 4.3. Our basic empirical conclusions, in accord with a growing literature, particularly with respect to the temperature statistics, are that β is mostly in the range 0.2–0.4 over land and $\approx$0.6 over the ocean.

3. CLIMATE CHANGE

3.1. What Is Climate Change?

We briefly surveyed the weather scaling, focusing on the transition to the low-frequency weather regime for time scales longer than the lifetimes of planetary-scale eddies, $\tau_w \approx 5$–20 days. This picture was complicated somewhat by the qualitatively similar (and nonlinearly coupled) transition from the analogous ocean "weather" to "low-frequency ocean weather" at $\tau_o \approx 1$ year. Using purely spectral analyses, we found that these low-frequency regimes continued until scales of the order of $\tau_c \approx 10$–100 years, after which the spectra started to steeply rise, marking the beginning of the true climate regime. While the high-frequency regime clearly corresponds to "weather," we termed the intermediate regime "low-frequency weather" since its statistics are not only well reproduced with (unforced) "control" runs of GCMs (Figure 1b) but also by (stochastic, turbulent) cascade models of the weather when these are extended to low frequencies. The term "climate regime" was thus reserved for the long times $\tau > \tau_c$, where the low-frequency weather regime gives way to a qualitatively different and much more variable regime. The new climate regime is thus driven either by new (internal) low-frequency nonlinear interactions or by appropriate low-frequency solar, volcanic, anthropogenic, or eventually orbital forcing at scales $\tau > \tau_c$.

This three-scale-range scaling picture of atmospheric variability leads to a clarification of the rough idea that the climate is nothing more than long-term averages of the weather. It allows us to precisely define a *climate state* as the average of the weather over the entire low-frequency weather regime up to τ_c (i.e., up to decadal or centennial scales). This paves the way for a straightforward definition of climate *change* as the long-term *changes* in this climate state, i.e., of the statistics of these climate states at scales $\tau > \tau_c$.

3.2. What Is τ_c?

In Figures 1a and 1b, we gave some evidence that τ_c was in the range (10 years)$^{-1}$ to (100 years)$^{-1}$; that is, it was near the extreme low-frequency limit of instrumental data. We now attempt to determine it more accurately. Up until now, we primarily used spectral analysis since it is a classical, straightforward technique, whose limitations are well known, and it was adequate for the purpose of determining the basic scaling regimes in time and in space. We now focus on the low frequencies corresponding to several years to $\approx$100 kyr so that it is convenient to study fluctuations in real rather than Fourier space. There are several reasons for this. The first is that we are focusing on the lowest instrumental frequencies, and so spectral analysis provides only a few useful data points; for example, on data 150 years long, the time scales longer than 50 years are characterized only by three discrete frequencies $\omega = 1, 2, 3$; Fourier methods are "coarse" at low frequencies. The second is that in order to extend the analysis to lower frequencies, it is imperative to use proxies, and these need calibration: the mean absolute amplitudes of fluctuations at a given scale enable us to perform a statistical calibration. The third is that the absolute amplitudes are also important for gauging the physical interpretation and hence significance of the fluctuations.

3.3. Fluctuations and Structure Functions

The simplest fluctuation is also the oldest, the difference: $(\Delta v(\Delta t))_{\text{diff}} = \Delta v(t + \Delta t) - \Delta v(t)$. According to equation (3), the fluctuations follow:

$$\Delta v = \varphi_{\Delta t} \Delta t^H, \tag{9}$$

where $\varphi_{\Delta t}$ is a resolution Δt turbulent flux. From this, we see that the statistical moments follow:

$$\langle \Delta v(\Delta t)^q \rangle = \langle \varphi_{\Delta t}^q \rangle \Delta t^{qH} \approx \Delta t^{\xi(q)}; \quad \xi(q) = qH - K(q); \tag{10}$$

$\xi(q)$ is the (generalized) structure function exponent, and $K(q)$ is the (multifractal, cascade) intermittency exponent, equation (4). The turbulent flux has the property that it is independent of scale Δt, i.e., the first-order moment $\langle \varphi_{\Delta t} \rangle$ is constant; hence, $K(1) = 0$ and $\xi(1) = H$. The physical significance of H is thus that it determines the rate at which fluctuations grow ($H > 0$) or decrease ($H < 0$) with scale Δt. Since the spectrum is a second-order moment, there is the following useful and simple relation between real space and Fourier space exponents:

$$\beta = 1 + \xi(2) = 1 + 2H - K(2). \tag{11}$$

A problem arises since the mean *difference* cannot decrease with increasing Δt; hence, differences are clearly inappropriate when studying scaling processes with $H < 0$: the differences simply converge to a spurious constant depending on the highest frequencies present in the sample. Similarly, when $H > 1$, fluctuations defined as differences saturate at a large Δt independent value; they depend on the lowest frequencies present in the sample. In both cases, the exponent $\xi(q)$ is no longer correctly estimated. The problem is that we need a definition of fluctuations such that $\Delta v(\Delta t)$ is dominated by frequencies $\approx \Delta t^{-1}$.

The need to more flexibly define fluctuations motivated the development of wavelets [e.g., *Bacry et al.*, 1989; *Mallat and Hwang*, 1992; *Torrence and Compo*, 1998], and the related DFA technique [*Peng et al.*, 1994; *Kantelhardt et al.*, 2001; *Kantelhardt et al.*, 2002] for polynomial and multifractal extensions, respectively. In this context, the classical difference fluctuation is only a special case, the "poor man's wavelet." In the weather regime, most geophysical H parameters are indeed in the range 0 to 1 (see, e.g., the review by *Lovejoy and Schertzer* [2010]) so that fluctuations tend to *increase* with scale, so that this classical difference structure function is generally adequate. However, a prime characteristic of the low-frequency weather regime is precisely that $H < 0$ (section 2.5) so that fluctuations *decrease* rather than increase with scale; hence, for studying this regime, difference fluctuations are inappropriate. To change the range of H over which fluctuations are usefully defined, one changes the shape of the defining wavelet, changing both its real and Fourier space localizations. In the usual wavelet framework, this is done by modifying the wavelet directly, e.g., by choosing the Mexican hat or higher-order derivatives of the Gaussian, etc., or by choosing them to satisfy some special criterion. Following this, the fluctuations are calculated as convolutions with fast Fourier (or equivalent) numerical techniques.

A problem with this usual implementation of wavelets is that not only are the convolutions numerically cumbersome, but the physical interpretation of the fluctuations is lost. In contrast, when $0 < H < 1$, the difference structure function is both simple and gives direct information on the typical difference ($q = 1$) and typical variations around this difference

($q = 2$) and even typical skewness ($q = 3$) or typical Kurtosis ($q = 4$) or, if the probability tail is algebraic, of the divergence of high-order moments of differences. Similarly, when $-1 < H < 0$, one can define the "tendency structure function" (below), which directly quantifies the fluctuation's deviation from zero and whose exponent characterizes the rate at which the deviations decrease when we average to larger and larger scales. These poor man's and tendency fluctuations are also very easy to directly estimate from series with uniformly spaced data and, with straightforward modifications, to irregularly spaced data.

The study of real space fluctuation statistics in the low-frequency weather regime therefore requires a definition of fluctuations valid at least over the range $-1 < H < 1$. Before discussing our choice, the Haar wavelet, let us recall the definitions of the difference and tendency fluctuations; the corresponding structure functions are simply the corresponding qth-order statistical moments. The difference/poor man's fluctuation is thus

$$(\Delta v(\Delta t))_{\text{diff}} \equiv |\delta_{\Delta t} v|; \quad \delta_{\Delta t} v = v(t + \Delta t) - v(t), \qquad (12)$$

where δ is the difference operator. Similarly, the "tendency fluctuation" [*Lovejoy and Schertzer*, 2012] can be defined using the series with overall mean removed: $v'(t) = v(t) - \overline{v(t)}$ with the help of the summation operator s by

$$(\Delta v(\Delta t))_{\text{tend}} = \left|\frac{1}{\Delta t}\delta_{\Delta t} s v'\right|; \quad s v' = \sum_{t' \le t} v'(t'), \qquad (13)$$

where $(\Delta v(\Delta t))_{\text{tend}}$ has a straightforward interpretation in terms of the mean tendency of the data but is useful only for $-1 < H < 0$. It is also easy to implement: simply remove the *overall* mean and then take the mean over intervals Δt: this is equivalent to taking the mean of the differences of the running sum.

We can now define the Haar fluctuation, which is a special case of the Daubechies family of orthogonal wavelets [see, e.g., *Holschneider*, 1995] (for a recent application, see *Ashok et al.* [2010] and for a comparison with the related DFA technique, see *Koscielny-Bunde et al.* [1998, 2006]). This can be done by

$$(\Delta v(\Delta t))_{\text{Haar}} = \left|\frac{2}{\Delta t}\delta^2_{\Delta t/2} s\right| = \left|\frac{1}{\Delta t}((s(t) + s(t + \Delta t)) - 2s(t + \Delta t/2))\right|$$
$$= \left|\frac{2}{\Delta t}\left[\sum_{t+\Delta t/2 \le t' \le t+\Delta t} v(t') - \sum_{t \le t' \le t+\Delta t/2} v(t')\right]\right|. \qquad (14)$$

From this, we see that the Haar fluctuation at resolution Δt is simply the first difference of the series degraded to resolution $\Delta t/2$. Although this is still a valid wavelet (but with the extra normalization factor Δt^{-1}), it is almost trivial to calculate, and (thanks to the summing) the technique is useful for series with $-1 < H < 1$.

For pure scaling functions, the difference ($1 > H > 0$) or tendency ($-1 < H < 0$) structure functions are adequate and have obvious interpretations. The real advantage of the Haar structure function is apparent for functions with two or more scaling regimes, one with $H > 0$, one with $H < 0$. From equation (11), we see that ignoring intermittency, this criterion is the same as $\beta < 1$ or $\beta > 1$; hence (see, e.g., Figure 1a), Haar fluctuations will be useful for the data analyzed, which straddle (either at high or low frequencies) the boundaries of the low-frequency weather regime.

Is it possible to "calibrate" the Haar structure function so that the amplitude of typical fluctuations can still be easily interpreted? To answer this, consider the definition of a "hybrid" fluctuation as the maximum of the difference and tendency fluctuations:

$$(\Delta T)_{\text{hybrid}} = max((\Delta T)_{\text{diff}}, (\Delta T)_{\text{tend}}); \qquad (15)$$

the "hybrid structure function" is thus the maximum of the corresponding difference and tendency structure functions and therefore has a straightforward interpretation. The hybrid fluctuation is useful if a calibration constant C can be found such that

$$\langle \Delta T(\Delta t)^q_{\text{hybrid}} \rangle \approx C^q \langle \Delta T(\Delta t)^q_{\text{Haar}} \rangle. \qquad (16)$$

In a pure scaling process with $-1 < H < 1$, this is clearly possible since the difference or tendency fluctuations yield the same scaling exponent. However, in a case with two or more scaling regimes, this equality cannot be exact, but as we see this in the next section, it can still be quite a reasonable approximation.

3.4. Application of Haar Fluctuations to Global Temperature Series

Now that we have defined the Haar fluctuations and corresponding structure function, we can use them to analyze a fundamental climatological series: the monthly resolution global mean surface temperature. At this resolution, the high-frequency weather variability is largely filtered out, and the statistics are dominated first by the low-frequency weather regime ($H < 0$) and then at low enough frequencies by the climate regime ($H > 0$).

Several such series have been constructed. The three we chose are the NOAA National Climatic Data Center (NCDC) merged land air and SST data set (from 1880 on a 5° × 5° grid) (see *Smith et al.* [2008] for details), the NASA Goddard Institute for Space Studies (GISS) data set (from 1880 on a

2° × 2°) [*Hansen et al.*, 2010], and the HadCRUT3 data set (from 1850 to 2010 on a 5° × 5° grid). HadCRUT3 is a merged product created out of the Climate Research Unit HadSST2 [*Rayner et al.*, 2006] SST data set and its companion data set CRUTEM3 of atmospheric temperatures over land. The NOAA NCDC and NASA GISS are both heavily based on the Global Historical Climatology Network [*Peterson and Vose*, 1997] and have many similarities including the use of sophisticated statistical methods to smooth and reduce noise. In contrast, the HadCRUTM3 data is less processed. Unsurprisingly, these series are quite similar, although analysis of the scale by scale differences between the spectra is interesting [see *Lovejoy and Schertzer*, 2012].

Each grid point in each data set suffered from missing data points so that here we consider the globally averaged series obtained by averaging over all the available data for the common 129 year period 1880–2008. Before analysis, each series was periodically detrended to remove the annual cycle; if this is not done, then the scaling near $\Delta t \approx 1$ year will be artificially degraded. The detrending was done by setting the amplitudes of the Fourier components corresponding to annual periods to the "background" spectral values.

Figure 5 shows the comparison of the difference, tendency, hybrid, and Haar root-mean-square (RMS) structure functions $\langle \Delta T(\Delta t)^2 \rangle^{1/2}$, the latter increased by a factor $C = 10^{0.35} \approx 2.2$. Before commenting on the physical implications, let us first make some technical remarks. It can be seen that the "calibrated" Haar and hybrid structure functions are very close; the deviations are ±14% over the entire range of nearly a factor 10^3 in Δt. This implies that the indicated amplitude scale of the calibrated Haar structure function in degrees K is quite accurate and that to a good approximation, the Haar structure function can preserve the simple interpretation of the difference and tendency structure functions: in regions where the logarithmic slope is between −1 and 0, it approximates the tendency structure function, whereas in regions where the logarithmic slope is between 0 and 1, the calibrated Haar structure function approximates the difference structure function. For example, from the graph, we can see that global-scale temperature fluctuations decrease from ≈0.3 K at monthly scales, to ≈0.2 K at 10 years and then increase to ≈0.8 K at ≈100 years. All of the numbers have obvious implications, although note that they indicate the mean overall range of the fluctuations so that, for example, the 0.8 K corresponds to ±0.4 K, etc.

From Figure 5, we also see that the global surface temperatures separate into two regimes at about $\tau_c \approx 10$ years, with negative and positive logarithmic slopes = $\xi(2)/2 \approx -0.1, 0.4$ for $\Delta t < \tau_c$, and $\Delta t > \tau_c$, respectively. Since $\beta = 1 + \xi(2)$ (equation (11)), we have $\beta \approx 0.8, 1.8$. We also analyzed the

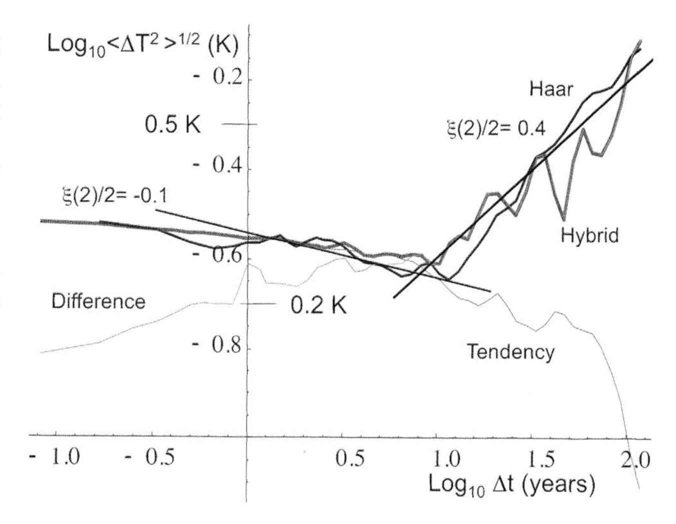

Figure 5. A comparison of the different structure function analyses (root-mean-square (RMS)) applied to the ensemble of three monthly surface series discussed in section 3.4 (NASA GISS, NOAA NCDC, and HadCRUT3), each globally averaged, from 1881 to 2008 (1548 points each). The usual (difference, poor man's) (bottom left) structure function (thin gray), (bottom right) the tendency structure function (thin gray), the maximum of the two ("Hybrid," thick, gray), and the Haar (in black) are shown. The latter has been increased by a factor $C = 10^{0.35} = 2.2$; the resulting RMS deviation with respect to the hybrid is ±14%. Reference slopes with exponents $\xi(2)/2 \approx 0.4$ and −0.1 are also shown (corresponding to spectral exponents $\beta = 1 + \xi(2) = 1.8$ and 0.8, respectively). In terms of difference fluctuations, we can use the global RMS $\langle \Delta T(\Delta t)^2 \rangle^{1/2}$ annual structure functions (fitted for 129 years > Δt > 10 years), obtaining $\langle \Delta T(\Delta t)^2 \rangle^{1/2} \approx 0.08 \Delta t^{0.33}$ for the ensemble. In comparison, *Lovejoy and Schertzer* [1986] found the very similar $\langle \Delta T(\Delta t)^2 \rangle^{1/2} \approx 0.077 \Delta t^{0.4}$ using Northern Hemisphere data (these correspond to $\beta_c = 1.66$ and 1.8, respectively).

first-order structure function whose exponent $\xi(1) = H$; at these scales, the intermittency ($K(2)$, equation (4)) ≈ 0.03 so that $\xi(2) \approx 2H$ so that $H \approx -0.1, 0.4$ confirming that fluctuations decrease with scale in the low-frequency weather regime but increase again at lower frequencies in the climate regime (more precise intermittency analyses are given in the work of *Lovejoy and Schertzer* [2012]). Note that ignoring intermittency, the critical value of β discriminating between growing and decreasing fluctuations (i.e., $H < 0$, $H > 0$) is $\beta = 1$.

Before pursuing the Haar structure function, let us briefly consider its sensitivity to nonscaling perturbations, i.e., to nonscaling external trends superposed on the data, which break the overall scaling. Even when there is no particular reason to suspect such trends, the desire to filter them out is commonly invoked to justify the use of special wavelets, or nearly equivalently, of various orders of the multifractal detrended fluctuation analysis technique (MFDFA) [*Kantelhardt et al.*, 2002]. A simple way to produce a higher-order Haar wavelet that eliminates polynomials of order n is simply to iterate ($n + 1$ times) the difference operator in equation (14). For example, iterating it three times yields the "quadratic Haar" fluctuation $(\Delta v(\Delta t))_{\text{Haarquad}} = \frac{3}{\Delta t}(s(t+\Delta t) - 3s(t+\Delta t/3) + 3s(t-\Delta t/3) - s(t-\Delta t))$. This fluctuation is sensitive to structures of size Δt^{-1} and, hence, useful over the range $-1 < H < 2$, and it is blind to polynomials of order 1 (lines). In comparison, the nth-order DFA technique defines fluctuations using the RMS deviations of the summed series $s(t)$ from regressions of nth-order polynomials so that quadratic Haar fluctuations are nearly equivalent to the quadratic MFDFA RMS deviations (although these deviations are not strictly wavelets, note that the MFDFA uses a scaling function $\approx \Delta v\, \Delta t$; hence, with DFA exponent, $\alpha_{\text{DFA}} = 1 + H$). Although at first sight the insensitivity of these higher-order wavelets to trends may seem advantageous, it should be recalled that, on the one hand, they only filter out polynomial trends (and not, for example, the more geophysically relevant periodic trends), while on the other hand, even for this, they are "overkill" since the trends they filter are filtered at all scales, not just the largest. Indeed, if one suspects the presence of external polynomial trends, it suffices to eliminate them over the whole series (i.e., at the largest scales) and then to analyze the resulting deviations using the Haar fluctuations.

Figure 6 shows the usual (linear) Haar RMS structure function (equation (14)) compared to the quadratic Haar and quadratic MFDFA structure functions. Unsurprisingly, the latter two are close to each other (after applying different calibration constants, see the figure caption), that the low and high-frequency exponents are roughly the same. However, the transition point has shifted by nearly a factor of 3 so that, overall, they are rather different from the Haar structure function, and it is clearly not possible to simultaneously "calibrate" the high- and low-frequency parts. The drawback with these higher-order fluctuations is thus that we lose the simplicity of interpretation of the Haar wavelet, and unless $H > 1$, we obtain no obvious advantage.

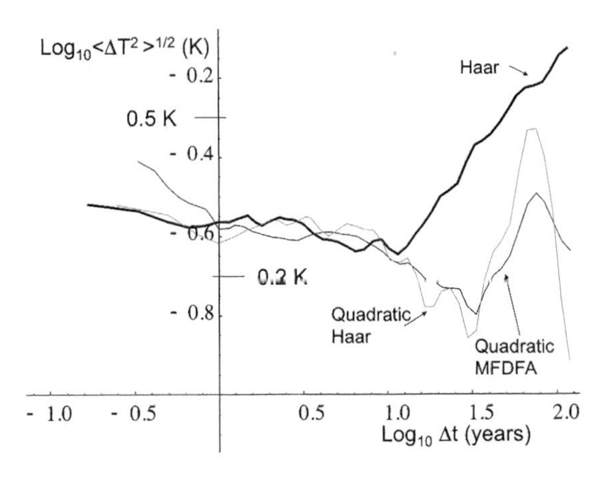

Figure 6. Same temperature data as Figure 5: a comparison of the RMS Haar structure function (multiplied by $10^{0.35} = 2.2$), the RMS quadratic Haar (multiplied by $10^{0.15} = 1.4$), and the RMS quadratic multifractal detrended fluctuation analysis (multiplied by $10^{1.5} = 31.6$).

4. THE TRANSITION FROM LOW-FREQUENCY WEATHER TO THE CLIMATE

4.1. Intermediate-Scale Multiproxy Series

In section 2, we discussed atmospheric variability over the frustratingly short instrumentally accessible range of time scales (roughly $\Delta t < 150$ years) and saw evidence that weakly variable low-frequency weather gives way to a new highly variable climate regime at a scale τ_c, somewhere in the range 10–30 years. In Figure 1a, we already glimpsed the much longer 1–100 kyr scales accessible primarily via ice core paleotemperatures (see also below); these confirmed that, at least when averaged over the last 100 kyr or so, the climate does indeed have a new scaling regime with fluctuations increasing rather than decreasing in amplitude with scale ($H > 0$).

Since the temporal resolution of the high-resolution Greenland Ice Core Project (GRIP) paleotemperatures was ≈ 5.2 years (and for the Vostok series $\approx$100 years), these paleotemperature resolutions do not greatly overlap the instrumental range; it is thus useful to consider other intermediates: the "multiproxy" series that have been developed following the work of *Mann et al.* [1998]. Another reason to use intermediate-scale data is because we are living in a climate epoch, which is exceptional in both its long- and short-term aspects. For example, consider the long stretch of relatively mild and stable conditions since the retreat of the last ice sheets about 11.5 kyr ago, the "Holocene." This epoch is claimed to be at least somewhat exceptional: it has even been suggested that such stability is a precondition for the invention of farming and thus for civilization itself [*Petit et al.*, 1999]. It is therefore possible that the paleoclimate statistics averaged over series 100 kyr or longer may not be as pertinent as we would like for understanding the current epoch. Similarly, at the high-frequency end of the spectrum, there is the issue of "twentieth century exceptionalism," a consequence of twentieth century warming and the probability that at least some of it is of anthropogenic, not natural origin. Since these affect a large part of the instrumental record, it is problematic to use the latter as the basis for extrapolations to centennial and millennial scales. In the following, we try to assess both "exceptionalisms" in an attempt to understand the natural variability in the last few centuries.

4.2. The Holocene Exception: Climate Variability in Time and in Space

The high-resolution GRIP core gives a striking example of the difference between the Holocene and previous epochs in central Greenland (Figure 7). Even a cursory visual inspection of the figure confirms the relative absence of low-frequency variability in the current 10 kyr section compared to previous 10 kyr sections. To quantify this, we can turn to Figure 8, which compares the RMS Haar structure functions for both GRIP (Arctic) and Vostok (Antarctic) cores for both the Holocene 10 kyr section and for the mean and spread of the eight earlier 10 kyr sections. The GRIP Holocene curve is clearly exceptional, with the fluctuations decreasing with scale out to $\tau_c \approx 2$ kyr in scale and with $\xi(2)/2 \approx -0.3$. This implies a spectral exponent near the low-frequency weather value $\beta \approx 0.4$, although it seems that as before $\xi(2)/2 \approx 0.4$ ($\beta \approx 1.8$) for larger Δt. The main difference however is that τ_c is much larger than for the other series (see Table 1 for quantitative comparisons). The exceptionalism is quantified by noting that the corresponding RMS fluctuation function ($S(\Delta t)$) is several standard deviations below the average of the previous eight 10 kyr sections. In comparison (to the right in Figure 8), the Holocene period of the Vostok core is also somewhat exceptional, although less so: up to $\tau_c \approx 1$ kyr, it has $\xi(2)/2 \approx -0.3$ ($\beta \approx 0.4$), and it is more or less within one standard deviation limits of its mean, although τ_c is still large. Beyond scales of ≈ 1 kyr, its fluctuations start to increase;

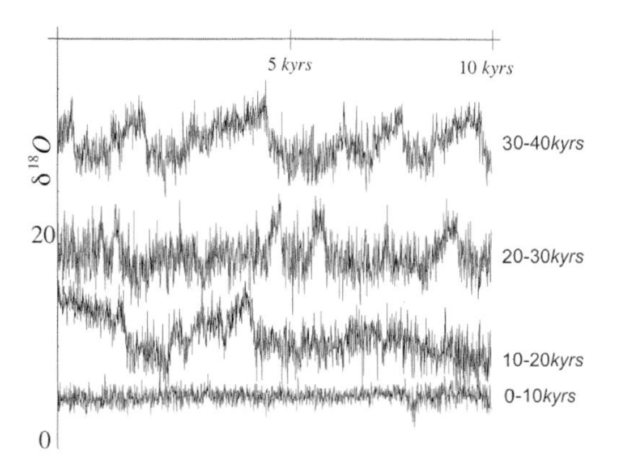

Figure 7. (bottom to top) Four successive 10 kyr sections of the high-resolution GRIP data, the most recent to the oldest. Each series is separated by 10 mils in the vertical for clarity (vertical units, mils, i.e., parts per thousand). The bottom Holocene series is indeed relatively devoid of low-frequency variability compared to the other 10 kyr sections, a fact confirmed by statistical analysis discussed in the text and Figure 8.

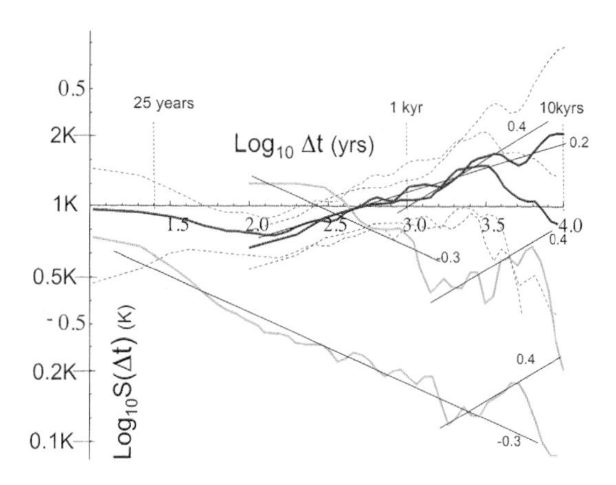

Figure 8. Comparison of the RMS Haar structure function ($S(\Delta t)$) for both Vostok and GRIP high-resolution cores (resolutions 5.2 and 50 years, respectively, over the last 90 kyr). The Haar fluctuations were calibrated and are accurate to ±20%. For Vostok, we used the *Petit et al.* [1999] calibration; for GRIP, we used 0.5 K mil^{-1}. The series were broken into 10 kyr sections. The thick gray lines show the most recent of these (roughly the Holocene, (top) Vostok, and (bottom) GRIP), whereas the long dark gray and short dark gray lines are the mean of the eight 10–90 kyr GRIP and Vostok $S(\Delta t)$ cores, respectively. The 1 standard deviation variations about the mean are indicated by dashed lines. Also shown are reference lines with slopes $\xi(2)/2 = -0.3$, 0.2, and 0.4 corresponding to $\beta = 0.4$, 1.4, and 1.8, respectively. Although the Holocene is exceptional for both series, for GRIP, it is exceptional by many standard deviations. For the Holocene, we can see that $\tau_c \approx 1$ kyr for Vostok and ≈ 2 kyr for GRIP, although for the previous 80 kyr, we see that $\tau_c \approx 100$ years for both.

Table 1 quantifies the differences. We corroborated this conclusion by an analysis of the 2 kyr long (yearly resolution) series from other (nearby) Greenland cores (as described by *Vinther et al.* [2008]) where *Blender et al.* [2006] also obtained $\beta \approx 0.2$–0.4 and also obtained similar low β estimates for the Greenland GRIP, GISP2 cores over the last 3 kyr.

Although these analyses convincingly demonstrate that the Greenland Holocene was exceptionally stable, nevertheless, their significance for the overall natural variations of Northern Hemisphere temperatures is doubtful. For example, on the basis of paleo-SST reconstructions just 1500 km southeast of Greenland [*Andersen et al.*, 2004; *Berner et al.*, 2008], it was concluded that the latter region was on the contrary "highly unstable." Using several ocean cores as proxies, Holocene SST reconstructions were produced, which included a difference between maximum and minimum of roughly 6 K and "typical variations" of 1–3 K. In comparison, from Figure 8, we see that the mean temperature fluctuation deduced from the GRIP core in the last 10 kyr is ≈ 0.2 K. However, also from Figure 8, we see that the mean

Table 1. Comparison of Various Paleoexponents Estimated Using the Haar Structure Function Over Successive 10 kyr Periods[a]

	H				β			
	Holocene		10–90 kyr		Holocene		10–90 kyr	
Range of regressions	100 years $< \Delta t$ < 2 kyr	2 kyr $< \Delta t$ < 10 kyr	100 years $< \Delta t$ < 10 kyr	100 years $< \Delta t$ < 10 kyr ensemble	100 years $< \Delta t$ < 2 kyr	2 kyr $< \Delta t$ < 10 kyr	100 years $< \Delta t$ < 10 kyr	100 years $< \Delta t$ < 10 kyr ensemble
GRIP	−0.25	0.21	0.14 ± 0.18	0.17	0.43	1.33	1.14 ± 0.33	1.20
Vostok	−0.40	0.38	0.19 ± 0.28	0.31	0.18	1.76	1.29 ± 0.51	1.49

[a]Vostok at 50 year resolution, Greenland Ice Core Project (GRIP) at 5.2 year resolution, all regressions over the scale ranges indicated. The Holocene is the most recent period (0–10 kyr). Note that while the Holocene exponents are estimates from individual series, the 10–90 kyr exponents are the means of the estimates from each 10 kyr section and (to the right) the exponent of the ensemble mean of the latter. Note that the mean of the exponents is a bit below the exponent of the mean indicating that a few highly variable 10 kyr sections can strongly affect the ensemble averages. For the Holocene, the separate ranges <2 kyr and $\Delta t > 2$ kyr were chosen because according to Figure 8, $\tau_c \approx 1$–2 kyr.

over the previous eight 10 kyr sections was ≈ 1–2 K, i.e., quite close to these paleo-SST variations (and about amount expected in order to explain the glacial/interglacial temperature swings, see section 5). These paleo-SST series thus underline the strong geographical climate variability effectively undermining the larger significance of the Greenland Holocene experience. At the same time, they lend support to the application of standard statistical stationarity assumptions to the variability over longer periods (e.g., to the relevance of spectra and structure functions averaged over the whole cores). Finally, *Lovejoy and Schertzer* [2012] argue that the spatial variability over both the low-frequency weather and climate regimes has very high intermittency and that this corresponds to the existence of climate zones.

4.3. Multiproxy Temperature Data, Centennial-Scale Variability, and the Twentieth Century Exception

The key to linking the long but geographically limited ice core series with the short but global-scale instrumental series are the intermediate category of "multiproxy temperature reconstructions." These series of Northern Hemisphere average temperatures, pioneered by *Mann et al.* [1998, 1999], have the potential of capturing "multicentennial" variability over at least the (data rich) Northern Hemisphere. These series are at typically annual resolutions and combine a variety of different data types ranging from tree rings, ice cores, lake varves, boreholes, ice melt stratigraphy, pollen, Mg/Ca variation in shells, ^{18}O in foraminifera, diatoms, stalactites (in caves), biota, and historical records. In what follows, we analyze eight of the longest of these (see Table 2 for some of statistical characteristics and descriptions).

Before reviewing the results, let us discuss some of the technical issues behind the continued development of new series. Consideration of the original series [*Mann et al.*, 1998] (extended back to 1000 A.D. by *Mann et al.* [1999]) illustrates both the technique and its attendant problems. The basic difficulty is in getting long series that are both temporally uniform and spatially representative. For example, the original six century long multiproxy series presented in the work of *Mann et al.* [1998] has 112 indicators going back to 1820, 74 to 1700, 57 to 1600, and only 22 to 1400. Since only a small number of the series go back more than two or three centuries, the series' "multicentennial" variability depends critically on how one takes into account the loss of data at longer and longer time intervals. When it first appeared, the Mann et al. series created a sensation by depicting a "hockey stick"-shaped graph of temperature, with the fairly flat "handle" continuing from 1000 A.D. until a rapid twentieth century increase. This lead to the famous conclusion, echoed in the IPPC Third Assessment Report [*Intergovernmental Panel on Climate Change*, 2001], that the twentieth century was the warmest century of the millennium, that the 1990s was the warmest decade, and that 1998 was the warmest year. This multiproxy success encouraged the development of new series using larger quantities of more geographically representative proxies [*Jones et al.*, 1998], by the introduction new types of data [*Crowley and Lowery*, 2000], to the more intensive use of pure dendrochronology [*Briffa et al.*, 2001], or to improved methodologies [*Esper et al.*, 2002].

However, the interest generated by reconstructions also attracted criticism, in particular, *McIntyre and McKitrick* [2003] pointed out several flaws in the *Mann et al.* [1998] data collection and in the application of the principal component analysis technique, which had been borrowed from econometrics. After correction, the same proxies yielded series with significantly larger low-frequency variability and included the reappearance of the famous "medieval warming" period at around 1400 A.D., which had disappeared in

Table 2. A Comparison of Parameters Estimated From the Multiproxy Data From 1500 to 1979 (480 Years)[a]

	β (High Frequency (4–10 years)$^{-1}$)	β (Lower Frequency Than (25 years)$^{-1}$)	H_{high} (4–10 years)	H_{low} (>25 years)
Jones et al. [1998]	0.52	0.99	−0.27	0.063
Mann et al. [1998, 1999]	0.57	0.53	−0.22	−0.13
Crowley and Lowery [2000]	2.28	1.61	0.72	0.31
Briffa et al. [2001]	1.19	1.18	0.15	0.13
Esper et al. [2002]	0.88	1.36	0.01	0.22
Huang [2004]	0.94	2.08	0.02	0.61
Moberg et al. [2005]	1.15	1.56	0.09	0.32
Ljundqvist [2010]	–	1.84	–	0.53

[a]The Ljundquist high-frequency numbers are not given since the series has decadal resolution. Note that the β for several of these series was estimated by *Rybski et al.* [2006], but no distinction was made between low-frequency weather and climate; the entire series was used to estimate single (hence generally lower) β.

the original. Later, an additional technical critique [*McIntyre and McKitrick*, 2005] underlined the sensitivity of the methodology to low-frequency red noise variability present in the calibration data (the latter modeled this with exponentially correlated processes probably underestimating that which would have been found using long-range correlated scaling processes). Other work in this period, notably by *von Storch et al.* [2004] using "pseudo proxies" (i.e., the simulation of the whole calibration process with the help of GCMs), similarly underlined the nontrivial issues involved in extrapolating proxy calibrations into the past.

Beyond the potential social and political implications of the debate, the scientific upshot was that increasing attention had to be paid to the preservation of the low frequencies. One way to do this is to use borehole data, which, when combined with the use of the equation of heat diffusion, has essentially no calibration issues whatsoever. *Huang* [2004] used 696 boreholes (only back to 1500 A.D., roughly the limit of this approach) to augment the original [*Mann et al.*, 1998] proxies so as to obtain more realistic low-frequency variability. Similarly, in order to give proper weight to proxies with decadal and lower resolutions (especially lake and ocean sediments), *Moberg et al.* [2005] used wavelets to separately calibrate the low- and high-frequency proxies. Once again, the result was a series with increased low-frequency variability. Finally, *Ljundqvist* [2010] used a more up to date, more diverse collection of proxies to produce a decadal-resolution series going back to 1 A.D. The low-frequency variability of the new series was sufficiently large that it even included a third century "Roman warm period" as the warmest century on record and permitted the conclusion that "the controversial question whether Medieval Warm Period peak temperatures exceeded present temperatures remains unanswered" [*Ljundqvist*, 2010].

With this context, let us quantitatively analyze the eight series cited above; we use the Haar structure function. We concentrate here on the period 1500–1979 because (1) it is common to all eight reconstructions, (2) being relatively recent, it is more reliable (it has lower uncertainties), and (3) it avoids the Medieval Warm Period and thus the possibility that the low-frequency variability is artificially augmented by the possibly unusual warming in the 1300s. The result is shown in Figure 9, where we have averaged the structure functions into the five pre-2003 and three post-2003 reconstructions. Up to $\Delta t \approx 200$ years, the basic shapes of the curves are quite similar to each other and, indeed, to the surface temperature $S(\Delta t)$ curves back to 1881 (Figure 5). However, quite noticeable for the pre-2003 constructions is the systematic drop in RMS fluctuations for $\Delta t \approx >200$ years, which contrasts with their continued rise in the post-2003 reconstructions. This confirms the above analysis to the effect that the post-2003 analyses were more careful in their treatments of multicentennial variability. Table 2 gives a quantitative intercomparison of the various statistical parameters.

Figure 9 compares the mean multiproxies with the ensemble average of the instrumental global surface series. This confirms the basic behavior: small Δt scaling with $\xi(2)/2 = -0.1$ ($\beta = 0.8$) followed by large Δt scaling with $\xi(2)/2 = 0.4$ ($\beta = 1.8$) is displayed by all the data, all the pre-2003 $S(\Delta t)$ functions drop off for $\Delta t \approx >200$ years. Notable are (1) the transition scale in the global instrumental temperature at $\tau_c \approx 10$ years, which is somewhat lower than that found in the multiproxy reconstructions ($\tau_c \approx 40$–100 years) and (2) over the common low-frequency part that the amplitudes of the reconstruction RMS fluctuations are about a factor of 2 lower than for the global instrumental series. The reason for the amplitude difference is not at all clear since, on the one hand, the monthly and annually averaged Haar structure functions of the instrumental series are very close to each other up to $\Delta t \approx 10$ years (the temporal resolution is not an issue), and similarly, the difference between the Northern Hemisphere

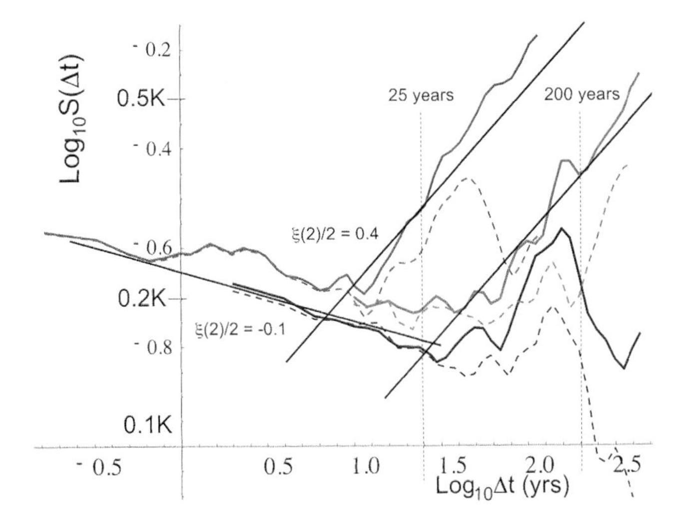

Figure 9. (bottom) RMS Haar fluctuation for the mean of the pre-2003 and post-2003 series from 1500 to 1979 (solid black and gray curves, respectively, and excluding the Crowley series due to its poor resolution), along with (top) the mean of the globally averaged monthly resolution surface series from Figure 5 (solid gray). In order to assess the effect of the twentieth century warming, the structure functions for the multiproxy data were recalculated from 1500 to 1900 only (the associated thin dashed lines) and for the instrumental surface series with their linear trends from 1880 to 2008 removed (the data from 1880 to 1899 are too short to yield a meaningful $S(\Delta t)$ estimate for the lower frequencies of interest). While the large Δt variability is reduced a little, the basic power law trend is robust, especially for the post-2003 reconstructions. Note that the decrease in $S(\Delta t)$ for the linearly detrended surface series over the last factor of 2 or so in lag Δt is a pure artifact of the detrending. We may conclude that the low-frequency rise is not an artifact of an external linear trend. Reference lines corresponding to $\beta = 0.8$ and 1.8 have been added.

and the Southern Hemisphere instrumental $S(\Delta t)$ functions is much smaller than this (only about ≈15%).

To get another perspective on this low-frequency variability, we can compare the instrumental and multiproxy structure functions with those from ice core paleotemperature discussed in more detail in the next section. In Figure 10, we have superposed the calibrated deuterium-based RMS temperature fluctuations with RMS multiproxy and RMS surface series fluctuations (we return to this in section 5). We see that extrapolating the latter out to 30–50 kyr is quite compatible with the Vostok core with the "interglacial window" (i.e., the rough quasiperiod and amplitude corresponding to the interglacials). Although the Vostok $S(\Delta t)$ curve is from the entire 420 kyr record (not just the Holocene), this certainly makes it plausible that while the surface series appear to have realistic low-frequency variability, the variability of the reconstructions is too small (although the post-2003 reconstructions are indeed more realistic than the contrasting relative lack of variability in the pre-2003 reconstructions).

4.4. Twentieth Century Exceptionalism

Although the reconstructions and instrumental series qualitatively agree, their quantitative disagreement is large and requires explanation. In this section, we consider whether this could be a consequence of the twentieth century "exceptionalism" discussed earlier, the fact that, irrespective of the cause, the twentieth century is somewhat warmer than the nineteenth and earlier centuries. It has been recognized that this warming causes problems for the calibration of the proxies [e.g., *Ljundqvist*, 2010], and it will clearly contribute to the RMS multicennial variability in Figure 9. In order to demonstrate that the basic type of statistical variability is not an artifact of the inclusion of exceptional twentieth century temperatures in Figure 9, we also show the corresponding Haar structure functions for the earlier period 1500–1900. Truncating the instrumental series at 1900 would result in a series only 20 years long; therefore, the closest equivalent for the surface series was to remove overall linear trends and then redo the analysis. As expected, the figure shows that all the large Δt fluctuations are reduced but that the basic scaling behaviors are apparently not affected. We conclude that both the type of variability as characterized by the scaling exponents and the transition scale τ_c are fairly robust, if difficult,

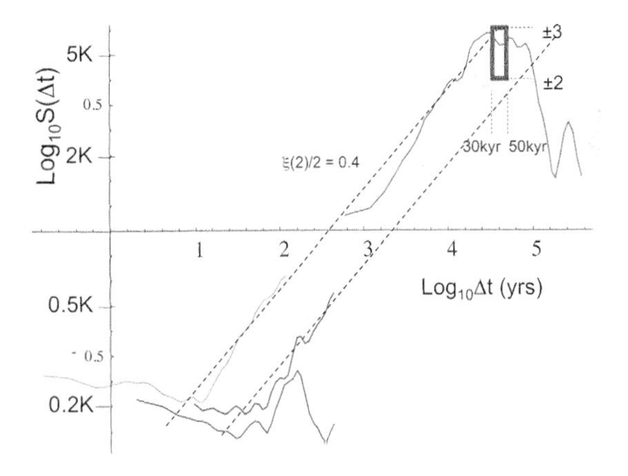

Figure 10. (bottom left) RMS Haar fluctuations for the mean monthly, global surface series (thin gray, from Figure 5) and (bottom right) the mean pre-2003 (medium gray) and (bottom middle) mean post-2003 proxies (dark gray, from Figure 9) as well as the mean (top right) Vostok $S(\Delta t)$ function over the last 420 kyr interpolated to 300 year resolution and using the *Petit et al.* [1999] calibration. Also shown (the rectangle) is the "interglacial window," the probable typical range of fluctuations and quasiperiods of the glacial-interglacials. The "calibration" of the fluctuation amplitudes is accurate to ±25%.

to accurately determine; they are not artifacts of external linear trends. The instrumental and reconstruction discrepancy in Figure 9 thus remains unexplained.

5. TEMPORAL SPECTRAL SCALING IN THE CLIMATE REGIME: 10–10^5 YEARS

5.1. Review of Literature

Thanks to several ambitious international projects, many ice cores exist, particularly from the Greenland and Antarctic ice sheets, which provide surrogate temperatures based on $\delta^{18}O$ or deuterium concentrations in the ice. The most famous cores are probably the GRIP and Vostok (Antarctica) cores, each of which are over 3 km long (limited by the underlying bedrock) and go back, respectively, 240 and 420 kyr. Near the top of the cores, individual annual cycles can be discerned (in some cases, going back over 10 kyr); below that, the shearing of ice layers and diffusion between the increasingly thin annual layers make such direct dating impossible, and models of the ice flow and compression are required. Various "markers" (such as dust layers from volcanic eruptions) are also used to help fix the core chronologies.

A problem with the surrogates is their highly variable temporal resolutions combined with strong depth dependences. For example, *Witt and Schumann* [2005] used wavelets, *Davidsen and Griffin* [2010] used (monofractal) fractional Brownian motion as a model, and *Karimova et al.* [2007] used (mono) fractal interpolation to attempt to handle this; *Lovejoy and Schertzer* [2012] found that the temporal resolution itself has multifractal intermittency. The main consequence is that the intermittency of the interpolated surrogates is a bit too high but that serious spectral biases are only present at scales of the order of the mean resolution or higher frequencies.

With these caveats, Table 3 summarizes some of the spectral scaling exponents, scaling ranges. It is interesting to note that the three main orbital ("Milankovitch") forcings at 19, 23 (precessional), and 41 kyr (obliquity) are indeed visible, but only barely, above the scaling "background" [see especially *Wunsch*, 2003].

5.2. A Composite Picture of Atmospheric Variability From 6 Hours to 10^5 Years

To clarify our ideas about the variability, it is useful to combine data over huge ranges of scale into a single composite analysis (such as the spectra shown in Figures 1a, 1b) but using real space fluctuations rather than spectra. Some time ago, such a composite already clarified the following points: (1) that there is a distinction between the variability of regional- and global-scale temperatures, (2) that global averages had particularly small transition scale τ_c, (3) that there was a scaling range for global averages between scales of about 3 years and 40–50 kyr (where the variability apparently "saturates") with a realistic exponent $\beta_c \approx 1.8$, and (4)

Table 3. An Intercomparison of Various Estimates of the Spectral Exponents β_c of the Low-Frequency Climate Regime and Scaling Range Limits[a]

Series	Authors	Series Length (kyr)	Resolution (years)	β_c
Composite ice cores, instrumental	*Lovejoy and Schertzer* [1986]	10^{-9} to 1000	1000	1.8
Composite (Vostok) (ice core, instrumental)	*Pelletier* [1998]	10^{-5} to 1000	0.1 to 500	2
$\delta^{18}O$ from GRIP Greenland	*Wunsch* [2003]	100	100	1.8
Planktonic $\delta^{18}O$ ODP677 (Panama basin)	*Wunsch* [2003]	1000	300	2.3
CO_2, Vostok (Antarctica)	*Wunsch* [2003]	420	300	1.5
$\delta^{18}O$ from GRIP Greenland	*Ditlevsen et al.* [1996]	91	5	1.6
$\delta^{18}O$ from GRIP Greenland	*Schmitt et al.* [1995]	123	200	1.4
$\delta^{18}O$ from GISP Greenland	*Ashkenazy et al.* [2003]	110	100	1.3
$\delta^{18}O$ from GRIP Greenland	*Ashkenazy et al.* [2003]	225	100	1.4
$\delta^{18}O$ from Taylor (Antarctica)	*Ashkenazy et al.* [2003]	103	100	1.8
$\delta^{18}O$ from Vostok	*Ashkenazy et al.* [2003]	420	100	2.1
Composite, midlatitude	*Huybers and Curry* [2006]	10^{-4} to 1000	0.1 to 10^3	1.6
Composite tropics	*Huybers and Curry* [2006]	10^{-4} to 1000	0.1 to10^3	1.3
$\delta^{18}O$ from GRIP Greenland	*Lovejoy and Schertzer* [2012]	91	5	1.4
$\delta^{18}O$ from Vostok	*Lovejoy and Schertzer* [2012]	420	300	1.7
$\delta^{18}O$ from GRIP, Greenland	*Blender et al.* [2006]	3	3	0.4
$\delta^{18}O$ from GISP2 Greenland	*Blender et al.* [2006]	3	3	0.7
$\delta^{18}O$ from GRIP Greenland (last 10 kyr only)	Figure 8	10	5	0.2

[a]For series with resolution ≈100 years, the last three rows are for the (anomalous) Holocene only [see *Lovejoy and Schertzer*, 2012].

that this scaling regime could potentially quantitatively explain the magnitudes of the temperature swings between interglacials [*Lovejoy and Schertzer*, 1986].

Similar scaling composites, but in Fourier space, were adopted by *Pelletier* [1998] and, more recently, by *Huybers and Curry* [2006], who made a more data-intensive study of the scaling of many different types of paleotemperatures collectively spanning the range of about 1 month to nearly 10^6 years. The results are qualitatively very similar, including the positions of the scale breaks; the main innovations are (1) the increased precision on the β estimates and (2) the basic distinction made between continental and oceanic spectra including their exponents. We could also mention the composite of *Fraedrich et al.* [2009], which is a modest adaptation of that of *Mitchell* [1976], although it does introduce a single scaling regime spanning only 2 orders of magnitude: from ≈3 to ≈100 years (with $\beta \approx 0.3$), at lower frequencies, the composite exhibits a *decrease* (rather than *increase*) in variability.

Figure 11 shows an updated composite where we have combined the 20CR reanalysis spectra (both local, single grid point and global) with the GRIP 55 cm and GRIP high-resolution spectra (both for the last 10 kyr and averaged over the last 90 kyr), and the three surface global temperature series. For reference, we have also included the 500 year control run of the Institut Pierre Simon Laplace (IPSL) GCM used in the IPCC Fourth Assessment Report. We use difference structure functions so that the interpretation is particularly simple, although a consequence (see section 3.3) is that all the logarithmic slopes are >0. In order to avoid this problem, compare this to the Haar structure functions (Figure 12).

Key points to note are (1) the use of annually averaged instrumental data in Figure 11 (differences), but of daily data in Figure 12 (Haar) and (2) the distinction made between globally and locally averaged quantities whose low-frequency weather have different scaling exponents. Also shown is the interglacial window (Δt is the half quasiperiod, and for a white noise, S is double the amplitude). The calibration of the paleotemperatures is thus constrained so that it goes through the window at large Δt but joins up to the *local* instrumental $S(\Delta t)$ at small Δt. In addition, as discussed in Section 4.2, since the last 10 kyr GRIP fluctuations are anomalously low (Figure 4 and see the nearly flat red curve compared with the full 91 kyr red curve), the calibration must be based on this flatter $S(\Delta t)$ (Figure 11). Starting at $\tau_c \approx 10$ years, one can plausibly extrapolate the global $S(\Delta t)$ using $H = 0.4$ ($\beta \approx 1.8$), all the way to the interglacial window (with nearly an identical $S(\Delta t)$ as given by *Lovejoy and Schertzer* [1986]), although the Northern Hemisphere reconstructions do not extrapolate as well, possibly because of their higher intermittency. The local temperatures extrapolate (starting at $\tau_c \approx$ 20 years) with a lower exponent corresponding to $\beta \approx 1.4$ (see Figure 1a), which is close to the other Greenland paleotemperature exponents (Table 3) presumably reflecting the fact that the Antarctic temperatures are better surrogates for global rather than local temperatures; these exponents are all averages over spectra of series of ≈100 kyr or more in length. An interesting feature of the Haar structure function (Figure 12) is that it shows that local (grid scale) temperature fluctuations are roughly the same amplitude for monthly as for the much longer glacial/interglacials periods. Not only can we make out the three scaling regimes discussed above but also for $\Delta t > 100$ kyr, we can start to discern a new "low-

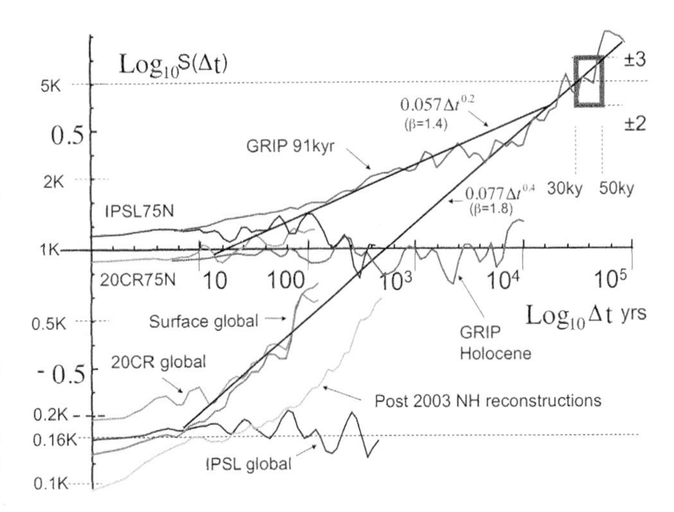

Figure 11. (top right) A comparison of the RMS structure function $S(\Delta t)$ of the high-resolution (5.2 year) GRIP (gray), IPSL (top left) 75°N and (bottom left) global, 20CR (top) 75°N (light gray) and (bottom) global (light gray), (bottom left) mean surface series (darker gray), (bottom left) mean of the three post-2003 Northern Hemisphere reconstructions (light gray) for globally averaged temperatures, and (top) the mean at Greenland latitudes, all using fluctuations defined as differences (poor man's wavelet) so that the vertical scale directly indicates typical changes in temperature. In addition, the GRIP data are divided into two groups: the Holocene (taken as the last 10 kyr, along the axis) and (top right) the entire 91 kyr of the high-resolution GRIP series (gray). The GRIP $\delta^{18}O$ data have been calibrated by lining up the Holocene structure function with the mean 75°N 20CR reanalysis structure function (corresponding to ≈0.65 K mil^{-1}). When this is done, the 20CR and surface mean global structure functions can be extrapolated with exponent $H \approx 0.4$ (see the corresponding line) to the "interglacial window" (box) corresponding to half pseudoperiods between 30 and 50 kyr with variations (= $\pm S/2$) between ±2 and ±3 K. This line corresponds to spectral exponent $\beta = 1.8$. Finally, we show a line with slope $\xi(2)/2 = 0.2$, corresponding to the GRIP $\beta = 1.4$ (see Figures 1a and 1b); we can see that extrapolating it to 50 kyr explains the local temperature spectra quite well.

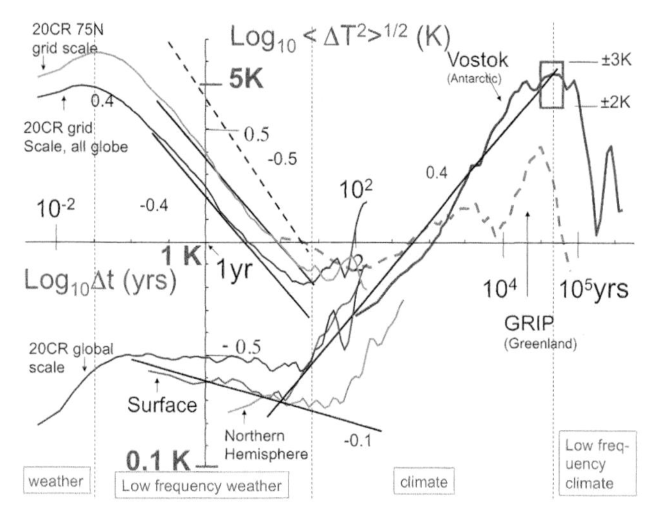

Figure 12. The equivalent of Figure 11 except for the RMS Haar structure function rather than the RMS difference structure function and including daily resolution 20CR data and monthly resolution surface temperatures. We show (top left) grid point scale (2° × 2°) daily scale fluctuations for both 75°N and globally averaged along with reference slope $\xi(2)/2 \approx H = -0.4$ (20CR, 700 mb). We see (lower left) at daily resolution, the corresponding globally averaged structure function. Also shown are the average of the three in situ surface series (Figure 5) as well as the post-2003 multiproxy structure function (from Figure 9). We show (right) both the GRIP (55 cm resolution, calibrated with 0.5 K mil^{-1}) and the Vostok paleotemperature series. All the fluctuations have been multiplied by 2.2 so that the calibration scale in degrees K is fairly accurate (compare with Figure 11). Also shown are the interglacial window and a reference line with slope −0.5 corresponding to Gaussian white noise.

frequency climate" regime. For more scaling paleoclimate analyses, including "paleocascades," see the work of *Lovejoy and Schertzer* [2012].

6. CONCLUSIONS

Just as the laws of continuum mechanics emerge from those of statistical mechanics when large enough numbers of particles are present, so do the laws of turbulence emerge at high-enough Reynold's numbers and at strong-enough nonlinearity. However, the classical turbulence laws were constrained by strong assumptions of homogeneity and isotropy and could not cover wide scale ranges in the atmosphere. By generalizing them to take into account anisotropy (especially vertical stratification) and intermittency, their range of applicability is vastly increased. In the last 5 years, thanks in part to the ready availability of huge global-scale data sets of all kinds, it has been possible to verify that these generalized emergent laws accurately hold up to planetary scales. For a recent review, see the work of *Lovejoy and Schertzer* [2010].

These "weather regime" laws show that the horizontal variability is fundamentally controlled by solar forcing via the energy flux. First principle calculations show that this accurately accounts for the large-scale winds and predicts a drastic "dimensional transition" at $\tau_w \approx 10$ days, the typical lifetime of planetary-scale structures. Beyond this time scale, spatial interactions are rapidly quenched so that the longer scales are driven by essentially temporal degrees of freedom, and the spectra of atmospheric fields display a shallow "spectral plateau" with small logarithmic slope. By making a third generalization of the classical laws, the statistical behavior in this "low-frequency weather" regime can be predicted. This qualitative change in the statistics at τ_w is just as expected; it is neither due to the action of new mechanisms of internal variability nor to external climate forcing; it is apparently nothing more than low-frequency weather. Similarly (forcing free), GCM control runs reproduce the same type of variability out to their low-frequency limits. The main complication is that due to similar effects from ocean turbulence, whose corresponding outer time scale is $\tau_o \approx 1$ year, there is enhanced intermittency up to that scale, with a slightly steeper "low-frequency ocean weather" regime beyond. Depending on the location and atmospheric variable of interest, this scaling continues up to scales of ≈10–100 years beyond which the type of variability drastically changes; new mechanisms of internal variability or of external climate forcing must come into play: the true climate regime begins. It is notable that if we consider pre-twentieth century Northern Hemisphere temperatures, the result is qualitatively similar so that, presumably, anthropogenic forcings are not responsible for this new regime.

In section 2, we discussed the weather and low-frequency weather regimes using spectral techniques. However, in later sections, we turned to real space fluctuation statistics. For these, the key parameter is the fluctuation exponent H, which determines the rate at which the mean fluctuations increase ($H > 0$) or decrease ($H < 0$) with scale Δt. We underlined the particularly simple interpretations afforded by the usual difference and tendency structure functions. However, the corresponding wavelets only usefully characterize the scaling of the fluctuations over narrow ranges in the exponent H: the differences ($0 < H < 1$) and the tendency ($-1 < H < 0$), i.e., useful only for fluctuations increasing in amplitude with scale or decreasing in amplitude with scale, respectively. Since both the weather and climate generally have $H > 0$ and the low-frequency weather regime has $H < 0$, we instead defined fluctuations using the Haar wavelet, which is thus useful over the entire range $-1 < H < 1$ and can also be "calibrated" to directly yield fluctuation amplitudes and can be easily implemented numerically.

In order to evaluate the statistical variability of the atmosphere over as wide a range as possible, we combined the Haar wavelet with temperature data from the 20CR (1871–2008), (2° × 2°, 6 hourly), three surface series (5° × 5°, monthly), eight intermediate-length resolution "multiproxy" series of the Northern Hemisphere from 1500 to 1980 (yearly), and GRIP and Vostok paleotemperatures at 5.2 and ≈100 year resolutions over lengths 91 and 420 kyr. The basic findings were that the key transition scale τ_c from low-frequency weather to climate, was somewhat variable, depending on the field and geographical location. For example, for surface global temperatures, we found $\tau_c \approx 10$ years, whereas for the more reliable post-2003 Northern Hemisphere reconstructions, $\tau_c \approx 30$ years, for the Holocene GRIP (Greenland) core, $\tau_c \approx 2$ kyr, the Holocene Vostok (Antarctica) core, $\tau_c \approx 1$ kyr, and the mean pre-Holocene paleotemperature value, $\tau_c \approx 100$ years. We also found $H_{lw} \approx -0.4$ and -0.1 for local (e.g., 2° × 2° resolution) and global series, respectively, and $H_c \approx 0.4$; although the GRIP value was a little lower, these values correspond to $\beta_{lw} \approx 0.2$, 0.8, 1.8, respectively (ignoring intermittency corrections $K(2)$, which ranged from $K_{lw}(2) \approx 0.05$ to $K_c(2) \approx 0.1$; a full characterization of the intermittency, i.e., $K(q)$ was been performed but was not discussed here).

Although this basic overall three-scaling regime picture is 25 years old, much has changed to make it more convincing. Obviously, an important element is the improvement in the quantity and quality of the data, but we have also benefited from advances in nonlinear dynamics as well as in data analysis techniques. In combination, these advances make the model a seductive framework for understanding atmospheric variability over huge ranges of space-time scales. It allows us to finally clarify the distinction between weather, its straightforward extension, without new elements, to low-frequency weather, and finally to the climate regime. It allows for objective definitions of the weather (scales $<\tau_w$), climate states (averages up to τ_c), and hence of climate change (scales $>\tau_c$). This new understanding of atmospheric variability is essential for evaluating the realism of both atmospheric and climate models. In particular, since without special external forcing, GCMs only model low-frequency weather, the question is posed as to what types of external forcing are required so that the GCM variability makes a transition to the climate regime with realistic scaling exponents and at a realistic time scales.

Acknowledgments. We thank P. Ditlevsen for providing the high-resolution GRIP data and A. Bunde for useful editorial and scientific comments.

REFERENCES

Andersen, C., N. Ko, and M. Moros (2004), A highly unstable Holocene climate in the subpolar North Atlantic: Evidence from diatoms, *Quat. Sci. Rev.*, *23*, 2155–2166.

Ashkenazy, Y., D. R. Baker, H. Gildor, and S. Havlin (2003), Nonlinearity and multifractality of climate change in the past 420,000 years, *Geophys. Res. Lett.*, *30*(22), 2146, doi:10.1029/2003GL018099.

Ashok, V., T. Balakumaran, C. Gowrishankar, I. L. A. Vennila, and A. Nirmalkumar (2010), The fast Haar wavelet transform for signal and image processing, *Int. J. Comput. Sci. Inf. Security*, *7*, 126–130.

Bacry, A., A. Arneodo, U. Frisch, Y. Gagne, and E. Hopfinger (1989), Wavelet analysis of fully developed turbulence data and measurement of scaling exponents, in *Turbulence and Coherent Structures*, edited by M. Lessieur and O. Metais, pp. 703–718, Kluwer Acad., Dordrecht, Netherlands.

Berner, K. S., N. Koç, D. Divine, F. Godtliebsen, and M. Moros (2008), A decadal-scale Holocene sea surface temperature record from the subpolar North Atlantic constructed using diatoms and statistics and its relation to other climate parameters, *Paleoceanography*, *23*, PA2210, doi:10.1029/2006PA001339.

Blender, R., and K. Fraedrich (2003), Long time memory in global warming simulations, *Geophys. Res. Lett.*, *30*(14), 1769, doi:10.1029/2003GL017666.

Blender, R., K. Fraedrich, and B. Hunt (2006), Millennial climate variability: GCM-simulation and Greenland ice cores, *Geophys. Res. Lett.*, *33*, L04710, doi:10.1029/2005GL024919.

Bolgiano, R., Jr. (1959), Turbulent spectra in a stably stratified atmosphere, *J. Geophys. Res.*, *64*(12), 2226–2229.

Briffa, K., T. Osborn, F. Schweingruber, I. Harris, P. Jones, S. Shiyatov, and E. Vaganov (2001), Low-frequency temperature variations from a northern tree ring density network, *J. Geophys. Res.*, *106*(D3), 2929–2941.

Brunt, D. (1939), *Physical and Dynamical Meteorology*, 2nd ed., 454 pp., Cambridge Univ. Press, New York.

Bunde, A., J. F. Eichner, S. Havlin, E. Koscielny-Bunde, H. J. Schellnhuber, and D. Vyushin (2004), Comment on "Scaling of atmosphere and ocean temperature correlations in observations and climate models," *Phys. Rev. Lett.*, *92*, 039801, doi:10.1103/PhysRevLett.92.039801.

Burgert, R., and W. W. Hsieh (1989), Spectral analysis of the AVHRR sea surface temperature variability off the west coast of Vancouver Island, *Atmos. Ocean*, *27*, 577–587.

Clayson, C. A., and L. H. Kantha (1999), Turbulent kinetic energy and its dissipation rate in the equatorial mixed layer, *J. Phys. Oceanogr.*, *29*, 2146–2166.

Compo, G. P., et al. (2011), The Twentieth Century Reanalysis Project, *Q. J. R. Meteorol. Soc.*, *137*, 1–28.

Corrsin, S. (1951), On the spectrum of isotropic temperature fluctuations in an isotropic turbulence, *J. Appl. Phys.*, *22*, 469–473.

Crowley, T. J., and T. S. Lowery (2000), How warm was the Medieval Warm period?, *Ambio*, *29*, 51–54.

Davidsen, J., and J. Griffin (2010), Volatility of unevenly sampled fractional Brownian motion: An application to ice core records, *Phys. Rev. E*, *81*, 016107, doi:10.1103/PhysRevE.81.016107.

Deschamps, P. Y., R. Frouin, and L. Wald (1981), Satellite determination of the mesoscale variability of the sea surface temperature, *J. Phys. Oceanogr.*, *11*, 864–870.

Deschamps, P., R. Frouin, and M. Crépon (1984), Sea surface temperatures of the coastal zones of France observed by the HCMM satellite, *J. Geophys. Res.*, *89*(C5), 8123–8149.

Ditlevsen, P. D., H. Svensmark, and S. Johson (1996), Contrasting atmospheric and climate dynamics of the last-glacial and Holocene periods, *Nature*, *379*, 810–812.

Eichner, J. F., E. Koscielny-Bunde, A. Bunde, S. Havlin, and H.-J. Schellnhuber (2003), Power-law persistence and trends in the atmosphere: A detailed study of long temperature records, *Phys. Rev. E*, *68*, 046133, doi:10.1103/PhysRevE.68.046133.

Esper, J., E. R. Cook, and F. H. Schweingruber (2002), Low-frequency signals in long tree-ring chronologies for reconstructing past temperature variability, *Science*, *295*(5563), 2250–2253.

Fraedrich, K., and K. Blender (2003), Scaling of atmosphere and ocean temperature correlations in observations and climate models, *Phys. Rev. Lett.*, *90*, 108501, doi:10.1103/PhysRevLett.90.108501.

Fraedrich, K., R. Blender, and X. Zhu (2009), Continuum climate variability: Long-term memory, scaling, and 1/f-noise, *Int. J. Mod. Phys. B*, *23*, 5403–5416.

Goldman, J. L. (1968), The power spectrum in the atmosphere below macroscale, *TRECOM 365G5-F*, Inst. of Desert Res., Univ. of St. Thomas, Houston, Tex.

Grant, H. L., R. W. Steward, and A. Moillet (1962), Turbulence spectra from a tidal channel, *J. Fluid Mech.*, *2*, 263–272.

Hansen, J., R. Ruedy, M. Sato, and K. Lo (2010), Global surface temperature change, *Rev. Geophys.*, *48*, RG4004, doi:10.1029/2010RG000345.

Holschneider, M. (1995), *Wavelets: An Analysis Tool*, 423 pp., Clarendon, New York.

Huang, S. (2004), Merging information from different resources for new insights into climate change in the past and future, *Geophys. Res. Lett.*, *31*, L13205, doi:10.1029/2004GL019781.

Huybers, P., and W. Curry (2006), Links between annual, Milankovitch and continuum temperature variability, *Nature*, *441*, 329–332.

Hwang, H. J. (1970), Power density spectrum of surface wind speed on Palmyra island, *Mon. Weather Rev.*, *98*, 70–74.

Inoue, E. (1951), On the turbulent diffusion in the atmosphere, *J. Meteorol. Soc. Jpn.*, *29*, 32.

Intergovernmental Panel on Climate Change (2001), *Climate Change 2001: The Scientific Basis: Contribution of Working Group I to the Third Assessment Report of the Intergovernmental Panel on Climate Change*, edited by J. T. Houghton et al., 881 pp., Cambridge Univ. Press, New York.

Intergovernmental Panel on Climate Change (2007), *Climate Change 2007: The Physical Science Basis: Working Group I Contribution to the Fourth Assessment Report of the IPCC*, edited by S. Solomon et al., Cambridge Univ. Press, New York.

Jones, P. D., K. R. Briffa, T. P. Barnett, and S. F. B. Tett (1998), High-resolution paleoclimatic records for the last Millenium: Interpretation, integration and comparison with General Circulation Model control-run temperatures, *Holocene*, *8*, 477–483.

Kantelhardt, J. W., E. Koscielny-Bunde, H. H. A. Rego, S. Havlin, and S. Bunde (2001), Detecting long range correlations with detrended fluctuation analysis, *Physica A*, *295*, 441–454.

Kantelhardta, J. W., S. A. Zschiegner, E. Koscielny-Bunde, S. Havlin, A. Bundea, and H. E. Stanley (2002), Multifractal detrended fluctuation analysis of nonstationary time series, *Physica A*, *316*, 87–114.

Karimova, L., Y. Kuandykov, N. Makarenko, M. M. Novak, and S. Helama (2007), Fractal and topological dynamics for the analysis of paleoclimatic records, *Physica A*, *373*, 737–746.

Kolesnikov, V. N., and A. S. Monin (1965), Spectra of meteorological field fluctuations, *Izv. Acad. Sci. USSR Atmos. Oceanic Phys., Engl. Transl.*, *1*, 653–669.

Kolmogorov, A. N. (1941), Local structure of turbulence in an incompressible liquid for very large Reynolds numbers, *Dokl. Akad. Nauk. SSSR*, *30*, 301–305. [English translation: *Proc. R. Soc. London, Ser. A*, *434*, 9–17, 1991]

Koscielny-Bunde, E., A. Bunde, S. Havlin, H. E. Roman, Y. Goldreich, and H.-J. Schellnhuber (1998), Indication of a universal persistence law governing atmospheric variability, *Phys. Rev. Lett.*, *81*, 729–732.

Koscielny-Bunde, E., J. W. Kantelhardt, P. Braund, A. Bunde, and S. Havlin (2006), Long-term persistence and multifractality of river runoff records: Detrended fluctuation studies, *J. Hydrol.*, *322*, 120–137.

Landau, L. D., and E. M. Lifschitz (1959), *Fluid Mechanics*, 536 pp., Pergamon, London, U. K.

Lanfredi, M., T. Simoniello, V. Cuomo, and M. Macchiato (2009), Discriminating low frequency components from long range persistent fluctuations in daily atmospheric temperature variability, *Atmos. Chem. Phys.*, *9*, 4537–4544.

Lennartz, S., and A. Bunde (2009), Trend evaluation in records with long-term memory: Application to global warming, *Geophys. Res. Lett.*, *36*, L16706, doi:10.1029/2009GL039516.

Le Traon, P. Y., P. Klein, B. L. Hua, and G. Dibarboure (2008), Do altimeter wavenumber spectra agree with the interior or surface quasigeostrophic theory?, *J. Phys. Oceanogr.*, *38*, 1137–1142.

Lien, R.-C., and E. A. D'Asaro (2006), Measurement of turbulent kinetic energy dissipation rate with a Lagrangian float, *J. Atmos. Oceanic Technol.*, *23*, 964–976.

Ljundqvist, F. C. (2010), A new reconstruction of temperature variability in the extra-tropical Northern Hemisphere during the last two millennia, *Geogr. Ann. Ser. A*, *92*, 339–351.

Lorenz, E. N. (1995), *Climate is What You Expect*, 55 pp., MIT Press, Cambridge, Mass.

Lovejoy, S., and D. Schertzer (1986), Scale invariance in climatological temperatures and the spectral plateau, *Ann. Geophys., Ser. B*, *4*, 401–410.

Lovejoy, S., and D. Schertzer (2010), Towards a new synthesis for atmospheric dynamics: Space-time cascades, *Atmos. Res.*, *96*, 1–52, doi:10.1016/j.atmosres.2010.01.004.

Lovejoy, S., and D. Schertzer (2012), *The Weather and Climate: Emergent Laws and Multifractal Cascades*, 660 pp., Cambridge Univ. Press, Cambridge, U. K.

Lovejoy, S., Y. Tessier, M. Claeredeboudt, W. J. C. Currie, J. Roff, E. Bourget, and D. Schertzer (2000), Universal multifractals and ocean patchiness phytoplankton, physical fields and coastal heterogeneity, *J. Plankton Res.*, *23*, 117–141.

Lovejoy, S., D. Schertzer, M. Lilley, K. B. Strawbridge, and A. Radkevitch (2008), Scaling turbulent atmospheric stratification. I: Turbulence and waves, *Q. J. R. Meteorol. Soc.*, *134*, 277–300, doi:10.1002/qj.201.

Mallat, S., and W. Hwang (1992), Singularity detection and processing with wavelets, *IEEE Trans. Inf. Theory*, *38*, 617–643.

Mann, M. E., R. S. Bradley, and M. K. Hughes (1998), Global-scale temperature patterns and climate forcing over the past six centuries, *Nature*, *392*, 779–787.

Mann, M. E., R. S. Bradley, and M. K. Hughes (1999), Northern hemisphere temperatures during the past millennium: Inferences, uncertainties, and limitations, *Geophys. Res. Lett.*, *26*(6), 759–762.

Matsuno, T., J.-S. Lee, M. Shimizu, S.-H. Kim, and I.-C. Pang (2006), Measurements of the turbulent energy dissipation rate ε and an evaluation of the dispersion process of the Changjiang Diluted Water in the East China Sea, *J. Geophys. Res.*, *111*, C11S09, doi:10.1029/2005JC003196.

McIntyre, S., and R. McKitrick (2003), Corrections to the Mann et al. (1998) "Proxy data base and Northern Hemispheric average temperature models," *Energy Environ.*, *14*, 751–771.

McIntyre, S., and R. McKitrick (2005), Hockey sticks, principal components, and spurious significance, *Geophys. Res. Lett.*, *32*, L03710, doi:10.1029/2004GL021750.

McLeish, W. (1970), Spatial spectra of ocean surface temperature, *J. Geophys. Res.*, *75*, 6872–6877.

Mitchell, J. M. (1976), An overview of climatic variability and its causal mechanisms, *Quat. Res.*, *6*, 481–493.

Moberg, A., D. M. Sonnechkin, K. Holmgren, and N. M. Datsenko (2005), Highly variable Northern Hemisphere temperatures reconstructed from low- and high-resolution proxy data, *Nature*, *433*, 613–617.

Monetti, R. A., S. Havlin, and A. Bunde (2003), Long-term persistence in the sea surface temperature fluctuations, *Physica A*, *320*, 581–589.

Monin, A. S. (1972), *Weather Forecasting as a Problem in Physics*, 216 pp., MIT Press, Boston, Mass.

Monin, A. S., and A. M. Yaglom (1975), *Statistical Fluid Mechanics*, 886 pp., MIT Press, Boston Mass.

Moum, J. N., M. C. Gregg, and R. C. Lien (1995), Comparison of turbulent Kinetic energy dissipation rate estimates from two ocean microstructure profiler, *J. Atmos. Oceanic Technol.*, *12*, 346–366.

Nakajima, H., and N. Hayakawa (1982), A cross-correlation analysis of tidal current, water temperature and salinity records, *J. Oceanogr. Soc. Jpn.*, *38*, 52–56.

Obukhov, A. (1949), Structure of the temperature field in a turbulent flow, *Izv. Akad. Nauk SSSR, Ser. Geogr. Geofiz*, *13*, 55–69.

Obukhov, A. (1959), Effect of archimedean forces on the structure of the temperature field in a turbulent flow, *Dokl. Akad. Nauk SSSR*, *125*, 1246–1248.

Palmén, E. (1959), *The Atmosphere and the Sea in Motion*, edited by B. Bolen, pp. 212–224, Oxford Univ. Press, New York.

Panofsky, H. A. (1969), The spectrum of temperature, *J. Radio Sci.*, *4*(12), 1143–1146.

Parisi, G., and U. Frisch (1985), A multifractal model of intermittency, in *Turbulence and Predictability in Geophysical Fluid Dynamics and Climate Dynamics*, edited by M. Ghil, R. Benzi and G. Parisi, pp. 84–88, North-Holland, Amsterdam, Netherlands.

Pelletier, J. D. (1998), The power spectral density of atmospheric temperature from time scales of 10^{-2} to 10^{6} yr, *Earth Planet. Sci. Lett.*, *158*, 157–164.

Peng, C.-K., S. V. Buldyrev, S. Havlin, M. Simons, H. E. Stanley, and A. L. Goldberger (1994), Mosaic organisation of DNA nucleotides, *Phys. Rev. E*, *49*, 1685–1689.

Penland, C. (1996), A stochastic model of IndoPacific sea surface temperature anomalies, *Physica D*, *98*, 534–558.

Peterson, T. C., and R. S. Vose (1997), An overview of the Global Historical Climatology Network temperature database, *Bull. Am. Meteorol. Soc.*, *78*, 2837–2849.

Petit, J. R., et al. (1999), Climate and atmospheric history of the past 420,000 years from the Vostok ice core, Antarctica, *Nature*, *399*, 429–436.

Pinus, N. Z. (1968), The energy of atmospheric macro-turbulence, *Izv. Acad. Sci. USSR Atmos. Oceanic Phys., Engl. Transl.*, *4*, 461.

Radkevitch, A., S. Lovejoy, K. B. Strawbridge, D. Schertzer, and M. Lilley (2008), Scaling turbulent atmospheric stratification. III: Space-time stratification of passive scalars from lidar data, *Q. J. R. Meteorol. Soc.*, *134*, 317–335, doi:10.1002/qj.1203.

Rayner, N. A., et al. (2006), Improved analyses of changes and uncertainties in marine temperature measured in situ since the mid-nineteenth century: The HadSST2 dataset, *J. Clim.*, *19*, 446–469.

Robert, C. W. (1976), Turbulent energy dissipation in the Atlantic equatorial undercurrent, Ph.D. thesis, Univ. of Br. Columbia, Vancouver, B. C., Canada.

Robinson, G. D. (1971), The predictability of a dissipative flow, *Q. J. R. Meteorol. Soc.*, *97*, 300–312.

Rybski, D., A. Bunde, S. Havlin, and H. von Storch (2006), Long-term persistence in climate and the detection problem, *Geophys. Res. Lett.*, *33*, L06718, doi:10.1029/2005GL025591.

Rybski, D., A. Bunde, and H. von Storch (2008), Long-term memory in 1000-year simulated temperature records, *J. Geophys. Res.*, *113*, D02106, doi:10.1029/2007JD008568.

Saunders, P. M. (1972), Space and time variability of temperature in the upper Ocean, *Deep Sea Res. Oceanogr. Abstr.*, *19*, 467–480.

Schertzer, D., and S. Lovejoy (1985a), Generalised scale invariance in turbulent phenomena, *Phys. Chem. Hydrodyn. J.*, *6*, 623–635.

Schertzer, D., and S. Lovejoy (1985b), The dimension and intermittency of atmospheric dynamics, in *Turbulent Shear Flow 4*, edited by B. Launder, pp. 7–33, Springer, Berlin.

Schertzer, D., and S. Lovejoy (1987), Physical modeling and analysis of rain and clouds by anisotropic scaling multiplicative processes, *J. Geophys. Res.*, *92*(D8), 9693–9714.

Schertzer, D., S. Lovejoy, F. Schmitt, Y. Chigirinskaya, and D. Marsan (1997), Multifractal cascade dynamics and turbulent intermittency, *Fractals*, *5*, 427–471.

Schertzer, D., I. Tchiguirinskaia, S. Lovejoy, and A. F. Tuck (2012), Quasi-geostrophic turbulence and generalized scale invariance, a theoretical reply, *Atmos. Chem. Phys.*, *12*, 327–336.

Schmitt, F., S. Lovejoy, and D. Schertzer (1995), Multifractal analysis of the Greenland Ice-Core Project climate data, *Geophys. Res. Lett.*, *22*(13), 1689–1692.

Seuront, L., F. Schmitt, D. Schertzer, Y. Lagadeuc, and S. Lovejoy (1996), Multifractal analysis of Eulerian and Lagrangian variability of physical and biological fields in the ocean, *Nonlin. Processes Geophys.*, *3*, 236–246.

Smith, T. M., R. W. Reynolds, T. C. Peterson, and J. Lawrimore (2008), Improvements to NOAA's historical merged land-ocean surface temperature analysis (1880–2006), *J. Clim.*, *21*, 2283–2293.

Talkner, P., and R. O. Weber (2000), Power spectrum and detrended fluctuation analysis: Application to daily temperatures, *Phys. Rev. E*, *62*, 150–160.

Torrence, T., and G. P. Compo (1998), A practical guide to wavelet analysis, *Bull. Am. Meteorol. Soc.*, *79*, 61–78.

Vallis, G. (2010), Mechanisms of climate variability from years to decades, in *Stochastic Physics and Climate Modelling*, edited by T. Palmer and P. Williams, pp. 1–34, Cambridge Univ. Press, Cambridge, U. K.

Van der Hoven, I. (1957), Power spectrum of horizontal wind speed in the frequency range from .0007 to 900 cycles per hour, *J. Meteorol.*, *14*, 160–164.

Vinnichenko, N. K. (1969), The kinetic energy spectrum in the free atmosphere for 1 second to 5 years, *Tellus*, *22*, 158–166.

Vinnichenko, N., and J. Dutton (1969), Empirical studies of atmospheric structure and spectra in the free atmosphere, *Radio Sci.*, *4*(12), 1115–1126.

Vinther, B. M., H. B. Clausen, D. A. Fisher, R. M. Koerner, S. J. Johnsen, K. K. Andersen, D. Dahl-Jensen, S. O. Rasmussen, J. P. Steffensen, and A. M. Svensson (2008), Synchronizing ice cores from the Renland and Agassiz ice caps to the Greenland Ice Core Chronology, *J. Geophys. Res.*, *113*, D08115, doi:10.1029/2007JD009143.

von Storch, H., E. Zorita, J. M. Jones, Y. Dimitriev, F. González-Rouco, and S. F. B. Tett (2004), Reconstructing past climate from noisy data, *Science*, *306*, 679–682.

Vyushin, D., I. Zhidkov, S. Havlin, A. Bunde, and S. Brenner (2004), Volcanic forcing improves Atmosphere-Ocean Coupled General Circulation Model scaling performance, *Geophys. Res. Lett.*, *31*, L10206, doi:10.1029/2004GL019499.

Witt, A., and A. Y. Schumann (2005), Holocene climate variability on millennial scales recorded in Greenland ice cores, *Nonlin. Processes Geophys.*, *12*, 345–352.

Wunsch, C. (2003), The spectral energy description of climate change including the 100 ky energy, *Clim. Dyn.*, *20*, 353–363.

S. Lovejoy, Department of Physics, McGill University, 3600 University St., Montreal, QC H3A 2T8, Canada. (lovejoy@physics.mcgill.ca)

D. Schertzer, LEESU, Ecole des Ponts ParisTech, Université Paris Est, F-77455 Paris, France.

Extreme Space Weather: Forecasting Behavior of a Nonlinear Dynamical System

D. N. Baker

Laboratory for Atmospheric and Space Physics, University of Colorado, Boulder, Colorado, USA

Vulnerability of modern societies to extreme space weather is an issue of increasing concern. Recent assessments are that individual severe space weather episodes can cause tens of millions to many billions of dollars worth of damage to space and ground-based assets. The most extreme events could cause disruptions to society lasting months to years and could cost more than $1 trillion. There are very good reasons to believe that efficient and effective forecasts of certain aspects of extreme space weather are possible and that the benefits of such forecasts would be immense. Data analysis and recent modeling show that the coupled Sun-Earth system is a complex, nonlinear dynamical system. Improved forecasts of severe space weather must take advantage of our increasing understanding of this dynamical behavior of the Earth's space environment. Such forecasts may hold the key to successful mitigation strategies for dealing with extreme conditions and thereby greatly benefit policy makers and society as a whole.

1. INTRODUCTION

Modern society depends heavily on a variety of technologies that are vulnerable to the effects of intense geomagnetic storms and solar energetic particle (SEP) events. Strong currents flowing in the ionosphere can disrupt and damage Earth-based electric power grids and contribute to the accelerated corrosion of oil and gas pipelines. Magnetic storm-driven ionospheric disturbances interfere with HF, VHF, and UHF radio communications and navigation signals from GPS satellites. Exposure of spacecraft to solar particles and radiation belt enhancements can cause temporary operational anomalies, damage critical electronics, degrade solar arrays, and blind optical systems such as imagers and star trackers. Moreover, intense solar energetic particle events present a significant radiation hazard for astronauts during the high-latitude segment of the International Space Station orbit as well as for future human explorers of the Moon and Mars who will be unprotected by the Earth's magnetic field [*National Research Council (NRC)*, 2008].

In addition to such direct effects as spacecraft anomalies or power grid outages, a thorough assessment of the impact of severe space weather events on present-day society must include the collateral effects of space-weather-driven technology failures. For example, polar cap absorption (PCA) events due to solar particles can degrade and, during severe events, completely black out, radio communications along transpolar aviation routes, requiring aircraft flying these routes to be diverted to lower latitudes. This can add considerable cost to the airlines and can greatly inconvenience passengers.

Modern technological society is characterized by a complex set of interdependencies among its critical infrastructures. A complete picture of the socioeconomic impact of severe space weather must include both direct, industry-specific effects and the collateral effects of space-weather-driven technology failures on dependent infrastructures and services. In this chapter, we describe possible extreme space weather impacts as well as the nonlinear aspects of our geospace environment. We conclude with a vision for specification and forecasting of hazardous space weather conditions in this complex, highly coupled system.

Extreme Events and Natural Hazards: The Complexity Perspective
Geophysical Monograph Series 196

10.1029/2011GM001075

2. AN EXTREME SPACE WEATHER EVENT AND ITS SOCIETAL IMPLICATIONS

From 28 August through 4 September of 1859, auroral displays of extraordinary brilliance were observed throughout North and South America, Europe, Asia, and Australia. Such displays were seen as far south as Hawaii, the Caribbean, and Central America in the Northern Hemisphere (see Figure 1) and in the Southern Hemisphere as far north as Santiago, Chile. Even after daybreak, when the intense auroras were no longer visible to human eyes, their presence continued to be felt through the effect of strong auroral currents. Magnetic observatories recorded disturbances in the Earth's field so extreme that magnetometer traces were driven off scale, and telegraph networks around the world, the "Internet" of that age, experienced severe disruptions and outages. "The electricity which attended this beautiful phenomenon took possession of the magnetic wires throughout the country," the Philadelphia Evening Bulletin reported, "and there were numerous side displays in the telegraph offices where fantastical and unreadable messages came through the instruments, and where the atmospheric fireworks assumed shape and substance in brilliant sparks" (quoted by *NRC* [2008]). In several locations, operators disconnected their systems from the batteries and sent messages using only the current induced by the aurora.

Intense auroral displays were the visible manifestation of two intense magnetic storms that occurred near the 1859–1860 peak of the then-current sunspot cycle. On 1 September, the day before the onset of the second storm, Richard Carrington, a British amateur astronomer, observed an outburst of intensely bright white light from a large and complex group of sunspots near the center of the Sun's disk [*Green and Boardsen*, 2006]. The dazzling auroral displays, magnetic disturbances, and disruptions of the telegraph network that occurred between 28 August and 4 September 1859 were recognized by contemporary observers as especially spectacular manifestations of a connection between sunspots and terrestrial magnetism. This connection had been noted earlier in the decade on the basis of the regular correspondence observed between changes in the Earth's magnetic field and the number of sunspots. The evident connection between the aurora borealis and terrestrial magnetism was well established by this time as well.

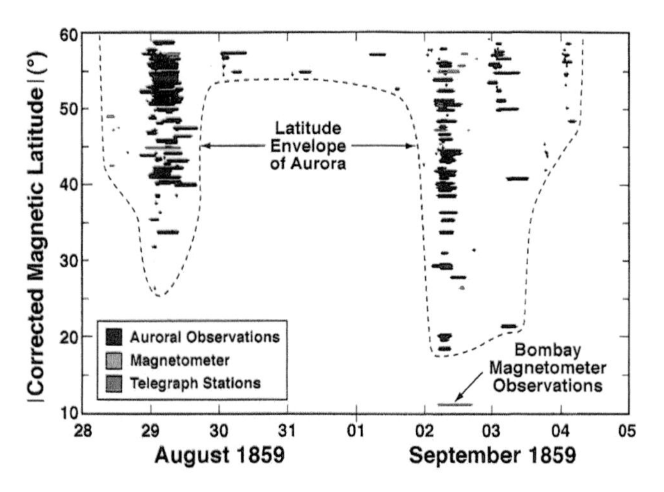

Figure 1. Magnetic latitudes at which strong auroral displays, magnetometer fluctuations, and telegraph station disruptions were observed during the August–September 1859 solar storms. Reprinted from *Green and Boardsen* [2006], with permission from Elsevier.

Although the existence of the links between solar, geomagnetic, and auroral phenomena was recognized by the time of the 1859 events, the nature of this link was not understood. The white light flare observations by Carrington furnished a critical clue. But it would not be until decades later that the significance of such observations was appreciated, and a full picture of the phenomena that constitute what we now call space weather would not emerge until well into the Space Age [see *Baker and Green*, 2011].

A key point of improved understanding of space weather came with the discovery of coronal mass ejections (CMEs) in the 1970s and with the recognition that these (rather than solar flares) are the cause of major nonrecurrent geomagnetic storms. Large-scale eruptions of plasma and magnetic fields from the Sun's corona, CMEs contain as much as 10^{16} g (10 billion t) of coronal material and travel at speeds as high as 3000 km s^{-1}, with a kinetic energy of up to 10^{32} ergs. Eruptive flares and CMEs occur most often around solar maximum and result from the release of energy stored in the Sun's magnetic field. CMEs and flares can occur independently of one another; however, both are generally observed at the start of a space weather event that leads to a large magnetic storm. To be highly geoeffective, i.e., to drive a strong geomagnetic storm, a CME must be launched from near the center of the Sun onto a trajectory that will cause it to impact the Earth's magnetic envelope. It also must be fast ($\geq$1000 km s^{-1}) and massive, thus having large kinetic energy, and it must possess a strong magnetic field whose orientation is opposite that of the Earth's.

Moving substantially faster than the surrounding solar wind plasma, fast CMEs create a shock wave that accelerates coronal and solar wind ions (predominantly protons) and electrons to relativistic and near-relativistic velocities. Particles are accelerated by solar flares as well, and large SEP events, although dominated by shock-accelerated particles, generally include flare-accelerated particles (some of which may be further accelerated by the shock). Traveling near the speed of light, SEPs begin arriving at Earth within less than an hour of the CME liftoff/flare eruption and are channeled along geomagnetic field lines into the upper atmosphere above the North and South Poles, where they enhance the ionization of the lower ionosphere over the entire polar regions, PCA events, and can initiate ozone-depleting chemistry in the middle atmosphere. SEP events can last for several days.

The mid-nineteenth century lacked the means to detect and measure SEPs, and the technologies of that age were essentially unaffected by them. Thus, in contrast to the widely observed auroral displays and magnetic disturbances, the radiation storm produced by the solar eruption on 1 September 1859 went unnoticed by contemporary observers. There is, however, a natural record of the storm that can be retrieved even today. Nitrates, produced by SEP impacts of the atmosphere above the poles, precipitate out of the atmosphere within weeks of a SEP event and are preserved in the polar ice. Analysis of anomalous nitrate concentrations in ice core samples allows the magnitude of pre-space-era SEP events to be estimated. Such an analysis (see Figure 2) indicates that the 1859 event is the largest SEP event known, with a total fluence of 1.9×10^{10} cm^{-2} for protons with energies greater than 30 MeV, four times that of the much discussed August 1972 event, the largest in the space era [*Baker and Green*, 2011].

The shock responsible for the radiation storm hit the Earth's magnetosphere at 04:50 UT on 2 September 1859. It dramatically compressed the geomagnetic field, producing a steep increase in the magnitude of the field's horizontal (H) component, which marked the onset of the geomagnetic storm. The compression of the field would also have triggered an almost instantaneous brightening of the entire auroral oval. A rough quantitative measure of its intensity is provided by the Colaba (India) data, which show a precipitous reduction (by 1600 nT) in H at the peak of the storm's main phase. Converted to 1 h averages, these data yield a proxy *Dst* index of approximately −850 nT. For comparison, the largest *Dst* index recorded since the International Geophysical Year (1957) is −548 nT for the superstorm of 14 March 1989.

The 1859 storm was at its most intense on 2 September, and the geomagnetic field required several days to recover. Balfour Stewart, the director of the Kew Observatory near London, reported that the magnetic elements "remained in a state of considerable disturbance until September 5, and scarcely attained their normal state even on September 7 or 8" [*NRC*, 2008]. As we will discuss below, an extreme event of the Carrington class, were it to occur in today's technological age, could and probably would produce devastating socioeconomic consequences [see *Baker and Green*, 2011].

3. REQUIREMENTS AND IMPLICATIONS OF FORECASTING EXTREME EVENTS

As noted in the Introduction, modern society is characterized by an amazing array of dependencies and interdependencies. This interconnectedness of elements of infrastructure is shown graphically by a diagram developed by the U.S. Department of Homeland Security (see Figure 3). Some of the main technological sectors of present-day society are shown by the variously colored panels. These range from fuels (oil/gas) and electric power at the top to financial and government services at the bottom. The colored arrows in the diagram are meant to portray some of the main connections and dependencies of one sector with another. As shown, the linkages form an intricate and complex web.

A clear and present danger for any modern society is that a disruption in one sector portrayed in Figure 3 would almost

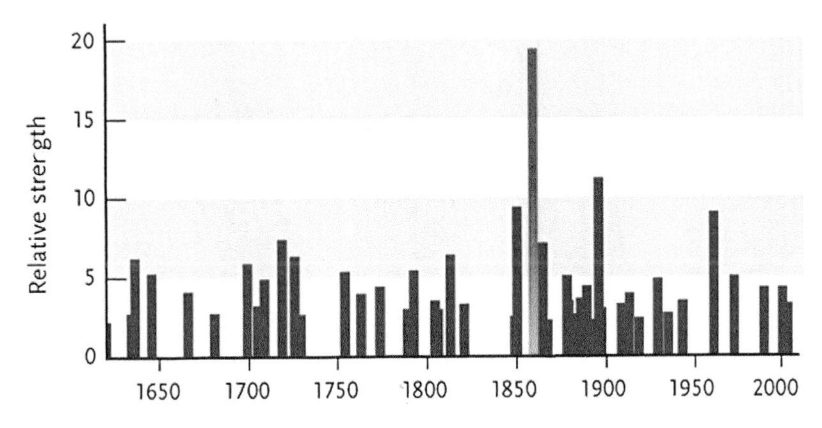

Figure 2. History of fluences of $E > 30$ MeV protons from the years ~1625 to 2000 as recorded in ice core measurements of nitrate compounds. The 1859 Carrington event was by far the largest event in the record. From *Baker and Green* [2011].

certainly cascade in negative ways into other sectors. As an illustration, consider the case of electric power. This part of our infrastructure has been called a "cornerstone" technology. Loss of the electric power grid would quickly mean that transportation systems, communications, pumping of oil and gas, provision of water, and ready (and reliable) availability of emergency services would be threatened. Brief outages of electrical power (hours to a day or two) might be tolerable, but extended outages of several days to weeks (or longer) would produce huge socioeconomic impacts [*NRC*, 2008].

Recent research examining the fundamental properties of interdependent networks [see *Buldyrev et al.*, 2010] has, indeed, shown that a failure of nodes in one network may lead to a catastrophic failure of dependent nodes in other linked networks. *Buldyrev et al.* [2010], in fact, studied such a cascade of failures in a real-world example of an electrical blackout in Italy in 2003. We suggest that during a large, highly stressed situation in an extreme space weather event, one would almost certainly see a progression of technological failures.

Analysis of an historic solar storm that occurred on 14–15 May 1921 (as reported by the *NRC* [2008] by J. Kappenman) demonstrates how widely and substantially a large geomagnetic storm could affect the modern power grid. Kappenman noted that changes made to increase operational efficiencies in the grid and to allow transport of power over vast distances (i.e., continental scales) have inadvertently escalated the risks associated with geomagnetic storms. Steps taken in the U.S. power system to boost operational reserves generally do not reduce risks associated with geomagnetically induced currents (GICs). For very large storms (those with rapid changes of magnetic field (dB/dt) associated with very large GICs), the scale and speed of developing problems could exceed anything experienced by grid operators in the modern era (including the famous HydroQuebec failure in March 1989).

As shown in Figure 4, the May 1921 storm was estimated to have dB/dt values of 4800 nT min^{-1}. These rates would be 10 times larger than those experienced in the March 1989 event. A great storm of the 1921 class (or the Carrington event of 1859), were it to occur today, would cause huge GICs in the extremely high voltage (EHV) transformers in the North American power grid backbone. This would have the potential to cause transformer failures throughout the eastern United States and could lead to power loss for >130 million citizens. Moreover, the transformer failures estimated in these cases would cost millions of dollars each to replace and would take many months to manufacture [*Baker and Green*, 2011].

Representatives of the electrical power companies have noted that accurate forecasts of powerful geomagnetic storms, especially tailored forecasts of regional dB/dt values,

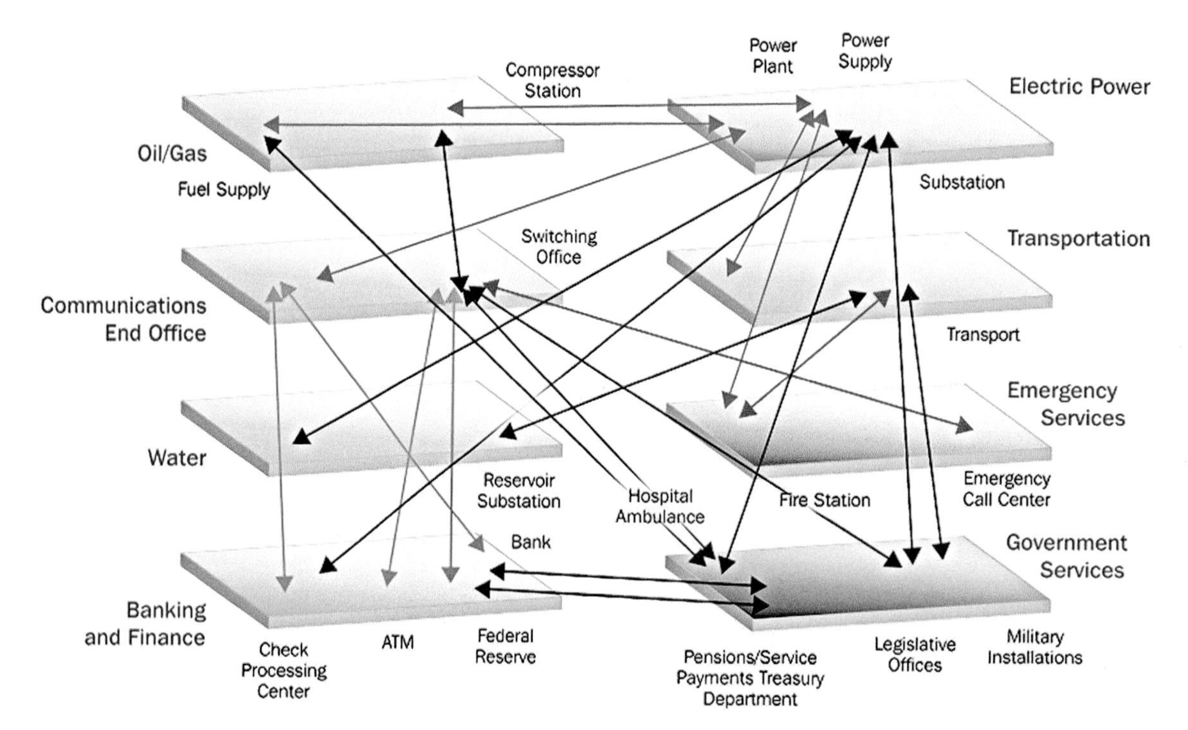

Figure 3. Connections and key interdependencies of various components across the economy. This schematic shows the highly interconnected infrastructures of modern societies and how effects in one sector might impact other sectors. Source is Department of Homeland Security, National Infrastructure Protection Plan, reprinted in the *National Research Council* [2008] work.

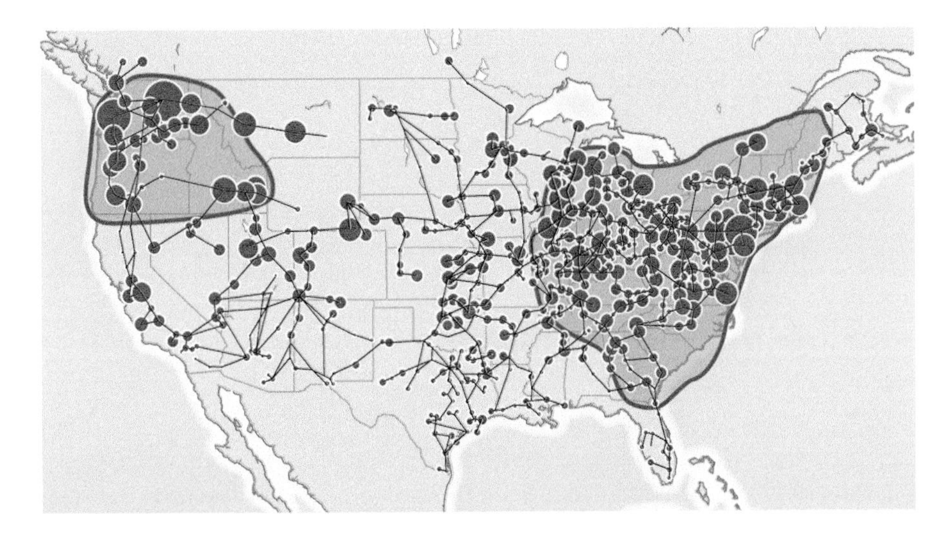

Figure 4. Illustration represents the continental United States with obvious outlines of individual states. Dark lines show routes of EHV transmission lines and major power substations. Geomagnetic electric currents from a solar storm will flow through these lines to major transformers, marked by red dots. The relative sizes of the dots indicate the magnitude of the current. Due to the powerful flows of the current, 300 large EHV transformers would be at risk of permanent damage or failure. The electric grid and transformers in the gray shaded regions could suffer a catastrophic collapse, leaving more than 130 million people without electricity. From *Baker and Greene* [2011] and courtesy of J. Kappenman, Metatech Corp.

would be of immense operational value to the power industry. Such forecasts could allow generation of extra power in regions where this was needed and could also allow shifting of power away from threatened sectors during the height of great storms. Such actions, while detrimental to customers in the short term, would allow the grid infrastructure to avert long-term damage.

Another economic sector greatly affected by severe space weather is the airline industry. Solar flares (X-rays) can cause HF radio disruptions at midlatitudes, and SEP events can be of concern at higher latitudes. One of the most recent concerns is the SEP effect for transpolar airline flight. The issues, as presented by United Airlines (UAL), are shown here in Figure 5. As described by UAL, the polar routes allow aircraft to avoid the strong wintertime headwinds and decrease travel time and, therefore, transport more passengers and cargo, thus offering a more economical and convenient service to its customers.

Federal regulations require flights to maintain communications with Air Traffic Control and their company operations center over the entire route of flight. UAL relies on SATCOM, which is communication via satellites in geosynchronous orbit (located about 22,300 miles (35,900 km) above the equator). Aircraft lose the ability to communicate with these satellites when they go above 82°N latitude (within the yellow circle shown toward the center of Figure 5). In this region, aircraft communications are reliant on HF radio links.

Strong solar activity causes HF radio blackouts in the polar region. If and when the Sun emits a shower of high-energy protons and other ions (SEP events as defined above), the protons hit the Earth's outermost atmosphere (the ionosphere). They can greatly increase the density of ionized gas, which, in turn, affects the ability of radio waves to propagate [*Patterson et al.*, 2001]. HF radio frequencies in the polar regions are particularly affected because the solar protons can directly reach the ionosphere in the polar parts of the Earth's magnetic field. The radio blackouts over the poles are called PCA events. When a solar event causes severe HF degradation in the polar region, aircraft that are dependent on SATCOM have to be diverted to latitudes below 82°N so that SATCOM satellite communication links can be used. UAL currently utilizes the NOAA Space Weather Prediction Center (SWPC) space weather scales and alerts to plan upcoming flights and to instruct planes in transit to divert from polar routes.

PCA blackouts can last up to several days, depending on the size and location of the disturbance on the Sun that triggers them [*Patterson et al.*, 2001]. For example, between 15 and 19 January 2005, five separate X-ray solar flares occurred that produced radio blackouts of extremely high intensity [*Patterson et al.*, 2001]. For four consecutive days, flights from Chicago to Hong Kong could not operate on polar routes. The longer nonpolar routes required an extra refueling stop in Anchorage, Alaska, which added delays ranging from 3 to

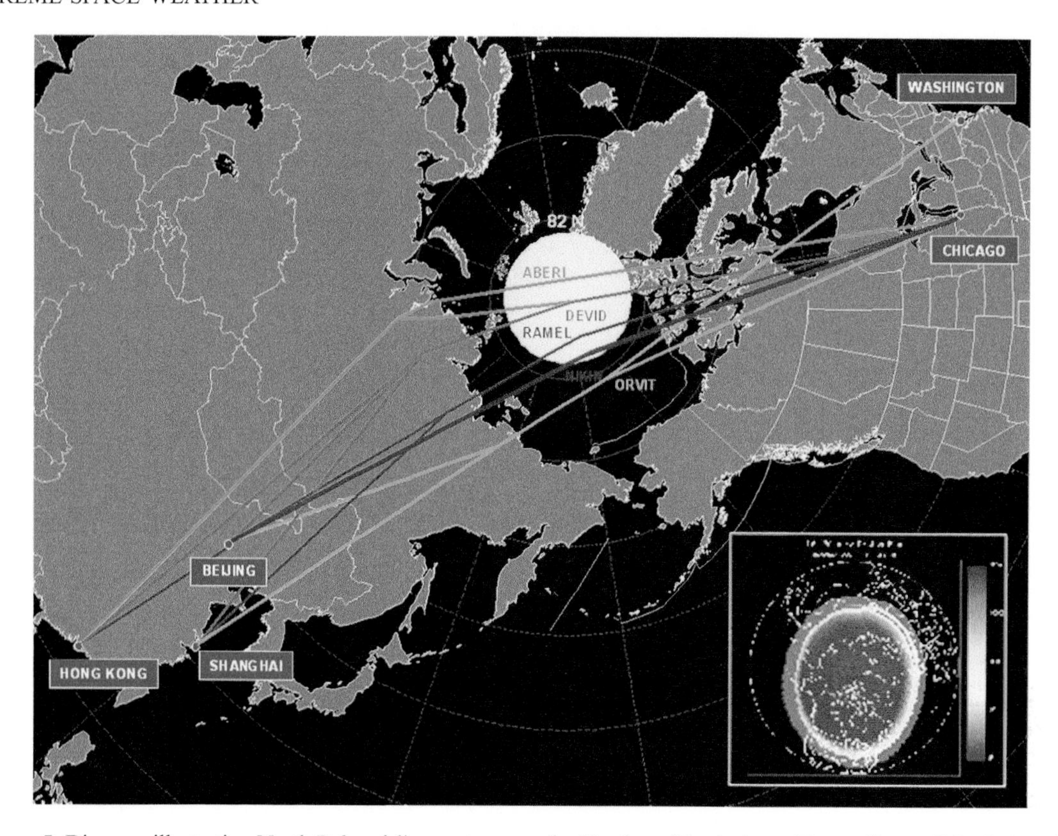

Figure 5. Diagram illustrating North Polar airline routes over the Northern Hemisphere. The outlines of North America, Asia, and Europe are obvious in the diagram. Colored lines show routes for air traffic that necessitate HF radio communication (shown in yellow circle near center of image). The inset shows that a powerful solar storm such as the Halloween storms of 2003 portrayed in the diagram can completely disrupt ionospheric radio propagation over this polar region. Figure courtesy of United Airlines.

3.5 h. In total, 26 flights operated on less than optimal polar routes or nonpolar routes. Increased flight time and extra landings and takeoffs increase fuel consumption and cost, and the delays disrupted connections to other flights.

There are a whole host of companies and industries that are concerned about, and are affected by, space weather in its many forms. Figure 6 shows the growth of space weather customers for the NOAA SWPC over the period ~1940 to present. It is, indeed, sobering to think about all the technological sectors that can be disrupted, or even totally disabled, by severe space storms. In a large number of cases, improved space weather forecasts could mitigate the duration, severity, and consequences of extreme space weather events [*NRC*, 2008].

4. NONLINEARITY IN THE SOLAR-TERRESTRIAL SYSTEM

Our current understanding is that energy extracted from solar wind sources is episodically stored in the Earth's magnetosphere, with a large fraction of the energy often being in the form of excess magnetic flux in the magnetotail lobes. This stored energy is then explosively dissipated in the near-Earth, nightside region [e.g., *Baker et al.*, 1996]. It is generally accepted that these episodes of energy release, called substorms, consist of both directly "driven" and "unloading" processes. Here we will describe nonlinear dynamical aspects of the global solar wind-magnetosphere interaction. We argue that it is important to include nonlinearity in our picture in order to replicate the main features of geomagnetic activity and to address issues of predictability.

For weak solar wind-magnetosphere coupling periods, the loading of flux into the Earth's magnetotail is relatively small, and corresponding magnetic reconnection at the distant magnetic neutral line [see *Hones*, 1979] can balance the dayside magnetic "merging" rate. However, for strongly southward interplanetary magnetic field (IMF) and/or high solar wind speed, the dayside magnetic merging completely overwhelms the reconnection rate and return of flux from the distant X line. This imbalance forces near-Earth neutral line

Growth of Space Weather Customers

Commercial Space Transportation
Airline Polar Flights
Microchip technology
Precision Guided Munitions
Cell phones
Atomic Clock
Satellite Operations
Carbon Dating experiments
GPS Navigation
Ozone Measurements
Aircraft Radiation Hazard
Commercial TV Relays
Communications Satellite Orientation
Spacecraft Charging
Satellite Reconnaissance & Remote Sensing Instrument Damage
Geophysical Exploration.
Pipeline Operations
Anti-Submarine Detection
Satellite Power Arrays
Power Distribution
Long-Range Telephone Systems
Radiation Hazards to Astronauts
Interplanetary Satellite experiments
VLF Navigation Systems (OMEGA, LORAN)
Over the Horizon Radar
Solar-Terres. Research & Applic. Satellites
Research & Operations Requirements
Satellite Orbit Prediction
Solar Balloon & Rocket experiments
Ionospheric Rocket experiments
Radar
Short-wave Radio Propagation

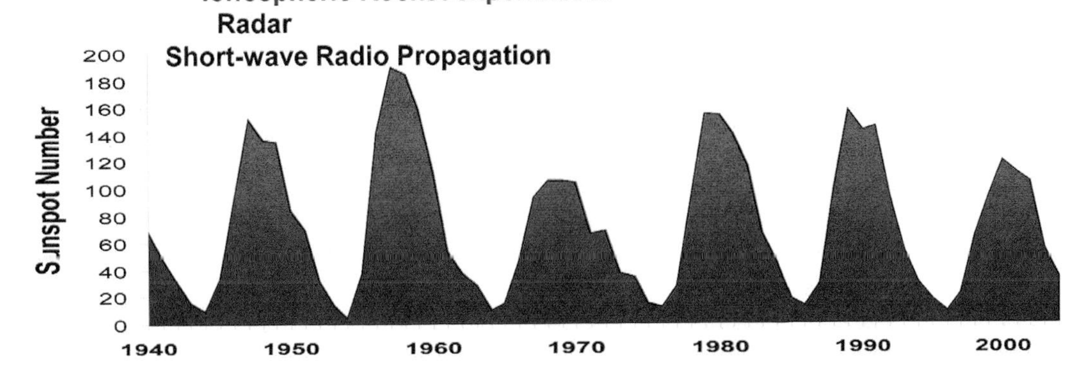

Sunspot Cycles

Figure 6. Growth of space weather customers from 1940 to present. Courtesy of NOAA Space Weather Prediction Center.

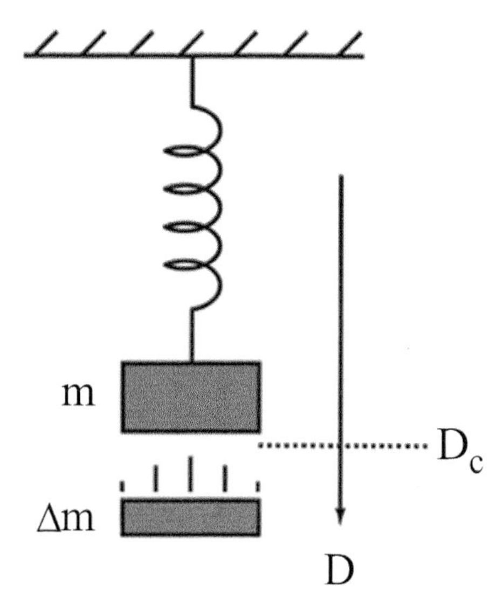

Figure 7. Dripping faucet mass-on-a-spring analog model [*Klimas et al.*, 1992].

development and magnetospheric substorm expansion phase onset. Reconnection in the near tail rapidly returns magnetic flux to the dayside and rids the tail of large amounts of energy by release of the substorm plasmoid [e.g., *Baker et al.*, 1996; *Hones*, 1979]. A major form of energy dissipation during substorms is the currents that flow from the magnetotail and close through the nightside ionosphere. These "field-aligned" currents produce high levels of Joule heating in the auroral zone near local midnight and can be monitored quite effectively by ground-based magnetic indices, such as the "auroral electrojet" indices, *AE* or *AL*. Modern multispacecraft missions such as Cluster [e.g., *Baker et al.*, 2002, 2005] and THEMIS [*Angelopoulos et al.*, 2008; *Pu et al.*, 2010], have confirmed as well as extended our understanding of magnetotail-to-ionosphere coupling.

In an effort to study the relative roles of driven and unloading behavior during substorms, *Bargatze et al.* [1985] assembled an extensive solar wind and geomagnetic activity data set. They examined several dozen different intervals when a monitoring spacecraft was in the interplanetary medium. These intervals of complete, continuous solar wind coverage were typically several days long. The concurrent geomagnetic activity, as measured by the *AL* index, was then examined for these intervals. The solar wind dawn-to-dusk electric field VB_s (the solar wind speed times the southward component of the IMF) was associated with the corresponding *AL* index to study temporal and causal relations.

The *Bargatze et al.* [1985] study used linear prediction filter analysis in order to characterize substorm responses to interplanetary drivers. The work showed two characteristic magnetospheric response timescales. A first response (lag time of 15–20 min) was interpreted as being related to the directly driven (loading) portion of the solar wind-magnetosphere interaction. The second characteristic timescale (~60 min) in the Bargatze et al. analysis was taken as the period of energy loading (growth phase) and the time to develop the plasma sheet instability (reconnection) leading to substorm expansion phase onset. A key point of the Bargatze et al. analysis was that a fundamental change occurred as solar wind driving progressed from relatively weak to relatively strong levels. The prediction filters showed that the magnetospheric response behavior progressed from periodic to much more complex variability.

For weak and moderate activity, as noted above, the magnetosphere generally exhibits both a driven and an unloading character; for stronger activity, the Bargatze et al. results show that the loading and the unloading responses tend to be commingled. *Baker et al.* [1990] suggested that this change in the nature of the magnetospheric response represents an evolution toward essentially chaotic internal magnetospheric dynamics during strong solar wind driving conditions. This nonlinear evolution of magnetospheric dynamics was likened to the "dripping faucet" model (Figure 7). Such behavior was first suggested in the plasmoid analogy described by *Hones* [1979] (see Figure 8).

Based upon the above interpretations and upon "phase-space reconstruction" methods [*Vassiliadis et al.*, 1995],

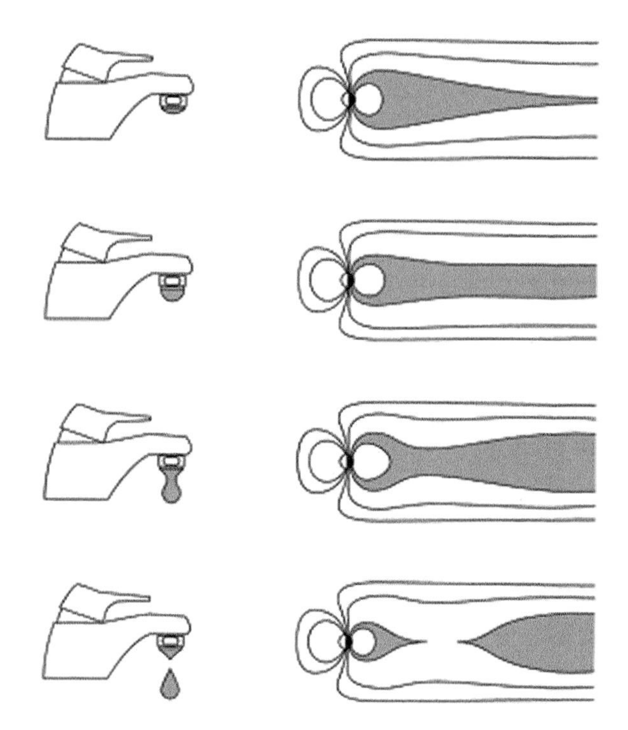

Figure 8. Analogy of plasmoid formation to a leaky faucet. Adapted from the work of *Hones* [1979].

nonlinear dynamical models of the magnetosphere have been developed. These models attempt to replicate the internal magnetospheric dynamical evolution including loading of energy from the solar wind and dumping of energy from the system by plasmoid formation. The Faraday loop model [*Klimas et al.*, 1992] in particular, provided a simple, but dynamically complete representation of the substorm cycle including energy storage in the magnetotail (growth phase), sudden unloading of tail energy (expansion phase), and a return toward the equilibrium or ground state (recovery phase).

The Faraday loop model has been used with highly time-dependent solar wind driving of the model [*Klimas et al.*, 1996]. The observed VB_s from the work of *Bargatze et al.* [1985] was used as a driver for the model input. The result of such work showed that the measured values of *AL* compare well with the values of *AL* computed from the Faraday loop model using VB_s as the driver. Without unloading in the model, one can fit reasonably well the low-frequency response of the magnetosphere to the solar wind, but one cannot at all fit the sharp substorm onsets or the large amplitude changes in *AL*. Thus, the analog model with realistic solar wind driving produces unloading episodes very similar to those seen in the real magnetosphere [*Bargatze et al.*, 1985; *Baker et al.*, 1990]. This supports the idea that, with both driven and unloading aspects included, the analog model provides a reasonably good replication of the magnetospheric dynamics as observed in the *AL* time series.

From observed dynamics and reflections on the complexity of the magnetospheric systems, one can conclude the following: (1) The solar wind-magnetosphere-ionosphere system clearly is characterized by nonlinear dynamics. (2) Methods borrowed from other branches of physics, chemistry, and engineering can offer useful, if imperfect, analogies on how to deal with nonlinearity in the Sun-Earth system. (3) Ultimately, space weather prediction probably must depart from idealized local "stability" analyses to consider true global system responses and must incorporate realistic, nonlinear aspects.

Therefore, there is strong motivation to go away from usual plasma physics techniques to consider more aggressive and comprehensive approaches. It may be that the magnetotail has a distributed set of interacting control agents, some of which are local plasma conditions and some of which are remote in the ionosphere or solar wind. Thus, it may appear that instabilities are spontaneous or even random [e.g., *Lewis*, 1991]. We need to use new analysis tools to address such issues. Figure 9 presents a time line progression of the increased appreciation of nonlinear physical aspects of solar wind-magnetospheric dynamics. Incorporating both traditional forecasting methods for solar storms and also using the array of methods embodied in the work portrayed in Figure 9 offers the best hope for dealing with extreme space weather events.

5. A VISION FOR SPACE WEATHER PREDICTION

As described above, severe space weather episodes can (and do) have devastating consequences for modern technological systems. A goal of the National Space Weather Program (NSWP) [see *Office of the Federal Coordinator for Meteorological Services and Supporting Research (OFCM)*, 2010] is to develop and provide accurate, tailored space weather information for all the societal sectors that need it. The objective is to work with (and coordinate) service providers to allow suitable warnings and forecasts of damaging space weather events.

As might be inferred from Figure 10, the vision of a space weather plan forecasting begins with understanding what is, in effect, space "climate." We know that the Sun undergoes an approximately 11 year cycle of increasing and then

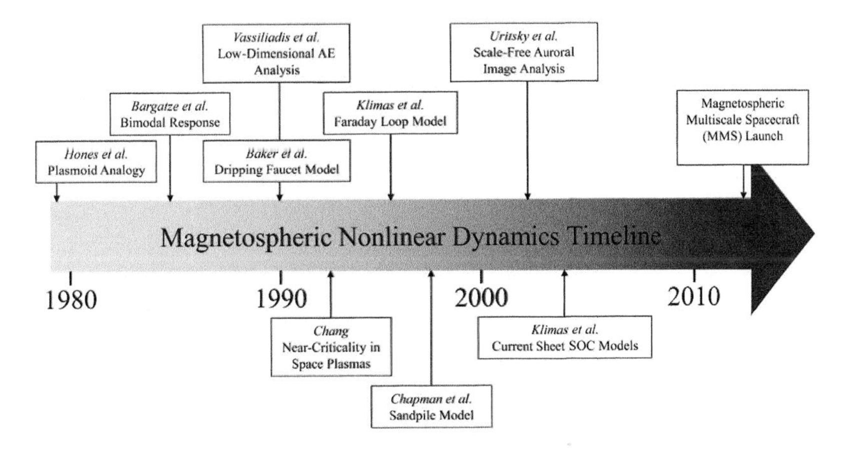

Figure 9. A time line showing the progression of studies describing the nonlinear aspects of solar-terrestrial coupling.

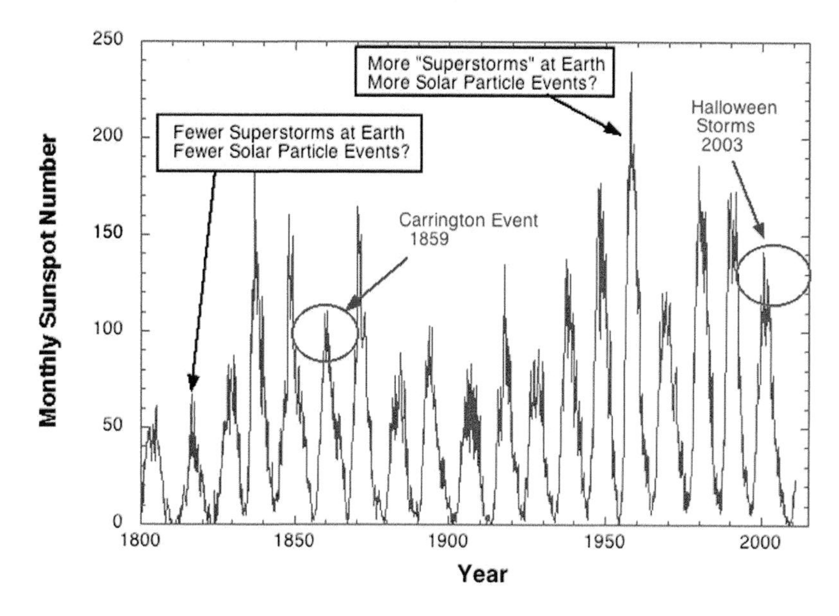

Figure 10. Record of average sunspot activity from 1800 to present. Sunspot cycles may have relatively high or low peak monthly numbers. The 1859 "Carrington" event period and the 2003 Halloween storm period are both indicated.

decreasing sunspot activity. Some solar cycles have been relatively weak, while others in the historical record have been very large. The stronger maxima tend to have more episodes of strong solar flares, intense outbursts of UV light, and (perhaps) greater chances of solar particle "superstorms." Conversely, weaker maxima may exhibit fewer of these extreme characteristics. However, even these general patterns of behavior are not ironclad. The 1859 Carrington event, for example, occurred during a relatively weak sunspot maximum period (see Figure 10).

Going to shorter temporal scales, we know that forecasting space weather events will require continuous and effective observations of the Sun and its corona. Being able to see the initiation of a powerful solar flare and detecting an Earthward-bound CME at its early stages give us the best chance of predicting a powerful geomagnetic storm. Such CME observations can give perhaps 18 h (or longer) alerts that a solar-induced geomagnetic storm is imminent.

When CME observations and modeling are combined with global solar wind models (see Figure 11), one begins to have the elements of an operational space weather forecasting capability. This has recently been attained by the National Centers for Environmental Prediction of NOAA [see *OFCM*, 2010].

By using models developed in the research community and driving such models with space-based and ground-based solar observations, the United States has moved an important

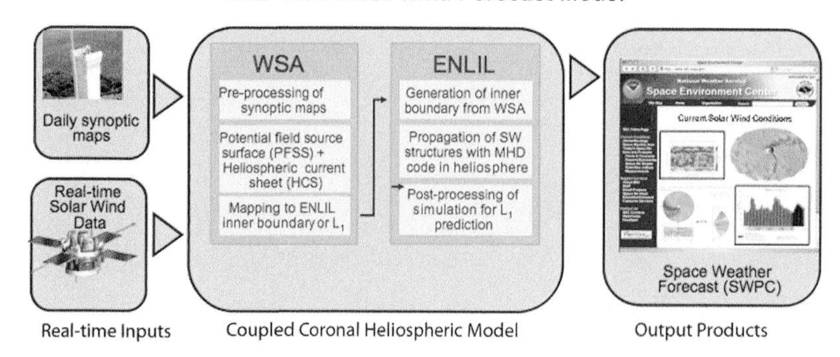

Figure 11. Elements of the real-time coupled coronal-heliospheric model used in NOAA near-operational forecasting. Global Oscillation Network Group data are used as input for the synoptic maps, and forecast outputs are given at the test bed site. This Wang-Sheeley-Arge solar coronal model is used with a physics-based heliospheric model, ENLIL From *Baker et al.* [2011].

step closer to providing tens of hours warning of impending geomagnetic storms. As shown in Figure 11, both ground-based and space-based observations of the Sun can be fed into heliospheric models to forecast solar wind conditions near the Earth. In the magnetosphere-ionosphere system, models of the complete, coupled geospace system can be used to provide forecasts ranging from tens of minutes to several days. The present NSWP still has a long way to go to achieve the levels of accuracy and reliability that are desired by the U.S. industrial, military, and commercial sectors. Moreover, knowing with accuracy the near-Earth solar and solar wind drivers of space weather still demands further understanding of the nonlinear responses of the geospace system that were described in the prior section of this chapter. The complex feedback within the solar wind-magnetosphere-ionosphere system may ultimately limit our ability to predict the size, duration, and precise location of magnetospheric current systems that can wreak havoc on human technologies.

Despite the immense challenges of understanding and modeling the geospace environment, we view the opportunities and rewards of space weather forecasting with great hope and enthusiasm. Remarkable strides have been made in the last decade [*OFCM*, 2006, 2010], and prospects are very good that observational and modeling improvements will continue to pick up steam. Thus, should an extreme solar disturbance occur in the near future, it is likely that some measure of warning will be offered by our present systems. The big challenge and the big question is how well emergency preparedness agencies, industries, and policy makers are positioned to use such forecasts to minimize impacts on human technological systems [*NRC*, 2008; *Baker and Green*, 2011].

Acknowledgments. The author thanks countless numbers of colleagues in the space physics and space weather communities for help and advice on the work reviewed here. Special thanks to those involved in the 2008 National Research Council report.

REFERENCES

Angelopoulos, V., et al. (2008), Tail reconnection triggering substorm onset, *Science*, *321*, 931–935.

Baker, D. N., and J. L. Green (2011), The perfect solar superstorm, *Sky Telescope*, February, 30–36.

Baker, D. N., A. J. Klimas, R. L. McPherron, and J. Büchner (1990), The evolution from weak to strong geomagnetic activity: An interpretation in terms of deterministic chaos, *Geophys. Res. Lett.*, *17*(1), 41–44.

Baker, D. N., T. Pulkkinen, V. Angelopoulos, W. Baumjohann, and R. McPherron (1996), Neutral line model of substorms: Past results and present view, *J. Geophys. Res.*, *101*(A6), 12,975–13,010.

Baker, D. N., et al. (2002), Timing of magnetic reconnection initiation during a global magnetospheric substorm onset, *Geophys. Res. Lett.*, *29*(24), 2190, doi:10.1029/2002GL015539.

Baker, D. N., R. L. McPherron, and M. W. Dunlop (2005), Cluster observations of magnetospheric substorm behavior in the near- and mid-tail region, *Adv. Space Res.*, *36*(10), 1809–1817.

Baker, D. N., et al. (2011), The space environment of Mercury at the times of the second and third MESSENGER flybys, *Planet. Space Sci.*, *59*(15), 2066–2074.

Bargatze, L., D. Baker, R. McPherron, and E. Hones Jr. (1985), Magnetospheric impulse response for many levels of geomagnetic activity, *J. Geophys. Res.*, *90*(A7), 6387–6394.

Buldyrev, S. V., R. Parshani, G. Paul, H. E. Stanley, and S. Havlin (2010), Catastrophic cascade of failures of interdependent networks, *Nature*, *64*(15), 1025–1028, doi:10.1038/nature08932.

Green, J. L., and S. Boardsen (2006), Duration and extent of the great auroral storm of 1859, *Adv. Space Res.*, *38*(2), 130–135.

Hones, E. W., Jr. (1979), Transient phenomena in the magnetotail and their relation to substorms, *Space Sci. Rev.*, *23*(3), 393–410, doi:10.1007/BF00172247.

Klimas, A., D. Baker, D. Roberts, D. Fairfield, and J. Büchner (1992), A nonlinear dynamical analogue model of geomagnetic activity, *J. Geophys. Res.*, *97*(A8), 12,253–12,266.

Klimas, A., D. Vassiliadis, D. Baker, and D. Roberts (1996), The organized nonlinear dynamics of the magnetosphere, *J. Geophys. Res.*, *101*(A6), 13,089–13,113.

Lewis, Z. V. (1991), On the apparent randomness of substorm onset, *Geophys. Res. Lett.*, *18*(8), 1627–1630.

National Research Council (NRC) (2008), *Severe Space Weather Events—Understanding Societal and Economic Impacts: A Workshop Report*, Natl. Acad. Press, Washington, D. C.

Office of the Federal Coordinator for Meteorological Services and Supporting Research (OFCM) (2006), Report of the Assessment Committee for the National Space Weather Program, *Rep. FCM-R24-2006*, Off. of the Fed. Coord. for Meteorol. Serv., Silver Spring, Md.

Office of the Federal Coordinator for Meteorological Services and Supporting Research (OFCM) (2010), The National Space Weather Program Strategic Plan, *Rep. FCM-P30-2010*, Off. of the Fed. Coord. for Meteorol. Serv., Washington, D. C.

Patterson, J., T. Armstrong, C. Laird, D. Detrick, and A. Weatherwax (2001), Correlation of solar energetic protons and polar cap absorption, *J. Geophys. Res.*, *106*(A1), 149–163.

Pu, Z. Y., et al. (2010), THEMIS observations of substorms on 26 February 2008 initiated by magnetotail reconnection, *J. Geophys. Res.*, *115*, A02212, doi:10.1029/2009JA014217.

Vassiliadis, D., A. Klimas, D. Baker, and D. Roberts (1995), A description of the solar wind-magnetosphere coupling based on nonlinear filters, *J. Geophys. Res.*, *100*(A3), 3495–3512.

D. N. Baker, Laboratory for Atmospheric and Space Physics, University of Colorado, 1234 Innovation Drive, Boulder, CO 80309, USA. (daniel.baker@lasp.colorado.edu)

Supermagnetic Storms: Hazard to Society

G. S. Lakhina and S. Alex

Indian Institute of Geomagnetism, Navi Mumbai, India

B. T. Tsurutani

NASA Jet Propulsion Laboratory, California Institute of Technology, Pasadena, California, USA

W. D. Gonzalez

Instituto Nacional de Pesquisas Espaciais, Sao Paulo, Brazil

Magnetic storms are an important component of space weather effects on Earth. Superintense magnetic storms (defined here as those with $Dst < -500$ nT, where Dst stands for the disturbance storm time index that measures the strength of the magnetic storm), although relatively rare, can be hazardous to technological systems in space as well as on the ground. Such storms can cause life-threatening power outages, satellite damage, communication failures, and navigational problems. The data for such magnetic storms during the last 50 years is rather scarce. Research on historical geomagnetic storms can help to create a good database for intense and superintense magnetic storms. The superintense storm of 1–2 September 1859 is analyzed in the light of new knowledge of interplanetary and solar causes of storms gained from the space-age observations. We will discuss the results in the context of some recent intense storms and also the occurrence probability of such superstorms.

1. INTRODUCTION

We all know about the extreme weather events, like thunderstorms, hurricanes, cyclones, typhoons, floods, earthquakes, and tsunamis. Depending on their severity, these events can cause loss of life and property. These are the familiar natural hazards to the society. We can feel them, for example, the force of winds and rains during cyclones, shaking of the ground during the earthquakes, etc. There is another natural hazard, which we cannot feel directly, but it can impose a great danger to the society, as it can damage the technological systems in space and on the ground. This is the space storm or magnetic storm, and it comes under the area of space weather and also space geomagnetism.

In addition to electromagnetic radiation, e.g., visible light, X-rays, UV, etc., the Sun emits continuously a stream of charged particles, called the solar wind, in all directions. Interaction of the solar wind with the geomagnetic field leads to the formation of the Earth's magnetosphere. When there are big solar eruptions, like solar flares and coronal mass ejections (CMEs), they hurl huge blobs of ionized matter in space. Since such solar ejecta move with supersonic speeds, higher than the solar wind speeds, they produce interplanetary (IP) shocks. When the IP shocks and the solar ejecta hit the magnetosphere, they produce worldwide geomagnetic

Extreme Events and Natural Hazards: The Complexity Perspective
Geophysical Monograph Series 196

10.1029/2011GM001073

field disturbances lasting for about half a day to several days, known as magnetic storms.

2. MAGNETIC STORMS

2.1. Historical Background

Geomagnetism, a new branch of physics at that time, was born with the publication of *De Magnete* by William Gilbert in 1600 A.D., proposing that the Earth acts as a great magnet [*Gilbert*, 1600]. That the Earth has a magnetic field was first realized by Petrus Peregrinus in 1259 A.D. Some ancient texts from Greece, China, and India mention the attracting properties of magnetic materials such as magnetite (Fe_3O_4: lodestone) as early as 800 B.C. The earliest form of magnetic compass using lodestone is believed to have been invented in China for navigation purposes around the first century A.D. The first map of field declination was made by Edmund Halley in the beginning of the eighteenth century. Alexander von Humboldt played an important role in the development of geomagnetism, in addition to his fundamental contributions to natural sciences.

Alexander von Humboldt and his colleague observed the local magnetic declination in Berlin, every half hour from midnight to morning, for the period starting from May 1806 until June 1807. On the night of 21 December 1806, von Humboldt observed strong magnetic deflections for six consecutive hours and noted the presence of correlated northern lights (aurora) overhead. When the aurora disappeared at dawn, the magnetic perturbations disappeared as well. From this, von Humboldt concluded that the magnetic disturbances on the ground and the auroras in the polar sky were two manifestation of the same phenomenon. Von Humbold gave this phenomenon the name "Magnetische Ungewitter" or magnetic storms [*von Humboldt*, 1808]. The worldwide network of magnetic observatories later confirmed that such "storms" were indeed worldwide phenomena [*Schröder*, 1997].

2.2. Sunspots, Solar Flares, and Geomagnetic Activity

Sunspots are the dark spots on the photosphere; they appear dark as their temperature is reduced compared to the surrounding area due to the strong magnetic fields embedded in them. Due to the invention of telescope and Galileo's interpretation of sunspots in 1612, a new era in solar observation started. An amateur German astronomer, S. Heinrich Schwabe, began observing the Sun and making counts of sunspots in 1826. He reported periodic behavior in spot counts of 10 years [*Schwabe*, 1843].

A decennial period in the daily variation of magnetic declination had been reported by Lamont from Munich in 1851, but he did not relate it to the sunspot cycle. From his extensive studies, *Sabine* [1851, 1852] discovered that geomagnetic activity paralleled the recently discovered sunspot cycle.

On 1 September 1859 (Thursday) morning, *Richard Carrington* [1859] was studying a big group of sunspots. He was surprised by the sudden appearance of two brilliant beads of blinding white light over the sunspots, which intensified with time. Carrington later wrote, "I hastily ran to call someone to witness the exhibition with me. On returning within 60 seconds, I was mortified to find that it was already much changed and enfeebled." He and his witness watched the white spots contract to mere pinpoints and disappear. This was the first observation of the white light (visible) solar flare on record. The 1 September 1859 solar flare was also observed by *R. Hodgson* [1859], but somehow, it came to be known as Carrington flare.

The very next day, a severe magnetic storm was recorded by the Kew, and some other observatories, especially Colaba, Bombay. Carrington knew about the occurrence of the magnetic storm but failed to connect it to the solar flare. It took nearly 100 years to gather sufficient statistics to make a convincing case for an association between large solar flares and severe magnetic storms [*Hale*, 1931; *Chapman and Bartels*, 1940; *Newton*, 1943].

3. MODERN OUTLOOK

The solar wind is essentially a highly conducting plasma, and it obeys the simple Ohm's law of the form

$$\mathbf{J} = \sigma(\mathbf{E} + \mathbf{V} \times \mathbf{B}), \tag{1}$$

where **J**, **E**, **B**, **V**, and σ are the electrical current density, the electric field, the magnetic field, the flow velocity, and the electrical conductivity, respectively. For infinite conductivity, i.e., $\sigma = \infty$,

$$\mathbf{E} + \mathbf{V} \times \mathbf{B} = 0, \quad \text{or} \quad \mathbf{V} = \mathbf{E} \times \mathbf{B}/B^2, \tag{2}$$

meaning that solar magnetic field is frozen in the solar wind, and it is dragged with the flow into the interplanetary space. This is the interplanetary magnetic field (IMF). The IMF plays an important role in the transfer of energy from the solar wind into the magnetosphere and in driving the magnetic storms.

The magnetopause boundary layer is the site where solar wind energy and momentum is transferred into the magnetosphere. Two main processes by which the solar wind plasma can cross the magnetopause are (1) direct entry involving magnetic reconnection [*Dungey*, 1961; *Gonzalez et al.*, 1989; *Lakhina*, 2000] and (2) the cross-field transport due to the scattering of particles by magnetopause boundary layer waves [*Gurnett et al.*, 1979; *Tsurutani et al.*, 1998; *Lakhina et al.*, 2000] across the closed magnetopause field lines. The boundary layer waves provide a specific mechanism for "viscous interaction" [*Axford and Hines*, 1961; *Tsurutani and Gonzalez*, 1995] in which the solar wind flow energy is transferred to the magnetosphere. Several other processes, like impulsive penetration of the magnetosheath plasma elements with an excess momentum density, plasma entry due to solar wind irregularities [*Schindler*, 1979], the Kelvin-Helmholtz instability [*Miura*, 1987], and plasma percolation due to overlapping of a large number of tearing islands at the magnetopause have been suggested for the plasma entry across the magnetopause [*Galeev et al.*, 1986]. All these processes, however, appear to play only a minor role in solar wind energy transfer.

Magnetic reconnection is recognized as a basic plasma process, which converts magnetic energy into plasma kinetic energy accompanied by topological changes in the magnetic field configuration. It prohibits the excessive buildup of magnetic energy in the current sheets [*Dungey*, 1961]. Magnetic reconnection is very effective when the IMF is directed southward leading to strong plasma injection from the tail toward the inner magnetosphere causing intense auroras at high-latitude nightside regions. About 5% to 10% of solar wind energy is transferred into the Earth's magnetosphere [*Gonzalez et al.*, 1989; *Weiss et al.*, 1992] during substorms and storms. During northward IMF intervals, however, magnetic reconnection is not very effective, and the wave-particle cross-field transport may become dominant. *Tsurutani and Gonzalez* [1995] have estimated that only about 0.1% to 0.3% of the solar wind energy gets transferred to the magnetosphere during northward IMFs.

A schematic of magnetic reconnection process responsible for the solar wind energy transfer in the magnetosphere is shown in Figure 1. When the IMFs are directed opposite to the Earth's fields, there is magnetic erosion on the dayside magnetosphere (by magnetic connection) and magnetic field accumulation on the nightside magnetotail region. Subsequent reconnection in the nightside magnetotail leads to plasma injection at these local times and auroras occurring at high-latitude nightside regions. As the magnetotail plasma get injected into the nightside magnetosphere, the energetic protons drift to the west and electrons to the east, forming a ring of current around the Earth. This current, called the "ring current," causes a diamagnetic decrease in the Earth's magnetic field measured at near-equatorial magnetic stations. The decrease in the equatorial magnetic field strength is directly related to the total energy of the ring current particles and, thus, is a good measure of the energetics of the magnetic storm [*Dessler and Parker*, 1959; *Sckopke*, 1966; *Carovillano and Siscoe*, 1973].

3.1. Geomagnetic Storms

A geomagnetic storm is characterized by a *main phase* during which the horizontal component of the Earth's low-latitude magnetic fields are significantly depressed over a time span of one to a few hours followed by its recovery, which may extend over several days [*Rostoker*, 1997]. The main phase is caused by the intensified ring current, which moves closer to the Earth producing a depression in geomagnetic field H component. The *recovery phase* begins with the

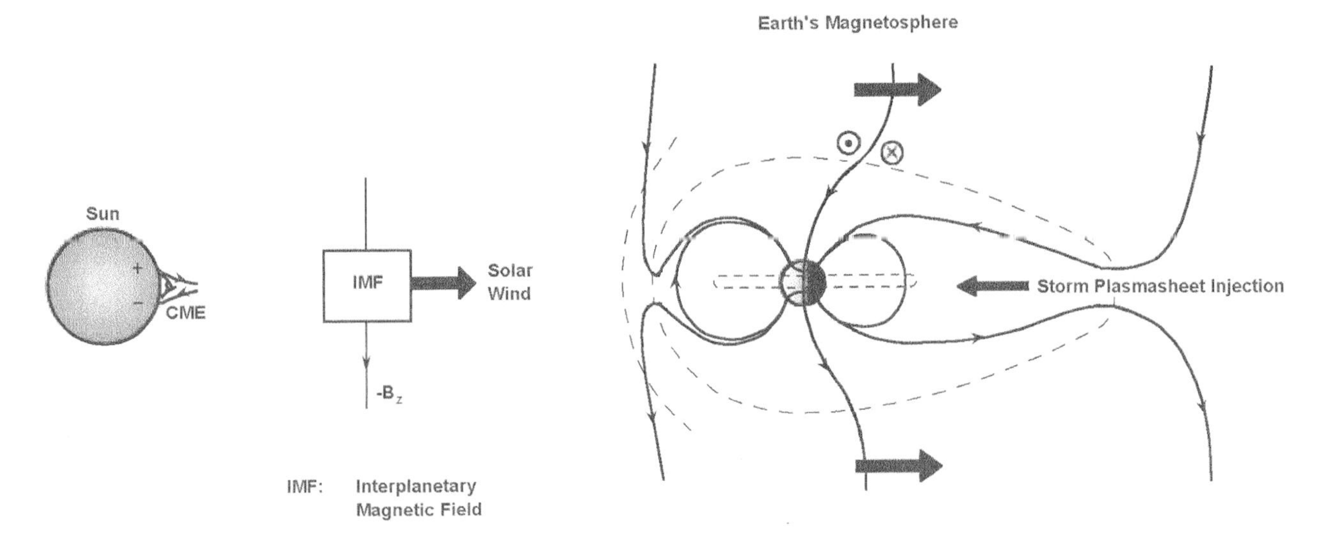

Figure 1. A schematic of the magnetic reconnection process. From *Tsurutani et al.* [2003].

decay of the ring current due to charge exchange, Coulomb collision, and wave-particle interaction processes. The auroral activity becomes intense, and auroras are not confined only to the auroral oval; rather, the auroras can be seen at the subauroral to midlatitudes. The intensity of a geomagnetic storm is expressed in terms of the disturbance storm time index (*Dst*) or symmetric-H (SYM-H) index, which is a measure of the intensity of the ring current. Both indices are basically the same but differ in time resolution of the magnetic data used, whereas *Dst* is an hourly index, the SYM-H has 1 min resolution.

There are two types of geomagnetic storms. The storms that are characterized by a sudden increase in the horizontal magnetic field intensity shortly before the main phase are called sudden commencement (SC)-type storms (see top panel of Figure 2). This sudden increase in magnetic field strength is caused by the interplanetary shock compression of the magnetosphere. The period between the SC and the storm main phase is called the initial phase. However, all magnetic storms do not have an initial phase. The SC-type magnetic storms are driven by the CMEs [*Gonzalez et al.*, 1994; *Taylor et al.*, 1994]. On the other hand, geomagnetic storms that are not accompanied by SCs are called gradual geomagnetic storm (SG) type (bottom panel of Figure 2). The SG-type geomagnetic storms are caused by the corotating interaction regions (CIRs) associated with fast streams emanating from coronal holes [*Taylor et al.*, 1994; *Tsurutani and Gonzales*, 1995; *Tsurutani et al.*, 2006].

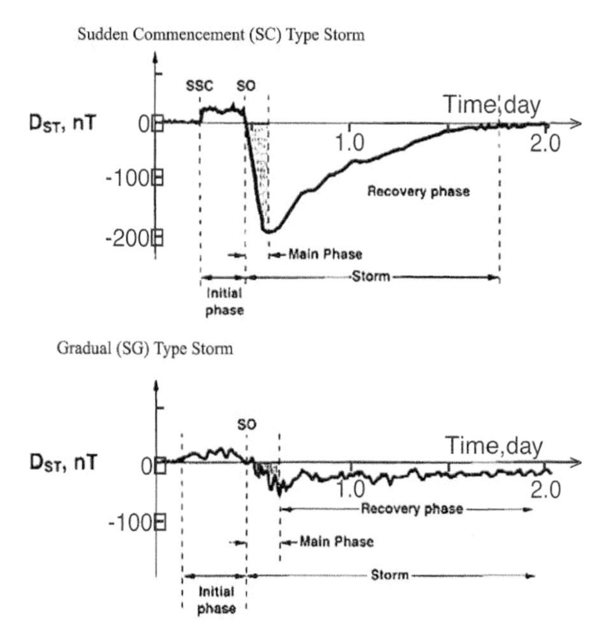

Figure 2. (top) A schematic of a sudden . . . driven by an ICME and (bottom) gradual (SG) type . . . by a CIR. All storms may not have initial phases. Slightly modified from *Tsurutani et al.* [2006] and *Lakhina and Tsurutani* [2011].

Both SC and SG storms are further classified into type 1 or type 2 storms. Magnetic storms having a single main phase, wherein the *Dst* decreases more or less continuously to a minimum value and then starts to recover, are called type 1 or one-step storms. Geomagnetic storms, where the main phase undergoes a two-step growth in the ring current [*Kamide et al.*, 1998] in a way that before the ring current had decayed to a significant prestorm level, a new major particle injection occurs, leading to further buildup of the ring current and further decrease of *Dst*, are called type 2 or two-step storms. Two-step storms are caused by the compressed IMF B_s (southward component) in the sheath region downstream of the interplanetary coronal mass ejection (ICME) shocks (first main phase) followed by the magnetic cloud field (second main phase) [*Tsurutani and Gonzalez*, 1997]. In general, even multistep (i.e., more than two steps) storms can sometime occur depending on the ring current injection events caused by the solar and interplanetary conditions. The intensity of the magnetic storm is measured by the *Dst* index at peak of the main phase. The magnetic storms are called weak when $Dst > -50$ nT, moderate when $-50 > Dst > -100$ nT, and intense when $Dst < -100$ nT [*Kamide et al.*, 1998] and superintense when $Dst < -500$ nT [*Tsurutani et al.*, 2003]. It is interesting to note that the number of intense ($Dst < -100$ nT) storms follow the solar cycle sunspot number, but for weak to moderate storms, there is a much smaller solar cycle dependence [*Tsurutani et al.*, 2006]. It was found that CIRs and high-speed streams are presumably responsible for most of the weaker storms.

Tsurutani et al. [2006] also showed that the CIR-generated magnetic storms appear to have very long recovery phases compared to those driven by ICMEs. Relativistic "killer" electrons appear during "recovery" phase of the magnetic storms. These electrons pose great danger for the spacecraft. The relativistic electrons are usually detected in the inner magnetosphere during intervals of high-speed solar wind streams. The exact mechanism for relativistic electron acceleration is presently unknown. However, there are two popular mechanisms. One mechanism entails electron radial diffusion due to ultralow frequency (ULF) waves that break the particles' third adiabatic invariant [*Mann et al.*, 2004, and references therein]. The second mechanism is based on energy diffusion by cyclotron resonant interactions of electron with chorus that breaks the particles' first adiabatic invariant [*Summers et al.*, 2004, and references therein].

The main interplanetary causes of geomagnetic storms are the ICMEs having unusually intense magnetic fields and high solar wind speeds near the Earth and the CIRs created by the interaction of fast streams emanating from solar

Table 1. List of Intense and Superintense Magnetic Storms From Colaba and Alibag Magnetic Observatories

Serial Number	Year	Month	Day	H (nT)	*Dst* (nT)	Remark
1	1847	Sep	24	471.0	-	Colaba
2	1847	Oct	23	535.0	-	Colaba
3	1848	Nov	17	404.0	-	Colaba
4	1857	Dec	17	306.0	-	Colaba
5	1859	Sep	1–2	1722.0	-	Colaba
6	1859	Oct	12	984.0	-	Colaba
7	1872	Feb	4	1023.0	-	Colaba
8	1872	Oct	15	430.0	-	Colaba
9	1882	Apr	17	477.0	-	Colaba
10	1882	Nov	17	445.0	-	Colaba
11	1882	Nov	19	446.0	-	Colaba
12	1892	Feb	13	612.0	-	Colaba
13	1892	Aug	12	403.0	-	Colaba
14	1894	Jul	20	525.0	-	Colaba
15	1894	Aug	10	607.0	-	Colaba
16	1903	Oct	31	819.0	-	Colaba
17	1909	Sep	25	>1500.0	-	Alibag
18	1921	May	13–16	>700.0	-	Alibag
19	1928	Jul	7	779.0	-	Alibag
20	1935	Jun	9	452.0	-	Alibag
21	1938	Apr	16	532.0	-	Alibag
22	1944	Dec	16	424.0	-	Alibag
23	1957	Jan	21	420.0	−250	Alibag
24	1957	Sep	4–5	419.0	−324	Alibag
25	1957	Sep	13	582.0	−427	Alibag
26	1957	Sep	29	483.0	−246	Alibag
27	1958	Feb	11	660.0	−426	Alibag
28	1958	Jul	8	610.0	−330	Alibag
29	1960	Apr	1	625.0	−327	Alibag
30	1972	Jun	18	230.0	−190	Alibag
31	1972	Aug	9	218.0	−154	Alibag
32	1972	Nov	1	268.0	−199	Alibag
33	1980	Dec	19	479.0	−240	Alibag
34	1981	Mar	5	406.0	−215	Alibag
35	1981	Jul	25	367.0	−226	Alibag
36	1982	Jul	13–14	410.0	−325	Alibag
37	1982	Sep	5–6	434.0	−289	Alibag
38	1986	Feb	9	342.0	−307	Alibag
39	1989	Mar	13	Loss	−589	Alibag
40	1989	Nov	17	425.0	−266	Alibag
41	1991	Mar	24	Loss	−298	Alibag
42	1992	Feb	9	225.0	−201	Alibag
43	1992	Feb	21	304.0	−171	Alibag
44	1992	May	10	503.0	−288	Alibag
45	1998	Sep	25	300.0	−207	Alibag
46	2000	Apr	6	384.0	−207	Alibag
47	2000	Jul	15	407.0	−288	Alibag
48	2001	Mar	31	480.0	−358	Alibag
49	2001	Apr	11	332.0	−256	Alibag
50	2001	Nov	6	359.0	−277	Alibag
51	2001	Nov	24	455.0	−213	Alibag
52	2003	Aug	18	254.0	−168	Alibag
53	2003	Oct	29	441.0	−345	Alibag
54	2003	Oct	30	506.0	−401	Alibag
55	2003	Nov	20	749.0	−472	Alibag
56	2004	Jul	27	342.0	−182	Alibag
57	2004	Nov	8	459.0	−383	Alibag
58	2005	May	15	352.0	−263	Alibag
59	2005	Aug	24	457.0	−216	Alibag

coronal holes with the slow solar wind streams [*Tsurutani et al.*, 1999, 2006]. Geomagnetic storms are caused by long intervals of southward IMFs [*Echer et al.*, 2008], which considerably enhance the transfer of energy from solar wind to the magnetosphere via magnetic reconnection process, leading to strong plasma injection from the magnetotail toward the inner magnetosphere. This leads to intense auroras at high-latitude nightside regions and at the same time intensifies the ring current, which causes a diamagnetic decrease in the Earth's magnetic field measured at near-equatorial magnetic stations [*Tsurutani and Gonzalez*, 1997].

Superintense magnetic storms (defined here as those with $Dst < -500$ nT), although relatively rare, have the largest societal and technological relevance. Such storms can cause life-threatening power outages, satellite damage, and loss of low Earth-orbiting (LEO) satellites, data loss, communication failures and navigational problems, damage to power transmission lines, and corrosion of long pipelines due to strong geomagnetically induced current (GIC). The data for superintense magnetic storms are rather scarce. For example, only one truly superintense magnetic storm has been recorded ($DST = -640$ nT, 13 March 1989) during the space age (since 1958).

There was a great media interest about the possible super-magnetic storms in October–November 2003. Though the solar flares on 28 and 29 October were of class X17 and X10, they failed to produce a superintense storm; they produced intense double storm of mere $Dst \sim -400$ nT [*Mannucci et al.*, 2005]. A much weaker solar flare (and CME) of class M3.2/2N on 18 November 2003 resulted in a near super-intense storm on 20 November with $Dst \sim -490$ nT. This clearly shows that it is not only the energy of the solar flare and speed of the ejecta, which control the strength of the geomagnetic storm, but also the solar magnetic field, too, plays a critical role.

Although there is record of only one or two superintense magnetic storms during the space age, many such storms may have occurred many times in the last ~160 years or so when the regular observatory network came into existence. Research on historical geomagnetic storms can help to create a good database for intense and superintense magnetic storms [*Lakhina et al.*, 2005]. From the application of knowledge of interplanetary and solar causes of storms gained from the space-age observations, to this superintense storm data set, one can deduce their possible causes and construct a database for solar ejecta.

Table 1 gives a partial chronological list of some large magnetic storms, which had occurred during the past 160 years or so. The list includes the "Remarkable Magnetic Storms" described in the works of *Moos* [1910] and *Chapman and Bartels* [1940] and "Large Magnetic Storms" in the work of *Tsurutani et al.* [2003]. One can see that some of the events fall under the category of superintense magnetic storms. Analysis of these events can form a very useful database for superintense storms.

3.2. *Case Study of Supermagnetic Storm of 1–2 September 1859*

The supermagnetic storm of 1–2 September 1859, which was caused by the Carrigton flare that occurred on 1 September 1859, has been discussed in detail by *Tsurutani et al.* [2003]. This created a lot of interest in studying the historical Carrington 1–2 September 1989 supermagnetic storm; a detailed account can be found in the special issue of *Advances in Space Research*, edited by *Clauer and Siscoe* [2006]. *Tsurutani et al.* [2003] used the reduced ground magnetometer data of Colaba Observatory, Mumbai, India, for the 1–3 September 1859, published papers [*Carrington*, 1859], auroral reports based on newspapers [*Kimball*, 1960], and recently obtained (space age) knowledge of interplanetary and solar causes of storms, to identify the probable causes of this superstorm. We will briefly discuss the main points concerning this superstorm.

3.2.1. Solar flare of 1 September 1859 and geomagnetic phenomena. The solar flare of 1 September 1859 was observed and reported by R. C. Carrington and R. Hodgson in 1859 in the *Monthly Notices of the Royal Astronomical Society* and became the best-known solar event of all times. Of particular note was the intensity of the event as quoted by *Carrington* [1859, p. 14], "For the brilliancy was fully equal to that of direct sunlight," and by *Hodgson* [1859, pp. 15–16], "I was suddenly surprised at the appearance of a very brilliant star of light, much brighter than the sun's surface, most dazzling to the protected eye."

The solar flare was followed by a magnetic storm at the Earth. The time delay was ~17 h and 40 min (stated in the Carrington paper). Although Carrington carefully noted this relationship, he was cautious in his appraisal: "and that towards four hours after midnight there commenced a great magnetic storm, which subsequent accounts established to have been as considerable in the southern as in the northern hemisphere". While the contemporary occurrence may deserve noting, he would not have it supposed that he even leans toward connecting them "one swallow does not make a summer" [*Carrington*, 1859].

The auroras occurred globally during the magnetic storm. *Kimball* [1960, p. 34] has provided the most complete indexing of auroral sightings: "Red glows were reported as visible from within 23° of the geomagnetic equator in both north and southern hemispheres during the display of September 1–2."

In both the United States and Europe, many fires were caused by arcing from currents induced in telegraph wires during this magnetic storm [*Loomis*, 1861].

3.2.2. Magnetic data of Colaba Observatory during 1–2 September 1859. During 1846–1867, the Colaba Observatory was using the magnetometers made by Thomas Grubb of Dublin for measuring declination and horizontal magnetic field components. The description of these magnetometers is provided by the *Royal Society* [1840, 1842]. Measurements were taken at hourly intervals 24 h a day. When a magnetic storm (main phase) was occurring, measurements were made at 15 min intervals or even less. The original readings were in units of grains and feet. The data was reduced and converted into nanotesla (nT). The final absolute values "H" plotted in Figure 3 are in nT (as converted from the cgs units) [*Tsurutani et al.*, 2003].

The magnetogram of 1–2 September 1859 from Colaba Observatory (Figure 3) indicates that the storm sudden commencement (SSC) was about 120 nT, the maximum H-component depression at Colaba was $\Delta H \approx -1600$ nT, and duration of the main phase of the storm (corresponding to the plasma injection) was ~1-1/2 h. The location at Colaba (~12 LT) was ideal to detect the maximum magnetic response to the storm as the ring current asymmetry effect is reduced at noon. However, based on observations from this one station, one can say that this is now the most intense magnetic storm on record. Magnetometers at high latitudes, e.g., Kew and others, were either saturated or nonoperational for this event.

We will apply the recently gained knowledge about Sun-Earth connection and use other related information to determine the cause of the superstorm of 1–2 September 1859.

The information that the lowest latitudes of the auroras were seen at 23° [*Kimball*, 1960] can be used to identify the plasmapause location, which in turn was used to

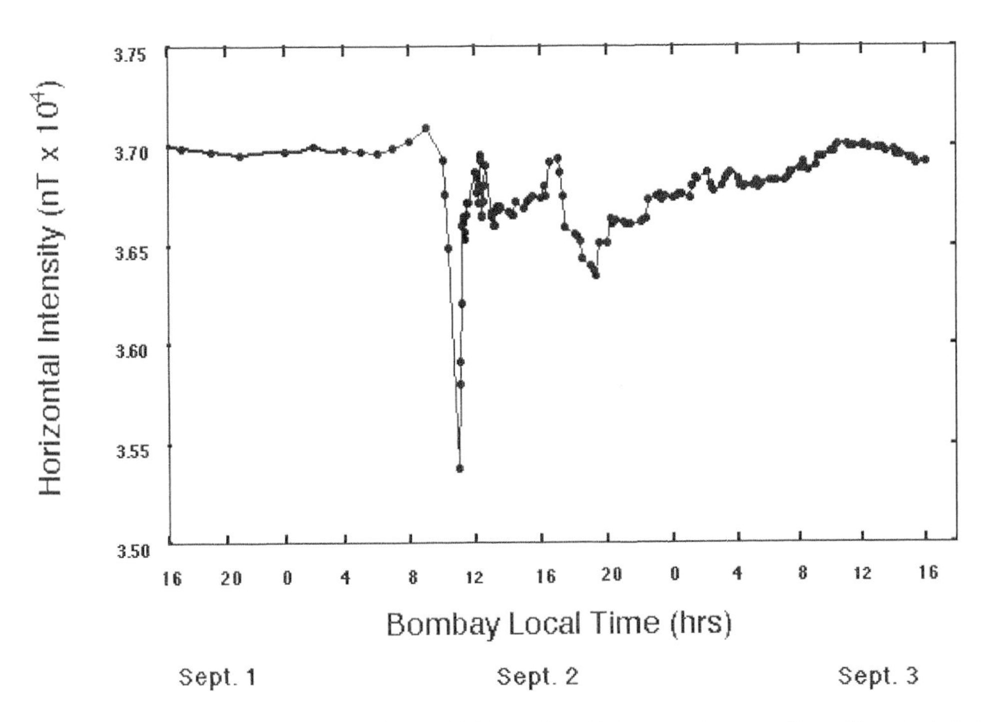

Figure 3. The Colaba (Bombay) magnetogram for the 12 September 1859 magnetic storm. From *Tsurutani et al.* [2003].

determine the magnetospheric convection electric fields, $E_C \sim 20$ mV m^{-1}.

Knowing that the transit time of the ICME from the Sun to the Earth was ~17 h and 40 min [*Carrington*, 1859], an average shock transit speed of $V_{\text{shock}} \approx 2380$ km s^{-1} is deduced. Then, using the relationships, $V_{SW} = 0.775\ V_{\text{shock}}$ [*Cliver et al.*, 1990] and B (nT) = 0.047 V_{SW} (km s^{-1}) [*Gonzalez et al.*, 1998], where B is the magnetic field and V_{SW} is the peak solar wind speed of the ejecta at 1 AU, the maximum possible electric field for this extremely fast interplanetary event can be expressed as [*Tsurutani et al.*, 2003]

$$E = 2.8 \times 10^{-5} V_{\text{shock}}^2 \quad \text{mV m}^{-1}. \tag{3}$$

On putting $V_{\text{shock}} = 2380$ km s^{-1} in the above expression, we get the interplanetary electric field, $E_I \sim 160$ mV m^{-1}. This estimates compares well with the estimated convection electric field, $E_C \sim 20$ mV m^{-1}, if a reasonable value of the penetration efficiency of ~8% of the interplanetary electric field, E_I, is assumed [*Gonzalez et al.*, 1989].

The peak intensity or *Dst* index for this magnetic storm can be obtained from an empirical relation for the evolution of the ring current [*Burton et al.*, 1975]:

$$\frac{\mathrm{d}Dst}{\mathrm{d}t} = Q - \frac{Dst}{\tau}, \tag{4}$$

where Q is the energy input and τ is the decay constant for the storm time ring current. From equation (4), it follows that the *Dst* will attain the maximum value when there is an energy balance of the ring current, thus, giving

$$Dst = \tau Q. \tag{5}$$

For very intense storms, the energy input, Q, can be expressed [*Burton et al.*, 1975] as

$$Q - 1.5 \times 10^{-3} (V_{SW} B_S) \text{ nT s}^{-1}, \tag{6}$$

where $(V_{SW}B_S)$ is expressed in mV m^{-1}. Considering $\tau = 1.5$ h (taken from Colaba magnetogram, Figure 3), we get from equations (5) and (6), $Dst = -1760$ nT, a value consistent with Colaba measurement of $\Delta H = -1600$ nT [*Tsurutani et al.*, 2003]. We must remark here that the ring current decay time of about 2 h (based on Colaba data) appears to be much shorter than most storms. This short time (until the level of *Dst* around -300 nT) is most probably related to the fast loss of particles at the magnetopause (short magnetopause distance) due to strong convection during the expected intense partial ring current phase. Then, from the data, it looks that a more slower and typical recovery started at about *Dst* of -300 nT.

The profile of the *Dst* index for this storm (the first main dip) indicates that it was due to a simple plasma injection, and there is no evidence for the possibility of a complex storm. Storm main phase "compound" events or "double storms" [*Kamide et al.*, 1998] due first to sheath fields and then to cloud fields [*Tsurutani et al.*, 1988] can be ruled out by the (simple) storm profile. The possibility of sheath fields alone causing this superstorm can be ruled out because the compression factor of magnetic fields following fast shocks is only approximately four times [*Kennel et al.*, 1985]. Since typical quiet interplanetary fields are ~3 to 10 nT, the compressed fields would be too low to generate the inferred interplanetary and magnetospheric electric fields for the storm. Therefore, the most likely mechanism for this intense, short duration storm would be a magnetic cloud with intense B_S (southward) magnetic fields. The second and third depressions in *Dst* are caused by the new injections near the end of the fast recovery phase of the main storm, thus prolonging the recovery.

Li et al. [2006] have modeled the 1–2 September 1859 supermagnetic storm using an updated *Dst* prediction model of *Temerin and Li* [2002]. According to their model prediction, a very fast solar wind with a very large negative IMF B_z can produce a supermagnetic storm with $Dst < -1600$ nT, and thus, such a storm is likely to occur again. However, in the *Li et al.* [2006] model, the extremely fast recovery of the *Dst* requires an extremely large pressure enhancement following the shock.

3.2.3. How rare was the solar flare of 1–2 September 1859? In addition to "white light," solar flares radiate at a variety of other wavelengths as well. The energy of 1–2 September 1859 flare for the white light portion based on the observation of *Carrington* [1859] is estimated to be about $\sim 2 \times 10^{30}$ ergs by D. Neidig (private communication, 2001), where as the total energy of this flare is estimated to be and about $\sim 10^{32}$ ergs by K. Harvey (private communication, 2001).

On the other hand, using general scalings, *Lin and Hudson* [1976] have estimated the total energy of August 1972 flare as $\sim 10^{32}$ to 10^{33} ergs. *Kane et al.* [1995] has estimated the 1 June 1991 flare energy as $\sim 10^{34}$ ergs. The comparison shows that the 1 September 1859 Carrington flare was not exceptional in terms of total energy released. However, analysis of thin nitrate layers in ice cores indicates that the flux of solar energetic particles (SEPs) for Carrington event was the biggest in the last 500 years [*McCracken et al.*, 2001a, 2001b].

3.3. What Is the Probability of Occurrence of Supermagnetic Storms Similar to or Higher Than 1–2 September 1859?

From the above arguments, it is clear that the 1859 flare/CME ejecta was not unique. The 4 August 1972 flare was definitely equally or even more energetic, and the average interplanetary ejecta speed of 2850 km s^{-1} (with a delay time of 14.6 h) was faster than that of 1859 flare ejecta [*Vaisberg and Zastenker*, 1976]. Actually, 4 August 1972 event had the highest transit speed on record [*Cliver et al.*, 1990]. Yet, the accompanying magnetic storm was a moderate one with a $Dst \sim -120$ nT. Unfortunately, there were no measurements for the IMF near the Earth (i.e., at 1 AU). *Tsurutani et al.* [1992] analyzed the Pioneer 10 (at about 2 AU) magnetic data and extrapolated it back to 1 AU. They observed that the magnetic cloud responsible for this event had its axis highly tilted from the ecliptic and had more or less northward B_z during its passage past the Earth. Had the magnetic cloud associated with 4 August 1972 solar flare ejecta had its axis on the ecliptic or had an opposite orientation (i.e., a large southward B_z), the resulting storm at the Earth would have been as intense as the 1–2 September 1859 event or perhaps even more [*Gonzalez et al.*, 2011].

Therefore, it is likely that 1859, like supermagnetic storms, can occur again in the near future. How often can they occur? Are even more intense events possible? Can one assign probabilities to the occurrence of a similar storm or to a greater intensity storm? At this stage, it is difficult to answer these questions. To answer the first question, the one big flare per solar cycle (11 years) has the potential for creating a storm with an intensity similar to the 1859 storm. However in reality, we know that this was the largest storm in the last 150 years (about 14 solar cycles). The predictability of similar or greater intensity events require knowledge of two things: full understanding of the physical processes involved in the phenomenon and good empirical statistics of the tail of the energy distribution. If one knows the physical processes causing solar flares or magnetic storms, then the high energy tail (extreme event) distributions could be readily ascertained. Since we do not fully understand these specific saturation processes, it is therefore not known whether flares with energy $>10^{34}$ ergs or magnetic storms with $Dst < -1760$ nT are possible or not.

Can one use statistics to infer the probabilities of flares with energies less than, but close to, 10^{34} ergs and storms with $Dst < -1760$ nT? Unfortunately, not with any accuracy. Even assuming that there are no major internal changes in the Sun or the magnetosphere ("stationarity," in a statistical sense), one easily notes that the statistics for extreme solar flares with energies greater than 10^{32} ergs and extreme magnetic storms with $Dst < -500$ nT are poor. The shapes of these high-energy tails are essentially unknown. One can therefore assign accurate probabilities to flares and storms for only the lower energies where the number of observed events is statistically significant.

Willis et al. [1997] have applied extreme-value statistics to the first, second, and third largest geomagnetic storms per solar cycle for 14 solar cycles (1844–1993) using the daily *aa* index. They predict a 99% probability that the daily *aa* index will satisfy the condition $aa < 550$ for the largest geomagnetic storm in the next 100 solar cycles.

Recently, it is found that the previously thought "upper limit" of 10^{32} ergs for the energy of a flare can be broken by a wide margin [*Kane et al.*, 1995]. Quite possibly, we may have not detected events at the saturation limit (either flares or magnetic storms) during the short span of only hundreds of years of observations.

It is debateable whether the Sun has had flares at superflare energy (approximately 10^{38}–10^{39} ergs) levels? Most probably not [*Lingenfelter and Hudson*, 1980], but perhaps 10^{35} ergs is feasible for our Sun. By using probabilistic approach, *Yazev and Spirina* [2009] have estimated the occurrence of superflares of 10^{36} erg (Wolf sunspot 400) to be once in every 10^4 years.

4. INTENSE MAGNETIC STORM AND SOCIETY

Modern society is becoming ever increasingly dependent on space technology for daily routine functions, such as communication, navigation, data transmission, global surveillance of resource surveys, atmospheric weather, etc. Space weather refers to conditions on the Sun and in the solar wind, magnetosphere, ionosphere, and thermosphere that can influence the performance and reliability of spaceborne and ground-based technological systems and can endanger human life or health.

Intense and superintense geomagnetic storms create hostile space weather conditions that can generate many hazards to the spacecraft as well as technological systems on the ground. Geomagnetic storms can cause life-threatening power outages such as the Hydro Quebec power failure during March 1989 magnetic storm. Orders of magnitude relativistic electron flux increases are observed in the radiation belts during intense magnetic storms; they can lead to malfunctioning and failure of satellite instruments due to deep dielectric charging. During the past decade alone, many magnetospheric satellites malfunctioned for several hours or even were permanently damaged due to adverse space weather conditions during intense geomagnetic storms. Strong GICs produced by sudden short-period variations in the geomagnetic field during intense magnetic storms can damage power transmission lines and corrode long pipelines. Intense and superintense geomagnetic storms produce disturbances in the ionosphere-thermosphere system that can cause communication failure and navigational errors. Heating and subsequent expansion of the thermosphere during such storms could produce extra drag [*Tsurutani et al.*, 2007] on the LEO satellites that could reduce their lifetimes significantly. Superintense geomagnetic storms like the 1–2 September 1859 event, if they were to occur today, would produce adverse space weather conditions on a much larger scale than the intense storm of March 1989 with catastrophic consequences for the society.

5. SUMMARY

Intense and superintense magnetic storms are caused by the solar ejecta (due mainly to CMEs) having unusually intense magnetic fields (with large southward component) and high solar wind speeds near the Earth. However, no strong relationship between the strengths of the flares and the speed and magnetic intensities of the ICMEs has been found, yet it is certainly noted that the most intense magnetic storms are indeed related to intense solar flares, i.e., the two phenomena have a common cause, that is, magnetic reconnection at the Sun. At this stage (see discussion under Section 3.3), it is not possible to make any accurate prediction of a supermagnetic storm having similar or higher intensity than that of 1–2 September 1859.

GLOSSARY

Chorus: A right-hand, circularly polarized electromagnetic plane whistler mode wave. Chorus is generated close to the geomagnetic equatorial plane or in minimum B field pockets in the Earth's magnetosphere by the loss cone instability due to anisotropic ~10 – 100 keV electrons. Dayside chorus is a bursty emission composed of rising frequency "elements" with duration of ~ 0.1 to 1.0 s. Each element is composed of coherent subelements with durations of ~1 to 100 ms or more. It is believed that cyclotron resonant interaction of high-energy electrons with chorus emission can accelerate them to relativistic energies.

Coronal mass ejection (CME): A transient outflow of plasma from or through the solar corona. CMEs are often, but not always, associated with erupting prominences, disappearing solar filaments, and solar flares. CMEs usually occur in large-scale closed coronal structures. The average mass and energy of the material ejected during CME can be a few times 10^{15} g and 10^{31} erg.

***Dst* index**: Disturbance storm time that measures variation in the geomagnetic field due to the equatorial ring current. It is computed from the H components at approximately four near-equatorial stations at hourly intervals.

Geomagnetically induced currents (GIC): Currents produced because of rapid temporal changes of the geomagnetic field during magnetic storms in technological conductor systems such as power grids or pipelines.

Interplanetary coronal mass ejection (ICME): Interplanetary counterpart of CME.

Photosphere: Lowest layer of the solar atmosphere. It is essentially the solar "surface" that we see when we look at the Sun in "white" (i.e., regular or visible) light. The photosphere is not a solid surface but consists of a layer of hydrogen and helium gas about 100 km thick. It is the site of sunspots and solar flares.

Solar ejecta: A transient outflow of material from the Sun, which propagates out from the Sun and generates major interplanetary and magnetospheric effects. The ejecta can be produced from solar flares, coronal mass ejections, erupting prominences, etc. The velocity of the ejecta in the solar atmosphere can be less than 100 km s^{-1} to greater than 1000 km s^{-1}.

Solar flare: A sudden eruption of magnetic energy release in the solar atmosphere lasting from minutes to hours, accompanied by bursts of electromagnetic radiation and charged particles. Solar flares occur near complex sunspots. Solar flares are classified according to their X-ray brightness in the wavelength range 1 to 8 Å. There are three categories: X-class flares are big with $I > 10^{-4}$ W m^{-2}, M-class flares are medium-sized with $10^{-5} \leq I < 10^{-4}$ W m^{-2}, and C-class flares are small with $10^{-6} \leq I < 10^{-5}$ W m^{-2}, where I denotes the peak X-ray burst intensity measured at the Earth. Each category for X-ray flares has nine subdivisions ranging from, e.g., X1 to X9, M1 to M9, and C1 to C9.

Symmetric-H (SYM-H) index: Same as *Dst* but computed at a higher resolution of 1 min instead of 1 h used for *Dst*.

Ultralow frequency (ULF) waves: That portion of the radio frequency spectrum from approximately 1 mHz to 30 Hz. They are produced by a variety of plasma processes occurring in the magnetosphere and the solar wind.

Acknowledgments. GSL thanks the Indian National Science Academy, New Delhi, for the support under the Senior Scientist scheme. Portions of this research were performed at the Jet Propulsion Laboratory, California Institute of Technology, under contract with NASA.

REFERENCES

Axford, W. I., and C. O. Hines (1961), A unifying theory of high-latitude geophysical phenomena and geomagnetic storms, *Can. J. Phys.*, *39*, 1433–1464.

Burton, R. K., R. L. McPherron, and C. T. Russell (1975), An empirical relationship between interplanetary conditions and *Dst*, *J. Geophys. Res.*, *80*(31), 4204–4214.

Carovillano, R. L., and G. L. Siscoe (1973), Energy and momentum theorems in magnetospheric processes, *Rev. Geophys.*, *11*(2), 289–353.

Carrington, R. C. (1859), Description of a singular appearance seen in the Sun on September 1, 1859, *Mon. Not. R. Astron. Soc.*, *20*, 13–15.

Chapman, S., and J. Bartels (1940), *Geomagnetism*, vol. 1, pp. 328–337, Oxford Univ. Press, New York.

Clauer, C. R., and G. Siscoe (Eds.) (2006), The Great Historical Geomagnetic Storm of 1859: A Modern Look, *Adv. Space Res.*, *38*, 274 pp.

Cliver, E., J. Feynman, and H. Garrett (1990), An estimate of the maximum speed of the solar wind, 1938–1989, *J. Geophys. Res.*, *95*(A10), 17,103–17,112.

Dessler, A. J., and E. N. Parker (1959), Hydromagnetic theory of magnetic storms, *J. Geophys. Res.*, *64*(12), 2239–2252.

Dungey, J. W. (1961), Interplanetary magnetic field and the auroral zones, *Phys. Res. Lett.*, *6*, 47–48.

Echer, E., W. D. Gonzalez, B. T. Tsurutani, and A. L. C. Gonzalez (2008), Interplanetary conditions causing intense geomagnetic storms ($Dst \leq -100$ nT) during solar cycle 23 (1996–2006), *J. Geophys. Res.*, *113*, A05221, doi:10.1029/2007JA012744.

Galeev, A. A., M. M. Kuznetsova, and L. M. Zelenyi (1986), Magnetopause stability threshold for patchy reconnection, *Space Sci. Rev.*, *44*, 1–41.

Gilbert, W. (1600), *De Magnete*, Chiswick, London, U. K. [English translation by P. F. Mottelay, Dover, New York, 1958.]

Gonzalez, W. D., B. T. Tsurutani, A. L. C. Gonzalez, E. J. Smith, F. Tang, and S. I. Akasofu (1989), Solar wind-magnetosphere coupling during intense magnetic storms (1978–1979), *J. Geophys. Res.*, *94*(A7), 8835–8851.

Gonzalez, W., J. Joselyn, Y. Kamide, H. Kroehl, G. Rostoker, B. Tsurutani, and V. Vasyliunas (1994), What is a geomagnetic storm?, *J. Geophys. Res.*, *99*(A4), 5771–5792.

Gonzalez, W. D., A. L. C. de Gonzalez, A. Dal Lago, B. T. Tsurutani, J. K. Arballo, G. S. Lakhina, B. Buti, C. M. Ho, and S.-T. Wu (1998), Magnetic cloud field intensities and solar wind velocities, *Geophys. Res. Lett.*, *25*(7), 963–966.

Gonzalez, W. D., E. Echer, A. L. Clade Gonzalez, B. T. Tsurutani, and G. S. Lakhina (2011), Extreme geomagnetic storms, recent Gleissberg cycles and space era-superintense storms, *J. Atmos. Sol. Terr. Phys.*, *73*(11–12), 1447–1453, doi:10.1016/j.jastp.2010.07.023.

Gurnett, D., R. Anderson, B. Tsurutani, E. Smith, G. Paschmann, G. Haerendel, S. Bame, and C. Russell (1979), Plasma wave turbulence at the magnetopause: Observations from ISEE 1 and 2, *J. Geophys. Res.*, *84*(A12), 7043–7058.

Hale, G. E. (1931), The spectrohelioscope and its work. Part III. Solar eruptions and their apparent terrestrial effects, *Astrophys. J.*, *73*, 379–412.

Hodgson, R. (1859), On a curious appearance seen in the Sun, *Mon. Not. R. Astron. Soc.*, *20*, 15–16.

Kamide, Y., N. Yokoyama, W. Gonzalez, B. Tsurutani, I. Daglis, A. Brekke, and S. Masuda (1998), Two-step development of geomagnetic storms, *J. Geophys. Res.*, *103*(A4), 6917–6921.

Kane, S. R., K. Hurley, J. M. McTiernan, M. Sommer, M. Boer, and M. Niel (1995), Energy release and dissipation during giant solar flares, *Astrophys. J.*, *446*, L47–L50.

Kennel, C. F., J. P. Edmiston, and T. Hada (1985), A quarter century of collisionless shock research, in *Collisionless Shocks in the Heliosphere: A Tutorial Review*, *Geophys. Monogr. Ser.*, vol. 34, edited by R. G. Stone and B. T. Tsurutani, pp. 1–36, AGU, Washington, D. C., doi:10.1029/GM034p0001.

Kimball, D. S. (1960), A study of the aurora of 1859, *Sci. Rep. 6, UAG-R109*, Univ. of Alaska Fairbanks, Fairbanks.

Lakhina, G. S. (2000), Magnetic reconnection, *Bull. Astron. Soc. India*, *28*, 593–646.

Lakhina, G. S., and B. T. Tsurutani (2011), Electromagnetic pulsations and magnetic storms, in *Encyclopedia of Solid Earth Geophysics*, 2nd ed., edited by H. K. Gupta, pp. 792–796, Springer, New York.

Lakhina, G. S., B. Tsurutani, H. Kojima, and H. Matsumoto (2000), "Broadband" plasma waves in the boundary layers, *J. Geophys. Res.*, *105*(A12), 27,791–27,831.

Lakhina, G. S., S. Alex, B. T. Tsurutani, and W. D. Gonzalez (2005), Research on historical records of geomagnetic storms, in *Coronal and Stellar Mass Ejections: Proceedings of the 226th Symposium of the International Astronomical Union held in Beijing, China, September 13–17, 2004*, edited by K. P. Dere, J. Wang and Y. Yan, pp. 3–15, Cambridge Univ. Press, Cambridge, U. K.

Li, X., M. Temerin, B. T. Tsurutani, and S. Alex (2006), Modeling of 1–2 September 1859 super magnetic storm, *Adv. Space Res.*, *38*, 273–279.

Lin, R. P., and H. S. Hudson (1976), Non-thermal processes in large solar flares, *Sol. Phys.*, *50*, 153–178.

Lingenfelter, R. E., and H. S. Hudson (1980), Solar particle fluxes and the ancient Sun, in *The Ancient Sun: Fossil Record in the Earth, Moon and Meteorites*, edited by R. O. Pepin, J. A. Eddy and R. B. Merrill, pp. 69–79, Pergamon, New York.

Loomis, E. (1861), On the great auroral exhibition of Aug. 28th to Sept. 4, 1859, and on auroras generally, *Am. J. Sci.*, *82*, 318–335.

Mann, I. R., T. P. O'Brien, and D. K. Milling (2004), Correlations between ULF wave power, solar wind speed, and relativistic electron flux in the magnetosphere: Solar cycle dependence, *J. Atmos. Sol. Terr. Phys.*, *66*, 187–198.

Mannucci, A. J., B. T. Tsurutani, B. A. Iijima, A. Komjathy, A. Saito, W. D. Gonzalez, F. L. Guarnieri, J. U. Kozyra, and R. Skoug (2005), Dayside global ionospheric response to the major interplanetary events of October 29–30, 2003 "Halloween Storms", *Geophys. Res. Lett.*, *32*, L12S02, doi:10.1029/2004GL021467.

McCracken, K., G. Dreschhoff, E. Zeller, D. Smart, and M. Shea (2001a), Solar cosmic ray events for the period 1561–1994 1. Identification in polar ice, 1561–1950, *J. Geophys. Res.*, *106*(A10), 21,585–21,598.

McCracken, K., G. Dreschhoff, D. Smart, and M. Shea (2001b), Solar cosmic ray events for the period 1561–1994 2. The Gleissberg periodicity, *J. Geophys. Res.*, *106*(A10), 21,599–21,609.

Miura, A. (1987), Simulation of the Kelvin-Helmholtz instability at the magnetopheric boundary, *J. Geophys. Res.*, *92*(A4), 3195–3206.

Moos, N. A. F. (1910), Magnetic observations made at the government observatory, Bombay, for the period 1846–1905, and their discussion. Part II: The phenomenon and its discussion, Gov. Cent. Press, Bombay, India.

Newton, H. W. (1943), Solar flares and magnetic storms, *Mon. Not. R. Astron. Soc.*, *103*, 244–257.

Rostoker, G., E. Friedrich, and M. Dobbs (1997), Physics of magnetic storms, in *Magnetic Storms*, *Geophys. Monogr. Ser.*, vol. 98, edited by B. T. Tsurutani et al., pp. 149–160, AGU, Washington, D. C., doi:10.1029/GM098p0149.

Royal Society (1840), *Report of the Committee of Physics, Including Meteorology on the Objects of Scientific Inquiry in Those Sciences*, J. E. Taylor, London, U. K.

Royal Society (1842), *Revised Instructions for the Use of the Magnetic Meteorological Observations and for Magnetic Surveys*, Comm. of Phys. and Meteorol., R. Soc., London, U. K.

Sabine, E. (1851), On periodical laws discoverable in mean effects on the larger magnetic disturbances, *Philos. Trans. R. Soc. London*, *141*, 123–139.

Sabine, E. (1852), On periodical laws discoverable in mean effects on the larger magnetic disturbances, II, *Philos. Trans. R. Soc. London*, *142*, 103–129.

Schindler, K. (1979), On the role of irregularities in plasma entry into the magnetosphere, *J. Geophys. Res.*, *84*(A12), 7257–7266.

Schröder, W. (1997), Some aspects of the earlier history of solar-terrestrial physics, *Planet. Space Sci.*, *45*, 395–400.

Schwabe, S. H. (1843), Solar observations during 1843, *Astron. Nachr.*, *20*(495), 234–235.

Sckopke, N. (1966), A general relation between the energy of trapped particles and the disturbance field near the Earth, *J. Geophys. Res.*, *71*, 3125–3130.

Summers, D., C. Ma, N. P. Meridith, R. B. Horne, R. M. Thorne, and R. R. Anderson (2004), Modeling outer-zone relativistic electron response to whistler mode chorus activity during substorms, *J. Atmos. Sol. Terr. Phys.*, *66*, 133–146.

Taylor, J. R., M. Lester, and T. K. Yeoman (1994), A superposed epoch analysis of geomagnetic storms, *Ann. Geophys.*, *12*, 612–624.

Temerin, M., and X. Li (2002), A new model for the prediction of *Dst* on the basis of the solar wind, *J. Geophys. Res.*, *107*(A12), 1472, doi:10.1029/2001JA007532.

Tsurutani, B. T., and W. D. Gonzalez (1995), The efficiency of "viscous interaction" between the solar wind and the magnetosphere during intense northward IMF events, *Geophys. Res. Lett.*, *22*(6), 663–666.

Tsurutani, B. T., and W. D. Gonzalez (1997), The Interplanetary causes of magnetic storms: A review, in *Magnetic Storms*, *Geophys. Monogr. Ser.*, vol. 98, edited by B. T. Tsurutani et al., pp. 77–89, AGU, Washington, D. C., doi:10.1029/GM098p0077.

Tsurutani, B. T., W. D. Gonzalez, F. Tang, S. I. Akasofu, and E. J. Smith (1988), Origin of interplanetary southward magnetic fields responsible for major magnetic storms near solar maximum (1978–1979), *J. Geophys. Res.*, *93*(A8), 8519–8531.

Tsurutani, B. T., W. D. Gonzalez, F. Tang, Y. T. Lee, M. Okada, and D. Park (1992), Reply to L. J. Lanzerotti: Solar wind RAM pressure corrections and an estimation of the efficiency of viscous interaction, *Geophys. Res. Lett.*, *19*(19), 1993–1994.

Tsurutani, B. T., G. S. Lakhina, C. M. Ho, J. K. Arballo, C. Galvan, A. Boonsiriseth, J. S. Pickett, D. A. Gurnett, W. K. Peterson, and R. M. Thorne (1998), Broadband plasma waves observed in the polar cap boundary layer: Polar, *J. Geophys. Res.*, *103*(A8), 17,351–17,366.

Tsurutani, B. T., Y. Kamide, J. K. Arballo, W. D. Gonzalez, and R. P. Lepping (1999), Interplanetary causes of great and superintense magnetic storms, *Phys. Chem. Earth*, *24*, 101–105.

Tsurutani, B. T., W. D. Gonzalez, G. S. Lakhina, and S. Alex (2003), The extreme magnetic storm of 1–2 September 1859, *J. Geophys. Res.*, *108*(A7), 1268, doi:10.1029/2002JA009504.

Tsurutani, B. T., et al. (2006), Corotating solar wind streams and recurrent geomagnetic activity: A review, *J. Geophys. Res.*, *111*, A07S01, doi:10.1029/2005JA011273.

Tsurutani, B. T., O. P. Verkhoglyadova, A. J. Mannucci, T. Araki, A. Sato, T. Tsuda, and K. Yumoto (2007), Oxygen ion uplift and satellite drag effects during the 30 October 2003 daytime superfountain event, *Ann. Geophys.*, *25*, 569–574.

Vaisberg, O. L., and G. N. Zastenker (1976), Solar wind and magnetosheath observations at Earth during August 1972, *Space Sci. Rev.*, *19*, 687–702.

von Humboldt, A. (1808), Die vollständigste aller bisherigen Beobachtungen über den Einfluss des Nordlichts auf die Magnetnadel, *Ann. Phys.*, *29*, 425–429.

Weiss, L. A., P. H. Reiff, J. J. Moses, and B. D. Moore (1992), Energy dissipation in substorms, in *Substorms 1*, *Eur. Space Agency Spec. Publ.*, *ESA SP-335*, 309–317 .

Willis, D. M., P. R. Stevens, and S. R. Crothers (1997), Statistics of the largest geomagnetic storms per solar cycle (1844–1993), *Ann. Geophys.*, *15*, 719–728.

Yazev, S. A., and E. A. Spirina (2009), The probabilistic approach to estimating the origination of certain types of extreme solar events, *Geomagn. Aeron.*, *49*, 898–903, doi:10.1134/S0016793209070123.

S. Alex and G. S. Lakhina, Indian Institute of Geomagnetism, Plot No. 5, Sector-18, New Panvel (W), Navi Mumbai 410 218, India. (lakhina@iigs.iigm.res.in; salex@iigs.iigm.res.in)

W. D. Gonzalez, Instituto Nacional de Pesquisas Espaciais, Caixa Postal 515, 12200 Sao Jose Dos Campos, Sao Paulo, Brazil. (Gonzalez@dge.inpe.br)

B. T. Tsurutani, NASA Jet Propulsion Laboratory, California Institute of Technology, Pasadena, CA 91109, USA. (bruce.tsurutani@jpl.nasa.gov)

Development of Intermediate-Scale Structure in the Nighttime Equatorial Ionosphere

A. Bhattacharyya

Indian Institute of Geomagnetism, Navi Mumbai, India

Equatorial plasma bubbles (EPBs), a phenomenon observed in the nighttime equatorial ionosphere, are an important component of space weather. EPBs are associated with irregularities of the ionospheric plasma that span a large range of scale sizes of which the intermediate-scale size (~100 m to a few kilometers) irregularities are capable of scattering incident radio waves of VHF and higher frequencies to produce a pattern of varying signal amplitude and phase, as the signal propagates to a ground receiver. Movement of the irregularities across the signal path causes the amplitude and phase of the signal recorded by a receiver to undergo temporal fluctuations or scintillations, which can adversely affect the operation of satellite-based communication and navigation systems. EPBs occur because of the growth of a generalized Rayleigh-Taylor instability on the bottomside of the nighttime equatorial ionosphere, and nonlinear development of the instability depends on ambient ionospheric conditions. As the severity of signal degradation depends on the strength as well as the spectrum of the irregularities, it is of great interest to identify the ambient ionospheric conditions that determine the complexity of structures generated by the instability. This chapter presents the current status of our understanding of this problem based on observations and theoretical models, with an emphasis on scintillation-producing intermediate-scale irregularities in the postsunset equatorial ionosphere, as intense scintillations are indicative of an extreme phenomenon in the Earth's ionosphere.

1. INTRODUCTION

A variety of plasma instability phenomena occur in the F region of the nighttime equatorial ionosphere, which together are known by the generic name of "equatorial spread F" (ESF). The plasma density structures that have been observed in ESF span a large range of scale sizes extending from submeter to hundreds of kilometers. Of these, the geomagnetic field-aligned structures identified as regions of depleted plasma called equatorial plasma bubbles (EPBs) [*Woodman and LaHoz*, 1976; *Kelley et al.*, 1976; *Rino et al.*, 1981; *Tsunoda et al.*, 1982] are considered to be an important component of space weather because electron density irregularities of intermediate-scale sizes (~100 m to a few kilometers) that develop within an EPB are capable of scattering incident radio waves of VHF and higher frequencies and thus can cause degradation and even disruption in the operation of satellite-based communication and navigation systems such as the GPS in the dip equatorial and low-latitude regions. It is well known that growth of the Rayleigh-Taylor (R-T) instability on the bottomside of the nighttime equatorial ionosphere, where a steep upward gradient in plasma density exists after sunset, involves perturbation electric fields that cause lower-density plasma to move upward into a region of higher-density plasma, and in the nonlinear phase of development of the R-T instability, the plasma-depleted region

Extreme Events and Natural Hazards: The Complexity Perspective
Geophysical Monograph Series 196

10.1029/2011GM001078

may push through to the topside of the F region where the plasma density gradient is downward such that the R-T instability would not grow there in the linear regime [*Ott*, 1978; *Keskinen et al.*, 1980; *Zalesak and Ossakow*, 1980; *Ossakow*, 1981]. Gravity need not be the only destabilizing factor, and a generalized R-T instability also includes the influence of an ambient electric field and neutral wind [*Sekar and Raghavarao*, 1987; *Sekar and Kelley*, 1998], and a seeding mechanism for the R-T instability such as gravity waves also plays a role in the occurrence of EPBs [*Keskinen and Vadas*, 2009].

The intermediate-scale-length structures, which develop in the course of evolution of the R-T instability, have been observed in situ by instruments on board rockets and satellites [*Dyson et al.*, 1974; *Rino et al.*, 1981; *Valladares et al.*, 1983; *Kil and Heelis*, 1998; *Sinha et al.*, 1999; *Hysell*, 2000; *Su et al.*, 2001; *Hysell et al.*, 2009; *Heelis et al.*, 2010]. Forward scattering of incident VHF and higher-frequency radio waves by these irregularities produces fluctuations or scintillations in phase and amplitude of the received radio wave signal. Hence, measurement of ionospheric scintillations on transionospheric radio wave signals received at equatorial and low-latitude locations have also provided information about the characteristics of these irregularities [*Yeh and Liu*, 1982; *Tsunoda*, 1985; *Bhattacharyya et al.*, 1992; *Basu et al.*, 2002; *Valladares et al.*, 2004]. The ESF irregularities have been the subject of numerous observational and theoretical studies for several decades [*Woodman*, 2009], and recently, several new studies have been carried out using instruments on board the low-inclination Communication Navigation Outage Forecast System (C/NOFS) satellite [*Burke et al.*, 2009; *de La Beaujardière et al.*, 2009; *Rodrigues et al.*, 2009; *Heelis et al.*, 2010], which is dedicated to the study of scintillation-producing ESF irregularities. Important questions related to the day-to-day variability in the occurrence pattern and characteristics of these irregularities still remain, since the basic condition for growth of the R-T instability, existence of an upward gradient in plasma density on the bottomside of the postsunset F region of the equatorial ionosphere at a time when the Pedersen conductivity in the conjugate E regions at the feet of the geomagnetic field lines connected to the equatorial F region is sufficiently low such that the perturbation electric field associated with the R-T instability is not rapidly short circuited by currents flowing through the E region as happens during daytime, is apparently present every night.

A majority of studies done on ESF irregularities, both observational and theoretical, have been concerned with the onset conditions for ESF, while a few have investigated the nature of the irregularity spectrum over different spatial scale regimes under different ambient conditions. As far as scintillations are concerned, evolution of the irregularity spectrum is a key issue. Current understanding of spectral characteristics of irregularities in the intermediate-scale range derived from in situ observations and computer simulations of the development of these irregularities based on nonlinear theories of an electrostatic R-T instability is described in section 2. Information about the evolution of intermediate-scale-length irregularities, derived from observations of ionospheric scintillations produced by these structures in the nighttime equatorial ionosphere, is the topic of discussion in section 3. Developments in theory, which offer an alternate way of considering the effect of ambient ionospheric conditions, such as the height of the F layer and Pedersen conductivity in the conjugate E regions on the nonlinear development of EPBs with complex spatial structure capable of scattering GPS signals, are discussed in section 4.

2. INTERMEDIATE-SCALE IRREGULARITY SPECTRUM

Early in situ data from satellite [*Dyson et al.*, 1974] and rocket [*Kelley et al.*, 1976] measurements suggested a 1-D irregularity spectrum of the form $P_{\Delta N}(k) \propto k^{-2}$. Analysis of data from AE satellites led *Valladares et al.* [1983] to suggest the existence of a new class of ESF irregularities: bottomside sinusoidal irregularities, which were observed in the bottomside of the equatorial F region, and had a central wavelength lying in the range 300 m to 3 km, with a steep power spectral slope (-5 or -6) at shorter wavelengths. Rocket data obtained in project Condor provided the first spectra for electric field fluctuations associated with the R-T instability, which showed that the spectra of plasma density and electric field fluctuations had the same power law form in the intermediate-scale range, while the ratio of the spectral densities of electric field and density fluctuations assumed a k^2 dependence for wavelengths shorter than 100 m, indicating a change in the physical process involved [*LaBelle et al.*, 1986]. Using ion density and vertical ion drift velocity data from ROCSAT1 satellite, *Su et al.* [2001] found that at scale lengths shorter than 100 m, power spectrum of velocity fluctuations has a spectral slope that is shallower than that of density fluctuations in a stagnated bubble, while the two slopes match in an active bubble. The first results from power spectral analysis of high sampling rate simultaneous measurements of ion densities and electric fields made by C/NOFS have also shown a break in the power law type of spectrum for ion density fluctuations at a scale length of around 70 m, and in the intermediate-scale range, spectra of ion density as well as electric field fluctuations have a nearly $-5/3$ spectral slope [*Rodrigues et al.*, 2009].

In VHF radar observations of ESF such as those carried out at the Jicamarca Radio Observatory, the radio signal coherently backscattered by meter-scale irregularities is measured, and hence, such observations do not yield direct information about intermediate-scale irregularities. However, the range-time-intensity (RTI) plots that show the variation of signal-to-noise ratio with altitude and local time have provided unique information about the evolution of EPBs, such as the vertical extent of the associated "plumes" as well as the formation and structure of bottomside irregularities [*Hysell*, 2000]. Figure 1 shows the RTI plots obtained with a radar operating at 53 MHz at the low-latitude station Gadanki (geomagnetic latitude 4°N): on a night when EPBs extended to a height of about 600 km above Gadanki (Figure 1a) and on another night when only a relatively thin layer of irregularities confined to heights below 300 km was seen after sunset (Figure 1b). Drawing upon satellite, sounding rocket, and coherent scatter radar data, *Hysell* [2000] concluded that during the growth of an interchange instability, such as the R-T instability in the postsunset equatorial ionosphere, bottomside irregularities often begin to emerge as the F region dynamo takes over from the E region dynamo in controlling the electrodynamics of the bottomside F region. A 2-D numerical simulation of the development of a collisional interchange instability on the bottomside of the equatorial ionosphere was carried out by *Hysell* [2000] to explain the appearance of steepened structures in rocket observations and solitary waves in AE-E satellite observations of bottomside irregularities. On the other hand, spectral analysis of plasma drift data from AE-E satellite during its passage through strong spread F irregularities, which would map along geomagnetic field lines to apex heights of ~600–900 km in the topside equatorial F region, showed that in the scale length range of about 1 to 100 km, the 1-D spectral slope was close to $-5/3$, which was indicative of an inertial regime [*Shume and Hysell*, 2004]. Their results could not be extended below 1 km on account of the instrument noise floor.

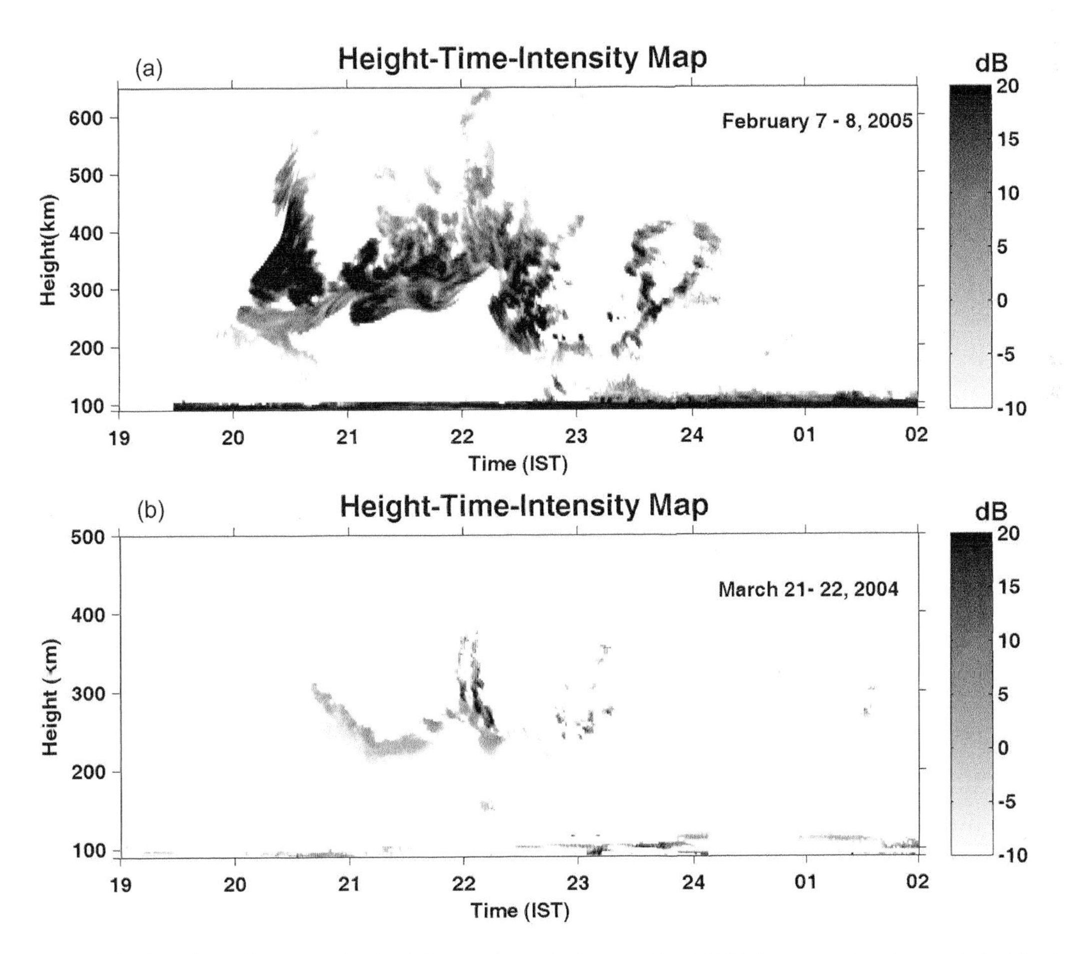

Figure 1. Height-time-intensity maps obtained for a coherently backscattered 53 MHz radar signal recorded at the low-latitude station Gadanki (geomagnetic latitude 4°N) on (a) 7–8 February 2005 and (b) 21–22 March 2004.

Conventionally, interchange instabilities in the ionosphere, such as the generalized R-T instability, which act to interchange high-plasma-density and low-plasma-density regions, have been treated theoretically under two different regimes: collisional and inertial. In the collisional limit, ion-neutral collisions dominate, and ion inertia is neglected, while in the inertial limit applicable at higher altitudes, effects of ion inertia are retained, while ion-neutral collisions may be neglected. It has also been the practice to consider the plasma fluid equations in the electrostatic limit since the ionospheric plasma is a low β plasma (the plasma pressure is much less than the magnetic field pressure). A 2-D numerical simulation of turbulence generated by the growth of the collisional interchange instability in the nighttime equatorial ionosphere yielded, in the plane perpendicular to the geomagnetic field **B**, a markedly anisotropic time-averaged 1-D spectrum of the electrostatic potential for length scales such that $kL_n >> 1$ (L_n is the scale length of the density gradient on the bottomside of the postsunset equatorial *F* region, typically ~10 km) [*Hassam et al.*, 1986]. In the inertial regime, these authors found the potential spectrum to be isotropic, and it exhibited Kolmogorov scaling. In their 2-D numerical simulation of the generalized R-T instability, *Zargham and Seyler* [1987] found that the collisional interchange instability evolves to an anisotropic state, which has nearly sinusoidal variations of plasma density along an effective electric field that includes an ambient eastward electric field as well as gravity and shocklike structures propagating in the perpendicular direction. In this environment, rocket measurements would yield a power law spectrum with a spectral slope between -3 and -2, while satellites would sample quasiperiodic structures with a kneelike spectrum having a steep spectral slope of -6 to -4 for large wavenumbers. This is in agreement with the observations of *Kelley et al.* [1976], *Valladares et al.* [1983], and *Hysell* [2000]. In a self-consistent simulation of the interchange instability in the inertial regime, where the time evolution of the source of the instability, i.e., the density gradient on the bottomside of the equatorial *F* region was also considered, *Zargham and Seyler* [1989] found the density power spectrum to be isotropic with a 1-D spectral slope ranging between -2 and -1.5. Thus, the present picture of intermediate-scale spectrum for bottomside irregularities that has emerged is that the two dimensions in the plane perpendicular to the geomagnetic field **B** are not equivalent [*Kelley*, 2009]. At higher heights, the spectrum is expected to be isotropic in the plane perpendicular to **B**, with the 1-D spectral slope approaching $-5/3$, indicative of Kolmogorov scaling. The C/NOFS results reported by *Rodrigues et al.* [2009] are from observations made around local midnight at an altitude of ~450 km, where the inertial regime results were unexpected. The authors explained their observation of an inertial subrange at this altitude to be due to the extended solar minimum conditions, which saw much lower than usual neutral densities and, hence, much lower ion-neutral collision frequencies at this altitude.

3. IONOSPHERIC SCINTILLATION OBSERVATIONS

In the presence of ionospheric irregularities, variations ΔN in electron density give rise to fluctuations $\Delta\mu$ in the refractive index of the ionosphere, which for VHF and higher-frequency radio waves are simply proportional to ΔN: $\Delta\mu = -\lambda^2 r_e \Delta N/2\pi$, where λ is the signal wavelength and r_e is the classical electron radius [*Yeh and Liu*, 1982]. Such radio signals, while propagating through a structured ionosphere to a ground receiver, are scattered by intermediate-scale size irregularities, essentially in the forward direction, to form a diffraction pattern on the ground. As the signal path sweeps across the irregularities due to the movement of the irregularities across the signal path, as in the case of signals transmitted from a geostationary satellite, or due to movement of the irregularities as well as the satellite as in the case of GPS signals, a receiver on the ground records fluctuations or scintillations in both the phase and amplitude of the received signal. While amplitude scintillations may cause signals to fade, rapid fluctuations in phase of the signal may cause loss of lock of GPS signals. The statistical characteristics of recorded scintillations depend on the strength and power spectrum of the irregularities and the speed with which irregularities drift across the signal path. It should be noted that all the irregularities in the path of the signal contribute to scintillations, with maximum contribution coming from the region of the peak plasma density in the *F* region where ΔN would be the largest for a given $\Delta N/N$.

A measure of the strength of amplitude scintillations on a radio signal is the S_4 index, which is the standard deviation of normalized intensity fluctuations. The power spectrum of weak amplitude scintillations ($S_4 < 0.5$) has a direct relationship with the irregularity power spectrum for scale sizes shorter than the Fresnel scale length d_F, which is given by $d_F = \sqrt{2\lambda z}$, where λ is the signal wavelength and z is the average distance of the irregularity layer from the receiver along the signal path. The temporal scale of the scintillation spectrum may be converted into a spatial scale for the irregularities using the drift speed V_0 of the ground scintillation pattern, provided that the drift velocity is constant and there are no changes in the characteristics of the irregularities as they drift across the signal path. However, estimation of the drift speed of the ground scintillation pattern in the magnetic east-west direction, using data from two receivers spaced along a magnetic east-west baseline, has shown that during the initial phase of EPB development, often extending up to

2 h after scintillations are first recorded, the maximum cross-correlation between the two spaced receiver signals is significantly less than unity, becoming close to unity at a later time. Hence the drift speed estimated from the time lag for maximum cross correlation is an apparent drift speed, V_A, which may exceed the true drift speed [*Spatz et al.*, 1988; *Bhattacharyya et al.*, 1989]. The true drift speed, V_0, is estimated using the full correlation technique originally suggested by *Briggs* [1984], which is based on the assumption that the space-time correlation function of the recorded intensity scintillations has the following form:

$$C_I(x,t) = f[(x - V_0 t)^2 + V_C^2 t^2], \tag{1}$$

where f is a monotonically decreasing function of its argument with maximum value $f(0) = 1$, and V_C is the so-called "random velocity," which accounts for the decorrelation of the signals. It can be seen from equation (1) that $V_A = (V_0^2 + V_C^2)/V_0$. Comparing the forms of the spatial correlation function (time lag $t = 0$) and temporal correlation function (spatial lag $x = 0$) as obtained from equation (1), it is seen that a spatial scale d is converted into a temporal scale τ through the relationship: $d = V\tau$, where $V = \sqrt{V_0 V_A}$. For a 1-D irregularity power spectrum of the form $P_{\Delta N}(k_x) \propto k_x^{-m+1}$, along the magnetic east-west direction, power spectrum of weak amplitude scintillations is of the form $P(f) \propto f^{-m}$, for frequencies f greater than the Fresnel frequency, $f_F = V/d_F$, with frequencies f related to the wavenumbers k_x through $f = k_x V/2\pi$. It should be pointed out that these theoretical results for weak scintillations hold whether the irregularities are confined to a thin layer, which may be described by a single phase-changing screen or a thick layer of irregularities is considered [*Yeh and Liu*, 1982]. Weak phase scintillations also yield the same power law form of spectrum except that for phase fluctuations, the spectrum may be extended to scale lengths longer than the Fresnel scale as Fresnel filtering does not reduce phase fluctuations at these longer scale lengths, which are directly related to the variations in total electron content along the signal path and are not caused by forward scattering of the signal by the irregularities [*Bhattacharyya et al.*, 2000].

For a possible interpretation of V_C in terms of irregularity characteristics, a space-time correlation function for a turbulent flowing plasma suggested by *Shkarofsky* [1968] has been introduced in the theories of weak scintillations produced by a phase screen [*Wernik et al.*, 1983], saturated scintillations due to a phase screen [*Franke et al.*, 1987], and scintillations of all strengths produced by a thick layer of irregularities [*Bhattacharyya et al.*, 1989]. Based on the assumption that the two characteristic scale lengths required for a realistic representation of the turbulence spectrum, characterizing, respectively, the scale sizes that contain maximum energy and the ones that undergo rapid dissipation, are independent of time, *Shkarofsky* [1968] suggested the following form for the wavenumber frequency-dependent turbulence spectrum:

$$S_{\Delta N}(\mathbf{k}, \omega) = P_{\Delta N}(\mathbf{k})\Psi(\mathbf{k}, \omega). \tag{2}$$

Here $P_{\Delta N}(\mathbf{k})$ is the power spectrum of irregularities simply convecting with the prevailing mean velocity, and different forms for the Fourier transform $\Psi(\mathbf{k},t)$ of $\Psi(\mathbf{k},\omega)$ were considered for different processes such as velocity fluctuations and diffusion, which could lead to loss of correlation. For modeling the space-time variation of the ground scintillation pattern produced by geomagnetic field-aligned irregularities, only the variation in the magnetic east-west (x) direction of ionospheric electron content, integrated along the signal path, need be considered. The spatial Fourier transform of the space-time correlation function of the irregularities could thus have the form:

$$S_{\Delta N}(k_x, t) = P_{\Delta N}(k_x)\Psi(k_x, t), \tag{3}$$

where $P_{\Delta N}(k_x)$ is the 1-D power spectrum of the irregularities as would be measured by a satellite for instance. In a situation where there is loss of correlation due to fluctuations of the drift velocity, $\Psi(k_x,t)$ for irregularities drifting across the signal path with an average speed V_0 in the x direction and with random fluctuations in the drift speed having standard deviation σ_V, may be taken as [*Shkarofsky*, 1968]:

$$\Psi(k_x, t) = \exp[-ik_x V_0 t - k_x^2 \sigma_V^2 t^2/2]. \tag{4}$$

Describing the space-time structure of ionospheric irregularities using equations (3) and (4), scintillations have been modeled to show that intensity scintillations could have a space-time correlation function of the form given in equation (1), where the random velocity V_C could now be identified with σ_V [*Wernik et al.*, 1983; *Franke*, 1987; *Bhattacharyya et al.*, 1989]. This picture of the irregularities also provides a possible explanation for the large decorrelation of spaced receiver signals during the period when scintillations are produced by irregularities in the growth phase of the EPBs, which is characterized by the presence of significant perturbation electric fields associated with the R-T instability [*Bhattacharyya et al.*, 1989, 2001]. Once these perturbation electric fields are eroded a couple of hours after the generation of the EPB, possibly due to downward movement of the background ionosphere [*Bhattacharyya et al.*, 1989], the two signals become well correlated, irrespective of the strength of the intensity scintillations. The fact that the scintillating signal generally has a path which is inclined to the vertical implies that the drift speed of the irregularities transverse to the signal path has a contribution from the vertical drift of the EPB as well, which is substantial in the growth phase of the EPBs. This led *Bhattacharyya et al.* [2001] to suggest that for a given time interval, the maximum cross correlation obtained at a time lag t_m between the signals recorded by two receivers spaced a distance X_0 apart along a magnetic east-west baseline, $C_I(X_0,t_m)$, which is significantly smaller than 1, often falling below 0.5, when an EPB with scintillation-producing irregularities is in its nascent stage, may be used to estimate the "age" of the irregularities that produce the scintillations. This is an important issue in the use of ionospheric scintillation observations to study the evolution of the spectrum of intermediate-scale-length ESF irregularities. It is also crucial for differentiating between EPBs that start growing in the postmidnight period as a result of a change in the ambient electric field from the normal westward at the time to an eastward electric field caused by magnetic activity and the EPBs that have developed earlier during postsunset hours and later drifted into the path of the signal recorded by a receiver [*Bhattacharyya et al.*, 2002]. A recent 3-D simulation of an EPB [*Krall et al.*, 2010] has demonstrated that bubbles become "fossilized"; that is, the perturbation electric field associated with the R-T instability becomes very small, and the upward movement of an EPB ceases when the flux tube-integrated ion mass density just inside the bubble is equal to that of the adjacent background. However, the EPB persists for hours after it has stopped rising. Thus, scintillations, even strong scintillations on a VHF signal, may persist long after the EPBs become "fossil" bubbles, and the strength of scintillations is not an adequate indicator of the "age" of the EPB.

Two examples of the evolution of irregularity spectrum as obtained from amplitude scintillation spectra are shown in Figures 2a and 2b for a magnetically quiet and a magnetically disturbed day. The data used in the analysis are of amplitude scintillations on a 251 MHz signal transmitted from a geostationary satellite located at 72°E and recorded by two receivers spaced 540 m apart along a magnetic east-west baseline at the equatorial station Tirunelveli (8.7°N, 77.8°E, geomagnetic latitude: 0.3°S). It is to be noted that the spectral indices m have been obtained for only those 3 min intervals, which have weak amplitude scintillations ($S_4 < 0.5$), where the irregularity spectrum may be directly related to the spectrum of amplitude scintillations. On the magnetically quiet day, temporal variation of the maximum cross correlation between the spaced receiver signals indicates that by 22 LT, the perturbation electric field associated with the R-T instability has nearly disappeared, and a gradual steepening of intermediate-scale irregularity spectrum is evident in Figure 2a. During the postmidnight period, irregularities encountered by the signal have a steep 1-D spectrum in the magnetic east-west direction with spectral indices ≥ 4, characteristic of bottomside irregularities. As long as irregularities with scale sizes close to the Fresnel scale d_F, which is approximately 1 km for a radio signal of frequency 251 MHz incident on an irregularity layer at an average distance of 400 km from the receiver along the signal path, are present with sufficient strength in a bottomside layer, they can cause strong scintillations on a VHF signal even though they do not produce significant scintillations on a GPS signal. On the magnetically disturbed day, the irregularities generated shortly after sunset initially have a 1-D power spectral index: $m - 1 \approx 5/3$, which is indicative of Kolmogorov scaling although the maximum cross correlation indicates that the perturbation electric field associated with the R-T instability is much reduced, so the bubble is "fossilized," and as time progresses, the incident signal encounters irregularities in a decay mode with the irregularity spectrum becoming progressively steeper. However, around midnight possibly as a result of the ambient electric field, which is normally westward at this time on a quiet day, turning eastward due to the effect of magnetic activity, fresh irregularities are generated as the nascent EPB reaches higher altitudes where inertial effects dominate over the effects of ion-neutral collisions [*Bhattacharyya et al.*, 2002]. The important point is that the postsunset equatorial ionosphere provides a dynamic background in which the generalized R-T instability evolves nonlinearly to produce the irregularities. Hence, there are several factors which contribute to the day-to-day variability of not only the occurrence pattern of ESF irregularities but also the development of scintillation-producing intermediate-scale structures in the nighttime equatorial ionosphere.

4. ROLE OF *E* REGION CONDUCTIVITY AND *F* LAYER HEIGHT

Huba et al. [1985] retained both ion inertia and ion-neutral collisions in a theoretical description of the nonlinear evolution of the electrostatic R-T instability in the ionosphere using a three-mode system. They showed that the nonlinear equations, which described the evolution of this system, were identical in form to the Lorenz equations that were used to describe approximately the evolution of the Rayleigh-Benard instability [*Lorenz*, 1963]. In the case of the electrostatic R-T instability, the condition for instability of the fixed states for the three-mode system was found to be: $\nu_{in} < \sqrt{g/2L_n}$, where ν_{in} is the ion-neutral collision frequency and L_n is the density gradient scale length on the bottomside of the equatorial F region [*Huba et al.*, 1985]. Hence, these

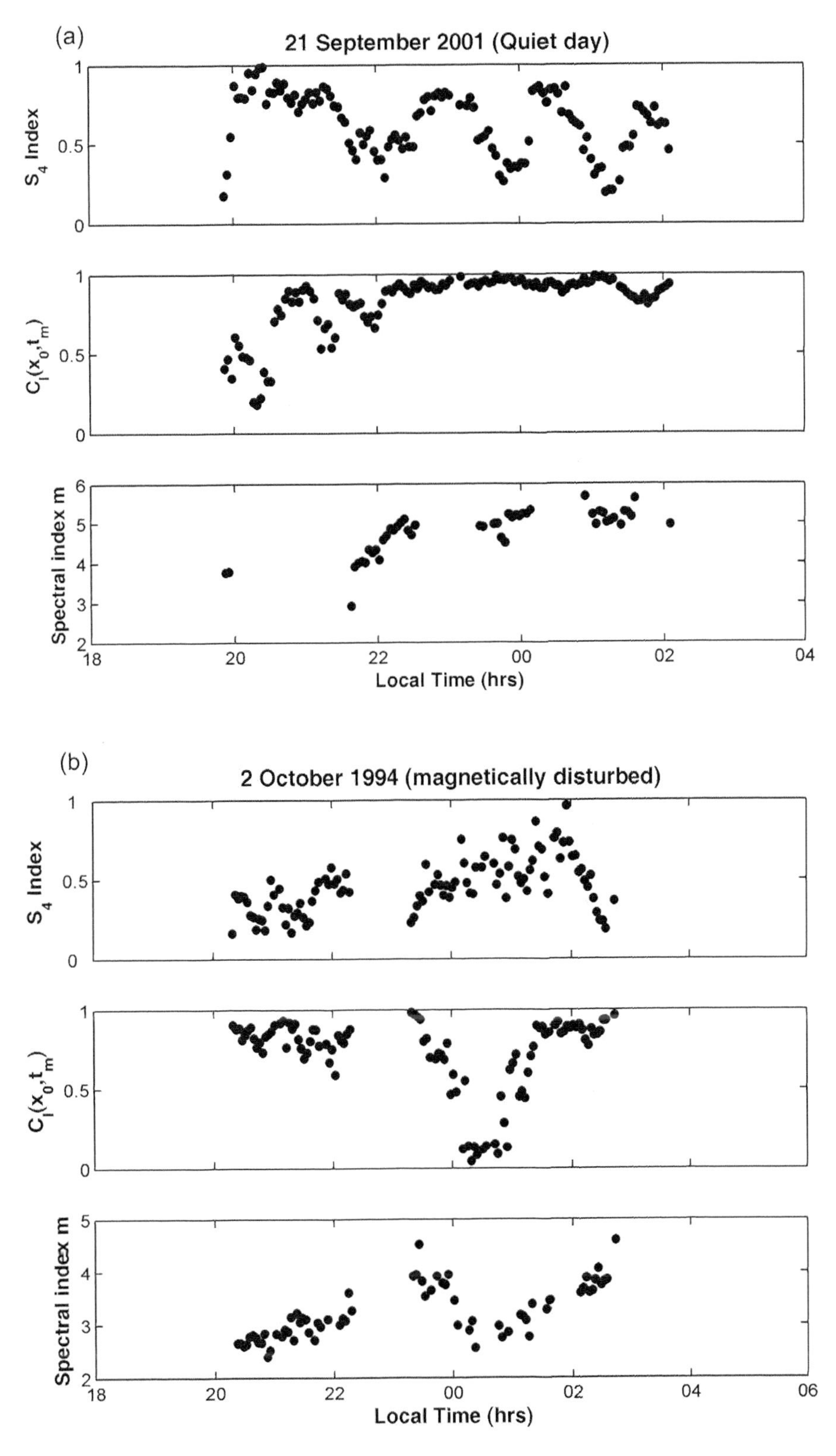

Figure 2. (top) S_4 index, (middle) maximum cross correlation between spaced receiver records of intensity scintillations on a 251 MHz signal, and (bottom) spectral index of a power law spectrum of intensity scintillations computed for every 3 min interval of the scintillation events recorded at Tirunelveli on (a) a magnetically quiet day and (b) a magnetically disturbed day.

authors concluded that chaotic behavior of the ionospheric fluid may result from the R-T instability at altitudes above 500 km where the above condition would be met. However, a many-mode study of the evolution of the electrostatic R-T instability in a 2-D approximation by *Hassam et al.* [1986] demonstrated that the behavior of large convection cells did not follow the pattern suggested by the three-mode study of *Huba et al.* [1985].

Theoretical descriptions of ionospheric scintillations have been based on a stochastic approach on account of the complexity of electron density structures that result from the growth of plasma instabilities in the ionosphere. In the previous section, it was shown how statistical parameters derived from scintillation data, such as the S_4 index or amplitude power spectrum, are used to characterize the ionospheric irregularities that produce the scintillations. Instead of the conventional spectral studies of scintillations, *Bhattacharyya* [1990] treated the irregularity layer as a fractal phase-changing screen in an attempt to deduce the fractal structure of ionospheric turbulence from amplitude and phase scintillation data. It was concluded in this paper that propagation effects dominate for amplitude scintillation data so that reconstruction of an attractor to describe the chaotic behavior of strong ionsopheric turbulence using amplitude scintillation data is not possible but phase scintillation data yield a low information dimension (<4) for the fractal phase screen. *Wernik and Yeh* [1994] simulated amplitude and phase scintillations that would be produced by a phase-changing screen with phase fluctuations represented by colored noise and calculated the information dimension from the scintillations for different spectral indices. They found that even for weak scintillations, propagation effects tend to dominate for amplitude scintillations, whereas phase scintillations yield the same information dimension irrespective of the strength of the phase fluctuations. Further studies on these lines, to understand the role of ambient ionospheric conditions in the development of ionospheric turbulence, have not been carried out since phase scintillation data on a signal transmitted by a geostationary satellite are now unavailable for the equatorial region.

One of the limitations of earlier theoretical studies of the spectrum of intermediate-scale-length ESF irregularities, mentioned in section 2, is that the models are 2-D so that the effect of including $k_{\|}$, which couples the equatorial *F* region with the conjugate *E* regions at the feet of the geomagnetic field lines that connect them, is absent. As the geomagnetic field-aligned EPBs are considered to form through the interchange of flux tubes containing low-density plasma and high-density plasma, the *E* regions at the feet of the flux tubes, which are stable, have a pivotal role to play, since *E* regions with significant electrical conductivity would impede the interchange of flux tubes. Hence, the effect of *E* regions needs to be considered separately from the *F* regions. Based on simultaneous measurements of electric and magnetic field fluctuations associated with EPBs updrafting at supersonic speeds, using DE 2 satellite, *Aggson et al.* [1992] suggested that field-aligned currents (FACs) that couple the equatorial *F* region with conjugate *E* regions may be carried by Alfvén waves launched in the equatorial *F* region when a bubble starts to grow there. In a seminal study of the electromagnetic interchange mode in a partially ionized collisional plasma, *Hudson and Kennel* [1975] pointed out the fundamental electromagnetic nature of the interchange mode and found that the coupling to the intermediate Alfvén mode has a stabilizing effect for finite parallel wavelengths, which would tend to restrict the interchange instability to the lowest-order flute perturbations of an entire flux tube. However, they did not consider the coupling with conducting *E* regions at the end of the flux tubes, which as discussed above are impediments to the interchange of flux tubes. In a transmission line analogy proposed to model the process of FACs being carried by Alfvén waves launched in the equatorial *F* region when a bubble starts to grow there [*Bhattacharyya and Burke*, 2000], the closure of FACs also through polarization currents associated with shear Alfvén waves, particularly in those regions along geomagnetic field lines connecting the equatorial *F* region with conjugate *E* regions where the Pedersen conductivity is much smaller than that in *E* and F regions, was considered. The profile of local Pedersen conductivity at sunset as a function of the apex altitude of a geomagnetic field line, obtained in a model calculation by *Sultan* [1996], is shown in Figure 3. It is seen from this figure that the *E* region conductivity may be large enough even at sunset to be necessarily taken into consideration in the modeling of the development of an EPB. In the

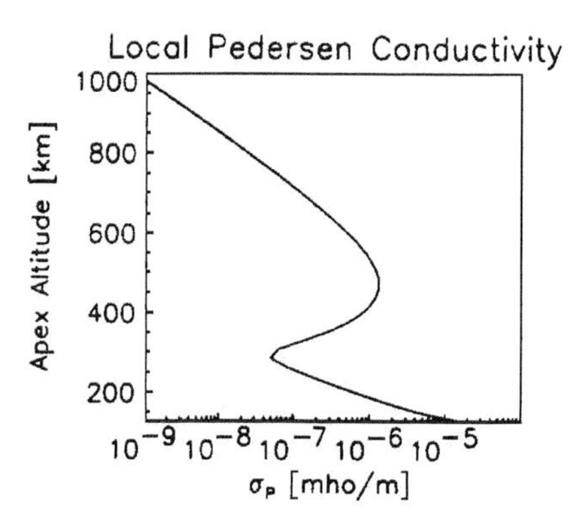

Figure 3. Profile of local Pedersen conductivity at sunset as a function of the apex altitude of a geomagnetic field line over the dip equator obtained in a model calculation. After the work of *Sultan* [1996].

transmission line analogy, $k_{\|}$ is small enough that the EPBs are still nearly geomagnetic field aligned. One of the important results to emerge from this model is that in the linear regime, the electromagnetic version of the R-T instability would give rise to nonvanishing density perturbation in the equatorial *F* region only if the field line-integrated *E* region Pedersen conductivities in the two hemispheres are identical and small. The requirement of identical *E* region conductivities in the two hemispheres provides a theoretical explanation for the observed seasonal and longitudinal occurrence pattern of equatorial scintillations, which showed that near alignment of the sunset terminator with the geomagnetic flux tubes provided a favorable condition for the growth of the intermediate-scale ESF irregularities [*Tsunoda*, 1985]. The same condition was also indicated by the seasonal and longitudinal occurrence pattern of EPBs based on electron density measurements by DMSP satellites during the period 1989–2002 [*Burke et al.*, 2004]. It should be mentioned that the presence of off-equatorial sporadic *E* layers could sometimes play quite a different role in the generation of EPB. It was suggested by *Tsunoda* [2006] that a polarization electric field generated by a sporadic *E* layer instability could map along geomagnetic field lines to the base of the equatorial *F* layer and produce a large-scale wave structure on the bottomside, the crests of which would be the locations for the growth of EPBs.

Magnetic field fluctuations associated with EPBs have been observed by CHAMP satellite [*Stolle et al.*, 2006], as well as DEMETER satellite [*Pottelette et al.*, 2007], and the characteristics of FACs associated with EPBs have also been studied in detail using CHAMP data [*Park et al.*, 2009]. An example of magnetic field fluctuations associated with EPBs as observed by CHAMP satellite at an altitude of about 400 km is shown in Figure 4. Simultaneous electric field measurements are not available from CHAMP satellite, but such measurements were carried out by the DEMETER satellite and also by the Extremely Low Frequency Wave Analyzer instrument on board CRRES [*Koons et al.*, 1997]. DEMETER observations showed the occurrence of electromagnetic fluctuations in regions with plasma density structures of 1–10 km scale sizes, whereas broadband electrostatic fluctuations were seen in association with shorter scale sizes

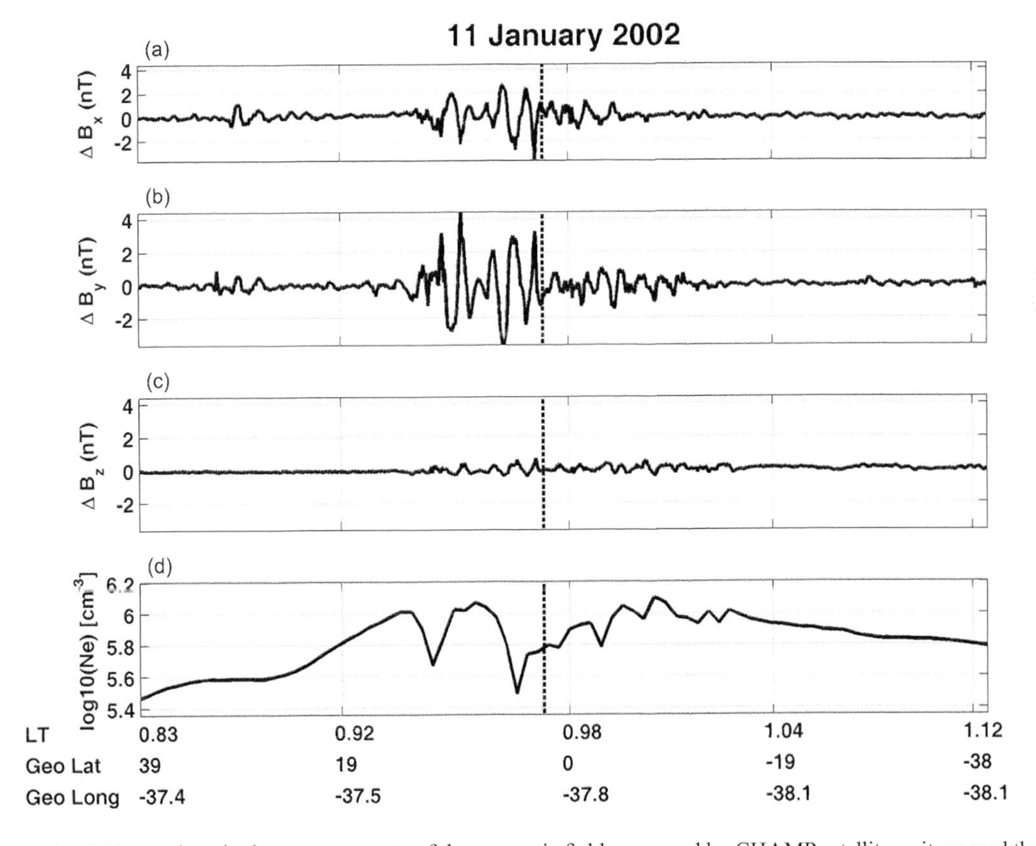

Figure 4. (a–c) Fluctuations in three components of the magnetic field measured by CHAMP satellite as it crossed the dip equator (indicated by the dashed vertical lines in each graph), where the geomagnetic field is taken to be in the *z* direction. (d) Plot of the electron density measured by CHAMP, which shows the presence of equatorial plasma bubbles (EPBs).

(10–100 m) [*Pottelette et al.*, 2007]. These authors also presented some results on the planarity of propagating electromagnetic waves and the polar angle of propagation of the wave vector with respect to the background magnetic field, which supported their identification of the observed electromagnetic bursts with kinetic Alfvén waves. *Koons et al.* [1997] also reported CRRES observations of both electrostatic waves and electromagnetic waves propagating in the extraordinary mode simultaneously with density depletions in the topside equatorial F region. In order to identify the wave modes, these authors also computed the refractive index, cB/E, which showed a variation as the square root of electron density obtained from Langmuir probe data [*Koons et al.*, 1997, Figure 4] for the electromagnetic modes associated with the density depletions, as would be expected in the case of Alfvén waves.

In order to determine how small the E region conductivity has to be for the growth of EPBs, the transmission line analogy was extended to study the nonlinear evolution of the electromagnetic R-T instability [*Bhattacharyya*, 2004]. For a three mode system, the nonlinear equations obtained in this paper are identical to those obtained by *Huba et al.* [1985] for an electrostatic interchange instability, except that the parameters now contain an additional term, which depends on $\sum_P^E$, the field line integrated E region Pedersen conductivity in either hemisphere. As these equations correspond exactly to the Lorenz equations, a condition for unstable fixed states is obtained:

$$\nu_i + \frac{\mu_0 V_A^2 \Sigma_P^E}{l} < \sqrt{\frac{g}{2L_n}}. \quad (5)$$

Here $\nu_i = \nu_{ie} + \nu_{in}$ includes ion collision frequency with both neutrals and electrons, $V_A (= B/\sqrt{\mu_0 n_0 m_i})$ is the Alfvén speed, and l is the length of the geomagnetic field line

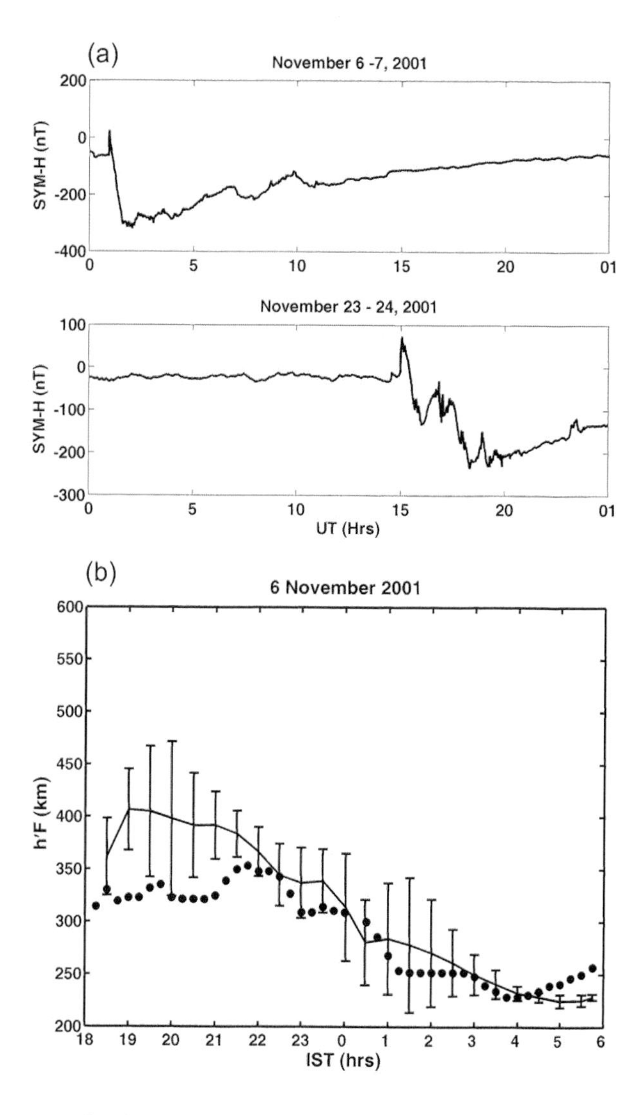

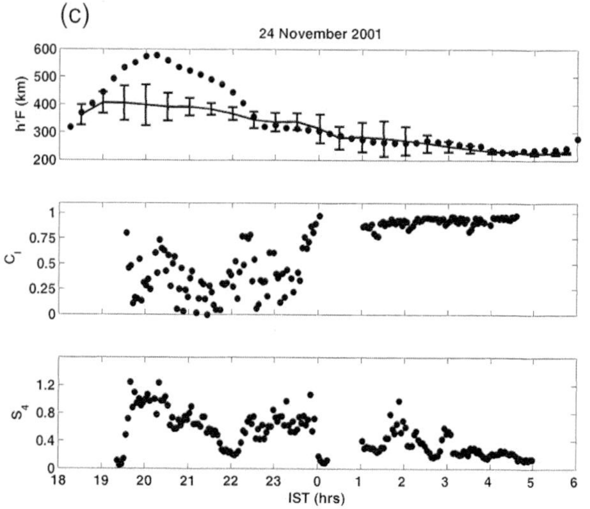

Figure 5. (a) SYM-H index as a function of UT (Indian standard time, IST = UT + 5.5 h) shows the occurrence of major magnetic storms on 6 and 24 November 2001. (b) Virtual height of the equatorial F layer, $h'F$, obtained from ionosonde observations at equatorial station, Trivandrum, during the night of 6 November 2001 (solid circles). The average of $h'F$ for five magnetically quiet International Quiet Days (IQD) of November 2001, together with their standard deviation, is also shown for comparison. (c) (top) The $h'F$ obtained from ionosonde observations at Trivandrum, during the night of 24 November 2001 (solid circles) and the average $h'F$ for the five IQD of November 2001. (middle) The maximum cross correlation of scintillations recorded on a 251 MHz signal by two spaced receivers at equatorial station Tirunelveli (0.8°E of Trivandrum) shows the generation of EPBs until 23:30 IST. (bottom) S_4 index indicating the strength of amplitude scintillations on the 251 MHz signal recorded at Tirunelveli.

from the equatorial F region to the conjugate E region in either hemisphere. The difference between this condition and that obtained by *Huba et al.* [1985] is that in equation (5), the E region conductivity, together with the equatorial F region polarizability, introduces a new time scale in the phenomenon, which is the time required to discharge the bubble by currents flowing through the E region. With $\sum_P^E \approx 1$ mho, $l \approx$ 1600 km and $V_A \approx 300$ km s^{-1}, this time is about 100 s, which is comparable to the other timescales involved in the phenomenon. From extensive studies based on radar observations, height of the nighttime equatorial F layer has emerged as one of the key factors in the generation of the EPBs [*Fejer et al.*, 1999]. According to equation (5), the larger the value of l, the lesser is the effect of E region. An example of the role of nighttime equatorial F region height in the development of scintillation-producing irregularities is given in Figure 5, where 2 days with major magnetic activity are shown. On 6 November 2001, the symmetric disturbance field in H component (SYM-H) index, which is a measure of the strength of the magnetospheric ring current, plotted in Figure 5a shows that the main phase of the magnetic storm started at around 02:00 UT (07:30 IST (Indian standard time)) and a disturbance dynamo setup in the ionosphere due to this magnetic activity earlier in the day resulted in suppression of the postsunset prereversal enhancement of the eastward electric field in the equatorial ionosphere, which would have otherwise raised the altitude of the equatorial F region and thus created a condition conducive to the growth of an EPB. As a result of this suppression, the virtual height of the bottom of the F layer, $h'F$, obtained from ionosonde observations at the equatorial station Trivandrum (8.5°N, 77°E, geomagnetic latitude: 0.3°S) did not go above 350 km as seen in Figure 5b, and on this day, no scintillations were observed on a VHF signal recorded at the equatorial station Tirunelveli, which is located 0.8°E of Trivandrum. On the other hand, on 24 November 2001, magnetic activity that started at around 06:00 UT enhanced the eastward electric field in the postsunset hours well above what might be expected from the average postsunset prereversal enhancement under quiet conditions, and $h'F$ reached a value of 600 km, and remained above 400 km for more than 3 h. Ionospheric conditions were thus conducive for the growth of new EPBs almost up to midnight, which produced scintillations on a 251 MHz signal recorded by two spaced receivers at Tirunelveli, and the spaced receiver signals showed significant decorrelation at least up to 23:30 IST as shown in Figure 5c. It should be pointed out that from equation (5), it may also be concluded that apart from a high altitude of the F layer and low E region Pedersen conductivity, a higher plasma density in the F region also increases the time required to discharge the plasma bubble and, hence, should be more favorable for the evolution of an EPB with complex structure. This is expected because a higher ion mass density hampers oscillations of the geomagnetic field lines, which are involved in the propagation of Alfvén waves, and thus reduces the coupling of the equatorial F region with the conjugate E regions. However, it is to be noted that equation (5) is based on a number of simplifying assumptions such as a three-mode system, and also, the effects of neutral winds have not been taken into consideration at all.

5. SUMMARY

One of the important space weather phenomena in the Earth's equatorial ionosphere is the development of complex structures in the postsunset equatorial ionosphere. The intermediate-scale ionospheric structures can cause strong scattering of incident VHF and higher-frequency radio waves, which affects the operation of satellite-based communication and navigation systems. A clear understanding of the factors involved in the day-to-day variability in the occurrence and complexity of the structures continues to be elusive. In recent years, it has been realized that 2-D simulations of the growth of the geomagnetic field-aligned structures are not adequate to answer the questions that remain and that 3-D models may be required to understand the development of scintillation-producing irregularities in the postsunset equatorial ionosphere through nonlinear evolution of R-T instability or interchange instability. This may be necessary because (1) the conjugate E regions at the feet of the geomagnetic field lines connecting them with the equatorial F region do not take part in the instability, as the ions in the E region are not magnetized, (2) geomagnetic FACs, which couple the equatorial F region with the conjugate E regions, need to be considered explicitly since currents that flow through conducting E regions at the feet of the field lines act to short-circuit the perturbation electric field associated with the plasma instability so that the E region conductivity is an impediment to the interchange of flux tubes, and (3) dynamics of the background plasma and the changing background plasma distribution all along the geomagnetic field lines plays an important role in the evolution of the EPB and associated electric and magnetic field fluctuations. Results of a number of 3-D simulations of the development of an EPB through nonlinear growth of the electrostatic R-T instability or interchange instability have been reported [*Keskinen et al.*, 2003; *Keskinen and Vadas*, 2009; *Retterer*, 2010; *Aveiro and Hysell*, 2010; *Krall et al.*, 2010; *Huba and Joyce*, 2010]. In some of these studies, finite parallel conductivity effects have been considered together with a time-dependent 3-D model of the background ionosphere, and in others, the background ionosphere has been modeled in a self-consistent manner including global neutral wind-driven dynamo electric field, while the geomagnetic field lines have been considered to be equipotentials. However, the longitudinal resolution in all these models is insufficient to allow a study of the development of intermediate-scale ESF irregularities. Although the magnetic field fluctuations that have been observed to be associated with EPBs [*Aggson et al.*, 1992; *Koons et al.*, 1997; *Stolle et al.*, 2006; *Pottelette et al.*, 2007; *Park et al.*, 2009] are small compared to the geomagnetic field strength, they indicate the possibility of the involvement of an electromagnetic version of the R-T instability in the development of an EPB, which introduces a new component of the phenomenon: FACs carried by Alfvén waves. Observations by C/NOFS of electric and magnetic field fluctuations associated with EPBs, together with 3-D simulation of the nonlinear evolution of an electromagnetic version of the R-T instability, could shed some further light on this question.

Acknowledgments. The author gratefully acknowledges the use of radar data obtained from NARL, Gadanki, India; CHAMP magnetic field data obtained from GFZ, Potsdam, Germany; and ionosonde data obtained from Space Physics Laboratory, VSSC, Thiruvananthapuram, India. The author also thanks D. Tiwari, B. Kakad, and S. Sripathi for their assistance in analysis of the data.

REFERENCES

Aggson, T., W. Burke, N. Maynard, W. Hanson, P. Anderson, J. Slavin, W. Hoegy, and J. Saba (1992), Equatorial bubbles updrafting at supersonic speeds, *J. Geophys. Res.*, *97*(A6), 8581–8590.

Aveiro, H. C., and D. L. Hysell (2010), Three-dimensional numerical simulation of equatorial *F* region plasma irregularities with bottomside shear flow, *J. Geophys. Res.*, *115*, A11321, doi:10.1029/2010JA015602.

Basu, S., K. M. Groves, S. Basu, and P. J. Sultan (2002), Specification and forecasting of scintillations in communication/navigation links: Current status and future plans, *J. Atmos. Sol. Terr. Phys.*, *64*, 1745–1754.

Bhattacharyya, A. (1990), Chaotic behaviour of ionospheric turbulence from scintillation measurements, *Geophys. Res. Lett.*, *17* (6), 733–736.

Bhattacharyya, A. (2004), Role of *E* region conductivity in the development of equatorial ionospheric plasma bubbles, *Geophys. Res. Lett.*, *31*, L06806, doi:10.1029/2003GL018960.

Bhattacharyya, A., and W. J. Burke (2000), A transmission line analogy for the development of equatorial ionospheric bubbles, *J. Geophys. Res.*, *105*(A11), 24,941–24,950.

Bhattacharyya, A., S. J. Franke, and K. C. Yeh (1989), Characteristic velocity of equatorial *F* region irregularities determined from spaced receiver scintillation data, *J. Geophys. Res.*, *94*(A9), 11,959–11,969.

Bhattacharyya, A., K. C. Yeh, and S. J. Franke (1992), Deducing turbulence parameters from transionospheric scintillation measurements, *Space Sci. Rev.*, *61*, 335–386.

Bhattacharyya, A., T. L. Beach, S. Basu, and P. M. Kintner (2000), Nighttime equatorial ionosphere: GPS scintillations and differential carrier phase fluctuations, *Radio Sci.*, *35*(1), 209–224.

Bhattacharyya, A., S. Basu, K. M. Groves, C. E. Valladares, and R. Sheehan (2001), Dynamics of equatorial *F* region irregularities from spaced receiver scintillation observations, *Geophys. Res. Lett.*, *28*(1), 119–122.

Bhattacharyya, A., S. Basu, K. M. Groves, C. E. Valladares, and R. Sheehan (2002), Effect of magnetic activity on the dynamics of equatorial *F* region irregularities, *J. Geophys. Res.*, *107*(A12), 1489, doi:10.1029/2002JA009644.

Briggs, B. H. (1984), The analysis of spaced sensor records by correlation techniques, in *Middle Atmosphere Program, Handbook for MAP*, vol. 13, edited by R. A. Vincent, pp. 166–186, ICSU, Paris.

Burke, W. J., L. C. Gentile, C. Y. Huang, C. E. Valladares, and S. Y. Su (2004), Longitudinal variability of equatorial plasma bubbles observed by DMSP and ROCSAT-1, *J. Geophys. Res.*, *109*, A12301, doi:10.1029/2004JA010583.

Burke, W. J., O. de La Beaujardière, L. C. Gentile, D. E. Hunton, R. F. Pfaff, P. A. Roddy, Y.-J. Su, and G. R. Wilson (2009), C/NOFS observations of plasma density and electric field irregularities at post-midnight local times, *Geophys. Res. Lett.*, *36*, L00C09, doi:10.1029/2009GL038879.

de La Beaujardière, O., et al. (2009), C/NOFS observations of deep plasma depletions at dawn, *Geophys. Res. Lett.*, *36*, L00C06, doi:10.1029/2009GL038884.

Dyson, P. L., J. P. McClure, and W. B. Hanson (1974), In situ measurements of the spectral characteristics of *F* region ionospheric irregularities, *J. Geophys. Res.*, *79*(10), 1497–1502.

Fejer, B. G., L. Scherliess, and E. R. de Paula (1999), Effects of the vertical plasma drift velocity on the generation and evolution of equatorial spread *F*, *J. Geophys. Res.*, *104*(A9), 19,859–19,869.

Franke, S. (1987), The space-time intensity correlation function of scintillation due to a deep random phase screen, *Radio Sci.*, *22* (4), 643–654.

Hassam, A. B., W. Hall, J. D. Huba, and M. J. Keskinen (1986), Spectral characteristics of interchange turbulence, *J. Geophys. Res.*, *91*(A12), 13,513–13,522.

Heelis, R. A., R. Stoneback, G. D. Earle, R. A. Haaser, and M. A. Abdu (2010), Medium-scale equatorial plasma irregularities observed by Coupled Ion-Neutral Dynamics Investigation sensors aboard the Communication Navigation Outage Forecast System in a prolonged solar minimum, *J. Geophys. Res.*, *115*, A10321, doi:10.1029/2010JA015596.

Huba, J. D., and G. Joyce (2010), Global modeling of equatorial plasma bubbles, *Geophys. Res. Lett.*, *37*, L17104, doi:10.1029/2010GL044281.

Huba, J. D., A. B. Hassam, I. B. Schwartz, and M. J. Keskinen (1985), Ionospheric turbulence: Interchange instabilities and chaotic fluid behavior, *Geophys. Res. Lett.*, *12*(1), 65–68.

Hudson, M. K., and C. F. Kennel (1975), The electromagnetic interchange mode in a partly-ionized collisional plasma, *J. Plasma Phys.*, *14*(1), 121–134.

Hysell, D. L. (2000), An overview and synthesis of plasma irregularities in equatorial spread *F*, *J. Atmos. Sol. Terr. Phys.*, *62*, 1037–1056.

Hysell, D. L., R. B. Hedden, J. L. Chau, F. R. Galindo, P. A. Roddy, and R. F. Pfaff (2009), Comparing *F* region ionospheric irregularity observations from C/NOFS and Jicamarca, *Geophys. Res. Lett.*, *36*, L00C01, doi:10.1029/2009GL038983.

Kelley, M. C. (2009), *The Earth's Ionosphere: Plasma Physics and Electrodynamics*, pp. 134–138, Academic, San Diego, Calif.

Kelley, M. C., G. Haerendel, H. Kappler, A. Valenzuela, B. B. Balsley, D. A. Carter, W. L. Ecklund, C. W. Carlson, B. Häusler, and R. Torbert (1976), Evidence for a Rayleigh-Taylor type instability and upwelling of depleted density regions during equatorial spread *F*, *Geophys. Res. Lett.*, *3*(8), 448–450.

Keskinen, M. J., and S. L. Vadas (2009), Three-dimensional nonlinear evolution of equatorial ionospheric bubbles with gravity wave seeding and tidal wind effects, *Geophys. Res. Lett.*, *36*, L12102, doi:10.1029/2009GL037892.

Keskinen, M. J., S. L. Ossakow, and P. K. Chaturvedi (1980), Preliminary report of numerical simulations of intermediate wavelength collisional Rayleigh-Taylor instability in equatorial spread *F*, *J. Geophys. Res.*, *85*(A4), 1775–1778.

Keskinen, M. J., S. L. Ossakow, and B. G. Fejer (2003), Three-dimensional nonlinear evolution of equatorial ionospheric spread-*F* bubbles, *Geophys. Res. Lett.*, *30*(16), 1855, doi:10.1029/2003GL017418.

Kil, H., and R. A. Heelis (1998), Global distribution of density irregularities in the equatorial ionosphere, *J. Geophys. Res.*, *103*, 407–417.

Koons, H. C., J. L. Roeder, and P. Rodriguez (1997), Plasma waves observed inside plasma bubbles in the equatorial *F* region, *J. Geophys. Res.*, *102*(A3), 4577–4583.

Krall, J., J. D. Huba, S. L. Ossakow, and G. Joyce (2010), Why do equatorial ionospheric bubbles stop rising?, *Geophys. Res. Lett.*, *37*, L09105, doi:10.1029/2010GL043128.

LaBelle, J., M. C. Kelley, and C. E. Seyler (1986), An analysis of the role of drift waves in equatorial spread *F*, *J. Geophys. Res.*, *91*, 5513–5525.

Lorenz, E. N. (1963), Deterministic nonperiodic flow, *J. Atmos. Sci.*, *20*, 130–141.

Ossakow, S. L. (1981), Spread-*F* theories—A review, *J. Atmos. Terr. Phys.*, *43*, 437–452.

Ott, E. (1978), Theory of Rayleigh-Taylor bubbles in the equatorial ionosphere, *J. Geophys. Res.*, *83*(A5), 2066–2070.

Park, J., et al. (2009), The characteristics of field-aligned currents associated with equatorial plasma bubbles as observed by the CHAMP satellite, *Ann. Geophys.*, *27*, 2685–2697.

Pottelette, R., M. Malingre, J. J. Berthelier, E. Seran, and M. Parrot (2007), Filamentary Alfvénic structures excited at the edges of equatorial plasma bubbles, *Ann. Geophys.*, *25*, 2159–2165.

Retterer, J. M. (2010), Forecasting low-latitude radio scintillation with 3-D ionospheric plume models: 2. Scintillation calculation, *J. Geophys. Res.*, *115*, A03307, doi:10.1029/2008JA013840.

Rino, C. L., R. T. Tsunoda, J. Petriceks, R. C. Livingston, M. C. Kelley, and K. D. Baker (1981), Simultaneous rocket-borne beacon and in situ measurements of equatorial spread *F*—Intermediate wavelength results, *J. Geophys. Res.*, *86*(A4), 2411–2420.

Rodrigues, F. S., M. C. Kelley, P. A. Roddy, D. E. Hunton, R. F. Pfaff, O. de La Beaujardière, and G. S. Bust (2009), C/NOFS observations of intermediate and transitional scale-size equatorial spread *F* irregularities, *Geophys. Res. Lett.*, *36*, L00C05, doi:10.1029/2009GL038905.

Sekar, R., and M. C. Kelley (1998), On the combined effects of vertical shear and zonal electric field patterns on nonlinear equatorial spread *F* evolution, *J. Geophys. Res.*, *103*(A9), 20,735–20,747.

Sekar, R., and R. Raghavarao (1987), Role of vertical winds on the Rayleigh-Taylor instabilities of the nighttime equatorial ionosphere, *J. Atmos. Terr. Phys.*, *49*, 981–985.

Shkarofsky, I. P. (1968), Turbulence functions useful for probes (space-time correlation) and for scattering (wave number-frequency spectrum) analysis, *Can. J. Phys.*, *46*, 2683–2702.

Shume, E. B., and D. L. Hysell (2004), Spectral analysis of plasma drift measurements from the AE-E satellite: Evidence of an inertial subrange in equatorial spread *F*, *J. Atmos. Sol. Terr. Phys.*, *66*, 57–65.

Sinha, H. S. S., S. Raizada, and R. N. Misra (1999), First simultaneous in situ measurement of electron density and electric field fluctuations during spread *F* in the Indian Zone, *Geophys. Res. Lett.*, *26*(12), 1669–1672.

Spatz, D. E., S. J. Franke, and K. C. Yeh (1988), Analysis and interpretation of spaced receiver scintillation data recorded at an equatorial station, *Radio Sci.*, *23*(3), 347–361.

Stolle, C., H. Lühr, M. Rother, and G. Balasis (2006), Magnetic signatures of equatorial spread *F* as observed by the CHAMP satellite, *J. Geophys. Res.*, *111*, A02304, doi:10.1029/2005JA011184.

Su, S.-Y., H. C. Yeh, and R. A. Heelis (2001), ROCSAT 1 ionospheric plasma and electrodynamics instrument observations of equatorial spread *F*: An early transitional scale result, *J. Geophys. Res.*, *106*(A12), 29,153–29,159.

Sultan, P. J. (1996), Linear theory and modeling of the Rayleigh-Taylor instability leading to the occurrence of equatorial spread *F*, *J. Geophys. Res.*, *101*(A12), 26,875–26,891.

Tsunoda, R. T. (1985), Control of the seasonal and longitudinal occurrence of equatorial scintillations by the longitudinal gradient in integrated *E* region Pedersen conductivity, *J. Geophys. Res.*, *90*(A1), 447–456.

Tsunoda, R. T. (2006), Day-to-day variability in equatorial spread *F*: Is there some physics missing?, *Geophys. Res. Lett.*, *33*, L16106, doi:10.1029/2006GL025956.

Tsunoda, R. T., R. C. Livingston, J. P. McClure, and W. B. Hanson (1982), Equatorial plasma bubbles: Vertically elongated wedges from the bottomside *F* layer, *J. Geophys. Res.*, *87*(A11), 9171–9180.

Valladares, C. E., W. B. Hanson, J. P. McClure, and B. L. Cragin (1983), Bottomside sinusoidal irregularities in the equatorial *F* region, *J. Geophys. Res.*, *88*(A10), 8025–8042.

Valladares, C. E., J. Villalobos, R. Sheehan, and M. P. Hagan (2004), Latitudinal extension of low-latitude scintillations measured with a network of GPS receivers, *Ann. Geophys.*, *22*, 3155–3175.

Wernik, A. W., and K. C. Yeh (1994), Chaotic behavior of ionospheric scintillation: Modeling and observations, *Radio Sci.*, *29*(1), 135–144.

Wernik, A. W., C. H. Liu, and K. C. Yeh (1983), Modeling of spaced-receiver scintillation measurements, *Radio Sci.*, *18*(5), 743–764.

Woodman, R. F. (2009), Spread *F*—An old equatorial aeronomy problem finally resolved?, *Ann. Geophys.*, *27*, 1915–1934.

Woodman, R. F., and C. LaHoz (1976), Radar observations of *F*-region equatorial irregularities, *J. Geophys. Res.*, *81*(31), 5447–5466.

Yeh, K. C., and C. H. Liu (1982), Radio wave scintillations in the ionosphere, *Proc. IEEE*, *70*(4), 324–360.

Zalesak, S. T., and S. L. Ossakow (1980), Nonlinear equatorial spread *F*: Spatially large bubbles resulting from large horizontal scale initial perturbations, *J. Geophys. Res.*, *85*, 2131–2142.

Zargham, S., and C. E. Seyler (1987), Collisional interchange instability 1. Numerical simulations of intermediate-scale irregularities, *J. Geophys. Res.*, *92*(A9), 10,073–10,088.

Zargham, S., and C. E. Seyler (1989), Collisional and inertial dynamics of the ionospheric interchange instability, *J. Geophys. Res.*, *94*(A7), 9009–9027.

A. Bhattacharyya, Indian Institute of Geomagnetism, New Panvel, Navi Mumbai 410 218, India. (abh@iigs.iigm.res.in)

Complex Analysis of Polar Auroras for 1996

James Wanliss and Joshua Peterson

Department of Physics and Computer Science, Presbyterian College, Clinton, South Carolina, USA

Geomagnetic and auroral fluctuations feature bursts of activity on multiple spatial and temporal scales. Statistics of these activity bursts, at both high- and low-latitudes, show strong scaling properties that indicate the magnetosphere tends to operate in a stable critical state. Recent work shows how this complex behavior is ubiquitous throughout the magnetosphere and possesses power law intermittency statistics implying global and/or local self-organized critical dynamics. Here we provide a detailed description of an analysis technique to study multiscale correlations of high-resolution ultraviolet images of the nighttime sector of the northern aurora from the Ultraviolet Imager onboard the Polar spacecraft. Each image is considered as a fractal surface, whose roughness varies and is reorganized during intense magnetospheric activity. We test to see how the fractal roughness varies as a function of other measures of activity, such as the geomagnetic indices *SYM-H* and *AE*.

1. INTRODUCTION

Scale-invariant dynamics have been observed in numerous physical systems, the magnetosphere included. Various attempts have been made to describe such driven and disordered systems in terms of the theory of nonequilibrium critical phenomena. In these theories, in the vicinity of the critical point, the system tends to produce long-range scale-free correlations over many spatial and/or temporal scales [*Bunde and Havlin*, 1996; *Takayasu*, 1990; *Turcotte*, 1997].

Statistical studies of high-latitude magnetospheric disturbances [*Consolini*, 2002; *Uritsky et al.*, 2003, 2006] concentrated on various measures of activity bursts in local and global geomagnetic variations as well as on spatiotemporal development of auroral emissions. These analyses obtained robust power law statistical relations consistent with the dynamics of nonequilibrium systems undergoing transitions between multiple metastable configurations.

Extreme Events and Natural Hazards: The Complexity Perspective
Geophysical Monograph Series 196

10.1029/2011GM001083

Since the physics of the inner magnetosphere of the Earth is significantly different from other magnetospheric regions, it is not clear that the apparent critical behavior at high latitudes should have any similar response in magnetospheric regions that map closer to the equator. It was found, however, that similar scaling laws and relations, but with a different set of power law exponents, govern bursty behavior of low-latitude and midlatitude geomagnetic fluctuations [*Wanliss and Uritsky*, 2010].

Different methods have been described in the literature that explores scaling behavior of the nonequilibrium magnetosphere. In this chapter, we will present a transparent explanation of the methodology employed by *Uritsky et al.* [2006], elucidating limitations and strengths, particularly related to the gridding of data. We will use this approach to explore the development of scaling in auroral emissions, and test for the presence of fractal correlations, and whether symmetry breaking in these correlations is related to changes in ground-based indices such as *AE* and *SYM-H*.

AE represents high-latitude effects in that all its component magnetometer stations are in or near the auroral oval, a region of intense auroral activity caused primarily by precipitating protons and electrons. At low latitudes, closer to the Earth's equator, the plasma and electrodynamic coupling between the Earth and magnetosphere map along magnetic field lines to the ring current and near-Earth plasma sheet. *SYM-H* provides information related to those regions; like *AE*, it is a global geomagnetic index but is constructed from data at low-latitude magnetometer stations. *SYM-H* thus measures the horizontal magnetic field fluctuation near the geomagnetic equator and is partly reflective of perturbations in the ring current. It is operationally interchangeable with *Dst* [*Iyemori*, 1990; *Wanliss and Showalter*, 2006], the traditional low time resolution (1 h) index used to study inner magnetosphere energetics.

2. ANALYSIS METHODOLOGY

Structure functions provide an ideal tool to investigate the possible turbulent behavior at auroral altitudes, with the field to investigate being the brightness of the auroral emissions. Structure functions have traditionally been used in studies of plasma turbulence [*Yu et al.*, 2003]. Compared to Fourier and many other analysis techniques that are designed to work with even and regularly sampled data, structure function analysis makes no assumptions about the missing data. This makes it ideally suited to analyze auroral data, which are patchy with many data gaps (not uniform and continuous).

The structure function $C_p(x, r)$ of order p of a field $h(x)$ is given by

$$C_p(x,r) = \langle |h(x+r) - h(x)|^p \rangle,$$

where x is the position, r is a lag with respect to x, and the angled braces, $\langle \cdot \rangle$, indicate an ensemble average. The second-order structure function C_2 is obtained by setting $p = 2$ in the equation. Structure functions of high order (large values of p) provide information regarding the spatial distribution of the large fluctuations of the field, while the smaller fluctuations generally dominate the lower-order structure functions.

If increments of the field $h(x + r) - h(x)$ are assumed to be statistically homogeneous, $C_p(x, r)$ is only a function of the lag r and may be written as $C_p(r)$. In this case, the ergodic theorem indicates that one may use a spatial average to approximate the ensemble average in the equation. In our analysis, the field is the luminosity of auroral emissions.

When the field has homogeneous increments and is statistically isotropic, $C_p(r)$ is a function of the distance between the locations of the field values. For such a field, contours of $C_p(r)$ will be concentric circles, and thus, an average can be taken about annuli of constant r to obtain a more reliable approximation of $C_p(r)$. The independent variables of the structure function, x and r, are vectors in the situation we consider here.

For white noise [*Hida et al.*, 1993], the second-order structure function scales trivially in time as $C_2 \sim c$, a constant. For diffusive processes, like Brownian motion, the scaling obeys as $C_2 \sim r$, and for fractional Brownian motion (having white noise and Brownian motion as its subclasses), the scaling obeys $C_2 \sim r^\alpha$. This is a result found for a system in a steady critical state and for spatial scales significantly smaller than the overall system size. Here the alpha parameter is equivalent to twice the well-known Hurst scaling exponent, H.

Thus, given an auroral image, the basic statistical characteristics of activity surface structure are given by C_2, described above, which is essentially the same as its height-height correlation function [*Barabasi and Stanley*, 1995; *Uritsky et al.*, 2006].

We will define the height h of the surface above each point as the Polar Ultraviolet Imager (UVI) auroral brightness in the Lyman-Birge-Hopfield long spectral band. We will focus on the nighttime auroral sector within the ranges 2000–0400 magnetic local time (MLT) and 55–75 MLT.

3. DATA PREPARATION

As mentioned above, in this chapter, we exploit the data set of global auroral images obtained by UVI [*Torr et al.*, 1995]. The UVI instrument was launched aboard NASA's Polar spacecraft in February 1996 into a 9.0 R_E by 1.8 R_E orbit with an inclination of 86°. With an orbit period of roughly 18 h, the Polar spacecraft spends upward of 9 h during each orbit at altitudes great than 6.0 R_E over the northern auroral zone, making it an ideal platform for observing the aurora continuously over time scales much longer than the characteristic time scales of auroral variations.

The Polar satellite allows scientists to study magnetospheric activity in terms of the particle precipitation and subsequent auroral activity that is observed. The Polar UVI obtains wide-field images of the auroral regions continuously over long time scales and with high time resolution. During its nominal periods of operation, UVI provides upward of 9 h of continuous observations of the aurora out of its 18 h orbit. Polar UVI obtains global observations of the aurora in four FUV band passes: two channels in the Lyman-Birge-Hopfield band of N2 emissions, O I 1304 Å, and O I 1356 Å. Polar UVI has collected over 10 years of auroral observations since 1996, and the data set includes observations through the last solar maximum period. This makes UVI an ideal instrument for characterizing the behavior of the aurora during times when the variations in the magnetosphere-ionosphere system span several temporal and spatial scales. Polar UVI images allow us to characterize auroral forms with scale sizes ranging from a few tens of kilometers to synoptic scale features spanning the entire auroral oval.

For the proposed analysis of auroral scale sizes, both good spatial coverage and spatial resolution are required. Global auroral imaging from space is a continual compromise between

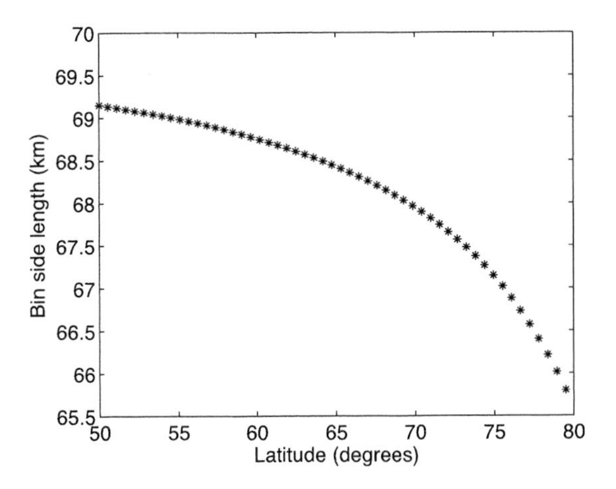

Figure 1. Box lengths (km) versus latitude.

these two factors: greater global coverage generally translates into coarser spatial resolution since a larger area is projected on a fixed number of pixels. The Polar UVI instrument strikes a good balance between spatial coverage and resolution. UVI's CCD detector comprises 200 × 228 pixels covering a circular 8° field of view. The spatial resolution is then 0.0363° × 0.0357° per pixel in each respective spatial dimension. When the equatorward edge of the auroral oval is near 60° magnetic latitude, good coverage of the auroral oval will be achieved for Polar altitudes greater than about 6 R_E. For the Polar orbit, this condition is satisfied for periods of roughly 9 h.

3.1. Exploring Different Data Binning Techniques

Each data point collected from the Polar satellite has a specific latitude and longitude associated with it. To compensate for the smearing effects caused by satellite wobble and also to ensure equal resolution at different satellite altitudes, images of the same region of the Earth were averaged together in roughly square bins with 70 km sides. Owing to the spherical nature of the Earth, it is impossible to produce nonoverlapping adjacent square bins with each side exactly equal to 70 km. We will discuss two different methods of binning the data, namely, the Peterson method and the Uritsky method. The detail is necessary since it is not apparent a priori that the method adopted by *Uritsky et al.* [2006] is appropriate. The Peterson methodology is described for the first time in this chapter.

For both the Peterson and Uritsky methods, a pseudo-square grid is drawn on the surface of the Earth. This 3-D grid is then transferred over to a 2-D plane. It is during this mapping to the 2-D plane where the two methods differ. To create the grid surrounding the Earth, the following method was used. Square bins were produced on the surface of the Earth with the edges of the bins parallel to the geographic latitude and longitude lines of the Earth. As a side note, magnetic coordinates could be used with only slight modifications to this methodology. A bin with the eastern and western edges having lengths of 70 km extends through a colatitude angle θ, given by the equation

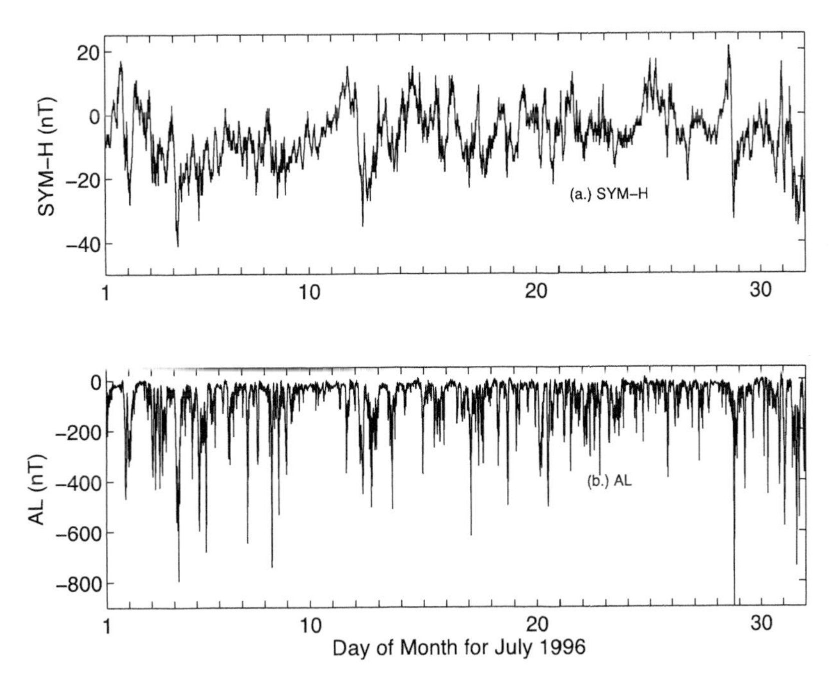

Figure 2. *SYM-H* and *AL* for July 1996.

$$\theta = \frac{70 \times 180}{\pi r}, \tag{1}$$

where r is the radius of the Earth given in kilometers. The colatitude angle is independent of the longitudinal position. A bin with the northern and southern edges having lengths of 70 km extends through a longitudinal angle ϕ, defined by the equation

$$\theta = \frac{70 \times 180}{\pi r \sin\left(\text{colat} \times \frac{\pi}{180}\right)}, \tag{2}$$

where colat is the colatitude position of the southern side of the bin. The longitudinal angle is dependent on the latitudinal location of the bin.

The lengths of the eastern and western edges of the each bin are 70 km as are the southern (nonpole) sides of each bin. The length of the northern side (pole side) of each bin is slightly smaller, ranging in distance from 69 km at a latitude of 50° to 66 km at a latitude of 80°. The lengths of each side of the all boxes are shown in Figure 1 versus the latitude of each box. Bin size is independent of longitudinal location of the bin.

In the Southern Hemisphere, the southern side of each bin (now the pole side) still has a length of 70 km, and the northern side (now the nonpole side) of each bin has a distance larger than 70 km, ranging from 71 km at latitudes of −50° to 74 km at latitudes of −80°.

3.2. Peterson Gridding

The corners of the bins described above will act as the corners for the bounding boxes in which the data will be course grained for a comparison of fractal scaling results from both the Peterson and Uritsky methods. In the Peterson method, the geographic Cartesian coordinate corner of each box is projected to $z = 0$ in the xy plane, producing a 2-D grid in which the data can be sorted. This is simply done by only using the x and y coordinates of each box. The system origin is at the center of the Earth. Since the geographical coordinates of each data point from the satellite are also known, determining the appropriate bin for each data point is trivial.

The Peterson method has several shortcomings. For instance, near the equator (latitudes below 20°), this method will not work as effectively because the orientation of the bins is nearly perpendicular to the xy plane; the projection of the boxes on the xy plane will have very little width. However, for the purposes of this study, we are primarily concerned with the polar region of the Earth; therefore, this complication will not affect our results, and it may make them more accurate than those used by *Uritsky et al.* [2006].

3.3. Uritsky Gridding

The *Uritsky et al.* [2006] method of binning the data uses the corners of the bounding boxes and does a 1-1 mapping to an xy plane where the North Pole is at the origin, the equator is at a radial distances of $r \sim 10{,}000$, and the South Pole is spread out over an entire circle at a radial distance of $r \sim 20{,}000$. The equations used for this mapping are

$$x = \frac{\pi r(90 - \text{lat})}{180} \sin\left(\frac{2\pi \text{LT}}{24}\right) \tag{3}$$

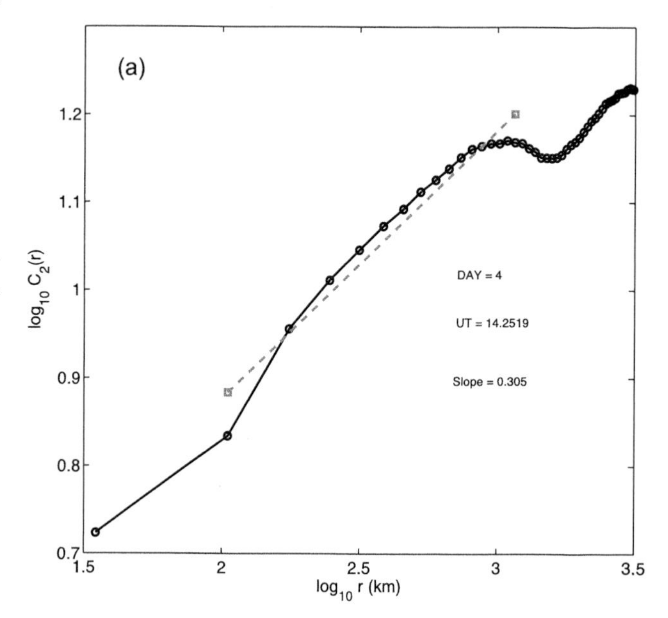

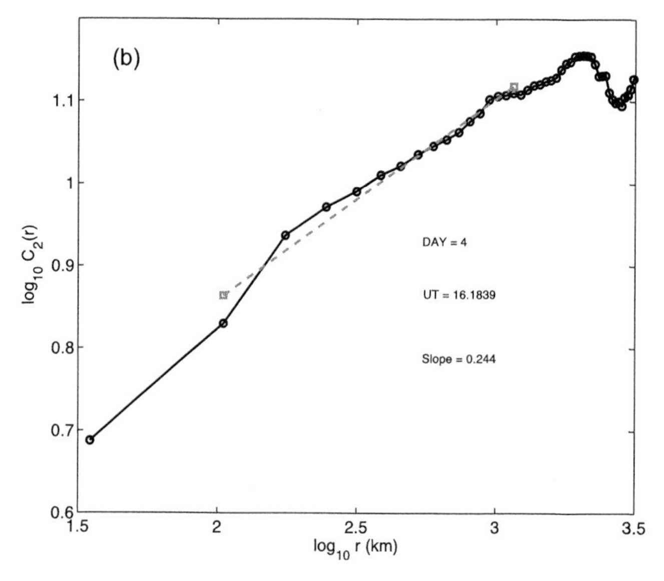

Figure 3. Representative samples of $C_2(r)$ for auroral activity surfaces.

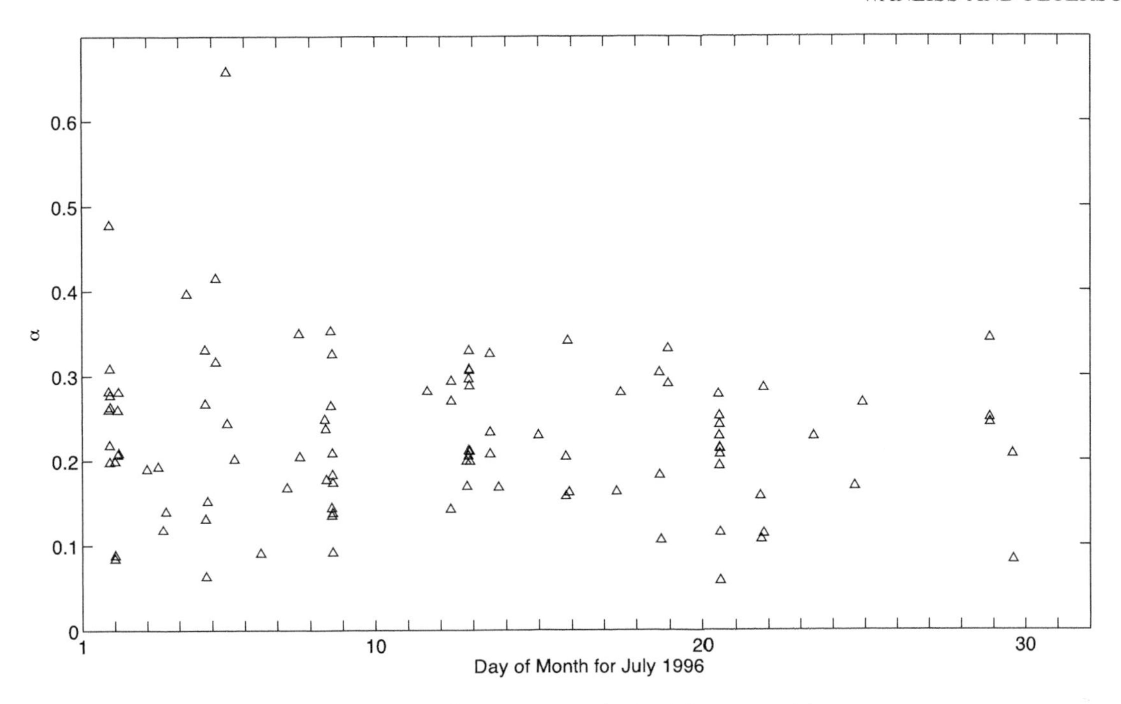

Figure 4. Fractal scaling exponent calculated from auroral images.

$$y = \frac{\pi r(90 - \text{lat})}{180}\cos\left(\frac{2\pi \text{LT}}{24}\right), \quad (4)$$

where LT is the local time of the point and lat is the latitude of the point in degrees. It should be mentioned that the x-y points are not geographic (or any other physical) coordinates.

To bin the data, the Uritsky method performs the transformations defined in equations (3) and (4) to every bounding box corner, in addition to each data point from the satellite. Because all points of interest must be mapped to a new space using the Uritsky method, significantly more calculations (~40,000) are performed for each image from Polar UVI.

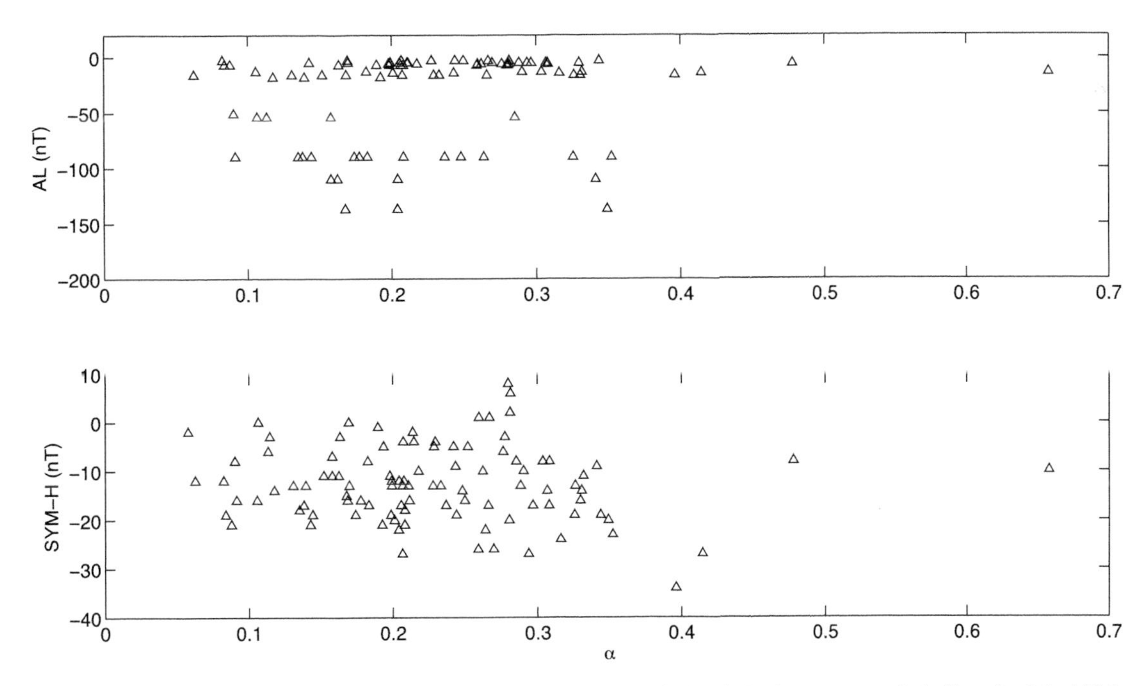

Figure 5. Comparison of scaling exponents against midlatitude and high-latitude geomagnetic indices for July 1996.

3.4. Grid Comparison

We tested the two methods against each other as follows. A sample set of data points consists of evenly spaced points every 0.05° in both latitude and longitude, over a range of latitudes from 50° to 80° and a range of longitudes from 120° to 240°. The sample data are passed to the two different grid algorithms. There were 205 differences out of 1,440,000, a 0.014% discrepancy between the two methods.

4. RESULTS

We will present the results of our analysis of 1 month of Polar UVI data for July 1996. Figure 2 shows the *AL* and *SYM-H* data for this interval. The *AL* data shows the presence of multiple substorms during the month. Several of these are clearly associated with activity at lower latitudes, as measured by *SYM-H*. For instance, substorms on 4 July and 28 July are clearly associated with sharp drops in *SYM-H*, which can be described as small space storms [*Wanliss and Showalter*, 2006].

Figure 3 shows several representative examples of $C_2(r)$ versus correlation length, r. Here the field $h(r)$ used is the luminosity of auroral emissions measured by Polar. To calculate roughness exponent values for the auroral activity surfaces, we have computed least squares linear regression slopes of $C_2(r)$ against r, plotted in double logarithmic coordinates. The accuracy of estimation is determined by an adaptive algorithm, which finds the largest ratio (r_{max}/r_{min}) that had a correlation coefficient over 0.99. If none of the possible combination of points had a correlation coefficient over 0.99, then the correlation coefficient cutoff was lowered to 0.98, and the largest ratio (r_{max}/r_{min}) was calculated again. As before, if none of the possible combination of points had a correlation coefficient over 0.98, then the correlation coefficient cutoff was lowered to 0.97 and the largest ratio (r_{max}/r_{min}) was again calculated. This process was continued for correlation coefficients given by 0.96, 0.95, 0.925, and 0.90. When none of the data had a correlation coefficient over 0.90, then the slope was not recorded. We also eliminate all calculated scaling exponents for which the uncertainty is larger than the value of the exponent.

Figure 4 shows the value of the remaining experimentally calculated scaling exponents, given as a function of time for the month of July 1996. In each case, as before, we calculate the scaling exponent from data in the nighttime auroral sector within the ranges 2000–0400 and 55–75 MLT. There is no meaningful correlation between these scaling exponents, and the ground-based data shown in Figure 2. Figure 5 compares the scaling exponents with the ground-based indices. The correlation coefficients for *SYM-H* and *AL* are 0.109 and 0.151, respectively.

5. CONCLUSIONS

We have explored the relationship between time-dependent fractal measures of auroral activity and ground-based measures of geomagnetic activity. Two different gridding methods were applied, but the results were found to be independent of method. Previous work suggested that measures such as these we have calculated can be used as sensitive indicators of early stages of the development of geomagnetic disturbances providing important auxiliary information on phases of the nonlinear magnetospheric response to the solar wind driver. Here, however, we have found no correlation between the auroral scaling response and geomagnetic activity.

REFERENCES

Barabasi, A.-L., and H. E. Stanley (1995), *Fractal Concepts in Surface Growth*, Cambridge Univ. Press, Cambridge, U. K.

Bunde, A., and S. Havlin (Eds.) (1996), *Fractals and Disordered Systems*, 408 pp., Springer, Berlin.

Consolini, G. (2002), Self-organized criticality: A new paradigm for the magnetotail dynamics, *Fractals*, *10*, 275–283, doi:10.1142/S0218348X02001397.

Hida, T., H.-H. Kuo, J. Potthoff, and L. Streit (1993), *White Noise: An Infinite Dimensional Calculus*, Kluwer Acad., Dordrecht, Netherlands.

Iyemori, T. (1990), Storm-time magnetospheric currents inferred from midlatitude geomagnetic field variations, *J. Geomagn. Geoelectr.*, *42*, 1249–1265.

Takayasu, H. (1990), *Fractals in the Physical Science*, 180 pp., Manchester Univ. Press, Manchester, U. K.

Torr, M. R., et al. (1995), A far ultraviolet imager for the International Solar-Terrestrial Physics Mission, *Space Sci. Rev.*, *71*, 329–383.

Turcotte, D. L. (1997), *Fractals and Chaos in Geology and Geophysics*, 398 pp., Cambridge Univ. Press, Cambridge, U. K.

Uritsky, V. M., A. J. Klimas, and D. Vassiliadis (2003), Evaluation of spreading critical exponents from the spatiotemporal evolution of emission regions in the nighttime aurora, *Geophys. Res. Lett.*, *30*(15), 1813, doi:10.1029/2002GL016556.

Uritsky, V. M., A. Klimas, and D. Vassiliadis (2006), Analysis and prediction of high-latitude geomagnetic disturbances based on a self-organized criticality framework, *Adv. Space Res.*, *37*, 539–546.

Wanliss, J. A., and K. M. Showalter (2006), High-resolution global storm index: *Dst* versus SYM-H, *J. Geophys. Res.*, *111*, A02202, doi:10.1029/2005JA011034.

Wanliss, J., and V. Uritsky (2010), Understanding bursty behavior in midlatitude geomagnetic activity, *J. Geophys. Res.*, *115*, A03215, doi:10.1029/2009JA014642.

Yu, Z.-G., V. V. Anh, and K. S. Lau (2003), Multifractal and correlation analyses of protein sequences from complete genomes, *Phys. Rev. E*, *68*(2), 021913-1, doi:10.1103/PhysRevE.68.021913.

J. Peterson and J. Wanliss, Department of Physics and Computer Science, Presbyterian College, 503 South Broad Street, Clinton, SC 29325, USA. (jawanliss@presby.edu)

On Self-Similar and Multifractal Models for the Scaling of Extreme Bursty Fluctuations in Space Plasmas

N. W. Watkins

British Antarctic Survey, Cambridge, UK

Centre for Fusion, Space and Astrophysics, University of Warwick, Coventry, UK

Centre for the Analysis of Time Series, London School of Economics and Political Science, London, UK

B. Hnat and S. C. Chapman

Centre for Fusion, Space and Astrophysics, University of Warwick, Coventry, UK

A direct inspiration for the investigation of scaling behavior and extreme fluctuations in space plasmas has come from inherently multiscale physical theories such as self-organized criticality and turbulence. An additional benefit, with "space weather" implications, is an ability to assess the likelihood of an extreme fluctuation of a given size. If it is present, however, scaling behavior may not be captured by a single self-similarity exponent H but might instead require a multifractal spectrum of scaling exponents. We believe that it is, nonetheless, useful to assess how well simple monofractal models can capture the "stylized facts" of the scaling behavior of auroral indices and solar wind quantities and here illustrate it by studying the use of linear fractional stable motion (LFSM) as a model for solar wind and ionospheric time series, an example that can be taken as a prototype for other possible models. By postulating such a description, we can heuristically explore how the previously experimentally measured scaling exponents for quantities like superposed epoch averaged activity, or the probability distribution of the differenced time series, depend on the model's parameters. We can then also derive predicted scaling exponents for the exponents of more complicated measurements that have also been made, such as size and duration of bursts above a threshold or the survival probability of a burst. Comparison of these predictions with data is then used to assess the usefulness of LFSM as a toy model for space physics time series.

1. INTRODUCTION

In many areas of the geosciences, we are concerned with two related questions: "How large will the next large event be," and "how long will we have to wait until it happens"? Very broadly speaking, the first question appears at first more of a statistical one, while the second seems more obviously a dynamical one, but in any system, more detailed consideration tends to reveal statistical and dynamical aspects of both questions. Space plasma physics has been no exception to this, and indeed, the presence of both fluctuations on many scales and coupling between these scales is an ever-present aspect of all plasmas [*Sagdeev*, 1993].

Since the pioneering work of *Bargatze et al.* [1985], many dynamical models of the solar wind-magnetosphere interaction

Extreme Events and Natural Hazards: The Complexity Perspective
Geophysical Monograph Series 196

10.1029/2011GM001084

have been proposed, which explicitly treat the magnetosphere on substorm space and time scales as an input-output system. Many of these models use the auroral indices *AU*, *AL*, and *AE* as proxies. The corresponding frameworks include conditional probability [*Ukhorskiy et al.*, 2004], cusp catastrophe [*Lewis*, 1991], and phase transition [*Sitnov et al.*, 2001] models, neural networks, nonlinear filters, and many others (see, e.g., the review of *Vassiliadis* [2006]). The observed multiscale properties may be both directly induced by the complexity present in the solar wind and also generated by the internal processes in the magnetosphere.

Another series of investigations has focused on statistical signatures of the physics of the system. These were initially inspired by models for the physics of fluctuations in complex systems, particularly self-organized criticality (SOC) [*Jensen*, 1998; *Aschwanden*, 2011] and fluid turbulence [*Frisch*, 1995], and so initially were less focused on the nonautonomous nature of the considered system, in this case, the Earth's magnetosphere driven by the solar wind. Instead, their primary purpose was to describe, and provide a physical underpinning for, two other so-called "fractal" (H-self-similar) [see *Embrechts and Maejima*, 2002] aspects of the system, which had hitherto received little attention. These were the long-tailed nature of the probability distributions of the auroral indices, and the "$1/f$" behavior of their power spectral densities (PSDs). Long-tailed probability density functions (PDFs) have a much greater chance of an extreme fluctuation than the Gaussian (normal) distribution, while "$1/f$" power spectra result in long-ranged correlation in time and long-lived spontaneous trends even without a driver. As with many other areas of physical science and economics in which SOC or turbulence ideas have been applied [*Jensen*, 1998; *Mantegna and Stanley*, 2000; *Aschwanden*, 2011], such approaches have already provided as a by-product new measurements of the scaling of extreme fluctuations. An additional motivation is the "space weather" question of how large, and long-lived, a given geoeffective plasma fluctuation is likely to be.

In many of the above studies, a widely used approach has been to assume a one-to-one match between a measured quantity and an observable that appears in a given complex physical model. In the works of *Watkins* [2002] and *Watkins et al.* [2005], some of the present authors argued for the parallel consideration of another approach, in which one first tries to capture the "stylized facts" of fluctuating data such as the auroral index *AE*, and some solar wind time series, by means of deliberately oversimplified phenomenological models. Our advocacy was inspired by "econophysics" [e.g., *Mantegna and Stanley*, 2000; *Simonson et al.*, 2003], which was itself influenced by the introduction of stylized facts into economics by *Kaldor* [1961]. *Kaldor* [1961, p. 178], argued that theory construction should begin with a summary of the relevant facts but noted a problem equally familiar to geoscientists: "[the] . . . facts as recorded by statisticians, are always subject to numerous snags and qualifications, and for that reason are incapable of being summarized." *Kaldor* [1961, p. 178] proposed that theorists "should be free to start off with a stylised view of the facts i.e. concentrate on broad tendencies, ignoring individual detail."

It might be thought that renormalization group [RG] theory and the study of phase transitions and critical phenomena in equilibrium systems actually give a sufficient mathematical foundation for relevant and irrelevant aspects. We would certainly agree that for a system at a scale-free "fixed" point [e.g., *Cardy*, 1996; *Goldenfeld*, 1992], RG does allow one to distinguish between scaled fluctuating variables, which vanish on coarse graining ("irrelevant") and those which do not (see also the very clear discussion by *Sethna et al.* [2001]). However, we are less convinced that such arguments will carry over to all complex natural systems. Despite the most optimistic early claims surrounding SOC, such systems need not be critical in the above sense, and their fluctuations need not be an additive process. The limited success of RG when applied in turbulence (which is often modeled as a multiplicative cascade) could perhaps be seen as supporting this view. Instead, we see the development of appropriate aggregation methods and limit theorems for fluctuations in systems, which are highly correlated but are not necessarily scale-free (or even multifractal) as a very important frontier for complexity science, with obvious relevance to extremes.

In the stochastic framework discussed earlier, two natural choices to adopt as stylized facts [cf., *Watkins et al.*, 2005] would seem to be the presence of long tails in the PDFs of fluctuations and the singular ("$1/f$") behavior of their PSDs. Such tendencies are (at least approximately) observed in data, so potentially observable parameters corresponding to them could be the exponent β of the power spectral density $S(f) \sim f^{-\beta}$ and the power law tail exponent α estimated from the fluctuation PDF $p(x) \sim x^{-(1+\alpha)}$.

In its turn, β estimates the long-range memory parameter d. In a monofractal approach, one considers an H-self-similar model, linear fractional stable motion (LFSM). This has only two independent scaling exponents, α, d, and has a third parameter, H, the self-similarity exponent, which depends on the other two additively via

$$H(\alpha, d) = d + 1/\alpha. \tag{1}$$

In physics, one tends to view H as a function of d and α, representing Mandelbrot's "Joseph" and "Noah" effects, respectively. Alternatively, one could take the fundamental exponents as H and α and regard d as dependent, as is frequently done in the stochastics literature. This latter viewpoint sees d as in some sense measuring the "distance" between the magnitude of self-similarity measured by H (see equation (3) below) and that of stability measured by α. In the special case of fractional Brownian motion, i.e., $\alpha = 2$ (Gaussian) fluctuations, the spectral index is $\beta = 1 + 2H = 2(1 + d)$, and we retrieve the familiar Brownian motion value of $S(f) = f^{-2}$ by further taking $H = 1/2$.

If a system is only approximately H-self-similar though, or is multifractal, the above arbitrariness in parametrization disappears. The property of long-range dependence d then becomes independent of self-similarity, requiring, in fact, only asymptotic self-similarity, and can still be defined directly via the singular low-frequency behavior of the spectral density (again $S(f) = f^{-\beta}$) or via the exponent $\zeta(2)$ of the second-order structure function (see below). The identity $\beta = 1 + \zeta(2)$ holds even in non-Gaussian cases as long as the power spectral density itself still exists, as it does for multifractals [e.g., see *Rodriguez-Iturbe and Rinaldo*, 1997, pp. 238–241].

In this chapter, we continue to work with essentially the same stylized facts that we did in the *Watkins et al.* [2005] work, in order to examine the possible utility of a model parameterized by H and α. We will, however, see that because the system is not self-similar, some of the evidence originally interpreted as being for self-similarity needs to be reconsidered.

In writing the *Watkins et al.* [2005] article, we felt, and would still argue, that an attractive aspect for physicists of model building using stylized facts is that models can first be built by using apparently robust tendencies, e.g., the scaling property, and existing measurements, such as α and d. They can then be tested against time series and other data. If they indeed do prove robust, they can then be used as an "intermediate layer," which captures phenomenology well enough for more detailed and physics-inspired models to be tested against. This avoids the property of "overfitting" a model to replicate one given data set, sometimes known as "wiggle matching." Then, even when the models fail, the manner in which they do so is likely to be instructive.

The constructive and stimulating dialog, which such a process should give rise to, is exemplified by several recent interesting papers [*Rypdal and Rypdal*, 2010; *Moloney and Davidsen*, 2010, 2011]. These authors give evidence suggesting that the behavior of AE and solar wind quantities is *not* just self-similar but essentially multifractal, and they claim that a monofractal description could only be of limited use, at best a "rough approximation" in the words of *Moloney and Davidsen* [2010]. Other work [*Abel et al.*, 2007, 2009] has indicated that the degree of intermittency and multifractality in the solar wind-ionosphere coupling process may be a function of IMF clock angle.

These new studies raise an interesting question, which cuts across many of the disciplines and themes represented in this volume. If a system is initially thought to be (mono)fractal, but an increasing balance of evidence favors multifractality, are any aspects of the monofractal description, such as the (now approximate) self-similarity exponent H or the parameters d and α, still useful? Can some problems still be made partially tractable in a monofractal approximation? Similar issues are faced when interpreting the scaling seen in solar wind fluctuations exceeding imposed high thresholds, a problem recently studied by *Moloney and Davidsen* [2011]. Although originally suggested in the context of monofractal paradigms like SOC, the results are of continuing interest in a multifractal context.

Two aspects of this question seem particularly timely to us and are the subject of this short contribution: (1) multiplicity of diagnostics and (2) estimating burst distributions. For multiplicity of diagnostics, we pose the following questions: Which of the many different scaling exponents that have been calculated by various workers are actually returning different information, such as the exponents d and α in the case of LFSM? Which, however, such as H in the example of LFSM, are, in fact, dependent on one or more of the others? Estimating burst distributions raises the following: Can analytical scaling arguments based on a monofractal approximation still give a useful estimate of the probability distribution of bursts above thresholds in multifractal data such as AE? If not, can they be adapted to the more complicated case of a multifractal spectrum rather than a single exponent H, or will such efforts have to start completely afresh [cf. *Bartolozzi*, 2007; *Uritsky et al.*, 2010]?

Although our discussion of both the above questions is focused around solar wind and ionospheric plasmas, which motivated our initial interest in the problem, our conclusions should be of potential relevance across the themes of this volume. Examples of heavy tails and long-range dependence are abundant in the geosciences. Notable examples include earthquake magnitudes for the former [e.g., *Sornette*, 2004] and atmospheric temperature for the latter [e.g., *Franzke et al.*, 2012].

The main diagnostics we discuss are integrated activity bursts above thresholds, and the spreading exponents η and δ, which have been used to measure activity in space plasma time series. Questions about their scaling are relatively simple to pose (conceptually at least) if a model time series is monofractal. The answers to the questions are subtle, though less obvious for multifractals, and even less obvious still for natural series, which, after all, need not have been purely fractal or multifractal in the first place.

We first (section 2.1) very briefly recap the original evidence that suggested that some solar wind and ionospheric time series are approximately self-similar and α-stable. We go on (section 2.2) to summarize the divergent paths, which research in this area has taken, exemplified by two influential papers by *Consolini et al.* [1996] and *Consolini* [1997], which concentrated, respectively, on the multifractal and the heavy-tailed burst distribution properties of the *AE* index. Inspired by the latter property, we then (section 2.3) briefly describe a self-similar α-stable toy model, LFSM.

In section 3, we summarize the *AE* and solar wind data sets used. We then, in section 4, consider the immediate consequences, which follow simply from assuming *H*-self-similarity. In section 4.1, we explore the implications of the fact that a self-similar curve after first crossing a threshold scales with increasing time τ as τ^H. This property is used, for example, by *Kearney and Majumdar* [2005] to calculate the area of the burst defined by such a curve from its upward crossing time until its first return to the threshold. We examine the conjecture that an approximately self-similar time series such as *AE* might still be expected to have an effective spreading exponent η, which would be close to *H*. If true, this would explain the observation of such an exponent by *Uritsky et al.* [2001] and the value they found for it. In section 4.2, we go on to note that *H*-self-similarity would also account for the observed power law dependence of burst size on duration, although there is an interesting discrepancy between the exponent predicted by LFSM and that reported by *Uritsky et al.* [2001], suggesting limitations to the LFSM model.

We then move from the general consequences of self-similarity to those depending on the assumption of a particular form of self-similar model or on its applicability to a given set of data. In section 5.1, we show that a stable self-similar LFSM implies a power law distribution for first-passage time (FPT). It follows immediately that a "survival exponent" δ defined in the manner of *Uritsky et al.* [2001] must then also exist. However, the specific values of the FPT exponent and survival exponent both depend on the details of the underlying process (section 5.2). Information about the FPT may then be folded in with the scaling of burst size versus duration discussed in section 4.2 to estimate a burst size distribution (section 5.3). Section 6 then discusses in more detail the limitations of LFSM as a model, in particular with respect to the phenomenon of "volatility clustering," seen not only in finance but also in turbulence and by *Rypdal and Rypdal* [2010] in the *AE* index, for example. Section 7 offers some conclusions and considers some possible ways forward.

2. MOTIVATION FOR A SELF-SIMILAR, STABLE, DESCRIPTION

2.1. Two Tentative "Stylized Facts" of Space Plasma Fluctuations

The initial work of *Watkins et al.* [2005] was based on two stylized facts: (1) stylized fact I, "Space plasma fluctuations are approximately self-similar in time" and (2) stylized fact II, "Space data have an approximately stable tail PDF shape for its increments."

Dealing first with *self-similarity*, approximate "*H*-self-similarity" here means that the first differences Δx of a time series $x(t)$ taken at a scale $t - t_0$

$$\Delta x(t - t_0) = x(t) - x(t_0) \tag{2}$$

have the same statistical properties (i.e., are equal in distribution) as the process on a time scale dilated by c, to within a scaling factor c^H, i.e.,

$$\Delta x(c(t - t_0)) \stackrel{d}{=} c^H \Delta x(t - t_0). \tag{3}$$

The first data *interpreted* as showing approximate self-similarity in some solar wind quantities came with the pioneering power spectral studies of *Coleman* [1968], while for auroral indices, the corresponding first study was that of *Tsurutani et al.* [1990]. The relevant scaling exponent in both cases was the spectral exponent β of the empirical estimated power spectral density $S(f)$, where $S(f) \sim f^{-\beta}$. A review of solar wind power spectral studies is given by *Goldstein and Roberts* [1999].

Subsequently, work on the auroral indices by *Takalo* [1994], also interpreted in terms of self-similarity, used the second-order structure function S_2. This is defined by $S_2(\tau) = \langle \|x(t+\tau) - x(t)\|^2 \rangle$. A corresponding scaling exponent $\zeta(2)$ exists if $S_2(\tau)$ is found to take a power law form $S_2(\tau) \sim \tau^{\zeta(2)}$. Other methods that are frequently interpreted as diagnostics for *H*-self-similarity were also employed [e.g., *Price and Newman*, 2001], for example, the Hurst "*R/S*" method.

Unfortunately, with the benefit of hindsight, one can now see that most of the methods used in this period were measuring long-range dependence via *d*, and not in fact *H*. They would only have been measuring *H* if the fluctuations in the time series had been Gaussian [*Mandelbrot and Wallis*, 1969]. For further detailed discussions of this point, see, e.g., *Mercik et al.* [2003] and *Franzke et al.* [2012].

With hindsight, most of the evidence supporting this first stylized fact has, thus, not proved to be very robust, and moreover, the approximate monofractality seen in scaling collapse of PDFs of differences of the time series has been convincingly interpreted as being a byproduct of multifractality by *Chang et al.* [2006] and *Rypdal and Rypdal* [2010]. We will, nonetheless, assume it in what follows, in order to explore how well a monofractal model may yet serve in the specific problem of estimating burst statistics.

Turning now to consider *stability*, a little later than the initial spectral density work described above, it was realized that the fluctuation PDF of some space plasma fluctuation data was, in fact, non-Gaussian, but rather was approximately α-stable, and that the tail PDF shape resembled a power law [*Consolini and De Michelis*, 1998; *Hnat et al.*, 2002a]. This suggested that the tail exponent α might also be a meaningful parameter for modeling the data. Some workers [e.g., *Hnat et al.*, 2002a] were, in part, inspired by the use of exponentially truncated α-stable distributions in econophysics by *Mantegna and Stanley* [2000].

There is also a more *physical motivation*: In addition to the above two phenomenological considerations, it has been realized [*Krishnamurthy et al.*, 2000] that LFSM may have more direct physical motivation, via the activity map of extremal processes.

2.2. Burst Distributions and Multifractals

The early observations of heavy-tailed PDFs for space plasma fluctuations, and further evidence interpreted as self-similarity, inspired interest in physical mechanisms to explain them. One notable example was the paradigm of SOC, one motivation for which [*Takalo*, 1993; *Chapman et al.*, 1998; *Consolini*, 1997; *Uritsky and Pudovkin*, 1998] was the results from thresholding the *AE* time series. This method was used [*Takalo*, 1993; *Consolini*, 1997] to define "bursts" during which the signal exceeds a given threshold level. The distributions of size and duration of these bursts were approximately of power law form. The resulting exponents could thus be interpreted as the scaling exponents predicted by SOC models.

At about the same time, though, some workers were struck by experimental evidence that, rather than a single self-similarity exponent, space plasmas might show a continuously varying fractal behavior with time and scale, i.e., multifractality. Notable early examples are from the works of *Consolini et al.* [1996] and *Vörös et al.* [1998], where the multifractality was seen as arising from a turbulent cascade. Both monofractal descriptions, particularly those inspired by SOC, and multifractal descriptions have remained subjects of active investigation in space physics, with their reconciliation remaining an incompletely solved problem, taken up by *Rypdal and Rypdal* [2010]. In this chapter, we are mainly concerned with self-similar monofractals, but we will briefly discuss multifractals in section 6.

2.3. LFSM: A Self-Similar, Stable, Model

By 2005, it was clear to us that estimates obtained by different authors using various fractal measures appeared contradictory, a paradox that had also been noted by other workers (M. Sitnov, private communication, 2005). We felt that a model was needed, which could combine the "superdiffusive" property of observed heavy-tailed PDFs with the "subdiffusive" values (i.e., less than the value of 0.5 corresponding to Brownian motion) observed for H. This evidence led *Watkins et al.* [2005] to suggest the use of LFSM.

One way to understand LFSM is to consider a random walk, which asymptotically tends to LFSM [*Krishnamurthy et al.*, 2000]. If activity measured by a time series was x_0 at time t_0, then the probability that it is x at t is [*Krishnamurthy et al.*, 2000]

$$P(|x - x_0|, t - t_0) = (t - t_0)^{-H} \phi_\alpha\left(\frac{|x - x_0|}{(t - t_0)^H}\right), \qquad (4)$$

with a power law scaling function

$$\phi_\alpha(r) \propto r^{-\alpha-1} \qquad (5)$$

for $r >> 1$ and

$$\phi_\alpha(r) \propto r^0 \qquad (6)$$

for $r << 1$.

Two more familiar limiting cases of LFSM are ordinary Lévy motion with $H = 1/\alpha$ and fractional Brownian motion for which $\phi_2(r)$ goes to a Gaussian.

A first interesting consequence of LFSM is that the asymptotic behavior (with r) of the time-independent function ϕ_α is controlled by α, while H controls propagation of the activity as a function of time t. H does not need to be $1/\alpha$; i.e., the tail exponent of collapse function need not be uniquely determined by the self-similarity exponent because α-stability is not generally the same as self-similarity. An α-stable random variable is named after its property of preserving its PDFs shape under the addition of many variables. This is nicely illustrated, for example, in Figures 3.1 and 3.2 of *Mantegna and Stanley* [2000]. The most famous example is the Gaussian, corresponding to $\alpha = 2$, but each value of α between 2 and 0 has a corresponding PDF, sometimes known in the physics literature as a "Lévy" or "Lévy-stable" distribution. In consequence, the time series of LFSM can indeed have power law tails (with $\alpha < 2$) for its PDF, while the measured H value can still be less than 1/2.

The LFSM expression for $P(x,t)$ is said to be of scaling form; i.e., it has a power law self-similar prefactor times a scaling function ϕ_α. The function ϕ_α can only be found from collapse rather than through dimensional analysis [*Goldenfeld*, 1992]. It is in this sense that LFSM is a simple special case of "finite size scaling" (FSS) [*Cardy*, 1996; *Sornette*, 2004]. In the more general case of FSS, the exponent H in the denominator of argument of ϕ need not be same as the negative exponent of the prefactor (see, e.g., the discussion of *Wilmott et al.* [1995], page 74).

3. DATA SETS USED

The auroral indices used in this study were *AU*, *AL*, and *AE*, for the period 1 January 1978 to 31 December 1978 obtained from the World Data Centre, Chilton, at the UK Solar System Data Centre. They are the same index time series as were studied by *Hnat et al.* [2002b, 2003]. Figure 1 illustrates the three curves of *AU* and *AL*, and their difference, *AE*, for part of 10 April 1978, contained within the above interval. The magnetograms from which this interval's *AU* and *AL* indices were derived can be seen in Figure D.1 of the work of *Parks* [1991].

The solar wind quantity studied was ε, given by

$$\varepsilon = v\frac{B^2}{\mu_0}l_0^2\sin^4(\theta/2). \quad (7)$$

This was constructed from measurements of solar wind speed v and magnetic field **B** from 1 January 1995 to 31 December 1995 by the SWE [*Lepping et al.*, 1995] and MFI experiments [*Ogilvie et al.*, 1995] on NASA's WIND spacecraft and, thus, corresponds to the first year of the period studied by *Hnat et al.*, [2002b, 2003].

We follow *Hnat et al.* [2002b] by first differencing time series of the indices *AE, AU, AL*, and ε at intervals τ of 1, 2, 3... times the fundamental sampling period (1 min for the indices and 46 s for ε). For further details of the data set and preprocessing, see the work of *Hnat et al.* [2002b, 2005] and references therein.

4. GENERAL CONSEQUENCES OF ANY SELF-SIMILAR MODEL

Some consequences of self-similarity are quite general and do not depend on the specific type of self-similar model chosen. We conjecture that these would also be somewhat robust even in the event that a system is only approximately self-similar. They are thus candidates for investigation as possible ingredients for a "stylized fact"-based model. An interesting example is the behavior of activity bursts with time. Inspired by SOC models, these were defined as excursions above a fixed threshold L by *Takalo* [1993] and

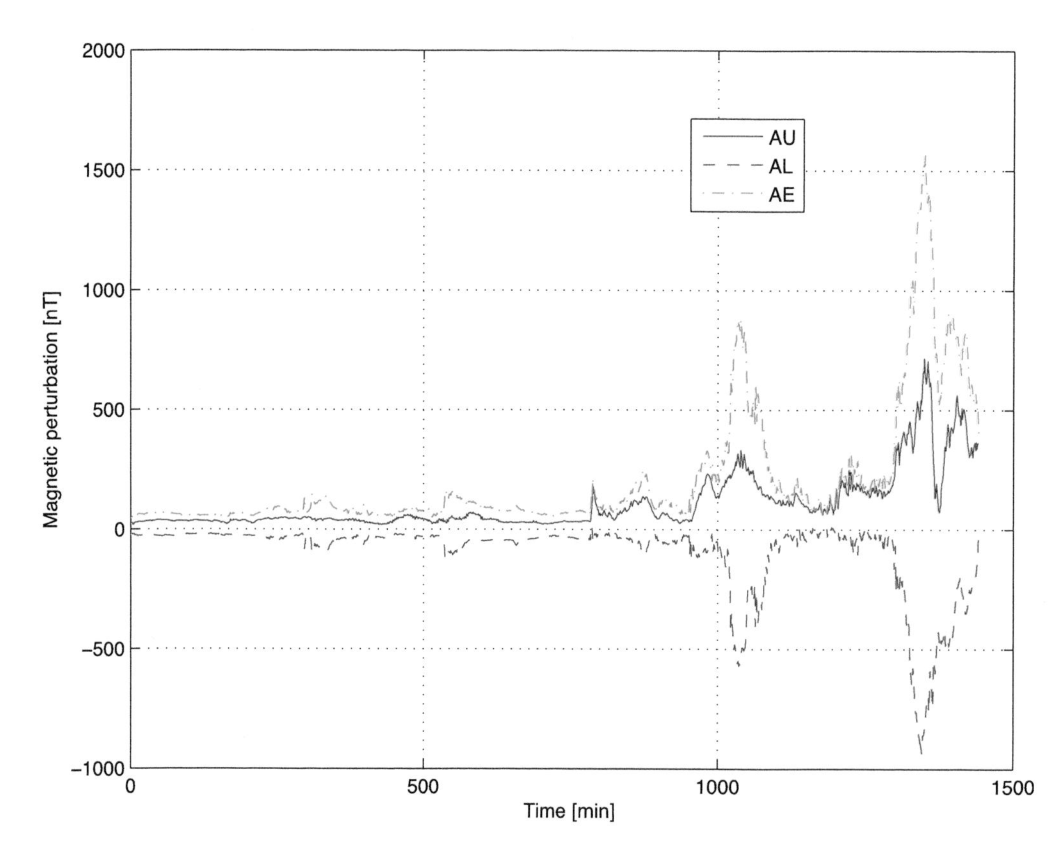

Figure 1. The *AE* (top trace, dash-dotted line) index and its constituent *AU* (middle trace, solid line) and *AL* (bottom trace, dashed line) indices for 10 April 1978.

Consolini [1997]. Their definition of burst size A_I is based on the isoset, i.e., the set of upcrossing and downcrossing times $\{t_i\}$ such that

$$A_I = \int_{t_i}^{t_{i+1}} (x(t') - L)\mathrm{d}t', \tag{8}$$

and is to be read in the same sense as the area A defined after equation (1) of the work of *Kearney and Majumdar* [2005]. It is estimated on data using a sum of the type used by *Carbone et al.* [2004] in their equation 7, but with a fixed threshold rather than the running one they use. A first-passage-based definition can only generate one burst for a given realization of a time series, while the number of bursts generated by an isoset is controlled by how often the curve exceeds the threshold.

In addition, *Uritsky et al.* [2001] introduced another new burst analysis method into space plasma physics, also from condensed matter theory, by using the two scaling exponents δ and η, known as "spreading exponents." When perturbed, critical systems, and other complex systems such as shell models of turbulence [*Mikkelsen*, 2000], show superposed epoch-averaged activity increasing with time as τ^{η} (taken as a signature of a scale-free growing "avalanche"). Conversely, event survival probability decays as $\tau^{-\delta}$.

The presence of these properties in activity across diverse SOC, nonequilibrium critical, and turbulent systems suggests to us that it may indeed be ubiquitous. However, it also seems important to us to also consider whether it may follow from the presence, in much natural data, of approximate H-self-similarity rather than a specific mechanism. We sketch some heuristic arguments for this below.

4.1. Conjecture 1

Conjecture 1 is as follows: In an H-self-similar time series, H acts as the dynamical exponent. The activity exponent η should thus be controlled only by H.

The activity exponent η was estimated by *Uritsky et al.* [2001] using a function N^*:

$$N^*(\tau) = \langle X(t_1 + \tau)\rangle_{t_1} - L, \tag{9}$$

where for any given τ, the brackets denote a superposed epoch average taken over all "uptimes" t_1 (at which X first exceeds a threshold L) of *surviving bursts*, i.e., those which have not returned to the threshold yet. We can rewrite this by bringing the constant L into the average and making the survival condition explicit by use of a Heaviside step function $\theta(x)$, which is 1 for positive argument x and 0 for negative argument:

$$N^*(\tau) \equiv \langle (X(t_1 + \tau) - L)[\theta(X(t_1 + \tau) - L)]\rangle_{t_1}. \tag{10}$$

To make an analytic estimate of how N^* should scale, for an arbitrary self-similar time series, provided only with the knowledge that a self-similarity exponent H exists, we first note that $L \approx X(t_1)$. This is less true in the case of very heavy-tailed series, where a crossing can substantially "overshoot" the threshold, but will be a reasonable approximation for AE. We can thus take

$$N^*(\tau) \approx \langle (X(t_1+\tau) - X(t_1))[\theta(X(t_1+\tau) - X(t_1))]\rangle_{t_1} = S_1^*(\tau), \tag{11}$$

where we have defined an auxiliary function S_1^*. It can be recognized as functionally similar to the first-order structure function S_1 but (1) estimated only at the points t_1 and, importantly, (2) for surviving (positive) bursts only. The normalization implicit in the bracket notation $\langle\ldots\rangle_{t_1}$ is thus not to all bursts but just surviving bursts.

For a symmetric series, S_1 does not scale, and restriction to upward-going bursts cannot change this. However, the further restriction to *surviving bursts* breaks the up-down symmetry, and so, S_1^* can behave more like the generalized variogram C_1 (where the absolute value is taken) than the (signed) structure function S_1. For a simple monofractal such as fractional Brownian motion (fBm), C_1 scales as τ^H, so we might expect a similar property to hold true in S_1^* and thus N^*.

This behavior is intriguingly close to what *Uritsky et al.* [2001] observed. Figure 2 of the work of *Uritsky et al.* [2001]

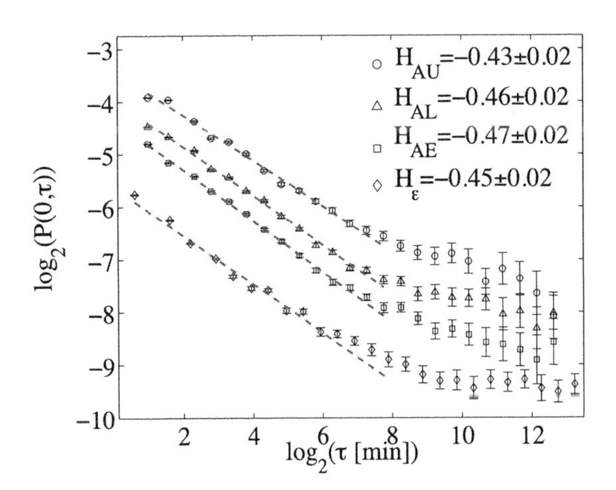

Figure 2. Estimation of a self-similarity exponent H via scaling of peaks $P(0)$ of probability density functions of differenced time series $X(t+\tau) - X(t)$ as a function of differencing interval τ. Plots are for (1) auroral indices: $X = AU$ (circle), $X = AL$ (triangle), and AE (square) and (2) solar wind: ε (diamond). Period covered is 1 January 1978 to 31 December 1978. Plots have been offset vertically for clarity.

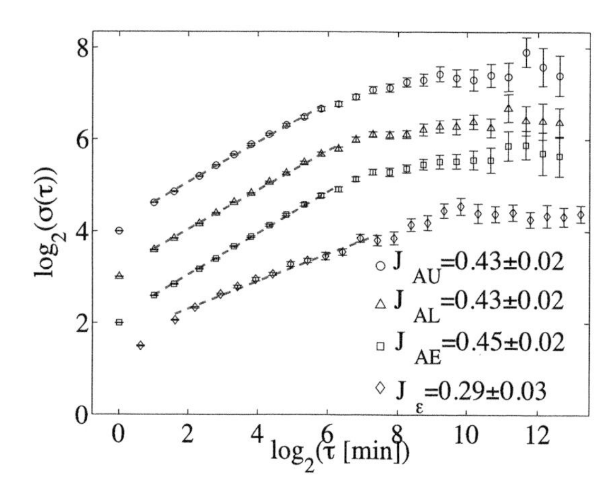

Figure 3. Estimation of Joseph exponent $J = d + 1/2$ for scaling of the standard deviation σ of the differenced series versus τ for the same quantities as Figure 2. Notation as in Figure 2.

shows N^* scaling like τ^η for small τ and then a scaling break at about τ = 130 min. One can, of course, make other estimates of H on the time series. One way to estimate H for a self-similar process is via the exponent returned from a log-log plot of peak scaling $P(0,\tau)$ versus τ. This is because it corresponds to setting $x = x_0$. Our Figure 2 shows how $P(0,\tau)$ behaves for solar wind ε and the *AE*, *AU*, and *AL* indices. The scaling range is small, but the range of inferred H values is 0.43 to 0.47 ± 0.02. The self-similarity exponent H and the 1-D spreading exponent η are seen to have essentially the same values if we accept that the measured H of 0.45 from our Figure 2 and their measured η of 0.41 are indistinguishable within plausible errors. Unlike published plots of C_1 and C_2 for *AE*, however [e.g., *Takalo and Timonen*, 1994], there is a falloff in the values of N^* for larger τ. This is a consequence of the average over surviving bursts in N_1, which is not present in C_1.

With hindsight, such a scaling seems to us to be at least plausible because H controls the propagation of the activity of a monofractal curve as a function of time t. This was a property used explicitly by *Kearney and Majumdar* [2005], for example. We find our argument above suggestive, and the simple simulations of Brownian motion and LFSM that we have performed so far (not shown) support it, but further detailed investigation would be needed to make it robust. It may now be an obsolete question because the clear evidence for volatility bunching in *AE* means such a study would be more usefully done with a multifractal model like that of *Rypdal and Rypdal* [2010]. We have presented the above arguments in the belief that they may have value to the more general question of extremes in scaling systems, some of which may still be better fitted by monofractal models.

As noted earlier, however, even when considering ideal self-similar, α-stable models, not all the measurements that are frequently referred to as "Hurst" exponents are actually of the self-similarity exponent H. This is because some methods are only sensitive to d rather than α, which also contributes to H [*Mandelbrot and Wallis*, 1969; *Mercik et al.*, 2003; *Watkins et al.*, 2005; *Franzke et al.*, 2012]. This effect is illustrated in Figure 3 for the same data as was used in our Figure 2. Instead of $P(0,\tau)$, we use a log-log plot of how the standard deviation of the four time series grow as a function of the differencing interval τ. In three cases, for the family of *AE* indices, one would estimate a similar value to that seen by $P(0,\tau)$, but the solar wind-derived series has a notably different and smaller value, around 1/3.

The fact that the scaling exponent measured by some diagnostics, such as the growth of standard deviation in Figure 3, or Mandelbrot's celebrated "*R*/*S*" method, is not the self-similarity exponent H but rather a Joseph exponent J, may come as a surprise. For example, if one's intuition was formed by considering equations (4) and (5) that describe LFSM, the appearance in these of H and α as the governing exponents might lead to a different expectation. Two particularly important cases of this effect are by now well documented in the literature.

The first case is for multifractal cascades [e.g., *Rodriguez-Iturbe and Rinaldo*, 1997] where the process retains finite moments but is no longer purely self-similar. The second-order moment has no privileged place relative to the other moments, but it is related to the power spectral density. Conversely, for LFSM, all moments of order α and above, including the variance, are infinite, but the process is still H-self-similar. In both cases, it is found that the empirical standard deviation grows like $J = d + 1/2$. In the multifractal case, d is directly related to the power spectrum, while for LFSM, the behavior is understood as a property of finite data series [cf. *Mandelbrot and Wallis*, 1969; *Mercik et al.*, 2003; *Watkins et al.*, 2005; *Rypdal and Rypdal*, 2010].

The chosen range of the fit will significantly alter the estimate of the exponent and increase the real error bars. In addition, if present, multifractality will make scaling exponents sensitive to bin size (see discussions of *Parkinson* [2006] and *Moloney and Davidsen* [2010]).

4.2. Conjecture 2

Conjecture is as follows: In an H-self-similar time series, burst area will scale with duration.

We can also invoke self-similarity to partially explain another interesting property remarked on by *Uritsky et al.* [2001] that the area of the bursts in auroral indices was found to scale with their duration. In this, and section 5, we follow the work of *Watkins et al.* [2009], who adapted the proofs of *Kearney and Majumdar* [2005] and *Carbone et al.* [2004]. These papers should be consulted for more detail.

Kearney and Majumdar [2005] considered zero-drift Wiener Brownian motion and were concerned with the time t_f at which a curve $x(t)$ returned to a threshold after it was first exceeded (by an infinitesimal amount) by a curve at time t_i. They first noted that $x(t) \sim t^{1/2}$ for large t. This allows definition of a first-passage-based area measure A_{FP} by

$$A_{FP} = \int_{t_i}^{t_f} x(t')\mathrm{d}t' \qquad (12)$$

so the integration implies that large A_{FP} would scale as $t_f^{3/2}$.

Simple inversion of this expression implies that t_f must scale as

$$t_f {\sim} A_{FP}^{2/3}. \qquad (13)$$

For a walk that is non-Brownian but still H-self-similar, we can still argue, as we did in section 4.1, that $x(t) \sim t^H$ for large t, and the same arguments as given above then predict that

$$t_I \sim A_I^{-(1+H)}, \qquad (14)$$

where we are also [*Watkins et al.*, 2009] using burst area A_I defined by the time t_I after which a curve that has exceeded a fixed threshold returns to it.

This behavior can be studied by numerically simulating LFSM. In the fBm limit, we have confirmed that area depends on duration to the power $1 + H$ (see Figure 5 of *Watkins et al.* [2009]), a similar behavior to that seen by *Carbone et al.* [2004] using a variable time-varying threshold (see their Figure 3b). LFSM simulations do not quantitatively change this (plots not shown). For parameters that are similar, typical of the *AE* "family," i.e., H = 0.4 to 0.45, we would thus expect an exponent of about 1.4. The observed exponent dependence of burst size on burst duration found by *Uritsky et al.* [2001] for *AE* was actually ≈1.8, however. This again may reflect the multifractal nature of the series.

We plan to investigate this further and note the interesting complementary studies of bursts in turbulence simulations of *Uritsky et al.* [2010] as well as the earlier work on bursts in simulated multifractals of *Bartolozzi* [2007] and in magnetometer data.

5. SPECIFIC CONSEQUENCES OF A PARTICULAR CHOICE OF SELF-SIMILAR MODEL

Other consequences of self-similarity are less general; i.e., the choice of *which* self-similar model is taken becomes relevant.

5.1. Conjecture 3

Conjecture 3 is as follows: The use of a self-similar α-stable model implies a power law FPT distribution. If this assumption is valid, or the data are anyway found to have a power law FPT distribution, it then also follows immediately that a survival exponent defined by *Uritsky et al.* [2001] must exist.

When applied to a natural time series, the "survival" exponent δ for bursts defined above a threshold L has interesting properties. Notably, a power law form was found for the survival exponent by *Uritsky et al.* [2001].

However, rather than being unique to (and thus diagnostic for) a critical system per se, these properties can be derived exactly from the "optimal investment horizon" discussed by *Simonsen et al.* [2003] and thus from the familiar scaling properties of the FPT of a time series. To see this, first recall the definition of the FPT as the time taken by a discrete-time stochastic process to return to a given threshold for the first time. A discrete-time stochastic process is essentially a random vector with components indexed by time and that the time series observed in our application can be seen as one realization of this random vector. The PDF of this time $f(t)$ is found to be a power law

$$f(t) \sim \tau^{-(1+\delta)} \qquad (15)$$

for many self-similar processes of interest, including, for example, Brownian motion, where we have defined an exponent $-(1 + \delta)$.

The survival probability $P_s(\tau)$ can then be found directly from the definition given by *Uritsky et al.* [2001] as

$$P_s(\tau) = \int_t^{\infty} f(t')\mathrm{d}t' \sim t^{-\delta}, \qquad (16)$$

and so we see that for *any* process that is already known to have a power law FPT, the result of defining survival probability as in the work of *Uritsky et al.* [2001] will be a power law with exponent $-\delta$.

The assumption of a power law form for the FPT is consistent with what *Freeman et al.* [2000a, 2000b] found in earlier work. These papers found that the isosets of *AU* and *AL* scaled as $\tau^{-(1+H)}$ where in each case H was the same value (typically in the range 0.4 to 0.45) as that extracted from peak scaling.

5.2. Specific Values of FPT and Survival Exponents Depend on the Underlying Process

Granted that we observe a power law form for the FPT, however, we still have the question of whether the experimentally determined exponent agrees with the prediction of

LFSM or other self-similar model. As noted in section 5.1, the form for δ found in *AE* data implies an FPT exponent of $1 + H$, rather than the predicted FPT exponent of $2 - H$ for LFSM [*Krishnamurthy et al.*, 2000]. We have found [*Watkins et al.*, 2009] that burst durations, as well as FPT, scale as $2 - H$ for simulated LFSM (see the findings of *Carbone et al.* [2004] for fBm). Further work is in progress to employ more accurate error estimates such as boxplots and improving ensemble sizes. We accept that all such discussions need to have an estimate of the extent to which such differences are detectable factored into them. We see this as part of the broader discussion [e.g., *Clauset et al.*, 2009] of the accuracy of scaling estimates in real-world data that is currently occurring in complexity science.

On balance, it seems that the survival exponent does not *detect* critical behavior in any sense other than the approximate H-self-similarity already known to be present in *AE*; its value, nonetheless, seems to be sensitive to more than just H and so it seems likely that it usefully further constrains the choice of possible fractal models for the series. Interestingly, in the case of *AE*, $\delta \approx H$; i.e., the estimate δ is found to be 0.41, i.e., the same as η.

5.3. Folding in Information About the FPT With the Scaling of Burst Size Versus Duration to Give a Burst Size Distribution

This step can again be demonstrated for the Brownian motion case, by following the methods of *Kearney and Majumdar* [2005]. We again use the approach of section 4.1 but independently also know the standard result for FPT for Brownian motion:

$$P(t_f) \sim t_f^{-3/2}. \tag{17}$$

To get $P(t_f)$ as a function of A_{FP}, i.e., $P(t_f(A_{FP}))$, we need to insert the expression for t_f as a function of A_{FP} in the above equation and, in addition, need a Jacobian to give

$$P(A) \approx A^{-4/3}. \tag{18}$$

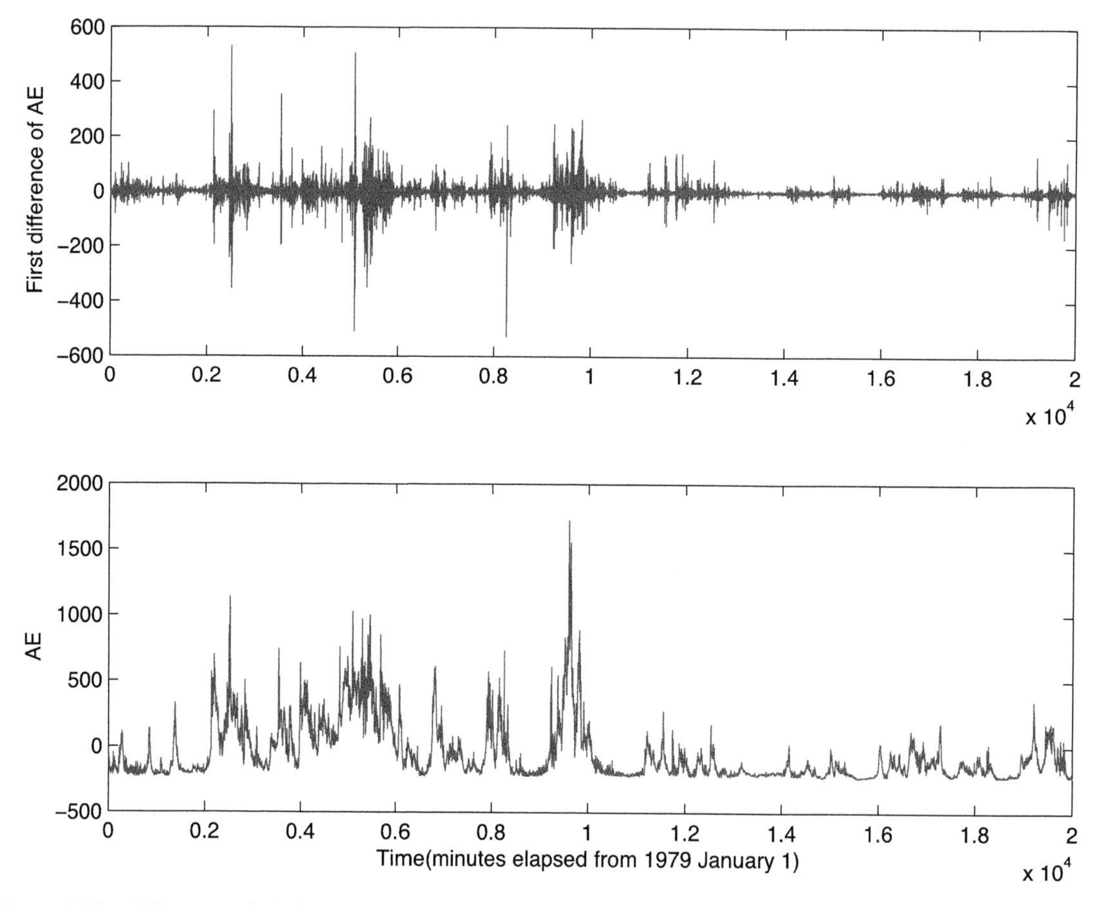

Figure 4. First differences of *AE* for 20,000 min from 1 January 1979 onward, illustrating the volatility clustering noted by *Consolini et al.* [1996].

A similar argument in the non-Brownian case gives

$$P(A) = A^{-2/(1+H)}, \tag{19}$$

which can be checked for the Brownian case, where $H = 1/2$, to retrieve $P(A) \sim A^{-4/3}$. Again, numerical simulations using LFSM show this scaling [*Watkins et al.*, 2009]. Typical estimates of burst size distribution exponents in *AE* are ~1.2 [cf. *Consolini*, 1997] and so closer to $2 - 2H$ than $-2/(1 + H)$, although the question deserves more detailed examination.

6. LIMITATIONS OF THE LFSM MODEL

LFSM was introduced into space physics by *Watkins et al.* [2005] mainly as a way of illustrating how an *H*-self-similar model could, nonetheless, present different scaling exponents when studied using different diagnostics. It has limitations as a faithful model of the complex dynamics of the *AE* family, some of which are discussed briefly below.

6.1. Total Amplitude of Series

The amplitude distribution of *AE* and *AL* has been shown to be approximately lognormal (more precisely bilognormal) [*Consolini and De Michelis*, 1998]. This suggests that a more appropriate model might involve a log transformation such as that employed by *Rypdal and Rypdal* [2010]. This would tend to reduce the need for the α-stable tails. An interesting example of the use of a log-transformed α-stable distribution in ionospheric physics was the fit to the spectral width of SuperDARN data performed by *Freeman and Chisham* [2004].

6.2. Sign

The illustrative simulations we have done of LFSM used a symmetric α-stable distribution and so took positive as well as negative values. *AE*, however, is a difference of *AU*, which is usually positive, and *AL*, which is usually negative, and so is nearly always positive. LFSM can mimic this property

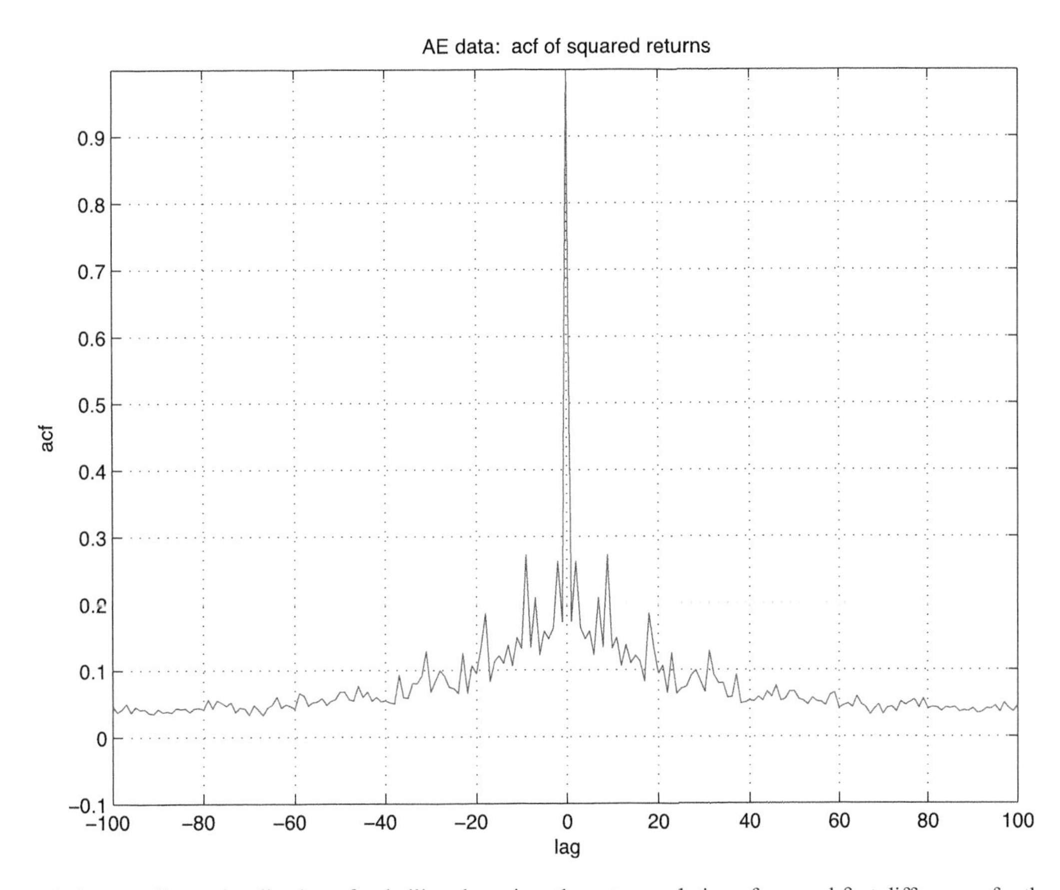

Figure 5. A more direct visualization of volatility clustering, the autocorrelation of squared first differences for the data presented in Figure 4, using the method of *Rypdal and Rypdal* [2010].

better than we have done so far, by using more skewed α-stable distributions or by using Pareto distributions instead, as *Mandelbrot and Wallis* [1969, Figure 9] did in their pioneering study of a "fractional hyperbolic" noise.

6.3. *Volatility*

A more profound limitation of LFSM as a model though is posed by a phenomenon, which has been noted in several natural and man-made time series, including some taken from fluid turbulence and ice cores. The first differences of *AE* seen in Figure 4 are "clumpy" as well as "bursty"; that is, the eye is showing that large jumps tend to be clustered together, as remarked on for *AE* by *Consolini et al.* [1996]. In finance, this is known as "volatility clustering," and it can be seen (Figures 5 and 6) by taking the autocorrelation of squared differences [cf. *Rypdal and Rypdal*, 2010] or absolute differences rather than that of differences, which for *AE* is essentially a delta function at high frequencies. Modeling this phenomenon is one of the key motivations for the multifractal model of *Rypdal and Rypdal* [2010].

7. CONCLUSIONS

The arguments above suggest that three better stylized facts for many solar wind and ionospheric quantities would be (1) the presence of heavy tails as a consequence of multifractality, (2) the appearance of long-range dependence also as a consequence of multifractality, and (3) the presence of volatility bunching manifested through correlations in the absolute values of differences.

Although the presence of volatility bunching, in particular, seems to be a fundamental limitation to complete success of a self-similar model in space plasmas, we have, nonetheless, shown how one may approach the study of some of the preexisting observations of burst size and duration using scaling arguments derived from monofractals. At least some of the scaling properties of extremes are qualitatively

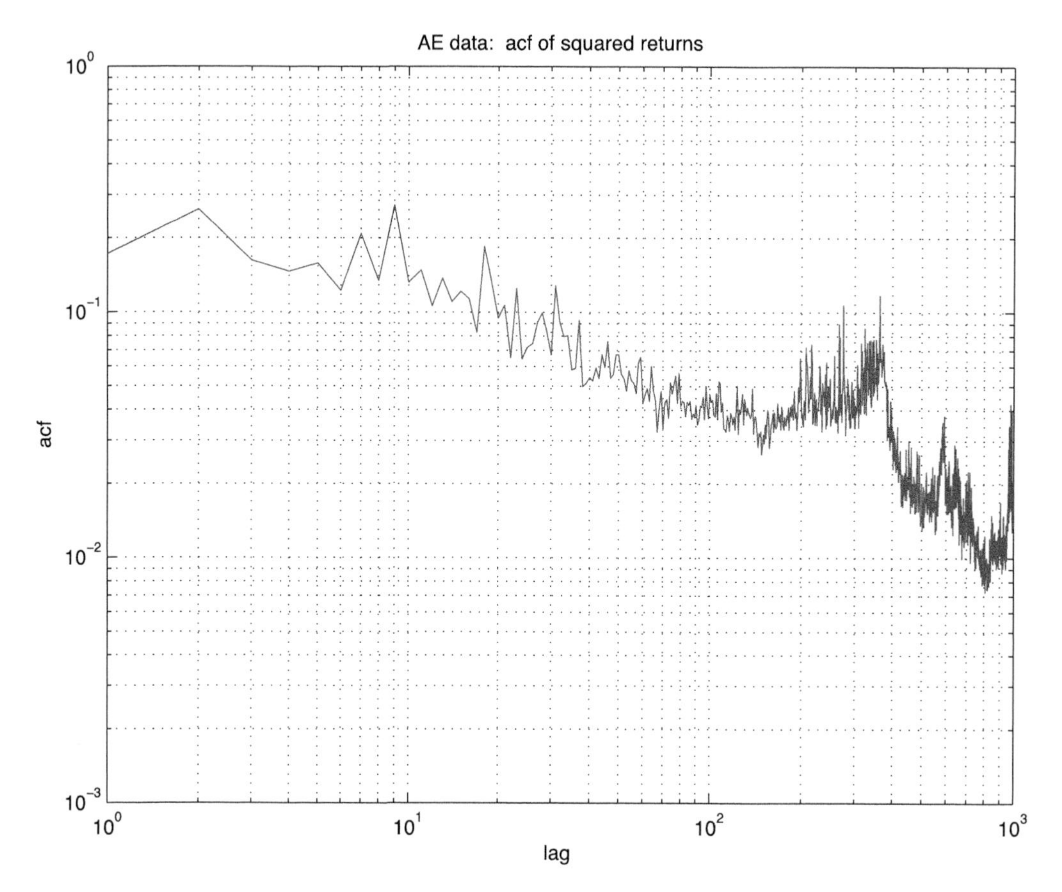

Figure 6. Again, the autocorrelation of squared first differences for the data presented in Figure 4 but on a log-log scale and with an additional decade. The feature at a few hundred minutes corresponds to substorm time scales.

reproduced, but quantitative comparison begins to make the need for better models clear (see also *Rypdal and Rypdal* [2010] and *Moloney and Davidsen* [2010], for example). We plan in the future to develop our preliminary arguments and to link our approach both to the contemporary understanding of first-passage processes in fractal noises and to that of extremes in multifractals, both themes discussed in the Hyderabad conference, in order to generate more realistic models of the wild fluctuations seen in space plasmas and elsewhere.

Acknowledgments. We thank the referees for valuable and constructive suggestions and Vadim Uritsky, Misha Sitnov, Mervyn Freeman, Nicola Longden, Tim Graves, Bobby Gramacy, Christian Franzke, Sam Rosenberg, Dan Credgington, Joern Davidsen, Martin and Kristoffer Rypdal, Tom Chang, Khurom Kiyani, John Greenhough, Mike Kearney, Satya Majumdar, and many others for their continuing interest and for valuable discussions over many years. It is a pleasure also to thank the AGU, the organizers, and the hosts of the 2010 Hyderabad Chapman Conference. S.C.C. and B.H. thank the UK EPSRC and STFC for support. The work at BAS is part of the British Antarctic Survey Polar Science for Planet Earth Programme and was funded by the UK NERC.

REFERENCES

Abel, G. A., M. P. Freeman, G. Chisham, and N. W. Watkins (2007), Investigating turbulent structure of ionospheric plasma velocity using the Halley SuperDARN radar, *Nonlinear Processes Geophys.*, *14*, 799–809.

Abel, G. A., M. P. Freeman, and G. Chisham (2009), IMF clock angle control of multifractality in ionospheric velocity fluctuations, *Geophys. Res. Lett.*, *36*, L19102, doi:10.1029/2009GL040336.

Aschwanden, M. (2011), *Self-Organized Criticality in Astrophysics: The Statistics of Nonlinear Processes in the Universe*, 416 pp., Springer, Berlin.

Bargatze, L. F., D. N. Baker, R. L. McPherron, and E. W. Hones Jr. (1985), Magnetospheric impulse response for many levels of geomagnetic activity, *J. Geophys. Res.*, *90*(A7), 6387–6394.

Bartolozzi, M., (2007), Scale-free avalanches in the multifractal random walk, *Eur. Phys. J. B*, *57*(3), 337–345, doi:10.1140/epjb/e2007-00178-3.

Carbone, A., G. Castelli, and H. E. Stanley (2004), Analysis of clusters formed by the moving average of a long-range correlated time series, *Phys. Rev. E*, *69*, 026105, doi:10.1103/PhysRevE.69.026105.

Cardy, J. (1996), *Scaling and Renormalization in Statistical Physics*, *Cambridge Lect. Notes Phys.*, vol. 5, 238 pp., Cambridge Univ. Press, Cambridge, U. K.

Chang, T. S., S. W. Y. Tam, and C. C. Wu (2006), Complexity in space plasmas – A brief review, *Space Sci. Rev.*, *122*, 281–291.

Chapman, S. C., N. W. Watkins, R. O. Dendy, P. Helander, and G. Rowlands (1998), A simple avalanche model as an analogue for magnetospheric activity, *Geophys. Res. Lett.*, *25*(13), 2397–2400.

Clauset, A., C. R. Shalizi, and M. E. J. Newman (2009), Power-law distributions in empirical data, *SIAM Rev.*, *51*, 661–703.

Coleman, P. J. (1968), Turbulence, viscosity and dissipation in the solar wind plasma, *Astrophys. J.*, *153*, 371–388.

Consolini, G. (1997), Sandpile cellular automata and magnetospheric dynamics, in *Proceedings of the 8th GIFCO Conference, Cosmic Physics in the Year 2000*, vol. 58, edited by S. Aiello et al., pp. 123–127, Soc. Ital. di Fis., Bologna, Italy.

Consolini, G., and P. De Michelis (1998), Non-Gaussian distribution function of *AE*-index fluctuations: Evidence for time intermittency, *Geophys. Res. Lett.*, *25*, 4087–4090.

Consolini, G., M. F. Marcucci, and M. Candidi (1996), Multifractal structure of auroral electrojet index data, *Phys. Rev. Lett.*, *76*, 4082–4085.

Embrechts, P., and M. Maejima (2002), *Selfsimilar Processes*, Princeton Univ. Press, Princeton, N. J.

Franzke, C. L. E., T. Graves, N. W. Watkins, R. B. Gramacy, and C. Hughes (2012), Robustness of estimators of long-range dependence and self-similarity under non-Gaussianity, *Philos. Trans. R. Soc. A*, *370*, 1250–1267.

Freeman, M. P., and G. Chisham (2004), On the probability distributions of SuperDARN Doppler spectral width measurements inside and outside the cusp, *Geophys. Res. Lett.*, *31*, L22802, doi:10.1029/2004GL020923.

Freeman, M. P., N. W. Watkins, and D. J. Riley (2000a), Evidence for a solar wind origin of the power law burst lifetime distribution of the *AE* indices, *Geophys. Res. Lett.*, *27*, 1087–1090.

Freeman, M. P., N. W. Watkins, and D. J. Riley (2000b), Power law distributions of burst duration and interburst interval in the solar wind: Turbulence or dissipative self-organized criticality?, *Phys. Rev. E.*, *62*, 8794–8797.

Frisch, U. (1995), *Turbulence: The Legacy of A. N. Kolmogorov*, 312 pp., Cambridge Univ. Press, Cambridge, U. K.

Goldenfeld, N. (1992), *Lectures on Phase Transitions and the Renormalization Group*, 394 pp., Perseus, Reading, Mass.

Goldstein, M., and D. A. Roberts (1999), Magnetohydrodynamic turbulence in the solar wind, *Phys. Plasmas*, *6*, 4154, doi:10.1063/1.873680.

Hnat, B., S. C. Chapman, G. Rowlands, N. W. Watkins, and W. M. Farrell (2002a), Finite size scaling in the solar wind magnetic field energy density as seen by WIND, *Geophys. Res. Lett.*, *29*(10), 1446, doi:10.1029/2001GL014587.

Hnat, B., S. C. Chapman, G. Rowlands, N. W. Watkins, and M. P. Freeman (2002b), Scaling of solar wind ε and the *AU*, *AL* and *AE* indices as seen by WIND, *Geophys. Res. Lett.*, *29*(22), 2078, doi:10.1029/2002GL016054.

Hnat, B., S. C. Chapman, G. Rowlands, N. W. Watkins, and M. P. Freeman (2003), Correction to "Scaling of solar wind ε and the *AU*, *AL* and *AE* indices as seen by WIND", *Geophys. Res. Lett.*, *30*(8), 1426, doi:10.1029/2003GL017194.

Hnat, B., S. C. Chapman, and G. Rowlands (2005), Scaling and a Fokker-Planck model for fluctuations in geomagnetic indices and comparison with solar wind as seen by Wind and ACE, *J. Geophys. Res.*, *110*, A08206, doi:10.1029/2004JA010824.

Jensen, H. J. (1998), *Self-Organized Criticality, Cambridge Lect. Notes Phys.*, vol. 10, 153 pp., Cambridge Univ. Press, Cambridge, U. K.

Kaldor, N. (1961), Capital accumulation and economic growth, in *The Theory of Capital*, edited by F. A. Lutz and D. C. Hague, pp. 177–222, St. Martin's Press, New York.

Kearney, M. J., and S. N. Majumdar (2005), On the area under a continuous time Brownian motion till its first-passage time, *J. Phys. A*, *38*, 4097–4104.

Krishnamurthy, S., A. Tanguy, P. Abry, and S. Roux (2000), A stochastic description of extremal dynamics, *Europhys. Lett.*, *51*(1), 1–7, doi:10.1209/epl/i2000-00330-9.

Lepping, R. P., et al. (1995), The WIND magnetic field investigation, *Space Sci. Rev.*, *71*, 207–229.

Lewis, Z. V. (1991), On the apparent randomness of substorm onset, *Geophys. Res. Lett.*, *18*, 1627–1630.

Mandelbrot, B. B., and J. R. Wallis (1969), Robustness of *R/S* in measuring noncyclic global statistical dependence, *Water Resour. Res.*, *5*, 967–988.

Mantegna, R., and H. E. Stanley (2000), *An Introduction to Econophysics: Correlations and Complexity in Finance*, 148 pp., Cambridge Univ. Press, Cambridge, U. K.

Mercik, S., K. Weron, K. Burnecki, and A. Weron (2003), Enigma of self-similarity of fractional Lévy stable motions, *Acta Phys. Pol. B*, *34*, 3773–3791.

Mikkelsen, R. (2000), Universality in transitions to spatio-temporal chaos, CSc thesis, Neils Bohr Inst., Copenhagen.

Moloney, N. R., and J. Davidsen (2010), Extreme value statistics in the solar wind: An application to correlated Lévy processes, *J. Geophys. Res.*, *115*, A10114, doi:10.1029/2009JA015114.

Moloney, N. R., and J. Davidsen (2011), Extreme bursts in the solar wind, *Geophys. Res. Lett.*, *38*, L14111, doi:10.1029/2011GL048245.

Ogilvie, K. W., et al. (1995), SWE, a comprehensive plasma instrument for the wind spacecraft, *Space Sci. Rev.*, *71*, 55–77.

Parks, G. K. (1991), *Physics of Space Plasmas*, 538 pp., Addison-Wesley, Redwood City, Calif.

Parkinson, M. L. (2006), Dynamical critical scaling of electric field fluctuations in the greater cusp and magnetotail implied by HF radar observations of *F*-region Doppler velocity, *Ann. Geophys.*, *24*, 689–705.

Price, C. P., and D. E. Newman (2001), Using the *R/S* statistic to analyze *AE* data, *J. Atmos. Sol. Terr. Phys.*, *63*, 1387–1397.

Rodriguez-Iturbe, I., and A. Rinaldo (1997), *Fractal River Basins: Chance and Self-Organization*, 547 pp., Cambridge Univ. Press, Cambridge, U. K.

Rypdal, M., and K. Rypdal (2010), Stochastic modeling of the *AE* index and its relation to fluctuations in B_z of the IMF on substorm time scales, *J. Geophys. Res.*, *115*, A11216, doi:10.1029/2010JA015463.

Sagdeev, R. Z. (Ed.) (1993), *Nonlinear Space Plasma Physics*, 505 pp., Am. Inst. of Physics, New York.

Sethna, J. P., K. A. Dahmen, and C. R. Myers (2001), Crackling noise, *Nature*, *410*, 242–250.

Simonsen, I., M. H. Jensen, and A. Johansen (2003), Optimal investment horizons, *Eur. Phys. J. B*, *27*, 583–586.

Sitnov, M. I., A. S. Sharma, K. Papdopoulos, and D. Vassiliadis (2001), Modeling substorm dynamics of the magnetosphere: From self-organization and self-organized criticality to nonequilibrium phase transitions, *Phys. Rev. E.*, *65*, 016116, doi:10.1103/PhysRevE.65.016116.

Sornette, D. (2004), *Critical Phenomena in Natural Sciences*, 2nd ed., 528 pp., Springer, Berlin.

Takalo, J. (1993), Correlation dimension of *AE* data, thesis for degree of Licentiate of Philosophy, Univ. of Jyväskylä, Jyväskyskylä, Finland.

Takalo, J. (1994), On the dynamics of the magnetosphere based on time series analysis of geomagnetic indices, PhD dissertation, Univ. of Jyväskylä, Jyväskylä, Finland.

Takalo, J., and J. Timonen (1994), Characteristic time scale of auroral electrojet data, *Geophys. Res. Lett.*, *21*, 617–620.

Tsurutani, B. T., M. Sugiura, T. Iyemori, B. E. Goldstein, W. D. Gonzalez, S. I. Akasofu, and E. J. Smith (1990), The nonlinear response of *AE* to the IMF B_s driver: A spectral break at 5 hours, *Geophys. Res. Lett.*, *17*, 279–282.

Ukhorskiy, A. Y., M. I. Sitnov, A. S. Sharma, and K. Papadopoulos (2004), Global and multi-scale features of solar wind-magnetosphere coupling: From modeling to forecasting, *Geophys. Res. Lett.*, *31*, L08802, doi:10.1029/2003GL018932.

Uritsky, V. M., and M. I. Pudovkin (1998), Low frequency $1/f$-like fluctuations of the AE-index as a possible manifestation of self-organized criticality in the magnetosphere, *Ann. Geophys.*, *28*, 1580–1588.

Uritsky, V. M., A. J. Klimas, and D. Vassiliadis (2001), Comparative study of dynamical critical scaling in the auroral electrojet index versus solar wind fluctuations, *Geophys. Res. Lett.*, *28*, 3809–3812.

Uritsky, V. M., A. Pouquet, D. Rosenberg, P. D. Mininni, and E. F. Donovan (2010), Structures in magnetohydrodynamic turbulence: Detection and scaling, *Phys. Rev. E.*, *82*, 056326, doi:10.1103/PhysRevE.82.056326.

Vassiliadis, D. (2006), Systems theory for geospace plasma dynamics, *Rev. Geophys.*, *44*, RG2002, doi:10.1029/2004RG000161.

Vörös, Z., P. Kovács, Á. Juhász, A. Körmendi, and A. W. Green (1998), Scaling laws from geomagnetic time series, *Geophys. Res. Lett.*, *25*(14), 2621–2624.

Watkins, N. W. (2002), Scaling in the space climatology of the auroral indices: Is SOC the only possible description?, *Nonlinear Processes Geophys.*, *9*, 389–397.

Watkins, N. W., D. Credgington, B. Hnat, S. C. Chapman, M. P. Freeman, and J. Greenhough (2005), Towards synthesis of solar wind and geomagnetic scaling exponents: A fractional Lévy motion model, *Space Sci. Rev.*, *121*, 271–284.

Watkins, N. W., D. Credgington, R. Sanchez, S. J. Rosenberg, and S. C. Chapman (2009), Kinetic equation of linear fractional stable motion and applications to modeling the scaling of intermittent bursts, *Phys. Rev. E*, *79*, 041124, doi:10.1103/PhysRevE.79.041124.

Wilmott, P., S. Howson, and J. Dewynne (1995), *The Mathematics of Financial Derivatives: A Student Introduction*, 336 pp., Cambridge Univ. Press, Cambridge, U. K.

S. C. Chapman and B. Hnat, CFSA, University of Warwick, Coventry CV4 7AL, UK.

N. W. Watkins, British Antarctic Survey, High Cross, Madingley Road, Cambridge CB3 0ET, UK. (nww@bas.ac.uk)

Extreme Value and Record Statistics in Heavy-Tailed Processes With Long-Range Memory

Aicko Y. Schumann

Complexity Science Group, Department of Physics and Astronomy, University of Calgary, Calgary, Alberta, Canada

Nicholas R. Moloney[1]

Max Planck Institute for the Physics of Complex Systems, Dresden, Germany

Jörn Davidsen

Complexity Science Group, Department of Physics and Astronomy, University of Calgary, Calgary, Alberta, Canada

Extreme events are an important theme in various areas of science because of their typically devastating effects on society and their scientific complexities. The latter is particularly true if the underlying dynamics does not lead to independent extreme events as often observed in natural systems. Here we focus on this case and consider stationary stochastic processes that are characterized by long-range memory and heavy-tailed distributions, often called fractional Lévy noise. While the size distribution of extreme events is not affected by the long-range memory in the asymptotic limit and remains a Fréchet distribution, there are strong finite-size effects if the memory leads to persistence in the underlying dynamics. Moreover, we show that this persistence is also present in the extreme events, which allows one to make a time-dependent hazard assessment of future extreme events based on events observed in the past. This has direct applications in the field of space weather as we discuss specifically for the case of the solar power influx into the magnetosphere. Finally, we show how the statistics of records, or record-breaking extreme events, is affected by the presence of long-range memory.

1. INTRODUCTION

History often turns on extreme events, rare occurrences of extraordinary nature, be they man-made or natural. Examples are global financial crises, military strikes, radical political events, or natural disasters such as floods, droughts, and earthquakes. Extreme events are often associated with catastrophes, and the word "extreme" is sometimes substituted by "freak" to suggest something unnatural and undesirable. Generally, the economic and social consequences of extreme events are a matter of enormous concern. In particular, the ever-increasing economic and human losses from natural hazards underscore the urgency for improved understanding of extreme events to develop effective strategies to reduce their impact.

The surprisingly high likelihood of extreme events is actually a key attribute of many complex systems, in both natural and man-made environments [see, e.g., *Albeverio*

[1]Formerly at Complexity Science Group, Department of Physics and Astronomy, University of Calgary, Calgary, Alberta, Canada.

Extreme Events and Natural Hazards: The Complexity Perspective
Geophysical Monograph Series 196

10.1029/2011GM001088

et al., 2006; *Bunde et al.*, 2002; *Embrechts et al.*, 2004; *Galambos et al.*, 1994; *Sornette*, 2006]. In particular, we know that records must be broken in the future, so if a flood design is based on the worst case of the past, then, we are still not prepared for all possible floods in the future. The classic approach to studying the probability of extreme events has been to assume independent and identically distributed (iid) event sizes. This has led to a powerful statistical theory [see, e.g., *Embrechts et al.*, 2004; *Coles*, 2007; *de Haan and Ferreira*, 2006], which has been successfully applied in many cases [see, e.g., *Easterling et al.*, 2000; *Glaser and Stangl*, 2004; *van den Brink and Können*, 2008]. The latter is also related to the fact that some parts of the theory, e.g., the limit distributions of block maxima, can be extended to a wide class of dependent stationary stochastic processes and their associated time series [*Berman*, 1964; *Leadbetter et al.*, 1983; *Leadbetter and Rootzén*, 1988; *Samorodnitsky*, 2004]. However, we are still far away from a general understanding of extreme events generated by such dependent processes, which are abundant in nature. Complicating factors typically include slow convergence and strong finite-size effects [*Györgyi et al.*, 2008], as well as nonlinear correlations [*Bogachev et al.*, 2007].

Strikingly, one often encounters dependence in the form of long-range persistence in recordings taken from natural systems. Persistence is defined as the tendency that subsequent values in a time series are similar: large values tend to be followed by large values, and small values tend to be followed by small values. Such behavior has been reported for water levels in rivers [*Hurst*, 1951] and in river runoff records [*Kantelhardt et al.*, 2003, 2006], in climatological temperature recordings [*Koscielny-Bunde et al.*, 1998; *Pelletier and Turcotte*, 1997; *Huybers and Curry*, 2006] and for temperature fluctuations in oceans [*Monetti et al.*, 2002], as well as in marine data [*Roman et al.*, 2008], to name only a few examples.

The theoretical studies of the extreme value statistics of stationary stochastic processes exhibiting long-range persistence have mainly focused on processes with Gaussian-distributed event sizes. In this and related cases, the limit distribution of the block maxima is the same as in the iid case: a Gumbel distribution [*Berman*, 1964; *Eichner et al.*, 2006]. Similar results have been obtained for beta-distributed random variables, in which case the Gumbel distribution is replaced by the Weibull distribution [*Moloney and Davidsen*, 2009]. Meanwhile, for distributions with power law tails the block maxima for iid events are Fréchet distributed [*Embrechts et al.*, 2004], and this is even true for a certain class of dependent sequences [*Leadbetter et al.*, 1983; *Leadbetter and Rootzén*, 1988]. Power law tails imply that there is no "typical" event size and that indeed event sizes vary over many orders of magnitude. Many natural systems even obey distributions of Pareto-like (power law–like) *heavy* tails, such that the second moment of the distribution is not defined. In space physics, such heavy-tailed distributions have been used to model, e.g., magnetic field line transport [*Pommois et al.*, 1998], fluctuations in various solar wind parameters [*Hnat et al.*, 2003; *Bruno et al.*, 2004; *Zaslavsky et al.*, 2008], and auroral indices [*Watkins et al.*, 2005; *Zaslavsky et al.*, 2008]. Other examples include, size distributions in geology [*Caers et al.*, 1999], meteorology [*Taqqu*, 1987], or sediments [*Painter and Paterson*, 1994].

Here we study the extreme value statistics of exactly those stationary stochastic processes that are characterized by persistent (or antipersistent) long-range memory and heavy tails. These processes are sometimes called fractional Lévy noise or linear stable fractional noise (see *Watkins et al.* [2009] and *Moloney and Davidsen* [2010] for discussions and further references). Specifically, we consider those symmetric α-stable (SαS)-distributed processes ($0 < \alpha < 2$) that are characterized by a self-similarity index H with $0 < H < 1$. These are particularly relevant in the context of the solar wind as discussed, for example, by *Moloney and Davidsen* [2010] for the energy influx into the magnetosphere, which is captured by the Akasofu ε parameter. While it is known that the limit distribution of the block maxima remains a Fréchet distribution for all α and H [*Samorodnitsky*, 2004], we systematically quantify the finite size corrections. As expected, they are particularly pronounced for strong persistence, but they also play a role in the presence of antipersistence. Moreover, we show that the conditional block maxima distributions deviate significantly from the unconditional distribution for small block sizes. This history dependence allows one to make a time-dependent hazard assessment of the value of the next block maximum based on the previous one. We also show how the statistics of records, or record-breaking extreme events, is affected by the presence of long-range memory and, thus, extend very recent results for the Gaussian-distributed case [*Newman et al.*, 2010]. Finally, we apply the time-dependent hazard assessment and the record analysis to the ε time series derived from ACE spacecraft measurements for the years 2000–2007.

The chapter is organized as follows: First, we review the basic properties and limit theorems of (maximum) extreme value statistics in section 2, followed by record statistics in section 3. We then describe α-stable (αS) or Lévy distributions and discuss the generation of stationary SαS-distributed processes with long-range memory in section 4. The presentation of our numerical results can be subdivided into two parts. While we focus on estimating the extreme value statistics of these processes including simple block maxima and conditional block maxima in sections 5.1 and 5.2, their record statistics are presented in section 5.3. Section 6 discusses the application of the introduced methodology to the solar power influx into the Earth's magnetosphere. Finally, we conclude in section 7.

2. CLASSICAL EXTREME VALUE STATISTICS

In classical extreme value statistics, one considers sets of independent identically distributed random variables $\{X_1, X_2, \ldots, X_n\}$, each drawn from the same cumulative distribution function $F(x)$. $F(x)$ can typically be represented by a unique probability density function $P(x)$. A time series $\{x_i\}_{i=1,\ldots,n}$ can then be understood as one possible realization of the set of random variables, $\{X_1 = x_1, X_2 = x_2, \ldots, X_n = x_n\}$, following this density function $P(x)$. The distribution of the (block) maximum or extreme value $M_n = \max\{X_1, \ldots, X_n\}$ is given by

$$\begin{aligned} Pr(M_n \leq m) &= Pr(X_1 \leq m, \ldots, X_n \leq m) \\ &= \prod_{i=1}^{n} Pr(X_i \leq m) = F^n(m). \end{aligned} \tag{1}$$

(Since $\min\{X_1, \ldots, X_n\} = \max\{-X_1, \ldots, -X_n\}$, the minimum can be mathematically treated in the same way.) The *Fisher-Tippett-Gnedenko theorem* states that if there exists a renormalization sequence $\{a_n, b_n\}$, $a_n \in \mathbb{R}^+$ and $b_n \in \mathbb{R}$, such that

$$\lim_{n\to\infty} Pr\left(\frac{M_n - b_n}{a_n} \leq m\right) \stackrel{d.}{=} G(m), \tag{2}$$

where $G(m)$ is nondegenerate and $d.$ means convergence in distribution, then, the asymptotic limit distribution $G(m)$ belongs to one of the following three function families [*Coles*, 2007]:

$$G_{\text{Gumbel}}(m) = \exp\left\{-\exp\left\{-\left(\frac{m-b}{a}\right)\right\}\right\}, \tag{3a}$$

$$G_{\text{Fréchet}}(m) = \begin{cases} 0 & : m \leq b \\ \exp\left\{-\left(\frac{m-b}{a}\right)^{-\alpha}\right\} & : m > b, \end{cases} \tag{3b}$$

$$G_{\text{Weibull}}(m) = \begin{cases} \exp\left\{-\left(-\left(\frac{m-b}{a}\right)\right)^{\alpha}\right\} & : m < b \\ 1 & : m \geq b \end{cases}. \tag{3c}$$

Here $a \in \mathbb{R}^+$, $b \in \mathbb{R}$, and $\alpha > 0$. As evident from equation (1), the exact limiting distribution is determined by $F(x)$ and, thus, $P(x)$.

If, for instance, $P(x)$ corresponds to the normal distribution or the exponential distribution, the limit distribution of the (block) maxima or extreme values follows a *Gumbel* distribution (equation (3a)). In cases where $P(x)$ exhibits power law right tails (Pareto-like tails), i.e., $\bar{P}(x) \sim L(x)x^{-1-\alpha}$ where $\bar{P}(x)$ is the tail of the distribution $P(x)$, $L(x)$ is some slowly varying function, the limit distribution of the (block) maximum or extreme value approaches the *Fréchet* distribution (equation (3b)) with the same $\alpha > 0$. Finally, the limit distribution of the maximum falls into the *Weibull* class for some distributions with bounded right tails, such as the beta distribution (equation (3c)). Note that the convergence to one of the three distributions in equation (3) requires the condition in equation (2) to be valid. For example, the condition in equation (2) is not generally satisfied for the Poisson distribution, the geometric distribution, or the negative binomial distribution [*de Haan and Ferreira*, 2006; *Embrechts et al.*, 2004].

All three extreme value distributions in equations (3) can be written in a compact form in terms of the generalized extreme value distribution (GEV)

$$G(z) = \exp\left\{-\left[1 + \xi\left(\frac{z-\mu}{\sigma}\right)\right]_+^{-1/\xi}\right\}, \tag{4}$$

where + indicates the constraint $\left[1 + \xi\left(\frac{z-\mu}{\sigma}\right)\right] > 0$. The parameters are the *shape parameter* ξ, $\xi \in \mathbb{R}$, the *scale parameter* σ, $\sigma \in \mathbb{R}^+$, and the *location parameter* μ, $\mu \in \mathbb{R}$. The shape parameter ξ determines the distribution family: $\xi = 0$ (interpreted as $\lim_{\xi\to 0}$) indicates the Gumbel class (equation (3a)), $\xi > 0$ the Fréchet class (equation (3b)), and $\xi < 0$ the Weibull class (equation (3c)).

In this chapter, we exclusively focus on SαS-distributed processes with long-range memory, for which the block maxima asymptotically converge in distribution to the Fréchet extreme value distribution in equation (3b) [*Samorodnitsky*, 2004].

3. CLASSICAL RECORD STATISTICS

The field of record statistics is closely related to extreme value statistics since records can be considered as a special type of extreme values: Records are simply record-breaking extreme values, i.e., the sequence of extreme values for increasing n. Studies of record values and record times started with the pioneering work by *Chandler* [1952] followed by important contributions by *Rény* [1962]. For a comprehensive overview and a historic summary, we refer to the works of *Glick* [1978] and *Nevzorov* [2001] and references therein. The classical theory of records is based on the assumption of iid random variables and has been successfully extended and applied to many natural systems, including earthquakes [*Davidsen et al.*, 2006, 2008; *Van Aalsburg et al.*, 2010; *Vasudevan et al.*, 2010; *Peixoto et al.*, 2010], climate dynamics [*Schmittmann and Zia*, 1999; *Benestad*, 2003; *Redner and Petersen*, 2006; *Benestad*, 2008;

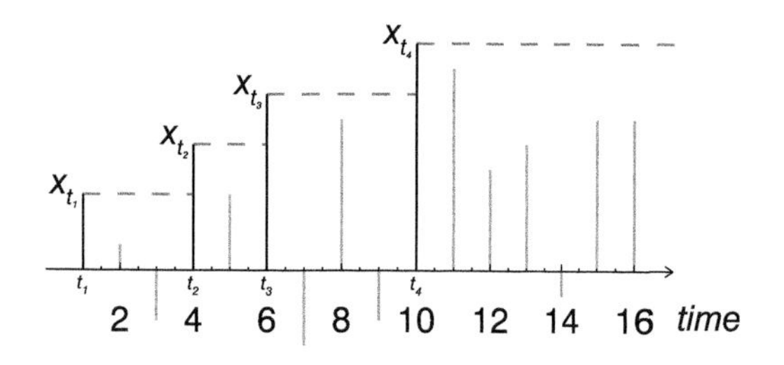

Figure 1. Illustration on the definition of records in a realization $\{x_i\}_{i=1,...,R}$ of the process $\{X_i\}_{i=1,...,R}$. Four records (black), $r_1 = x_{t_1=1}$, $r_2 = x_{t_2=4}$, $r_3 = x_{t_3=6}$, $r_4 = x_{t_4=10}$, are identified in the fragment shown.

Newman et al., 2010; *Wergen and Krug*, 2010], hydrology [*Vogel et al.*, 2001], and evolution [*Sibani and Jensen*, 2009].

Specifically, let again $\{X_1, X_2, ..., X_R\}$ be a sequence of iid random variables with a common continuous distribution $F(x)$. One possible realization of the set of random variables may then be given by the time series $\{x_i\}_{i=1,...,R} = \{x_1, x_2, ..., x_R\}$. We denote X_m as an (upper) record if $X_m = \max\{X_1, ..., X_m\}$ for $m \leq R$. (Similarly, one can define a lower record $X_m = \min\{X_1, ..., X_m\}$.) As illustrated in Figure 1, the set of records for a given realization is trivially ordered, and the index m at which the kth record occurs is defined as the kth record time, T_k. Obviously, $T_1 = 1$ and $T_{k\geq 2} = \min\{j \leq R : X_j > X_{T_{k-1}}\}$. Thus, $T_{k>1}$ as well as the number of records after a time $t \leq R$ in a block of size R, N_t, are random variables themselves.

Interestingly, many statistical properties of records are identical for all sequences of iid random variables $\{X_i\}_{i=1,...,R}$ and, thus, independent of the distribution $F(x)$. This includes the expected number of records and their variance. To see this, it is important to realize that each $X_{k\leq m}$ has the same probability of being the maximum, which is simply given by

$$P(X_k = \max\{X_1, ..., X_m\}) = 1/m. \tag{5}$$

Thus, this is, in particular, the probability that X_m is a record, which allows one immediately to calculate the expected number of records after a time t and their variance [*Glick*, 1978]:

$$\begin{aligned} E(N_t) &= \sum_{m=1}^{t} P(X_m = \max\{X_1, ..., X_m\}) \\ &= \sum_{m=1}^{t} 1/m = \ln(t) + \gamma + \mathcal{O}(1/t) \end{aligned} \tag{6a}$$

$$\begin{aligned} \mathrm{Var}(N_t) &= \sum_{m=1}^{t} \frac{1}{m} - \sum_{m=1}^{t} \frac{1}{m^2} \\ &= \ln(t) + \gamma + \mathcal{O}(1/t) - \pi^2/6, \end{aligned} \tag{6b}$$

where $\gamma = 0.577...$ is the Euler-Mascheroni constant.

Note, that there are other properties of records, as for instance the distribution of occurrence times of the kth records [*Glick*, 1978], that are also independent of the distribution of $\{X_i\}_{i=1,...,R}$ but will not be further discussed here. In contrast to the above distribution-independent properties, results involving actual record values do generally depend on the distribution $F(x)$ [*Nevzorov*, 2001; *Redner and Petersen*, 2006].

4. HEAVY-TAILED PROCESSES WITH LONG-RANGE MEMORY

A prototype of stationary stochastic processes with heavy-tailed distributions are α-stable (αS) processes. A random variable X is called stable if any set of independent copies of X, $\{X_1, ..., X_n\}$, obeys

$$X_1 + ... + X_n \stackrel{d.}{=} c_n X + d_n \tag{7}$$

for some $c_n \in \mathbb{R}^+$ and $d \in \mathbb{R}$ [*Embrechts and Schmidli*, 1994]. This property implies that the shape of the distribution remains invariant if sums over the random variable are considered. If $d_n = 0$, X is referred to as strictly stable, and X is symmetric if $X \stackrel{d.}{=} -X$. A generalized central limit theorem proves that all distributions that have a probability density function $P(x)$ with an associated Fourier transform or characteristic function of the form

$$\varphi_{\alpha,\beta,\delta,\gamma}(t) = \begin{cases} \exp\left\{it\delta - \gamma^\alpha|t|^\alpha\left(1 - i\beta\,\mathrm{sgn}(t)\tan\left(\frac{\pi\alpha}{2}\right)\right)\right\} & : \alpha \neq 1 \\ \exp\left\{it\delta - \gamma|t|\left(1 + \frac{2}{\pi}i\beta\,\mathrm{sgn}(t)\ln|t|\right)\right\} & : \alpha = 1 \end{cases} \tag{8}$$

are stable [*Samorodnitsky and Taqqu*, 1994]. In particular, the constants c_n are not arbitrary but scale as $n^{1/\alpha}$ with $\alpha \in (0,2]$, and hence, the corresponding distributions are called αS distributions. In equation (8), the *characteristic exponent* or *index of stability* $\alpha \in (0,2]$, the *skewness parameter* $\beta \in [-1,1]$, and the *scale parameter* $\gamma \in \mathbb{R}^+$ describe the shape of the distribution, while $\delta \in \mathbb{R}$ is the *location parameter.* Since many of these αS distributions do not possess a finite mean and standard deviation, these quantities are typically not used to characterize these distributions.

For symmetric α-stable (SαS) distributions, the skewness parameter obeys $\beta = 0$. In particular, the Gaussian distribution $\mathcal{N}(\mu,\sigma^2)$ is SαS with $(\alpha = 2, \beta = 0, \gamma = \sigma/\sqrt{2}, \delta = \mu)$. Throughout this chapter, we will only focus on SαS-distributed time series and set without loss of generality the location parameter δ to zero. We will further normalize our time series to ensure $\gamma = 1$ (see section 4.2). Hence, equation (8) reduces to the simple form

$$\varphi_\alpha(t) = \exp\{-|t|^\alpha\}. \tag{9}$$

While even in this case an analytical form of $P(x)$ is typically not known, the asymptotic behavior for $\alpha \in (0,2)$ is characterized by *Pareto-like* power law tails with $\bar{P}(x) \sim x^{-1-\alpha}$ [*Sornette*, 2006]. Consequently, the variance of all non-Gaussian αS distributions with $\alpha < 2$ is undefined. In addition, the distribution's mean value is undefined for $\alpha < 1$.

4.1. Long-Range Memory in SαS Processes

For a sequence of identically but *not* independently distributed random variables $\{X_1, X_2, \ldots, X_n\}$ with a common continuous distribution $F(x)$, the associated realizations or stationary time series $\{x_i\}_{i=1,\ldots,n} = \{x_1, x_2, \ldots, x_n\}$ are typically characterized by nontrivial correlations. One particular example is the presence of long-range linear autocorrelations. These can be described by three equivalent definitions, namely, a power law behavior in (1) the autocorrelation function with correlation exponent $0 < \gamma < 1$, $C(s) \sim s^{-\gamma}$, (2) the power spectrum with spectral exponent $0 < \beta < 1$, $P(f) \sim f^{-\beta}$, as well as (3) the fluctuation function with (monofractal) fluctuation exponent $h(2)$, $F(s) \sim s^{h(2)}$ for scales s, frequencies f, and $1 - \beta = \gamma = 2 - 2h(2)$ (for a comprehensive definition and discussion, see the work of *Schumann* [2011] and references therein).

While such long-range linear autocorrelations imply long-range persistence, the reverse is not necessarily true. In particular, the above definitions require that a finite variance exists, which is not the case for αS processes with $\alpha < 2$ as discussed above. Thus, one has to use an alternative way to quantify persistence and long-range memory, in general, for such processes. One possibility is to define persistence based on the closely related phenomenon of self-similarity. A (continuous) stochastic process $X = \{X(t), t \in \mathbb{R}\}$ is called self-similar with a self-similarity parameter $H > 0$ if $\forall \kappa > 0, t \in \mathbb{R}$ [*Embrechts and Makoto*, 2002],

$$\{X(\kappa t)\} \stackrel{d.}{=} \{\kappa^H X(t)\}. \tag{10}$$

It can be further shown that for self-similar SαS-distributed processes $X_{H,\alpha}$ with $0 < H < 1$, long-range persistence is achieved for $H > 1/\alpha$, while $H = 1/\alpha$ corresponds to the uncorrelated case, and antipersistence emerges for $H < 1/\alpha$ [*Stoev and Taqqu*, 2004]. Thus, persistent behavior can only be observed if $1 < \alpha < 2$, and we focus on this range of α values in the following.

4.2. Numerical Simulation of Self-Similar SαS Processes

There are two distinct algorithms typically used to generate stationary self-similar SαS-distributed processes characterized by the exponents α and H. One approach utilizes fractional calculus and derives the fractional integral or fractional derivative $X_{\nu,\alpha}(t)$ of an uncorrelated SαS process $X_\alpha(t)$ in Fourier space, $\hat{X}_\alpha(\omega) = (2\pi)^{-1}\int_R X_\alpha(t')exp\{-it'\omega\}\mathrm{d}t'$, [*Chechkin and Gonchar*, 2000],

$$\hat{X}_{\nu,\alpha}(\omega) = \frac{\hat{X}_\alpha(\omega)}{(i\omega)^\nu}, \quad \nu = H - 1/\alpha. \tag{11}$$

Hence, this approach is very similar to the established Fourier-filtering technique often used for generating correlated Gaussian noises [see, e.g., *Schumann and Kantelhardt*, 2011]. Note that *Chechkin and Gonchar* [2000] restrict their algorithm to $1 \le \alpha \le 2$. (This limitation mainly comes from the loss of validity of the Minkowski inequality for $\alpha < 1$, which is used in the work of *Chechkin and Gonchar* [2000] to estimate the upper boundary of H. It should nevertheless be possible to generalize the algorithm to arbitrary $\alpha \in (0,2]$. Moreover, one possible pitfall of the algorithm is that the probability density of the generated self-similar SαS process is not explicitly renormalized to be the same as for the uncorrelated noise it is initialized with. This has to be done manually.) They further discuss why high-frequency inaccuracies of fractional integration disturb numerical simulations in the antipersistent case (see also the work of *Chechkin and Gonchar* [2001] for their comprehensive study of fractional Gaussian integration).

The other algorithm is based on discretizing and numerically solving a stochastic integral that defines a *cumulative* self-similar SαS process $Y_{H,\alpha}(t)$, often called fractional Lévy motion,

$$Y_{H,\alpha}(t) = C_{H,\alpha}^{-1}\int_R\left((t-s)_+^{H-1/\alpha} - (-s)_+^{H-1/\alpha}\right)\mathrm{d}Y_\alpha(s), \tag{12}$$

where $Y_\alpha(s)$ is (standard) SαS motion and $C_{H,\alpha}$ is some norming constant. The increments $X_{H,\alpha}(k) = Y_{H,\alpha}(k) - Y_{H,\alpha}(k-1)$ in the discretized version of equation (12) then define a self-similar SαS process $X_{H,\alpha}$ that we study here. Note that the norming constant is chosen such that (1) the SαS scaling parameter is $\gamma = 1$ as previously discussed for deriving equation (9) from equation (8) and (2) the probability density functions are equal for the independent and self-similar SαS process. A stochastic integral such as in equation (12) can be discretized and efficiently be solved by exploiting the convolution theorem and employing a fast Fourier transform (for further details, we refer to the work of *Stoev and Taqqu* [2004]). (A similar supposedly faster algorithm was suggested later [*Wu et al.*, 2004], but not tested by us.)

In the following, we present results for self-similar SαS processes using the latter algorithm [*Stoev and Taqqu*, 2004], but we obtain very similar results using the former one [*Chechkin and Gonchar*, 2000] over the range of its established validity. These processes are parametrized by the two parameters H and α ($\gamma = 1$, $\delta = \beta = 0$ in equations (8) and (9)). Two further parameters m and M are required by the used algorithm. They control the mesh size (intermediate points) and the kernel of the fast Fourier transform, respectively. We chose the parameters $m = 64$ and $M = 48,576$ and generate time series of length $n = 1,000,000$. This choice of m and M with $mM \ll n$ is motivated by our observation that an undesired crossover in the scaling behavior of different parameters appears at scales of the order mM. (This crossover is not present in the algorithm by *Chechkin and Gonchar* [2000].) To the best of our knowledge, this has not been noticed before. Indeed, much shorter time series with $n + M = 2^{14}$, $m_{\max} = 256$, and $M_{\max} = 6000$ were studied by *Stoev and Taqqu* [2004] who concluded that larger values of m should be chosen for $0 < \alpha \leq 1$, while smaller m are preferable for $1 < \alpha \leq 2$ and small M. This is clearly not supported by our simulations for much larger values of n, which are necessary to estimate extreme value and record statistics of self-similar SαS processes reliably. (We also noted a number of artifacts in the generated time series for values $\alpha < 1$, which we do not consider in this chapter. One might be able to resolve these artifacts by significantly increasing the mesh size parameter m. However, doubling m results in practically doubling the memory requirements unless the series length n is reduced accordingly. We plan on testing this more systematically in the future. Note further that the values were chosen to ensure that $m(M + n)$ is an integer power of 2.)

For each parameter set (α,H), we generate $N_{\text{conf}} = 1500$ long-range persistent data sets. (Note that significant computational resources are required. For a single parameter set (α, H), about 15 GB are necessary to store the "double" raw data. Since the computation of the involved Fourier transform requires to allocate arrays of size $m(M + n)$ and there is a noticeable overhead for computation, main memory requirements become an issue very quickly as m and M grow.) For surrogate testing, we destroy correlations by shuffling those data sets using a Mersenne-Twister-19937 pseudorandom number generator [*Matsumoto and Nishimura*, 1998], which was also used to generate white Gaussian noise being the basis for the uncorrelated SαS noise. (Uncorrelated SαS-distributed random numbers can be generated from uncorrelated Gaussian-distributed random numbers following the work of *Chambers et al.* [1976, 1987].)

5. EFFECTS OF LONG-RANGE MEMORY ON THE STATISTICAL PROPERTIES OF SαS PROCESSES

5.1. Extreme Value Statistics

Since the theoretical analysis of extreme value statistics is mostly concerned with *asymptotic* distributions as discussed in section 2, these findings are not immediately applicable to time series $\{x_i\}_{i=1,...,n}$ of finite length n, which are measured in natural systems. In particular, the exact cumulative distribution function $F(x)$ is generally unknown and, hence, neither is $F^n(m)$ in equation (1). Thus, one typically has to assume that (1) the given time series is a single realization of an underlying stationary stochastic process and (2) the underlying dynamics is ergodic such that one can use time averages instead of, or in addition to, ensemble averages. This allows one to estimate the distribution of the (block) maximum or extreme value $M_R = \max\{X_1, ..., X_R\}$ from the given time series by considering nonoverlapping blocks of size $R \ll n$. To be more specific, the sequence of maximum values in disjunct blocks of length R is given by $\{m_j\}_{j=1,...,[n/R]}$, where $m_j = \max\{x_{(j-1)R+1}, x_{(j-1)R+2}, ..., x_{jR}\}$ is the maximum in block j, and [.] denotes the integer division (see Figure 2 for an illustration). Under the above assumptions, the estimated distribution $P_R(m)$ of the block maxima should converge toward a specific $G(m)$ as $R \to \infty$ if a suitable sequence $\{a_n, b_n\}$ exists as in equation (2). If this convergence is sufficiently fast, one can estimate $G(m)$ reliably even for finite R (and n) and a finite number of realizations N_{conf} provided that $[n/R] \times N_{\text{conf}} \gg 1$.

For independent SαS-distributed processes, it is well-known that the asymptotic limit distribution of its extreme

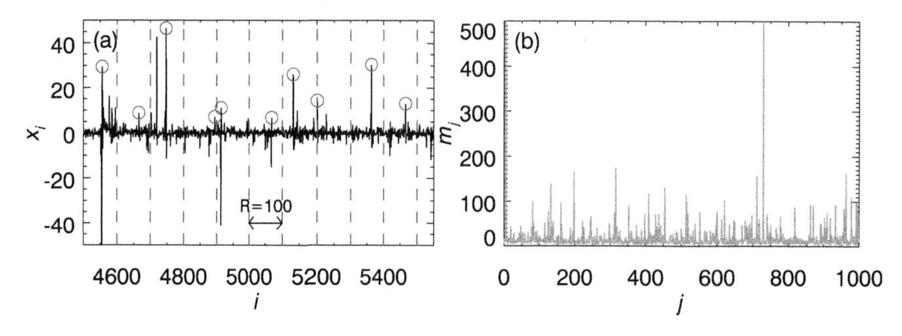

Figure 2. (a) By considering disjunct blocks of size $R = 100$ in the underlying time series $\{x_i\}_{i=0,1,...,N-1}$ (b) a sequence of block maxima $m_j = \max\{x_{jR}, ..., x_{j(R+1)-1}\}_{j=0,1,...,[N/R]-1}$ is defined. Only parts of both series are shown.

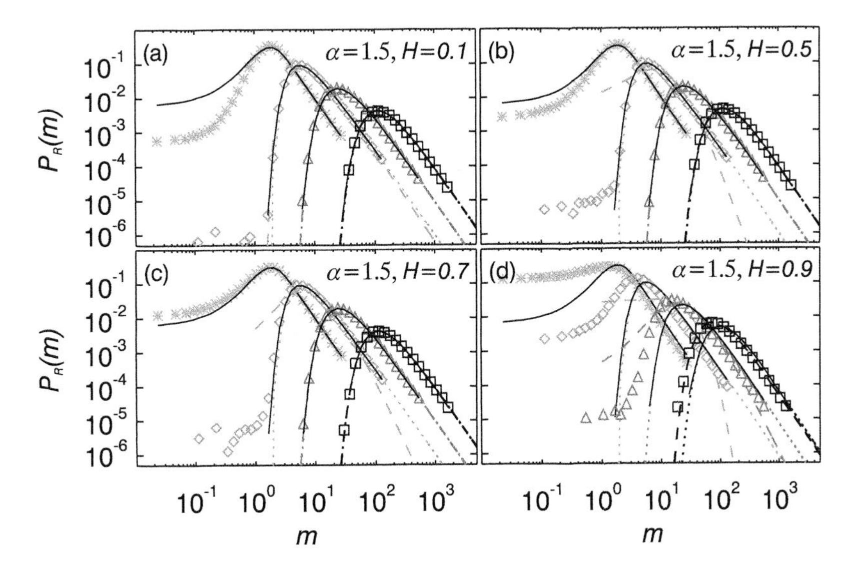

Figure 3. Probability density functions of block maxima for $\alpha = 1.5$ and different values of H and R. (a and b) Examples of antipersistence. (c and d) Examples of persistent behavior since $H = 1/\alpha = 2/3$ is the boundary between the two regimes. Shown are four different block sizes $R = 10$ (stars, light gray), $R = 100$ (diamonds, gray), $R = 1000$ (triangles, dark gray), and $R = 10000$ (squares, black). The surrogate data for the corresponding independent SαS processes are indicated by solid black lines. Maximum-likelihood generalized extreme value distribution (GEV) fits are plotted as dotted curves (surrogate data) and dashed curves (original data) in the same gray tones for comparison.

values is the Fréchet distribution in equation (3b) [*Embrechts et al.*, 2004] since $\bar{P}(x) \sim x^{-1-\alpha}$. More recently, it was proven mathematically that all self-similar SαS processes with $\alpha < 2$ also converge in distribution to the Fréchet extreme value distribution [*Samorodnitsky*, 2004]. Thus, the presence of long-range memory in the form of persistence or antipersistence does not have any significant influence on the asymptotic behavior in this case of stationary processes. However, we find that there are significant finite size effects, i.e., significant differences in the extreme value distribution for finite R between the cases with and without long-range memory.

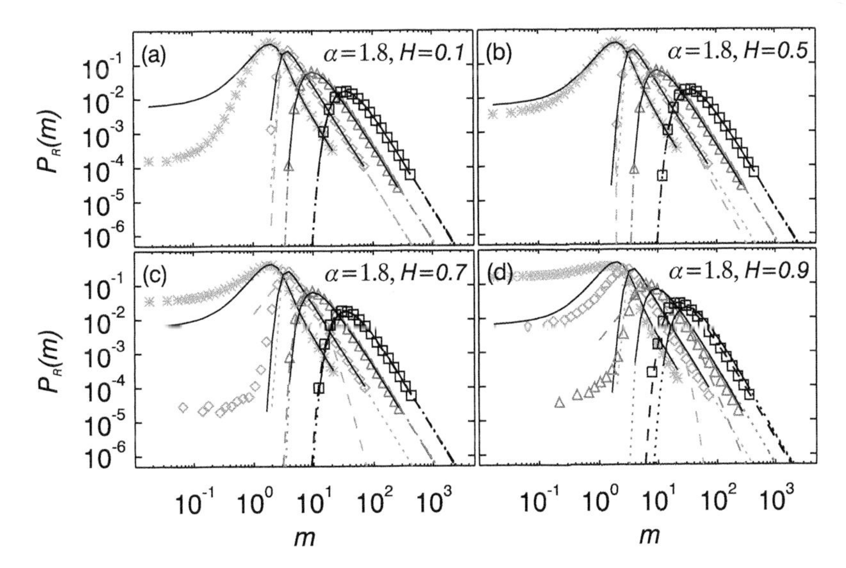

Figure 4. Same as Figure 3 but for $\alpha = 1.8$. Note that now, the independent case corresponds to $H = 1/\alpha = 5/9$, and we have antipersistence for $H < 5/9$ and persistence for $H > 5/9$. Thus, the persistence for $H = 0.9$ is stronger, and the antipersistence for $H = 0.1$ is weaker than in the case with $\alpha = 1.5$ shown in Figure 3.

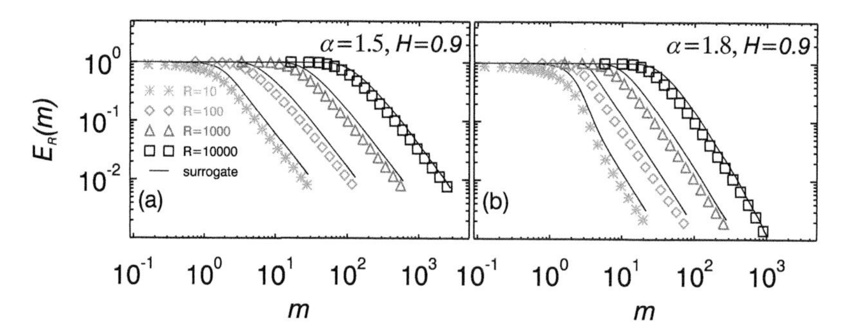

Figure 5. Exceedance probability, i.e., the probability to find an extreme event larger than m. Shown are results for self-similar SαS time series with $H = 0.9$ and the corresponding independent surrogates (black solid lines) for different block sizes R (symbols) and for different characteristic exponents (a) $\alpha = 1.5$ and (b) $\alpha = 1.8$.

To see this, we estimate the probability density function $P_R(m)$ of the maxima or extreme values in nonoverlapping blocks of length R, which are shown in Figures 3 and 4 for self-similar SαS processes with different parameters H and α including persistent as well as antipersistent cases. There are clear differences between the original data and the independent surrogate data, both of which were generated as described in section 4.2. These differences are particularly

Table 1. Shape Parameters ξ and Corresponding Maximum Error Obtained From 95% Confidence Intervals of a Maximum Likelihood Fit of Equation (4) to the Data[a]

	α =1.5		α =1.8	
	Self-Similar SαS Shape ξ	Shuffled SαS Shape ξ	Self-Similar SαS Shape ξ	Shuffled SαS Shape ξ
$H = 0.1$				
$R = 1000$	0.6679 ± 0.0017	0.6657 ± 0.0017	0.5751 ± 0.0017	0.5734 ± 0.0017
$R = 10,000$	0.6582 ± 0.0054	0.6532 ± 0.0054	0.5489 ± 0.0051	0.5448 ± 0.0050
$H = 0.2$				
$R = 1000$	0.6682 ± 0.0017	0.6653 ± 0.0017	0.5756 ± 0.0017	0.5738 ± 0.0017
$R = 10,000$	0.6575 ± 0.0054	0.6510 ± 0.0054	0.5491 ± 0.0051	0.5443 ± 0.0050
$H = 0.3$				
$R = 1000$	0.6682 ± 0.0017	0.6647 ± 0.0017	0.5763 ± 0.0017	0.5738 ± 0.0017
$R = 10,000$	0.6573 ± 0.0054	0.6509 ± 0.0054	0.5498 ± 0.0051	0.5451 ± 0.0051
$H = 0.4$				
$R = 1000$	0.6683 ± 0.0017	0.6649 ± 0.0017	0.5764 ± 0.0017	0.5739 ± 0.0017
$R = 10,000$	0.6584 ± 0.0054	0.6501 ± 0.0054	0.5501 ± 0.0051	0.5473 ± 0.0051
$H = 0.5$				
$R = 1000$	0.6685 ± 0.0017	0.6654 ± 0.0017	0.5761 ± 0.0017	0.5756 ± 0.0016
$R = 10,000$	0.6589 ± 0.0054	0.6513 ± 0.0054	0.5515 ± 0.0051	0.5496 ± 0.0051
$H = 0.6$				
$R = 1000$	0.6684 ± 0.0017	0.6667 ± 0.0017	0.5751 ± 0.0017	0.5756 ± 0.0017
$R = 10,000$	0.6608 ± 0.0054	0.6556 ± 0.0054	0.5533 ± 0.0051	0.5488 ± 0.0051
$H = 0.7$				
$R = 1000$	0.6681 ± 0.0017	0.6675 ± 0.0017	0.5742 ± 0.0017	0.5714 ± 0.0017
$R = 10,000$	0.6625 ± 0.0054	0.6568 ± 0.0054	0.5530 ± 0.0051	0.5418 ± 0.0051
$H = 0.8$				
$R = 1000$	0.6656 ± 0.0017	0.6586 ± 0.0017	0.5496 ± 0.0016	0.5653 ± 0.0017
$R = 10,000$	0.6622 ± 0.0054	0.6431 ± 0.0054	0.5519 ± 0.0051	0.5296 ± 0.0051
$H = 0.9$				
$R = 1000$	0.4199 ± 0.0017	0.6410 ± 0.0006	0.2716 ± 0.0005	0.5510 ± 0.0017
$R = 10,000$	0.6613 ± 0.0054	0.6202 ± 0.0054	0.5484 ± 0.0051	0.5108 ± 0.0050

[a]Note that for $R = 100$, there was often no convergence in the estimation, meaning the probability density cannot be properly approximated by the generalized extreme value distribution in equation (4). Positive values of ξ indicate a Fréchet distribution. Asymptotically, ξ should approach 1/a based on theoretical results [*Samorodnitsky*, 2004].

pronounced for small values of m and R and for strongly persistent cases (H = 0.9). For small m, we observe that persistence and antipersistence lead to different behaviors: the probability of finding a smaller m is enhanced in the case of persistence when compared to the surrogate data, while in the case of antipersistence, the behavior depends sensitively on R. Nevertheless, in the latter case, there seems to be a clear and rapid convergence to the distribution of the independent surrogate data with increasing R. Specifically, there are no significant differences for $R > 1000$. This observation can be readily explained. Antipersistence implies that in an associated time series, large values tend to be followed by small values, and small values are rather followed by large values, which is similar to an alternating behavior. This alternating behavior does not significantly constrain the maximum value found in a given block if its size is sufficiently large. Thus, long-range memory in the form of antipersistence does not play an important role for the extreme value distribution even for finite R.

The situation is very different for the case with long-range persistence. Since persistence corresponds to a clustering of large and small values, respectively, more blocks will contain only smaller values such that the block maximum will also be smaller than what would be expected based on independent values. Thus, the distribution of block maxima will indicate a higher probability to find smaller values compared to the independent surrogate data. This is exactly what Figures 3 and 4 show for H = 0.9, independent of R. As expected, the deviations from the extreme value distribution of the surrogate data are larger for stronger persistence as a comparison of α = 1.5 and α = 1.8 shows. Thus, long-range memory in the form of persistence reduces the probability of observing very large block maxima for finite R. This is quantified by Figure 5, which shows the exceedance probability defined as

$$E_R(m) = Pr(M_R > m) = \int_m^\infty P_R(m')\mathrm{d}m'. \quad (13)$$

This phenomenon is not only true for self-similar SαS processes considered here but also for stationary long-range correlated Gaussian and exponentially distributed processes [*Eichner et al.*, 2006].

In order to test how closely the probability density function $P_R(m)$ of the block maxima for finite R resembles its asymptotic limit given by the GEV distribution in equation (4), we employ a maximum likelihood estimation method, already implemented in MathWorks's MATLAB™ Statistics Toolbox (software package, version R2009b), to estimate its parameters and to quantify the goodness of fit [*Aldrich*, 1997]. The values of the shape parameters ξ determining the distribution class are reported for all considered values of H and α in Table 1. Figure 6 illustrates the convergence of ξ_R to $1/\alpha$ as $R \to \infty$ for selected values. The fits, themselves, are shown in Figures 3 and 4. While there are significant deviations from the GEV distribution for small R and strong persistence, we find relative agreement for $R \geq 10000$ in all cases considered. However, we observe systematic deviations from the theoretical value $1/\alpha$ for very large R. Although errors increase, the systematic underestimation of the shape parameter is possibly not exclusively caused by finite size effects but might be due to shortcomings in the GEV fitting algorithm as well as the data generation algorithm.

Further evidence for the rapid convergence to the asymptotic GEV distribution comes from the scaling of the mean, $\langle m \rangle_R$, the median, med_R, and the estimated scale parameter, σ_R, with R (see equation (4)), which is shown in Figure 7. For large R, the scaling approaches the same limit for the self-similar SαS processes and the independent surrogate data. (Note that the differences between the different surrogate data are partly statistical and partly due to the algorithm discussed in section 4.2.) For the rescaling coefficient a_R in equation (2), it is known that $a_R \propto R^{1/\alpha}$ asymptotically [*Embrechts et al.*, 2004]. This follows from studies of the domain of attraction of the Fréchet class, which have proven that if the cumulative distribution function $F(x)$ of the stochastic process decays asymptotically as $\bar{F}(x) \sim L(x)x^{-\alpha}$, then the rescaling coefficient $a_R = R^{1/\alpha}L(R)$ for some slowly varying function L [*Embrechts et al.*, 2004]. $L(R) \equiv$ *constant* for SαS processes. It was also shown that centering is not necessary, i.e., $b_R = 0$ [*Embrechts et al.*, 2004].

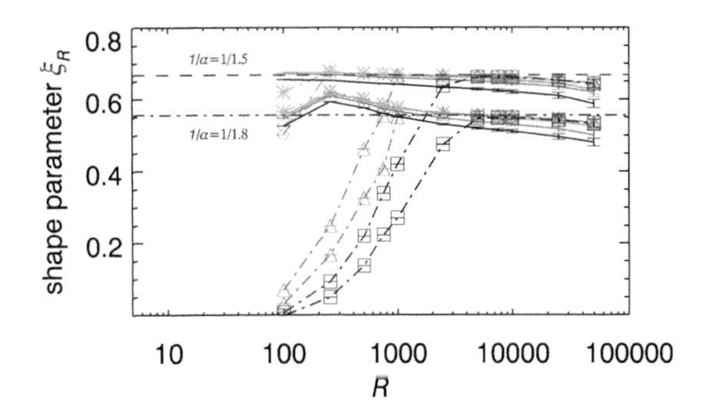

Figure 6. Convergence of the shape parameter ξ_R toward the theoretical value $1/\alpha$ as $R \to \infty$ obtained from GEV fits for four different self-similar SαS time series (grayscale coded dash-dotted lines and symbols) and corresponding independent surrogates (solid lines in same gray): H = 0.1 (light gray, stars), H = 0.2 (gray, diamonds), H = 0.8 (dark gray, triangles), and H = 0.9 (black, squares). Upper curves correspond to α = 1.5, while lower curves belong to α = 1.8. Error bars are 95% confidence intervals of ξ_R obtained from the GEV fit after combining the data of all 1500 configurations.

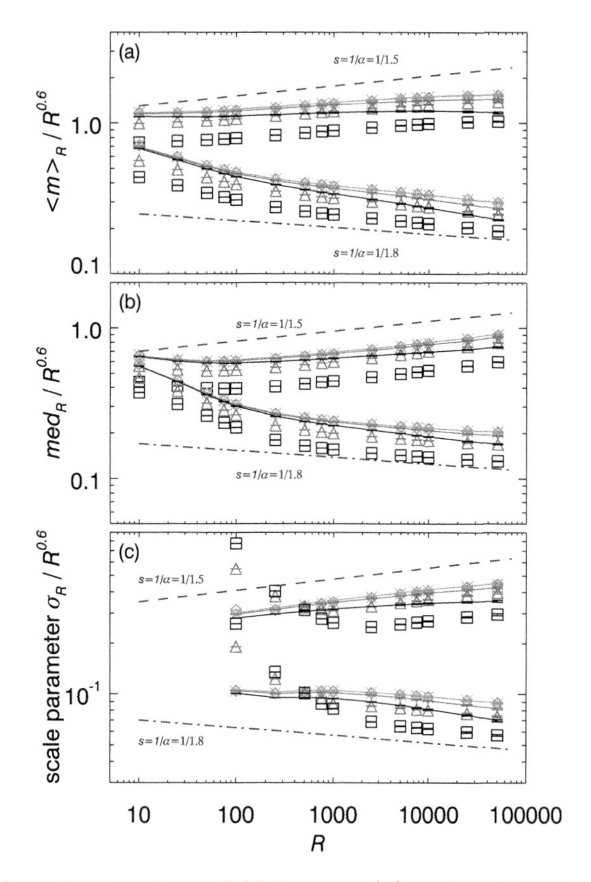

Figure 7. Dependence of (a) the mean $\langle m \rangle_R$ and (b) the median med_R block maximum together with (c) the scale parameter σ_R from a GEV fit on the block size R. Shown are results for four different self-similar SαS time series (grayscale-coded symbols) and corresponding independent surrogates (solid lines in same gray tones): $H = 0.1$ (light gray, stars), $H = 0.2$ (gray, diamonds), $H = 0.8$ (dark gray, triangles), and $H = 0.9$ (black, squares). Upper curves belong to $\alpha = 1.5$ and lower curves to $\alpha = 1.8$. Shown are averages of $\langle m \rangle_R$ and med_R obtained separately for each configuration ($N_{\text{conf}} = 1500$) together with the ensemble standard deviations as error bars in Figures 7a and 7b. Expected rescaling exponents are indicated by dashed and dash-dotted lines, respectively. Note that the y axis was rescaled by $R^{0.6}$ to enhance differences. Error bars in Figure 7c are 95% confidence intervals of σ_R obtained from a GEV fit after combining all N_{conf} configurations to increase statistics for large R.

Nevertheless, there exist to our best knowledge no analytical derivations on the convergence of $\langle m \rangle_R$, med_R, and σ_R. We expect a similar behavior in the limit $R \to \infty$ where a unique Fréchet distribution is approached. While we observe a scaling similar to a_R for the median in Figure 7b and with reservations for the scale σ_R in Figure 7c, there are still noticeable deviations for the mean $\langle m \rangle_R$, and for that reason, we exclusively focus on medians in our discussion of conditional extremes (see section 5.2). Besides this, Figure 7 provides further evidence that the asymptotic scaling of the renormalization sequence $\{a_n, b_n\}$, which ensures the convergence in equation (2), is not affected by the presence of long-range memory. Yet, we also see clear deviations from the asymptotic scaling for small R as expected. In particular, we observe a crossover in the medians at scales somewhat below $R_\times = 100$ for $\alpha = 1.5$ and around $R_\times = 600$–700 for $\alpha = 1.8$ in the medians. Note that this crossover is apparently independent of H. This indicates that long-range memory does not play a role for the deviations from the asymptotic scaling. Instead, the "heavier" the heavy-tailed distribution, the smaller the $R_\times$.

Based on the observed scaling behavior, it is straightforward to identify a suitable sequence $\{a_R, b_R\}$ to ensure that equation (2) holds. Namely, we chose

$$a_R = R^{1/\alpha} \text{ and } b_R = 0. \tag{14}$$

Using this sequence, we can rescale the probability density functions in Figures 3 and 4 according to equation (2) in order to obtain a scaling collapse resembling the asymptotic distribution. This is shown in Figures 8 and 9. While the collapse in the tails is excellent in all considered cases, the collapse is often not as good for small arguments. This is particularly true for large α and large H and confirms the findings above.

5.2. Conditional Extrema and Hazard Assessment

As a consequence of long-range memory in self-similar SαS processes, the value of $X(t)$ for a given t depends on the values at all earlier times. Thus, the probability of finding a certain value m_{j+1} as the block maximum also depends on the history and, thus, the preceding block maxima m_k with $k \leq j$. This directly allows one to make a time-dependent hazard assessment and potentially even predict the size of future extreme values based on the history of block maxima. In the absence of memory as for the iid case, only *time-independent* hazard assessment is possible.

As one possible approach to time-dependent hazard assessment, we study the statistics of all block maxima m_{j+1} in the sequence of maxima $\{m_j\}_{j=1,...,[N/R]}$ that follow a block maximum $m_j = m_0$ within a certain range (see *Schweigler and Davidsen* [2011] for a related but different approach). For example, Figures 10a and 10b display the conditional probability density functions of block maxima m that follow a block maximum m_0 larger than or equal to a threshold. The thresholds were chosen to obtain 15,000 conditional maxima when combining all realizations. (This corresponds to the upper 0.999 quantile or 1 per mille of all maxima.) As can be seen, long-range persistence shifts the distributions to the right compared to the independent surrogate case. The latter

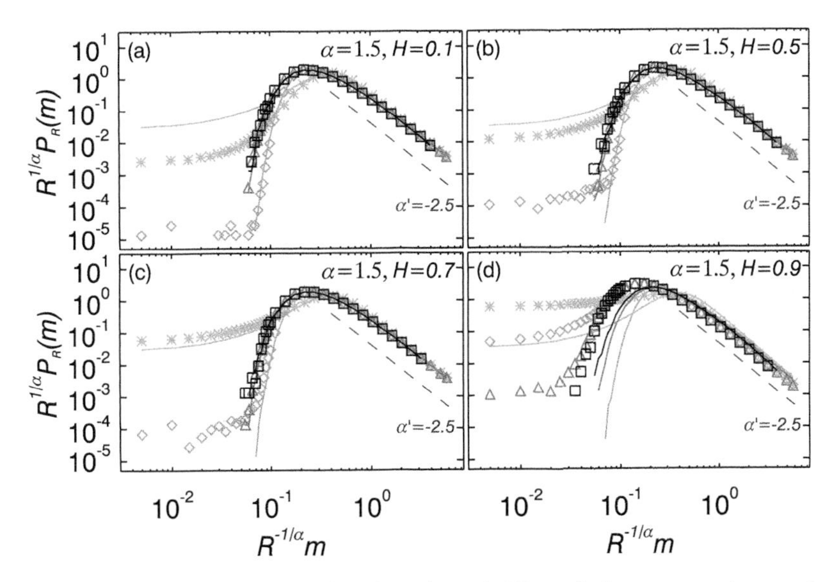

Figure 8. Rescaled versions of probability density functions shown in Figure 3; the same gray tones and symbols are used. The expected asymptotic scaling of the tails is indicated by the dark gray dashed lines $\alpha' = -1 - \alpha = -2.5$.

is equivalent to the iid case. Note that the results for the inverse condition (m_0 is smaller than the threshold) are also indistinguishable from the iid case. Figures 10c and 10d report the corresponding exceedance probabilities, clearly indicating clustering of extremes in the presence of long-range persistence compared with the time-independent case, i.e., larger events are more likely followed by large events in the time-dependent case.

For a more complete understanding, we now define the conditions m_0 as the geometric averages of nonoverlapping and exponentially growing ranges and combine bins at both ends to ensure at least 1000 conditional maxima per bin. (We start with a set of N_{bin}=20 nonoverlapping logarithmic bins, $\{\mathfrak{M}_k\}_{k=1,\ldots,N_{\text{bin}}} = \{[m_{k,min}, m_{k,max})\}_k$ with $m_{k,\text{min}} \leq m_{k,\text{max}}$ and $m_{1,\text{min}} > 0$, and associate the geometric mean $m_{k,0} = (m_{k,\text{min}} m_{k,\text{max}})^{1/2}$ with the condition $m_{k,0}$. Owing to the heavy tails in the Fréchet distribution, outer bins are much less populated than bins in the center. We therefore combine bins on both edges, starting from $k = 1$ and $k = N_{\text{bin}}$ and proceeding toward the center, until at least 1000 maxima m_j are contributing to each bin.) Figures 11 and 12 report the full results for conditional medians $\text{med}_R(m_{k,0}) = \text{median}\{m_{j,j=2,...,n} \mid m_{j-1} = m_{k,0}\}$ where the conditions $m_{k,0}$ are understood as ranges defined by the kth bin. In addition to the conditional

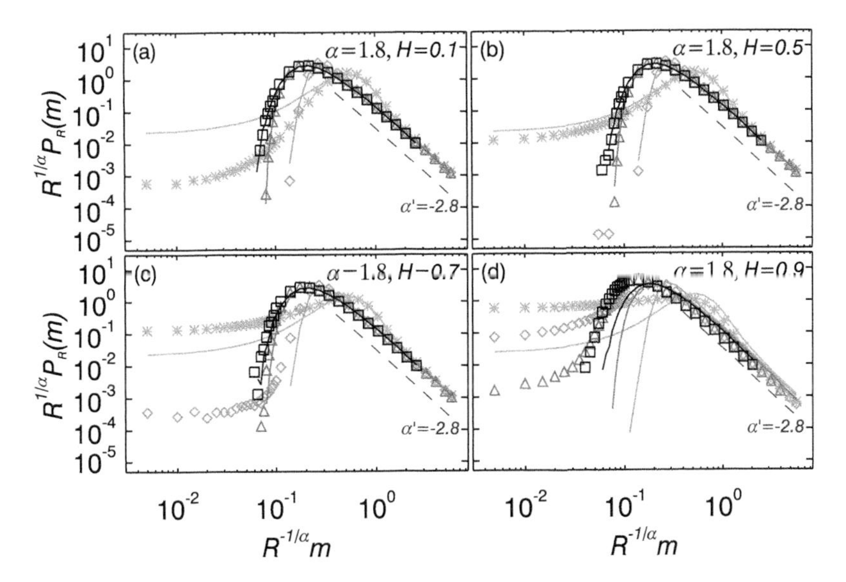

Figure 9. Same as Figure 8 but for $\alpha = 1.8$.

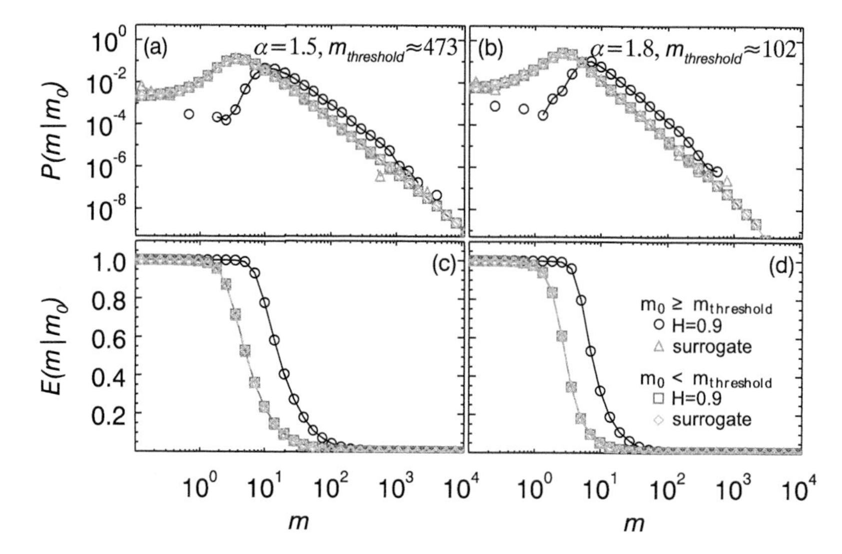

Figure 10. Examples of (a and b) the conditional probability density functions for block maxima and (c and d) corresponding exceedances for $R = 100$; (left) $\alpha = 1.5$ and (right) $\alpha = 1.8$. Shown are results for block maxima m that are preceded by a block maximum $m_0 \geq m_{\text{threshold}}$ for a (large) self-similarity index $H = 0.9$ (black circles) and for the corresponding randomized sequence of block maxima (gray triangles). For comparison results for the complementary condition, $m_0 < m_{\text{threshold}}$ are also displayed (self-similar SαS processes (dark gray squares), independent surrogate (light gray diamonds)). The threshold has been chosen to ensure 15,000 conditional maxima (Figures 10a and 10c $m_{\text{threshold}} \approx 473$, Figures 10b and 10d $m_{\text{threshold}} \approx 102$).

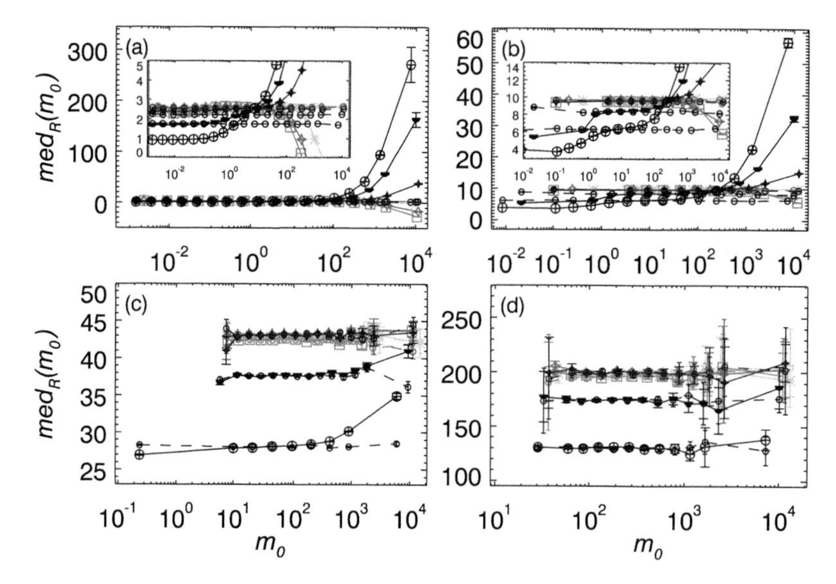

Figure 11. Median of the conditional distribution of block maxima, $\text{med}_R(m_0)$, given a preceding block maximum of m_0 for $\alpha = 1.5$. Results for different block sizes (a) $R = 10$, (b) $R = 100$, (c) $R = 1000$, and (d) $R = 10000$ are shown. Different self-similarity exponents are both gray tone and symbol coded; $H = 0.1$ (very light gray, stars), $H = 0.5$ (light gray, squares), $H = 0.6$ (gray, open diamonds), $H = 0.7$ (dark gray, solid diamonds), $H = 0.8$ (very dark gray, solid lower half circles), and $H = 0.9$ (black, circles with plus). The values expected in the absence of memory effects between block maxima are shown in the same grayscale coding but with smaller open circles (see text for details). Note that both corresponding curves for data with and without memory must intersect. Error bars are obtained from bootstrapping [*Efron*, 1979; *Efron and Tibshirani*, 1993] within each bin: We draw the same number of elements as in the bin with a cap at 50,000 and consider 10,000 realizations per bin. The insets in Figures 11a and 11b show magnifications of the intersection area.

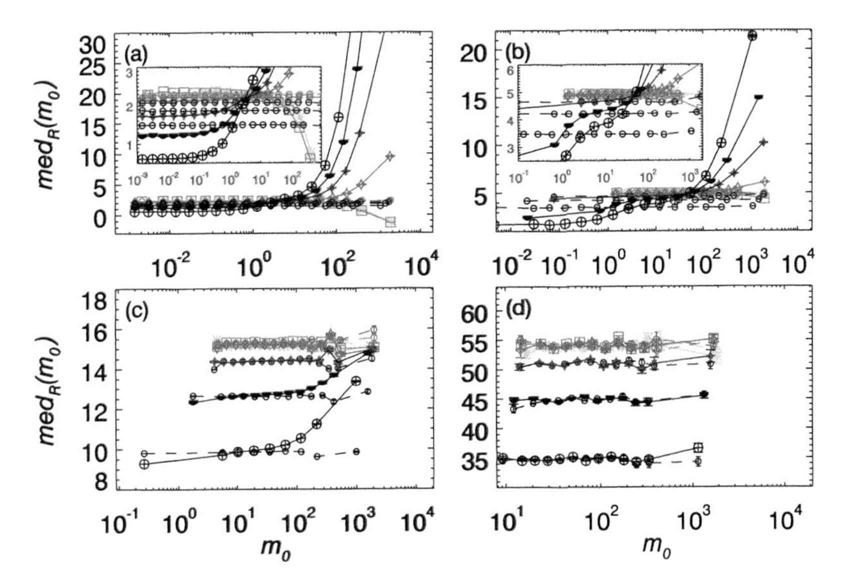

Figure 12. Same as Figure 11 but for a characteristic exponent $\alpha = 1.8$. The main plots have the same x axes as in Figure 11 to allow a better comparison.

median of self-similar SαS-distributed time series, we also show the conditional medians for the randomized sequence of block maxima for comparison. Shuffling the raw data affects the distribution of block maxima depending on the block size R. Hence, for generating independent surrogates, it is here important to shuffle the sequence of block maxima and *not* the underlying time series $\{x_i\}_i$ to preserve the distribution of block maxima. Note that shuffling the block maxima is equivalent to destroying long-range memory in $\{x_i\}$ on scales larger than the block size but preserving memory on smaller scales [*Schumann and Kantelhardt*, 2011]. Considering that the surrogates correspond to a time-independent hazard assessment, we observe that med_R (m_0) increases monotonically with m_0 in the case of strong persistence, again a clear indication of the clustering of block maxima. This dependence becomes less pronounced for larger R, and indeed, it should vanish in the limit $R \longrightarrow \infty$. This is also true for other processes with long-range persistence [*Eichner et al.*, 2006]. In the case of antipersistence, the deviations from the expected behavior for independent maxima are much smaller and only significant for small R. As expected, the behavior is opposite to the case of persistence: $\text{med}_R(m_0)$ decreases monotonically with m_0.

5.3. Record Statistics

While the theoretical results for record statistics presented in section 3 are all based on the assumption of iid processes, not much is known about the case of stationary processes with long-range memory [*Newman et al.*, 2010]. For self-similar SαS processes, we first analyze the probability $P(N_R)$ to find N_R records given a sequence of length R. As Figure 13 shows for $R = 10000$, this probability only deviates significantly from the iid prediction, which is independent of the underlying

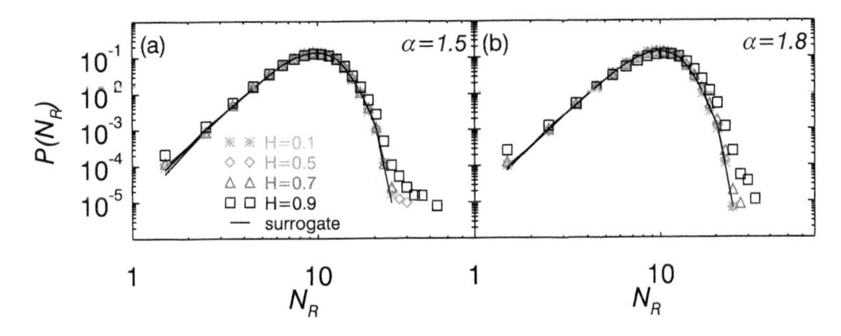

Figure 13. Probability of finding a given total number of records in sequences of block size $R = 10000$ for self-similar SαS processes with different parameters α and H (symbols). Results for corresponding surrogate data are shown for comparison (lines).

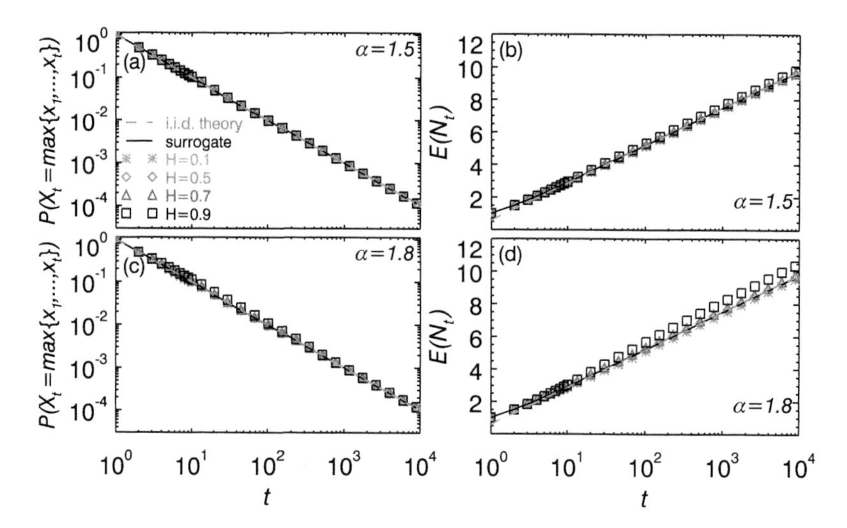

Figure 14. (a and c) Probability to break a record at a time t for different values of H and α. Black solid lines correspond to independent surrogate data ($H = 1/\alpha$), which do not deviate significantly from the theoretically expected $P(X_t = \max\{x_1, \ldots, x_t\}) = 1/t$ (see equation (5)). (b and d) Expected total number of records $E(N_t)$ that occur up to time t in the same color and symbol coding as in Figures 14a and 14c. The theoretical independent and identically distributed (iid) behavior is $E(N_t) = \ln|t| + \gamma$ (see equation (6a)). Significant deviations from the iid behavior are visible in the presence of strong persistence ($H = 0.9$, $\alpha = 1.8$).

distribution as follows from the discussion in section 3, in the case of strong persistence ($H > 1/\alpha$). Since persistence implies that large values tend to follow large values and small values rather follow small values, one expects that there is higher probability to encounter a small number of records as well as a larger number of records compared to the iid prediction. This is exactly what Figure 13 shows. While the above results could be specific to the chosen R, this is not confirmed by the dependence of the mean number of records $E(N_t)$ on t with $t \leq R$. Indeed, Figure 14 indicates that strong deviations from the iid behavior described by equation (6a) only occur in the presence of strong persistence. While the influence of persistence on record statistics is easily detectable in $E(N_t)$ and $P(N_R)$, the effect is more subtle if one considers the probability to break a record at a time t, $P(X_t = \max\{x_1, \ldots, x_t\})$ (see Figure 14). This is related to the fact that the former are cumulative measures of the latter as follows directly from the discussion in section 3.

Our results are consistent with earlier numerical findings for stationary Gaussian processes with persistent long-range memory [*Newman et al.*, 2010]. The authors found that the expected number of records $E(N_t)$ for fixed t increased monotonically with increasing persistence.

6. APPLICATION: SOLAR POWER INPUT INTO THE EARTH'S MAGNETOSPHERE

The solar wind is a prime example of plasma turbulence at low-frequency magnetohydrodynamic scales as evident from its power law energy distribution [*Zhou et al.*, 2004], the magnetic field correlations [*Matthaeus et al.*, 2005], and its intermittent dynamics [*Burlaga*, 2001]. While understanding this type of turbulence is an important challenge by itself, understanding the interplay between the solar wind and the Earth's magnetosphere is another open problem of considerable interest [*Baker*, 2000]. The difficulty of understanding the magnetospheric response to the solar wind variations is intimately related to the observed range of mechanisms of energy release and multiscale coupling phenomena [*Angelopoulos et al.*, 1999; *Lui et al.*, 2000; *Uritsky et al.*, 2002; *Borovsky and Funsten*, 2003; *D'Amicis et al.*, 2007; *Uritsky and Pudovkin*, 1998; *Chang*, 1999; *Chapman et al.*, 1999; *Klimas et al.*, 2000; *Sitnov et al.*, 2001; *Consolini*, 2002; *Aschwanden*, 2011]. Here we focus on the Akasofu ε parameter, which is a solar wind proxy for the energy input into the Earth's magnetosphere (see *Koskinen and Tanskanen* [2002] for a more recent discussion). In SI units, it is defined as

$$\varepsilon = v\frac{B^2}{\mu_0}\ell_0^2\sin^4(\theta/2), \tag{15}$$

where v is the solar wind velocity, B is the magnetic field, $\mu_0 = 4\pi \times 10^{-7}$ is the permeability of free space, $\ell_0 \approx 7R_E$, and $\theta = \arctan(|B_y|/B_z)$. Geocentric solar magnetospheric (GSM) coordinates are used.

The ACE spacecraft [*Stone et al.*, 1998] orbits the Earth-Sun L1 libration point approximately 1.5×10^9 m from the

Earth and monitors solar wind, interplanetary magnetic fields, and high-energy particles. The data can be downloaded from http://cdaweb.gsfc.nasa.gov. Specifically, for the years 2000–2007, we extracted the magnitude of the x component of the solar wind, and the y and z components of the magnetic fields, as seen, respectively, by the Solar Wind Electron, Proton, and Alpha Monitor (SWEPAM) and magnetometer (MAG) instruments (level 2 72 data) of the ACE spacecraft, all in GSM coordinates. The choice of components reflects the Poynting flux interpretation of the ε parameter. For the most part, measurements are available every 64 and 16 s for the wind velocity and magnetic fields, respectively. We calculated the ε parameter given by equation (15) every 64 s. Since the wind velocity and magnetic field measurements are not synchronized, we linearly interpolated the magnetic field measurements toward the time of the nearest wind velocity measurement. For these 8 years, the ε time series consisted of 3,944,700 points (see Figure 15a). Measurements for wind velocities or magnetic fields are sometimes unavailable. Approximately 9% of the points comprising the ε series are missing. As done by *Moloney and Davidsen* [2011], we set missing data points to the value of the last valid recording preceeding them (irrespective of the size of the data gap), thereby creating plateaus of constant intensity. This minimizes artifacts associated with points missing at regular experiment-specific frequencies. We have checked that nothing changes crucially in our statistical analyses by adopting other schemes.

Watkins et al. [2005] have suggested to model the ε parameter by a fractional Lévy motion as defined by equation (12), which implies that the series of changes in ε, $\Delta_\tau\varepsilon(t_k) = \varepsilon(t_k + \tau) - \varepsilon(t_k)$ with $t_k = t_0 + k\tau$ for $k \geq 0$, should form a self-similar SαS process. For the ε time series studied here and shown in Figure 15b for τ = 64s, *Moloney and Davidsen* [2010] have found that the properties of self-similarity and α stability both roughly hold for 64 s $\leq \tau \leq$ 4 h. Specifically, they found $\alpha = 1.55$ and $H = 0.40$ indicating the presence of weak antipersistence. Similar values have been observed for other ε series [see *Moloney and Davidsen* [2010], and references therein] always indicting the presence of antipersistence. Given these properties, one might ask to which extent the numerical results for time-dependent hazard assessment and record statistics of self-similar SαS processes presented in section 5 apply to the $\Delta_\tau\varepsilon$ series. This will be addressed in the following.

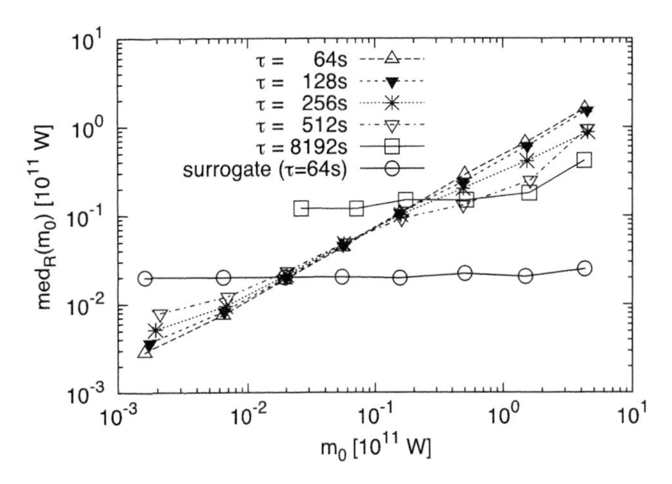

Figure 16. Median of the conditional distribution of block maxima $\text{med}_R(m_0)$ given a preceding block maximum of m_0 within a logarithmic bin for the $\Delta_\tau\varepsilon$ series with different values of τ (symbols and line style coded). The block size is $R = 100$. Surrogate data (obtained by shuffling the $\Delta_\tau\varepsilon$ = 64 s series' block maxima) corresponding to the iid case are shown for comparison (solid curve, circles). Surrogates for other values of τ (not shown) are similar but increase monotonically with τ.

6.1. Conditional Extrema and Hazard Assessment

Figure 16 shows that the median of the conditional distribution of the block maxima, $\text{med}_R(m_0)$, increases monotonically and approximately as a power law with m_0 for τ = 64 s. As τ is increased, the power law's exponent decreases, eventually leaving the power law regime and approaching the iid

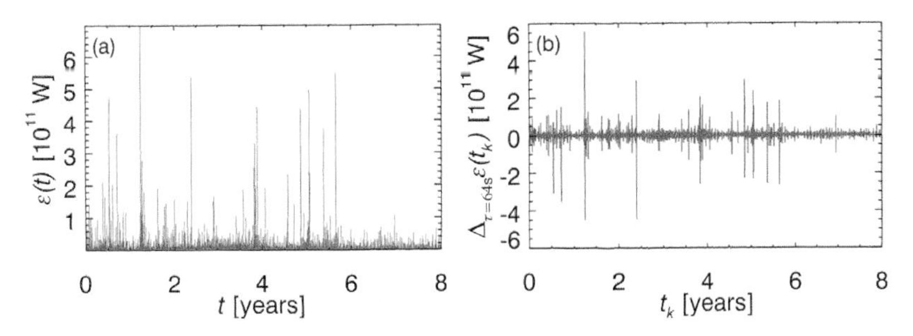

Figure 15. (a) Akasofu's $\varepsilon(t)$ time series derived from ACE data according to equation (15) for the years 2000 to 2007 with a sampling of 64 s. (b) $\Delta_{\tau=64s}\varepsilon(t_k)$ time series obeying a symmetric heavy-tailed distribution as shown by *Moloney and Davidsen* [2010].

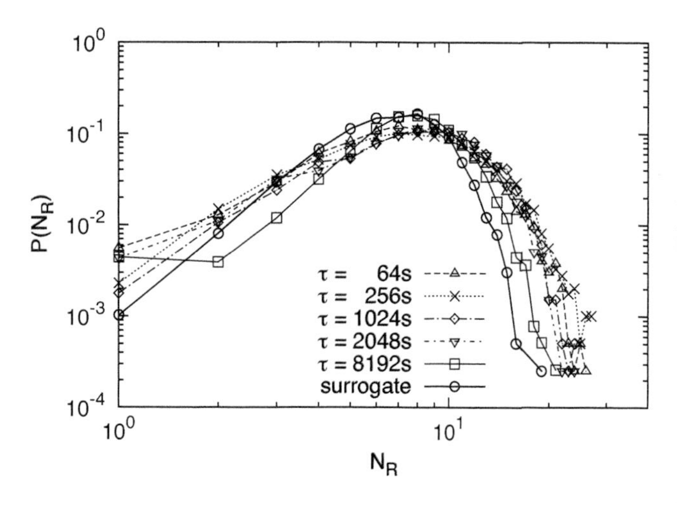

Figure 17. Probability density function for the number of records in nonoverlapping blocks of size $R = 1000$ of the $\Delta_\tau\varepsilon$ series for different values of τ. Surrogate data (obtained by shuffling the $\Delta_\tau\varepsilon$ series) corresponding to the iid case are shown for comparison.

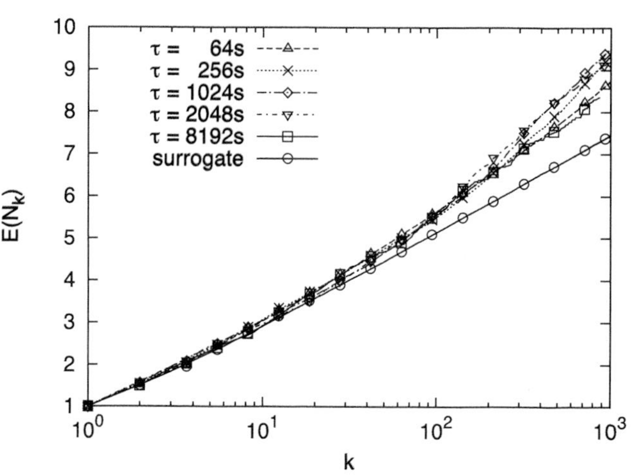

Figure 18. Expected number of records as a function of values k for the $\Delta_\tau\varepsilon$ series for different values of τ. Estimates are based on nonoverlapping blocks of size $R = 1000$. Surrogate data (obtained by shuffling the $\Delta_\tau\varepsilon$ series) corresponding to the iid case are shown for comparison.

case (median independent of condition) in the limit of very large τ. Note that the median values of the block maxima and their corresponding surrogates, obtained by shuffling the block maxima, naturally increase for larger block sizes since the fewer blocks are more and more dominated by the largest extreme values in the time series (compare $\tau = 8192$ s). Our findings indicate clustering of extreme values for not too large τ and, thus, persistence, in sharp contrast to what is expected for antipersistent self-similar SαS processes (see Figures 11 and 12). This contradiction provides clear evidence that only some properties of the $\Delta_\tau\varepsilon$ series can be (roughly) characterized by a self-similar SαS process. In particular, the memory reflected in the $\Delta_\tau\varepsilon$ is not reliably captured by a self-similar SαS process, potentially due to nonstationarities. A similar conclusion has been reached by *Moloney and Davidsen* [2010], who investigated the distributions of block maxima of $\Delta_\tau\varepsilon$. Nevertheless, the findings summarized by Figure 16 show unambiguously that time-dependent hazard assessment can be successfully applied to the considered time series and clearly outperforms any time-independent hazard assessment of future extreme values.

6.2. Record Statistics

The observed clustering in the $\Delta_\tau\varepsilon$ series discussed in section 6.1 is also present in the record statistics. Figure 17 shows that the probability $P(N_R)$ to find N_R records in a sequence of length R is much higher for large N_R than what is expected for the iid case. This is further confirmed by Figure 18, which shows that the expected number of records grows much faster than in the iid case. Both figures also suggest a dependence on τ. In particular, the deviations from the iid case for large times increase monotonically with τ until they reach a maximum around $\tau = 1024$ s. As τ is increased further, the deviations decrease monotonically with τ, and finally, they approach the iid case in the limit of very large τ. This is supported by the probability to observe a record at an elapsed time t shown in Figure 19. Note that in contrast to these results found for records, the conditional

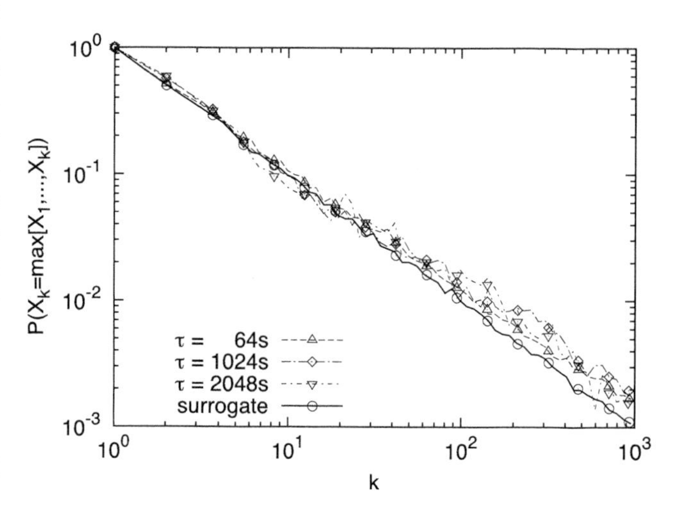

Figure 19. Probability to observe a record at values k for the $\Delta_\tau\varepsilon$ series for different values of τ. Estimates are based on nonoverlapping blocks of size $R = 1000$. Surrogate data (obtained by shuffling the $\Delta_\tau\varepsilon$ series) corresponding to the iid case are shown for comparison.

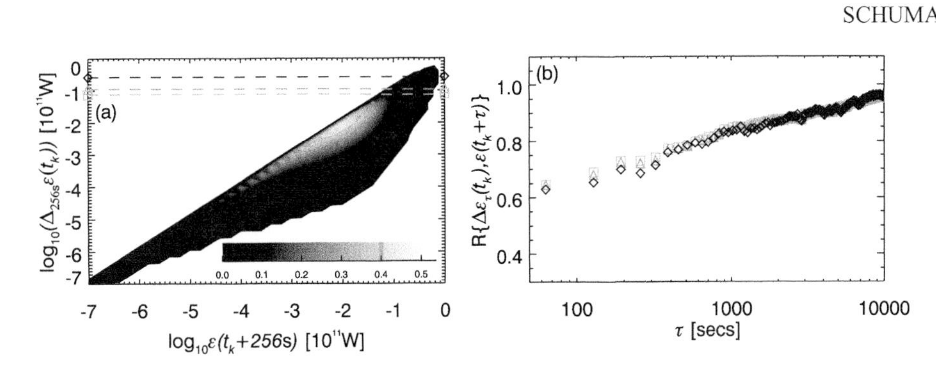

Figure 20. (a) Probability density function to observe a given pair of values $\varepsilon(t_k + \tau)$ and $\Delta_\tau\varepsilon(t_k) = \varepsilon(t_k + \tau) - \varepsilon(t_k)$ simultaneously for a fixed value of $\tau = 256$ s and only for pairs, where $\Delta_\tau\varepsilon(t_k) > 0$. Logarithmic bins were used. The dashed lines indicate from top to bottom the 99.9% (black, diamonds), 99.5% (gray, triangles), and the 99% (light gray, squares) quantile of the $\Delta_{256s}\varepsilon$ distribution. (b) Pearson's correlation coefficient calculated for the same upper quantiles as in Figure 20a (same symbols and gray tones) for all time series pairs $(\Delta_\tau\varepsilon(t_k), \varepsilon(t_k + \tau))$ as a function of the time shift τ.

maxima in Figure 16 apparently showed a monotonic dependence on τ.

6.3. Discussion

While we have focused here on $\Delta_\tau\varepsilon$ series, our results with respect to time-dependent hazard assessment are directly applicable to the ε series itself. As Figure 20 shows, there is a strong correlation between $\Delta_\tau\varepsilon(t_k)$ and $\varepsilon(t_k + \tau)$, as measured by the Pearson correlation coefficient, if one considers the largest values in $\Delta_\tau\varepsilon(t_k)$. This implies that a large extreme value in $\Delta_\tau\varepsilon$ typically indicates a large value in ε, while a smaller extreme value in $\Delta_\tau\varepsilon$ tends to correspond to a smaller value in ε, independent of the exact value of τ. Thus, one can translate our findings from section 6.1: Extreme values in ε tend to be clustered, which makes time-dependent hazard assessment in the context of space weather feasible. These results are just a first step and clearly need to be investigated in more detail in future work.

7. SUMMARY

In summary, our numerical analysis of self-similar SαS processes has shown that the presence of long-range memory can lead to significant finite size effects in extreme value and record statistics despite the fact that some of the asymptotic behavior is not affected. The finite size effects are particularly pronounced if the long-range memory leads to persistent behavior. We also found that long-range memory allows a time-dependent hazard assessment of the size of future extreme events based on extreme events observed in the past. Time-dependent hazard assessment based on the value of the previous block maximum significantly outperforms time-independent hazard assessment and, thus, provides a significant improvement.

Moreover, we showed that such a time-dependent hazard assessment is directly applicable in the context of the solar power influx into the magnetosphere. Since many processes in nature are characterized by long-range memory and heavy-tailed distributions, especially in space physics [*Watkins et al.*, 2005; *Moloney and Davidsen*, 2010], our findings are of general interest. Further examples outside geophysics include communications traffic [*Laskin et al.*, 2002].

Acknowledgments. We thank A. Rasmussen for helpful discussions. This project was supported by Alberta Innovates (formerly Alberta Ingenuity).

REFERENCES

Albeverio, S., V. Jentsch, and H. Kantz (Eds.) (2006), *Extreme Events in Nature and Society*, 352 pp., Springer, Berlin.

Aldrich, J. (1997), R.A. Fisher and the making of maximum likelihood 1912–1922, *Stat. Sci.*, *12*(3), 162–176, doi:10.1214/ss/1030037906.

Angelopoulos, V., T. Mukai, and S. Kokubun (1999), Evidence for intermittency in Earth's plasma sheet and implications for self-organized criticality, *Phys. Plasmas*, *6*, 4161–4168.

Aschwanden, M. (2011), *Self-Organized Criticality in Astrophysics*, 420 pp., Springer, Berlin.

Baker, D. N. (2000), Effects of the sun on the earth's environment, *J. Atmos. Sol. Terr. Phys.*, *62*(17–18), 1669–1681.

Benestad, R. E. (2003), How often can we expect a record event?, *Clim. Res.*, *23*, 3–13.

Benestad, R. E. (2008), A simple test for changes in statistical distributions, *Eos Trans. AGU*, *89*(41), 389, doi:10.1029/2008EO410002.

Berman, S. M. (1964), Limit theorems for the maximum term in stationary sequences, *Ann. Math. Stat.*, *33*, 502–516.

Bogachev, M. I., J. F. Eichner, and A. Bunde (2007), Effect of nonlinear correlations on the statistics of return intervals in multifractal data sets, *Phys. Rev. Lett.*, *99*, 240601, doi:10.1103/PhysRevLett.99.240601.

Borovsky, J. E., and H. O. Funsten (2003), Role of solar wind turbulence in the coupling of the solar wind to the Earth's magnetosphere, *J. Geophys. Res.*, *108*(A6), 1246, doi:10.1029/2002JA009601.

Bruno, R., L. Sorriso-Valvo, V. Carbone, and B. Bavassano (2004), A possible truncated-Lévy-flight statistics recovered from interplanetary solar-wind velocity and magnetic-field fluctuations, *Europhys. Lett.*, *66*(1), 146, doi:10.1209/epl/i2003-10154-7.

Bunde, A., J. Kropp, and H. J. Schellnhuber (Eds.) (2002), *The Science of Disasters: Climate Disruptions, Heart Attacks and Market Crashes*, 453 pp., Springer, Berlin.

Burlaga, L. F. (2001), Lognormal and multifractal distributions of the heliospheric magnetic field, *J. Geophys. Res.*, *106*, 15,917–15,927.

Caers, J., J. Beirlant, and M. A. Maes (1999), Statistics for modeling heavy tailed distributions in geology: Part II. Applications, *Math. Geol.*, *31*(4), 411–434.

Chambers, J. M., C. L. Mallows, and B. W. Stuck (1976), A method for simulating stable random variables, *J. Am. Stat. Assoc.*, *71* (354), 340–344.

Chambers, J. M., C. L. Mallows, and B. W. Stuck (1987), Corrections: A method for simulating stable random variables, *J. Am. Stat. Assoc.*, *82*(398), 704.

Chandler, K. N. (1952), The distribution and frequency of record values, *J. R. Stat. Soc., Ser. B*, *14*(2), 220–228.

Chang, T. (1999), Self-organized criticality, multi-fractal spectra, sporadic localized reconnections and intermittent turbulence in the magnetotail, *Phys. Plasmas*, *6*(11), 4137–4145.

Chapman, S. C., R. O. Dendy, and G. Rowlands (1999), A sandpile model with dual scaling regimes for laboratory, space, and astrophysical plasmas, *Phys. Plasmas*, *6*(11), 4169–4177.

Chechkin, A., and V. Gonchar (2000), A model for persistent Lévy motion, *Physica A*, *277*(3–4), 312–326.

Chechkin, A., and V. Gonchar (2001), Fractional Brownian motion approximation based on fractional integration of a white noise, *Chaos Solitons Fractals*, *12*(2), 391–398, doi:10.1016/S0960-0779(99)00183-6.

Coles, S. (2007), *An Introduction to Statistical Modeling of Extreme Values*, 4th ed., Springer, London, U. K.

Consolini, G. (2002), Self-organized criticality: A new paradigm for the magnetotail dynamics, *Fractals*, *10*(3), 275–283.

D'Amicis, R., R. Bruno, and B. Bavassano (2007), Is geomagnetic activity driven by solar wind turbulence?, *Geophys. Res. Lett.*, *34*, L05108, doi:10.1029/2006GL028896.

Davidsen, J., P. Grassberger, and M. Paczuski (2006), Earthquake recurrence as a record breaking process, *Geophys. Res. Lett.*, *33*, L11304, doi:10.1029/2006GL026122.

Davidsen, J., P. Grassberger, and M. Paczuski (2008), Networks of recurrent events, a theory of records, and an application to finding causal signatures in seismicity, *Phys. Rev. E*, *77*(6), 066104, doi:10.1103/PhysRevE.77.066104.

de Haan, L., and A. Ferreira (2006), *Extreme Value Theory: An Introduction*, 418 pp., Springer, New York.

Easterling, D. R., G. A. Meehl, C. Parmesan, S. A. Changnon, T. R. Karl, and L. O. Mearns (2000), Climate extremes: Observations, modeling, and impacts, *Science*, *289*, 2068–2074.

Efron, B. (1979), 1977 Bootstrap methods: Another look at the jackknife, *Ann. Stat.*, *7*(1), 1–26.

Efron, B., and R. J. Tibshirani (1993), *An Introduction to the Bootstrap*, 456 pp., Chapman and Hall, New York.

Eichner, J. F., J. W. Kantelhardt, A. Bunde, and S. Havlin (2006), Extreme value statistics in records with long-term persistence, *Phys. Rev. E*, *73*, 016130, doi:10.1103/PhysRevE.73.016130.

Embrechts, P., and M. Makoto (2002), *Selfsimilar Processes*, 152 pp., Princeton Univ. Press, Princeton, N. J.

Embrechts, P., and H. Schmidli (1994), Modelling of extremal events in insurance and finance, *Math. Methods Oper. Res.*, *39*(1), 1–34, doi:10.1007/BF01440733.

Embrechts, P., C. Klüppelberg, and T. Mikosch (2004), *Modelling Extremal Events for Insurance and Finance*, Springer, Berlin.

Galambos, J., J. Lechner, and E. Simin (Eds.) (1994), *Extreme Value Theory and Applications*, 536 pp., Kluwer, Dordrecht, Netherlands.

Glaser, R., and H. Stangl (2004), Climate and floods in central Europe since AD 1000: Data, methods, results and consequences, *Surv. Geophys.*, *25*, 485–510.

Glick, N. (1978), Breaking records and breaking boards, *Am. Math. Mon.*, *85*(1), 2–26.

Györgyi, G., N. R. Moloney, K. Ozogány, and Z. Rácz (2008), Finite-size scaling in extreme statistics, *Phys. Rev. Lett.*, *100*, 210601, doi:10.1103/PhysRevLett.100.210601.

Hnat, B., S. C. Chapman, and G. Rowlands (2003), Intermittency, scaling, and the Fokker-Planck approach to fluctuations of the solar wind bulk plasma parameters as seen by the WIND spacecraft, *Phys. Rev. E*, *67*(5), 056404, doi:10.1103/PhysRevE.67.056404.

Hurst, H. E. (1951), Long-term storage capacity of reservoirs, *Trans. Am. Soc. Civ. Eng.*, *116*, 770–799.

Huybers, P., and W. Curry (2006), Links between annual, Milankovitch, and continuum temperature variability, *Nature*, *441*, 329–332.

Kantelhardt, J. W., D. Rybski, S. A. Zschiegner, P. Braun, E. Koscielny-Bunde, V. Livina, S. Havlin, and A. Bunde (2003), Multifractality of river runoff and precipitation: Comparison of fluctuation analysis and wavelet methods, *Physica A*, *330*, 240–245.

Kantelhardt, J. W., E. Koscielny-Bunde, D. Rybski, P. Braun, A. Bunde, and S. Havlin (2006), Long-term persistence and multifractality of precipitation and river runoff records, *J. Geophys. Res.*, *111*, D01106, doi:10.1029/2005JD005881.

Klimas, A. J., J. A. Valdivia, D. Vassiliadis, D. N. Baker, M. Hesse, and J. Takalo (2000), Self-organized criticality in the substorm phenomenon and its relation to localized reconnection in the magnetospheric plasma sheet, *J. Geophys. Res.*, *105*(A8), 18,765–18,780.

Koscielny-Bunde, E., H. Roman, A. Bunde, S. Havlin, and H. Schellnhuber (1998), Long-range power-law correlations in local daily temperature fluctuations, *Philos. Mag. B*, *77*(5), 1331–1340.

Koskinen, H. E. J., and E. I. Tanskanen (2002), Magnetospheric energy budget and the epsilon parameter, *J. Geophys. Res.*, *107*(A11), 1415, doi:10.1029/2002JA009283.

Laskin, N., I. Lambadaris, F. C. Harmantzis, and M. Devetsikiotis (2002), Fractional Lévy motion and its application to network traffic modeling, *Comput. Networks*, *40*, 363–375.

Leadbetter, M. R., and H. Rootzén (1988), Extremal theory for stochastic processes, *Ann. Probab.*, *16*, 431–478.

Leadbetter, M. R., G. Lindgren, and H. Rootzen (1983), *Extremes and Related Properties of Random Sequences and Processes*, 336 pp., Springer, New York.

Lui, A. T. Y., S. C. Chapman, K. Liou, P. T. Newell, C. I. Meng, M. Brittnacher, and G. K. Parks (2000), Is the dynamic magnetosphere an avalanching system?, *Geophys. Res. Lett.*, *27*(7), 911–914.

Matsumoto, M., and T. Nishimura (1998), Mersenne twister: A 623-dimensionally equidistributed uniform pseudo-random number generator, *Trans. Model. Comput. Simul.*, *8*(1), 3–30.

Matthaeus, W. H., S. Dasso, J. M. Weygand, L. J. Milano, C. W. Smith, and M. G. Kivelson (2005), Spatial correlations of solar-wind turbulence from two-point measurements, *Phys. Rev. Lett.*, *95*, 231101, doi:10.1103/PhysRevLett.95.231101.

Moloney, N. R., and J. Davidsen (2009), Extreme value statistics and return intervals in long-range correlated uniform deviates, *Phys. Rev. E*, *79*, 041131, doi:10.1103/PhysRevE.79.041131.

Moloney, N. R., and J. Davidsen (2010), Extreme value statistics in the solar wind: An application to correlated Lévy processes, *J. Geophys. Res.*, *115*, A10114, doi:10.1029/2009JA015114.

Moloney, N. R., and J. Davidsen (2011), Extreme bursts in the solar wind, *Geophys. Res. Lett.*, *38*, L14111, doi:10.1029/2011GL048245.

Monetti, R. A., S. Havlin, and A. Bunde (2002), Long-term persistence in the sea surface temperature fluctuations, *Physica A*, *320*, 581–589.

Nevzorov, V. B. (2001), *Records: Mathematical Theory*, *Trans. Math. Monogr.*, vol. 194, 170 pp., Am. Math. Soc., Providence, R. I.

Newman, W. I., B. D. Malamud, and D. L. Turcotte (2010), Statistical properties of record-breaking temperatures, *Phys. Rev. E*, *82*(6), 066111, doi:10.1103/PhysRevE.82.066111.

Painter, S., and L. Paterson (1994), Fractional Lévy motion as a model for spatial variability in sedimentary rock, *Geophys. Res. Lett.*, *21*(25), 2857–2860.

Peixoto, T. P., K. Doblhoff-Dier, and J. Davidsen (2010), Spatiotemporal correlations of aftershock sequences, *J. Geophys. Res.*, *115*, B10309, doi:10.1029/2010JB007626.

Pelletier, J., and D. L. Turcotte (1997), Long-range persistence in climatological and hydrological time series: Analysis, modeling and application to drought hazard assessment, *J. Hydrol.*, *203*, 198–208.

Pommois, P., G. Zimbardo, and P. Veltri (1998), Magnetic field line transport in three dimensional turbulence: Lévy random walk and spectrum models, *Phys. Plasmas*, *5*(5), 1288–1297.

Redner, S., and M. R. Petersen (2006), Role of global warming on the statistics of record-breaking temperatures, *Phys. Rev. E*, *74*(6), 061114, doi:10.1103/PhysRevE.74.061114.

Rény, A. (1962), Theory des éléments saillants d'une suite d'observations [with summary in English], in *Colloquium on Combinatorial Methods in Probability Theory*, pp. 104–117, Mat. Inst., Aarhus Univ., Denmark.

Roman, H. E., A. Celi, and G. De Filippi (2008), Fluctuation analysis of meteo-marine data, *Eur. Phys. J. Spec. Top.*, *161*, 195–205, doi:10.1140/epjst/e2008-00761-4.

Samorodnitsky, G. (2004), Extreme value theory, ergodic theory and the boundary between short memory and long memory for stationary stable processes, *Ann. Probab.*, *32*, 1438–1468.

Samorodnitsky, G., and M. S. Taqqu (1994), *Stable Non-Gaussian Random Processes: Stochastic Models with Infinite Variance*, 632 pp., Chapman and Hall, London, U. K.

Schmittmann, B., and R. K. P. Zia (1999), "Weather" records: Musings on cold days after a long hot Indian summer, *Am. J. Phys.*, *67*, 1269–1276.

Schumann, A. Y. (2011), *Fluctuations and Synchronization in Complex Physiological Systems*, 245 pp., Logos, Berlin.

Schumann, A. Y., and J. W. Kantelhardt (2011), Multifractal moving average analysis and test of multifractal model with tuned correlations, *Physica A*, *390*(14), 2637–2654.

Schweigler, T., and J. Davidsen (2011), Clustering of extreme and recurrent events in deterministic chaotic systems, *Phys. Rev. E*, *84*, 016202, doi:10.1103/PhysRevE.84.016202.

Sibani, P., and H. J. Jensen (2009), Record statistics and dynamics, in *Encyclopedia of Complexity and System Science*, edited by R. A. Meyers, pp. 7583–7591, Springer, New York.

Sitnov, M. I., A. S. Sharma, K. Papadopoulos, and D. Vassiliadis (2001), Modeling substorm dynamics of the magnetosphere: From self-organization and self-organized criticality to nonequilibrium phase transitions, *Phys. Rev. E*, *65*, 016116, doi:10.1103/PhysRevE.65.016116.

Sornette, D. (2006), *Critical Phenomena in Natural Sciences*, 2nd ed., 550 pp., Springer, Berlin.

Stoev, S., and M. S. Taqqu (2004), Simulation methods for linear fractional stable motion and FARIMA using the Fast Fourier Transform, *Fractals*, *12*(1), 95–121.

Stone, E. C., A. M. Frandsen, R. A. Mewaldt, E. R. Christian, D. Margolies, J. F. Ormes, and F. Snow (1998), The advanced composition explorer, *Space Sci. Rev.*, *86*, 1–23.

Taqqu, M. S. (1987), Random processes with long-range dependence and high variability, *J. Geophys. Res.*, *92*(D8), 9683–9686.

Uritsky, V. M., and M. I. Pudovkin (1998), Low-frequency 1/*f*-like fluctuations of the AE-index as a possible manifestation of self-organized criticality in the magnetosphere, *Ann. Geophys.*, *16*, 1580–1588.

Uritsky, V. M., A. J. Klimas, D. Vassiliadis, D. Chua, and G. Parks (2002), Scale-free statistics of spatiotemporal auroral emissions as depicted by POLAR UVI images: Dynamic magnetosphere is an avalanching system, *J. Geophys. Res.*, *107*(A12), 1426, doi:10.1029/2001JA000281.

Van Aalsburg, J., W. I. Newman, D. L. Turcotte, and J. B. Rundle (2010), Record-breaking earthquakes, *Bull. Seismol. Soc. Am.*, *100*(4), 1800–1805, doi:10.1785/0120090015.

van den Brink, H. W., and G. P. Können (2008), The statistical distribution of meteorological outliers, *Geophys. Res. Lett.*, *35*, L23702, doi:10.1029/2008GL035967.

Vasudevan, K., D. W. Eaton, and J. Davidsen (2010), Intraplate seismicity in Canada: A graph theoretic approach to data analysis and interpretation, *Nonlinear Processes Geophys.*, *17*, 513–527.

Vogel, R. M., A. Zafirakou-Koulouris, and N. C. Matalas (2001), Frequency of record-breaking floods in the United States, *Water Resour. Res.*, *37*(6), 1723–1731.

Watkins, N. W., D. Credgington, B. Hnat, S. C. Chapman, M. P. Freeman, and J. Greenhough (2005), Towards synthesis of solar wind and geomagnetic scaling exponents: A fractional Lévy motion model, *Space Sci. Rev.*, *121*, 271–284.

Watkins, N. W., D. Credgington, R. Sanchez, S. J. Rosenberg, and S. C. Chapman (2009), Kinetic equation of linear fractional stable motion and applications to modeling the scaling of intermittent bursts, *Phys. Rev. E*, *79*(4), 041124, doi:10.1103/PhysRevE.79.041124.

Wergen, G., and J. Krug (2010), Record-breaking temperatures reveal a warming climate, *Europhys. Lett.*, *92*, 30008, doi:10.1209/0295-5075/92/30008.

Wu, W., G. Michailidis, and D. Zhang (2004), Simulating sample paths of linear fractional stable motion, *IEEE Trans. Inf. Theory*, *50*(6), 1086–1096, doi:10.1109/TIT.2004.828059.

Zaslavsky, G. M., P. N. Guzdar, M. Edelnman, M. I. Sitnov, and A. S. Sharma (2008), Multiscale behavior and fractional kinetics from the data of solar wind-magnetosphere coupling, *Commun. Nonlinear Sci. Numer. Simul.*, *13*, 314–330.

Zhou, Y., W. H. Matthaeus, and P. Dmitruk (2004), Magnetohydrodynamic turbulence and time scales in astrophysical and space plasmas, *Rev. Mod. Phys.*, *76*, 1015–1035.

J. Davidsen and A. Y. Schumann, Complexity Science Group, Department of Physics and Astronomy, University of Calgary, 2500 University Drive NW, Calgary, AB T2N 1N4, Canada. (davidsen@phas.ucalgary.ca; ay.schumann@ucalgary.ca)

N. R. Maloney, Max Planck Institute for the Physics of Complex Systems, Nöthnitzer Str. 38, D-01187 Dresden, Germany.

Extreme Event Recurrence Time Distributions and Long Memory

M. S. Santhanam

Indian Institute of Science Education and Research, Pune, India

The distribution of return intervals for extreme events is an important quantity of interest in many fields ranging from physics and geophysical sciences to finance and physiology. Extreme events are understood to imply exceedance of a discretely sampled time series above a prescribed threshold q. The recurrence or return interval is the elapsed time interval between successive extreme events. The question is if it is possible to determine the recurrence or return interval distribution from the knowledge of the temporal dependencies in a time series. The return interval distribution for uncorrelated time series is known to have an exponentially decaying form. What is the behavior of return interval distribution for processes that display long memory? This question assumes importance because most of the naturally occurring physical processes display long-memory property. Long memory implies slow decay of autocorrelations. In recent times, considerable research effort has been directed toward studying the distribution of return intervals for such time series. In this chapter, we review the work done in this area in the last few years and discuss the results arising from simulations as well as from analytical approaches while paying special attention to the recent developments in the recurrence time statistics for seismic activity. We also point out many parallel results in areas ranging from computing to finance.

1. INTRODUCTION

Extreme events in nature can be thought of as emergent phenomena in which many different physical processes working on disparate time scales conspire to ultimately produce events of magnitude much larger or smaller than the average. For instance, tremors result from complex spatio-temporal organization of seismic processes and take place almost all the time; however, it qualifies to be an extreme event called earthquake when the magnitude of tremors are far greater than its usual average. Floods, drought, cyclones or hurricanes, extreme space weather, forest fires, and tsunami are all examples of extreme events, which happen rarely but have tremendous impact on life and property [*Albeverio et al.*, 2005]. In general, similar extreme events could take place in the socioeconomic and technological domains, such as stock market crashes or network congestion and power grid failures. One widely reported event is the power blackout in the Northeastern United States in the year 2003 and is believed to have been caused due to cascading grid failures (see *Buldyrev et al.* [2010] for an analysis of cascading failures in interdependent networks). The extreme space weather event during October–November 2003 was another one that led to disruptions in power transmission and communications [*Gopalswamy et al.*, 2005a, 2005b]. From a statistical physics point of view, all these systems, whether natural or otherwise, belong to a class of driven, nonequilibrium phenomena [*Sornette*, 2005; *Majumdar*, 1999; *Mandelbrot and van Ness*, 1968]. Hence, considering the impact of extreme events on society, it is important to study and understand various properties related to extreme events and their occurrences. This will also further help understand the general phenomenology of nonequilibrium physics as well. It is indeed desirable to study the physical origins of extreme

Extreme Events and Natural Hazards: The Complexity Perspective
Geophysical Monograph Series 196

10.1029/2011GM001145

events using models that have a basis in physics of the system. In the case of most natural processes, this is a nontrivial and long-term research pursuit and is currently vigorously pursued by many groups. In this chapter, however, we take a different point of view. We assume that the measured time series of physical variables (such as the wind speed, tremors, temperature, etc.) encapsulates all the information about the physical origins including the spatiotemporal correlations that characterize the system. Hence, we can also study the extreme events by synthetically generating time series with specified spatial and temporal correlations. Invariably, the measured variables are always contaminated by noise from various sources. In this chapter, we do not attempt to account for these additional sources of noise in the measured variables. Thus, the implicit assumption is that the strength of these noise sources is sufficiently weak that they do not affect the temporal correlations inherent in the measured time series of the variables.

The study of extreme events in univariate time series has a long and distinguished history with significant results [*Gumbel*, 2004; *Sornette*, 2005]. Consider an independent and identically distributed sequence $X_1, X_2, X_3, \ldots X_n$. Then, after suitable rescaling, the limiting cumulative distribution for the maxima of this sequence for large n can be one of Fréchet, Gumbel, or Weibull distribution depending on the behavior of the tail of the probability density function (pdf) $f(x)$ [*Gumbel*, 2004]. Gumbel distribution is obtained for the cases in which the pdf is unbounded and decays faster than power law. Fréchet distribution is obtained in the case when pdf is bounded from below and falls as power law as $x \to \infty$. Weibull distribution corresponds to the case of pdf, which has an upper bound and decays slowly near the upper bound. This has been amply verified in many cases of interest. For a detailed exposition on the subject, the reader is referred to the work of *Kotz and Nadarajah* [2001] and *Gumbel* [2004]. In any application of extreme value theory, one talks of extreme value distributions in the context of block maxima, i.e., distributions of maxima occurring in a suitable time interval block. For instance, in many geophysical applications, 1 year might be a suitable time interval block. This implies that we reject all other measured data points except the maxima in a given block. One possible way to use all the available information is to model the exceedance over a chosen threshold, the so-called peaks-over-threshold approach. The distribution of these exceedance belongs to a class of generalized Pareto distribution [*Embrechts et al.*, 1997]. *Pisarenko and Sornette* [2003] provide an introduction to generalized Pareto distribution with emphasis on application to extreme seismic events. In this chapter, we do not pursue these questions about the distribution of extreme events, their properties, and prediction (see *Ghil et al.* [2011] for a detailed review) but rather study the distribution of intervals between the successive occurrences of the extreme events.

2. LONG MEMORY

It is by now well established that many of the naturally occurring processes display long-term memory [*Koscielny-Bunde et al.*, 1996, 1998; *Wang et al.*, 2006; *Vyushin and Kushner*, 2009; *Yamasaki et al.*, 2005; *Corral et al.*, 2008; *Wheatland*, 2003; *Mandelbrot and Wallis*, 1968; *Rybski et al.*, 2008; *Pelletier and Turcotte*, 1997; *Riemann-Campe et al.*, 2010; *Turcotte et al.*, 1997] in the sense described below (see also the work of *Lanfredi et al.* [2009] for a critical discussion on the presence of long-range memory in daily atmospheric temperature variability). Many computing and network-related processes, too, display similar memory features [*Falaki and Sorensen*, 1999; *Antoniou et al.*, 2002; *Cai et al.*, 2009]. This property implies slowly decaying autocorrelations in stark contrast to the fast decay of correlations when the system displays nearly no memory at all. For a stationary, discretely sampled time series $x(t)$, $t = 0,1,2\ldots N$ (with zero mean) this is measured by the two-point correlation function as $N \to \infty$ and is given by

$$C(\tau) = \frac{\langle x(t)x(t+\tau)\rangle}{\langle x^2(t)\rangle}, \tag{1}$$

where $\langle . \rangle$ represents the mean, and τ is the lag. In typical situations of statistical physics, such as the case of ideal gas in equilibrium at temperature T, the system forgets its initial state quickly, and the autocorrelation is $C(\tau) \sim \exp(-\alpha\tau)$, where $\alpha > 0$ is some constant. In particular, for systems that exhibit such exponentially decaying autocorrelations, we get

$$\sum_{\tau} C(\tau) < \infty, \tag{2}$$

and it is termed short memory or short-range memory. Physically, this property implies that in order to model such processes, we need not consider its long temporal history except possibly the last few time steps. In contrast, as $\tau \to \infty$, if

$$C(\tau) \sim \tau^{-\gamma} \quad 0 < \gamma < 1, \tag{3}$$

where γ is called the autocorrelation exponent, then, we obtain a slowly decaying autocorrelation of power law type that its correlation integral diverges, i.e., $\sum_{\tau} C(\tau) = \infty$. This implies that the system in principle does not forget where it came from, and in this sense, displays long-memory or long-range correlations. Further, systems with long memory do not have a typical time scale, while the systems with short memory possess typical time scale. For instance, some of the cases that display long-term memory are the daily temperature,

DNA sequences, river runoff, earth quakes and stock market volatility [*Bunde et al.*, 2002], rock fractures [*Davidsen et al.*, 2007], solar flares [*de Arcangelis et al.*, 2006]. Often, experimentally observed data do not necessarily exhibit such well-demarcated behavior as long- and short-range memory. For example, the estimated autocorrelation for the observed hourly wind data in Canadian coast exhibited exponential decay for short lags followed by an oscillatory behavior for large lags [*Brett and Tuller*, 1991]. Such results imply that more than one time scale might be present, and this has consequences for modeling efforts.

3. EXTREME EVENTS AND RETURN INTERVALS

3.1. Stretched Exponential Distribution

We consider a stationary time series $x(t)$, sampled at discrete time points $t = 1, 2, 3...N + 1$, with mean $\langle x \rangle = 0$. We denote extreme events in terms of exceedance above a prescribed threshold. In the spirit of this definition, we denote the threshold for extreme events by q, and an extreme event is said to occur at $t = t_0$ if $x(t_0) > q$. The return intervals are defined to be the time intervals between two successive occurrences of extreme events. From $x(t)$, we obtain a series for return intervals denoted by r_k, $k = 1, 2, 3...N$. Note that if the time series were to be continuous, then there would be infinite number of extreme events between successive threshold crossings. Hence, discretely sampled time series is suitable for properly defining return intervals. We show this schematically for an uncorrelated series in Figure 1. For such uncorrelated time series, the return intervals are also uncorrelated. Hence, the distribution of return intervals in this case is given by [*Ross*, 1996],

$$f_q(r) = \frac{1}{\langle r \rangle_q} e^{-r/\langle r \rangle_q}, \qquad (4)$$

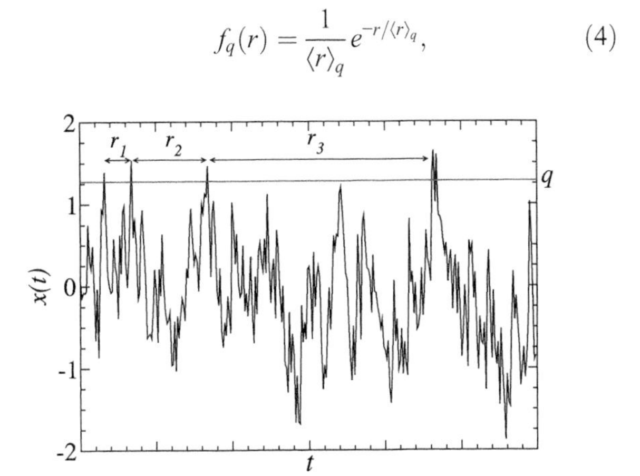

Figure 1. A schematic of an uncorrelated time series is shown. The value q indicates the threshold to define extreme events. Three sample return intervals r_1, r_2, and r_3 are also shown.

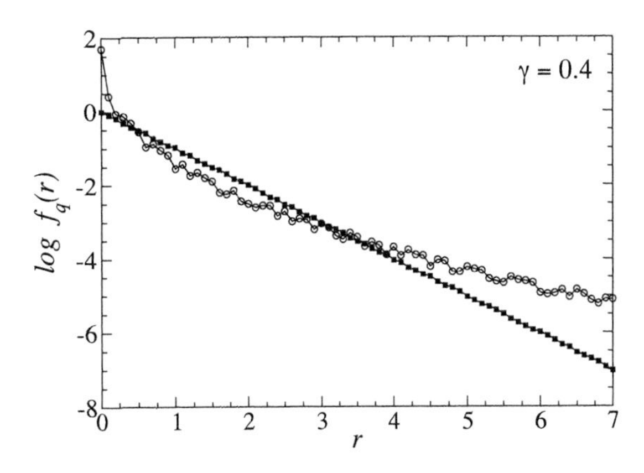

Figure 2. The stretched exponential distribution (circles) for $\gamma = 0.4$ obtained from simulations and the exponential distribution (squares) plotted in semilog scale. Note that the stretched exponential predicts higher probability of return intervals as $r \to 0$.

where the mean return interval is defined as

$$\langle r \rangle_q = \lim_{N\to\infty} \frac{1}{N} \sum_{k=1}^{N} r_k. \qquad (5)$$

As pointed out earlier, many natural systems, particularly in geophysical sciences, display long-range memory. Hence, the question of interest is how does the long memory in a time series affect its return interval distribution? Note that the long memory implies that the initial conditions will continue to affect the system even after a very long time, and this feature makes it analytically difficult to handle. This problem was addressed using synthetically generated linear long-range correlated data [*Bunde et al.*, 2003]. Long-range correlated noises with the desired autocorrelation exponent can be generated by a host of techniques. One of the popular methods is the Fourier filtering technique [*Rangarajan and Ding*, 2000], and many improved techniques exist as well [*Makse et al.*, 1996]. Let $x(t)$ be one realization of long-range correlated noise with autocorrelation exponent γ. Extreme events are the ones for which $x(t) > q$. Based on the numerical evidence, *Bunde et al.* [2003] proposed a stretched exponential form for the return interval distribution. In effect, this implies a distribution of the form

$$f_q(r) = Ae^{-Br^{\gamma'}}, \qquad (6)$$

where A and B are constants. The exponent γ' was nearly identical (within numerical accuracy) to the autocorrelation exponent γ of the original time series. In addition, the return intervals are long-range correlated as well.

In Figure 2, we show the return interval distribution $f_q(r)$ for long-range correlated (circles in Figure 2) and for uncorrelated time series (squares in Figure 2). One principal feature is that $f_q(r)$ predicts larger probability of extreme events for return intervals $r \to 0$. Physically, this could be understood as follows. A long-range correlated time series, by construction, displays persistence. This implies that positive (negative) increments are more likely to be followed by positive (negative) increments. Hence, due to persistence, if $x(t)$ crosses the threshold q for extreme events once, it is likely to remain above q for some time. This leads to an enhanced probability for recurrence interval as $r \to 0$ in comparison with that for an uncorrelated time series. The constants in equation (6) can be determined by performing the following integrals:

$$\int_0^\infty f_q(r)dr = 1 \tag{7}$$

$$\int_0^\infty rf_q(r)dr = 1. \tag{8}$$

Equation (7) is the normalization integral, and equation (8) requires that the mean return interval be unity. After transforming the variable as $r \to R = r/\langle r \rangle$, *Altmann and Kantz* [2005] determine the constants A and B to be

$$A = B\frac{\gamma}{\Gamma(1/\gamma)} \quad B = \frac{2}{(2\sqrt{\pi})^\gamma}\Gamma\left(\frac{2+\gamma}{2\gamma}\right)^\gamma. \tag{9}$$

Equations (6) and (9) provide a reasonably good representation for the return interval distribution obtained from empirical data. A good agreement with the empirically computed return interval distribution in the case of measured wind [*Santhanam and Kantz*, 2005] and temperature data [*Bunde et al.*, 2004] was also observed. It was shown that equation (6) provides a good fit to the recurrence distribution for several observed geophysical records such as the water level of River Nile, and long memory was proposed as a mechanism behind the clustering of extreme events [*Bunde et al.*, 2005]. The form of recurrence distribution in equation (6) derives theoretical support, for a choice of threshold $q = 0$, from a theorem due to *Newell and Rosenblatt* [1962]. Recently, a more general analytical treatment, similar to quantum field theoretic methods, by *Olla* [2007] also supports the stretched exponential form.

It is worth noting that power law autocorrelations of the form in equation (3) are not the only ones that would lead to stretched exponential in equation (6) for recurrence distribution. An instance of this is seen in the extreme events in the fluctuations of electrical resistors in nonequilibirum steady state [*Pennetta*, 2006] whose autocorrelation function is neither exponential nor a power law. The caveat is that our understanding of long memory and its recurrence statistics is still incomplete. Even within the constraint of power law autocorrelation as in equation (3), one could create a list of processes/systems that agree with stretched exponential form and a list that would not. For instance, for the long-range correlated random variables with the pdf that is uniform and bounded, the recurrence distribution is not of stretched exponential form [*Moloney and Davidsen*, 2009]. Nevertheless, in the last decade, several other examples of stretched exponential form for recurrence statistics have been reported, in disparate scenarios ranging from Internet traffic [*Cai et al.*, 2009] to intervals between wars in China [*Tang et al.*, 2010]. In general, we remark that extreme events could also be defined with respect to an interval $x(t) \in [q - \delta q, q + \delta q]$ instead of the standard choice $x(t) > q$. Equation (6) describes the recurrence distribution even when the extremes are defined with respect to an interval [*Altmann and Kantz*, 2005].

3.2. *Stretched Exponential to Weibull*

However, there are reasons to believe that $f_q(r)$ in equation (6) is not the complete result for recurrence statistics. For example, deviations from the stretched exponential distribution in equation (6) have been noted for return intervals shorter ($R < 1$) than the average, where $R = r/\langle r \rangle$. For short return intervals, i.e., $R < 1$, empirical results are better described by a power law with the exponent $\sim (\gamma - 1)$ [*Eichner et al.*, 2007], which is not explained by equation (6). Further, recent studies [*Blender et al.*, 2008; A. Witt and B. D. Malamud, E2C2: Extreme events, causes and consequences, open conference, Paris, 2008, e2c2.ipsl.jussieu.fr, hereinafter referred to as Witt and Malamud, open conference, 2008] have concluded that a Weibull distribution provides a good representation of the empirical results. In general, the form in equation (6) has been conjectured based on numerical simulations, and there are not many analytical results for this problem.

In this section, we discuss an analytical approach for return interval distribution [*Santhanam and Kantz*, 2008] that reproduces most of the reported empirical features discussed above. To begin with, we assume that the linear time series $x(t)$ is a Gaussian distributed, stationary, fractional noise process and is sampled at discrete intervals of τ such that $x_k = x(k\tau)$. We implicitly assume that the return intervals are much greater than sampling interval τ. With these assumptions, return intervals will be finite and its mean $\langle r \rangle > 0$.

As pointed out earlier, for stationary long-memory processes, physically, we anticipate that one extreme event is most likely to be followed by several other extreme events in quick succession. Then, the unconditional probability that the extreme event would take place after an interval r from the preceding occurrence is given by

$$P_{ex}(r) = ar^{-(1-\gamma)}, \tag{10}$$

where $0 < \gamma < 1$, and a is the normalization constant that will be fixed later. For an uncorrelated time series, we have $\gamma = 1$. In this case, $P_{ex}(r) = a$, a constant which concurs with the result that for the uncorrelated time series, extreme events are also uncorrelated, and hence, the probability of occurrence should be uniform. Further, the form of the probability model $P_{ex}(r)$ implies that it is a renewal process. After every extreme event, the time is reset, and the process starts again irrespective of the length of the preceding return interval. The values of γ being considered ensure that the process stops after a finite return interval. For further discussion on the validity of equation (10), the reader is directed to the work of *Santhanam and Kantz* [2008].

Given the unconditional probability $P_{ex}(r)$ for the occurrence of extreme events at time $t = r$, the general strategy is to first evaluate the probability $P_{noex}(r)$ that no extreme event occurs in the interval $[0, r]$. The probability that an extreme event occurs in the infinitesimal interval dr beyond r is $P_{ex}(r)\,dr$. Thus, the required probability distribution of the return intervals turns out to be

$$P(r)dr = P_{noex}(r)P_{ex}(r)dr. \qquad (11)$$

By equation (11), we have implicitly assumed that the return intervals are independent. This is strictly not true in many of the observed phenomena such as, for instance, the earthquakes. However, this allows for analytical treatment that captures the essential features of the recurrence distribution we seek to calculate. Both the probabilities $P_{noex}(r)$ and $P_{ex}(r)$ can be calculated and the normalized return interval distribution $P(R)$ is [*Santhanam and Kantz*, 2008],

$$P(R) = \gamma\left[\Gamma\left(\frac{1+\gamma}{\gamma}\right)\right]^{\gamma} R^{-(1-\gamma)} e^{-\left[\Gamma\left(\frac{1+\gamma}{\gamma}\right)\right]^{\gamma} R^{\gamma}}, \qquad (12)$$

where $R = r/\langle r\rangle$ and $\Gamma(.)$ is the gamma function [*Abromowitz and Stegun*, 1970].

First, we note that equation (12) has implicit dependence on the threshold q, and this has been empirically studied by *Santhanam and Kantz* [2008]. By substituting $\lambda = \frac{1}{\Gamma\left(\frac{1+\gamma}{\gamma}\right)}$, we obtain the Weibull distribution given by

$$P(R) = \frac{\gamma}{\lambda}\left(\frac{R}{\lambda}\right)^{-(1-\gamma)} e^{-\left(\frac{R}{\lambda}\right)^{\gamma}} \quad R \geq 0. \qquad (13)$$

It must be noted that even though Weibull distribution is one of the limiting forms for the distribution of extreme values [*Gumbel*, 2004; *Sornette*, 2005] for an uncorrelated series, $P(R)$ is unrelated to the distribution of extreme events itself.

The value $\gamma = 1$, being a crossover to uncorrelated time series, substituted in equation (12) leads to $P(R) = \exp(-R)$. This concurs with the known result for the uncorrelated time series shown in equation (4). If $R \ll 1$ implying that the return intervals are far less than the mean, the power law term with exponent $(\gamma - 1)$ dominates the return interval distribution $P(R)$, and we get

$$P(R) \propto R^{-(1-\gamma)} \quad (R \ll 1). \qquad (14)$$

This is in agreement with the simulations reported by *Eichner et al.* [2007]. On the other hand, for $R \gg 1$, we obtain an exponential distribution of the form,

$$P(R) \propto e^{-g_\gamma R^{\gamma}} \quad (R \gg 1), \qquad (15)$$

where g_γ is a constant dependent on γ. Clearly, most of the earlier simulations [*Bunde et al.*, 2005; *Santhanam and Kantz*, 2005; *Altmann and Kantz*, 2005; *Bunde et al.*, 2003] have focused on this regime of $R \gg 1$, which explains the good agreement with stretched exponential distribution. In the simulation results shown in Figure 3, we also account for the fact that in sampled time series, there cannot be zero return interval, $r = 0$ [*Santhanam and Kantz*, 2008].

The numerical simulations use long sequences ($2^{25} \approx 3 \times 10^7$) of time series [*Santhanam and Kantz*, 2008] obtained through Fourier filtering method for several values of γ. The return interval distribution for $q = 3$ is displayed in Figure 3, and Figure 3 reveals a good agreement between the analytical and the numerical results. In Figure 4, we focus on the regime of short return intervals, i.e., $R \ll 1$. We expect this regime to display a power law. These results show a good agreement with the theoretically expected slope of $(-1 + \gamma)$. Further, it

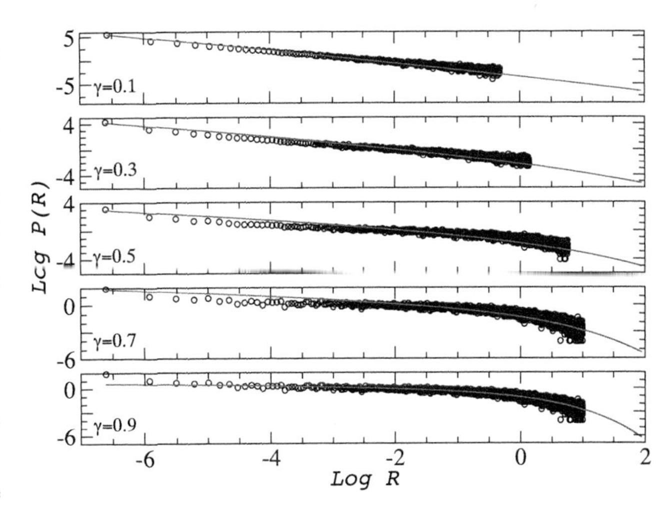

Figure 3. Distribution of return intervals of extreme events $P(R)$ in log-log scale compared with the analytical results given by equation (12). From *Santhanam and Kantz* [2008].

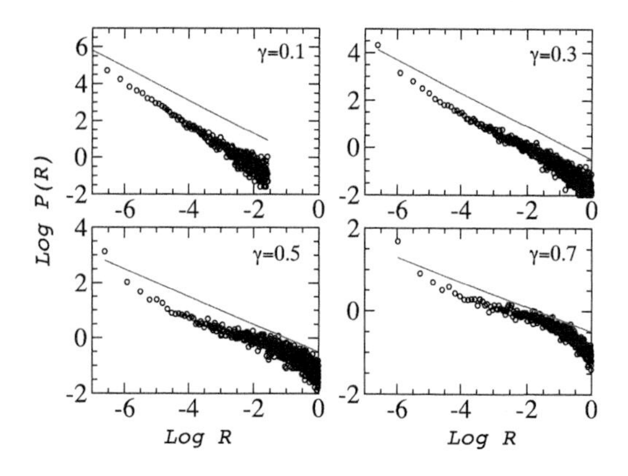

Figure 4. The return interval distribution focused on the regime $R << 1$. The open circles are the results from simulations. The solid line is a straight line with a slope $\gamma - 1$ to serve as a guide to the eye. Note that the simulated results are parallel to the solid line to a good approximation. From *Santhanam and Kantz* [2008].

must be noted that the agreement is good for large threshold q, which is not surprising considering that the theoretical result has been obtained under the assumption of independence of return intervals, which is almost true for $q \gg 1$.

Finally, we consider the extreme event statistics in the case of measured boundary layer wind speeds [*Ragwitz and Kantz*, 2000] whose autocorrelation exponent is $\gamma \approx 0.23$ [*Santhanam and Kantz*, 2005]. The empirical return interval distribution shown in Figure 5 agrees with the analytical result in the limit of large mean return interval or equivalently large values of threshold q. There have been other empirical results in agreement with the result in equation (12). The distribution of return intervals for landslides have revealed that Weibull distribution provides the best possible fit to the empirical results [*Blender et al.*, 2008; Witt and Malamud, open conference, 2008]. The distribution of interval between successive occurrences of the same word in online USENET group can also be obtained on the basis of equation (11) and finally results in Weibull distribution [*Altmann et al.*, 2009]. After rescaling seismic data, Weibull distribution is claimed to best represent the recurrence statistics for microrepeaters, i.e., microearthquakes that appear in the same location [*Goltz et al.*, 2009]. The recurrence distribution of the "stiff" slider-block model, being a relevant one for the earthquakes, is also Weibull distributed since the waiting time for the next event has scale invariant distribution [*Abaimov et al.*, 2007], similar to the assumption made in equation (10).

4. EARTHQUAKES AND RETURN INTERVALS

One of the significant results [*Bak et al.*, 2002; *Christensen et al.*, 2002; *Corral*, 2003, 2004] proposed about 10 years ago is that the distribution of recurrence times for stationary seismic activity, after suitable rescaling, is universal and is independent of the region and the magnitude considered. Based on an extensive study of seismic catalogues, *Corral* [2004] proposed that the recurrence distribution is

$$P(\tau) = \frac{1}{\langle\tau\rangle} f\left(\frac{\tau}{\langle\tau\rangle}\right), \tag{16}$$

where $f(\tau/\tau)$ has the universal form, a gamma distribution given by

$$f(x) = Ax^{-(1-\gamma)} \exp(-x^{\delta}/B), \tag{17}$$

where $A \approx 0.5$, $B \approx 1.58$, $\delta \approx 0.98$, and $\gamma \approx 0.67$. Figure 6, taken from the work of *Corral* [2004], depicts the unified scaling for earthquake data from different catalogs, and by appropriate scaling, they collapse on to one curve given by $f(x)$ in equation (17). Physically, this distribution in equation (17) implies that the seismic activity is strongly clustered in time and, as some authors have argued, possibly points to a complex process of spatiotemporal organization in a driven, nonequilibrium system. Thus, there have been suggestions that it might be a case of self-organized criticality [*Christensen et al.*, 2002]. However, the work of *Lennartz et al.* [2008a] suggests that the existence of long-range correlations in seismic data can explain the scaling form in equation (17). In this work, they estimate the correlation exponent to be $\gamma \approx 0.4$ for stationary seismic activity in northern and southern California. A synthetic sequence of Gutenberg-Richter distributed time series with $\gamma \approx 0.4$ provides a good parameter-free fit to the seismic data. Irrespective of the physical mechanisms for the origin of universality, there is

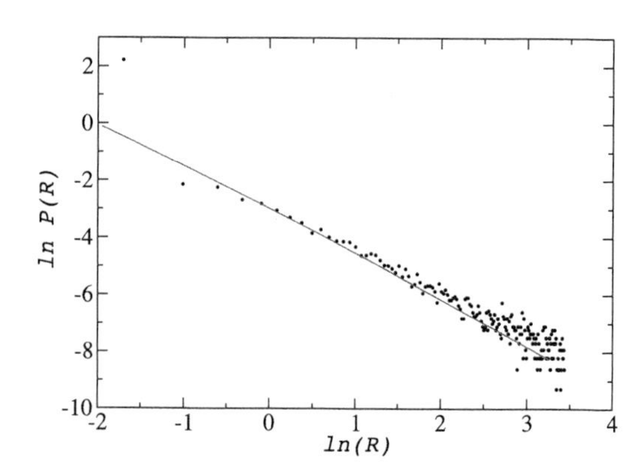

Figure 5. The return interval distribution for measured boundary layer wind data. The symbols correspond to distribution obtained from measured data. The solid line is the analytical result. The estimated value of $\gamma = 0.23$ has been used [*Santhanam and Kantz*, 2005]. The threshold for extreme events is taken to be $q = 3$ after standardizing the data to zero mean and unit variance.

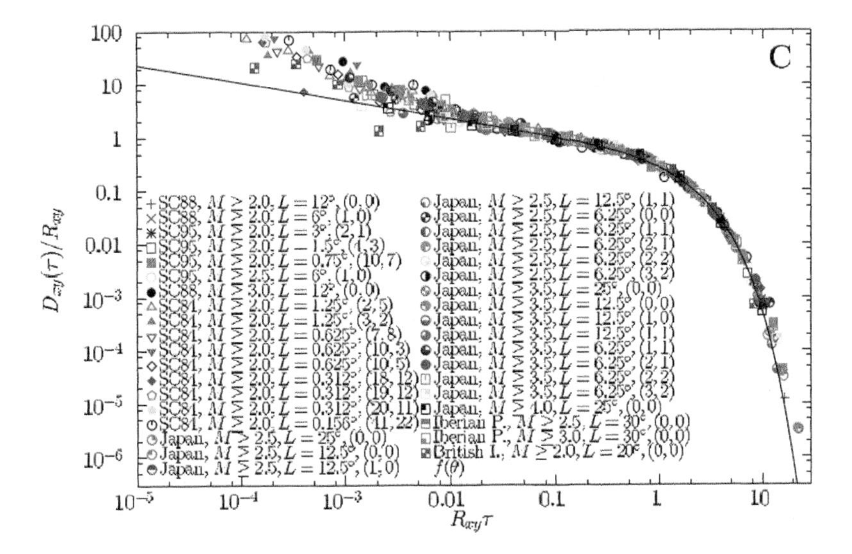

Figure 6. The return interval distribution for earthquakes from different catalogs, after appropriate scaling by their mean return interval, displays data collapse exhibiting a universal behavior. For details, see the work of *Corral* [2004]. Reprinted with permission from *Corral* [2004] (http://prl.aps.org/abstract/PRL/v92/i10/e108501). Copyright 2004 by the American Physical Society.

considerable debate on the very validity of claims of universality [*Lindman et al.*, 2005; *Corral and Christensen*, 2006; *Lindman et al.*, 2006]. For instance, *Saichev and Sornette*, 2006, 2007] have derived a universal form for $f(x)$, which differs from equation (17) and is based on the laws of seismicity (Gutenberg-Richter relation, Omori-Utsu law, and productivity law) within the framework of epidemic-type aftershock sequence model [see also *Sornette et al.*, 2008]. They show that the universal recurrence time distribution proposed by Corral can be obtained from the laws of seismicity, and universality does not strictly hold as well. In addition, $P(\tau)$ in equation (16) shows strong deviations from empirical data for $\tau \ll \langle\tau\rangle$, whereas the form proposed by *Saichev and Sornette* [2006, 2007] and *Sornette et al.* [2008] displays power law behavior in this regime in reasonable agreement with the data.

The claim of universality of recurrence distributions for earthquakes is based on the stationary windows of seismic data in which shortest interevent times (less than 10 s) are discarded. If the nonstationary sequences are taken into account and further all aftershocks on short time scales are also considered, once again, universality appears to break down [*Touati et al.*, 2009]. This work essentially points out that it might be too restrictive to describe earthquakes with a single time scale, namely, the mean recurrence interval. Indeed, it is argued that if the four characteristic time scales relating to the rate of independent events, rate of correlated events, time parameter in Omori law, and catalog duration are carefully taken into account, the universal scaling in equation (17) can be realized [*Bottiglieri et al.*, 2010]. Further, deviations from this scaling form can also be understood in terms of these time scales. Another alternative explanation is that the deviations from universality can be attributed to missing data in the aftershock sequences. It is known that after a main shock, many aftershocks, especially the small ones, are not registered by the measuring apparatus. Based on simulations with branching aftershock sequence model and four measured Californian aftershock sequences up to 365 days after the main shock, the authors argue that the missing data are responsible for the deviations from universality in return time statistics [*Lennartz et al.*, 2011]. Finally, we make two remarks. Notice that most of the arguments presented here are based on approximate analysis or synthetic data or observed seismic catalogs. There is some evidence to show that simple ab initio earthquake models can approximately reproduce interevent time distribution in equation (17) [*Weatherley and Abe*, 2004], but most of the cellular automaton-based seismicity models are not able to do so [*Weatherley*, 2006]. Finally, it must be noted that while equations (16) and (17) were proposed in the specific context of tectonic seismic activity, there is evidence to suggest that it might be more generally valid [*Verma et al.*, 2006; *de Arcangelis et al.*, 2006; *Geist and Parsons*, 2008; *Corral et al.*, 2008], including for the recurrence distribution of volcanic earthquakes [*Bottiglieri et al.*, 2009].

5. DISCUSSION

All these results discussed in sections 2–4, in summary, point to the fact that the recurrence interval distribution for long-memory series (including for earthquakes) is not easily susceptible to generalizations. In fact, a simple exercise involving the recurrence time distribution for a dichotomic noise illustrates this point. Dichotomic noise process [*Van Den Broeck*, 1983] consists of two discrete states +1 or −1 with mean zero. The noise process switches back and forth between these states at certain switching times t_i. If t_i represents a stationary point process with the time intervals between successive changes of states being independent, the distribution of interevent times has an asymptotic form $f(\tau) \sim \tau^{-\gamma-2}$ provided the autocorrelation of the original process is given by $\tau^{-\gamma}$. This is certainly different from the result anticipated by equation (12). Similar examples abound in the literature, well beyond the domain of geophysical sciences. In the case of Poincare recurrences in Hamiltonian dynamics [*Zaslavsky*, 2007], completely chaotic systems are characterized by exponential decay of autocorrelations. On the other hand, most realistic systems display mixed-phase space, i.e., contain regimes of regularity and chaos. In such cases, it is known that the autocorrelations display long memory, but the return interval distribution is given by a power law of the form $P(\tau) \sim \tau^{-\gamma}$ [*Zaslavsky*, 2007]. Again, this differs from the result in equation (12). We might remark here that the distribution of coronal mass ejection waiting times [*Wheatland*, 2003] and the return intervals between daily returns in financial data from stock and currency markets [*Bogachev et al.*, 2007] are also best described by a power law form.

One important reason for all these apparent "anomalies" is that the specification of the autocorrelation as a power law, as assumed in equation (3), does not completely specify all the temporal dependencies. The return interval distribution, in the ultimate analysis, is determined by the details of temporal dependencies in the time series. Hence, it is not entirely surprising that even within the family of power law correlated time series, the return interval distribution is not uniquely determined. In many cases, it is seen that the return intervals are strongly correlated as well, and it is so for earthquakes [*Livina et al.*, 2005]. Thus, the memory between successive return intervals modulates the recurrence distribution. This is certainly the case for seismic activity. Even though we have not discussed this aspect in this chapter, it is sufficient to remark that this makes analysis difficult to handle. It is also pertinent to note that the correlations in the magnitude of earthquakes are vigorously debated by several authors [*Lippiello et al.*, 2007, 2008; *Davidsen and Green*, 2011; *Lennartz et al.*, 2008b] since they have implications for the predictability of earthquakes. All these results illustrate the nuances of recurrence distribution in long-memory time series and provide a cautionary note on generalizing the empirical and analytical results disregarding the details of temporal correlations.

REFERENCES

Abaimov, S. G., D. L. Turcotte, R. Shcherbakov, and J. B. Rundle (2007), Recurrence and interoccurrence behavior of self-organized complex phenomena, *Nonlinear Processes Geophys.*, *14*, 455–464.

Abromowitz, M., and I. A. Stegun (Eds.) (1970), *Handbook of Mathematical Functions*, 1046 pp., Dover, New York.

Albeverio, S., V. Jentsch, and H. Kantz (Eds.) (2005), *Extreme Events in Nature and Society*, 352 pp., Springer, Berlin.

Altmann, E. G., and H. Kantz (2005), Recurrence time analysis, long-term correlations, and extreme events, *Phys. Rev. E*, *71*, 056106, doi:10.1103/PhysRevE.71.056106.

Altmann, E. G., J. B. Pierrehumbert, and A. E. Motter (2009), Beyond word frequency: Bursts, lulls and scaling in the temporal distributions of words, *PLoS ONE*, *4*, e7678, doi:10.1371/journal.pone.0007678.

Antoniou, I., V. V. Ivanov, V. V. Ivanov, and P. V. Zrelov (2002), On the log-normal distribution of network traffic, *Physica D*, *167*, 72–85.

Bak, P., K. Christensen, L. Danon, and T. Scanlon (2002), Unified scaling law for earthquakes, *Phys. Rev. Lett.*, *88*, 178501, doi:10.1103/PhysRevLett.88.178501.

Blender, R., K. Fraedrich, and F. Sienz (2008), Extreme event return times in long-term memory processes near 1/*f*, *Nonlinear Processes Geophys.*, *15*, 557–565.

Bogachev, M. I., J. F. Eichner, and A. Bunde (2007), Effect of nonlinear correlations on the statistics of return intervals in multifractal data sets, *Phys. Rev. Lett.*, *99*, 240601, doi:10.1103/PhysRevLett.99.240601.

Bottiglieri, M., C. Godano, and L. D'Auria (2009), Distribution of volcanic earthquake recurrence intervals, *J. Geophys. Res.*, *114*, B10309, doi:10.1029/2008JB005942.

Bottiglieri, M., L. de Arcangelis, C. Godano, and E. Lippiello (2010), Multiple-time scaling and universal behavior of the earthquake interevent time distribution, *Phys. Rev. Lett.*, *104*, 158501, doi:10.1103/PhysRevLett.104.158501.

Brett, A. C., and S. E. Tuller (1991), The autocorrelation of hourly wind speed observations, *J. Appl. Meteorol.*, *30*, 823–833.

Buldyrev, S. V., R. Parshani, G. Paul, H. E. Stanley, and S. Havlin (2010), Catastrophic cascade of failures in interdependent networks, *Nature*, *464*, 1025–1028.

Bunde, A., J. Kropp, and H.-J. Schellnhuber (Eds.) (2002), *The Science of Disasters: Climate Disruptions, Heart Attacks and Market Crashes*, 453 pp., Springer, Berlin.

Bunde, A., J. F. Eichner, S. Havlin, and J. W. Kantelhardt (2003), The effect of long-term correlations on the return periods of rare events, *Physica A*, *330*, 1–7.

Bunde, A., J. F. Eichner, S. Havlin, and J. W. Kantelhardt (2004), Return intervals of rare events in records with long-term persistence, *Physica A*, *342*, 308–314.

Bunde, A., J. F. Eichner, J. W. Kantelhardt, and S. Havlin (2005), Long-term memory: A natural mechanism for the clustering of extreme events and anomalous residual times in climate records, *Phys. Rev. Lett.*, *94*, 048701, doi:10.1103/PhysRevLett.94.048701.

Cai, S.-M., Z.-Q. Fu, T. Zhou, J. Gu, and P.-L. Zhou (2009), Scaling and memory in recurrence intervals of Internet traffic, *Europhys. Lett.*, *87*, 68001, doi:10.1209/0295-5075/87/68001.

Christensen, K., L. Danon, T. Scanlon, and P. Bak (2002), Unified scaling law for earthquakes, *Proc. Natl. Acad. Sci. USA.*, *99*, 2509–2513.

Corral, A. (2003), Local distributions and rate fluctuations in a unified scaling law for earthquakes, *Phys. Rev. E.*, *68*, 035102, doi:10.1103/PhysRevE.68.035102.

Corral, A. (2004), Long-term clustering, scaling, and universality in the temporal occurrence of earthquakes, *Phys. Rev. Lett.*, *92*, 108501, doi:10.1103/PhysRevLett.92.108501.

Corral, A., and K. Christensen (2006), Comment on "Earthquakes descaled: On waiting time distributions and scaling laws," *Phys. Rev. Lett.*, *96*, 109801, doi:10.1103/PhysRevLett.96.109801.

Corral, A., L. Telesca, and R. Lasaponara (2008), Scaling and correlations in the dynamics of forest-fire occurrence, *Phys. Rev. E*, *77*, 016101, doi:10.1103/PhysRevE.77.016101.

Davidsen, J., and A. Green (2011), Are earthquake magnitudes clustered?, *Phys. Rev. Lett.*, *106*, 108502, doi:10.1103/PhysRevLett.106.108502.

Davidsen, J., S. Stanchits, and G. Dresen (2007), Scaling and universality in rock fracture, *Phys. Rev. Lett.*, *98*, 125502, doi:10.1103/PhysRevLett.98.125502.

de Arcangelis, L., C. Godano, E. Lippiello, and M. Nicodemi (2006), Universality in solar flare and earthquake occurrence, *Phys. Rev. Lett.*, *96*, 051102, doi:10.1103/PhysRevLett.96.051102.

Eichner, J. F., J. W. Kantelhardt, A. Bunde, and S. Havlin (2007), Statistics of return intervals in long-term correlated records, *Phys. Rev. E*, *75*, 011128, doi:10.1103/PhysRevE.75.011128.

Embrechts, P., C. P. Kluppelberg, and T. Mikosh (1997), *Modelling Extremal Events*, 648 pp., Springer, Berlin.

Falaki, S. O., and S. A. Sorensen (1999), Traffic measurements on a local area computer network, *Comput. Commun.*, *15*, 192–197.

Geist, E. L., and T. Parsons (2008), Distribution of tsunami interevent times, *Geophys. Res. Lett.*, *35*, L02612, doi:10.1029/2007GL032690.

Ghil, P., et al. (2011), Extreme events: Dynamics, statistics and prediction, *Nonlinear Proc. Geophys.*, *18*, 295–350.

Goltz, C., D. L. Turcotte, S. G. Abaimov, R. M. Nadeau, N. Uchida, and T. Matsuzawa (2009), Rescaled earthquake recurrence time statistics: Application to microrepeaters, *Geophys. J. Int.*, *176*, 256–264.

Gopalswamy, N., L. Barbieri, E. W. Cliver, G. Lu, S. P. Plunkett, and R. M. Skoug (2005a), Introduction to violent Sun-Earth connection events of October–November 2003, *J. Geophys. Res.*, *110*, A09S00, doi:10.1029/2005JA011268.

Gopalswamy, N., S. Yashiro, Y. Liu, G. Michalek, A. Vourlidas, M. L. Kaiser, and R. A. Howard (2005b), Coronal mass ejections and other extreme characteristics of the 2003 October–November solar eruptions, *J. Geophys. Res.*, *110*, A09S15, doi:10.1029/2004JA010958.

Gumbel, A. (2004), *Statistics of Extremes*, 400 pp., Dover, New York.

Koscielny-Bunde, E., A. Bunde, S. Havlin, and Y. Goldreich (1996), Analysis of daily temperature fluctuations, *Physica A*, *231*, 393–396.

Koscielny-Bunde, E., A. Bunde, S. Havlin, H. E. Roman, Y. Goldreich, and H.-J. Schellnhuber (1998), Indication of a universal persistence law governing atmospheric variability, *Phys. Rev. Lett.*, *81*, 729–732.

Kotz, S., and S. Nadarajah (2001), *Extreme Value Distributions: Theory and Applications*, 185 pp., World Sci., Singapore.

Lanfredi, M., T. Simoniello, V. Cuomo, and M. Macchiato (2009), Discriminating low frequency components from long range persistent fluctuations in daily atmospheric temperature variability, *Atmos. Chem. Phys.*, *9*, 4537–4544, doi:10.5194/acp-9-4537-2009.

Lennartz, S., V. N. Livina, A. Bunde, and S. Havlin (2008a), Long-term memory in earthquakes and the distribution of interoccurrence times, *Europhys. Lett.*, *81*, 69001, doi:10.1209/0295-5075/81/69001.

Lennartz, S., A. Bunde, and D. L. Turcotte (2008b), Missing data in aftershock sequences: Explaining the deviations from scaling laws, *Phys. Rev. E*, *78*, 041115, doi:10.1103/PhysRevE.78.041115.

Lennartz, S., A. Bunde, and D. L. Turcotte (2011), Modelling seismic catalogues by cascade models: Do we need long-term magnitude correlations?, *Geophys. J. Int.*, *184*, 1214–1222.

Lindman, M., K. Jonsdottir, R. Roberts, B. Lund, and R. Bödvarsdvarsson (2005), Earthquakes descaled: On waiting time distributions and scaling laws, *Phys. Rev. Lett.*, *94*, 108501, doi:10.1103/PhysRevLett.94.108501.

Lindman, M., K. Jonsdottir, R. Roberts, B. Lund, and R. Bödvarsdvarsson (2006), Reply to the comment on "Earthquakes descaled: On waiting time distributions and scaling laws," *Phys. Rev. Lett*, *96*, 109802, doi:10.1103/PhysRevLett.96.109802.

Lippiello, E., C. Godano, and L. de Arcangelis (2007), Dynamical scaling in branching models for seismicity, *Phys. Rev. Lett.*, *98*, 098501, doi:10.1103/PhysRevLett.98.098501.

Lippiello, E., L. de Arcangelis, and C. Godano (2008), Influence of time and space correlations on earthquake magnitude, *Phys. Rev. Lett.*, *100*, 038501, doi:10.1103/PhysRevLett.100.038501.

Livina, V. N., S. Havlin, and A. Bunde (2005), Memory in the occurrence of earthquakes, *Phys. Rev. Lett.*, *95*, 208501, doi:10.1103/PhysRevLett.95.208501.

Majumdar, S. N. (1999), Persistence in nonequilibrium systems, *Curr. Sci.*, *77*, 370–375.

Makse, H. A., S. Havlin, M. Schwartz, and H. E. Stanley (1996), Method for generating long-range correlations for large systems, *Phys. Rev. E*, *53*, 5445–5449.

Mandelbrot, B. B., and J. W. van Ness (1968), Fractional Brownian motions, fractional noises and applications, *SIAM Rev.*, *10*, 422–437.

Mandelbrot, B. B., and J. R. Wallis (1968), Noah, Joseph and operational hydrology, *Water Resour. Res.*, *4*, 909–918.

Moloney, N. R., and J. Davidsen (2009), Extreme value statistics and return intervals in long-range correlated uniform deviates, *Phys. Rev. E*, *79*, 041131, doi:10.1103/PhysRevE.79.041131.

Newell, G. F., and M. Rosenblatt (1962), Zero crossing probabilities for Gaussian stationary processes, *Ann. Math. Stat.*, *33*, 1306–1313.

Olla, P. (2007), Return times for stochastic processes with power-law scaling, *Phys. Rev. E.*, *76*, 011122, doi:10.1103/PhysRevE.76.011122.

Pelletier, J. D., and D. L. Turcotte (1997), Long-range persistence in climatological and hydrological time series: Analysis, modeling and application to drought hazard assessment, *J. Hydrol.*, *203*, 198–208.

Pennetta, C. (2006), Distribution of return intervals of extreme events, *Eur. Phys. J. B*, *50*, 95–98.

Pisarenko, V. F., and D. Sornette (2003), Characterization of the frequency of extreme earthquake events by the generalized Pareto distribution, *Pure Appl. Geophys.*, *160*, 2343–2364.

Ragwitz, M., and H. Kantz (2000), Detecting non-linear structure and predicting turbulent gusts in surface wind velocities, *Europhys. Lett.*, *51*, 595–601.

Rangarajan, G., and M. Ding (2000), Integrated approach to the assessment of long range correlation in time series data, *Phys. Rev. E*, *61*, 4991–5001.

Riemann-Campe, K., R. Blender, and K. Fraedrich (2010), Global memory analysis in observed and simulated CAPE and CIN, *Int. J. Climatol.*, *31*, 1099–1107, doi:10.1002/joc.2148.

Ross, S. M. (1996), *Stochastic Processes*, 528 pp., John Wiley, New York.

Rybski, D., A. Bunde, and H. von Storch (2008), Long-term memory in 1000-year simulated temperature records, *J. Geophys. Res.*, *113*, D02106, doi:10.1029/2007JD008568.

Saichev, A., and D. Sornette (2006), "Universal" distribution of interearthquake times explained, *Phys. Rev. Lett.*, *97*, 078501, doi:10.1103/PhysRevLett.97.078501.

Saichev, A., and D. Sornette (2007), Theory of earthquake recurrence times, *J. Geophys. Res.*, *112*, B04313, doi:10.1029/2006JB004536.

Santhanam, M. S., and H. Kantz (2005), Long-range correlations and rare events in boundary layer wind fields, *Physica A*, *345*, 713–721.

Santhanam, M. S., and H. Kantz (2008), Return interval distribution of extreme events and long-term memory, *Phys. Rev. E.*, *78*, 051113, doi:10.1103/PhysRevE.78.051113.

Sornette, D. (2005), *Critical Phenomena in Natural Sciences, Chaos, Fractals, Self-organization and Disorder: Tools and Concepts*, 550 pp., Springer, Berlin.

Sornette, D., S. Utkin, and A. Saichev (2008), Solution of the nonlinear theory and tests of earthquake recurrence times, *Phys. Rev. E.*, *77*, 066109, doi:10.1103/PhysRevE.77.066109.

Tang, D., X.-P. Han, and B.-H. Wang (2010), Stretched exponential distribution of recurrent time of wars in China, *Physica A*, *389*, 2637–2641.

Touati, S., M. Naylor, and I. G. Main (2009), Origin and nonuniversality of the earthquake interevent time distribution, *Phys. Rev. Lett.*, *102*, 168501, doi:10.1103/PhysRevLett.102.168501.

Turcotte, D. L. (1997), *Fractals and Chaos in Geology and Geophysics*, 416 pp., Cambridge Univ. Press, Cambridge, U. K.

Van Den Broeck, C. (1983), On the relation between white shot noise, Gaussian white noise, and the dichotomic Markov process, *J. Stat. Phys.*, *31*, 467–483.

Verma, M. K., S. Manna, J. Banerjee, and S. Ghosh (2006), Universal scaling laws for large events in driven nonequilibrium systems, *Europhys. Lett.*, *76*, 1050–1056.

Vyushin, D. I., and P. J. Kushner (2009), Power-law and long-memory characteristics of the atmospheric general circulation, *J. Clim.*, *22*, 2890–2904.

Wang, F., K. Yamasaki, S. Havlin, and H. E. Stanley (2006), Scaling and memory of intraday volatility return intervals in stock markets, *Phys. Rev. E*, *73*, 026117, doi:10.1103/PhysRevE.73.026117.

Weatherley, D. (2006), Recurrence interval statistics of cellular automaton seismicity, *Pure Appl. Geophys.*, *163*, 1933–1947.

Weatherley, D., and S. Abe (2004), Earthquake statistics in a block slider model and a fully dynamic fault model, *Nonlinear Proc. Geophys.*, *11*, 553–560.

Wheatland, M. S. (2003), The coronal mass ejection waiting-time distribution, *Solar Phys.*, *214*, 361–373.

Yamasaki, K., L. Muchnik, S. Havlin, A. Bunde, and H. E. Stanley (2005), Scaling and memory in volatility return intervals in financial markets, *Proc. Natl. Acad. Sci. U. S. A.*, *102*, 9424–9428.

Zaslavsky, G. M. (2007), *Physics of Chaos in Hamiltonian Systems*, 328 pp., Imperial College Press, London, U. K.

M. S. Santhanam, Indian Institute of Science Education and Research, Pashan Road, Pune 411 008, India. (santh@iiserpune.ac.in)

Dealing With Complexity and Extreme Events Using a Bottom-Up, Resource-Based Vulnerability Perspective

Roger A. Pielke Sr.,[1] Rob Wilby,[2] Dev Niyogi,[3] Faisal Hossain,[4] Koji Dairuku,[5] Jimmy Adegoke,[6] George Kallos,[7] Timothy Seastedt,[8] and Katharine Suding[9]

We discuss the adoption of a bottom-up, resource-based vulnerability approach in evaluating the effect of climate and other environmental and societal threats to societally critical resources. This vulnerability concept requires the determination of the major threats to local and regional water, food, energy, human health, and ecosystem function resources from extreme events including those from climate but also from other social and environmental issues. After these threats are identified for each resource, then the relative risks can be compared with other risks in order to adopt optimal preferred mitigation/adaptation strategies. This is a more inclusive way of assessing risks, including from climate variability and climate change, than using the outcome vulnerability approach adopted by the Intergovernmental Panel on Climate Change (IPCC). A contextual vulnerability assessment using the bottom-up, resource-based framework is a more inclusive approach for policy makers to adopt effective mitigation and adaptation methodologies to deal with the complexity of the spectrum of social and environmental extreme events that will occur in the coming decades as the range of threats are assessed, beyond just the focus on CO_2 and a few other greenhouse gases as emphasized in the IPCC assessments.

1. INTRODUCTION

Rial et al. [2004, p. 11] state the following:

> The Earth's climate system is highly nonlinear: inputs and outputs are not proportional, change is often episodic and abrupt, rather than slow and gradual, and multiple equilibria are the norm . . . there is a relatively poor understanding of the different types of nonlinearities, how they manifest under various conditions, and whether they reflect a climate system driven by astronomical forcings, by internal feedbacks, or by a combination of both . . . [We] suggest a robust alternative to prediction that is based on using integrated assessments within the framework of vulnerability studies . . . It is imperative that the Earth's climate system research community embraces this nonlinear paradigm if we are to move forward in the assessment of the human influence on climate.

[1]Cooperative Institute for Research in Environmental Sciences, University of Colorado, Boulder, Colorado, USA.

[2]Department of Geography, Loughborough University, Loughborough, UK.

[3]Department of Agronomy and Department of Earth and Atmospheric Sciences, Purdue University, West Lafayette, Indiana, USA.

[4]Department of Civil and Environmental Engineering, Tennessee Technological University, Cookeville, Tennessee, USA.

[5]Disaster Prevention System Research Center, National Research Institute for Earth Science and Disaster Prevention, Tsukuba, Japan.

[6]Natural Resources and the Environment, CSIR, Pretoria, South Africa.

[7]School of Physics, University of Athens, Athens, Greece.

[8]Institute of Arctic and Alpine Research, University of Colorado, Boulder, Colorado, USA.

[9]Department of Environmental Science, Policy, and Management, University of California, Berkeley, California, USA.

Extreme Events and Natural Hazards: The Complexity Perspective
Geophysical Monograph Series 196

10.1029/2011GM001086

The concept of spatiotemporal chaos (e.g., discussed by T. Milanovic (Spatio-temporal chaos, Climate Etc., Weblog, available at http://judithcurry.com/2011/02/10/spatio-temporal-chaos, 2011)) reinforces this view of the complexity of the climate system and applies more generally to all components of society and the environment. Milanovic (paragraph 12) defines spatiotemporal chaos as dealing with the dynamics of spatial patterns:

> Weather and climate are manifestations of spatio temporal chaos of staggering complexity because there is not only Navier Stokes equations, but there are many more coupled fields. ENSO is an example of a quasi standing wave of the system.

The dominant scientific perspective is top-down and carbon dioxide centric. It focuses on multidecadal global climate model (GCM) predictions involving quasilinear responses dominated by the increases in greenhouse gases, which are downscaled to societal and environmental impacts (i.e., following the progression from the Working Group 1 [*Solomon et al.*, 2007] to the Working Group 2 reports [*Parry et al.*, 2007], which culminate in the Working Group 3 report [*Metz et al.*, 2007] of the Intergovernmental Panel on Climate Change (IPCC)). This narrow approach, however, has serious limitations in assessing risks of extreme events to key resources as is discussed below. An overview of these limitations, presented in Figure 1, reproduced from the work of *Kabat et al.* [2004], includes the spatial averaging of climate predictions over relatively large areas, the focus on single stressors, and gradual, near-linear predictions of climate change.

An additional limitation of the top-down approach is that if the ensemble of IPCC projections and actual climate trajectory differ significantly in coming decades, recognition of this error may occur too late for policy makers to realign the adaptation/mitigation strategy in order to respond to the actual state of climate at the local/regional scale. In contrast, if the adaptation strategy had considered more scenarios, then it could handle a larger margin of error than the constrained top-down approach.

This chapter begins by overviewing the limitations of the top-down approach to assess risk from extreme events as well as the difficulty in detecting changes in the threat of extreme events over time. We then discuss a bottom-up resource-based approach, which we conclude is a more robust tool to provide policy makers and the impact community with a

Approach	Scenario	Vulnerability
Assumed dominant stress	Climate, recent greenhouse gas emissions to the atmosphere, ocean temperatures, aerosols, etc .	Multiple Stresses: Climate (historical climate variability, land use and water use, altered disturbance regimes invasive species, contaminants/pollutants, habitat loss, etc.
Usual timeframe of concern	Long-term, doubled CO_2 30 to 100 years in the future.	Short-term (0-30 years) and long-term research.
Usual scale of concern	Global,sometimes regional. Local scale needs downscaling techniques. However, there is little evidence to suggest that present models provide realistic, accurate, or precise climate scenarios at local or regional scales.	Local, regional, national, and global scales.
Major parameters of concern	Spatially averaged changes in mean temperatures and precipitation in fairly large grid cells with some regional scenarios for drought.	Potential extreme values in multiple parameters (temperature, precipitations, frost-free days) and additional focus on extreme events (floods, fires, droughts, etc.) measures of uncertainty.
Major limitations for developing coping strategies	Focus on single stress limits preparedness for other stresses. Results often show gradual ramping of climate change-limiting preparedness for extreme events. Results represent only a limited subset of all likely future outcomes – usually unidirectional trends. Results are accepted by many scientists, the media, and the public as actual "predictions". Lost in the translation of results is that all models of the distant future have unstated (presently unknowable) levels of certainty or probability.	Approach requires detailed data on multiple stresses and their interactions at local, regional, national, and global scales – and many areas lack adequate information. Emphasis on short-term issues may limit preparedness for abrupt "threshold" changes in climate sometime in the short or long term. Requires preparedness for a far greater variation of possible futures, including abrupt changes in any direction – this is probably more realistic, yet difficult.

Figure 1. Contrast between a top-down versus bottom-up assessment of the vulnerability of resources to climate variability and change. From the work of *Kabat et al.* [2004].

much better estimate of the threats faced by key resources in the future. We conclude the chapter with examples illustrating why we need a bottom-up approach to assess the threats to water, food, energy, human health, and ecosystem function.

2. USE OF TOP-DOWN DOWNSCALING TO DETERMINE RISKS FROM EXTREME EVENTS

IPCC climate change projections are at relatively coarser resolution [*Solomon et al.*, 2007], whereas the impacts and potential mitigation policies of interest to stakeholders are mostly at local to regional scales. For example, climate models may project increasing drought at a regional scale. The resilience to such increased occurrence as well as changes in the intensity of droughts is, however, dependent on the local-scale environmental conditions (such as moisture storage and convective rainfall) and farming approaches (access to irrigation, timing of rain or stress, etc.). According to *Adger* [1996, p. 10] an important issue for IPCC-like global reports is to assess whether the top-down approach can incorporate the "aggregation of individual decision-making in a realistic way, so that results of the modelling are applicable and policy relevant."

There are also unresolved issues both for generating and applying IPCC-type model predictions to climate risk assessments for policy makers and other users [e.g., *Holman et al.*, 2005]. They are often presented as "projections" yet are actually forecasts (predictions) of the future climate based on different assumptions of greenhouse gas emissions. Such terminology has been debated before by *Pielke* [2002] and *MacCracken* [2002]. In this chapter, we use the terms projection, prediction, and forecast interchangeably [*Bray and von Storch*, 2009].

Multidecadal IPCC-type forecasts, if used without consideration of regional and local vulnerabilities, can lead to misleading outcomes and actions for the impacts and adaptation community as well as for policy makers [*Patt et al.*, 2010; *Pielke Jr. et al.*, 2007].

There are several reasons why top-down IPCC-type multidecadal global climate change model predictions are not able to accurately predict changes in the climate system over this time period. First, as a necessary condition for an accurate prediction, the multidecadal GCM simulations must include all first-order climate forcings and feedback. However, they do not [see, e.g., *National Research Council (NRC)*, 2005; *Pielke et al.*, 2009]. Natural climate forcings, such as large volcanic eruptions or long-term changes in solar irradiance, cannot be forecast skillfully. Omission of these natural forcings, as well as human climate forcings that are excluded or poorly understood, introduces large uncertainty in the local and regional estimates of impact on the atmospheric and oceanic circulations [e.g., *Myhre and Myhre*, 2003; *Matsui and Pielke*, 2006; *Davin et al.*, 2007].

Pielke et al. [2009, p. 413] state,

> In addition to greenhouse gas emissions, other first-order human climate forcings are important to understanding the future behavior of Earth's climate. These forcings are spatially heterogeneous and include the effect of aerosols on clouds and associated precipitation [e.g., *Rosenfeld et al.*, 2008], the influence of aerosol deposition (e.g., black carbon (soot) [*Flanner et al.*, 2007] and reactive nitrogen [*Galloway et al.*, 2004]), and the role of changes in land use/land cover [e.g., *Takata et al.*, 2009]. Among their effects is their role in altering atmospheric and ocean circulation features away from what they would be in the natural climate system [*NRC*, 2005]. As with CO_2, the lengths of time that they affect the climate are estimated to be on multidecadal time scales and longer.

Perhaps, at least partly for this reason, these global multidecadal predictions are unable to skillfully simulate major atmospheric circulation features such the Pacific Decadal Oscillation (PDO), the North Atlantic Oscillation (NAO), El Niño and La Niña, and the South Asian monsoon [*Pielke Sr.*, 2010; *Annamalai et al.*, 2007]. However, these large-scale atmospheric/ocean climate features determine the particular weather pattern for a region [e.g., *Otterman et al.*, 2002; *Chase et al.*, 2006]. Proposed decadal prediction efforts seek to address some of these deficiencies but are still under development [*Hurrell et al.*, 2009].

Dynamic and statistical regional downscaling yield higher spatial resolution; however, the regional climate models are strongly dependent on the lateral boundary conditions and interior nudging by their parent global models [e.g., see *Rockel et al.*, 2008]. Large-scale climate errors in the global models are retained and could even be amplified by the higher spatial-resolution regional models. Most downscaling methods also suffer from the inability to mimic second- or higher-order moments of climate variables on the regional and local scales and are typically conditioned to preserve the mean [*Salathe*, 2005]. In particular, the spatial gradient of precipitation may not be physically modeled well enough by downscaling methods to allow the accurate assessment of streamflow and other environmental features in regions of complex terrain [*Ferraris et al.*, 2003; *Salathe*, 2005; *Rahman et al.*, 2009; *Yang et al.*, 2009].

Moreover, since as reported above, the global multidecadal climate model predictions cannot accurately predict circulation features such as PDO, NAO, El Niño, and La Niña [*Compo et al.*, 2011], they cannot provide accurate lateral boundary conditions and interior nudging to the regional climate models. On the other hand, regional models themselves do not have the domain scale (or two-way interaction) to skillfully predict these larger-scale atmospheric features.

There is also only one-way interaction between regional and global models, which is not physically consistent. If the regional model significantly alters the atmospheric and/or ocean circulations, there is no way for this information to alter the larger-scale circulation features, which are being fed into the regional model through the lateral boundary conditions and nudging. Also, while there is information added when higher spatial analyses of land use and other forcings are considered in the regional domain, the errors and uncertainty from the larger model still persists, thus rendering the added complexity and details ineffective [*Ray et al.*, 2010; *Mishra et al.*, 2010].

In addition, lateral boundary conditions for input to regional downscaling require regional-scale information from a global forecast model. However, the global model does not have this regional-scale information because of its limited spatial resolution. This is, however, a logical paradox since the regional model needs something that can only be acquired by a regional model (or regional observations). Therefore, the acquisition of lateral boundary conditions with the needed spatial resolution becomes logically impossible.

There is sometimes an incorrect assumption that although GCMs cannot predict future climate change as an initial value problem, they can predict future climate statistics as a boundary value problem [*Palmer et al.*, 2008]. With respect to weather patterns, for the downscaling regional (and global) models to add value over and beyond what is available from the historical, recent paleorecord, and worse-case sequence of days, however, they must be able to skillfully predict the *changes* in the regional weather statistics. There is only value for predicting climate change *if* they could skillfully predict the *changes* in the statistics of the weather and other aspects of the climate system. There is no evidence, however, that the models can predict changes in these climate statistics even in hindcast. As highlighted by *Dessai et al.* [2009], the finer and time-space-based downscaled information can be "misconstrued as accurate," but the ability to get this finer-scale information does not necessarily translate into increased confidence in the downscaled scenario [*Fowler and Wilby*, 2010].

Statistical downscaling from the parent global model can be used as the benchmark (control) against which dynamic downscaling should improve [e.g., *Wilby et al.*, 1998; *Mearns et al.*, 1999]. If, however, the statistical relationship(s) between predictor(s) and predictants changes in the future, the method will not provide the actual real-world response. Under climate change, the statistical relationship between the climate and impacts would be expected to change [*Milly et al.*, 2008]. The same premise of stationarity also applies to the parameterized schemes within regional climate models.

There has also been a move toward higher spatial resolution and more complex GCMs. However, this added detail does not assure more skillful predictions of impacts to key resources decades from now. As concluded by *Landsea and Knaff* [2000, p. 2117], with respect to El Niño predictions, an increase in model complexity can, in fact, compound the input errors and downgrade the model skill. They write

> . . . the use of more complex, physically realistic dynamical models does not automatically provide more reliable forecasts. Increased complexity can increase by orders of magnitude the sources for error, which can cause degradation in skill.

Thus, neither dynamic downscaling nor statistical downscaling from multidecadal global model projections add *proven* value to spatial or temporal accuracy that can assist the impact community in ways beyond what is already available from historical records, paleorecords, or analog records [*Rajagopalan et al.*, 2009; *Parson et al.*, 2003]. The global and regional multidecadal climate change models are providing a level of confidence in forecast skill of the coming decades that is not warranted.

3. DETECTION TIME OF EXTREME EVENTS

Historically, changes in exposure and the value of capital at risk have been much more important drivers of economic losses from weather-related hazards than anthropogenic climate change [*Bouwer*, 2011; *Pielke Jr.*, 2010]. Nonetheless, our ability to detect *future* changes in extreme events depends on several additional factors: the strength of the

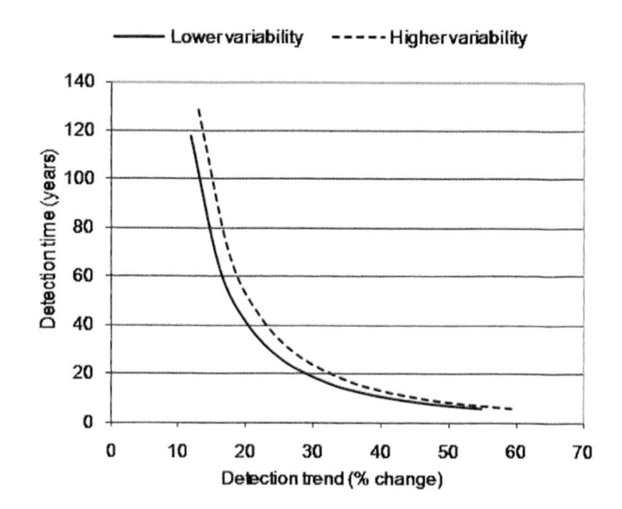

Figure 2. An example of the relationship between detection time (in years from 1990), the assumed strength of the climate change signal (percent change in mean), and interannual variability (variance) for summer flows in the River Itchen, southern England. This river had the shortest detection times (among the 15 basins studied) because of the relatively large climate changes projected for the region, combined with a large damping effect of groundwater on the flow regime. Adapted from the work of *Wilby* [2006].

predicted trend (signal) relative to the sample variance (noise), the length of time over which the trend persists, the choice of extreme index, the power of the statistical test, and the level of confidence required in the outcome of that statistical test (Figure 2). Quantitative predictions of extremes by climate models are highly uncertain due to the choice of model(s); unknown future changes in radiative and other climate forcing (by anthropogenic emissions, land surface modifications, and natural events (e.g., solar and volcanic)); and the random, internal variability of climate.

When taking all of these factors into account, it is hardly surprising that detection of robust anthropogenic signals in regional climate predictions is seldom possible within decision-making time scales of a few decades. For example, *Ziegler et al.* [2005] find that time series of 50–350 years are required to detect plausible trends in annual precipitation, evaporation, and discharge in the Missouri, Ohio, and Upper Mississippi River Basins. Likewise, *Wilby* [2006] showed that, under widely assumed climate change scenarios, expected trends in U.K. summer river flows are seldom detectable within typical planning horizons (i.e., by the 2020s). Again, depending on the climate model and underlying uncertainty of the regional projections, emergence time scales for U.S. tropical cyclone losses range between 120 and 550 years [*Crompton et al.*, 2011].

Hawkins and Sutton [2010] consider the extent to which the signal-to-noise ratio in future temperature and precipitation might vary in space and time, as well as the scope for improving predictive power by decreasing climate model uncertainties. Using the Coupled Model Intercomparison Project (CMIP3) ensemble, they show that the tropics have the highest S/N for temperature but the lowest for precipitation (which is greatest at the poles). Even when model uncertainty is set to zero, the gains in S/N for regional precipitation are only modest, especially for predictions over the next few decades. However, other model experiments suggest that changes in indices of extreme precipitation may be stronger than corresponding changes in mean precipitation [*Hegerl et al.*, 2004]. This view is supported by *Fowler and Wilby* [2010], who found that significant changes in multiday heavy rainfall accumulations could emerge in some parts of the United Kingdom within a decade or so (if the regional climate scenarios of the PRUDENCE ensemble are realized). Others assert that an attributable human fingerprint is already evident in the *risk* of flood occurrence at the scale of the United Kingdom [*Pall et al.*, 2011].

So what is the utility of top-down climate model prediction and detection of extreme events? Taken at face value, poorly discerned and attributed changes in extreme events imply either that adaptation decisions will have to be taken ahead of tangible evidence of the need to act or that those anticipatory measures should simply be deferred. The latter argument is sometimes supported by naïve mismatching of trends in historic weather extremes with regional climate model projections [see *Wilby et al.*, 2008].

Rather than an excuse for inaction, long emergence time scales reinforce the need for bottom-up, vulnerability-based responses. Anthropogenic climate change trends may already be underway but statistically undetectable for many more decades. This does not exclude the possibility that the same trends could have much earlier *practical* significance. For example, a rise in maximum temperatures of just a few tenths of a degree coinciding with lower river flows could result in abrupt changes in freshwater ecosystems that are already stressed by river regulation and pollution.

At least three steps can be taken to better detect complex, highly uncertain, and potentially dynamic patterns of extreme events. First, climate model outputs can be used to highlight potential "hot spots" of emerging risk (i.e., high S/N), thereby guiding a more targeted approach to environmental monitoring and assessment. For example, a strong signal is predicted for heavy rainfall in western England, particularly in the uplands. Early signs are that the expected trend may be emerging in the winter precipitation and streamflow record [*Dixon et al.*, 2006; *Fowler and Wilby*, 2010]. Of course, there is always a danger of making type I errors in such cases (i.e., erroneous trend detection when there is none), but this risk diminishes as the trend remains and the record grows. We should, therefore, be safeguarding lengthy, homogeneous records, while being mindful of other factors that can confound trend analysis. These include changes in instrument, location, observing/ recording practice, site characteristics, and sampling regime [*Pielke Sr. et al.*, 2007].

Second, regions with relatively low certainty in predicted extremes should be the focus for intensive field campaigns to improve understanding of regional climate forcing and representation in models. For example, large model uncertainty exists with respect to the future behavior of the South Asian monsoon. Rigorous scrutiny of the GCMs underpinning the IPCC reports revealed that just 6 of the 18 models have a plausible representation of monsoon precipitation climatology [*Annamalai et al.*, 2007]. Of these six GCMs, only four exhibited a robust El Niño–Southern Oscillation (ENSO)-monsoon correlation, including the well-known inverse relationship between ENSO and rainfall anomalies over India.

Another comprehensive assessment reviewed 79 GCM simulations from 12 different climate models and 6 different emission scenarios to ascertain whether any consensus can be reached about predicted changes in the main features of ENSO and the monsoon climates of South Asia [*Paeth et al.*, 2008]. Although most models project La Nina-like anomalies, and thus an intensification of the summer monsoon precipitation in India by the end of the twenty-first century, the response is barely distinguishable from natural climate variability. Early detection is unlikely in this case.

Third, more judicious selection of indices could increase S/N, as in the case of long-duration precipitation extremes. We should also recognize that some types of extreme (such as droughts linked to persistent Atlantic blocking or intense summer convective downpours and associated flash flooding) are not adequately resolved by the present generation of climate models, even under present conditions [e.g., *Fowler and Ekström*, 2009] or is there any guarantee that higher-resolution models will lead to reduced uncertainty, particularly if additional Earth system feedback are incorporated [*Hawkins and Sutton*, 2010]. However, by optimizing the choice of detection index, season and domain, it should be possible to identify a network of "sentinel" regions for earliest detection. But these hazard indices should not be so sophisticated that they lose societal relevance.

4. A BOTTOM-UP, RESOURCE-BASED VULNERABILITY PERSPECTIVE

4.1. Definitions of Vulnerability

In general, "vulnerability" may be defined as the concept of "threats" from potential hazards to the population, to key resources, and to the infrastructure. According to the IPCC Working Group 2 report [*Parry et al.*, 2007, p. 21]

> Vulnerability is the degree to which a system is susceptible to, and unable to cope with, adverse effects of climate change, including climate variability and extremes. Vulnerability is a function of the character, magnitude, and rate of climate change and variation to which a system is exposed, its sensitivity, and its adaptive capacity.

Bravo de Guenni et al. [2004] provides a useful summary below of the concept of vulnerability. Risk can be defined as a measure that combines, over a given time, the likelihoods and the consequences of a set of natural hazard scenarios [*Beer and Ismail-Zadeh*, 2003]. As summarized by A. Ismail-Zadeh (personal communication, 2011), the risk can be estimated as the probability of harmful consequences or expected losses (of lives and property) and damages (e.g., people injured, economic activity disrupted, environment damaged) due to a natural event resulting from interactions between hazards (*Hs*), vulnerability (*V*), and exposure (*E*). Conventionally, risk (*Rs*) is expressed quantitatively by the convolution of these three parameters: $Rs = Hs \times V \times E$. Such events can disrupt the human and/or the natural environment. A *hazard* is the combination of both the active physical exposure to a natural process and the vulnerability of the human and/or environmental system with which it is interacting. A hazard is commonly described as the "potential to do harm." The physical exposure is a function of both its intensity and duration. It has a magnitude and a probability of occurrence and takes place with respect to a particular resource at specified locations. The natural process becomes a hazard when it produces an event that exceeds a coping threshold, i.e., *an extreme value*. An extreme event, according to A. Ismail-Zadeh (personal communication, 2011), also could be more clearly defined as an occurrence that, with respect to other occurrences, is either notable, rare, unique, profound, or otherwise significant in terms of its impacts, effects, or outcomes. Hazard describes a phenomenon associated with a natural event (i.e., ground motion, ocean wave, atmospheric motion, etc.) that could cause harm and can be quantified by three parameters: a level of severity (expressed, for example, in terms of magnitude), and its occurrence frequency, and location. Hazard duration is determined by the length of time the threshold is exceeded. Resilience is the capacity of a system below which the thresholds of vulnerability are not exceeded [*Vogel*, 1998]. Figure 3 from the work of *Bravo de Guenni et al.* [2004] schematically illustrates the relationship between threshold and duration under different scenarios of threat and how they can change over time.

4.2. Two Approaches to Assessing Vulnerability Approach

The IPCC Fourth Assessment Report Working Groups (2 and 3) discuss vulnerability [*Pielke and Niyogi*, 2010; *Schneider et al.*, 2007]. The IPCC identifies seven criteria for "key" vulnerabilities: magnitude of impacts, timing of

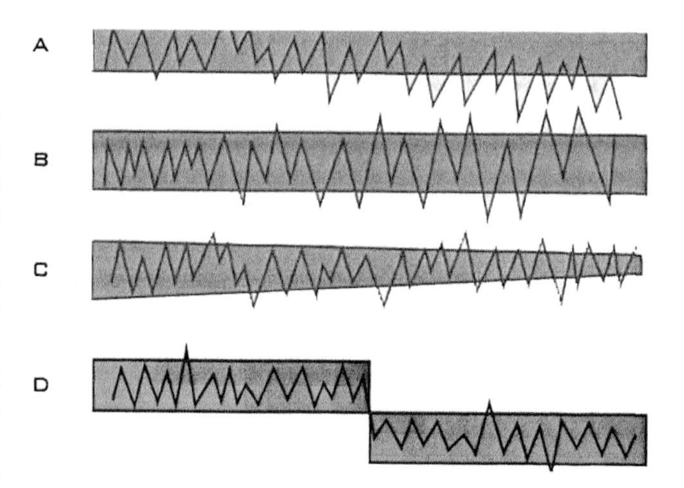

Figure 3. A schematic illustration in which risk changes because of variations in the physical system and the socioeconomic system. In all the cases, risk increases over time (with modifications after the work of *Smith* [1996]). From the work of *Kabat et al.* [2004].

impacts, persistence and reversibility of impacts, likelihood (estimates of uncertainty) of impacts and vulnerabilities and confidence in those estimates, potential for adaptation, distributional aspects of impacts and vulnerabilities, and the importance of the system(s) at risk.

The IPCC also refers to "outcome vulnerability" as illustrated in Figures 4 (left side) and 5 from the works of *O'Brien et al.* [2007] and *Füssel* [2007]. This is clearly a top-down driven perspective. The "contextual vulnerability" is, however, the more inclusive approach to assess risks to key resources since; rather than limiting to subset of threats, the entire spectrum of risks are considered.

For policy makers to develop resilient strategies, it is necessary to consider a multidimensional perspective as illustrated in Figure 6 (from the work of *Hossain et al.* [2011]) and Figure 4 (right side) (from the work of *Füssel* [2007]). *Klein et al.* [1999], for example, sought to determine whether the IPCC guidelines for assessing climate change impacts as well as adaptive strategies can be applied to the example of coastal adaptation. They recommend that a broader approach is needed, which has more local-scale information and input for assessing as well as monitoring the options. The missing link between local-scale features with global-scale projections becomes obvious.

The expanded eight-step approach of *Schroter et al.* [2005], designed to assess vulnerability to climate change, highlights the need to consider multiple interacting stresses. They assume that climate change can be a result of greenhouse gas changes, which are coupled to socioeconomic developments, which, in turn, are coupled to land use changes and that all of these drivers are expected to interactively affect the human, environmental system (such as crop yields). *Metzger et al.* [2006] concluded that most existing assessment studies cannot provide needed information on regional vulnerability.

5. EXAMPLES OF VULNERABILITY THRESHOLDS FOR KEY RESOURCES

There are five broad areas that we can use to define the need for contextual vulnerability assessments: *water*, *food*, *energy*, *human health*, and *ecosystem function*. Each sector is critical societal well-being. The vulnerability concept requires the determination of the major threats to these resources from extreme events including climate but also from other social and environmental pressures. After these threats are identified for each resource, relative risks can be compared in order to shape the preferred mitigation/adaptation strategy.

The questions to be asked for each key resource are as follows:

1. Why is this resource important? How is it used? To what stakeholders is it valuable?

2. What are the key environmental and social variables that influence this resource?

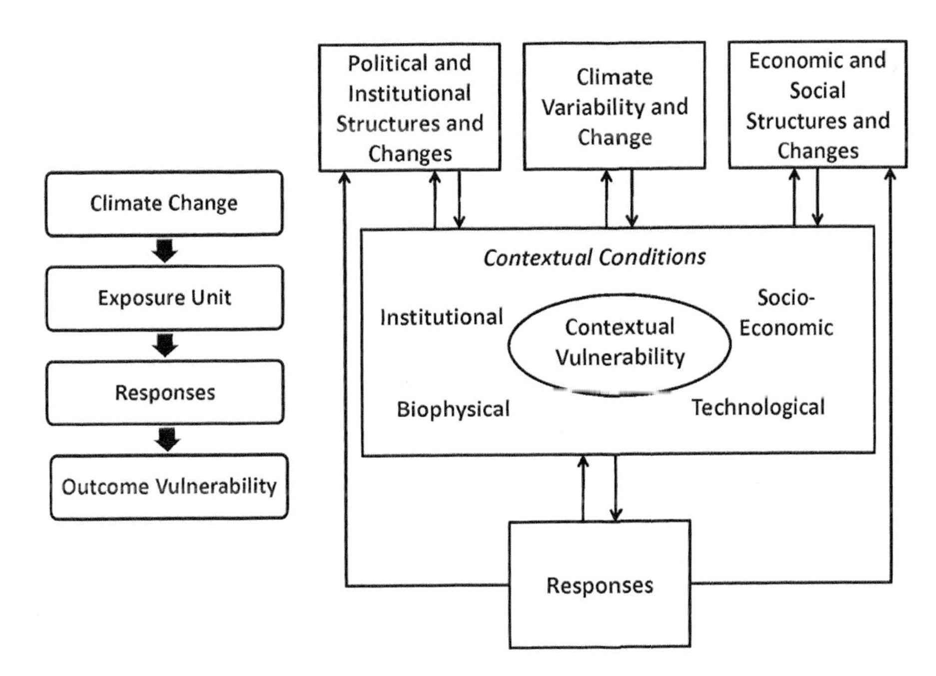

Figure 4. Framework depicting two interpretations of vulnerability to climate change: (left) outcome vulnerability and (right) contextual vulnerability. Adapted by D. Staley from the works of *Füssel* [2009] and *O'Brien et al.* [2007].

	End-Point Interpretation	Starting-Point Interpretation
Root problem	Climate change	Social vulnerability
Policy context	Climate change mitigation, compensation, technical adaptation	Social adaptation, sustainable development
Illustrative policy question	What are the benefits of climate change mitigation?	How can the vulnerability of societies to climatic hazards be reduced?
Illustrative research question	What are the expected net impacts of climate change in different regions?	Why are some groups more affected by climatic hazards than others?
Vulnerability and adaptive capacity	Adaptive capacity determines vulnerability	Vulnerability determines adaptive capacity
Reference for adaptive capacity	Adaptation to future climate change	Adaptation to current climate variability
Starting point of analysis	Scenarios of future climate hazards	Current vulnerability to climate stimuli
Analytical function	Descriptive, positivist	Explanatory, normative
Main discipline	Natural sciences	Social sciences
Meaning of "vulnerability"	Expected net damage for a given level of global climate change	Susceptibility to climate change and variability as determined by socioeconomic factors
Qualification according to the terminology from Section 2	Dynamic cross-scale integrated vulnerability [of a particular system] to a global climate change	Current internal socioeconomic vulnerability [of a particular social unit] to all climatic stressors
Vulnerability approach	Integrated, risk-hazard	Political economy
Reference	*McCarthy et al.* [2001]	*Adger* [1999]

Figure 5. Two interpretations of vulnerability in climate change research. From the work of *Füssel* [2007, 2009].

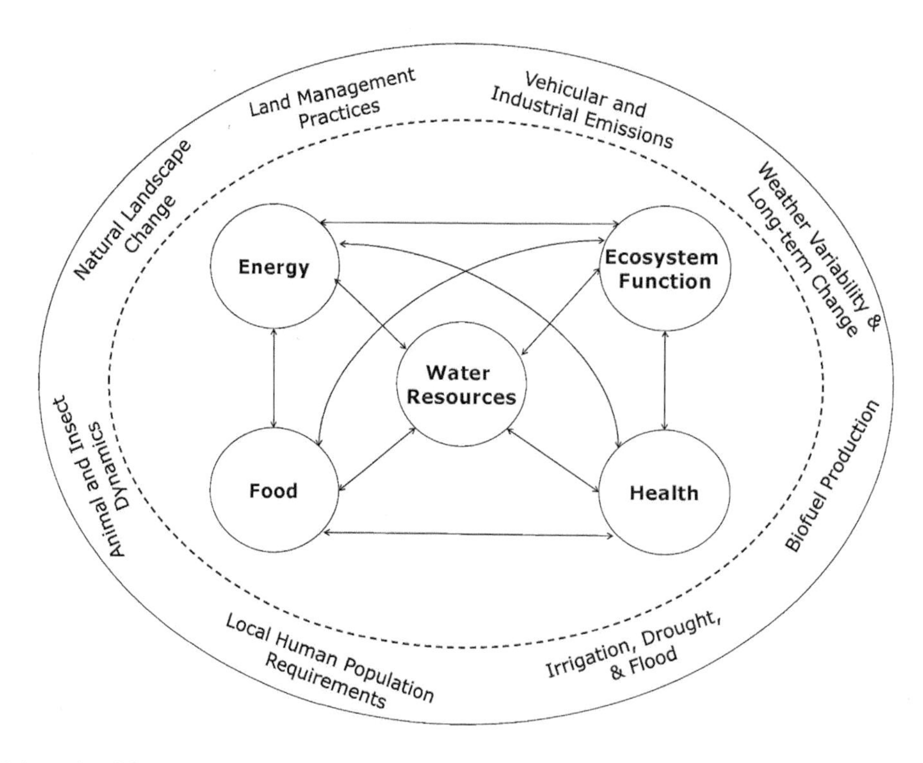

Figure 6. Schematic of the spectrum of risks to water resources. Other key resources associated with food, energy, human health, and ecosystem function can replace water resources in the central circle. From the work of *Hossain et al.* [2011].

3. What is the sensitivity of this resource to changes in each of these key variables? (This may include, but is not limited to, the sensitivity of the resource to climate variations and change on short (days), medium (seasons), and long (multidecadal) time scales.)

4. What changes (thresholds) in these key variables would have to occur to result in a negative (or positive) outcome for this resource?

5. What are the best estimates of the probabilities for these changes to occur? What tools are available to quantify the effect of these changes? Can these estimates be skillfully predicted?

6. What actions (adaptation/mitigation) can be undertaken in order to minimize or eliminate the negative consequences of these changes (or to optimize a positive response)?

7. What are specific recommendations for policy makers and other stakeholders?

Each of these concerns is explored in more detail in the following sections.

5.1. Water

To understand the vulnerability of water resources, we first need to recognize that the water that is usable can occur in various forms such as rainfall, surface water, rechargeable and fossil groundwater, snow, natural lakes, artificial reservoirs, and through state compacts and international treaties. The threats to these water resources are many, such as through health and contamination, changes in precipitation extremes, population demand, industrial and agricultural demand, contamination, national water policies, and climate [see *Vörösmarty et al.*, 2010]. There may also be "competition" between different applications (resource production). For example, most of today's agriculture and fossil fuel-based energy production is water intensive [*Jones*, 2008]. Population and industrialization have continued to increase over the last century, which results in more competition for available water resources between direct consumption (for public and industrial water supply) and resource production (for crops and energy).

The resilience to known threats to water availability can be region specific and vary due to a multiplicity of factors. The factors affecting availability of water in most parts of the world are many, and at least more than a few key issues are involved [*Vörösmarty et al.*, 2010]. The assessment of vulnerability of water resources requires an inherent recognition of these multiple threats (including from climate change and variability) to prioritize high-risk threats and plan adaptation strategies based on such multiple high-risk scenarios.

For example, let us consider for a country that the 50 year water availability is dictated overwhelmingly by rapid population growth and accompanying environmental degradation of water quality when compared to climate change (IPCC)-based projections [e.g., see *Vörösmarty et al.*, 2000]. A 50 year effective adaptation strategy that incorporates the 50 year population growth and expected water quality crises must therefore be resilient to any reasonably possible climate change. This is the inherent strength of a bottom-up approach versus the limited top-down counterpart.

5.2. Food

Agriculture, crop based as well as animal driven, is a risk-prone entity. For example, assuming a global model projection for a future climate is accurate for a particular region, one could ascribe a range of climatic changes. These could include higher temperatures, greater propensity for more intense rainfalls, and higher CO_2 levels. Each of these can positively affect the crop yield by promoting enhanced photosynthesis rates [*Curtis et al.*, 2003; *Jablonski et al.*, 2002] and taller and more robust crop and forest growth. Conversely, depending on the local conditions, the same changes could translate into increasing pest risk, higher ozone-related damages, increasing soil erosion risk, hail and frost damage, and reduced work days suitable for farm activities.

To extract the significance of the individual versus multiple climatic stressors on crop yields, *Mera et al.* [2006] developed a crop modeling study with over 25 different input scenarios of temperature, rainfall, and radiation changes at a farm scale for two crops that assimilate carbon differently (e.g., soybean and maize). As seen in many crop yield studies, the results suggested that yields were most sensitive to the amount of effective precipitation (estimated as rainfall minus physical evaporation/transpiration loss from the land surface). Changes in radiation had a nonlinear effect with crops showing an increased productivity for some reduction in the radiation as a result of cloudiness and increased diffuse radiation and a decline in yield with further reduction in radiation amounts. The impact of temperature changes, which has been at the heart of many climate projections, however, was quite limited, particularly if the soils did not have moisture stress. The analysis from the multiple climate change settings do not agree with those from individual changes, making a case for multivariable, ensemble approaches to identify the vulnerability and feedback when estimating climate-related impacts [cf. *Turner et al.*, 2003].

A big unknown in food security, however, is the so-called nonclimatic risks. This could include agricultural policies such as those permitting genetic versus organic farming standards for the region as practiced in some European Union countries or the ethanol blending mandated in the Midwest United States. Even when considering the climatic factors alone, a large number of if-then probable scenarios can be developed that can have positive or negative impacts on crop yield and agricultural sustainability.

Niyogi and Mishra [2012] assess a number of stresses including a temperature increase, which can lead to increased yield for an initial period and can affect fertility, graining, and future generations and the timing of the temperature and rainfall, both of which would be a significant source of uncertainties. For example, reduced rainfall during the 2 week sowing period can translate into reduced yield even if the rainfall was adequate for the entire growing season. The stress on the plants, particularly from heavy rain and even frost during the young stage, would be much higher than during the mature period when the roots would be much deeper. The uncertainties also include pathogen and weed stress due to increased humidity and temperature interactions. Weeds are expected to be at significant advantage and not currently considered in conventional crop yield impact studies. From an adaptation perspective, if the farmers have information about possible droughts and sowed the seed deeper into the soil, which requires extra energy and time investment, the negative impacts could be alleviated.

Assessing the adaptation and mitigation approaches therefore requires a much broader view on the production processes and the life cycle of the entity than CO_2-driven global model predictions can provide. Current crop impact studies adopt a typical approach in which the GCM scenario, often one or two extreme members instead of the ensemble, as input to simple process-based or statistical crop models. The bottom-up perspective provides a wider range of scenarios for the adaptive and mitigative strategies that individual growers, regional economies, and policy makers need to be able to respond to.

5.3. Energy

Two large categories of energy resources, namely, nuclear and renewable, are considered as unlimited, but this is practically untrue. The metals and other basic components used for producing the energy converters (e.g., nuclear reactors, solar cells, wind/wave generators, and farming of plants for biomass production) are limited and therefore vulnerable to human intervention. A more characteristic case is the growing need of materials with unique characteristics (rare metals). Climatic variability can influence all these energy resources in many ways. The increase of energy consumption requires intense mining of fossil fuels, increasing the areas covered by on/offshore renewable energy parks and platforms with a resultant substantial influence of local climate (e.g., due to changes in local wind and/or wave conditions from the physical presence of these structures).

Renewable energy is potentially vulnerable to climate variability and longer-term change. For example, biomass production involving land use change can alter the regional climate. *Costa et al.* [2007] found a significant reduction in rainfall when the land was converted to soybeans as contrasted with a conversion to grassland as a consequence of the larger albedo of the soybean fields. Wind turbines and solar panels are obviously strongly influenced by weather, and if they cover a large-enough area, it has been stated that they alter regional and even larger-scale climate patterns [see, e.g., *Wang and Prinn*, 2009]. Hydropower with its dependence on precipitation is obviously significantly affected by climate.

5.4. Human Health

The link between environmental conditions and human health is well established. Changes in weather and climate conditions on times scales ranging from days to decades can directly impact the conditions allowing certain diseases to flourish, on the one hand, while also affecting the exposure of human populations to disease, on the other.

For example, ranges and pathogen incubation periods of various vector-borne and waterborne diseases are directly linked to changes in climatic conditions (as is the case for malaria [see *Githeko*, 2009]). Similarly, heat-related morbidity and mortality associated with hot and cold waves are well documented [e.g., *Keatinge et al.*, 2000]. Changes in the precipitation regimes, length of growing seasons, and increased dust from drought all contribute to respiratory allergies, asthma, and airway diseases in vulnerable populations. The challenge of addressing the effects of climate on human health is very complex because local or regional cultural, political, and economic factors can exacerbate environmental stressors, and the decisions that people make also influence health.

A host of factors such as biological susceptibility, socioeconomic status, cultural norms, and the quality of infrastructure often come into play in determining the vulnerability to climate-related disease conditions. Effective response strategies have to necessarily be region specific, and these must include defining environmental risk factors, identifying vulnerable populations, and developing effective risk communication and prevention strategies [*Portier et al.*, 2010].

5.5. Ecosystem Function

Feedback from human activities have become directional drivers of change in both human and nonhuman-dominated ecosystems. These factors may be independent of climate change forcing, amplify, or attenuate the climate effects. For example, the fertilization of plants from enhanced atmospheric

CO_2, the positive and negative effects of nongreenhouse forms of inorganic nitrogen in the atmosphere, and the "wild card" effect of human-facilitated species introductions and extirpations are potentially changing local and regional landscapes as fast, or faster, than climate drivers [e.g., *Vitousek et al.*, 1997; *Hobbs et al.*, 2009]. Land use change can "trump" all of the above.

Thus, in considering how ecosystems will respond and interact in the twenty-first century, two points need to be emphasized. First, ecosystem function is vulnerable to human activities that are tangential to the climate drivers. Human activities have induced local and regional "tipping points" such as lake eutrophication [*Carpenter and Lathrop*, 2008] and desertification of rangelands [*Schlesinger et al.*, 1990]. These events are occurring because of factors independent of climate forcings. Clear evidence of climate variations and longer-term change is also occurring in many areas; thus, scenario planning, mitigation, and adaptation require that we understand how these different facets of global environmental change interact with the climate system. How do these other anthropogenic activities alter outcomes? How will they influence the vulnerability of these systems to change?

The above questions lead to the second point. The response functions of ecosystems to the climate drivers are determined by the net effect of past drivers on the current structure of the ecosystem. Ecosystems can respond to climate forcings by exhibiting resilience, a phenomenon well exemplified by the relatively benign response of the Great Plains grasslands to the drought of the 1930s. Conversely, the same areas can experience transformation, i.e., the dust storms and destruction of millions of hectares of agricultural lands caused by the same 1930s drought. Clearly, climate alone was not the causal mechanism for the dust bowl, and we know now that the subsequent feedback to the regional climate from either a vegetated or barren landscape were substantial [e.g., *Cook et al.*, 2009]. Resiliency and adaptive capacity is often associated with healthy diverse ecosystems; restoring ecosystem function of degraded ecosystems can convey resilience to future climate [*McAlpine et al.*, 2010]. Thus, current decisions about land management will affect the ecosystem response functions that influence subsequent global climate change drivers.

6. CONCLUSIONS

The adoption of a vulnerability assessment approach to evaluate the effect of climate and other environmental and societal threats to key resources is an inclusive way of assessing risks, including from climate variability and longer-term climate change. In contrast to the *outcome vulnerability* adopted by the IPCC, the *contextual vulnerability* discussed by *Füssel* [2009] is more inclusive and provides a more robust framework for policy makers to adopt mitigation and adaptation methodologies to deal with the spectrum of social and environmental issues in the coming decades.

The concept of contextual vulnerability enables the determination of major threats to water, food, energy, human health, and ecosystem function from extreme events including those arising from climate but also other social and environmental pressures (as given by *Pielke Jr.* [2010], *Wallace* [2010], *Webster and Hoyos* [2010], J. Curry and P. Webster (Pakistan flood follow-up, Climate Etc., Weblog, available at http://judithcurry.com/2011/01/05/pakistan-flood-follow-up/, 2010), and G. R. Carmichael (What goes around comes around: The globalization of air pollution and the implications for the quality of the air we breathe, the water we drink, and the food we eat, CIRES Distinguished Lecture Series, University of Colorado, Boulder, 6 March 2009)]. After these threats are identified for each resource, then relative risks can be determined in order to prioritize individual response measures and to shape the preferred mitigation/adaptation strategy.

Acknowledgments. The final editing of the paper was handled in the standard outstanding manner by Dallas Staley. Ray Taylor is thanked for alerting us to the Füssel article. Dev Niyogi acknowledges support from NSF CAREER ATM 0847472. Roger Pielke Sr. acknowledges support from NSF grant 0831331. The authors thank Alik Ismail-Zadeh for his valuable suggested edits in the final version.

REFERENCES

Adger, W. N. (1996), Approaches to vulnerability to climate change, *CSERGE Work. Pap. GEC 96-05*, Cent. for Soc. and Econ. Res. on the Global Environ., Univ. of East Anglia, Norwich, U. K.

Adger, W. N. (1999), Social vulnerability to climate change and extremes in coastal Vietnam, *World Dev.*, *27*, 249–269.

Annamalai, H., K. Hamilton, and K. R. Sperber (2007), The South Asian summer monsoon and its relationship with ENSO in the IPCC AR4 simulations, *J. Clim.*, *20*, 1071–1092, doi:10.1175/JCLI4035.1.

Beer, T., and A. T. Ismail-Zadeh (Eds.) (2003), *Risk Science and Sustainability*, 256 pp., Kluwer Acad., Dordrecht, Netherlands.

Bray, D., and H. von Storch (2009), 'Prediction' or 'Projection'? The nomenclature of climate science, *Sci. Comm.*, *30*, 534–543, doi:10.1177/1075547009333698.

Bouwer, L. M. (2011), Have disaster losses increased due to anthropogenic climate change?, *Bull. Am. Meteorol. Soc.*, *92*, 39–46, doi:10.1175/2010BAMS3092.1.

Bravo de Guenni, L., R. E. Schulze, R. A. Pielke Sr., and M. F. Hutchinson (2004), The vulnerability approach, in *Vegetation, Water, Humans and the Climate: A New Perspective on an*

Interactive System, edited by P. Kabat et al., chap. E.5, pp. 499–514, Springer, New York.

Carpenter, S., and R. C. Lathrop (2008), Probabilistic estimate of a threshold for eutrophication, *Ecosystems*, *11*, 601–613, doi:10.1007/s10021-008-9145-0.

Chase, T. N., K. Wolter, R. A. Pielke Sr., and I. Rasool (2006), Was the 2003 European summer heat wave unusual in a global context?, *Geophys. Res. Lett.*, *33*, L23709, doi:10.1029/2006GL027470.

Compo, G. P., et al. (2011), The Twentieth Century Reanalysis Project, *Q. J. R. Meteorol. Soc.*, *137*, 1–28, doi:10.1002/qj.776.

Cook, B. I., R. L. Miller, and R. Seager (2009), Amplification of the North American Dust Bowl drought through human-induced land degradation, *Proc. Natl. Acad. Sci. U. S. A.*, *106*(13), 4997–5001, doi:10.1073/pnas.0810200106.

Costa, M. H., S. N. M. Yanagi, P. J. O. P. Souza, A. Ribeiro, and E. J. P. Rocha (2007), Climate change in Amazonia caused by soybean cropland expansion, as compared to caused by pastureland expansion, *Geophys. Res. Lett.*, *34*, L07706, doi:10.1029/2007GL029271.

Crompton, R. P., R. A. Pielke Jr., and K. J. McAneney (2011), Emergence timescales for detection of anthropogenic climate change in US tropical cyclone loss data, *Environ. Res. Lett.*, *6*, 014003, doi:10.1088/1748-9326/6/1/014003.

Curtis, P. S., L. M. Jablonski, and X. Wang (2003), Assessing elevated CO_2 responses using meta-analysis, *New Phytol.*, *160*, 6–7, doi:10.1046/j.1469-8137.2003.00886.x.

Davin, E. L., N. de Noblet-Ducoudré, and P. Friedlingstein (2007), Impact of land cover change on surface climate: Relevance of the radiative forcing concept, *Geophys. Res. Lett.*, *34*, L13702, doi:10.1029/2007GL029678.

Dessai, S., M. Hulme, R. Lempert, and R. Pielke Jr. (2009), Do we need better predictions to adapt to a changing climate?, *Eos Trans. AGU*, *90*(13), 111, doi:10.1029/2009EO130003.

Dixon, H., D. M. Lawler, and A. Y. Shamseldin (2006), Streamflow trends in western Britain, *Geophys. Res. Lett.*, *33*, L19406, doi:10.1029/2006GL027325.

Ferraris, L., S. Gabellani, N. Rebora, and A. Provenzale (2003), A comparison of stochastic models for spatial rainfall downscaling, *Water Resour. Res.*, *39*(12), 1368, doi:10.1029/2003WR002504.

Flanner, M. G., C. S. Zender, J. T. Randerson, and P. J. Rasch (2007), Present-day climate forcing and response from black carbon in snow, *J. Geophys. Res.*, *112*, D11202, doi:10.1029/2006JD008003.

Fowler, H. J., and M. Ekström (2009), Multi-model ensemble estimates of climate change impacts on UK seasonal rainfall extremes, *Int. J. Climatol.*, *29*, 385–416, doi:10.1002/joc.1827.

Fowler, H. J., and R. L. Wilby (2010), Detecting changes in seasonal precipitation extremes using regional climate model projections: Implications for managing fluvial flood risk, *Water Resour. Res.*, *46*, W03525, doi:10.1029/2008WR007636.

Füssel, H.-M. (2007), Vulnerability: A generally applicable conceptual framework for climate change research, *Global Environ. Change*, *17*, 155–167.

Füssel, H.-M. (2009), Review and quantitative analysis of indices of climate change exposure, adaptive capacity, sensitivity, and impacts, background note, in *World Development Report 2010: Development and Climate Change*, report, 35 pp., World Bank, Washington, D. C. [Available at http://siteresources.worldbank.org/INTWDR2010/Resources/5287678-1255547194560/WDR2010_BG_Note_Fussel.pdf.]

Galloway, J. N., et al. (2004), Nitrogen cycles: Past, present, and future, *Biogeochemistry*, *70*(2), 153–226, doi:10.1007/s10533-004-0370-0.

Githeko, A. K. (2009), Malaria and climate change, in *Commonwealth Health Minister's Update 2009*, pp. 40–43, Commonw. Secr., London, U. K. [Available online at http://www.thecommonwealth.org/files/190385/FileName/Githeko_2009.pdf.]

Hawkins, E., and R. Sutton (2010), The potential to narrow uncertainty in projections of regional precipitation change, *Clim. Dyn.*, *37*(12), 407–418, doi:10.1007/s00382-010-0810-6.

Hegerl, G. C., F. W. Zwiers, P. A. Stott, and V. V. Kharin (2004), Detectability of anthropogenic changes in temperature and precipitation extremes, *J. Clim.*, *17*, 3683–3700, doi:10.1175/1520-0442(2004)017<3683:DOACIA>2.0.CO;2.

Hobbs, R. J., E. Higgs, and J. A. Harris (2009), Novel ecosystems: Implications for conservation and restoration, *Trends Ecol. Evol.*, *24*, 599–605, doi:10.1016/j.tree.2009.05.012.

Holman, I. P., M. D. A. Rounsevell, S. Shackley, P. A. Harrison, R. J. Nicholls, P. M. Berry, and E. Audsley (2005), A regional, multi-sectoral and integrated assessment of the impacts of climate and socio-economic change in the UK, Part I. Methodology, *Clim. Change*, *71*, 9–41, doi:10.1007/s10584-005-5927-y.

Hossain, F., D. Niyogi, J. Adegoke, G. Kallos, and R. Pielke Sr. (2011), Making sense of the water resources that will be available for future use, *Eos Trans. AGU*, *92*(17), 144, doi:10.1029/2011EO170005.

Hurrell, J., G. A. Meehl, D. Bader, T. L. Delworth, B. Kirtman, and B. Wielicki (2009), A unified modeling approach to climate system prediction, *Bull. Am. Meteorol. Soc.*, *90*, 1819–1832.

Jablonski, L. M., X. Wang, and P. S. Curtis (2002), Plant reproduction under elevated CO_2 conditions: A meta-analysis of reports on 79 crop and wild species, *New Phytol.*, *156*, 9–26, doi:10.1046/j.1469-8137.2002.00494.x.

Jones, W. D. (2008), How much water does it take to make electricity?, *IEEE Spectrum*, 23 April. [Available at http://spectrum.ieee.org/energy/environment/how-much-water-does-it-take-to-make-electricity.]

Kabat, P., et al. (Eds.) (2004), *Vegetation, Water, Humans and the Climate: A New Perspective on an Interactive System*, 566 pp., Springer, Berlin.

Keatinge, W. R., G. C. Donaldson, E. Cordioli, M. Martinelli, A. E. Kunst, J. P. Mackenbach, S. Nayha, and I. Vuori (2000), Heat related mortality in warm and cold regions of Europe: Observational study, *Br. Med. J.*, *321*, 670–673, doi:10.1136/bmj.321.7262.670.

Klein, R. J. T., R. J. Nicholls, and N. Mimura (1999), Coastal adaptation to climate change: Can the IPCC technical guidelines be applied?, *Mitigat. Adapt. Strat. Global Change*, *4*, 239–252, doi:10.1023/A:1009681207419.

Landsea, C. W., and J. A. Knaff (2000), How much skill was there in forecasting the very strong 1997–98 El Niño?, *Bull. Am. Meteorol. Soc.*, *81*(9), 2107–2120, doi:10.1175/1520-0477(2000)081<2107:HMSWTI>2.3.CO;2.

MacCracken, M. (2002), Do the uncertainty ranges in the IPCC and U.S. National Assessments account adequately for possibly overlooked climatic influences, *Clim. Change*, *52*, 13–23.

Matsui, T., and R. A. Pielke Sr. (2006), Measurement-based estimation of the spatial gradient of aerosol radiative forcing, *Geophys. Res. Lett.*, *33*, L11813, doi:10.1029/2006GL025974.

McCarthy, J. J., et al. (Eds.) (2001), *Climate Change 2001: Impacts, Adaptation, and Vulnerability: Contribution of Working Group II to the Third Assessment Report of the Intergovernmental Panel on Climate Change*, 1042 pp., Cambridge Univ. Press, Cambridge, U. K.

McAlpine, C. A., W. F. Laurance, J. G. Ryan, L. Seabrook, J. I. Syktus, A. E. Etter, P. M. Fearnside, P. Dargusch, and R. A. Pielke Sr. (2010), More than CO_2: A broader picture for managing climate change and variability to avoid ecosystem collapse, *Curr. Opin. Environ. Sustainability*, *2*, 334–336, doi:10.1016/j.cosust.2010.10.001.

Mearns, L. O., I. Bogardi, F. Giorgi, I. Matyasovszky, and M. Palecki (1999), Comparison of climate change scenarios generated from regional climate model experiments and statistical downscaling, *J. Geophys. Res.*, *104*(D6), 6603–6621, doi:10.1029/1998JD200042.

Mera, R. J., D. Niyogi, G. S. Buol, G. G. Wilkerson, and F. Semazzi (2006), Potential individual versus simultaneous climate change effects on soybean (C_3) and maize (C_4) crops: An agrotechnology model based study, *Global Planet. Change*, *54*, 163–182, doi:10.1016/j.gloplacha.2005.11.003.

Metz, B., et al. (Eds.) (2007), *Climate Change 2007: Mitigation of Climate Change: Contribution of Working Group III to the Fourth Assessment Report of the Intergovernmental Panel on Climate Change*, 890 pp., Cambridge Univ. Press, Cambridge, U. K.

Metzger, M. J., M. D. A. Rounsevell, L. Acosta-Michlik, R. Leemans, and D. Schroter (2006), The vulnerability of ecosystem services to land use change, *Agric. Ecosyst. Environ.*, *114*, 69–85, doi:10.1016/j.agee.2005.11.025.

Milly, P. C. D., J. Betancourt, M. Falkenmark, R. M. Hirsch, Z. W. Kundzewicz, D. P. Lettenmaier, and R. J. Stouffer (2008), Stationarity is dead: Whither water management?, *Science*, *319* (5863), 573–574, doi:10.1126/science.1151915.

Mishra, V., K. A. Cherkauer, D. Niyogi, M. Lei, B. C. Pijanowski, D. K. Ray, L. C. Bowling, and G. Yang (2010), A regional scale assessment of land use/land cover and climatic changes on water and energy cycle in the upper Midwest United States, *Int. J. Climatol.*, *30*, 2025–2044, doi:10.1002/joc.2095.

Myhre, G., and A. Myhre (2003), Uncertainties in radiative forcing due to surface albedo changes caused by land-use changes, *J. Clim.*, *16*, 1511–1524.

National Research Council (NRC) (2005), *Radiative Forcing of Climate Change: Expanding the Concept and Addressing Uncertainties*, 208 pp., Natl. Acad. Press, Washington, D. C.

Niyogi, D., and V. Mishra (2012), Climate-agriculture vulnerability assessment for the midwestern United States, in *Climate Change in the Midwest: Impacts, Risks, Vulnerability and Adaptation*, edited by S. C. Pryor, Indiana Univ. Press, Bloomington, in press.

O'Brien, K. L., S. Eriksen, L. Nygaard, and A. Schjolden (2007), Why different interpretations of vulnerability matter in climate change discourses, *Clim. Policy*, *7*(1), 73–88.

Otterman, J., et al. (2002), Are stronger North-Atlantic southwesterlies the forcing to the late winter warming in Europe?, *Int. J. Climatol.*, *22*, 743–750, doi:10.1002/joc.681.

Paeth, H., A. Scholten, P. Friederichs, and A. Hense (2008), Uncertainties in climate change prediction: El Niño-Southern Oscillation and monsoons, *Global Planet. Change*, *60*, 265–288, doi:10.1016/j.gloplacha.2007.03.002.

Pall, P., T. Aina, D. A. Stone, P. A. Stott, T. Nozawa, A. G. Hilberts, D. Lohmann, and M. R. Allen (2011), Anthropogenic greenhouse gas contribution to flood risk in England and Wales in autumn 2000, *Nature*, *470*, 382–386, doi:10.1038/nature09762.

Palmer, T. N., F. J. Doblas-Reyes, A. Weisheimer, and M. J. Rodwell (2008), Toward seamless prediction: Calibration of climate change projections using seasonal forecasts, *Bull. Am. Meteorol. Soc.*, *89*, 459–470, doi:10.1175/BAMS-89-4-459.

Parry, M. L., et al. (Eds.) (2007), *Climate Change 2007: Impacts, Adaptation and Vulnerability: Contribution of Working Group II to the Fourth Assessment Report of the Intergovernmental Panel on Climate Change*, 1000 pp., Cambridge Univ. Press, Cambridge, U. K.

Parson, E. A., et al. (2003), Understanding climatic impacts, vulnerabilities, and adaptation in the United States: Building a capacity for assessment, *Clim. Change*, *57*, 9–42, doi:10.1023/A:1022188519982.

Patt, A. G., D. P. van Vuuren, F. Berkhout, A. Aaheim, A. F. Hof, M. Isaac, and R. Mechler (2010), Adaptation in integrated assessment modeling: Where do we stand?, *Clim. Change*, *99*, 383–402, doi:10.1007/s10584-009-9687-y.

Pielke, R. A., Jr. (2010), *The Climate Fix: What Scientists and Politicians Won't Tell You About Global Warming*, 288 pp., Basic Books, New York.

Pielke, R. A., Jr., G. Prins, S. Rayner, and D. Sarewitz (2007), Lifting the taboo on adaptation, *Nature*, *445*(7128), 597–598, doi:10.1038/445597a.

Pielke, R. A., Sr. (2002), Overlooked issues in the U.S. National Climate and IPCC assessments, *Clim. Change*, *52*, 1–11.

Pielke, R. A., Sr. (2010), Comment on "A unified modeling approach to climate system prediction", *Bull. Am. Meteorol. Soc.*, *91*, 1699–1701, doi:10.1175/2010BAMS2975.1.

Pielke, R. A., Sr., and D. Niyogi (2010), The role of landscape processes within the climate system, in *Landform-Structure, Evolution, Process Control*, edited by J.-C. Otto and R. Didkau, *Lect. Notes Earth Sci.*, *115*, 67–85, doi:10.1007/978-3-540-75761-0_5.

Pielke, R. A., Sr., et al. (2007), Unresolved issues with the assessment of multidecadal global land surface temperature trends, *J. Geophys. Res.*, *112*, D24S08, doi:10.1029/2006JD008229.

Pielke, R. A., Sr., et al. (2009), Climate change: The need to consider human forcings besides greenhouse gases, *Eos Trans. AGU*, *90*(45), 413, doi:10.1029/2009EO450008.

Portier, C. J., et al. (2010), A human health perspective on climate change: A report outlining the research needs on the human health effects of climate change, report, Environ. Health Perspect./Natl. Inst. of Environ. Health Sci., Research Triangle Part, N. C., doi:10.1289/ehp.1002272. [Available at www.niehs.nih.gov/climatereport.]

Rahman, S., A. C. Bagtzoglou, F. Hossain, L. Tang, L. Yarbrough, and G. Easson (2009), Investigating spatial downscaling of satellite rainfall data for flood prediction, *J. Hydrometeorol.*, *10*, 1063–1079, doi:10.1175/2009JHM1072.1.

Rajagopalan, B., K. Nowak, J. Prairie, M. Hoerling, B. Harding, J. Barsugli, A. Ray, and B. Udall (2009), Water supply risk on the Colorado River: Can management mitigate?, *Water Resour. Res.*, *45*, W08201, doi:10.1029/2008WR007652.

Ray, D. K., R. A. Pielke Sr., U. S. Nair, and D. Niyogi (2010), Roles of atmospheric and land surface data in dynamic regional downscaling, *J. Geophys. Res.*, *115*, D05102, doi:10.1029/2009JD012218.

Rial, J., et al. (2004), Nonlinearities, feedbacks and critical thresholds within the Earth's climate system, *Clim. Change*, *65*, 11–38.

Rockel, B., C. L. Castro, R. A. Pielke Sr., H. von Storch, and G. Leoncini (2008), Dynamical downscaling: Assessment of model system dependent retained and added variability for two different regional climate models, *J. Geophys. Res.*, *113*, D21107, doi:10.1029/2007JD009461.

Rosenfeld, D., U. Lohmann, G. B. Raga, C. D. O'Dowd, M. Kulmala, S. Fuzzi, A. Reissell, and M. O. Andreae (2008), Flood or drought: How do aerosols affect precipitation?, *Science*, *321*(5894), 1309–1313, doi:10.1126/science.1160606.

Salathe, E. P. (2005), Downscaling simulations of future global climate with application to hydrologic modelling, *Int. J. Climatol.*, *25*, 419–436, doi:10.1002/joc.1125.

Schlesinger, W. H., J. F. Reynolds, G. L. Cunningham, L. F. Huenneke, W. M. Jarrell, R. A. Virginia, and W. G. Whitford (1990), Biological feedbacks in global desertification, *Science*, *247*, 1043–1048, doi:10.1126/science.247.4946.1043.

Schneider, S. H., et al. (2007), Assessing key vulnerabilities and the risk from climate change, in *Climate Change 2007: Impacts, Adaptation and Vulnerability: Contribution of Working Group II to the Fourth Assessment Report of the Intergovernmental Panel on Climate Change*, edited by M. L. Parry et al., pp. 779–810, Cambridge Univ. Press, Cambridge, U. K.

Schroeter, D., C. Polsky, and A. G. Patt (2005), Assessing vulnerabilities to the effects of global change: An eight step approach, *Mitigat. Adapt. Strat. Global Change*, *10*, 573–595.

Smith, K. (1996), *Environmental Hazards*, 376 pp., Routledge, London, U. K.

Solomon, S., et al. (Eds.) (2007), *Climate Change 2007: The Physical Science Basis: Contribution of Working Group I to the Fourth Assessment Report of the Intergovernmental Panel on Climate Change*, 1056 pp., Cambridge Univ. Press, Cambridge, U. K.

Takata, K., K. Saito, and T. Yasunari (2009), Changes in the Asian monsoon climate during 1700–1850 induced by preindustrial cultivation, *Proc. Natl. Acad. Sci. U. S. A.*, *106*, 9586–9589, doi:10.1073/pnas.0807346106.

Turner, B. L., et al. (2003), A framework for vulnerability analysis in sustainability science, *Proc. Natl. Acad. Sci. U. S. A.*, *100*, 8074–8079, doi:10.1073/pnas.1231335100.

Vitousek, P., H. A. Mooney, J. Lubchenco, and J. M. Melillo (1997), Human domination of Earth's ecosystems, *Science*, *277*, 494–499, doi:10.1126/science.277.5325.494.

Vogel, C. (1998), Vulnerability and global environmental change, *LUCC Newsl.*, *3*, 15–19.

Vörösmarty, C. J., P. Green, J. Salisbury, and R. B. Lammers (2000), Global water resources: Vulnerability from climate change and population growth, *Science*, *289*, 284–288, doi:10.1126/science.289.5477.284.

Vörösmarty, C. J., et al. (2010), Global threats to human water security and river biodiversity, *Nature*, *467*, 555–561, doi:10.1038/nature09440.

Wallace, J. M. (2010), Beyond climate change: Reframing the dialogue over environmental issues, *Seattle Times*, 26 March. [Available at http://seattletimes.nwsource.com/html/opinion/2011453141_guest28wallace.html.]

Wang, C., and R. G. Prinn (2009), Potential climatic impacts and reliability of very large-scale wind farm, *Rep. 175*, 21 pp., MIT Joint Program on the Sci. and Policy of Global Change, Cambridge, Mass. [Available at http://globalchange.mit.edu/files/document/MITJPSPGC_Rpt175.pdf.]

Webster, P. J., and C. Hoyos (2004), Prediction of monsoon rainfall and river discharge on 15–30-day time scales, *Bull. Am. Meteorol. Soc.*, *85*, 1745–1767, doi:10.1175/BAMS-85-11-1745.

Wilby, R. L. (2006), When and where might climate change be detectable in UK river flows?, *Geophys. Res. Lett.*, *33*, L19407, doi:10.1029/2006GL027552.

Wilby, R. L., T. M. L. Wigley, D. Conway, P. D. Jones, B. C. Hewitson, J. Main, and D. S. Wilks (1998), Statistical downscaling of general circulation model output: A comparison of methods, *Water Resour. Res.*, *34*(11), 2995–3008, doi:10.1029/98WR02577.

Wilby, R. L., K. J. Beven, and N. S. Reynard (2008), Climate change and fluvial flood risk in the UK: More of the same?, *Hydrol. Processes*, *22*, 2511–2523, doi:10.1002/hyp.6847.

Yang, G., L. C. Bowling, K. A. Cherkauer, B. C. Pijanowski, and D. Niyogi (2009), Hydroclimatic response of watersheds to urban intensity—An observational and modeling based analysis for the White River Basin, Indiana, *J. Hydrometeorol.*, *11*, 122–138, doi:10.1175/2009JHM1143.1.

Ziegler, A. D., E. P. Maurer, J. Sheffield, B. Nijssen, E. F. Wood, and D. P. Lettenmaier (2005), Detection time for plausible changes in annual precipitation, evapotranspiration, and streamflow in three Mississippi river sub-basins, *Clim. Change*, *72*, 17–36, doi:10.1007/s10584-005-5379-4.

J. Adegoke, Natural Resources and the Environment, CSIR, Pretoria 0001, South Africa. (jadegoke@csir.co.za)

K. Dairuku, Disaster Prevention System Research Center, National Research Institute for Earth Science and Disaster Prevention, 3-1 Tennodai, Tsukuba, Ibaraki 305-0006, Japan. (dairaku@bosai.go.jp)

F. Hossain, Department of Civil and Environmental Engineering, Tennessee Technological University, Cookeville, TN 38505-0001, USA. (FHossain@tntech.edu)

G. Kallos, School of Physics, University of Athens, 15784 Athens, Greece. (kallos@mg.uoa.gr)

D. Niyogi, Department of Earth and Atmospheric Sciences, Purdue University, West Lafayette, IN 47907, USA. (dniyogi@-dniyogi@purdue.edu)

R. A. Pielke Sr., Cooperative Institute for Research in Environmental Sciences, University of Colorado, Boulder, CO 80309, USA. (pielkesr@cires.colorado.edu)

T. Seastedt, Institute of Arctic and Alpine Research, University of Colorado, Boulder, CO 80309, USA. (timothy.seastedt@colo rado.edu)

K. Suding, Department of Environmental Science, Policy, and Management, University of California at Berkeley, Berkeley, CA 94720-3114, USA. (suding@berkeley.edu)

R. Wilby, Department of Geography, Loughborough University, Loughborough LE11 3TU, UK. (R.L.Wilby@lboro.ac.uk)

AGU Category Index

Index

Note: Page numbers with italicized *f* and *t* refer to figures and tables

F

T